Analysis of Messy Data

**VOLUME III:
ANALYSIS OF COVARIANCE**

Analysis of Messy Data

VOLUME III:
ANALYSIS OF COVARIANCE

George A. Milliken
Dallas E. Johnson

CHAPMAN & HALL/CRC

A CRC Press Company
Boca Raton   London   New York   Washington, D.C.

Library of Congress Cataloging-in-Publication Data

Milliken, George A., 1943–
　　Analysis of messy data / George A. Milliken, Dallas E. Johnson.
　　　　2 v. : ill. ; 24 cm.
　　Includes bibliographies and indexes.
　　Contents: v. 1. Designed experiments -- v. 2. Nonreplicated
experiments.
　　Vol. 2 has imprint: New York : Van Nostrand Reinhold.
　　ISBN 0-534-02713-X (v. 1) : $44.00 -- ISBN 0-442-24408-8 (v. 2)
　　1. Analysis of variance. 2. Experimental design. 3. Sampling
(Statistics)　I. Johnson, Dallas E., 1938– .　II. Title.
QA279 .M48 1984
519.5′352--dc19　　　　　　　　　　　　　　　　　　　　　　　　84-000839

This book contains information obtained from authentic and highly regarded sources. Reprinted material is quoted with permission, and sources are indicated. A wide variety of references are listed. Reasonable efforts have been made to publish reliable data and information, but the author and the publisher cannot assume responsibility for the validity of all materials or for the consequences of their use.

Apart from any fair dealing for the purpose of research or private study, or criticism or review, as permitted under the UK Copyright Designs and Patents Act, 1988, this publication may not be reproduced, stored or transmitted, in any form or by any means, electronic or mechanical, including photocopying, microfilming, and recording, or by any information storage or retrieval system, without the prior permission in writing of the publishers, or in the case of reprographic reproduction only in accordance with the terms of the licenses issued by the Copyright Licensing Agency in the UK, or in accordance with the terms of the license issued by the appropriate Reproduction Rights Organization outside the UK.

All rights reserved. Authorization to photocopy items for internal or personal use, or the personal or internal use of specific clients, may be granted by CRC Press LLC, provided that $1.50 per page photocopied is paid directly to Copyright Clearance Center, 222 Rosewood Drive, Danvers, MA 01923 USA. The fee code for users of the Transactional Reporting Service is ISBN 1-584-88083-X/02/$0.00+$1.50. The fee is subject to change without notice. For organizations that have been granted a photocopy license by the CCC, a separate system of payment has been arranged.

The consent of CRC Press LLC does not extend to copying for general distribution, for promotion, for creating new works, or for resale. Specific permission must be obtained in writing from CRC Press LLC for such copying.

Direct all inquiries to CRC Press LLC, 2000 N.W. Corporate Blvd., Boca Raton, Florida 33431.

Trademark Notice: Product or corporate names may be trademarks or registered trademarks, and are used only for identification and explanation, without intent to infringe.

Visit the CRC Press Web site at www.crcpress.com

© 2002 by Chapman & Hall/CRC

No claim to original U.S. Government works
International Standard Book Number 1-584-88083-X
Library of Congress Card Number 84-000839
Printed in the United States of America　1　2　3　4　5　6　7　8　9　0
Printed on acid-free paper

Table of Contents

Chapter 1 Introduction to the Analysis of Covariance 1

1.1 Introduction.. 1
1.2 The Covariate Adjustment Process 1
1.3 A General AOC Model and the Basic Philosophy 7
References.. 10

Chapter 2 One-Way Analysis of Covariance — One Covariate in a
Completely Randomized Design Structure 11

2.1 The Model 11
2.2 Estimation 12
2.3 Strategy for Determining the Form of the Model 14
2.4 Comparing the Treatments or Regression Lines 17
    2.4.1 Equal Slopes Model 18
    2.4.2 Unequal Slopes Model-Covariate by Treatment Interaction 21
2.5 Confidence Bands about the Difference of Two Treatments..................... 25
2.6 Summary of Strategies 25
2.7 Analysis of Covariance Computations via the SAS® System 26
    2.7.1 Using PROC GLM and PROC MIXED 26
    2.7.2 Using JMP® 31
2.8 Conclusions... 38
References.. 39
Exercise.. 39

Chapter 3 Examples: One-Way Analysis of Covariance — One Covariate
in a Completely Randomized Design Structure 41

3.1 Introduction.. 41
3.2 Chocolate Candy — Equal Slopes... 41
    3.2.1 Analysis Using PROC GLM 42
    3.2.2 Analysis Using PROC MIXED 47
    3.2.3 Analysis Using JMP® 50
3.3 Exercise Programs and Initial Resting Heart Rate — Unequal Slopes 54
3.4 Effect of Diet on Cholesterol Level: An Exception to the Basic
Analysis of Covariance Strategy... 66
3.5 Change from Base Line Analysis Using Effect of Diet on Cholesterol
Level Data... 70
3.6 Shoe Tread Design Data for Exception to the Basic Strategy 74

3.7　Equal Slopes within Groups of Treatments and Unequal Slopes between Groups 78
3.8　Unequal Slopes and Equal Intercepts — Part 1 83
3.9　Unequal Slopes and Equal Intercepts — Part 2 85
References 90
Exercises 90

Chapter 4　Multiple Covariates in a One-Way Treatment Structure in a Completely Randomized Design Structure 93

4.1　Introduction 93
4.2　The Model 93
4.3　Estimation 95
4.4　Example: Driving A Golf Ball with Different Shafts 95
4.5　Example: Effect of Herbicides on the Yield of Soybeans — Three Covariates 99
4.6　Example: Models That Are Quadratic Functions of the Covariate 105
4.7　Example: Comparing Response Surface Models 112
Reference 121
Exercises 121

Chapter 5　Two-Way Treatment Structure and Analysis of Covariance in a Completely Randomized Design Structure 123

5.1　Introduction 123
5.2　The Model 123
5.3　Using the SAS® System 127
　　　5.3.1　Using PROC GLM and PROC MIXED 128
　　　5.3.2　Using JMP® 129
5.4　Example: Average Daily Gains and Birth Weight — Common Slope 130
5.5　Example: Energy from Wood of Different Types of Trees — Some Unequal Slopes 136
5.6　Missing Treatment Combinations 144
5.7　Example: Two-Way Treatment Structure with Missing Cells 147
5.8　Extensions 158
Reference 160
Exercises 160

Chapter 6　Beta-Hat Models 163

6.1　Introduction 163
6.2　The Beta-Hat Model and Analysis 163
6.3　Testing Equality of Parameters 165
6.4　Complex Treatment Structures 166
6.5　Example: One-Way Treatment Structure 167
6.6　Example: Two-Way Treatment Structure 171

6.7	Summary	174
Exercises		174

Chapter 7 Variable Selection in the Analysis of Covariance Model 175

7.1	Introduction	175
7.2	Procedure for Equal Slopes	175
7.3	Example: One-Way Treatment Structure with Equal Slopes Model	177
7.4	Some Theory	184
7.5	When Slopes are Possibly Unequal	185
References		186
Exercises		186

Chapter 8 Comparing Models for Several Treatments 189

8.1	Introduction	189
8.2	Testing Equality of Models for a One-Way Treatment Structure	190
8.3	Comparing Models for a Two-Way Treatment Structure	191
8.4	Example: One-Way Treatment Structure with One Covariate	193
8.5	Example: One-Way Treatment Structure with Three Covariates	195
8.6	Example: Two-Way Treatment Structure with One Covariate	197
8.7	Discussion	200
References		201
Exercises		201

Chapter 9 Two Treatments in a Randomized Complete Block Design Structure 203

9.1	Introduction	203
9.2	Complete Block Designs	203
9.3	Within Block Analysis	204
9.4	Between Block Analysis	206
9.5	Combining Within Block and Between Block Information	207
9.6	Determining the Form of the Model	209
9.7	Common Slope Model	211
9.8	Comparing the Treatments	214
	9.8.1 Equal Slopes Models	215
	9.8.2 Unequal Slopes Model	215
9.9	Confidence Intervals about Differences of Two Regression Lines	215
	9.9.1 Within Block Analysis	216
	9.9.2 Combined Within Block and Between Block Analysis	216
9.10	Computations for Model 9.1 Using the SAS® System	217
9.11	Example: Effect of Drugs on Heart Rate	221
9.12	Summary	226
References		231
Exercises		231

Chapter 10 More Than Two Treatments in a Blocked Design Structure 233

10.1 Introduction 233
10.2 RCB Design Structure — Within and Between Block Information 233
10.3 Incomplete Block Design Structure — Within and Between Block Information 234
10.4 Combining Between Block and Within Block Information 236
10.5 Example: Five Treatments in RCB Design Structure 240
10.6 Example: Balanced Incomplete Block Design Structure with Four Treatments 247
10.7 Example: Balanced Incomplete Block Design Structure with Four Treatments Using JMP® 251
10.8 Summary 254
References 256
Exercises 256

Chapter 11 Covariate Measured on the Block in RCB and Incomplete Block Design Structures 259

11.1 Introduction 259
11.2 The Within Block Model 260
11.3 The Between Block Model 261
11.4 Combining Within Block and Between Block Information 261
11.5 Common Slope Model 263
11.6 Adjusted Means and Comparing Treatments 264
    11.6.1 Common Slope Model 264
    11.6.2 Non-Parallel Lines Model 264
11.7 Example: Two Treatments 265
11.8 Example: Four Treatments in RCB 269
11.9 Example: Four Treatments in BIB 277
11.10 Summary 282
References 284
Exercises 284

Chapter 12 Random Effects Models with Covariates 287

12.1 Introduction 287
12.2 The Model 287
12.3 Estimation of the Variance Components 292
12.4 Changing Location of the Covariate Changes the Estimates of the Variance Components 297
12.5 Example: Balanced One-Way Treatment Structure 299
12.6 Example: Unbalanced One-Way Treatment Structure 304
12.7 Example: Two-Way Treatment Structure 309
12.8 Summary 315
References 320
Exercises 321

Chapter 13 Mixed Models 325

13.1   Introduction 325
13.2   The Matrix Form of the Mixed Model 325
13.3   Fixed Effects Treatment Structure 329
13.4   Estimation of Fixed Effects and Some Small Sample Size
       Approximations 329
13.5   Fixed Treatments and Locations Random 331
13.6   Example: Two-Way Mixed Effects Treatment Structure in a CRD 332
13.7   Example: Treatments are Fixed and Locations are Random with a
       RCB at Each Location 337
References 350
Exercises 351

Chapter 14 Analysis of Covariance Models with Heterogeneous Errors 353

14.1   Introduction 353
14.2   The Unequal Variance Model 353
14.3   Tests for Homogeneity of Variances 354
       14.3.1   Levene's Test for Equal Variances 354
       14.3.2   Hartley's F-Max Test for Equal Variances 355
       14.3.3   Bartlett's Test for Equal Variances 355
       14.3.4   Likelihood Ratio Test for Equal Variances 356
14.4   Estimating the Parameters of the Regression Model 356
       14.4.1   Least Squares Estimation 356
       14.4.2   Maximum Likelihood Methods 357
14.5   Determining the Form of the Model 357
14.6   Comparing the Models 359
       14.6.1   Comparing the Nonparallel Lines Models 359
       14.6.2   Comparing the Parallel Lines Models 361
14.7   Computational Issues 362
14.8   Example: One-Way Treatment Structure with Unequal Variances 362
14.9   Example: Two-Way Treatment Structure with Unequal Variances 369
14.10 Example: Treatments in Multi-location Trial 381
14.11 Summary 389
References 389
Exercises 389

**Chapter 15 Analysis of Covariance for Split-Plot and Strip-Plot Design
             Structures** 391

15.1   Introduction 391
15.2   Some Concepts 392
15.3   Covariate Measured on the Whole Plot or Large Size of Experimental
       Unit 392
15.4   Covariate is Measured on the Small Size of Experimental Unit 395

15.5	Covariate is Measured on the Large Size of Experimental Unit and a Covariate is Measured on the Small Size of Experimental Unit	398
15.6	General Representation of the Covariate Part of the Model	399
	15.6.1 Covariate Measured on Large Size of Experimental Unit	401
	15.6.2 Covariate Measured on the Small Size of Experimental Units	403
	15.6.3 Summary of General Representation	405
15.7	Example: Flour Milling Experiment — Covariate Measured on the Whole Plot	406
15.8	Example: Cookie Baking	414
15.9	Example: Teaching Methods with One Covariate Measured on the Large Size Experimental Unit and One Covariate Measured on the Small Size Experimental Unit	426
15.10	Example: Comfort Study in a Strip-Plot Design with Three Sizes of Experimental Units and Three Covariates	432
15.11	Conclusions	444
	References	446
	Exercises	446

Chapter 16 Analysis of Covariance for Repeated Measures Designs..............451

16.1	Introduction	451
16.2	The Covariance Part of the Model — Selecting R	453
16.3	Covariance Structure of the Data	456
16.4	Specifying the Random and Repeated Statements for PROC MIXED of the SAS® System	457
16.5	Selecting an Adequate Covariance Structure	458
16.6	Example: Systolic Blood Pressure Study with Covariate Measured on the Large Size Experimental Unit	459
16.7	Example: Oxide Layer Development Experiment with Three Sizes of Experimental Units Where the Repeated Measure is at the Middle Size of Experimental Unit and the Covariate is Measured on the Small Size Experimental Unit	470
16.8	Conclusions	479
	References	487
	Exercises	487

Chapter 17 Analysis of Covariance for Nonreplicated Experiments 493

17.1	Introduction	493
17.2	Experiments with A Single Covariate	495
17.3.	Experiments with Multiple Covariates	499
17.4	Selecting Non-null and Null Partitions	501
17.5	Estimating the Parameters	502
17.6	Example: Milling Flour Using Three Factors Each at Two Levels	503
17.7	Example: Baking Bread Using Four Factors Each at Two Levels	508
17.8	Example: Hamburger Patties with Four Factors Each at Two Levels	511

17.9	Example: Strength of Composite Material Coupons with Two Covariates	512
17.10	Example: Effectiveness of Paint on Bricks with Unequal Slopes	520
17.11	Summary	527
References		529
Exercises		530

Chapter 18 Special Applications of Analysis of Covariance 533

18.1	Introduction	533
18.2	Blocking and Analysis of Covariance	533
18.3	Treatments Have Different Ranges of the Covariate	543
18.4	Nonparametric Analysis of Covariance	552
	18.4.1 Heart Rate Data from Exercise Programs	552
	18.4.2 Average Daily Gain Data from a Two-Way Treatment Structure	555
18.5	Crossover Design with Covariates	559
18.6	Nonlinear Analysis of Covariance	564
18.7	Effect of Outliers	572
References		590
Exercises		590

Index 597

Preface

Analysis of covariance is a statistical procedure that enables one to incorporate information about concomitant variables into the analysis of a response variable. Sometimes this is done in an attempt to reduce experimental error. Other times it is done to better understand the phenomenon being studied. The approach used in this book is that the analysis of covariance model is described as a method of comparing a series of regression models — one for each of the levels of a factor or combinations of levels of factors being studied. Since covariance models are regression models, analysts can use all of the methods of regression analysis to deal with problems such as lack of fit, outliers, etc. The strategies described in this book will enable the reader to appropriately formulate and analyze various kinds of covariance models. When covariates are measured and incorporated into the analysis of a response variable, the main objective of analysis of covariance is to compare treatments or treatment combinations at common values of the covariates. This is particularly true when the experimental units assigned to each of the treatment combinations may have differing values of the covariates. Comparing treatments is dependent on the form of the covariance model and thus care must be taken so that mistakes are not made when drawing conclusions.

The goal of this book is to present the structure and philosophy for using the analysis of covariance by including descriptions of methodologies, illustrating the methodologies by analyzing numerous data sets, and occasionally furnishing some theory when required. Our aim is to provide data analysts with tools for analyzing data with covariates and to enable them to appropriately interpret the results.

Some of the methods and techniques described in this book are not available in other books, but two issues of *Biometrics* (1957, Volume 13, Number 3, and 1982, Volume 38, Number 3) were dedicated to the topic of analysis of covariance. The topics presented are among those that we, as consulting statisticians, have found to be most helpful in analyzing data when covariates are available for possible inclusion in the analysis.

Readers of this book will learn how to:

- Formulate appropriate analysis of covariance models
- Simplify analysis of covariance models
- Compare levels of a factor or of levels of combinations of factors when the model involves covariates
- Construct and analyze a model with two or more factors in the treatment structure
- Analyze two-way treatment structures with missing cells
- Compare models using the beta-hat model
- Perform variable selection within the analysis of covariance model

- Analyze models with blocking in the design structure and use combined intra-block and inter-block information about the slopes of the regression models
- Use random statements in PROC MIXED to specify random coefficient regression models
- Carry out the analysis of covariance in a mixed model framework
- Incorporate unequal treatment variances into the analysis
- Specify the analysis of covariance models for split-plot, strip-plot and repeated measures designs both in terms of the regression models and the covariance structures of the repeated measures
- Incorporate covariates into the analysis of nonreplicated experiments, thus extending some of the results in *Analysis of Messy Data, Volume II*

The last chapter consists of a collection of examples that deal with (1) using the covariate to form blocks, (2) crossover designs, (3) nonparametric analysis of covariance, (4) using a nonlinear model for the covariate model, and (5) the process of examining mixed analysis of covariance models for possible outliers.

The approach used in this book is similar to that used in the first two volumes. Each topic is covered from a practical viewpoint, emphasizing the implementation of the methods much more than the theory behind the methods. Some theory has been presented for some of the newer methodologies. The book utilized the procedures of the SAS® system and JMP® software packages to carry out the computations and few computing formulae are presented. Either SAS® system code or JMP® menus are presented for the analysis of the data sets in the examples. The data in the examples (except for those using chocolate chips) were generated to simulate real world applications that we have encountered in our consulting experiences.

This book is intended for everyone who analyzes data. The reader should have a knowledge of analysis of variance and regression analysis as well as basic statistical ideas including randomization, confidence intervals, and hypothesis testing. The first four chapters contain the information needed to form a basic philosophy for using the analysis of covariance with a one-way treatment structure and should be read by everyone. As one progresses through the book, the topics become more complex by going from designs with blocking to split-plot and repeated measures designs. Before reading about a particular topic in the later chapters, read the first four chapters. Knowledge of Chapters 13 and 14 from *Analysis of Messy Data, Volume I: Designed Experiments* would be useful for understanding the part of Chapter 5 involving missing cells. The information in Chapters 4 through 9 of *Analysis of Messy Data, Volume II: Nonreplicated Experiments* is useful for comprehending the topics discussed in Chapter 17.

This book is the culmination of more than 25 years of writing. The earlier editions of this manuscript were slanted toward providing an appropriate analysis of split-plot type designs by using fixed effects software such as PROC GLM of the SAS® system. With the development of mixed models software, such as PROC MIXED of the SAS® system and JMP®, the complications of the analysis of split-plot type designs disappeared and thus enabled the manuscript to be completed without including the difficult computations that are required when using fixed effects software. Over the years, several colleagues made important contributions. Discussions with Shie-Shien Yang were invaluable for the development of the variable selection process described in Chapter 7. Vicki Landcaster and Marie Loughin read some of the earlier versions and provided important feedback. Discussions with James Schwenke, Kate Ash, Brian Fergen, Kevin Chartier, Veronica Taylor, and Mike Butine were important for improving the chapters involving combining intra- and inter-block information and the strategy for the analysis of repeated measures designs. Finally, we cannot express enough our thanks to Jane Cox who typed many of the initial versions of the chapters. If it were not for Jane's skills with the word processor, the task of finishing this book would have been much more difficult.

We dedicate this volume to all who have made important contributions to our personal and professional lives. This includes our wives, Janet and Erma Jean, our children, Scott and April and Kelly and Mark, and our parents and parents in-law who made it possible for us to pursue our careers as statisticians. We were both fortunate to study with Franklin Graybill and we thank him for making sure that we were headed in the right direction when our careers began.

1 Introduction to the Analysis of Covariance

1.1 INTRODUCTION

The statistical procedure termed analysis of covariance has been used in several contexts. The most common description of analysis of covariance is to adjust the analysis for variables that could not be controlled by the experimenter. For example, if a researcher wishes to compare the effect that ten different chemical weed control treatments have on yield of a specific wheat variety, the researcher may wish to control for the differential effects of a fertility trend occurring in the field and for the number of wheat plants per plot that happen to emerge after planting. The differential effects of a fertility trend can possibly be removed by using a randomized complete block design structure, but it may not be possible to control the number of wheat plants per plot (unless the seeds are sewn thickly and then the emerging plants are thinned to a given number of plants per plot). The researcher wishes to compare the treatments as if each treatment were grown on plots with the same average fertility level and as if every plot had the same number of wheat plants. The use of a randomized complete block design structure in which the blocks are constructed such that the fertility levels of plots within a block are very similar will enable the treatments to be compared by averaging over the fertility levels, but the analysis of covariance is a procedure which can compare treatment means after first adjusting for the differential number of wheat plants per plot. The adjustment procedure involves constructing a model that describes the relationship between yield and the number of wheat plants per plot for each treatment, which is in the form of a regression model. The regression models, one for each level of the treatment, are then compared at a predetermined common number of wheat plants per plot.

1.2 THE COVARIATE ADJUSTMENT PROCESS

To demonstrate the type of adjustment process that is being carried out when the analysis of covariance methodology is applied, the set of data in Table 1.1 is used in which there are two treatments and five plots per treatment in a completely randomized design structure. Treatment 1 is a chemical application to control the growth of weeds and Treatment 2 is a control without any chemicals to control the weeds. The data in Table 1.1 consist of the yield of wheat plants of a specific variety from plots of identical size along with the number of wheat plants that emerged

TABLE 1.1
Yield and Plants per Plot Data for the Example in Section 1.2

Treatment 1		Treatment 2	
Yield per plot	Plants per plot	Yield per plot	Plants per plot
951	126	930	135
957	128	790	119
776	107	764	110
1033	142	989	140
840	120	740	102

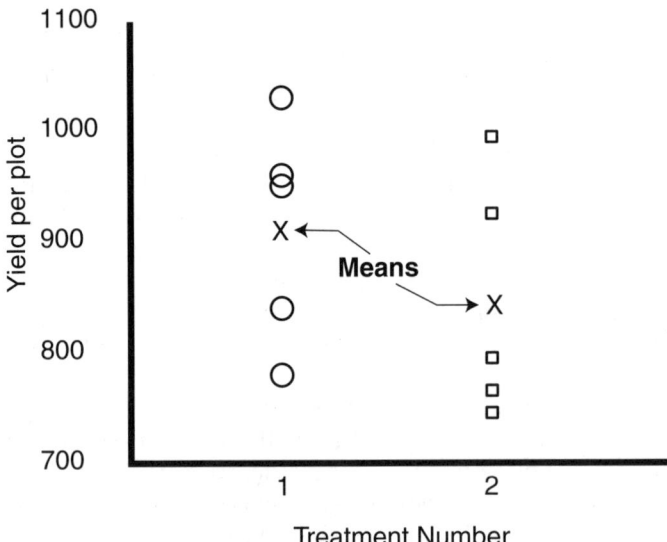

FIGURE 1.1 Plot of the data for the two treatments, with the "X" denoting the respective means.

after planting per plot. The researcher wants to compare the yields of the two treatments for the condition when there are 125 plants per plot.

Figure 1.1 is a graphical display of the plot yields for each of the treatments where the circles represent the data points for Treatment 1 and the boxes represent the data points for Treatment 2. An "X" is used to mark the means of each of the treatments.

If the researcher uses the two-sample t-test or one-way analysis of variance to compare the two treatments without taking information into account about the number of plants per plot, a t statistic of 1.02 or a F statistic of 1.05 is obtained, indicating the two treatment means are not significantly different ($p = 0.3361$). The results of the analysis are in Table 1.2 in which the estimated standard error of the difference of the two treatment means is 67.23.

TABLE 1.2
Analysis of Variance Table and Means for Comparing the Yields of the Two Treatments Where No Information about the Number of Plants per Plot is Used

Source	df	SS	MS	FValue	ProbF
Model	1	11833.60	11833.60	1.05	0.3361
Error	8	90408.40	11301.05		
Corrected total	9	102242.00			

Source	df	SS (type III)	MS	FValue	ProbF
TRT	1	11833.60	11833.60	1.05	0.3361

Parameter	Estimate	StdErr	tValue	Probt
Trt 1 – Trt 2	68.8	67.23	1.02	0.3361

TRT	LSMean	ProbtDiff
1	911.40	0.3361
2	842.60	

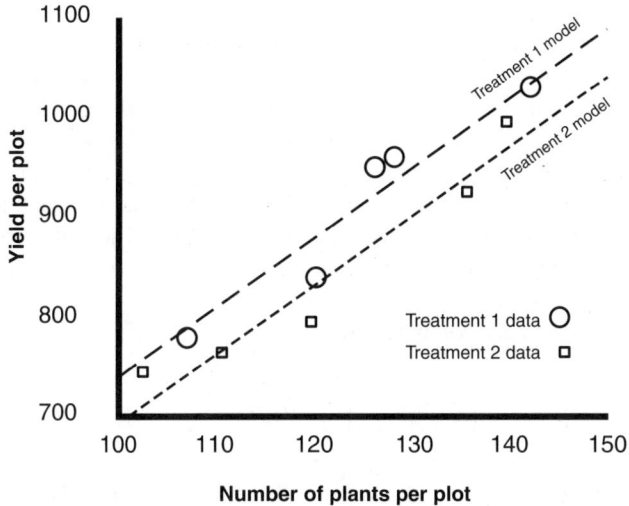

FIGURE 1.2 Plot of the data and the estimated regression models for the two treatments.

The next step is to investigate the relationship between the yield per plot and the number of plants per plot. Figure 1.2 is a display of the data where the number of plants is on the horizontal axis and the yield is on the vertical axis. The circles denote the data for Treatment 1 and the boxes denote the data for Treatment 2. The two lines on the graph, denoted by Treatment 1 model and Treatment 2 model, were computed from the data by fitting the model $y_{ij} = \alpha_i + \beta x_{ij} + \varepsilon_{ij}$, $i = 1, 2$ and $j = 1$,

TABLE 1.3
Analysis of Covariance to Provide the Estimates of the Slope and Intercepts to be Used in Adjusting the Data

Source	df	SS	MS	FValue	ProbF
Model	3	7787794.74	2595931.58	3167.28	0.0000
Error	7	5737.26	819.61		
Uncorr Total	10	7793532.00			

Source	df	SS(Type III)	MS	FValue	ProbF
TRT	2	4964.18	2482.09	3.03	0.1128
Plants	1	84671.14	84671.14	103.31	0.0000

Parameter	Estimate	StdErr	tValue	Probt
Trt 1 – Trt 2	44.73	18.26	2.45	0.0441

Parameter	Estimate	StdErr	tValue	Probt
TRT 1	29.453	87.711	0.34	0.7469
TRT 2	−15.281	85.369	−0.18	0.8630
Plants	7.078	0.696	10.16	0.0000

2, ..., 5, a model with different intercepts and common or equal slopes. The results are included in Table 1.3.

Now analysis of covariance is used to compare the two treatments when there are 125 plants per plot. The process of the analysis of covariance is to slide or move the observations from a given treatment along the estimated regression model (parallel to the model) to intersect the vertical line at 125 plants per plot. This sliding is demonstrated in Figure 1.3 where the solid circles represent the adjusted data for Treatment 1 and the solid boxes represent the adjusted data for Treatment 2.

The lines join the open circles to the solid circles and join the open boxes to the solid boxes. The lines indicate that the respective data points slid to the vertical line at which there are 125 plants per plot.

The adjusted data are computed by

$$y_{Aij} = y_{ij} - \left(\hat{\alpha}_i + \hat{\beta} x_{ij}\right) + \left(\hat{\alpha}_i + \hat{\beta} 125\right) = y_{ij} + \hat{\beta}\left(125 - x_{ij}\right)$$

The terms $y_{ij} - (\hat{\alpha}_i + \hat{\beta} x_{ij})$ i = 1,2 and j = 1,2,...,5 are the residuals or deviations of the observations from the estimated regression models. The preliminary computations of the adjusted yields are in Table 1.4. These adjusted yields are the predicted yields of the plots as if each plot had 125 plants.

The next step is to compare the two treatments through the adjusted yield values by computing a two-sample t statistic or the F statistic from a one-way analysis of variance. The results of these analyses are in Table 1.5.

A problem with this analysis is that it assumes the adjusted data are not adjusted data and so there is no reduction in the degrees of freedom for error due to estimating the slope of the regression lines. Hence the final step is to recalculate the statistics

Introduction to the Analysis of Covariance

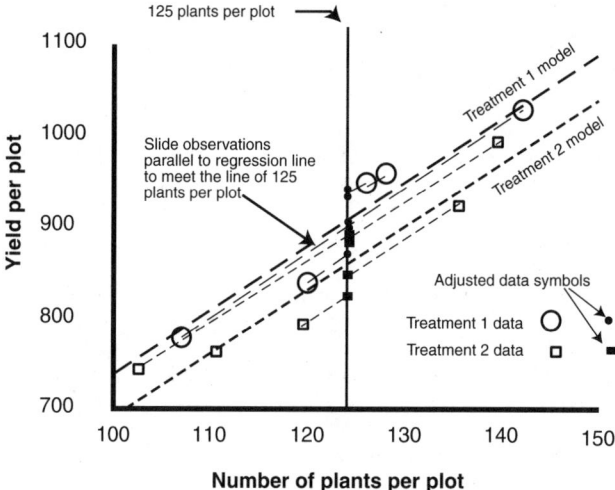

FIGURE 1.3 Plot of the data and estimated regression models showing how to compute adjusted yield values at 125 plants per plot.

TABLE 1.4
Preliminary Computations Used in Computing Adjusted Data for Each Treatment as If All Plots Had 125 Plants per Plot

Treatment	Yield Per Plot	Plants Per Plot	Residual	Adjusted Yield
1	951	126	29.6905	943.922
1	957	128	21.534	935.765
1	776	107	−10.8232	903.408
1	1033	142	−1.5611	912.67
1	840	120	−38.8402	875.391
2	930	135	−10.2795	859.218
2	790	119	−37.0279	832.469
2	764	110	0.6761	870.173
2	989	140	13.3294	882.827
2	740	102	33.3019	902.799

by changing the degrees of freedom for error in Table 1.5 from 8 to 7 (the cost of estimating the slope). The sum of squares error is identical for both Tables 1.3 and 1.5, but the error sum of squares from Table 1.5 is based on 8 degrees of freedom instead of 7. To account for this change in degrees of freedom in Table 1.5, the estimated standard error for comparing the two treatments needs to be multiplied by $\sqrt{8/7}$, the t statistic needs to be multiplied by $\sqrt{7/8}$, and the F statistic needs to be multiplied by 7/8. The recalculated statistics are presented in Table 1.6. Here the estimated standard error of the difference between the two means is 18.11, a 3.7-fold reduction over the analysis that ignores the information from the covariate. Thus,

TABLE 1.5
Analysis of the Adjusted Yields (Too Many Degrees of Freedom for Error)

Source	df	SS	MS	FValue	ProbF
Model	1	5002.83	5002.83	6.98	0.0297
Error	8	5737.26	717.16		
Corrected Total	9	10740.09			

Source	df	SS (Type III)	MS	FValue	ProbF
TRT	1	5002.83	5002.83	6.98	0.0297

Parameter	Estimate	StdErr	tValue	Prob$t$
Trt 1 − Trt 2	44.734	16.937	2.641197	0.0297

TRT	LSMean	ProbtDiff
1	914.231	0.0297
2	869.497	

TABLE 1.6
Recalculated Statistics to Reflect the Loss of Error Degrees of Freedom Due to Estimating the Slope before Computing the Adjusted Yields

Recalculated estimated standard error	18.11
Recalculated t-statistic	2.47
Recalculated F-statistic	6.10
Recalculated significance level	0.0428

by taking into account the linear relationship between the yield of the plot and the number of plants in that plot, there is a tremendous reduction in the variability of the data. In fact, the analysis of the adjusted data shows there is a significant difference between the yields of the two treatments when adjusting for the unequal number of plants per plot ($p = 0.0428$), when the analysis of variance in Table 1.2 did not indicate there is a significant difference between the treatments ($p = 0.3361$). The final issue is that since this analysis of the adjusted data overlooks the fact the slope has been estimated, the estimated standard error of the difference of two means is a little small as compared to the estimated standard error one gets from the analysis of covariance. The estimated standard error of the difference of the two means as computed from the analysis of covariance in Table 1.3 is 18.26 as compared to 18.11 for the analysis of the adjusted data. Thus the two analyses are not quite identical.

This example shows the power of being able to use information about covariates or independent variables to make decisions about the treatments being included in

Introduction to the Analysis of Covariance 7

the study. The analysis of covariance uses a model to adjust the data as if all the observations are from experimental units with identical values of the covariates.

A typical discussion of analysis of covariance indicates that the analyst should include the number of plants as a term in the model so that term accounts for variability in the observed yields, i.e., the variance of the model is reduced. If including the number of plants in the model reduces the variability enough, then it is used to adjust the data before the variety means are compared. It is important to remember that there is a model being assumed when the covariate or covariates are included in a model.

1.3 A GENERAL AOC MODEL AND THE BASIC PHILOSOPHY

In this text, the analysis of covariance is described in more generality than that of adjusting for variation due to uncontrollable variables. The analysis of covariance is defined as a method for comparing several regression surfaces or lines, one for each treatment or treatment combination, where a different regression surface is possibly used to describe the data for each treatment or treatment combination.

A one-way treatment structure with t treatments in a completely randomized design structure (Milliken and Johnson, 1992) is used as a basis for setting up the definitions for the analysis of covariance model. The experimental situation involves selecting N experimental units from a population of experimental units and measuring k characteristics $x_{1ij}, x_{2ij}, \ldots, x_{kij}$ on each experimental unit. The variables $x_{1ij}, x_{2ij}, \ldots, x_{kij}$ are called covariates or independent variables or concomitant variables. It is important to measure the values of the covariates before the treatments are applied to the experimental units so that the levels of the treatments do not effect the values of the covariates. At a minimum, the values of the covariate should not be effected by the applied levels of the treatments. In the chemical weed treatment experiment, the number of plants per plot occur after applying a particular treatment on a plot, so the value of the covariate (number of plants per plot) could not be determined before the treatments were applied to the plots. If the germination rate is affected by the applied treatments, then the number of plants per plot cannot be used as a covariate in the conventional manner (see Chapter 2 for further discussion). After the set of experimental units is selected and the values of the covariates are determined (when possible), then randomly assign n_i experimental units to treatment i, where $N = \sum_i n_i$. One generally assigns equal numbers of experimental units to the levels of the treatment, but equal numbers of experimental units per level of the treatment are not necessary. After an experimental unit is subjected its specific level of the treatment, then measure the response or dependent variable which is denoted by y_{ij}. Thus the variables used in the discussions are summarized as:

y_{ij} is the dependent measure
x_{1ij} is the first independent variable or covariate
x_{2ij} is the second independent variable or covariate
x_{kij} is the kth independent variable or covariate

At this point, the experimental design is a one-way treatment structure with t treatments in a completely randomized design structure with k covariates. If there is a linear relationship between the mean of y for the ith treatment and the k covariates or independent variables, an analysis of covariance model can be expressed as:

$$y_{ij} = \beta_{oi} + \beta_{1i}x_{1ij} + \beta_{2i}x_{2ij} + \ldots + \beta_{ki}x_{kij} + \varepsilon_{ij} \tag{1.1}$$

for $i = 1, 2, \ldots, t$, and $j = 1, 2, \ldots, n_i$, and the $\varepsilon_{ij} \sim$ iid $N(0, \sigma^2)$, i.e., the ε_{ij} are independently identically distributed normal random variables with mean 0 and variance σ^2. The important thing to note about this model is that the mean of the y values from a given treatment depends on the values of the x's as well as on the treatment applied to the experimental units.

The analysis of covariance is a strategy for making decisions about the form of the covariance model through testing a series of hypotheses and then making treatment comparisons by comparing the estimated responses from the final regression models. Two important hypotheses that help simplify the regression models are

H_{01}: $\beta_{h1} = \beta_{h2} = \ldots = \beta_{ht} = 0$ vs. H_{a1}: (not H_{01}:), that is, all the treatments' slopes for the hth covariate are zero, $h = 1, 2, \ldots, k$, or

H_{02}: $\beta_{h1} = \beta_{h2} = \ldots = \beta_{ht}$ vs. H_{a2}: (not H_{02}:), that is, the slopes for the hth covariate are equal across the treatments, meaning the surfaces are parallel in the direction of the hth covariate, $h = 1, 2, \ldots, k$.

The analysis of covariance model in Equation 1.1 is a combination of an analysis of variance model and a regression model. The analysis of covariance model is part of an analysis of variance model since the intercepts and slopes are functions of the levels of the treatments. The analysis of covariance model is also part of a regression model since the model for each treatment is a regression model.

An experiment is designed to purchase a certain number of degrees of freedom for error (generally without the covariates) and the experimenter is willing to sell some of those degrees of freedom for good or effective covariates which will help reduce the magnitude of the error variance. The philosophy in this book is to select the simplest possible expression for the covariate part of the model before making treatment comparisons.

This process of model building to determine the simplest adequate form of the regression models follows the principle of parsimony and helps guard against foolishly selling degrees of freedom for error to retain unnecessary covariate terms in the model. Thus the strategy for analysis of covariance begins with testing hypotheses such as H_{01} and H_{02} to make decisions about the form of the covariate or regression part of the model. Once the form of the covariate part of the model is finalized, the treatments are compared by comparing the regression surfaces at predetermined values of the covariates.

The structure of the following chapters leads one through the forest of analysis of covariance by starting with the simple model with one covariate and building through the complex process involving analysis of covariance in split-plot and

Introduction to the Analysis of Covariance 9

repeated measures designs. Other topics discussed are multiple covariates, experiments involving blocks, and graphical methods for comparing the models for the various treatments.

Chapter 2 discusses the simple analysis of covariance model involving a one-way treatment structure in a completely randomized design structure with one covariate and Chapter 3 contains several examples demonstrating the strategies for situations involving one covariate. Chapter 4 presents a discussion of the analysis of covariance models involving more than one covariate which includes polynomial regression models. Models involving two-way treatment structures, both balanced and unbalanced, are discussed in Chapter 5. A method of comparing parameters via beta-hat models is described in Chapter 6. Chapter 7 describes a method for variable selection in the analysis of covariance where many possible covariates were measured. Chapter 8 discusses methods for testing the equality of several regression models.

The next set of chapters (9 through 11) discuss analysis of covariance in the randomized complete block and incomplete block design structures. The analysis of data where the values of a characteristic are used to construct blocks is described, i.e., where the value of the covariate is the same for all experimental units in a block. In the analysis of covariance context, inter- or between block information about the intercepts and slopes is required to extract all available information about the regression lines or surfaces from the data. Usual analysis methods extract only the intra-block information from the data. A mixed models analysis involving methods of moments and maximum likelihood estimation of the variance components provides combined estimates of the parameters and should be used for blocked experiments. Chapter 12 describes models where the levels of the treatments are random effects (Littell et al., 1996). The models in Chapter 12 include random coefficient models. Chapter 13 provides a discussion of mixed models with covariates and Chapter 14 presents a discussion of unequal variance models.

Chapters 15 and 16 discuss problems with applying the analysis of covariance to experiments involving repeated measures and split-plot design structures. One has to consider the size of experimental unit on which the covariate is measured. Cases are discussed where the covariate is measured on the large size of an experimental unit and when the covariate is measured on the small size of an experimental unit. Several examples of split-plot and repeated measures designs are presented. A process of selecting the simplest covariance structure for the repeated measures part of the model and the simplest covariate (regression model) part of the model is described. The analysis of covariance in the nonreplicated experiment is discussed in Chapter 17. The half-normal plot methodology (Milliken and Johnson, 1989) is used to determine the form of the covariate part of the model and to determine which effects are to be included in the intercept part of the model.

Finally, several special applications of analysis of covariance are presented in Chapter 18, including using the covariate to construct blocks, crossover designs, nonlinear models, nonparameteric analysis of covariance, and a process for examining mixed models for possible outliers in the data set.

The procedures of the SAS® system (1989, 1996, and 1997) and JMP® (2000) are used to demonstrate how to use software to carry out the analysis of covariance computations. The topic of analysis of covariance has been the topic of two volumes of *Biometrics*, Volume 13, Number 3, in 1957 and Volume 38, Number 3, in 1982. The collection of papers in these two volumes present discussions of widely diverse applications of analysis of covariance.

REFERENCES

Littell, R. C., Milliken, G. A., Stroup, W. W., and Wolfinger, R. D. (1996) *SAS® System for Mixed Models*, SAS Institute Inc., Cary, NC.

Milliken, G. A. and Johnson, D. E. (1989) *Analysis of Messy Data, Volume II: Nonreplicated Experiments*, Chapman & Hall, London.

Milliken, G. A. and Johnson, D. E. (1992) *Analysis of Messy Data, Volume I: Design Experiments*, Chapman & Hall, London.

SAS Institute Inc. (1989) *SAS/STAT® User's Guide, Version 6, Fourth Edition, Volume 2*, Cary, NC.

SAS Institute Inc. (1996) *SAS/STAT® Software: Changes and Enhancements Through Release 6.11*, Cary, NC.

SAS Institute Inc. (1997) *SAS/STAT® Software: Changes and Enhancements Through Release 6.12*, Cary, NC.

SAS Institute Inc. (2000) *JMP® Statistics and Graphics Guide, Version 4*, Cary, NC.

2 One-Way Analysis of Covariance — One Covariate in a Completely Randomized Design Structure

2.1 THE MODEL

Suppose you have N homogeneous experimental units and you randomly divide them into t groups of n_i units each where $\sum_{i=1}^{t} n_i = N$. Each of the t treatments of a one-way treatment structure is randomly assigned to one group of experimental units, providing a one-way treatment structure in a completely randomized design structure. It is assumed that the experimental units are subjected to their assigned treatments independently of each other. Let y_{ij} (dependent variable) denote the jth observation from the ith treatment and x_{ij} denote the covariate (independent variable) corresponding to the $(i,j)^{th}$ experimental unit. As in Chapter 1, the values of the covariate are not to be influenced by the levels of the treatment. The best case is where the values of the covariate are determined before the treatments are assigned. In any case, it is a good strategy to use the analysis of variance to check to see if there are differences among the treatment covariate means (see Chapter 18).

Assume that the mean of y_{ij} can be expressed as a linear function of the covariate, x_{ij}, with possibly a different linear function being required for each treatment. It is important to note that the mean of an observation from the ith treatment group depends on the value of the covariate as well as the treatment. In analysis of variance, the mean of an observation from the ith treatment group depends only on the treatment.

The analysis of covariance model for a one-way treatment structure with one covariate in a completely randomized design structure is

$$y_{ij} = \alpha_i + \beta_i X_{ij} + \varepsilon_{ij},$$
$$i = 1, 2, \ldots, t. \quad j = 1, 2, \ldots, n_i$$

(2.1)

where the mean of y_i for a given value of X is $\mu_{Y_i|X} = \alpha_i + \beta_i X$. For making inferences, it is assumed that $\varepsilon_{ij} \sim$ iid $N(0, \sigma^2)$. Model 2.1 has t intercepts $(\alpha_1, \ldots, \alpha_t)$, t slopes $(\beta_1, \ldots, \beta_t)$, and one variance σ^2, i.e., the model represents a collection of simple linear regression models with a different model for each level of the treatment.

Before analyzing this model, make sure that the data from each treatment can in fact be described by a simple linear regression model. Various regression diagnostics should be run on the data before continuing. The equal variance assumption should also be checked (see Chapter 14). If the simple linear regression model is not adequate to describe the data for each treatment, then another model must be selected before continuing with the analysis of covariance.

The analysis of covariance is a process of comparing the regression models and then making decisions about the various parameters of the models. The process involves comparing the t slopes, comparing the distances between the regression lines (surfaces) at preselected values of X, and possibly comparing the t intercepts. The analysis of covariance computations are typically presented in summation notation with little emphasis on interpretations. In this and the following chapters, the various covariance models are expressed in terms of matrices (see Chapter 6 of Milliken and Johnson, 1992) and their interpretations are discussed. Software is used as the mode of doing the analysis of covariance computations. The matrix form of Model 2.1 is

$$\begin{bmatrix} y_{11} \\ \vdots \\ y_{1n_1} \\ y_{21} \\ \vdots \\ y_{2n_2} \\ \vdots \\ y_{t1} \\ \vdots \\ y_{tn_t} \end{bmatrix} = \begin{bmatrix} 1 & x_{11} & 0 & 0 & \cdots & 0 & 0 \\ \vdots & \vdots & \vdots & \vdots & & \vdots & \vdots \\ 1 & x_{1n_1} & 0 & 0 & \cdots & 0 & 0 \\ 0 & 0 & 1 & x_{21} & \cdots & 0 & 0 \\ \vdots & \vdots & \vdots & \vdots & & \vdots & \vdots \\ 0 & 0 & 1 & x_{2n_2} & \cdots & 0 & 0 \\ \vdots & \vdots & \vdots & \vdots & & \vdots & \vdots \\ 0 & 0 & 0 & 0 & \cdots & 1 & x_{t1} \\ \vdots & \vdots & \vdots & \vdots & & \vdots & \vdots \\ 0 & 0 & 0 & 0 & \cdots & 1 & x_{tn_t} \end{bmatrix} \begin{bmatrix} \alpha_1 \\ \beta_1 \\ \alpha_2 \\ \beta_2 \\ \vdots \\ \alpha_t \\ \beta_t \end{bmatrix} + \varepsilon. \quad (2.2)$$

which is expressed in the form of a linear model as $y = X\beta + \varepsilon$. The vector y denotes the observations ordered by observation within each treatment, the $2t \times 1$ vector β denotes the collection of slopes and intercepts, the matrix X is the design matrix, and the vector ε represents the random errors.

2.2 ESTIMATION

The least squares estimator of the parameter vector β is $\hat{\beta} = (X'X)^{-1}X'y$, but the least squares estimator of β can also be obtained by fitting the simple linear regression model to the data from each treatment and computing the least squares estimator of each pair of parameters (α_i, β_i). For data from the ith treatment, fit the model

$$\begin{bmatrix} y_{i1} \\ \vdots \\ y_{in_i} \end{bmatrix} = \begin{bmatrix} 1 & x_{i1} \\ \vdots & \vdots \\ 1 & x_{in_i} \end{bmatrix} \begin{bmatrix} \alpha_i \\ \beta_i \end{bmatrix} + \varepsilon_i, \quad (2.3)$$

One-Way Analysis of Covariance

which is expressed as $y_i = X_i\beta_i + \varepsilon_i$. The least squares estimator of β_i is $\hat{\beta}_i = (X_i'X_i)^{-1}X_i'y$, the same as the estimator obtained for a simple linear regression model. The estimates of β_i and α_i in summation notation are

$$\hat{\beta}_i = \frac{\sum_{j=1}^{n_i} x_{ij}y_{ij} - n_i \bar{x}_{i.} \bar{y}_{i.}}{\sum_{j=1}^{n_i} x_{ij}^2 - n_i \bar{x}_{i.}^2}$$

and

$$\hat{\alpha}_i = \bar{y}_{i.} - \hat{\beta}_i \bar{x}_{i.}.$$

The residual sum of squares for the ith model is

$$SSRes_i = \sum_{j=1}^{n_i} \left(y_{ij} - \hat{\alpha}_i - \hat{\beta}_i x_{ij}\right)^2.$$

There are $n_i - 2$ degrees of freedom associated with $SSRes_i$ since the ith model involves two parameters. After testing the equality of the treatment variances (see Chapter 14) and deciding there is not enough evidence to conclude the variances are unequal, the residual sum of squares for Model 2.1 can be obtained by pooling residual sums of the squares for each of the t models, i.e., sum the $SSRes_i$ together to obtain

$$SSRes = \sum_{i=1}^{t} SSRes_i. \quad (2.4)$$

The pooled residual sum of squares, SSRes, is based on the pooled degrees of freedom, computed and denoted by

$$d.f._{SSRes} = \sum_{i=1}^{t}(n_i - 2) = \sum_{i=1}^{t} n_i - 2t = N - 2t.$$

The best estimate of the variance of the experimental units is $\hat{\sigma}^2 = SSRes/(N - 2t)$. The sampling distribution of $(N - 2t)\hat{\sigma}^2/\sigma^2$ is central chi-square with $(N - 2t)$ degrees of freedom. The sampling distribution of the least squares estimator, $\hat{\beta}' = (\hat{\alpha}_1, \hat{\beta}_1, \ldots, \hat{\alpha}_t, \hat{\beta}_t)$ is normal with mean $\beta' = (\alpha_1, \beta_1, \ldots, \alpha_t, \beta_t)$ and variance-covariance matrix $\sigma^2 (X'X)^{-1}$, which can be written as

$$\sigma^2(X'X)^{-1} = \sigma^2 \begin{bmatrix} (X_1'X_1)^{-1} & & 0 \\ \vdots & \ddots & \vdots \\ 0 & \cdots & (X_t'X_t)^{-1} \end{bmatrix} \quad (2.5)$$

where

$$X_i = \begin{bmatrix} 1 & x_{i1} \\ \vdots & \vdots \\ 1 & x_{in_i} \end{bmatrix}$$

2.3 STRATEGY FOR DETERMINING THE FORM OF THE MODEL

The main objective of an analysis of covariance is to compare the t regression lines at several predetermined fixed values of the covariate, X. Depending on the values of the slopes, β_i, there are various strategies one can use to compare the regression lines.

The first question that needs to be answered is, does the mean of y given X depend on the value of X? If the data have been plotted for each of the treatment groups, and there seems to be a linear relationship between the values of y and x, then the question can be subjectively answered. That question can be answered statistically by testing the hypothesis

$$H_{01}: E(y_{ij}|X=x) = \alpha_i \text{ vs. } H_{a1}: E(y_{ij}|X=x) = \alpha_i + \beta_i x \text{ for } i = 1, 2, \ldots, t.$$

This hypothesis is equivalent to testing the hypothesis that the slopes are all zero, i.e.,

$$H_{01}: \beta_1 = \beta_2 = \ldots = \beta_t = 0 \text{ vs. } H_{a1}: (\text{not } H_0). \tag{2.6}$$

The null hypothesis states that none of the treatments' means depend linearly on the value of the covariate, X. The notation $E(y_{ij}|X=x)$ denotes the mean of the distribution of y for a given value of X, X = x.

The principle of conditional error (Milliken and Johnson, 1992) or model comparison method (Draper and Smith, 1981) provides an excellent way of obtaining the desired test statistic. The model restricted by the conditions of the null hypothesis, H_{01}, is

$$y_{ij} = \alpha_i + \varepsilon_{ij} \quad i = 1, 2, \ldots, t, \quad j = 1, 2, \ldots, n_i. \tag{2.7}$$

Model 2.7 is the usual analysis of variance model for the one-way treatment structure in a completely randomized design structure. The residual sum of squares for Model 2.7 is

$$SSRes(H_{01}) = \sum_{i=1}^{t} \sum_{j=1}^{n_i} (y_{ij} - \bar{y}_{i\cdot})^2 \tag{2.8}$$

One-Way Analysis of Covariance

which is based on d.f.$_{SSRes(H_{01})}$ = N − t degrees of freedom (where the mean of the model under H_{01} has t parameters, the intercepts). Using the principle of conditional error, the sum of squares due to deviations from H_{01}, denoted by SSH_{01}, is computed as,

$$SSH_{01} = SSRes(H_{01}) - SSRes, \qquad (2.9)$$

which is based on d.f.$_{SSRes(H_{01})}$ − d.f.$_{SSRes}$ = (N − t) − (N − 2t) = t degrees of freedom. The degrees of freedom associated with SSH_{01} is equal to t since the hypothesis being tested is that the t slope parameters are all equal to zero, i.e., the values of t parameters are specified; thus there are t degrees of freedom associated with the sum of squares. The sampling distribution of SSH_{01}/σ^2 is a noncentral chi-square distribution with t degrees of freedom where the noncentrality parameter is zero if and only if H_{01} is true, i.e., all slopes are equal to zero. A statistic for testing H_{01} vs. H_{a1} is

$$F_{H_{01}} = \frac{SSH_{01}/t}{\hat{\sigma}^2} \qquad (2.10)$$

and, when H_{01} is true, the sampling distribution of is that of a central $F_{H_{01}}$ distribution with t and N − 2t degrees of freedom.

If you fail to reject H_{01}, then you can conclude that the means of the treatments do not depend linearly on the value of the covariate, X. In this case, the next step in the analysis is to use analysis of variance to make comparisons among the treatments' means, i.e., compare the α_i, i = 1, 2, ..., t (as is discussed in Chapter 1, Milliken and Johnson, 1992). Recall you have already determined that the simple linear regression model adequately describes the data. Thus if the slopes are zero, then you conclude the models are of the form

$$y_{ij} = \alpha_i + \varepsilon_{ij}, \quad i = 1, 2, ..., t, \quad j = 1, 2, ..., n_i.$$

If H_{01} is rejected, then you conclude that the mean of y does depend linearly on the value of the covariate X for at least one of the treatments. In this case, the next step in the analysis of covariance is to determine whether or not the means of the treatments depend on the covariate X differently (as represented by unequal slopes which provide nonparallel lines). A test for homogeneity or equality of the slopes answers that question. The appropriate null hypothesis stating the slopes are equal is

$$H_{02}: E(y_{ij}|X = x) = \alpha_i + \beta x \text{ vs. } H_{a2}: E(y_{ij}|X = x) = \alpha_i + \beta_i x \text{ or equivalently} \qquad (2.11)$$

$$H_{02}: \beta_1 = \beta_2 = ... = \beta_t = \beta \text{ vs. } H_{a2} \text{ (not } H_{02}\text{)}$$

where β is unspecified and represents the common slope of the t parallel regression lines.

The model in the form of Model 2.1 that satisfies the conditions of H_{02} is

$$y_{ij} = \alpha_i + \beta x_{ij} + \varepsilon_{ij} \quad i = 1, 2, \ldots, t, \quad j = 1, 2, \ldots, n_i \tag{2.12}$$

which represents t parallel lines each with slope β and intercepts $\alpha_1, \ldots, \alpha_t$. The matrix form of Model 2.12 is

$$\begin{bmatrix} y_{11} \\ \vdots \\ y_{1n_1} \\ y_{21} \\ \vdots \\ y_{2n_2} \\ y_{t1} \\ \vdots \\ y_{tn_t} \end{bmatrix} = \begin{bmatrix} 1 & 0 & \cdots & 0 & x_{11} \\ \vdots & \vdots & & & \vdots \\ 1 & 0 & \cdots & 0 & x_{1n_1} \\ 0 & 1 & \cdots & 0 & x_{21} \\ \vdots & \vdots & & & \vdots \\ 0 & 1 & \cdots & 0 & x_{2n_2} \\ \vdots & \vdots & & & \vdots \\ 0 & 0 & \cdots & 1 & x_{t1} \\ \vdots & & & & \vdots \\ 0 & 0 & \cdots & 1 & x_{tn_t} \end{bmatrix} \begin{bmatrix} \alpha_1 \\ \vdots \\ \alpha_t \\ \beta \end{bmatrix} + \varepsilon. \tag{2.13}$$

The residual sum of squares for Model 2.13 is

$$SSRes(H_{02}) = \sum_{i=1}^{t} \sum_{j=1}^{n_i} \left(y_{ij} - \hat{\alpha}_i - \hat{\beta} x_{ij}\right)^2 \tag{2.14}$$

where $\hat{\alpha}_i = 1, 2, \ldots, t$ and $\hat{\beta}$ denote the least squares estimators of the corresponding parameters from Model 2.13. The residual sum of squares in Model 2.14 is based on d.f.$_{SSRes(H_{02})} = N - t - 1$ degrees of freedom as the mean of Model 2.12 has $t + 1$ parameters, i.e., t intercepts and one slope. Using the principle of conditional error, the sum of squares due to deviations from H_{02} is

$$SSH_{02} = SSRes(H_{02}) - SSRes \tag{2.15}$$

which is based on d.f.$_{SSRes(H_{02})}$ – d.f.$_{SSRes} = t - 1$ degrees of freedom. There are $t - 1$ degrees of freedom associated with SSH_{02} since t parameters (testing equality) are being compared and there are $t - 1$ linearly independent comparisons of the t slopes whose values are specified to be zero. The sampling distribution of SSH_{02}/σ^2 is noncentral chi-square with $t - 1$ degrees of freedom where the noncentrality parameter is zero if and only if H_{02} is true. The statistic used to test H_{02} is

$$F_{H_{02}} = \frac{SSH_{02}/(t-1)}{\hat{\sigma}^2} \tag{2.16}$$

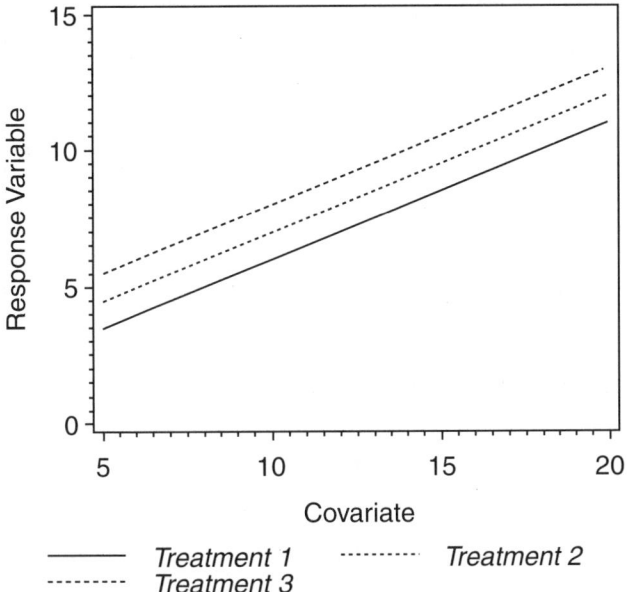

FIGURE 2.1 Graph of parallel lines models: common slopes.

which has a noncentral F sampling distribution with $t - 1$ and $N - 2t$ degrees of freedom. If you fail to reject H_{02}, then conclude that the lines are parallel (equal slopes) and proceed to compare the distances among the parallel regression lines by comparing their intercepts, α_i's (the topic of Section 2.4). Figure 2.1 displays the relationships among the treatment means as a function of the covariate X when the lines are parallel. Since the lines are parallel, i.e., the distance between any two lines is the same for all values of X, a comparison of the intercepts is a comparison of the distances between the lines.

If you reject H_{02}, then conclude that at least two of the regression lines have unequal slopes and hence, the set of lines are not parallel. Figure 2.2 displays a possible relationship among the means of treatments as a linear function of the covariate for the nonparallel lines case. When the lines are not parallel, the distance between two lines depends on the value of X; thus the nonparallel lines case is called covariate by treatment interaction.

2.4 COMPARING THE TREATMENTS OR THE REGRESSION LINES

An appropriate method for comparing the distances among the regression lines depends on the decision you make concerning the slopes of the models. If you reject H_{01} in Model 2.6 and fail to reject H_{02} in Model 2.11, the resulting model is a set of parallel lines (equal slopes as in Figure 2.1). A property of two parallel lines is that they are the same distance apart for every value of X. Thus, the distance between any two lines can be measured by comparing the intercepts of the two lines. When

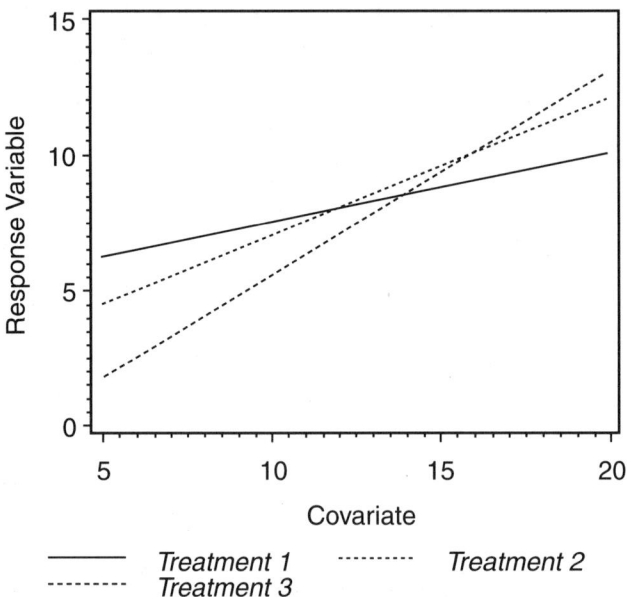

FIGURE 2.2 Graph of nonparallel lines models: unequal slopes.

the lines are parallel, contrasts between the intercepts are used to compare the treatments. When the slopes are unequal, there are two types of comparisons that are of interest, namely, comparing the distances between the various regression lines at several values of the covariate X and comparing specific parameters, such as comparing the slopes or comparing the models evaluated at the same or different values of X.

2.4.1 Equal Slopes Model

At this step in the analysis, you must remember that H_{02} was not rejected; thus the model used to describe the mean of y as a function of the covariate is

$$y_{ij} = \alpha_i + \beta X_{ij} = \varepsilon_{ij}. \tag{2.17}$$

The residual sum of squares for Model 2.17 is $SSRes(H_{02})$, the residual sum of squares to which you will compare other models, which was given in Equation 2.14. The first hypothesis to be tested is that the distances between the lines evaluated at a given value of X, say X_0, are equal,

$$H_{03}: \alpha_1 + \beta X_0 = \alpha_2 + \beta X_0 = \ldots = \alpha_t + \beta X_0 = \alpha + \beta X_0$$

$$\text{vs. } H_{a3}: \left(\text{not } H_{05}\right)$$

One-Way Analysis of Covariance

where α and β are unspecified. An equivalent hypothesis is that the intercepts are equal and is expressed as

$$H_{03}: \alpha_1 = \alpha_2 = \ldots = \alpha_t = \alpha \text{ vs. } H_{a3}: (\text{not } H_{03}) \qquad (2.18)$$

where α is unspecified. The model in Equation 2.17 restricted by the conditions of H_{03} is

$$y_{ij} = \alpha + \beta X_{ij} + \varepsilon_{ij}. \qquad (2.19)$$

Model 2.19 is a single simple linear regression model that is to be fit to the data from all of the treatments, i.e., one model is used to describe all of the data. The residual sum of squares corresponding to Model 2.19 is

$$SSRes(H_{03}) = \sum_{i=1}^{t} \sum_{j=1}^{n_i} \left(y_{ij} - \hat{\alpha} - \hat{\beta} X_{ij} \right)^2 \qquad (2.20)$$

where $\hat{\alpha}$ and $\hat{\beta}$ are least squares estimators of α and β from Model 2.19. Since Model 2.19 consists of two parameters, $SSRes(H_{03})$ is based on d.f.$_{SSRes(H_{03})} = N - 2$ degrees of freedom.

Applying the model comparison method, the sum of squares due to deviation from H_{03}, given that the slopes are equal (H_{02} is true), is

$$SSH_{03} = SSRes(H_{03}) - SSRes(H_{02}) \qquad (2.21)$$

which is based on d.f.$_{SSRes(H_{03})}$ − d.f.$_{SSRes(H_{02})}$ = $t - 1$ degrees of freedom. The appropriate test statistic is

$$F_{H_{03}} = \frac{SSH_{03}/(t-1)}{SSRes(H_{02})/(n-t-1)}. \qquad (2.22)$$

The sampling distribution of $F_{H_{03}}$ is that of a noncentral F distribution with $t - 1$ and $N - t - 1$ degrees of freedom. If H_{03} is not rejected, conclude that all of the data come from a single simple linear regression model with slope β and intercept α, i.e., there are no treatment differences. If H_{03} is rejected, then conclude that the distances between one or more pairs of lines are different from zero. Since the lines are parallel, the distance between any two lines can be compared at any chosen value of X. If the distance between any two lines is compared at $X = 0$, it is a comparison of the difference between two intercepts as $\alpha_i - \alpha_{i'}$. A multiple comparison procedure can be used to compare the distances between pairs of regression lines. An LSD type of multiple comparison procedure can be used for controlling the comparison-wise error rate and a Fisher protected LSD, Bonferroni, or Scheffe type of multiple comparison procedure can be used to control the experiment-wise error rate (see Chapter 3 of Milliken and Johnson, 1992). To use a multiple comparison procedure, you must compute the estimated standard error of the difference between two α's or intercepts as

$$S_{\hat{\alpha}_i - \hat{\alpha}_{i'}} = \sqrt{S_{\hat{\alpha}_i}^2 + S_{\hat{\alpha}_{i'}}^2 - 2\text{Cov}(\hat{\alpha}_i, \hat{\alpha}_{i'})} \qquad (2.23)$$

where $S_{\hat{\alpha}_i}^2$ is the estimated variance of $\hat{\alpha}_i$ and $\text{Cov}(\hat{\alpha}_i, \hat{\alpha}_{i'})$ is the estimated covariance between $\hat{\alpha}_i$ and $\hat{\alpha}_{i'}$. If there are specific planned comparisons (such as linear or quadratic effects for treatments with quantitative levels or comparing a set of treatment to a control, etc.) between the treatments, those comparisons would be made by constructing the necessary contrasts between the intercepts.

For example, if you are interested in the linear combination $\theta = \sum_{i=1}^{t} c_i \alpha_i$, the estimate is $\hat{\theta} = \sum_{i=1}^{t} c_i \hat{\alpha}_i$ and the estimated standard error of this estimate is

$$S_{\hat{\theta}} = \sqrt{\sum_{i=1}^{t} c_i^2 S_{\hat{\alpha}_i}^2 + \sum_{i \neq i'} c_i c_{i'} \text{Cov}(\hat{\alpha}_i, \hat{\alpha}_{i'})}$$

The statistic to test H_0: $\theta = 0$ vs. H_a: $\theta \neq 0$ is $t_{\hat{\theta}} = \hat{\theta}/S_{\hat{\theta}}$ which is distributed as a Student t distribution with $N - t - 1$ degrees of freedom. A $(1 - \alpha)$ 100% confidence interval about

$$\theta = \sum_{i=1}^{t} c_i \alpha_i \text{ is } \hat{\theta} \pm t_{\frac{\alpha}{2}, N-t-1} S_{\hat{\theta}}$$

If you have four equally spaced quantitative levels of the treatment, the linear contrast of the four levels is $\theta = 3\alpha_1 + 1\alpha_2 - 1\alpha_3 - 3\alpha_4$.

When analysis of covariance was first developed, it was mainly used to adjust the mean of y for a selected value of the covariate. The value usually selected was the mean of the covariate from all t treatments. Thus the term adjusted means was defined as the mean of y evaluated at $X = \bar{x}..$, where $\bar{x}..$ is the mean value of all the x_{ij}'s. The estimators of the means of the treatments evaluated at $X = \bar{x}..$, called the adjusted means, are

$$\hat{\mu}_{Y_i|\beta X = \beta \bar{x}} = \hat{\alpha}_i + \hat{\beta}\bar{x}.. \quad i = 1, 2, \ldots, t. \qquad (2.24)$$

where $\hat{\alpha}_i$ and $\hat{\beta}$ are least squares estimators of α_i and β from Model 2.17. The covariance matrix of the adjusted means can be constructed from the elements of the covariance matrix of $\hat{\alpha}_i, \ldots, \hat{\alpha}_t$ and $\hat{\beta}$.

One-Way Analysis of Covariance

The standard errors of the adjusted means are computed as

$$S_{\hat{\mu}_{Y_i|\beta X=\beta\bar{x}}} = \left[S_{\hat{\alpha}_i}^2 + \bar{X}^2 S_{\hat{\beta}}^2 + 2\bar{X}\operatorname{cov}(\hat{\alpha}_i,\hat{\beta})\right]^{1/2}, \quad i = 1, 2, \ldots, t.$$

One hypothesis of interest is that the expected value of the adjusted means are equal. This hypothesis can be expressed as

$$H_{04}: \mu_{Y_1|\beta X=\beta\bar{x}} = \ldots = \mu_{Y_t|\beta X=\beta\bar{x}} \quad \text{vs.} \quad H_a(\text{not } H_{04}).$$

However, since the lines are parallel, the difference between two adjusted means is the difference between intercepts as $\mu_{Y_1|\beta x = \beta\bar{x}} - \mu_{Y_2|\beta X = \beta\bar{x}} = \alpha_1 - \alpha_2$; thus H_{04} is equivalent to H_{03}.

Preplanned treatment comparisons and multiple comparison procedures can be carried out to compare the adjusted means by computing the standard error of the difference between pairs of adjusted means. Since the difference of two such adjusted means is $\hat{\alpha}_i - \hat{\alpha}_j$, the estimated standard error in Equation 2.23 can be used with any selected multiple comparison procedure. Contrasts among adjusted means, which are also contrasts among the intercepts, measuring linear, quadratic, etc. trends, should be used when appropriate.

2.4.2 Unequal Slopes Model-Covariate by Treatment Interaction

When you reject H_{02}, then conclude that the nonparallel lines Model 2.1 is necessary to adequately describe the data. The graph of such a possible situation was given in Figure 2.2. Since the lines are not parallel, the distance between any two lines depends on which value of the covariate is selected (see Figure 2.3). This is called covariate by treatment interaction. In the nonparallel lines case, a comparison of the intercepts is only a comparison of the regression lines at $X = 0$. That will generally be a meaningful comparison only when $X = 0$ is included in or is close to the range of X values in the experiment. The equal intercept hypothesis given the slopes are unequal is expressed as

$H_{05}: E(y_{ij}|X = x) = \alpha + \beta_i x$ vs. $H_{a5}: E(y_{ij}|X = x) = \alpha_i + \beta_i x$ or, equivalently,
$H_{05}: \alpha_1 = \alpha_2 = \ldots = \alpha_t = \alpha$ given $\beta_i \neq \beta'_i$ vs. $H_{a5}(\text{not } H_0)$

The model comparison method or principle of conditional error is used to compute the statistic for testing this hypothesis. Model 2.1 restricted by the conditions of H_{05} is

$$y_{ij} = \alpha + \beta_i X_{ij} + \varepsilon_{ij} \tag{2.25}$$

and the corresponding residual sum of squares is

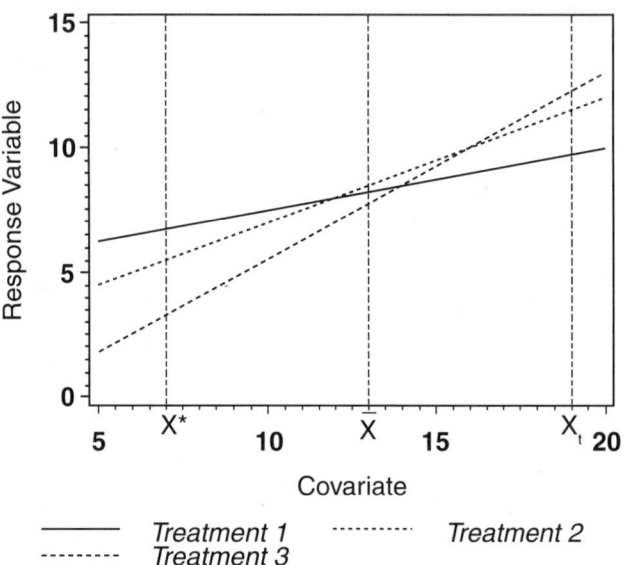

FIGURE 2.3 Graph demonstrating that for nonparallel lines model, the comparisons between the treatments depend on the value of X.

$$SSRes(H_{05}) = \sum_{i=1}^{t}\sum_{j=1}^{n_i}\left(y_{ij} - \hat{\alpha} - \hat{\beta}_i X_{ij}\right)^2 \quad (2.26)$$

which is based on d.f.$_{SSRes(H_{05})} = N - t - 1$ degrees of freedom. The values of $\hat{\alpha}$ and $\hat{\beta}_i$ in Equation 2.26 are the least squares estimators of the parameters of Model 2.25. Using the principle of conditional error, the sum of squares due to deviations from H_{05} is computed as

$$SSH_{05} = SSRes(H_{05}) - SSRes,$$

which is based on d.f.$_{SSRes(H_{05})}$ − d.f.$_{SSRes}$ = $t - 1$ degrees of freedom. The statistic to test H_{05} is

$$F_{H_{05}} = \frac{SSH_{05}/(t-1)}{SSRes/(N-2t)} \quad (2.27)$$

which is distributed as a noncentral F distribution based on $t - 1$ and $N - 2t$ degrees of freedom. The conclusion you make at X = 0 may be different from that you make at X = $\bar{x}_{..}$ or X = X_0, where X_0 is some other preselected or fixed value of the covariate. Suppose the you want to compare the distances between the regression lines at a selected value of X, say X = X_0. The hypothesis to be tested is

One-Way Analysis of Covariance

$$H_{06}: \mu_{Y_1|\beta_1 X=\beta_1 X_0} = \mu_{Y_2|\beta_2 X=\beta_2 X_0} = \ldots \mu_{Y_t|\beta_t X=\beta_t X_0} = \mu_{X_0}, \quad (2.28)$$

where $\mu_{Y_i|\beta_i X = \beta_i X_0} = \alpha_i + \beta_i X_0$ and μ_{X_0} is the unspecified common parameter. The model in Equation 2.1 can be equivalently expressed as

$$y_{ij} = \alpha_i + \beta_i X_0 - \beta_i X_0 + \beta_i X_{ij} + \varepsilon_{ij}$$
$$= \mu_{Y_i|\beta_i X=\beta_i X_0} + \beta_i (X_{ij} - X_0) + \varepsilon_{ij}. \quad (2.29)$$

The model restricted by H_{06} is

$$y_{ij} = \mu_{X_0} + \beta_i (X_{ij} - X_0) + \varepsilon_{ij} \quad (2.30)$$

which is a set of linear regression models with a common intercept (μ_{X_0}) and t slopes. The residual sum of squares corresponding to Model 2.30 is

$$SSRes(H_{06}) = \sum_{i=1}^{t} \sum_{j=1}^{n_i} \left(y_{ij} - \hat{\mu}_{X_0} - \beta_i (X_{ij} - X_0) \right)^2, \quad (2.31)$$

which is based on d.f.$_{SSRes(H_{06})}$ = N – t – 1 degrees of freedom. Using the principle of conditional error, the sum of squares due to deviations from H_{06} is computed as $SSH_{06} = SSRes(H_{06}) - SSRes$, which is based on d.f.$_{SSRes(H_{06})}$ – d.f.$_{SSRes}$ = t – 1 degrees of freedom. The resulting statistic to test H_{06},

$$F_{H_{06}} = \frac{SH_{06}/(t-1)}{SSRes/(N-2t)} \quad (2.32)$$

which is distributed as a noncentral F distribution based on t – 1 and N – 2t degrees of freedom. It is important for you to make comparisons among the regression lines at several different values of the covariate. The usual comparison of adjusted means, i.e., at $X = \bar{x}..$ is only one of many comparisons that are probably of interest. Figure 2.3 shows three possible values of X (covariate) at which you might make comparisons. The statistics for testing H_{06} with $X_0 = X_t$, $X_0 = \bar{x}..$ and $X_0 = X^*$ can be computed by using Model 2.30 and the sum of squares residual in Equation 2.31 for each of the three values of X. If you reject a hypothesis corresponding to a selected value of X_0, then a multiple comparison procedure could be used to compare the distances between pairs of regression lines or to make preplanned treatment comparisons among the regression lines at X_0.

The difference between two simple linear regression lines at $X = X_0$, is $\hat{\mu}_{y_i|\beta_i X=\beta_i X_0} - \hat{\mu}_{Y_{i'}|\beta_{i'} X = \beta_{i'} X_0}$, and the estimated standard error of the difference is

$$S_{\hat{\mu}_{Y_i|\beta_iX=\beta_iX_0}-\hat{\mu}_{Y_{i'}|\beta_{i'}X=\beta_{i'}X_0}} = \sqrt{S^2_{\hat{\mu}_{Y_i|\beta_iX=\beta_iX_0}} + S^2_{\hat{\mu}_{Y_{i'}|\beta_{i'}X=\beta_{i'}X_0}}} \qquad (2.33)$$

where the standard errors $S\hat{\mu}_{Y_i|\beta_iX = \beta_iX_0}$ can be obtained from the standard errors of the intercept parameters in Model 2.29 or can be computed as from the linear combination of the parameter estimates $\hat{\alpha}_i + \hat{\beta}_i X_0$ to obtain

$$S_{\hat{\mu}_{Y_i|\beta_iX=\beta_iX_0}} = \sqrt{S^2_{\hat{\alpha}_i} + X^2_0 S^2_{\hat{\beta}_i} + 2X_0 \operatorname{cov}(\hat{\alpha}_i, \hat{\beta}_i)}, \quad i = 1, 2, \ldots, t.$$

These estimated standard errors are computed with the assumption that the two models have no common parameters so the covariance between the two adjusted means is zero. Again, for preplanned comparisons, LSD, Bonferroni, or Scheffe types of multiple comparison procedures can be used to help interpret the results of comparing the regression lines at each selected value of the covariate. For most experiments, comparisons should be made for at least three values of X, one in the lower range, one in the middle range, and one in the upper range of the X's obtained for the experiment. There are no set rules for selecting the values of X at which to compare the models, but, depending on your objectives, the lines could be compared at:

1. The mean of the X values, the mean of the X values minus one standard deviation of the X values and the mean of the X values, plus one standard deviation of the X values
2. The median of the X values, the 25th percentile of the X values, and the 75th percentile of the X values
3. The median of the X values, the γ percentile of the X values, and the $1 - \gamma$ percentile of the X values for some choice of γ (such as .01, .05, .10, or .20)

You might be interested in determining which treatment mean responds the most to a change in the value of the covariate. In this case an LSD approach can be used to make size α comparison-wise tests about the linear combinations of the β_i's. Alternatively, you could also use a Fisher's protected LSD or Bonferroni or Scheffe-type approach to control the experiment-wise error rate. In any case, the standard error of the difference between two slopes is

$$S_{\hat{\beta}_i-\hat{\beta}_{i'}} = \sqrt{S^2_{\hat{\beta}_i} + S^2_{\hat{\beta}_{i'}}}$$

where $S_{\hat{\beta}_i}$ denotes the standard error associated with $\hat{\beta}_i$ and it is assumed the covariance between the two slope parameters is zero (which is the case here since the two models do not have common parameters.) The degrees of freedom for the appropriate percentage point corresponding to the selected test is $N - 2t$. Preplanned treatment comparisons can be made by comparing the slopes of the various models. Section 2.7 shows how to carry out the computations via the procedures of the SAS® system and JMP®.

One-Way Analysis of Covariance

2.5 CONFIDENCE BANDS ABOUT THE DIFFERENCE OF TWO TREATMENTS

When the slopes are unequal, it is often useful to determine the region of the covariate where two treatments produce significantly different responses. A confidence band can be constructed about the differences of models at several values of X in the region of the covariate. Then the region of the covariate where the confidence band does not contain zero is the region where the treatments produce significantly different responses. A Scheffe-type confidence statement should be used to provide experiment-wise error rate protection for each pair of treatments. The estimated difference between the two regression lines for Treatments 1 and 2 at $X = X_0$ is

$$\hat{\mu}_{Y_1|\beta_1 X = \beta_1 X_0} - \hat{\mu}_{Y_2|\beta_2 X = \beta_2 X_0} = \hat{\alpha}_1 + \hat{\beta}_1 X_0 - \hat{\alpha}_2 - \hat{\beta}_2 X_0$$

which has estimated standard error

$$S_{1-2|\beta_i X_0} = \sqrt{S^2_{\hat{\mu}_{Y_1|\beta_1 X = \beta_1 X_0}} + S^2_{\hat{\mu}_{Y_2|\beta_2 X = \beta_2 X_0}}}$$

where $S^2_{\hat{\mu}_{Y_1|\beta_1 X = \beta_1 X_0}} = S^2_{\hat{\alpha}_i} + X_0^2 S^2_{\hat{\beta}_i} + 2X_0 \text{Cov}(\hat{\alpha}_i, \hat{\beta}_i)$. The difference between the two models is

$$\hat{\mu}_{Y_1|\beta_1 X = \beta_1 X_0} - \hat{\mu}_{Y_2|\beta_2 X = \beta_2 X_0} = \alpha_1 + \beta_1 X_0 - \alpha_2 - \beta_2 X_0$$

$$= (\alpha_1 - \alpha_2) + (\beta_1 - \beta_2) X_0$$

which is a straight line with two parameters, i.e., the intercept is $\alpha_1 - \alpha_2$ and the slope is $\beta_1 - \beta_2$. Thus to construct a confidence band about the difference of the two models at $X = X_0$ for a range of X values based on Scheffe percentage points, use

$$\hat{\mu}_{Y_1|\beta_1 X = \beta_1 X_0} - \hat{\mu}_{Y_2|\beta_2 X = \beta_2 X_0} \pm \sqrt{2 F_{(2, N-2t)}} \, S_{1-2|\beta_i X_0}$$

where the number 2 corresponds to the two parameters in the difference of the two models. An example of the construction and use of the confidence band about the difference of two regression lines is presented in Section 3.3

2.6 SUMMARY OF STRATEGIES

Sections 2.3 and 2.4 describe the strategies for determining the form of the model and the resulting analyses. Those strategies are summarized in Table 2.1 which lists the paths for model determination. The first step is to make sure that a straight line is an adequate model to describe the data for each of the treatments. If the straight line does

TABLE 2.1
Strategy for Determining the Form of the Analysis of Covariance Model Involving One Covariate Assuming a Simple Linear Regression Model Will Describe Each Treatment Data

a. Test the hypothesis that the slopes are zero:
   i. If fail to reject, compare the treatments via analysis of variance.
   ii. If reject go to (b).
b. Test the hypothesis that the slopes are equal:
   i. If fail to reject, use a parallel lines model and compare the treatments by comparing the intercepts or adjusted means (LSMEANS).
   ii. If reject go to (c).
c. Use the unequal slope model and
   i. Compare the slopes of the treatments to see if treatments can be grouped into groups with equal slopes.
   ii. Compare the models of at least three values of the covariate, low, middle, and high value.
   iii. Construct confidence bands about the difference of selected pairs of models.

not fit the data, then the analysis must not continue until an adequate model is obtained. The strategies described are not all encompassing as there are possible exceptions.

There are at least two possible exceptions to the strategy. First, it is possible to reject the equal slopes hypothesis and fail to reject the slopes equal to zero hypothesis when there are both positive and negative slopes (see Section 3.4). In this case, use the nonparallel lines model. Second, it is possible to fail to reject the slopes equal to zero hypothesis when in fact the common slope of a parallel lines model is significantly different from zero. Many experiments have a very few observations per treatment. In that case, there is not enough information from the individual treatments to say the individual slopes are different from zero, but the combining the information into a common slope does detect the linear relationship (see Section 3.6). In this case, use the common slope or parallel lines model. There apparently can be other exceptions that can be constructed, but at this moment they are not evident.

2.7 ANALYSIS OF COVARIANCE COMPUTATIONS VIA THE SAS® SYSTEM

The SAS® system can be used to compute the various estimators and tests of hypotheses discussed in the previous sections. The SAS® system statements required for each part of the analysis are presented in this section. Section 2.7.1 describes the syntax needed for using PROC GLM or PROC MIXED (through Version 8). Section 2.7.2 describes the process of using JMP® Version 4. Detailed examples are discussed in Chapter 3.

2.7.1 USING PROC GLM AND PROC MIXED

All the models will be fit assuming that the data were read in by the following statements:

One-Way Analysis of Covariance

```
DATA ONECOV;INPUT TRT Y X;
DATALINES;
```

The required SAS System statements needed to fit Model 2.1 using PROC GLM (same for PROC MIXED) are

```
PROC GLM; CLASSES TRT;
MODEL Y = TRT X*TRT/NOINT SOLUTION;
```

The term TRT with the no intercept (NOINT) option generates the part of the design matrix corresponding to the intercepts and enables one to obtain the estimators of the intercepts. The term X*TRT generates the part of the design matrix corresponding to the slopes. The SOLUTION option is used so that the estimators and their standard errors are printed. (PROC GLM and PROC MIXED do not automatically provide the estimators when there is a CLASS variable unless SOLUTION is specified.) The sum of squares corresponding to ERROR is SSRes of Equation 2.4 and the MEAN SQUARE ERROR is $\hat{\sigma}^2$, the estimate of the sampling variance. For PROC MIXED, the value of $\hat{\sigma}^2$ is obtained from the Residual line from the covariance parameter estimates.

The type III sum of squares corresponding to X*TRT tests H_{01} of Model 2.6, i.e., $\beta_1 = \beta_2 = \cdots = \beta_t = 0$ given that the unequal intercepts are in the model. The Type III sum of squares corresponding to TRT tests H_{05}, i.e., $\alpha_1 = \alpha_2 = \ldots = \alpha_t = 0$ given that the unequal slopes are in the model. The intercepts equal to zero hypothesis is equivalent to testing that all of the treatment regression lines are equal to zero at X = 0. This hypothesis is often not of interest, but a test is available in case you have a situation where a zero intercept hypothesis is interpretable.

Next, to test H_{02} of Model 2.11, the required SAS® system statements are

```
PROC GLM; CLASSES TRT;
MODEL Y = TRT X X*TRT/SOLUTION;
```

The Type III sum of squares corresponding to X*TRT tests H_{02}. The Type III sum of squares corresponding to X tests if the average value of the slopes is zero and the Type III sums of squares corresponding to TRT tests H_{05}. By including X and/or removing the NOINT option, the model is singular and the provided least squares solutions is not directly interpretable. The least squares solution satisfies the set-to-zero restrictions (see Chapter 6 of Milliken and Johnson, 1984). If one uses the model statement Y = TRT X*TRT X, where X*TRT is listed before X, the Type I sum of squares corresponding to X*TRT tests H_{01} while the type III sum of squares tests H_{02}. A list of Type I and III estimable functions can be obtained and used to verify the hypothesis being tested by each sum of squares.

If one fails to reject H_{02}, the parallel lines or equal slope model of Equation 2.12 should be fit to the data. The appropriate SAS® system statements are

```
PROC GLM; CLASS TRT;
MODEL Y = TRT X/SOLUTION; .
```

The Type III sum of squares corresponding to TRT is SSH_{03} of Equation 2.21, and the resulting F ratio tests that the distances between the lines are zero given that the parallel line model is adequate to describe the data.

Estimates of the mean of y given $X = \bar{x}$. for each treatment, which are often called adjusted means, can be obtained by including the statement

```
LSMEANS TRT/STDERR PDIFF;
```

after the MODEL statement. The estimated adjusted means $\hat{\mu}_{y_i|\beta_i x} = \hat{\alpha}_i + \hat{\beta}_i \bar{x}$ are the estimates of the treatment means at $X = \bar{x}$ and are called least squares means. These adjusted means are predicted values computed from the respective regression models evaluated at $X = \bar{x}$. The option STDERR provides the corresponding standard errors of the adjusted means. The PDIFF option provides significance levels for t-tests of H_0: $\mu_{Y_i|\beta_i X = \beta_i \bar{x}} = \mu_{Y_{i'}|\beta_{i'} X = \beta_{i'} \bar{x}}$ for each pair of adjusted means. (The TDIFF option can be included and it provides the values of the t-tests from which the PDIFF values are obtained.) A comparison of adjusted means is also a comparison of the α_i's for the parallel lines model. The significance probabilities can be used to construct a LSD or a Bonferroni multiple comparison procedure for comparing the distances between pairs of lines.

Estimates of the mean of y given $X = A$ for each treatment, which are also called adjusted means, can be obtained by including the statement

```
LSMEANS TRT/STDERR PDIFF AT X = A;

(where A is a numerical constant)
```

after the MODEL statement. This provides the adjusted means $\hat{\mu}_{Y_i|\beta_i X = \beta_i A} = \hat{\alpha}_i + \hat{\beta}_i A$, which are the predicted values from the regression lines at $X = A$ or the estimates of the treatment means at $X = A$. The PDIFF option provides significance levels for t-tests of H_0: $\mu_{Y_i|\beta_i X = \beta_i A} = \mu_{Y_{i'}|\beta_{i'} X = \beta_{i'} A}$ for each pair of adjusted means.

Any comparisons among parameters can be made by using the ESTIMATE or CONTRAST statement in the GLM procedure. There are two situations where such statements are needed.

First, if the conclusion is that the slopes are not equal, then one can apply a multiple comparison procedure in order to compare some or all pairs of slopes. This is easily done by including an ESTIMATE statement following the MODEL statement for each comparison of interest. For example, if there are three treatments and it is of interest to compare all pairs of slopes, then the following statements would be used:

```
PROC GLM; CLASSES TRT;

MODEL Y = TRT X*TRT/NOINT SOLUTION;

ESTIMATE 'B1-B2' X*TRT 1 -1 0;

ESTIMATE 'B1-B3' X*TRT 1 0 -1;

ESTIMATE 'B2-B3' X*TRT 0 1 -1;
```

Each ESTIMATE statement produces an estimate of the linear combination of parameters, a computed t-value, and it's significance level using the residual mean square as the estimate of σ^2. The significance levels obtained from these comparisons can be used to construct a LSD or Bonferroni multiple comparison procedure.

One-Way Analysis of Covariance

For the unequal slope model, adjusted means need to be obtained at $X = \bar{x}$ and at other specified values of X. Such adjusted means can be used to make comparisons between the treatments. The SAS® system code for using ESTIMATE statements to obtain adjusted means at, say, $X = 7.3$ is

```
ESTIMATE  'T1 AT 7.3'  TRT 1 0 0  X*TRT 7.3 0 0;
ESTIMATE  'T3 AT 7.3'  TRT 0 1 0  X*TRT 0 7.3 0;
ESTIMATE  'T3 AT 7.3'  TRT 0 0 1  X*TRT 0 0 7.3;.
```

Contrasts among the adjusted means at a predetermined value of X, say $X = 7.3$, can be obtained by subtracting the respective estimate statement values from the adjusted mean estimate statements as

```
ESTIMATE  'T1 - T2 AT 7.3'  TRT 1 -1 0  X*TRT 7.3 -7.3 0;
ESTIMATE  'T1 - 2T2 + T3 AT 7.3'  TRT 1 -2 1  X*TRT
           7.3 -14.6 7.3;.
```

The estimates of the treatments means at $X = 7.3$ and comparisons of those means can also be achieved by using

```
LSMEANS TRT/PDIFF At X = 7.3;
```

The SAS® system can be used to construct confidence bands about the difference of two models. Since it is unlikely that the values of the covariates will provide a uniform coverage of the covariate region, one must add to the data set additional observations for each treatment corresponding to the values of the covariate at which the confidence intervals are to be constructed (y is assigned a missing value for these observations). The following SAS® system code generates a set of values to be added to the data set; fits the three models and constructs the confidence band about the difference between the models of Treatments 1 and 2.

First read in the data set where Y is the response variable and X is the covariate, using:

```
DATA RAW; INPUT TRT Y X;
DATALINES;
(THE DATA)
```

For each treatment, generate a grid over the values of X where the confidence intervals are to be constructed. For this example, the range of X is from 1 to 10. Set the value of Y to missing.

```
DATA GENERATE; Y = .;
   DO TRT = 1 TO 3;
      DO X = 1 TO 10;
         OUTPUT;
      END;
   END;
```

Combine the two data sets together, by:

```
DATA ALL; SET RAW GENERATE;
```

Fit the model to the combined data set where predicted values and their standard errors are computed and output data set is generated (VALUE). The ODS statements would be used to construct the output data set for PROC MIXED. Predicted values will be computed for each observation in RAW as well for each of the grid points in GENERATE.

```
PROC GLM; CLASS TRT;
MODEL Y = TRT X*TRT/SOLUTION;
OUTPUT OUT = VALUE P = PY
   STDP = STD;
```

Next construct two data sets, one with the predicted values for Treatment 1 (ONE) and one for the predicted values for Treatment 2 (TWO). New variables are constructed where PY1 and STD1 are the predicted values and estimated standard errors for Treatment 1 and PY2 and STD2 are the predicted values and estimated standard errors for Treatment 2. Sort the two data sets by the value of X and then merge the two data sets by the value of X.

```
DATA ONE; SET VALUE; IF TRT = 1;
PY1 = PY; STD1 = STD;
PROC SORT; BY X;
DATA TWO; SET VALUE; IF TRT = 2;
PY2 = PY; STD2 = STD;
PROC SORT; BY X;
```

The merged data set is used to compute the quantities needed for constructing the confidence intervals at each value of X. The value of Q is the square root of $2*F_{\alpha,2,27}$ where 27 is the pooled degrees of freedom for error accumulated from all three models. The value of Q corresponds to a Scheffe percentage point for simultaneous confidence intervals. DIF is the difference between predicted values and STEDIF is the estimated standard error of the difference. LOW and HIGH are the upper and lower confidence limits about the difference of the two models at the respective value of X. The plot statements provide a line printer plot, but PROC GPLOT can be used to provide an excellent looking graphic.

```
DATA ONETWO; MERGE ONE TWO; BY X;
Q = SQRT(2*FINV(.05,2,27));
*  compute the ci's;
DIF = PY1 - PY2;
```

```
STEDIF = SQRT(STD1**2 + STD2**2);
LOW = DIFF - Q*STEDIF;
HIGH = DIFF + Q*STEDIF;
PROC PLOT; PLOT
  DIFF*X = '*' LOW*X = '-' HIGH*X = '-'
  /OVERLAY;
```

The analysis and the SAS® System computations described in this section are demonstrated in detail by examples in Chapter 3.

2.7.2 Using JMP®

The process of using JMP® is through a series of point and click movements, although there is the possibility of generating a script of the process and then replaying the script for new variables. The data is in the form of a spread sheet as shown in Figure 2.4. The variables are denoted as nominal for the levels of the treatments and continuous for time and cctime. In this case, the levels of treatment correspond to types of chocolate candy, time is the time it takes a person to dissolve a piece of candy in their mouth, and cctime is the time it took the person to dissolve a chocolate chip. The response variable is time and the covariate is cctime. The Fit Model option from the Analyze menu is selected to provide the selections to fit the analysis of covariance model to the data set. Figure 2.5 displays the model specification screen.

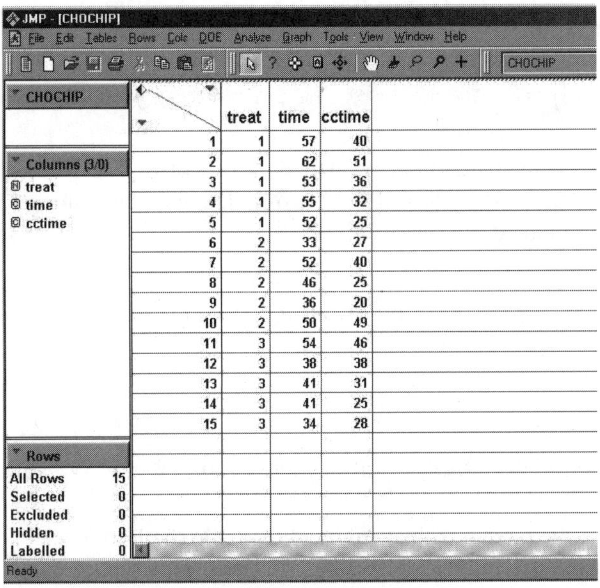

	treat	time	cctime
1	1	57	40
2	1	62	51
3	1	53	36
4	1	55	32
5	1	52	25
6	2	33	27
7	2	52	40
8	2	46	25
9	2	36	20
10	2	50	49
11	3	54	46
12	3	38	38
13	3	41	31
14	3	41	25
15	3	34	28

FIGURE 2.4 Data screen for JMP®.

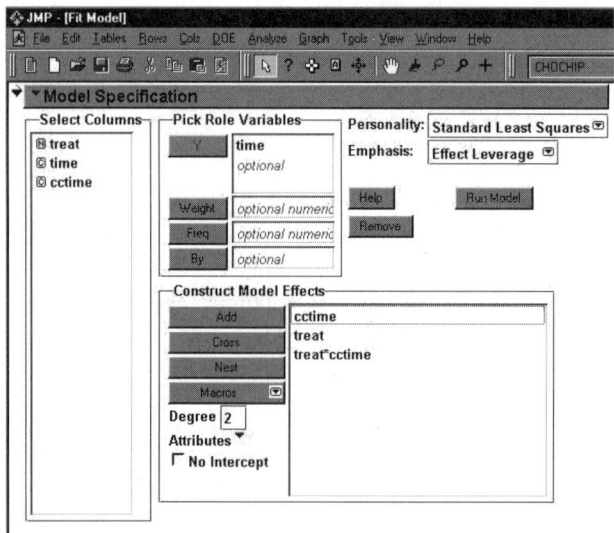

FIGURE 2.5 Model Specification screen to fit unequal slopes model.

Select the arrow in the Model Specification area and deselect the Polynomial Centering option. Highlight the variable time and click on the "Y" button to declare time as the response variable. To construct the model, highlight cctime and click the Add button, highlight treat and click the Add button and highlight treat in the Selected Columns area, highlight cctime in the Construct Model Effects area and click Cross to include cctime, treat, and cctime*treat in the model.

The constructed model fits a full rank model using the sum-to-zero restrictions as

$$y_{ij} = \theta + \tau_i + \phi x_{ij} + \delta_i x_{ij} + \varepsilon_{ij}, \quad i=1, \ldots, t, \; j=1, \ldots, n_i$$

where $\theta = (\bar{\alpha})_\bullet$, $\alpha_i - (\bar{\alpha})_\bullet$, $\phi = (\bar{\beta})_\bullet$, and $\delta_i = \beta_i - (\bar{\beta})_\bullet$. The results provided are estimates of θ, τ_1, ..., τ_{t-1}, ϕ, and δ_1, ..., δ_{t-1}. The estimates of the parameters of Model 2.1 are obtained by

$$\hat{\alpha}_1 = \hat{\theta} + \hat{\tau}_1, \; \hat{\alpha}_2 = \hat{\theta} + \hat{\tau}_2, \ldots, \hat{\theta}_{t-1} = \hat{\theta} - \hat{\tau}_1 - \hat{\tau}_2 - \ldots - \hat{\tau}_{t-1}$$

and

$$\hat{\beta}_1 = \hat{\phi} + \hat{\delta}_1, \; \hat{\beta}_2 = \hat{\phi} + \hat{\delta}_2, \ldots, \hat{\beta}_{t-1} = \hat{\phi} + \hat{\delta}_{t-1}, \; \hat{\beta}_t = \hat{\phi} - \hat{\delta}_1 - \hat{\delta}_2 - \ldots - \hat{\delta}_{t-1}$$

The analysis of the model is obtained by clicking on the Run Model button, and the results are displayed in Figure 2.6. The analysis of variance table contains the estimate of σ^2 denoted as the Mean Square Error. The effect tests provide tests of H_{05} for source Treat, H_{02} for source Treat*Cctime and H_0: $\theta = \bar{\beta}_\bullet = 0$ for source

One-Way Analysis of Covariance

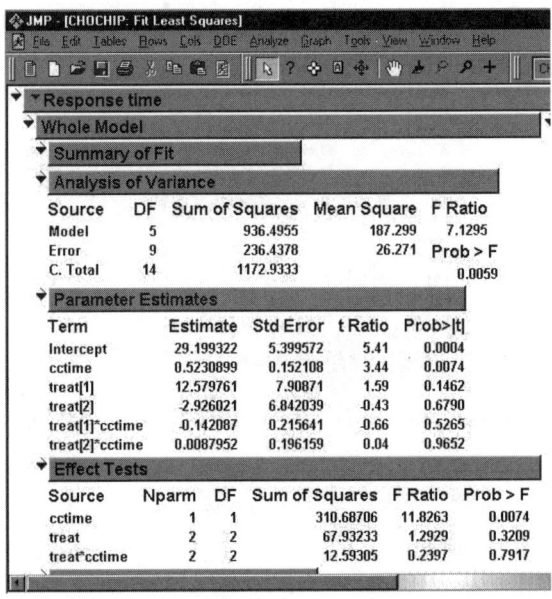

FIGURE 2.6 Estimates of the parameters and analysis of variance table for the unequal slopes model.

Cctime. The parameter estimates part provide estimates of the parameters by using the θ, τ_1, ..., τ_{t-1}, ϕ, and δ_1, ..., δ_{t-1}, corresponding to Intercept, treat[1], treat[2], cctime, treat[1]*cctime, and treat[2]*cctime, respectively.

Least squares means are provided which are estimates of the regression models at $\bar{X}_{..}$ and are computed by

$$\text{LSMEAN}(\text{Treat } i) = (\hat{\theta} + \hat{\tau}_i) + (\hat{\phi} + \hat{\delta}_i)\bar{X}_{..}, \quad i = 1, 2, ..., t-1$$

and

$$\text{LSMEAN}(\text{Treat } t) = (\hat{\theta} - \hat{\tau}_i - ... - \hat{\tau}_{t-1}) + (\hat{\phi} + \hat{\delta}_i - ... - \hat{\delta}_{t-1})\bar{X}_{..}$$

which are LSMEAN(Treat i) = $\hat{\alpha}_i + \hat{\beta}_i \bar{X}_{..}$, $i = 1, 2, ..., t$. The least squares means are displayed in Figure 2.7, where the estimated standard error and the unadjusted or raw means are also exhibited. Included in Figure 2.7 is the Contrast option that can be used to construct estimates and tests of hypotheses about contrasts of the $\alpha_i + \beta_i \bar{X}_{..}$, $i = 1, 2, ..., t$.

An option available with the Parameter Estimates is the Custom Test. The Custom Test allows one to estimate and construct a test of hypothesis about any function of the parameters of the model. Figure 2.8 contains the Custom Test window filled out to obtain estimates for the intercepts α_1, α_2, and α_3 and relationships

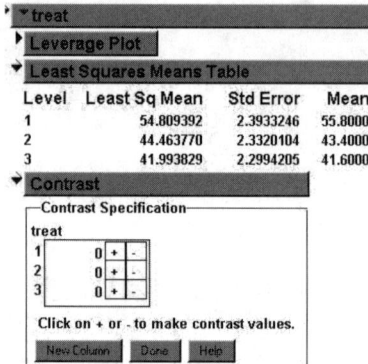

FIGURE 2.7 Least squares means and contrast window for unequal slopes model.

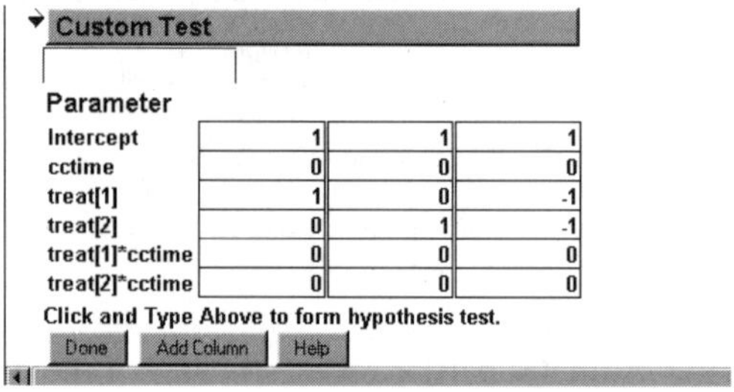

FIGURE 2.8 Custom Test window to obtain estimates of the model intercepts for the unequal slopes model.

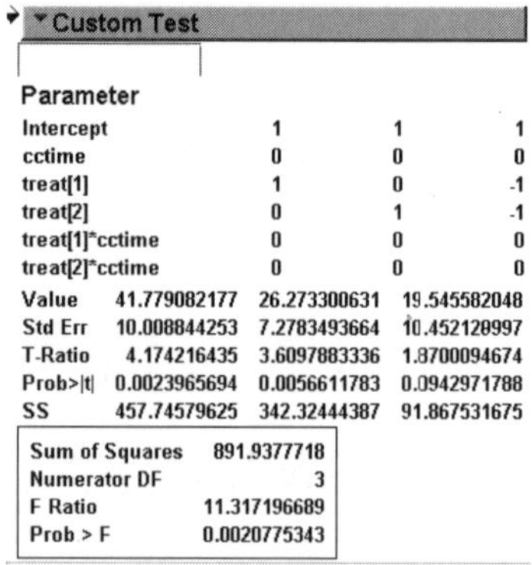

FIGURE 2.9 Estimates of the model intercepts for unequal slopes model with test of the intercepts equal to zero.

between θ, τ_1, and τ_2 and Figure 2.9 displays the estimates, estimated standard errors, and t statistics for testing H_0: $\alpha_i = 0$. The sum of squares and corresponding F ratio. The results of the Custom Test for the intercepts are in Figure 2.9 which provides a test of the hypothesis H_0: $\alpha_1 = \alpha_2 = \alpha_3 = 0$. Similarly, Figures 2.10 and 2.11 contain

Custom Test

Parameter			
Intercept	0	0	0
cctime	1	1	1
treat[1]	0	0	0
treat[2]	0	0	0
treat[1]*cctime	1	0	-1
treat[2]*cctime	0	1	-1

Click and Type Above to form hypothesis test.

Done | Add Column | Help

FIGURE 2.10 Custom Test window to obtain estimates of the model slopes for the unequal slopes model.

Custom Test

Parameter			
Intercept	0	0	0
cctime	1	1	1
treat[1]	0	0	0
treat[2]	0	0	0
treat[1]*cctime	1	0	-1
treat[2]*cctime	0	1	-1
Value	0.3810032017	0.5318850736	0.6563814867
Std Err	0.2647508893	0.2145335584	0.3035026295
T-Ratio	1.439100744	2.479262813	2.1626879735
Prob>|t|	0.1839781135	0.0350337886	0.0588095548
SS	54.407257204	161.48030834	122.87461431

Sum of Squares	338.76217985
Numerator DF	3
F Ratio	4.2983247726
Prob > F	0.0385340965

FIGURE 2.11 Estimates of the model slopes for unequal slopes model with test of the slopes equal to zero.

the results from using the Custom Test window to provide estimates of the slopes of the models β_1, β_2, β_3 from the relationships with ϕ, δ_1, δ_2, the corresponding estimated standard errors, and t statistics for testing H_0: $\beta_i = 0$. The sum of squares in Figure 2.11 provides a test of H_{01} of Equation 2.6.

The remaining discussion relates to using JMP® to fit the equal slopes model of Equation 2.12. Figure 2.12 is the Model Specification window with selections identical to those in Figure 2.5, except the treat*cctime term is not included. Clicking on the Run Model button provides the analysis of variance table and estimates of the parameters as displayed in Figure 2.13. The constructed model fits a full rank model using the sum-to-zero restrictions as

$$y_{ij} = \theta + \tau_i + \beta x_{ij} + \varepsilon_{ij}, \quad i = 1, \ldots, t, \quad j = 1, \ldots, n_i$$

where $\theta = (\bar{\alpha})._{}$, and $\tau_i = \alpha_i - (\bar{\alpha})..$ The results provided are estimates of θ, τ_1, …, τ_{t-1}, and β. The estimates of the intercepts of Model 2.12 are obtained by

$$\hat{\alpha}_1 = \hat{\theta} + \hat{\tau}_1,\ \hat{\alpha}_2 = \hat{\theta} + \hat{\tau}_2, \ldots, \hat{\alpha}_{t-1} = \hat{\theta} + \hat{\tau}_{t-1},\ \hat{\alpha}_t = \hat{\theta} - \hat{\tau}_1 - \hat{\tau}_2 - \ldots - \hat{\tau}_{t-1}$$

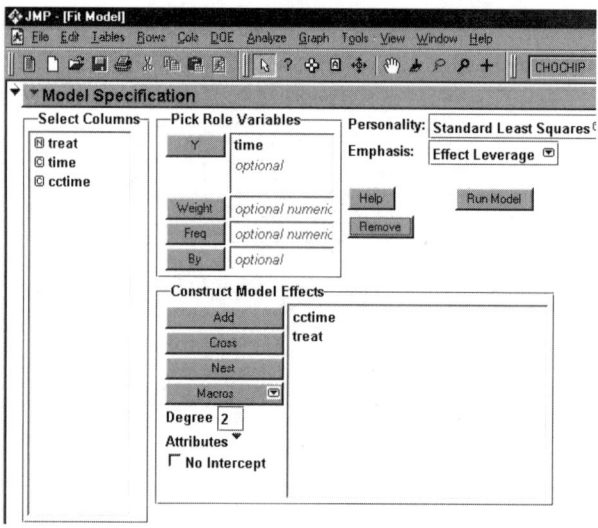

FIGURE 2.12 Model Specification window for the equal slopes model.

Analysis of Variance

Source	DF	Sum of Squares	Mean Square	F Ratio
Model	3	923.9025	307.967	13.6033
Error	11	249.0309	22.639	Prob > F
C. Total	14	1172.9333		0.0005

Parameter Estimates

Term	Estimate	Std Error	t Ratio	Prob>|t|
Intercept	29.327613	4.798279	6.11	<.0001
cctime	0.5147871	0.135624	3.80	0.0030
treat[1]	7.5282201	1.772822	4.25	0.0014
treat[2]	-2.503759	1.758446	-1.42	0.1822

Effect Tests

Source	Nparm	DF	Sum of Squares	F Ratio	Prob > F
cctime	1	1	326.16913	14.4073	0.0030
treat	2	2	426.14123	9.4116	0.0041

FIGURE 2.13 Estimates of the parameters and analysis of variance table for the equal slopes model.

The analysis of variance table contains the estimate of σ^2 denoted as the Mean Square Error. The effect tests provide tests of H_{03} of Equation 2.18 for source Treat and H_0: $\beta = 0$ for source cctime. The parameter estimates provide estimates of the parameters θ, τ_1, ..., τ_{t-1}, β, corresponding in this case to Intercept, treat[1], treat[2], and cctime, respectively.

Least squares means are provided which are estimates of the regression models evaluated at $\bar{X}_{..}$ and are computed by

$$\text{LSMEAN}(\text{Treat } i) = \left(\hat{\theta} + \hat{\tau}_i\right) + \hat{\beta}\bar{X}_{..}, \quad i = 1, 2, ..., t-1$$

and

One-Way Analysis of Covariance

$$\text{LSMEAN}(\text{Treat t}) = \left(\hat{\theta} - \hat{\tau}_1 - \ldots - \hat{\tau}_{t-1}\right) + \hat{\beta}\overline{X}_{..},$$

which are LSMEAN(Treat i) = $\hat{\alpha}_i = \hat{\beta}\overline{X}_{..}$, $i = 1, 2, \ldots, t$. The least squares means are displayed in Figure 2.14, where the estimated standard error and the unadjusted or raw means are also exhibited. The Custom Test option of the Parameter Estimates is used to provide estimates of the intercepts of the equal slopes models using the relationship between α_1, α_2, and α_3, and θ, τ_1, and τ_2, which are exhibited in Figure 2.15. Estimates of the intercepts, estimated standard errors and t-statistics to test hypotheses H_0: $\alpha_i = 0$ vs. H_a: $\alpha_i \neq 0$ $i = 1, 2, 3$ are provided. The results are displayed in Figure 2.16. The sum of squares provides a test of H_0: $\alpha_1 = \alpha_2 = \alpha_3 = 0$ vs. H_a:(notH_0:) for the equal slopes model.

Least Squares Means Table

Level	Least Sq Mean	Std Error	Mean
1	54.461553	2.1568905	55.8000
2	44.429574	2.1450896	43.4000
3	41.908872	2.1294262	41.6000

FIGURE 2.14 Least squares means.

Custom Test

Parameter			
Intercept	1	1	1
cctime	0	0	0
treat[1]	1	0	-1
treat[2]	0	1	-1

Click and Type Above to form hypothesis test.

FIGURE 2.15 Custom Test window providing coefficients to estimate the intercepts for the equal slopes model.

Custom Test

Parameter			
Intercept	1	1	1
cctime	0	0	0
treat[1]	1	0	-1
treat[2]	0	1	-1
Value	36.855833604	26.823854404	24.303152421
Std Err	5.4256375931	4.8579141814	5.029291514E
T-Ratio	6.7929036859	5.5216814052	4.8323212824
Prob>|t|	0.0000298181	0.0001802597	0.0005254095
SS	1044.6514752	690.24488571	528.65471539

Sum of Squares	1100.8021139
Numerator DF	3
F Ratio	16.207927753
Prob > F	0.0002379687

FIGURE 2.16 Estimates of the model intercepts for equal slopes model with test of the intercepts equal to zero.

```
         Custom Test

         Parameter
         Intercept          1            1            1
         cctime            50           50           50
         treat[1]           1            0           -1
         treat[2]           0            1           -1
         Value     62.59519012  52.56321092  50.042508937
         Std Err   2.7807878928  3.2180345879   3.07815664
         T-Ratio   22.50987581  16.333948403  16.257297724
         Prob>|t|  0.0000000001  0.0000000046  0.0000000049
         SS         11471.143322  6040.0824473  5983.5266036

         Sum of Squares   12783.428875
         Numerator DF                3
         F Ratio          188.21992529
         Prob > F         0.000000001
```

FIGURE 2.17 Custom test slopes window for evaluating adjusted means at cctime = 50 sec.

```
         Custom Test

         Parameter
         Intercept     |  1 |  1 |  1
         cctime        | 50 | 50 | 50
         treat[1]      |  1 |  0 | -1
         treat[2]      |  0 |  1 | -1
         Click and Type Above to form hypothesis test.
            Done    Add Column   Help
```

FIGURE 2.18 Least squares means evaluated at cctime = 50.

JMP® does not have a statement like LSMEANS TRT/at X = A, but the adjusted means evaluated at some value of the covariate can be obtained by using the Custom Test window. Figure 2.17 contains the Custom Test window with entries to compute the adjusted means at cctime = 50 sec. The first column evaluates the adjusted mean for treat = 1, as

$$\text{LSMEAN}_1 \text{ at } 50 = \hat{\mu}_{\text{time/cctime}=50}$$

$$= \hat{\theta} + \hat{\tau}_1 + \hat{\beta}\, 50$$

$$= \hat{\alpha}_1 + \hat{\beta}\, 50$$

which is a predicted value from the Treatment 1 regression line evaluated at 50 sec. Figure 2.18 contains the results, providing adjusted means of 62.6, 52.6, and 50.0 sec for Treatments 1, 2, and 3, respectively. The estimated standard errors and t statistics are displayed where the t statistic tests $H_0: \alpha_i + \beta 50 = 0$ vs. $H_a: \alpha_i + \beta 50 \neq 0$ for i = 1, 2, 3. Some examples in Chapter 3 were analyzed using the JMP® software.

2.8 CONCLUSIONS

The discussion in this chapter is concerned with the simple linear regression model being fit to the data set. It is important that one makes sure that a simple linear regression model does indeed fit the data for each of the treatment groups. Chapter 3 contains a set of examples where the simple linear regression models are adequate to describe the data. Chapter 4 contains an example where a quadratic model is the adequate model.

REFERENCES

Draper, N. and Smith, H. (1981) *Applied Regression Analysis,* 2nd ed., Wiley, New York..
JMP® and SAS® are registered trademarks of SAS Institute Inc.
Milliken, G. A. and Johnson, D. E. (1992) *Analysis of Messy Data, Volume 1: Design Experiments*, Chapman & Hall, London.

EXERCISE

Exercise 2.1: Find two articles in which the authors used analysis of covariance in analyzing their data set. Write a brief description of the situation when analysis of covariance was used and discuss the appropriateness of using the analysis. Include the reference citation. Do not use articles from statistical journals.

3 Examples: One-Way Analysis of Covariance — One Covariate in a Completely Randomized Design Structure

3.1 INTRODUCTION

This chapter provides a collection of examples to demonstrate the use of the strategies of analysis of covariance and to provide guidance in using software and interpreting the resulting output. The first example presents the results of the amount of time in seconds it takes to dissolve chocolate candy pieces by mouth as it relates to the amount of time in seconds it took to dissolve a butterscotch chip. The analysis of this data set is presented in detail using PROC GLM, PROC MIXED, and JMP®. The remaining examples are analyzed in less detail and with only one of the computational procedures. All of the examples are included to demonstrate particular features of the analysis of covariance, including how to prevent making some terrible mistakes. Examples of exceptions to the strategy in Section 2.6 are presented.

3.2 CHOCOLATE CANDY — EQUAL SLOPES

The data in Table 3.1 are from an experiment where 35 persons were randomly assigned to 1 of 6 types of chocolate candy: blue M&M®, Button, chocolate chip, red M&M®, small M&M®, and a Sno-Cap®. Eight plastic bags were prepared for each of the types of candy and each bag included the respective type of candy plus a butterscotch chip. After randomly assigning the plastic bags to persons, the process was to record the amount of time in seconds it took to dissolve the butterscotch chip in their mouth, denoted on Table 3.1 by "bstm." Then each participant was to record the amount of time in seconds it took to dissolve the specific type of chocolate candy in their mouth. The feeling was that there were differences among persons as to their abilities to dissolve chocolate candy, so collecting the time to dissolve the butterscotch chip to be used as a possible covariate was an attempt to reduce the variation among persons before comparing the mean times to dissolve the six candy types. The strategy described in Chapter 2 was used to provide an analysis of the data set.

TABLE 3.1
Data for the Chocolate Chip Experiment Where bstm is the Time in Seconds to Dissolve the Butterscotch Chip and Time is the Time in Seconds to Dissolve the Respective Type of Chocolate.

Blue M&M		Button		Choc chip		Red M&M		Small M&M		Sno-Cap	
bstm	time	bstm	time	bstm	time	bstm	time	bstm	time	bstm	time
28	60	27	53	17	36	20	30	30	25	15	20
30	45	16	47	29	51	35	45	32	25	21	29
19	38	19	39	20	40	16	32	30	33	16	23
33	48	35	90	14	34	19	47	22	26	40	44
19	34	34	65					32	30	19	26
24	42	40	58							21	29
25	48	28	72								
28	48	23	45								

3.2.1 Analysis Using PROC GLM

The first step in the analysis is to plot the data for each level of type to make sure that a simple linear regression line with equal variances does in fact adequately describe the data (plots not shown). Next, fit the unequal slopes model of Equation 2.1. The SAS® system statements required for PROC GLM to fit the model are in Table 3.2. The model contains type with the NOINT option to provide estimates of the intercepts and bstime*type to provide estimates of the slopes. The lower part of Table 3.2 displays the estimates of the intercepts ($\hat{\alpha}_i$'s corresponding to "type") and the slopes ($\hat{\beta}_i$'s corresponding to "bstime*type") of the regression lines for each of the levels of type. For example, the estimated equation to describe the mean time to dissolve a blue M&M for a given time to dissolve a butterscotch chip is 17.97 + 1.064 (bstime). The Error Mean Square is computed by pooling the variances across the levels of type and provides the estimate of the variance as 67.17.

The Type III sum of squares and corresponding F statistic for bstime*type provides a test of the hypothesis that all slopes are equal to zero or providing information about the quesiton" "Do I need the covariate in the model?" (see Equation 2.6). In this case, the significance level is 0.0066, indicating there is strong evidence that at least one slope is not equal to zero. Thus conclude the time models for the levels of type are linearly related to bstime for at least one type. The Type III sum of squares and resulting F statistic for source type provides a test that all of the intercepts are equal to zero, i.e., testing to see if all of the regression lines have a zero intercept. The significance level is 0.1075, indicating that there is not strong evidence to conclude that the intercepts are not all zero.

After concluding the slopes are not all zero, the next step is to determine if a common slope model is adequate to describe all of the data. The PROC GLM code in Table 3.3 has a model statement consisting of type, bstime, and bstime*type. Including these terms and not using the NOINT option results in a less than full

Examples: One-Way Analysis of Covariance

TABLE 3.2
Proc GLM Code, Analysis of Variance Table and Parameter Estimates for Full Rank Unequal Slopes Model.

```
proc glm data=mmacov; class type;
model time=type bstime*type/noint solution ss3;
```

Source	df	SS	MS	FValue	ProbF
Model	12	66741.99	5561.83	82.80	0.0000
Error	23	1545.01	67.17		
Uncorrected Total	35	68287.00			

Source	df	SS (Type III)	MS	FValue	ProbF
type	6	805.13	134.19	2.00	0.1075
bstime*type	6	1628.79	271.46	4.04	0.0066

Parameter	Estimate	StdErr	tValue	Probt
type Blue M&M	17.97	16.19	1.11	0.2784
type Button	21.57	10.78	2.00	0.0574
type Choc chip	16.92	15.17	1.12	0.2762
type Red M&M	26.58	13.17	2.02	0.0555
type Small M&M	22.20	29.08	0.76	0.4531
type Sno-Cap	8.70	9.41	0.92	0.3650
bstime*type Blue M&M	1.064	0.619	1.72	0.0989
bstime*type Button	1.335	0.374	3.57	0.0016
bstime*type Choc chip	1.167	0.730	1.60	0.1237
bstime*type Red M&M	0.530	0.556	0.95	0.3507
bstime*type Small M&M	0.192	0.988	0.19	0.8477
bstime*type Sno-Cap	0.900	0.400	2.25	0.0343

TABLE 3.3
PROC GLM Code and Analysis of Variance for the Singular Model with Unequal Slopes

```
proc glm data=mmacov; class type;
model time=type bstime bstime*type/solution ss3;
```

Source	df	SS	MS	FValue	ProbF
Model	11	6089.16	553.56	8.24	0.0000
Error	23	1545.01	67.17		
Corrected Total	34	7634.17			

Source (Type III)	df	SS (Type III)	MS	FValue	ProbF
type	5	101.99	20.40	0.30	0.9056
bstime	1	722.63	722.63	10.76	0.0033
bstime*type	5	155.16	31.03	0.46	0.8004

TABLE 3.4
PROC GLM Code and Analysis of Variance with Type I and Type III Sums of Squares Testing Different Hypotheses

```
proc glm data=mmacov; class type;
model time=type bstime*type bstime/ss1 ss3 solution;
```

Source	df	SS	MS	FValue	ProbF
Model	11	6089.16	553.56	8.24	0.0000
Error	23	1545.01	67.17		
Corrected Total	34	7634.17			

Source	df	SS (Type I)	MS	FValue	ProbF
type	5	4460.37	892.07	13.28	0.0000
bstime*type	6	1628.79	271.46	4.04	0.0066
bstime	0	0.00			

Source	df	SS (Type III)	MS	FValue	ProbF
type	5	101.99	20.40	0.30	0.9056
bstime*type	5	155.16	31.03	0.46	0.8004
bstime	1	722.63	722.63	10.76	0.0033

rank or singular model. This less than full rank model provides tests of interesting hypotheses.

Since the model contains both bstime and bstime*type, the Type III sum of squares corresponding to bstime*type tests the equal slope hypothesis, H_{02} of Equation 2.11. The significance level of the corresponding F statistic is 0.8004, indicating there is not enough evidence to conclude the slopes are not equal; thus conclude that a common slope model adequately describes the data set. Without the NOINT option the Type III sum of squares corresponding to type provides a test of the hypothesis that all of the intercepts are equal (H_{05} of Equation 2.25), i.e., tests that the regression lines intersect at bstime = 0. This test is not powerful since bstime = 0 is not close to the range of bstime.

Statistics to test hypotheses H_{01} and H_{02} can be obtained from one set of PROC GLM or Mixed code by writing the model statement as displayed in Table 3.4. The model has bstime*type occurring before bstime and the Type I and Type III sum of squares are requested. The Type I sum of squares corresponding to bstime*type tests H_{01} and the Type III sum of squares corresponding to bstime*type tests H_{02}. The results are in Table 3.4 and the relationships between the test statistics are obtained by comparing the results to those in Tables 3.2 and 3.3.

The PROC GLM code to fit parallel lines model of (2.12) is in Table 3.5. The model statement contains type and bstime with the NOINT option. This is a full rank model and the parameter estimates provide the estimate of the intercepts for each level of type ($\hat{\alpha}_i$ corresponding to type) and the estimate of the common slope ($\hat{\beta}$ corresponding to bstime). The estimated regression model for the mean time to dissolve a blue M&M for a given value of bstime is $\hat{\mu}_{BlueM\&M|bstime} = 19.74 + bstime \times 0.996$.

Examples: One-Way Analysis of Covariance

TABLE 3.5
PROC GLM Code, Analysis of Variance and Estimates of the Model's Parameters for the Full Rank Equal Slopes Model

```
proc glm data=mmacov; class type;
model time=type bstime/noint solution ss3;
```

Source	df	SS	MS	FValue	ProbF
Model	7	66586.83	9512.40	156.66	0.0000
Error	28	1700.17	60.72		
Uncorrected Total	35	68287.00			

Source	df	SS(Type III)	MS	FValue	ProbF
type	6	4652.48	775.41	12.77	0.0000
bstime	1	1473.63	1473.63	24.27	0.0000

Parameter	Estimate	StdErr	tValue	Probt
type Blue M&M	19.74	5.89	3.35	0.0023
type Button	31.00	6.25	4.96	0.0000
type Choc chip	20.34	5.61	3.62	0.0011
type Red M&M	16.10	5.99	2.69	0.0119
type Small M&M	−1.27	6.85	−0.19	0.8543
type Sno-Cap	6.60	5.47	1.21	0.2376
bstime	0.996	0.202	4.93	0.0000

The Type III sum of squares for type tests the hypothesis that the intercepts are all equal to zero and for bstime it tests the hypothesis that the common slope is equal to zero.

The model used to construct the analysis in Table 3.6 consists of type and bstime without the NOINT option. This is a less than full rank model, and in this case the Type III sum of squares corresponding to type tests the hypothesis that all of the intercepts are equal.

Since the model has a common slope, if all of the intercepts are equal, then a single simple linear regression model would describe the data for all of the levels of type. Thus, testing the hypothesis that the intercepts are equal for the common slope model is testing to see if the treatments are equal, i.e., testing to see if one simple linear regression will describe the data from all of the treatments. The F statistic corresponding to type has a value of 13.14 with significance level equal to 0.0000. Thus there is sufficient evidence to conclude that the regression models are not identical since they have different intercepts.

In order to compare the regression models, adjusted means were computed. The PROC GLM code in Table 3.7 provides the computation of the least squares means for each level of type. These least squares means are predicted values from the regression models evaluated at the average value of bstime from the whole data set or at 25.03 sec. The least squares mean for the blue M&M in Table 3.7 is computed as $\hat{\mu}_{\text{BlueM\&M}|\text{bstime}} = 19.74 + 25.03 \times 0.996 = 44.66$ sec.

TABLE 3.6
PROC GLM Code, Analysis of Variance Parameter Estimates for the Less Than Full Rank Equal Slopes Model

```
proc glm data=mmacov; class type;
model time=type bstime/solution ss3;
```

Source	df	SS	MS	FValue	ProbF
Model	6	5934.00	989.00	16.29	0.0000
Error	28	1700.17	60.72		
Corrected Total	34	7634.17			

Source	df	SS (Type III)	MS	FValue	ProbF
type	5	3988.09	797.62	13.14	0.0000
bstime	1	1473.63	1473.63	24.27	0.0000

Parameter	Estimate	Biased	StdErr	tValue	Probt
Intercept	6.60	1	5.47	1.21	0.2376
type Blue M&M	13.14	1	4.28	3.07	0.0047
type Button	24.40	1	4.37	5.59	0.0000
type Choc chip	13.74	1	5.05	2.72	0.0110
type Red M&M	9.50	1	5.03	1.89	0.0693
type Small M&M	−7.87	1	4.94	−1.59	0.1223
type Sno-Cap	0.00	1			
bstime	0.996	0	0.202	4.93	0.0000

The pdiff and adjust=simulate(report) options on the LSMEANS statement provide a multiple comparison procedure using the simulate method (Westfall et al., 1999). The t-statistics to compare adjusted means for types i and i′ are computed as

$$t_c = \frac{\text{Difference between two adjusted means}}{\text{Standard error of the difference of two adjusted means}}$$

$$= \frac{\hat{\mu}_{\text{type}_i|\text{bstime}=25.03} - \hat{\mu}_{\text{type}_{i'}|\text{bstime}=25.03}}{\hat{S}_{\hat{\mu}_{\text{type}_i|\text{bstime}=25.03} - \hat{\mu}_{\text{type}_{i'}|\text{bstime}=25.03}}}$$

where the standard error is computed by using Equation 2.23. A comparison of two least squares means is a comparison of the intercepts for the common slope model since $\hat{\mu}_{\text{type}_i|\text{bstime}=25.03} - \hat{\mu}_{\text{type}_{i'}|\text{bstime}=25.03} = \hat{\alpha}_i - \hat{\alpha}_{i'}$, i.e., the difference of two least squares means is the same no matter the value of bstime used to compute the adjusted means.

The probability values in the lower part of Table 3.7 correspond to carrying out the simulate multiple comparison for making pairwise comparisons among the means. The results indicate that the small M&Ms dissolve in less time than all,

Examples: One-Way Analysis of Covariance

TABLE 3.7
Adjusted or Least Squares Means and Pairwise Comparisons Using the Simulate Multiple Comparison Method for Equal Slopes Model

```
lsmeans type/pdiff stderr adjust=simulate(report);
```

type	LSMean	StdErr	LSMean Number
Blue M&M	44.66	2.76	1
Button	55.92	2.81	2
Choc chip	45.26	4.03	3
Red M&M	41.02	3.93	4
Small M&M	23.65	3.59	5
Sno-Cap	31.52	3.24	6

RowName	_1	_2	_3	_4	_5	_6
1		0.0734	1.0000	0.9729	0.0006	0.0477
2	0.0734		0.3014	0.0507	0.0000	0.0000
3	1.0000	0.3014		0.9704	0.0078	0.0995
4	0.9729	0.0507	0.9704		0.0351	0.4236
5	0.0006	0.0000	0.0078	0.0351		0.6011
6	0.0477	0.0000	0.0995	0.4236	0.6011	

except the Sno-Caps ($p < 0.05$). Sno-Caps dissolve in less time than the blue M&Ms and the Buttons ($p < 0.05$). A graph of the data and estimated regression lines is displayed in Figure 3.1 where the letters denote the data points and the lines denote the estimated regression models.

3.2.2 Analysis Using PROC MIXED

The PROC MIXED code is very similar to that of PROC GLM, but since PROC MIXED has several methods of estimating the variance components in the model (Littell et al., 1996), some of the results are presented in a different form than PROC GLM. Table 3.8 contains the PROC MIXED code and the results corresponding to fitting the full rank model with PROC GLM in Table 3.2.

The estimation method is selected by method=type3, indicating that the Type III sums of squares are used to provide estimates of the variance components. This model has only one variance component and its estimate is 67.174 as denoted by Residual of the Cov Parm set. The method=type3 requests that PROC MIXED provide an analysis of variance table with Type III sums of squares. The Type III analysis of variance table from Table 3.8 provides the same information as provided by PROC GLM in Table 3.2. PROC MIXED also displays an analysis of variance table for testing hypotheses about the fixed effects in the mixed model. The corresponding table contains the effect, numerator and denominator degrees of freedom, the value of the F statistic, and its significance level. The fixed effects analysis of variance table does not contain sums of squares nor mean squares.

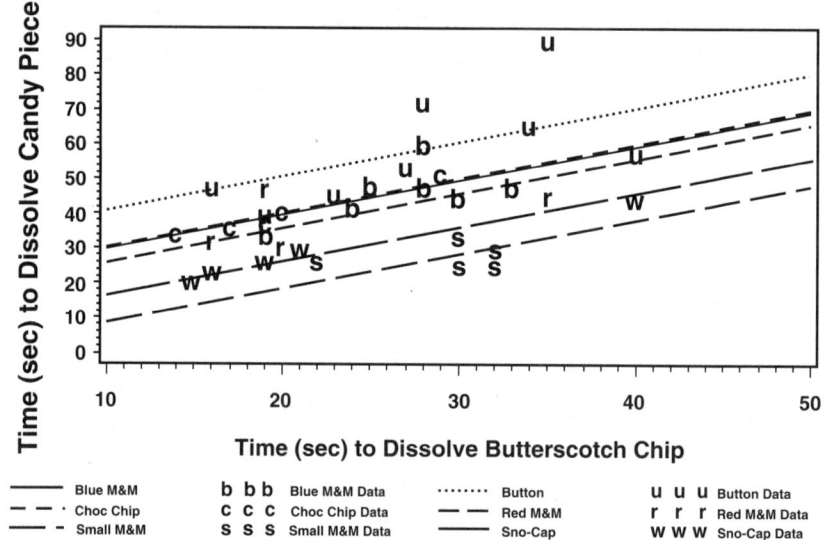

FIGURE 3.1 Plot of the data and regression lines for the equal slope or parallel lines model.

The estimates of the intercepts and slopes are identical to those from PROC GLM. The slopes all equal to zero hypothesis are tested by the effect bstime*type in Table 3.8. Table 3.9 contains the PROC MIXED code to fit the less than full rank model to provide a test of the equal slopes hypothesis corresponding to that carried out in Table 3.3. The F-statistic corresponding to bstime*type has a value of 0.46.

Table 3.10 contains the PROC MIXED code and results for fitting the less than full rank common slopes model to the data set as was done in Table 3.6 with PROC GLM. The estimate of the variance is 60.720. For the source corresponding to type, the Type III analysis and the Fixed Effects analysis provide the test statistic for testing the equality of the intercepts with an F-value of 13.14 with significance level of 0.0000.

The PROC MIXED code in Table 3.11 provides the least squares means, the predicted values at bstime=25.03, for each of the levels of type. The simulate method was used to carry out the pairwise multiple comparisons as in Table 3.7. PROC MIXED uses the diff option rather than the pdiff option used by PROC GLM. PROC MIXED provides a table that includes the estimated differences between the adjusted means, the estimated standard errors, denominator degrees of freedom, computed t-value, the significance level corresponding to the t-statistic, and the adjusted significance level from the simulate method.

Finally, Table 3.12 contains the PROC MIXED code to have the software compute predicted values or adjusted means for bstime=10 sec and bstime=50 sec.

These predicted values were used to provide the plot of the estimated regression lines in Figure 3.1.

TABLE 3.8
PROC MIXED Code, Analysis of Variance Tables and Parameter Estimates for the Full Rank Unequal Slopes Model

```
proc mixed data=mmacov method=type3; class type;
model time=type bstime*type/solution noint;
```

CovParm	Estimate
Residual	67.174

Source	df	SS (Type III)	MS	FValue	ProbF
type	6	805.13	134.19	2.00	0.1075
bstime*type	6	1628.79	271.46	4.04	0.0066
Residual	23	1545.01	67.17		

Effect	NumDF	DenDF	FValue	ProbF
type	6	23	2.00	0.1075
bstime*type	6	23	4.04	0.0066

Effect	Type	Estimate	StdErr	df	tValue	Probt
type	Blue M&M	17.97	16.19	23	1.11	0.2784
type	Button	21.57	10.78	23	2.00	0.0574
type	Choc chip	16.92	15.17	23	1.12	0.2762
type	Red M&M	26.58	13.17	23	2.02	0.0555
type	Small M&M	22.20	29.08	23	0.76	0.4531
type	Sno-Cap	8.70	9.41	23	0.92	0.3650
bstime*type	Blue M&M	1.064	0.619	23	1.72	0.0989
bstime*type	Button	1.335	0.374	23	3.57	0.0016
bstime*type	Choc chip	1.167	0.730	23	1.60	0.1237
bstime*type	Red M&M	0.530	0.556	23	0.95	0.3507
bstime*type	Small M&M	0.192	0.988	23	0.19	0.8477
bstime*type	Sno-Cap	0.900	0.400	23	2.25	0.0343

Oftentimes one wonders if the additional work required to go through the analysis of covariance strategy is worthwhile. To address this point, an analysis of variance was conducted. Table 3.13 contains the PROC MIXED code to fit the one-way treatment structure in a completely randomized design structure model in Equation 2.7. The estimate of the variance without the covariate is 109.441, compared to 60.720 when the covariate was included in the model. The F statistic for testing the equality of the type means has a value of 8.15 with a significance level of 0.0001, which is strong evidence that the type means are not equal.

Table 3.14a contains the least squares means for the levels of type along with their estimated standard errors. The estimated standard errors for the analysis of variance in Tables 3.14a are about 30% larger than the estimated standard errors for the analysis of covariance model from Table 3.11. So, by including information

TABLE 3.9
PROC MIXED Code and Analysis of Variance Tables for the Less Than Full Rank Unequal Slopes Model

```
proc mixed data=mmacov method=type3; class type;
    model time=type bstime bstime*type/solution;
```

CovParm	Estimate				
Residual	67.174				

Source	df	SS (Type III)	MS	FValue	ProbF
type	5	101.99	20.40	0.30	0.9056
bstime	1	722.63	722.63	10.76	0.0033
bstime*type	5	155.16	31.03	0.46	0.8004
Residual	23	1545.01	67.17		

Effect	NumDF	DenDF	FValue	ProbF
type	5	23	0.30	0.9056
bstime	1	23	10.76	0.0033
bstime*type	5	23	0.46	0.8004

about the time to dissolve a butterscotch chip, there is a large reduction in the estimate standard errors of the means.

A similar reduction in the estimated standard errors of the differences of means is achieved by incorporating the covariate into the model. The estimated standard errors of the differences of the least squares means in Table 3.11 are considerably smaller than those in Tables 3.14a and 3.14b.

Finally, using the simulate adjusted p-values to make pairwise comparisons between means, the Button takes significantly more time to dissolve than either the red M&M, small M&M, and Sno-Cap. More significant differences were detected when the time to dissolve the butterscotch chip was included in the model.

3.2.3 ANALYSIS USING JMP®

The JMP® software can be used to carry out the above analyses. The data set is entered into a worksheet where type is a nominal variable and time and bstime are continuous variables. Figure 3.2 contains the fit model dialog window where time has been designated as the "Y" variable. To fit the unequal slopes model to the data set, bstime, type, and bstime*type have been selected as model effects. After clicking on the "Run Model" button, the analysis of variance table, estimates of the models parameters and least squares means are computed. Figure 3.3 contains the analysis of variance table and tests for the effects. The Mean Square Error is the estimate of the variance, 67.174. The F-statistic corresponding to type*bstime provides a test of the equal slope hypothesis of Equation 2.11. The significance level of the test is 0.8004, indicating there is not enough evidence to conclude the slopes are unequal. There is no automatic test of the all slopes equal to zero hypothesis of Equation 2.6.

TABLE 3.10
PROC MIXED Code, Analysis of Variance Tables and Parameter Estimates for the Less Than Full Rank Equal Slopes Model

```
proc mixed data=mmacov method=type3; class type;
model time=type bstime/solution;
```

CovParm	Estimate
Residual	60.720

Source	df	SS (Type III)	MS	FValue	ProbF
type	5	3988.09	797.62	13.14	0.0000
bstime	1	1473.63	1473.63	24.27	0.0000
Residual	28	1700.17	60.72		

Effect	NumDF	DenDF	FValue	ProbF
type	5	28	13.14	0.0000
bstime	1	28	24.27	0.0000

Effect	Type	Estimate	StdErr	df	tValue	Probt
Intercept		6.60	5.47	28	1.21	0.2376
type	Blue M&M	13.14	4.28	28	3.07	0.0047
type	Button	24.40	4.37	28	5.59	0.0000
type	Choc chip	13.74	5.05	28	2.72	0.0110
type	Red M&M	9.50	5.03	28	1.89	0.0693
type	Small M&M	−7.87	4.94	28	−1.59	0.1223
type	Sno-Cap	0.00				
bstime		0.996	0.202	28	4.93	0.0000

The "Custom Test" selection from the "Parameter Estimates" can be used to provide estimates of the slopes as well as provide the statistic to test the hypothesis of Equation 2.6. For the first five columns, put a "1" corresponding to bstime, and a "1" for each one of the type[..]*bstime. The slope of the last level of type, Sno-Cap, is obtained by putting a "1" in for bstime and a "−1" for each of the type[..]*bstime. The results of using the Custom Test are in Figure 3.4. The line corresponding to "Value" provides the estimates of the slopes followed with estimated standard errors and a t-statistic to test the hypothesis that an individual slope is zero. The box in Figure 3.4 contains the test statistic, F-Ratio, and significance level corresponding to testing the slopes are all equal to zero hypothesis. The significance level is 0.0066, indicating there is sufficient evidence to conclude that the slopes are not all equal to zero. In this case, the covariate is useful in helping describe the mean time to dissolve a piece of candy as a linear function of bstime.

The next step is to fit a common slope model to the data set. The model specification window is displayed in Figure 3.5 where the model effects only involve bstime and type. In Figure 3.6 the F statistic corresponding to type, tests the hypothesis that the intercepts are equal for the equal slopes model, i.e., it tests Hypothesis 2.18.

TABLE 3.11
Least Squares Means and Results of Using the Simulate Multiple Comparison Method for the Equal Slopes Model

```
lsmeans type/diff adjust=simulate(report);
```

Type	bstime	Estimate	StdErr	df	tValue	Probt
Blue M&M	25.03	44.66	2.76	28	16.19	0.0000
Button	25.03	55.92	2.81	28	19.90	0.0000
Choc chip	25.03	45.26	4.03	28	11.24	0.0000
Red M&M	25.03	41.02	3.93	28	10.44	0.0000
Small M&M	25.03	23.65	3.59	28	6.60	0.0000
Sno-Cap	25.03	31.52	3.24	28	9.73	0.0000

Type	_Type	Estimate	StdErr	df	tValue	Probt	Adjp
Blue M&M	Button	−11.26	3.92	28	−2.87	0.0076	0.0685
Blue M&M	Choc chip	−0.60	4.91	28	−0.12	0.9037	1.0000
Blue M&M	Red M&M	3.64	4.82	28	0.76	0.4562	0.9712
Blue M&M	Small M&M	21.01	4.50	28	4.67	0.0001	0.0009
Blue M&M	Sno-Cap	13.14	4.28	28	3.07	0.0047	0.0436
Button	Choc chip	10.66	5.02	28	2.12	0.0428	0.3004
Button	Red M&M	14.90	4.89	28	3.05	0.0050	0.0457
Button	Small M&M	32.27	4.45	28	7.25	0.0000	0.0000
Button	Sno-Cap	24.40	4.37	28	5.59	0.0000	0.0000
Choc chip	Red M&M	4.24	5.53	28	0.77	0.4500	0.9695
Choc chip	Small M&M	21.61	5.55	28	3.89	0.0006	0.0063
Choc chip	Sno-Cap	13.74	5.05	28	2.72	0.0110	0.0988
Red M&M	Small M&M	17.37	5.40	28	3.22	0.0033	0.0324
Red M&M	Sno-Cap	9.50	5.03	28	1.89	0.0693	0.4251
Small M&M	Sno-Cap	−7.87	4.94	28	−1.59	0.1223	0.6079

The significance level is very small, <0.0001, which indicates the intercepts are all not equal. Since the lines are parallel, the unequal intercepts mean the predicted values from the regression lines at a selected value of bstime are all not equal. The least squares means in Figure 3.7 provide predicted values at bstime = 25.03 sec. Figure 3.7 presents the least squares or adjusted means as well as the unadjusted means (Means). It is interesting to note that the unadjusted mean for blue M&M is larger than the unadjusted means for chocolate chip, but after the adjustment process of the analysis of covariance, the order of the two means reverses. This change occurs because the bstime means are not identical. The bstime means are 25.75, 27.75, 20.00, 22.50, 29.70, and 22.0 for blue M&M, Button, chocolate chip, red M&M, small M&M and Sno-Cap, respectively, indicating that all types did not have persons with the same average times of dissolving the butterscotch chips. This is part of the process of analysis of covariance in that the levels of the treatment do not need to have sets of experimental units with identical values of the covariate. The process of describing the mean of each treatment by a regression model enables the prediction of mean values at a selected value of the covariate (adjusted means)

TABLE 3.12
Least Squares Means for the Types at Butterscotch Times of 10 sec and 50 sec

```
lsmeans type/at bstime=10;
lsmeans type/at bstime=50;
```

type	bstime	Estimate	StdErr	df
Blue M&M	10	29.69	4.21	28
Button	10	40.95	4.52	28
Choc chip	10	30.29	4.39	28
Red M&M	10	26.06	4.64	28
Small M&M	10	8.69	5.22	28
Sno-Cap	10	16.55	4.00	28
Blue M&M	50	69.52	5.62	28
Button	50	80.78	5.27	28
Choc chip	50	70.12	7.21	28
Red M&M	50	65.88	6.79	28
Small M&M	50	48.51	5.46	28
Sno-Cap	50	56.38	6.49	28

TABLE 3.13
PROC MIXED Code and Analysis of Variance Table for Model without the Covariate

```
proc mixed data=mmacov method=type3; class type;
model time=type;
```

CovParm	Estimate		
Residual	109.441		

Effect	NumDF	DenDF	FValue	ProbF
type	5	29	8.15	0.0001

as if all experimental units had that selected value as their value of the covariate. So those for types with a mean bstime > 25.03 the adjusted means are less than the unadjusted means and for those types with a mean bstime < 25.03 the adjusted means are greater than the unadjusted means. The mean of the bstime values for the complete experiment was 25.03.

The general conclusions made about the analysis of covariance modeling of the chocolate candy data set are

1. The equal slopes model adequately describes the data
2. The times to dissolve the various types of candy are not equal
3. The covariate was useful in reducing the variability of the data and reduced the estimate of the variance and the estimated standard errors of the least squares means and the differences of the least squares means

TABLE 3.14a
Least Squares Means and Estimates Standard Errors for the Chocolate Candy Study without the Covariate

Type	Estimate	StdErr	df
Blue M&M	45.38	3.70	29
Button	58.63	3.70	29
Choc chip	40.25	5.23	29
Red M&M	38.50	5.23	29
Small M&M	27.80	4.68	29
Sno-Cap	28.50	4.27	29

TABLE 3.14b
Pairwise Multiple Comparisons of Means Using the Simulate Method for the Model without the Covariate

Type	_Type	Estimate	StdErr	df	tValue	Probt	Adjp
Blue M&M	Button	−13.25	5.23	29	−2.53	0.0170	0.1465
Blue M&M	Choc chip	5.13	6.41	29	0.80	0.4302	0.9667
Blue M&M	Red M&M	6.88	6.41	29	1.07	0.2920	0.8931
Blue M&M	Small M&M	17.58	5.96	29	2.95	0.0063	0.0628
Blue M&M	Sno-Cap	16.88	5.65	29	2.99	0.0057	0.0578
Button	Choc chip	18.38	6.41	29	2.87	0.0076	0.0728
Button	Red M&M	20.13	6.41	29	3.14	0.0039	0.0390
Button	Small M&M	30.83	5.96	29	5.17	0.0000	0.0002
Button	Sno-Cap	30.13	5.65	29	5.33	0.0000	0.0001
Choc chip	Red M&M	1.75	7.40	29	0.24	0.8147	1.0000
Choc chip	Small M&M	12.45	7.02	29	1.77	0.0866	0.4967
Choc chip	Sno-Cap	11.75	6.75	29	1.74	0.0925	0.5192
Red M&M	Small M&M	10.70	7.02	29	1.52	0.1382	0.6543
Red M&M	Sno-Cap	10.00	6.75	29	1.48	0.1494	0.6816
Small M&M	Sno-Cap	−0.70	6.33	29	−0.11	0.9128	1.0000

3.3 EXERCISE PROGRAMS AND INITIAL RESTING HEART RATE — UNEQUAL SLOPES

An exercise physiologist structured three types of exercise programs (EPRO) and conducted an experiment to evaluate and compare the effectiveness of each program. The experiment consisted of subjecting an individual to a given exercise program for 8 weeks. At the end of the training program, each individual ran for 6 min after which their heart rate was measured. An exercise program is deemed to be more effective if individuals on that program have lower heart rates after the 6-min run than individuals on another exercise program. Since individuals entered the experiment at differing degrees of fitness, the resting heart rate before beginning training

Examples: One-Way Analysis of Covariance 55

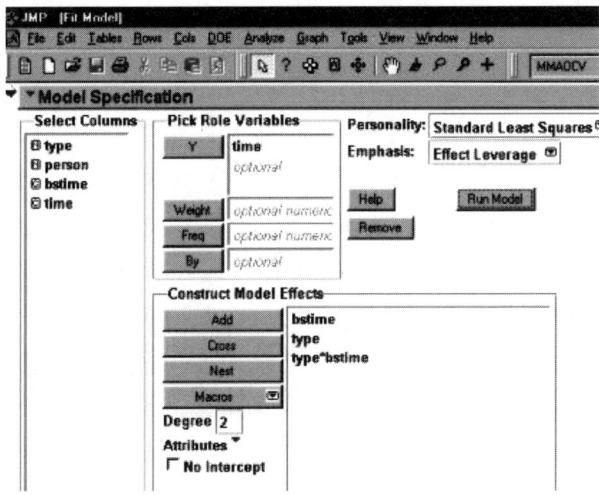

FIGURE 3.2 Fit model dialog window for unequal slopes model.

Summary of Fit	
RSquare	0.797619
RSquare Adj	0.700828
Root Mean Square Error	8.195998
Mean of Response	41.62857
Observations (or Sum Wgts)	35

Analysis of Variance					
Source	DF	Sum of Squares	Mean Square	F Ratio	Prob > F
Model	11	6089.1605	553.560	8.2406	<.0001
Error	23	1545.0109	67.174		
C. Total	34	7634.1714			

Effect Tests					
Source	Nparm	DF	Sum of Squares	F Ratio	Prob > F
bstime	1	1	722.62532	10.7575	0.0033
type	5	5	101.99054	0.3037	0.9056
type*bstime	5	5	155.15976	0.4620	0.8004

FIGURE 3.3 Analysis of variance table for unequal slopes model.

was recorded and was used as a covariate. The object of the study is to be able to compare exercise programs at a common initial resting heart rate. To carry out the experiment, 24 males between 28 and 35 years of age were selected. Then eight males were randomly assigned to each of the 3 EPRO treatments. The exercise program (EPRO), heart rate (HR) after the 6-min run at the completion of 8 weeks of training, and the initial resting heart rate (IHR) are given in Table 3.15.

After plotting the data to determine there is a linear relationship between HR and IHR for each of the exercise programs, the next step in the analysis is to fit Model 2.1 and test the slopes equal to zero hypothesis of Equation 2.6. Table 3.16

Custom Test

Parameter						
Intercept	0	0	0	0	0	0
bstime	1	1	1	1	1	1
type[Blue M&M]	0	0	0	0	0	0
type[Button]	0	0	0	0	0	0
type[Choc chip]	0	0	0	0	0	0
type[Red M&M]	0	0	0	0	0	0
type[Small M&M]	0	0	0	0	0	0
type[Blue M&M]*bstime	1	0	0	0	0	-1
type[Button]*bstime	0	1	0	0	0	-1
type[Choc chip]*bstime	0	0	1	0	0	-1
type[Red M&M]*bstime	0	0	0	1	0	-1
type[Small M&M]*bstime	0	0	0	0	1	-1
Value	1.0641025641	1.3352450469	1.1666666667	0.5299539171	0.1918604651	0.9
Std Err	0.618676047	0.3742894247	0.730157566	0.5563806217	0.9881153438	0.3999237694
T.Ratio	1.7199672903	3.5674132337	1.5978286345	0.9525024712	0.1941680861	2.2504288787
Prob> t	0.0988681854	0.0016354367	0.1237301331	0.350745808	0.8477490205	0.0342847305
SS	198.72115385	854.89064129	171.5	60.944700461	2.5325581395	340.2

Sum of Squares	1628.7890537
Numerator DF	6
F Ratio	4.0411955577
Prob > F	0.006556934

FIGURE 3.4 Results of using Custom Test to provide test of all slopes equal to zero for Example 3.2.

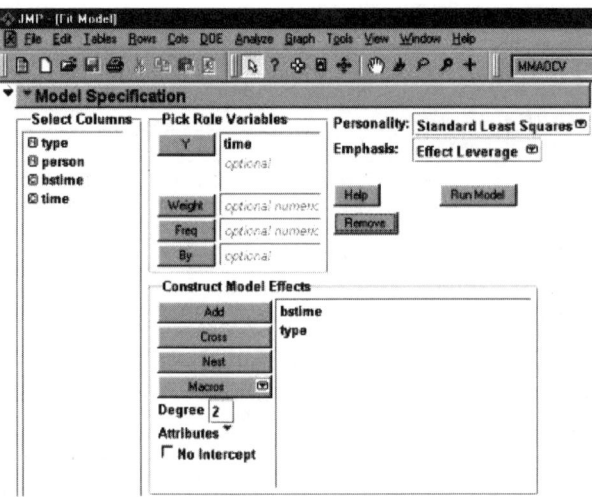

FIGURE 3.5 Model specification window for the equal slopes model for Example 3.2.

contains the PROC MIXED code, analysis of variance table, and estimates of the parameters. Since the model does not contain IHR as a separate term, the F statistic corresponding to IHR*EPRO provides a test of H_0: $\beta_1 = \beta_2 = \beta_3 = 0$ vs. H_a: (not H_0:). The significance level of the test statistic is less than 0.0001, indicating there is strong evidence that the null hypothesis is not true. The estimates of the intercepts, $\hat{\alpha}_i$ corresponding to EPRO, and slopes, $\hat{\beta}_i$ corresponding to IHR*EPRO, are in Table 3.16 where the model for EPRO1 is estimated by $\hat{\mu}_{EPRO1|IHR} = 46.538 + 1.4852$ IHR.

Examples: One-Way Analysis of Covariance

Analysis of Variance

Source	DF	Sum of Squares	Mean Square	F Ratio
Model	6	5934.0007	989.000	16.2878
Error	28	1700.1707	60.720	Prob > F
C. Total	34	7634.1714		<.0001

Effect Tests

Source	Nparm	DF	Sum of Squares	F Ratio	Prob > F
bstime	1	1	1473.6293	24.2691	<.0001
type	5	5	3988.0887	13.1359	<.0001

FIGURE 3.6 Analysis of variance table with tests for model effects for the equal slope model of Example 3.2.

Least Squares Means Table

Level	Least Sq Mean	Std Error	Mean
Blue M&M	44.656774	2.7588590	45.3750
Button	55.915652	2.8093617	58.6250
Choc chip	45.256249	4.0265107	40.2500
Red M&M	41.017347	3.9295306	38.5000
Small M&M	23.647089	3.5853479	27.8000
Sno-Cap	31.515127	3.2395455	28.5000

FIGURE 3.7 Least squares means and unadjusted means from the equal slopes model of Example 3.2.

TABLE 3.15
Data for Exercise Program Experiment with Eight Persons Assigned to Each Program for Example 3.3

EPRO 1		EPRO 2		EPRO 3	
HR	IHR	HR	IHR	HR	IHR
118	56	148	60	153	56
138	59	159	62	150	58
142	62	162	65	158	61
147	68	157	66	152	64
160	71	169	73	160	72
166	76	164	75	154	75
165	83	179	84	155	82
171	87	177	88	164	86

Next, test to see if a common slope model can be used to describe the data for the three treatments, i.e., test H_{02} of Equation 2.11 or H_0: $\beta_1 = \beta_2 = \beta_3$ vs. H_a: (not H_0:). Table 3.17 contains the PROC MIXED code and analysis of variance table for the model that includes both IHR and IHR*EPRO. The F statistic corresponding to

TABLE 3.16
Analysis of Variance Table with a Test for Slopes Equal to Zero and Estimates of the Intercepts and Slopes for Example 3.3

```
PROC MIXED data=heart; CLASSES EPRO;
MODEL HR=EPRO IHR*EPRO / NOINT SOLUTION OUTP=PRED;
```

CovParm	Estimate
Residual	28.203

Effect	NumDF	DenDF	FValue	ProbF
EPRO	3	18	60.38	0.0000
IHR*EPRO	3	18	31.33	0.0000

Effect	EPRO	Estimate	StdErr	df	tValue	Probt
EPRO	1	46.538	12.719	18	3.66	0.0018
EPRO	2	97.191	14.128	18	6.88	0.0000
EPRO	3	137.485	12.528	18	10.97	0.0000
IHR*EPRO	1	1.4852	0.1791	18	8.29	0.0000
IHR*EPRO	2	0.9380	0.1955	18	4.80	0.0001
IHR*EPRO	3	0.2638	0.1789	18	1.47	0.1576

TABLE 3.17
Analysis of Variance Table with Test of the Equal Slopes Hypothesis for Example 3.3

```
PROC mixed data=heart; CLASSES EPRO;
MODEL HR=EPRO IHR IHR*EPRO/ SOLUTION;
```

CovParm	Estimate
Residual	28.203

Effect	NumDF	DenDF	FValue	ProbF
EPRO	2	18	13.01	0.0003
IHR	1	18	70.59	0.0000
IHR*EPRO	2	18	11.68	0.0006

IHR*EPRO provides the test of the equal slope hypothesis. The significance level is 0.0006, indicating there is strong evidence to believe that the slopes are not equal.

Thus, continue the analysis assuming the unequal slopes model adequately describes the data for the three EPROs. Table 3.18 contains the results of testing the following special hypotheses:

1. $\beta_1 = \beta_2$, $\beta_1 = \beta_3$, and $\beta_2 = \beta_3$
2. $\mu_{Y_1|x=60} - \mu_{Y_2|x=80}$
3. $\alpha_1 = \alpha_2$, $\alpha_1 = \alpha_3$, and $\alpha_2 = \alpha_3$

TABLE 3.18
Results of the Estimate Statements for Example 3.3

```
ESTIMATE  'B1-B2'  IHR*EPRO  1 -1  0;
ESTIMATE  'B1-B3'  IHR*EPRO  1  0 -1;
ESTIMATE  'B2-B3'  IHR*EPRO  0  1 -1;
ESTIMATE  '1 AT 60 - 2 AT 80' EPRO 1 -1 0 IHR*EPRO 60 -80 0;
ESTIMATE  'A1-A2'  EPRO  1 -1  0;
ESTIMATE  'A2-A3'  EPRO  0  1 -1;
ESTIMATE  'A1-A3'  EPRO  1  0 -1;
```

Label	Estimate	StdErr	df	tValue	Probt
B1-B2	0.547	0.265	18	2.06	0.0537
B1-B3	1.221	0.253	18	4.83	0.0001
B2-B3	0.674	0.265	18	2.54	0.0203
1 AT 60 - 2 AT 80	−36.579	3.619	18	−10.11	0.0000
A1-A2	−50.653	19.010	18	−2.66	0.0158
A2-A3	−40.294	18.883	18	−2.13	0.0469
A1-A3	−90.947	17.853	18	−5.09	0.0001

The first hypothesis investigates the relationship among the slopes. This is reasonable, as one might wish to simplify the model by grouping treatments with "equal" slopes into groups where a parallel lines model is fit to the data within a group and nonparallel lines are fit between groups (see Section 3.7 for an example). When such a simplification can occur, the comparison process is easier as one can compare lines within a group with the LSMEANS.

The second hypothesis is a little strange in that it compares two treatments at different values of the covariate [(2) from above]. Hypotheses like this one are not usually tested, but there are some circumstances where such comparisons are warranted (comparing EPRO1 at IHR = 60 to EPRO2 at IHR = 80, is most likely not meaningful here), but such comparisons could be of interest if, for example, it costs the same to operate Process 1 at X_1 units of the covariate as it does to operate Process 2 at X_2 units of the covariate. In this case it is reasonable to compare the models at equal cost levels which means comparing the models at different values of X.

Comparison of the intercepts [(3) from above] compares the distances between the regression lines at IHR = 0. There are experiments when it is of interest to compare the regression lines when the value of the covariate is zero, but in this example, it is ridiculous to compare the intercepts since an IHR value of zero is not possible (the subjects would be dead and none of the exercise programs would help improve their fittness).

The results in Table 3.18 show that the slopes of EPRO1 and EPRO2 are significantly different from that of EPRO3 (p = 0.0001 and 0.0203, respectively). The significance level for comparing the slopes for EPRO1 and EPRO2 is 0.0537, but if a common slope is fit to these two data sets, a graph shows that it is not a good fit to the data (graph not shown). So, the process is to continue assuming the slopes are unequal for all of the three levels of EPRO. EPRO1 mean at IHR = 60 is significantly less than the EPRO2 mean at IHR = 80 with significance level less than 0.0001. Finally, if it were meaningful to compare the intercepts (but it is not the case here), all intercepts are significantly different ($p < 0.05$).

The final step is to compare the regression models. Since the slopes are not equal, it is necessary to compare the EPROs at several values of IHR. The experimenter chose to compare the three exercise programs at IHR = 55, 70, and 85. The first step in the process is to provide a test of the hypothesis that the models intersect (or are equal) at a selected value of the covariate or IHR. For IHR = IHR_0, test H_0: $\mu_{EPRO_1|IHR = IHR0} = \mu_{EPRO_2|IHR = IHR0} = \mu_{EPRO_3|IHR = IHR0}$ vs. H_a: (not H_0:) which is equivalent to testing

$$H_0: \alpha_1 + \beta_1 IHR_0 = \alpha_2 + \beta_2 IHR_0 = \alpha_3 + \beta_3 IHR_0 \text{ vs. } H_a:(\text{not } H_0:)$$

The test statistic can be obtained by fitting the model

$$y_{ij} = \alpha_i^* + \beta_i (IHR - IHR_0) + \varepsilon_{ij} \text{ where } \alpha_i^* = \alpha_i + \beta_i IHR_0.$$

The new intercepts of the models are the values of the regression models evaluated at IHR_0. By subtracting the value of IHR_0 from each value of IHR and using the adjusted values as a covariate in the model, the test corresponding to EPRO is a test that the intercepts of the new models are equal or a test that the models are equal at IHR_0. Tables 3.19 to 3.21 contain the results of testing the equality of the regression models at IHR = 55, 70, and 85 beats/sec, respectively, where the significance levels of the test corresponding to EPRO are 0.0001, 0.0009, and 0.0053. These tests indicate that the regression lines are not equal at 55, 70, and 85 beats/sec. To finish the investigation, adjusted means were obtained for the mean IHR (70.38) and for IHR of 55, 70, 85. Table 3.22 contains the PROC MIXED code to obtain the adjusted means. Many scientists just evaluate the adjusted means at the mean value of the covariate, but when the slopes are not equal, the adjusted means need to be evaluated at more than one place. The range of the data was the guide for selecting 55 to 85,

TABLE 3.19
Analysis of Variance Table with Test of Equal Models at IHR = 55 beats/sec for Example 3.3

```
PROC mixed data=heart; CLASSES EPRO;
MODEL HR=EPRO IHR55 IHR55*EPRO;
```

CovParm	Estimate			
Residual	28.203			

Effect	NumDF	DenDF	FValue	ProbF
EPRO	2	18	15.15	0.0001
IHR55	1	18	70.59	0.0000
IHR55*EPRO	2	18	11.68	0.0006

TABLE 3.20
Analysis of Variance Table with Test of Equal Models at IHR=70 beats/sec for Example 3.3

```
PROC mixed data=heart; CLASSES EPRO;
MODEL HR=EPRO IHR70 IHR70*EPRO;
```

CovParm	Estimate			
Residual	28.203			

Effect	NumDF	DenDF	FValue	ProbF
EPRO	2	18	10.69	0.0009
IHR70	1	18	70.59	0.0000
IHR70*EPRO	2	18	11.68	0.0006

TABLE 3.21
Analysis of Variance Table with Test of Equal Models at IHR = 85 beats/sec for Example 3.3

```
PROC mixed data=heart; CLASSES EPRO;
MODEL HR=EPRO IHR85 IHR85*EPRO;
```

CovParm	Estimate			
Residual	28.203			

Effect	NumDF	DenDF	FValue	ProbF
EPRO	2	18	7.12	0.0053
IHR85	1	18	70.59	0.0000
IHR85*EPRO	2	18	11.68	0.0006

with 70 occurring in the middle. The adjusted means can be computed at the desired value of IHR by including at "IHR = 55" as an option with the LSMEANS statement. Finally, pairwise comparisons of the adjusted means within a value of IHR were obtained by including the diff option with the LSMEANS statements in Table 3.22. The differences of the adjusted means, estimate standard errors, and t-statistics for testing H_0: $\mu_{EPRO_i|IHR = IHR0} = \mu_{EPRO_j|IHR = IHR0}$ vs. H_a: (not H_0:), $i \neq j$ are in Table 3.23.

Since there are three comparisons to be made for each value of IHR, a .05 Bonferroni approach was used to make the pairwise comparisons where a difference between means was declared significant within a IHR value if the t-test had a significance level less that or equal to $.05/3 = .01667$. At IHR = 70.35, the means for EPRO1 and EPRO3 are significantly less than the mean for EPRO2. At IHR = 55, the mean for EPRO1 is significantly less than the means for EPRO2 and EPRO3. At IHR = 70, the mean for EPRO1 is significantly less than the mean for EPRO2. At IHR = 85, the mean for EPRO3 is significantly less than the means for EPRO1

TABLE 3.22
Least Squares Means for the Exercise Programs at IHR = 70.38, 55, 70, and 85 beats/sec for Example 3.3

```
LSMEANS EPRO/ DIFF at means;
LSMEANS EPRO/ DIFF at IHR=55;
LSMEANS EPRO/ DIFF at IHR=70;
LSMEANS EPRO/ DIFF at IHR=85;
```

EPRO	IHR	Estimate	StdErr
1	70.38	151.06	1.88
2	70.38	163.20	1.89
3	70.38	156.05	1.89
1	55.00	128.23	3.31
2	55.00	148.78	3.75
3	55.00	151.99	3.17
1	70.00	150.50	1.88
2	70.00	162.85	1.90
3	70.00	155.95	1.88
1	85.00	172.78	3.24
2	85.00	176.92	3.22
3	85.00	159.90	3.39

TABLE 3.23
Differences between Least Squares Means for the Exercise Programs at IHR = 70.38, 55, 70, and 85 beats/sec for Example 3.3

EPRO	_EPRO	IHR	Estimate	StdErr	DF	tValue	Probt
1	2	70.38	−12.14	2.67	18	−4.55	0.0002
1	3	70.38	−4.99	2.66	18	−1.87	0.0775
2	3	70.38	7.16	2.67	18	2.68	0.0154
1	2	55.00	−20.56	5.01	18	−4.11	0.0007
1	3	55.00	−23.77	4.58	18	−5.19	0.0001
2	3	55.00	−3.21	4.91	18	−0.65	0.5215
1	2	70.00	−12.35	2.67	18	−4.62	0.0002
1	3	70.00	−5.44	2.66	18	−2.05	0.0555
2	3	70.00	6.90	2.68	18	2.58	0.0190
1	2	85.00	−4.14	4.57	18	−0.91	0.3769
1	3	85.00	12.88	4.69	18	2.75	0.0132
2	3	85.00	17.02	4.67	18	3.64	0.0019

Examples: One-Way Analysis of Covariance

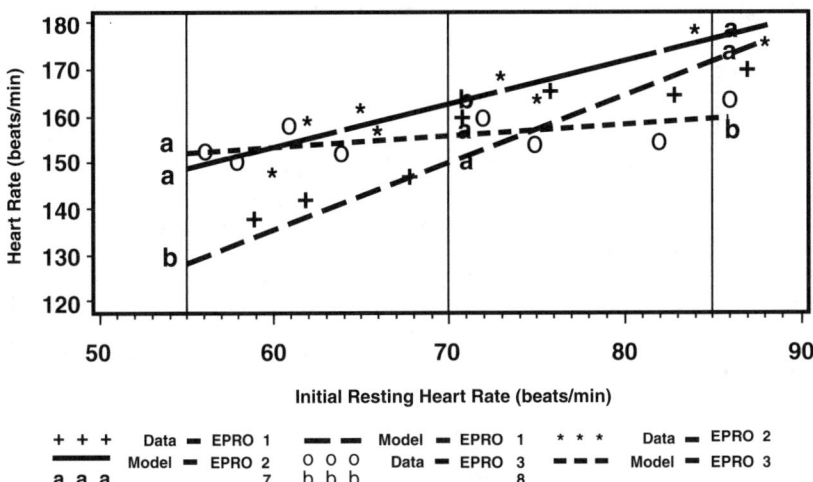

FIGURE 3.8 Plot of data and estimated regression models for the levels of EPRO for Example 3.3.

and EPRO2. The relationships among the adjusted means, plots of the data, and graphs of the estimated regression lines are displayed in Figure 3.8. The letters on the graph, "a" and "b," are such that if two means at the same value of IHR are followed by the same letter, the means are not significantly different via the Bonferroni adjustment, while means that do not have the same letter indicate a significant difference.

Another way to compare the regression lines is to construct confidence bands about the difference of each pair of models and determine the values of the independent variable where the bands exclude zero. To provide control over the error rates, Scheffe percentage points were used to construct a set of simultaneous confidence intervals (Milliken and Johnson, 1992) at several values of IHR in the range of 55 to 85 beats/min and then connecting them to provide the confidence bands. The difference between two models (say, 1 and 2) at IHR = IHR_0 is

$$(\alpha_1 - \alpha_2) + (\beta_1 - \beta_2)IHR_0$$

which depends on two parameters $(\alpha_1 - \alpha_2)$ and $(\beta_1 - \beta_2)$. Thus the Scheffe percentage point is $(2 F_{\alpha,2,\upsilon})^{1/2}$ where υ is the degrees of freedom associated with the MSERROR. Since the estimates of the parameters for the models are independent between the models, the variance of the difference of two models is the sum of the variances of the two predicted values. Most computer codes provide the predicted values and the standard errors of the predicted values which can be combined to construct the set of confidence intervals about the difference of two models at IHR_0. The end points of the confidence interval at a IHR = IHR_0 are

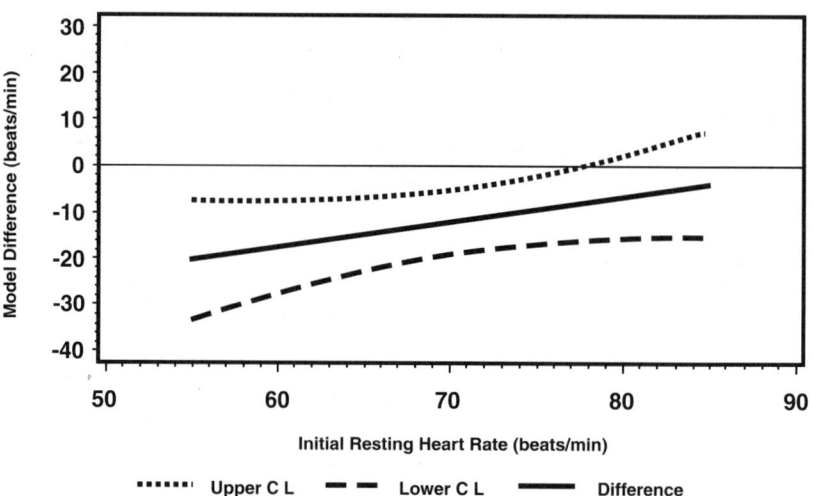

FIGURE 3.9 Confidence band about the difference of the models for EPRO1 minus EPRO2.

$$\left(\hat{\alpha}_1 + \hat{\beta}_1 \text{IHR}_0\right) - \left(\hat{\alpha}_2 + \hat{\beta}_2 \text{IHR}_0\right) \pm$$

$$\left(2\, F_{\alpha/2,2,\upsilon}\right)^{1/2} \left(\text{var}\left(\hat{\alpha}_1 + \hat{\beta}_1 \text{IHR}_0\right) + \text{var}\left(\hat{\alpha}_2 + \hat{\beta}_2 \text{IHR}_0\right)\right)^{1/2}$$

Computer code similar to that in Section 2.7.1 was used to generate the confidence bands about the differences of all pairs of models. The confidence bands are displayed in Figures 3.9 to 3.11. A horizontal line was inserted where the model difference is zero so that one can conclude that the regressions are not significantly different for the range where the horizontal line penetrates the confidence bands and that they are different for the range where the horizontal line does not penetrate the bands. The mean of EPRO1 is significantly less than the mean of EPRO2 for IHR < 77 beats/min, as illustrated in Figure 3.9. The mean of EPRO1 is significantly less than the mean of EPRO3 for IHR < 68 beats/min and is significantly greater than the means of EPRO3 for IHR > 84, as displayed in Figure 3.10. Finally, by using Figure 3.11, EPRO3 has a mean that is significantly less than that of EPRO2 for IHR > 72. There is one general conclusion that can be made by looking at Figure 3.8. Since the object of the study is to determine which EPRO reduces the heart rate after the 6-min run, it is easy to see that EPRO2 is not in consideration as a combination of EPRO1 and EPRO3 has a significantly lower mean heart rate for the whole range of IHR. The exercise scientist can use EPRO1 for IHR < 67, EPRO3 for IHR > 84, and use either EPRO1 or EPRO3 for 67 ≤ IHR ≤ 84.

Examples: One-Way Analysis of Covariance

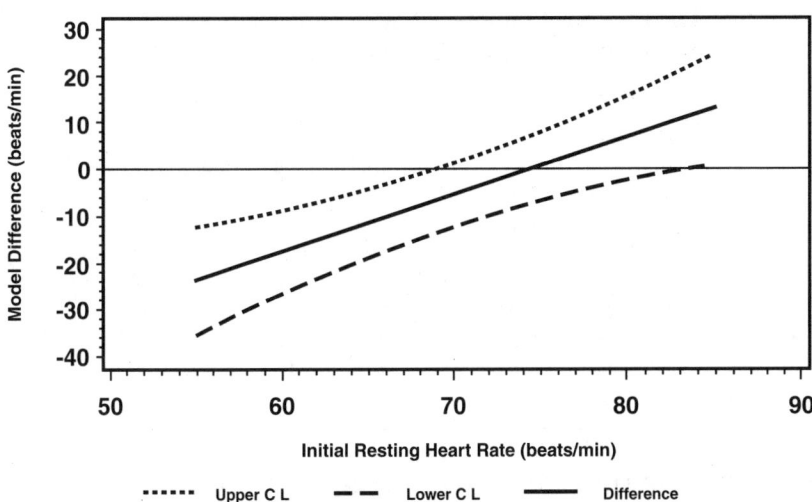

FIGURE 3.10 Confidence band about the difference of models for EPRO1 minus EPRO3.

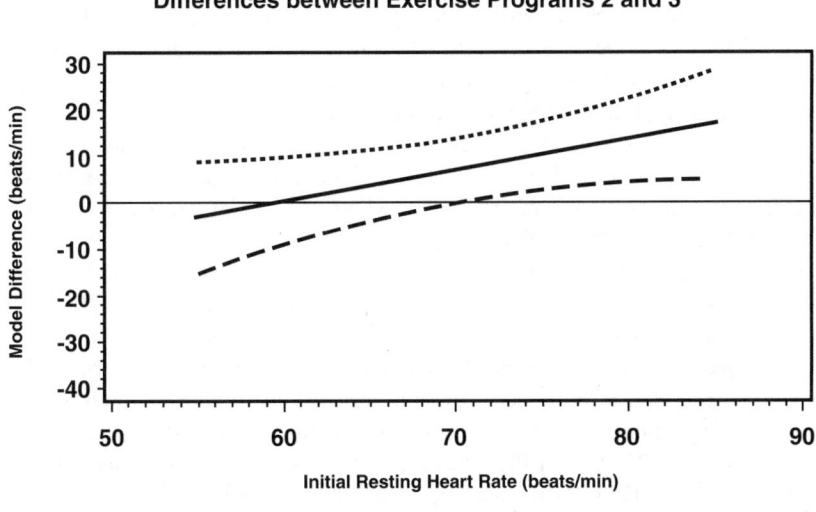

FIGURE 3.11 Confidence band about the difference of models for EPRO2 minus EPRO3.

TABLE 3.24
Data Set for Example 3.4 with Four Diets and Measurements of Pre and Post Cholesterol Levels

Diet 1		Diet 2		Diet 3		Diet 4	
Post Chol	Pre Chol	Post Chol	Pre Chol	Post Chol	Pre Chol	Post Chol	Pre Chol
174	221	211	203	199	249	224	297
208	298	211	223	229	178	209	279
210	232	201	164	198	166	214	212
192	182	199	194	233	223	218	192
200	258	209	248	233	274	253	151
164	153	172	268	221	234	246	191
208	293	224	249	199	271	201	284
193	283	222	297	236	207	234	168

3.4 EFFECT OF DIET ON CHOLESTEROL LEVEL: AN EXCEPTION TO THE BASIC ANALYSIS OF COVARIANCE STRATEGY

The data in Table 3.24 are the cholesterol levels (Y) of female human subjects after being on a specific diet for 8 weeks. The cholesterol levels of the subjects were determined before they started their diets. The objective of the study was to determine if any of the diets could lower cholesterol levels. The 32 subjects were allocated to the diets completely at random, 8 to each diet. Because the participants in the study had differing levels of cholesterol before the start of the study, the baseline cholesterol level was measure as a possible covariate before they started their assigned diet. Thus the design of the experiment was a one-way treatment structure with one covariate in a completely randomized design structure.

The first step in the analysis of covariance strategy is to determine if all of the slopes are equal to zero. The PROC GLM code and corresponding analysis of variance table are displayed in Table 3.25 where the model consists of DIET and PRE_CHOL*DIET with the NOINT option.

The F statistic corresponding to PRE_CHOL*DIET provides a test of the hypothesis that all slopes are zero (Equation 2.6). The significance level is 0.0785, thus when operating at $\alpha = 0.05$, one would conclude there is not enough evidence to reject the hypothesis. However, before stopping and concluding the covariate is not needed in the model, fit an effects model to test the equal slopes hypothesis.

Table 3.26 contains the PROC GLM code to fit the less than full rank model with DIET, PRE_CHOL, and PRE_CHOL*DIET. In this case the F statistic corresponding to PRE_CHOL*DIET provides a test of the hypothesis that the slopes are all equal (Equation 2.11). The significance level of the test statistic is 0.0417. In this case with $\alpha = 0.05$, one would conclude that the slopes are not all equal. This phenomenon occurs when the estimates of some slopes are positive and some are negative. The estimates of the intercepts and slopes (Table 3.27) for the regression

Examples: One-Way Analysis of Covariance

TABLE 3.25
PROC GLM Code and Analysis of Variance Table to Provide a Test of the Hypothesis That All of the Slopes are Zero for Example 3.4

```
proc glm data=one; class diet;
model post_chol = diet pre_chol*diet/noint ss3 solution;
```

Source	df	SS	MS	FValue	ProbF
Model	8	1428650.09	178581.26	732.28	0.0000
Error	24	5852.91	243.87		
Uncor Total	32	1434503.00			

Source	df	SS (Type III)	MS	FValue	ProbF
diet	4	59368.57	14842.14	60.86	0.0000
Pre_Chol*diet	4	2336.71	584.18	2.40	0.0785

TABLE 3.26
PROC GLM Code and Analysis of Variance for the Less Than Full Rank Model to Test the Equal Slopes Hypothesis for Example 3.4

```
proc glm; class diet;
model Post_chol = diet pre_chol
  Pre_chol*diet/noint;
```

Source	df	SS	MS	FValue	ProbF
Model	8	1428650.09	178581.26	732.28	0.0000
Error	24	5852.91	243.87		
Uncorrected Total	32	1434503.00			

Source	df	SS (Type III)	MS	FValue	ProbF
diet	4	59368.57	14842.14	60.86	0.0000
Pre_Chol	1	1.82	1.82	0.01	0.9318
Pre_Chol*diet	3	2334.81	778.27	3.19	0.0417

lines reveals that the slopes for diets 1 and 2 are positive while the slopes for diets 3 and 4 are negative. The negative slopes and the positive slopes are not quite significantly different from zero (using a Bonferroni adjustment), but the positive slopes are significantly different from the negative slopes. Thus, a model with unequal slopes is needed to adequately describe the data set. Comparisons among the diets are accomplished by using the unequal slopes model. Since the slopes are unequal, the diets need to be compared at least at three values of PRE_CHOL. For this study, the three values are the 75th percentile, median, and 25th percentile of the studies PRE_CHOL data, which are 281, 227.5, and 180. The least squares means computed at the three above values are in Table 3.28. Pair- wise comparisons among the levels of DIET were carried using the Tukey method for multiple comparisons

TABLE 3.27
Estimates of the Intercepts and Slopes for Full Rank Model of Table 3.25

Parameter	Estimate	StdErr	tValue	Probt
diet 1	137.63	27.27	5.05	0.0000
diet 2	195.74	32.03	6.11	0.0000
diet 3	223.73	33.71	6.64	0.0000
diet 4	276.60	23.67	11.69	0.0000
Pre_Chol*diet 1	0.2333	0.1113	2.10	0.0467
Pre_Chol*diet 2	0.0450	0.1367	0.33	0.7448
Pre_Chol*diet 3	–0.0232	0.1476	–0.16	0.8763
Pre_Chol*diet 4	–0.2333	0.1038	–2.25	0.0341

TABLE 3.28
PROC GLM Code and Corresponding Least Squares or Adjusted Means Evaluated at Three Values of PRE_CHOL of Example 3.4

```
lsmeans diet/pdiff stderr at pre_chol=281
  adjust=Tukey;  ***75th percentile;
lsmeans diet/pdiff stderr at pre_chol=227.5
  adjust=Tukey;  ***median or 50th percentile;
lsmeans diet/pdiff stderr at pre_chol=180
  adjust=Tukey;  ***25th percentile;
```

Post Chol	Diet	LSMean	StdErr	LSMean Number
281	1	203.19	7.16	1
	2	208.39	8.81	2
	3	217.20	9.91	3
	4	211.05	8.26	4
227.5	1	190.71	5.69	1
	2	205.98	5.54	2
	3	218.45	5.53	3
	4	223.53	5.55	4
180	1	179.63	8.66	1
	2	203.84	8.87	2
	3	219.55	8.67	3
	4	234.61	7.02	4

within each level of PRE_CHOL. The significance levels of the Tukey comparisons are in Table 3.29.

There were no significant differences among the DIET's means of POST_CHOL at PRE_CHOL=281: the mean of DIET 1 is significantly lower than the means of

TABLE 3.29
Tukey Significance Levels for Comparing the Levels of Diet at Each Value of Pre_Chol

Pre Chol	Row Name	1	2	3	4
281	1		0.9675	0.6653	0.8886
	2	0.9675		0.9092	0.9961
	3	0.6653	0.9092		0.9635
	4	0.8886	0.9961	0.9635	
227.5	1		0.2456	0.0094	0.0020
	2	0.2456		0.4013	0.1416
	3	0.0094	0.4013		0.9149
	4	0.0020	0.1416	0.9149	
180	1		0.2334	0.0164	0.0003
	2	0.2334		0.5918	0.0541
	3	0.0164	0.5918		0.5411
	4	0.0003	0.0541	0.5411	

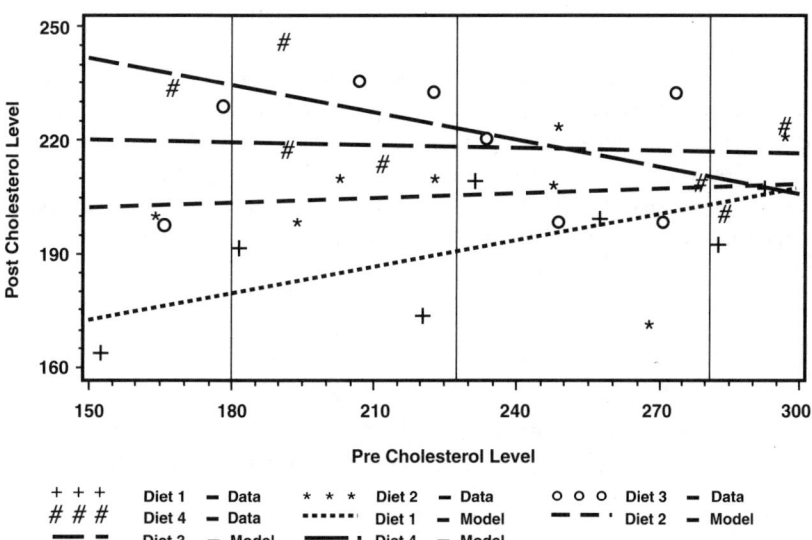

FIGURE 3.12 Graph of the diet models and data as a function of the pre-diet values for Example 3.4.

DIETs 3 and 4 at PRE_CHOL=227.5 and 180. A graph of the data and of the estimated models is in Figure 3.12, indicating there are large diet differences at low pre-diet cholesterol levels and negligible differences between the diets at high pre-diet cholesterol levels.

3.5 CHANGE FROM BASE LINE ANALYSIS USING EFFECT OF DIET ON CHOLESTEROL LEVEL DATA

There is a lot of confusion about the analysis of change from base line data. It might be of much interest to the dietician to evaluate the change in cholesterol level from the base line measurement or pre-diet cholesterol level discussed in Section 3.4. Some researchers think that by calculating the change of base line and then using analysis of variance to analyze that change, there is no need to consider base line as a covariate in the modeling process. The data set from Example 3.4 is used in the following to shed some light on the analysis of change from base line data. Table 3.30 contains the analysis of variance of the change from base line data calculated for each person as post cholesterol minus pre cholesterol. The estimate of the variance from Table 3.30 is 2695.58 compared to 243.87 for the analysis of covariance model from Table 3.15. In fact, the estimate of the variance based on the analysis of variance of just the post cholesterol values (ignoring the pre measurements) is 292.49 (analysis is not shown). So the change from base line data has tremendously more variability than the post diet cholesterol data. The analysis in Table 3.30 provides an F statistic for comparing diet means with a significance level of 0.2621. The analysis of variance on the post cholesterol (without the covariate) provides an F statistic with a significance level of 0.0054 and using the multiple comparisons, one discovers that the mean cholesterol level of diet 1 is significantly less than the means of diets 3 and 4. Therefore, the analysis of change from base line data is not necessarily providing appropriate information about the effect of diets on a person's cholesterol level.

TABLE 3.30
PROC GLM Code and Analysis of Variance of Change from Baseline, Pre Minus Post without the Covariate

```
proc glm data=one;class diet;
model change=diet;
```

Source	df	SS	MS	FValue	ProbF
Model	3	11361.09	3787.03	1.40	0.2621
Error	28	75476.13	2695.58		
Corrected Total	31	86837.22			

Source	df	SS(Type III)	MS	FValue	ProbF
diet	3	11361.09	3787.03	1.40	0.2621

Examples: One-Way Analysis of Covariance

What happens if the covariate is also used in the analysis of the change from base line data? Assume there are t treatments where y represents the response variable or post measurement and x denotes the covariate or pre measurement. Also, assume the simple linear regression model describes the relationship between the mean of y given x and x for each of the treatments. Then the model is

$$y_{ij} = \alpha_i + \beta_i x_{ij} + \varepsilon_{ij}, \quad i = 1, \ldots, t, \quad j = 1, \ldots, n_i$$

Next compute the change from base line data as $c_{ij} = y_{ij} - x_{ij}$. The corresponding model for c_{ij} is

$$c_{ij} = y_{ij} - x_{ij} = \alpha_i + (\beta_i - 1)x_{ij} + \varepsilon_{ij}, \quad i = 1, \ldots, t, \quad j = 1, \ldots, n_i$$

$$= \alpha_i + \gamma_i x_{ij} + \varepsilon_{ij},$$

where the slope of the model for c_{ij} is equal to the slope for the y_{ij} model minus 1. Thus testing $H_0: \gamma_1 = \ldots = \gamma_t = 0$ vs. H_a: (not H_0:) is equivalent to testing $H_0: \beta_1 = \ldots = \beta_t = 1$ vs. H_a: (not H_0:). Therefore, in order for the analysis on the change from base line data without the covariate to be appropriate is for the slopes of the models for y_{ij} to all be equal to 1. The following analyses demonstrate the importance of using the pre value as a covariate in the analysis of the change from base line. Tables 3.31 and 3.32 contain the results of fitting Model 2.1 to the change=post cholesterol minus pre cholesterol values where the full rank model is fit to get the results in Table 3.31 and the less than full rank model is fit to get the results in Table 3.32. The estimate of the variance from Table 3.31 is 243.87, the same as the estimate of the variance obtained from the analysis of covariance model in Table 3.25. The F statistic for source Pre_Chol*diet tests the equal slopes hypothesis of Equation 2.11. The significance level is 0.0417, the same as in Table 3.26.

TABLE 3.31
PROC GLM Code and Analysis of Variance Table for Change from Base Line Data with the Covariate to Test Slopes Equal Zero

```
proc glm data=one;class diet;
model change=diet Pre_chol*diet/solution noint;
```

Source	df	SS	MS	FValue	ProbF
Model	8	92122.09	11515.26	47.22	0.0000
Error	24	5852.91	243.87		
Uncorrected Total	32	97975.00			

Source	df	SS (Type III)	MS	FValue	ProbF
diet	4	59368.57	14842.14	60.86	0.0000
Pre_Chol*diet	4	69623.21	17405.80	71.37	0.0000

TABLE 3.32
PROC GLM Code and Analysis of Variance for the Change from Base Line Data to Provide the Test of the Slopes Equal Hypothesis

```
proc glm data=one;class diet;
model change=diet Pre_chol Pre_chol*diet;
```

Source	df	SS	MS	FValue	ProbF
Model	7	80984.31	11569.19	47.44	0.0000
Error	24	5852.91	243.87		
Corrected Total	31	86837.22			

Source	df	SS (Type III)	MS	FValue	ProbF
diet	3	3718.77	1239.59	5.08	0.0073
Pre_Chol	1	60647.40	60647.40	248.69	0.0000
Pre_Chol*diet	3	2334.81	778.27	3.19	0.0417

TABLE 3.33
Estimates of the Parameter from Full Rank Analysis of Covariance Model for Change from Base Line

Parameter	Estimate	StdErr	tValue	Probt
diet 1	137.63	27.27	5.05	0.0000
diet 2	195.74	32.03	6.11	0.0000
diet 3	223.73	33.71	6.64	0.0000
diet 4	276.60	23.67	11.69	0.0000
Pre_Chol*diet 1	−0.767	0.111	−6.89	0.0000
Pre_Chol*diet 2	−0.955	0.137	−6.98	0.0000
Pre_Chol*diet 3	−1.023	0.148	−6.93	0.0000
Pre_Chol*diet 4	−1.233	0.104	−11.88	0.0000

The estimates of the intercepts and slopes for the model in Table 3.31 are displayed in Table 3.33. The intercepts are identical to those in Table 3.27 and the slopes are the slopes in Table 3.27 minus 1. Just like in the analysis of the post cholesterol data, an unequal slopes model is needed to adequately describe the data set. The adjusted means or least squares means are computed at pre cholesterol levels of 281, 227.5, and 180. Those least squares means are presented in Table 3.34. Most of the time it is not of interest to consider the t-tests associated with the least squares means. The t-statistic is a test of the hypothesis that the respective population mean is equal to zero. The adjusted means are changed from base line values and diets 3 and 4 at Pre_chol=227.5 and diet 1 at Pre_chol=180 are not significantly different from zero. Table 3.35 consists of the significance levels for Tukey adjusted multiple comparisons for all pairwise comparisons of the diets means within the Pre_chol values of 281, 227.5, and 180. These significance levels are identical to those computed from Table 3.29.

Examples: One-Way Analysis of Covariance

TABLE 3.34
PROC GLM Code and Least Squares Means for the Change from Base Line Analysis

```
lsmeans diet/pdiff stderr at pre_chol=281
adjust=Tukey;***75th percentile;
lsmeans diet/pdiff stderr at pre_chol=227.5
adjust=Tukey;***median or 50th percentile;
lsmeans diet/pdiff stderr at pre_chol=180
adjust=Tukey;***25th percentile;
```

Pre_CHOL	Diet	LSMean	StdErr	Probt	LSMeanNumber
281	1	−77.81	7.16	0.0000	1
	2	−72.61	8.81	0.0000	2
	3	−63.80	9.91	0.0000	3
	4	−69.95	8.26	0.0000	4
227.5	1	−36.79	5.69	0.0000	1
	2	−21.52	5.54	0.0007	2
	3	−9.05	5.53	0.1148	3
	4	−3.97	5.55	0.4820	4
180	1	−0.37	8.66	0.9660	1
	2	23.84	8.87	0.0128	2
	3	39.55	8.67	0.0001	3
	4	54.61	7.02	0.0000	4

TABLE 3.35
Tukey Adjusted Significance Levels for Pairwise Comparisons of the Diets' Means at Three Levels of Pre Cholesterol

PRE_CHOL	RowName	_1	_2	_3	_4
281	1		0.9675	0.6653	0.8886
	2	0.9675		0.9092	0.9961
	3	0.6653	0.9092		0.9635
	4	0.8886	0.9961	0.9635	
227.5	1		0.2456	0.0094	0.0020
	2	0.2456		0.4013	0.1416
	3	0.0094	0.4013		0.9149
	4	0.0020	0.1416	0.9149	
180	1		0.2334	0.0164	0.0003
	2	0.2334		0.5918	0.0541
	3	0.0164	0.5918		0.5411
	4	0.0003	0.0541	0.5411	

In summary, when it is of interest to evaluate change from base line data, do the analysis, but still consider the base line values as possible covariates. The only time the analysis of variance of change from base line data is appropriate is when the slopes are all equal to 1. As this example shows, change from base line values can have considerable effect on the estimate of the variance and thus on the resulting conclusions one draws from the data analysis. So, carry out the analysis on the change from base line variables, but also consider the base line values as possible covariates.

3.6 SHOE TREAD DESIGN DATA FOR EXCEPTION TO THE BASIC STRATEGY

The data in Table 3.36 are the times it took males to run an obstacle course with a particular tread design on the soles of their shoes (Tread Time, sec). To help remove the effect of person-to-person differences, the time required to run the same course while wearing a slick-soled shoe was also measured (Slick Time, sec). Fifteen subjects were available for the study and were randomly assigned to one of three tread designs, five per design. The data are from a one-way treatment structure with one covariate in a completely randomized design structure. It is of interest to compare mean times for tread designs for a constant time to run the course with the slick sole shoes. Table 3.37 contains the analysis to test the hypothesis that the slopes are all equal to zero (Equation 2.6), which one fails to reject ($p = 0.2064$). None of the individual slopes are significantly different from zero, but they are all in the magnitude of 0.3.

The main problem is that there are only five observations per treatment group and it is difficult to detect a non-zero slope when the sample size is small. The basic

TABLE 3.36
Obstacle Course Time Data for Three Shoe Tread Designs

Tread Design					
1		2		3	
Slick Time	Tread Time	Slick Time	Tread Time	Slick Time	Tread Time
34	36	37	29	58	38
40	36	50	40	57	32
48	38	38	35	36	29
35	32	52	34	55	34
42	39	45	29	48	31

Note: Tread Time (sec) denotes time to run the course with the assigned tread and Slick Time (sec) denotes time to run the same course using a slick-soled shoe to be considered as a possible covariate for Example 3.6.

Examples: One-Way Analysis of Covariance

TABLE 3.37
PROC GLM Code, Analysis of Variance Table, and Estimates of the Parameters for the Full Rank Model for Example 3.6

```
proc glm data=two; class tread_ds;
model Tread_time = tread_ds Slick_Time*tread_ds/noint solution ss3;
```

Source	df	SS	MS	FValue	ProbF
Model	6	17620.18	2936.70	282.65	0.0000
Error	9	93.51	10.39		
Uncorrected Total	15	17713.69			

Source	df	SS (TypeIII)	MS	FValue	ProbF
tread_ds	3	122.92	40.97	3.94	0.0476
Slick_Time*tread_ds	3	58.04	19.35	1.86	0.2064

Parameter	Estimate	StdErr	tValue	Probt
tread_ds 1	23.51	11.75	2.00	0.0765
tread_ds 2	20.07	10.55	1.90	0.0897
tread_ds 3	18.24	8.88	2.05	0.0703
Slick_Time*tread_ds 1	0.325	0.294	1.11	0.2975
Slick_Time*tread_ds 2	0.301	0.236	1.28	0.2342
Slick_Time*tread_ds 3	0.286	0.173	1.65	0.1324

TABLE 3.38
PROC GLM Code and the Analysis of Variance Table for the Analysis of Tread Time without the Covariate for Example 3.6

```
proc glm data=two; class tread_ds;where tread_time ne .;
model Tread_time = tread_ds /solution;
```

Source	df	SS	MS	FValue	ProbF
Model	2	38.05	19.03	1.51	0.2608
Error	12	151.55	12.63		
Corrected Total	14	189.60			

Source	df	SS	MS	FValue	ProbF
tread_ds	2	38.05	19.03	1.51	0.2608

strategy says to continue the analysis of the shoe tread designs via analysis of variance, i.e., without the covariate. The analysis of variance of the time to run the obstacle course to compare the tread designs without using the covariate is in Table 3.38. The analysis of variance indicates there are no significant differences among the shoe tread design means. Table 3.39 displays the means for each of the tread designs and the significance levels indicate the means of the tread designs are not significantly different. If the basic strategy is ignored and other models are used (such as a common slope model), it becomes evident that the covariate is important

TABLE 3.39
PROC GLM Code, Least Squares Means and *p*-Values for Making Pairwise Comparisons among Shoe Tread Design Means for Tread Time (sec) of Example 3.6 without the Covariate

```
lsmeans tread_ds/stderr pdiff;
```

tread_ds	LSMean	StdErr	Probt	LSMeanNumber
1	36.40	1.59	0.0000	1
2	33.40	1.59	0.0000	2
3	32.74	1.59	0.0000	3

RowName	_1	_2	_3
1		0.2067	0.1294
2	0.2067		0.7740
3	0.1294	0.7740	

TABLE 3.40
PROC GLM Code, Analysis of Variance Table and Parameter Estimates for the Common Slope Model of Example 3.6

```
proc glm data=two; class tread_ds;
model Tread_time = tread_ds Slick_Time/solution;
```

Source	df	SS	MS	FValue	ProbF
Model	3	95.96	31.99	3.76	0.0444
Error	11	93.64	8.51		
Corrected Total	14	189.60			

Source	df	SS (Type III)	MS	FValue	ProbF
tread_ds	2	85.95	42.97	5.05	0.0278
Slick_Time	1	57.91	57.91	6.80	0.0243

Parameter	Estimate	StdErr	tValue	Probt
Intercept	17.67	5.92	2.98	0.0125
tread_ds 1	6.92	2.23	3.10	0.0100
tread_ds 2	2.54	1.98	1.28	0.2261
tread_ds 3	0.00			
Slick_Time	0.298	0.114	2.61	0.0243

in the comparison of the tread designs. The common slope model analysis is displayed in Table 3.40, which indicates there is a significant effect due to the covariate ($p = 0.0243$), i.e., indicating the common slope is significantly different from zero. Thus while there is not enough information from the individual tread design's data to conclude their slopes are different from zero, the combined data sets for a common slope does provide an estimate of the slope which is significantly different than zero.

Examples: One-Way Analysis of Covariance

TABLE 3.41
PROC GLM Code, Least Squares Means, and *p*-Values for Comparing the Shoe Tread Design Means

```
lsmeans tread_ds/stderr pdiff e;
```

tread_ds	LSMean	StdErr	Probt	LSMeanNumber
1	37.94	1.43	0.0000	1
2	33.57	1.31	0.0000	2
3	31.03	1.46	0.0000	3

RowName	_1	_2	_3
1		0.0436	0.0100
2	0.0436		0.2261
3	0.0100	0.2261	

The overall test of TREAD_DS in Table 3.40 indicates there is enough information to conclude that the tread designs yield different times, a different decision than from the analysis without the covariate (Table 3.38).

The adjusted means (LSMEANS at Slick Time = 44.88667) in Table 3.41 indicate that designs 2 and 3 are possibly better than design 1. If a Bonferroni adjustment is used, then runners using design 3 run significantly faster than design 1 ($\alpha = 0.05$). The graph of the estimated regression lines with a common slope is in Figure 3.13.

This example shows two important aspects of analysis of covariance. First, there could be enough evidence to conclude that a common slope model is necessary to

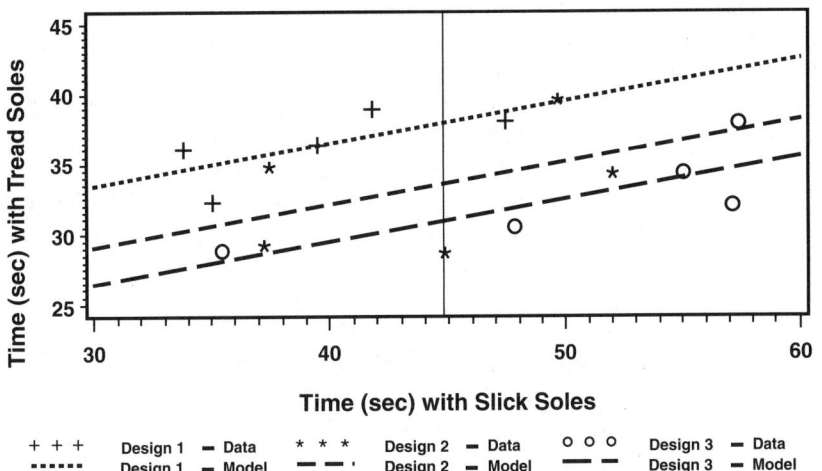

FIGURE 3.13 Plot of data and estimated models for each of the three tread designs for the common slope model of Example 3.6.

describe the data when the overall test of the hypothesis that the slopes are all equal to zero is not rejected. Second, the inclusion of the covariate in the model drastically changes the conclusions about the tread designs. Without the covariate, there were no significant differences among the design means, but with the covariate, significant differences appear.

3.7 EQUAL SLOPES WITHIN GROUPS OF TREATMENTS AND UNEQUAL SLOPES BETWEEN GROUPS

The data in Table 3.42 are the weights of potato bulbs harvested per plot (lb), y, after treated with one of six insecticides (TRT). To adjust for possible plot to plot differences in the initial infestation, the average number of insects per plant (from 10 plants) (cov) were measured to be considered as a possible covariate before the application of the insecticides.

The data come from a completely randomized design structure with a one-way treatment structure and one covariate. The SAS® system code to fit the unequal slopes model to the data set and provide the test of the slopes equal to zero hypothesis is in Table 3.43. The significance level of the statistic associated with COV*TRT is 0.0000, indicating the slopes are not all equal to zero. The SAS® system code for fitting the model to provide a test of the equality of the slopes is in Table 3.44.

The significance level associated with COV*TRT when COV is also in the model is 0.0000, indicating the slopes are not all equal. The statements in the upper part of Table 3.45 were used to provide multiple comparisons of the slopes of the regression lines of the insecticides. The results of the analysis are in the lower part of Table 3.45. Since the analysis in Table 3.44 indicates the slopes are significantly different, the comparisons between the slopes can be viewed as a Fisher Protected LSD type of multiple comparison (see Milliken and Johnson, 1992). The comparisons of the slopes indicate that the insecticides can be classified into three groups with equal slopes within a group and unequal slopes between groups. The three groups are group one contains insecticides 1, 2, and 3; group two contains insecticides 4 and 5; and group three contains insecticide 6.

A model with three slopes can be used to describe this data, which makes the comparisons between the insecticides a little simpler. To compare insecticides with in a group, one need only compare adjusted means evaluated at a single value of the covariate, while treatment mean comparisons between groups must be compared at several values of the covariate. The model with a different slope for each group is

$$y_{ij} = \mu_i + \beta_1 x_{ij} + \varepsilon_{ij} \text{ for } i = 1, 2, 3 \text{ and } j = 1, 2, \ldots, 10$$

$$y_{ij} = \mu_i + \beta_2 x_{ij} + \varepsilon_{ij} \text{ for } i = 4, 5 \text{ and } j = 1, 2, \ldots, 10$$

$$y_{6j} = \mu_6 + \beta_3 x_{6j} + \varepsilon_{6j} \text{ for } j = 1, 2, \ldots, 10.$$

TABLE 3.42
Data for Example 3.7, for Studying the Effect of Insecticide Treatments on Bulb Weight with Initial Infestation Used as Covariate

Treat 1		Treat 2		Treat 3		Treat 4		Treat 5		Treat 6	
Bulb Weight	Insects/Plant	Bulb Weight	Insects/Plant	Bulb Weight	Insects/Plant	Bulb Weight	Insects/Plant	Bulb Weight	Insects/Plant	Bulb Weight	Insects/Plant
47.93	2.9	61.00	2.5	59.53	2.6	58.33	4.2	64.62	8.5	68.07	6.0
55.68	6.0	55.93	3.5	40.33	2.8	58.19	1.7	50.19	8.1	72.64	8.0
52.97	4.2	54.66	3.8	62.67	2.8	62.12	4.0	62.76	2.7	42.69	4.4
47.92	4.8	63.52	4.6	69.83	4.7	50.68	5.0	55.88	7.8	42.17	3.6
62.99	7.6	72.23	7.1	66.59	5.2	63.45	1.2	80.64	1.1	45.15	4.8
50.53	1.9	55.29	3.3	55.06	4.9	59.77	2.7	61.34	7.1	73.88	8.0
50.98	4.9	64.08	6.5	74.92	7.7	53.79	7.4	50.88	4.8	43.11	3.8
49.63	1.2	63.98	7.1	67.43	3.5	51.94	6.5	57.83	8.3	45.87	4.0
47.77	2.5	52.04	4.6	80.32	6.9	62.35	1.9	59.92	5.1	29.23	2.6
55.35	6.1	69.71	6.3	68.29	6.6	50.95	3.8	63.84	2.9	45.18	3.0

TABLE 3.43
PROC GLM Code to Fit the Full Rank Model and Analysis of Variance Table with Test for Slopes All Equal to Zero for Example 3.7

```
proc glm data = one; classes trt;
model y = trt cov*trt / noint solution ss3;
```

Source	df	SS	MS	FValue	ProbF
Model	12	204831.54	17069.29	512.17	0.0000
Error	48	1599.70	33.33		
Uncorrected Total	60	206431.24			

Source	df	SS (Type III)	MS	FValue	ProbF
trt	6	25390.42	4231.74	126.98	0.0000
cov*trt	6	3073.47	512.25	15.37	0.0000

TABLE 3.44
PROC GLM Code to Fit Less Than Full Rank Model and Analysis of Variance Table to Test Slopes Equal for Example 3.7

```
proc glm data = one; classes trt;
model y = trt cov cov*trt / solution ss3;
```

Source	df	SS	MS	FValue	ProbF
Model	11	4540.63	412.78	12.39	0.0000
Error	48	1599.70	33.33		
Corrected Total	59	6140.34			

Source	df	SS (Type III)	MS	FValue	ProbF
trt	5	2828.42	565.68	16.97	0.0000
cov	1	949.84	949.84	28.50	0.0000
cov*trt	5	2610.44	522.09	15.67	0.0000

This type of modeling process is meaningful only if the comparisons between the slopes separate the treatments into non-overlapping groups, as is the case with this example.

A group variable was constructed to take on three values to correspond to the above classification of the slopes. Table 3.46 contains the SAS® system code and the analysis of variance table. The terms GROUP and TRT(GROUP) are used to describe the intercepts of the model and COV*GROUP is used to describe the slopes.

The residual mean square for the model with six slopes (see Table 3.43) is 33.33, while the residual mean square for the model with three slopes is 33.24. Thus, reducing the number of slope parameters from six to three did not distract from the model's ability to fit the data. Adjusted means for the insecticides at cov = 2, 5, and 8

TABLE 3.45
PROC GLM Estimate Statements to Make All Pairwise Comparisons among the Slopes for Example 3.7

```
estimate 'slope1-slope2' cov*trt 1 -1 0 0 0 0;
estimate 'slope1-slope3' cov*trt 1 0 -1 0 0 0;
estimate 'slope1-slope4' cov*trt 1 0 0 -1 0 0;
estimate 'slope1-slope5' cov*trt 1 0 0 0 -1 0;
estimate 'slope1-slope6' cov*trt 1 0 0 0 0 -1;
estimate 'slope2-slope3' cov*trt 0 1 -1 0 0 0;
estimate 'slope2-slope4' cov*trt 0 1 0 -1 0 0;
estimate 'slope2-slope5' cov*trt 0 1 0 0 -1 0;
estimate 'slope2-slope6' cov*trt 0 1 0 0 0 -1;
estimate 'slope3-slope4' cov*trt 0 0 1 -1 0 0;
estimate 'slope3-slope5' cov*trt 0 0 1 0 -1 0;
estimate 'slope3-slope6' cov*trt 0 0 1 0 0 -1;
estimate 'slope4-slope5' cov*trt 0 0 0 1 -1 0;
estimate 'slope4-slope6' cov*trt 0 0 0 1 0 -1;
estimate 'slope5-slope6' cov*trt 0 0 0 0 1 -1;
```

Parameter	Estimate	StdErr	tValue	Probt
slope1-slope2	−1.00	1.47	−0.68	0.4996
slope1-slope3	−2.35	1.40	−1.68	0.1001
slope1-slope4	3.52	1.32	2.66	0.0106
slope1-slope5	3.79	1.17	3.22	0.0023
slope1-slope6	−5.60	1.37	−4.09	0.0002
slope2-slope3	−1.35	1.54	−0.87	0.3868
slope2-slope4	4.52	1.47	3.07	0.0035
slope2-slope5	4.79	1.34	3.57	0.0008
slope2-slope6	−4.60	1.51	−3.04	0.0038
slope3-slope4	5.87	1.40	4.19	0.0001
slope3-slope5	6.13	1.26	4.86	0.0000
slope3-slope6	−3.25	1.44	−2.25	0.0288
slope4-slope5	0.26	1.18	0.22	0.8243
slope4-slope6	−9.12	1.37	−6.65	0.0000
slope5-slope6	−9.39	1.23	−7.64	0.0000

are in Table 3.47 and the comparisons between pairs of the adjusted means are in Table 3.48. The significance levels for all pairwise comparisons between the insecticide means are displayed at the top of Table 3.48 where COV = 2.

Only comparisons between insecticides from different groups are included in the lower two parts of Table 3.48. By grouping the insecticides, only 11 comparisons need be made at each value of COV, whereas without grouping, 15 comparisons would need to be made. The advantage is greater with larger group sizes and fewer groups. The other aspect of this process, although not necessarily appropriate here, is that those insecticides with similar slopes may be similar acting compounds, i.e.,

TABLE 3.46
PROC GLM Code to Fit a Model with Three Slopes, One for Each Level of GROUP, to the Data of Example 3.6

```
proc glm data = one; classes trt group;
model y = group trt(group) cov*group/solution ss3;
```

Source	df	SS	MS	FValue	ProbF
Model	8	4445.26	555.66	16.72	0.0000
Error	51	1695.07	33.24		
Corrected Total	59	6140.34			

Source	df	SS	MS	FValue	ProbF
group	2	2776.52	1388.26	41.77	0.0000
trt(group)	3	789.20	263.07	7.91	0.0002
cov*group	3	2978.10	992.70	29.87	0.0000

TABLE 3.47
Lsmeans Statements with Adjusted Means for the Insecticides Evaluated at COV = 2, 5, and 8, for Example 3.7

```
lsmeans trt(group)/pdiff at cov=2 stderr;
lsmeans trt(group)/pdiff at cov=5 stderr;
lsmeans trt(group)/pdiff at cov=8 stderr;
```

trt	Group	COV	LSMEAN	StdErr	LSMEAN Number
1	1	2	45.82	2.24	1
2	1	2	52.82	2.52	2
3	1	2	56.54	2.45	3
4	2	2	60.56	2.10	4
5	2	2	67.53	2.75	5
6	3	2	29.82	3.36	6
1	1	5	54.45	1.88	1
2	1	5	61.45	1.82	2
3	1	5	65.16	1.83	3
4	2	5	55.01	1.94	4
5	2	5	61.98	1.86	5
6	3	5	52.14	1.83	6
1	1	8	63.07	2.89	1
2	1	8	70.07	2.57	2
3	1	8	73.78	2.64	3
4	2	8	49.45	2.98	4
5	2	8	56.42	2.26	5
6	3	8	74.46	3.67	6

Examples: One-Way Analysis of Covariance

TABLE 3.48
Significance Levels of t-Statistics for Comparing Pairs of Means at COV=2, 5, and 8, for Example 3.7

RowName	COV	_1	_2	_3	_4	_5	_6
1	2		0.0099	0.0001	0.0000	0.0000	0.0002
2	2	0.0099		0.1562	0.0220	0.0002	0.0000
3	2	0.0001	0.1562		0.2178	0.0044	0.0000
4	2	0.0000	0.0220	0.2178		0.0152	0.0000
5	2	0.0000	0.0002	0.0044	0.0152		0.0000
6	2	0.0002	0.0000	0.0000	0.0000	0.0000	
1	5				0.8356	0.0064	0.3838
2	5				0.0192	0.8396	0.0007
3	5				0.0004	0.2277	0.0000
4	5	0.8356	0.0192	0.0004			0.2867
5	5	0.0064	0.8396	0.2277			0.0004
6	5	0.3838	0.0007	0.0000	0.2867	0.0004	
1	8				0.0019	0.0760	0.0182
2	8				0.0000	0.0002	0.3317
3	8				0.0000	0.0000	0.8815
4	8	0.0019	0.0000	0.0000			0.0000
5	8	0.0760	0.0002	0.0000			0.0001
6	8	0.0182	0.3317	0.8815	0.0000	0.0001	

by comparing the slopes, one might be able to make important decisions about the grouping of treatments. A graph of the data and resulting models are in Figure 3.14.

The last two examples are to help the data analyst make good decisions about the model. Too often one only looks at the summary statistics and fails to observe what is really happening with the models or how the models relate to the phenomenon being described.

3.8 UNEQUAL SLOPES AND EQUAL INTERCEPTS — PART 1

The data in Table 3.49 were generated to provide an example of where the intercepts are not different but the slopes are different. The range of the x values is 40 to 70 so there are not observations close to $x = 0$. The modeling process provides the conclusion that the slopes are not equal, so an unequal slopes model is used to describe the data set. Table 3.50 contains the SAS® system code, the analysis of variance table, and comparisons of the models at $x = 0$, 40, and 70. The analysis of variance provides a test that the treatment intercepts are equal though the source, TREAT. The significance level corresponding to the TREAT F statistic is 0.4171, indicating there is not enough evidence to believe that the treatment's intercepts are unequal.

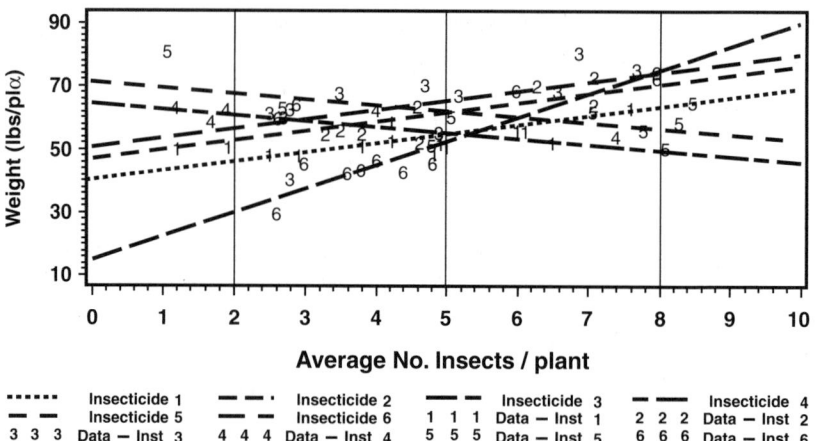

FIGURE 3.14 Graph of groups of lines indicating comparisons at 2, 6, and 8 insects per plant.

TABLE 3.49
Data for Example 3.8

Treat 1		Treat 2		Treat 3		Treat 4	
X	Y	X	Y	X	Y	X	Y
63	29	55	33	47	36	69	44
48	29	58	34	58	37	50	42
44	29	69	34	42	35	53	41
48	31	43	35	62	39	44	40
47	27	68	31	42	36	65	43
41	31	44	32	58	37	67	42
47	30	60	34	55	38	56	40
47	29	58	32	50	37	44	38
67	31	70	35	58	39	54	38
43	29	65	33	51	35	49	41
58	29	69	32			50	39
41	30	64	36				
68	30						

The question is, if the intercepts are not unequal, does that imply that the treatments are not different? Remember that the test provided by the TREAT F statistic is that the models are equal at x = 0. Since x = 0 is not near the range of x in the data set, it is unreasonable to compare the intercepts. The contrast statements in Table 3.50 and the corresponding results show that the regression lines are significantly different at X = 40 and X = 70. The contrast statement for comparing the regression models at X = 0 is provide to demonstrate that the same result is obtained as provided by

Examples: One-Way Analysis of Covariance

TABLE 3.50
PROC GLM Code and Analysis of Variance Table with Comparisons of the Regression Models at 0, 40, and 70 for Example 3.8

```
proc glm DATA=SET1; class treat;
model y=treat x*treat/solution ss3;
contrast "lines = at 0" treat 1 -1 0 0, treat 1 0 -1 0,
  treat 1 0 0 -1;
contrast "lines = at 40" treat 1 -1 0 0 x*treat 40 -40 0 0,
  treat 1 0 -1 0 x*treat 40 0 -40 0,
  treat 1 0 0 -1 x*treat 40 0 0 -40;
contrast "lines = at 70" treat 1 -1 0 0 x*treat 70 -70 0 0,
  treat 1 0 -1 0 x*treat 70 0 -70 0,
  treat 1 0 0 -1 x*treat 70 0 0 -70;
```

Source	df	SS	MS	FValue	ProbF
Model	7	845.26	120.75	70.21	0.0000
Error	38	65.35	1.72		
Corrected Total	45	910.61			

Source	df	SS (Type III)	MS	FValue	ProbF
treat	3	5.00	1.67	0.97	0.4171
x*treat	4	31.88	7.97	4.63	0.0038

Source	df	SS	MS	FValue	ProbF
lines = at 0	3	5.00	1.67	0.97	0.4171
lines = at 40	3	159.17	53.06	30.85	0.0000
lines = at 70	3	275.82	91.94	53.46	0.0000

the TREAT F statistic. So, just because the models have equal intercepts as measured by the TREAT F statistic does not mean that the models are equal for every value of X.

Figures 3.15 and 3.16 provide plots of the data and the models where Figure 3.15 uses 0 to 70 for the range of X and Figure 3.16 uses 40 to 70 (the range of the data) for the range of X. The straight lines approximate the relationship between Y and X in the range of the data, but since there are no data in the vicinity of X = 0, one should not attempt to interpret the F-test corresponding to TREAT.

3.9 UNEQUAL SLOPES AND EQUAL INTERCEPTS — PART 2

The data in Table 3.51 were generated to provide another example of where the intercepts are not different, the slopes are different, but the models are not significantly different through the range of the covariate. The range of the x values is 40 to 70, so there are no observations close to x = 0. The modeling process provides that the slopes are not equal, so an unequal slopes model is used to describe the data set. Table 3.52 contains the SAS® system code, the analysis of variance table, and comparisons of the models at x = 0, 40, and 70. The analysis of variance provides a test that the treatment intercepts are equal though the source, TREAT. The significance

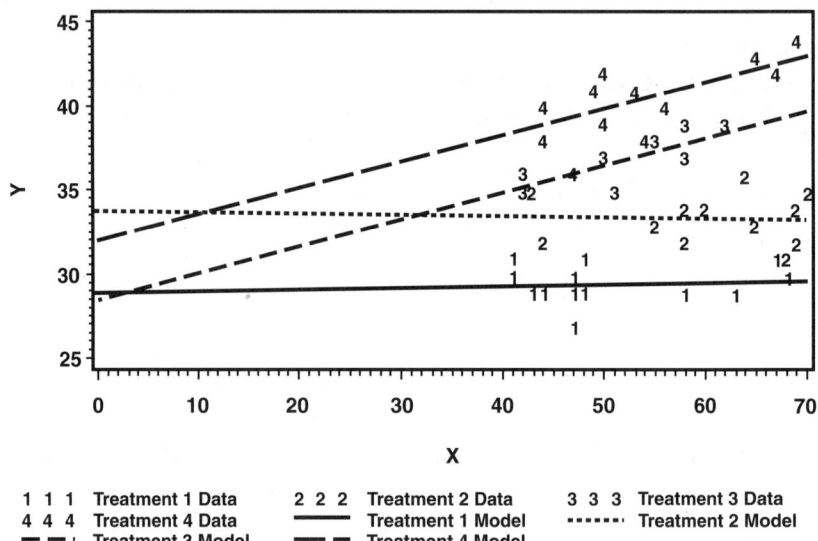

FIGURE 3.15 Plots of the data and predicted models over the range of X from 0 to 70 for Example 3.8.

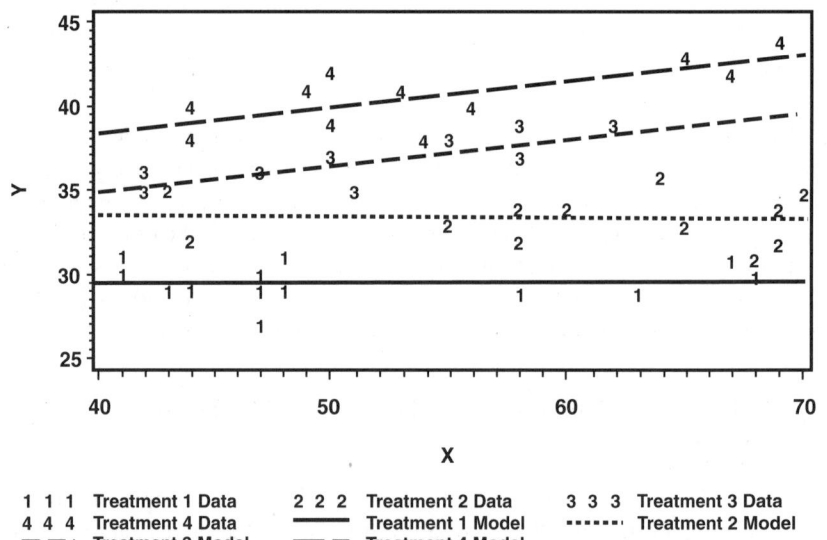

FIGURE 3.16 Plots of the data and predicted models over the range of X = 40 to 70 for Example 3.8.

Examples: One-Way Analysis of Covariance

TABLE 3.51
Data for Example 3.9

Treat 1		Treat 2		Treat 3		Treat 4	
X	Y	X	Y	X	Y	X	Y
63	7	51	5	44	4	50	9
48	9	45	10	60	6	64	4
64	6	63	7	44	11	58	7
61	1	67	2	61	5	43	11
41	10	65	3	57	8	57	5
54	10	52	8	57	7	64	9
68	5	54	9	60	6	45	9
54	5	50	7	45	9	41	8
48	7	62	3	49	7	45	8
51	9	68	4	56	7	47	8
60	3			67	5	54	10
70	4			42	10	42	7
55	9			63	7		
63	3						
43	7						

level corresponding to the TREAT F statistic is 0.4450, indicating there is not enough evidence to believe that the treatment's intercepts are unequal. The question is, if the intercepts are not unequal, does that imply that the treatments are not different? Remember that the test provided by the TREAT F statistic is that the models are equal at x = 0. The contrast statements in Table 3.52 and the corresponding results show that the regression lines are not significantly different at X = 40 and X = 70. The contrast statement for comparing the regression models at X = 0 is provided to demonstrate that the same result is obtained as obtained by the TREAT F statistic.

So, here is a case where the intercepts are not unequal, the slopes are unequal, but the models within the range of the data are not significantly different. Figures 3.17 and 3.18 provide plots of the data and the models where Figure 3.17 uses 0 to 70 for the range of X and Figure 3.18 uses 40 to 70 (the range of the data) for the range of X. Again, as for Example 3.8, the straight lines approximate the relationship between Y and X in the range of the data, but since there are no data in the vicinity of X = 0, one should not attempt to interpret the F-test corresponding to TREAT.

TABLE 3.52
PROC GLM Code and Analysis of Variance with Comparisons of the Regression Models at 0, 40, and 70 for the Example 3.9

```
proc glm DATA=SET1; class treat;
model y=treat x*treat/solution ss3;
contrast "lines = at 0" treat 1 -1 0 0, treat 1 0 -1 0,
  treat 1 0 0 -1;
contrast "lines = at 40" treat 1 -1 0 0 x*treat 40 -40 0 0,
  treat 1 0 -1 0 x*treat 40 0 -40 0,
  treat 1 0 0 -1 x*treat 40 0 0 -40;
contrast "lines = at 70" treat 1 -1 0 0 x*treat 70 -70 0 0,
  treat 1 0 -1 0 x*treat 70 0 -70 0,
  treat 1 0 0 -1 x*treat 70 0 0 -70;
```

Source	df	SS	MS	FValue	ProbF
Model	7	149.04	21.29	5.89	0.0001
Error	42	151.77	3.61		
Corrected Total	49	300.82			

Source	df	SS	MS	FValue	ProbF
treat	3	9.85	3.28	0.91	0.4450
x*treat	4	116.77	29.19	8.08	0.0001

Source	df	SS	MS	FValue	ProbF
lines = at 0	3	9.85	3.28	0.91	0.4450
lines = at 40	3	3.69	1.23	0.34	0.7960
lines = at 70	3	18.26	6.09	1.68	0.1849

Examples: One-Way Analysis of Covariance

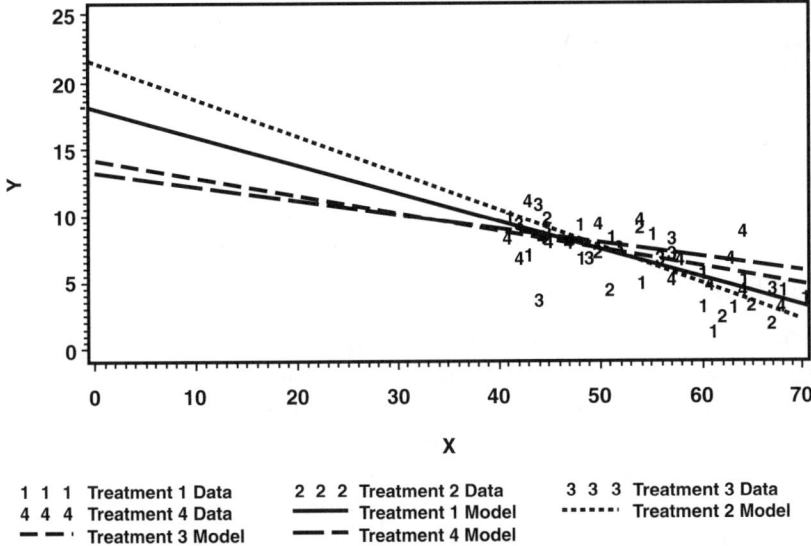

FIGURE 3.17 Plot of data and predicted models over the range of X = 0 to 70 for Example 3.9.

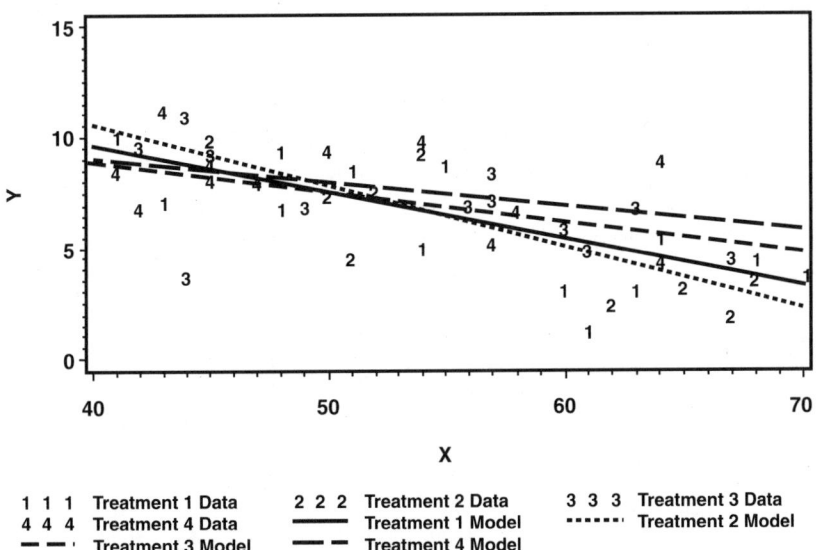

FIGURE 3.18 Plot of data and predicted models over the range of X = 40 to 70 for Example 3.9.

REFERENCES

Littell, R. C., Milliken, G. A., Stroup, W. W., and Wolfinger, R. D. (1996) *SAS® System for Mixed Models,* SAS Institute Inc., Cary, NC.

Milliken, G.A. and Johnson, D.E. (1992) *Analysis of Messy Data, Volume I: Design Experiments,* Chapman & Hall, London.

Westfall, P. H., Tobias, R. D., Rom, D., Wolfinger, R. D., and Hochberg, Y. (1999) *Multiple Comparisons and Multiple Tests Using the SAS® System,* SAS Institute Inc., Cary, NC.

EXERCISES

EXERCISE 3.1: In a study of growth regulators for sugar beets, it was determined that there was substantial plot to plot variation in the level of available nitrogen. The amount of nitrogen in the soil can affect the yield of the beets in addition to an effect due to the growth regulators. After planting the sugar beets, the available nitrogen was measured from soil samples obtained from each plot. The available nitrogen in the soil (pounds/acre) is to be used as the covariate and the dependent variable is the yield of the sugar beet roots per plot in pounds. The experimental design is a one-way treatment structure with three growth regulators (TRT) and ten plots per treatment in a completely randomized design structure. The data are in the following table. Use the strategy of analysis of covariance to determine if there are differences in the growth regulators ability to increase the amount of sugar beets per plot.

Sugar Beet Yields (lb) and Nitrogen Levels per plot (lb/acre) for Exercise 3.1

Treat 1		Treat 2		Treat 3	
YIELD	NIT	YIELD	NIT	YIELD	NIT
210	100	155	65	155	55
150	50	150	55	160	55
200	105	170	65	165	65
180	75	175	75	185	70
190	80	185	75	185	80
220	110	195	90	200	85
170	60	205	85	205	90
170	70	210	100	215	105
190	90	215	95	220	105
220	100	220	100	220	100

EXERCISE 3.2: Carry out an analysis of covariance for the following data set by determining the appropriate model and then making the needed treatment comparisons. Y is the response variable and X is the covariate. The design is a one-way treatment structure in a completely randomized design structure.

Data for Exercise 3.2

Treatment A		Treatment B		Treatment C		Treatment D	
X	Y	X	Y	X	Y	X	Y
42.6	33.4	27.7	43.7	23.3	40.4	45.8	27.3
30.6	40.8	25.3	47.0	36.5	34.4	47.2	23.2
46.3	33.7	46.1	35.1	47.8	21.5	35.7	34.0
41.5	39.3	43.1	39.9	41.9	32.6	46.1	31.0
20.3	50.5	27.2	53.0	33.9	39.4	25.0	44.8
		35.2	37.1	49.0	27.0	23.9	47.6
		49.5	25.2				

EXERCISE 3.3: Carry out an analysis of covariance for the following data set by determining the appropriate model and then making the needed treatment comparisons. Y is the response variable and X is the covariate. The experimental design is a one-way treatment structure in a completely randomized design structure.

Data for Exercise 3.3

Treatment A		Treatment B		Treatment C		Treatment D	
X	Y	X	Y	X	Y	X	Y
3.7	15.4	3.8	16.9	6.6	15.2	6.6	18.3
5.8	16.0	5.3	17.5	3.1	14.1	4.2	15.2
5.5	17.6	7.7	16.5	6.7	15.0	3.5	15.3
6.7	16.8	2.4	15.6	6.2	15.3	4.5	15.7
4.7	16.9	7.2	16.1	6.7	15.4	3.2	15.3
6.3	16.7	2.1	15.9	4.4	14.7	4.8	16.8

4 Multiple Covariates in a One-Way Treatment Structure in a Completely Randomized Design Structure

4.1 INTRODUCTION

Multiple covariates occur when more than one characteristic is needed to help account for the variability among the experimental units. If the relationship between the mean of y given the set of x values is a linear function of those x values, the resulting regression model is a plane or hyper-plane. Extensions of this model include polynomial regression models in one or more covariates. A recommended strategy is to determine the simplest relationship for the covariate part of the model and then compare the treatments or populations by comparing the regression models at selected values of the covariates. For additional information see the special issues on Analysis of Covariance in *Biometrics* 1957 and 1982.

4.2 THE MODEL

There are many experimental situations where more than one covariate is measured on each experimental unit. For data collected from a one-way treatment structure in a completely randomized design structure with k covariates, a model for each treatment can be expressed as

$$y_{ij} = \alpha_i + \beta_{1i}X_{1ij} + \beta_{2i}X_{2ij} + \ldots + \beta_{ki}X_{kij} + \varepsilon_{ij} \tag{4.1}$$

$$i = 1, 2, \ldots, t, \quad j = 1, 2, \ldots, n_i$$

where x_{pij} denotes the value of the pth covariate on the ij^{th} experimental unit, α_i denotes the intercept of the ith treatment's regression surface, β_{pi} denotes the slope in the direction of the pth covariate for treatment i, and it is assumed that ε_{ij}, j = 1,

$2, \ldots, n$, $i = 1, 2, \ldots, t$ are iid$N(0,\sigma^2)$. The model in Equation 4.1 is the equation of a k-dimensional plane in $k + 1$ dimensions which is assumed to adequately describe the data for each treatment. Thus, each treatment's model is a multiple linear regression model. The usual regression diagnostics should be applied to each treatment's data to check for model adequacy. The treatment variances should be tested for equality. When there is only one covariate, the analysis of covariance consists of comparing several regression lines. In the multiple covariates case, the analysis of covariance consists of comparing several regression planes or hyper-planes.

There are many forms these models can take on just as there are many forms of multiple linear regression models. The models for the treatments could also be a polynomial function of one independent variable as

$$y_{ij} = \alpha_i + \beta_{1i}X_{ij} + \beta_{2i}X_{ij}^2 + \ldots + \beta_{ki}X_{ij}^k + \varepsilon_{ij},$$

where the analysis of covariance would be comparing these polynomial regression models from several treatments.

The models for the treatments could be a quadratic function of two variables (like a quadratic response surface) as

$$y_{ij} = \alpha_i + \beta_{1i}X_{1ij} + \beta_{2i}X_{2ij} + \beta_{3i}X_{1ij}^2 + \beta_{4i}X_{2ij}^2 + \beta_{5i}X_{1ij}X_{2ij} + \varepsilon_{ij}.$$

The analysis of covariance would compare the t response surfaces. Four different examples are discussed in Sections 4.4 through 4.7. The matrix form of the multiple covariate model in Equation 4.1 is

$$\begin{bmatrix} y_{11} \\ y_{12} \\ \vdots \\ y_{1n_1} \\ y_{21} \\ y_{22} \\ \vdots \\ y_{2n_2} \\ \vdots \\ y_{t1} \\ y_{t2} \\ \vdots \\ y_{tn_t} \end{bmatrix} = \begin{bmatrix} 1 & X_{111} \cdots X_{k11} & 0 & 0 & \cdots & 0 & \cdots & 0 & 0 & \cdots & 0 \\ 1 & X_{112} \cdots X_{k12} & 0 & 0 & \cdots & 0 & \cdots & 0 & 0 & \cdots & 0 \\ \vdots & \vdots & \vdots & \vdots & & \vdots & & \vdots & \vdots & & \vdots \\ 1 & X_{11n_1} \cdots X_{k1n_1} & 0 & 0 & \cdots & 0 & \cdots & 0 & 0 & \cdots & 0 \\ 0 & 0 \cdots 0 & 1 & X_{121} & \cdots & X_{k21} & \cdots & 0 & 0 & \cdots & 0 \\ 0 & 0 \cdots 0 & 1 & X_{122} & \cdots & X_{k22} & \cdots & 0 & 0 & \cdots & 0 \\ \vdots & \vdots & \vdots & \vdots & & \vdots & & \vdots & \vdots & & \vdots \\ 0 & 0 \cdots 0 & 1 & X_{12n_2} & \cdots & X_{k2n_2} & \cdots & 0 & 0 & \cdots & 0 \\ \vdots & \vdots & \vdots & \vdots & & \vdots & & \vdots & \vdots & & \vdots \\ 0 & 0 \cdots 0 & 0 & 0 & \cdots & 0 & \cdots & 1 & X_{1t1} & \cdots & X_{kt1} \\ 0 & 0 \cdots 0 & 0 & 0 & \cdots & 0 & \cdots & 1 & X_{1t2} & \cdots & X_{kt2} \\ \vdots & \vdots & \vdots & \vdots & & \vdots & & \vdots & \vdots & & \vdots \\ 0 & 0 \cdots 0 & 0 & 0 & \cdots & 0 & \cdots & 1 & X_{1tn_t} & \cdots & X_{ktn_t} \end{bmatrix} \begin{bmatrix} \alpha_1 \\ \beta_{11} \\ \vdots \\ \beta_{k1} \\ \alpha_2 \\ \beta_{21} \\ \vdots \\ \beta_{k2} \\ \vdots \\ \alpha_t \\ \beta_{1t} \\ \vdots \\ \beta_{kt} \end{bmatrix} + \varepsilon \quad (4.2)$$

or $y = X\beta + \varepsilon$, where y is the data vector, X is the design matrix, β is the vector of parameters, and ε is the vector of errors. The next section presents the process of estimating the parameters of Model 4.1.

4.3 ESTIMATION

The least squares estimates of the parameters of Model 4.2 can be obtained by fitting the model in Equation 4.2 to the data as $\hat{\beta} = (\mathbf{X'X})^{-1}\mathbf{X'y}$ or by separately fitting t models of the form $\mathbf{y}_i = \mathbf{X}_i\beta_i + \varepsilon_i$, where

$$\mathbf{y}_i = \begin{bmatrix} y_{i1} \\ y_{i2} \\ \vdots \\ y_{in_i} \end{bmatrix} \quad \mathbf{X}_i = \begin{bmatrix} 1 & X_{1i1} & X_{2i1} & \cdots & X_{ki1} \\ 1 & X_{1i2} & X_{2i2} & \cdots & X_{2ik} \\ \vdots & \vdots & & & \vdots \\ 1 & X_{1in_i} & X_{2in_i} & \cdots & X_{kin_i} \end{bmatrix} \text{ and } \beta = \begin{bmatrix} \alpha_i \\ \beta_{1i} \\ \beta_{2i} \\ \vdots \\ \beta_{ki} \end{bmatrix}.$$

Either way, the least squares estimate of β_i is $\hat{\beta}_i = (\mathbf{X}'_i\mathbf{X}_i)^{-1}\mathbf{X}'_i\mathbf{y}_i$.

The estimate of the variance is obtained by pooling residual sum of squares from the t treatments or t models (one should test for equality of variances before pooling). Since the model for each treatment has $k + 1$ parameters, the pooled residual sum of squares, denoted by SSRes, is based on $N - t(k + 1)$ degrees of freedom where $N = \sum_{i=1}^{t} n_i$. The estimate of σ^2 is $\hat{\sigma}^2 = RSS/(N - t(k + 1))$. A recommended method of analysis has three parts, which are analogous to those given for the single covariate problem discussed in Chapter 2. First, determine which of the covariates are necessary to adequately describe the mean of y given the values of the covariates. Second, test for parallelism between the planes in the direction of each covariate. Third, compare the distances between the planes at various combinations of the covariates and carry out preplanned comparisons between the slopes of each covariate. If the treatment slopes are equal for each of the covariates, then the planes will be parallel. In that case, a comparison between the intercepts (α's) is also a comparison of the distances between the planes. Four examples are used to demonstrate some of the principles involved with multiple covariates.

4.4 EXAMPLE: DRIVING A GOLF BALL WITH DIFFERENT SHAFTS

A study was performed in which golfers were randomly assigned to one of three types of shafts put into a meter driver. The golfer hit five golf balls with the assigned shaft (with club head attached, of course) and the median distance was recorded as the response variable (DIST). The height (Ht in inches) and weight (Wt in pounds) of each golfer were recorded is case they might be useful as possible covariates in the analysis of the distance data. The data are in Table 4.1.

The model used to describe the Dist data is

$$\text{Dist}_{ij} = \alpha_i + \beta_{1i}\text{Ht}_{ij} + \beta_{2i}\text{Wt}_{ij} + \varepsilon_{ij}, \text{ where } \varepsilon_{ij} \sim \text{iid } N(0, \sigma^2)$$

This model assumes that each golfer is independent of the other golfers in the study (no identical twins or siblings). The matrix form of the model is

TABLE 4.1
Distances Golfers Hit Golf Balls with Different Types of Driver Shafts

Steel1			Graphite			Steel2		
Weight	Height	Distance	Weight	Height	Distance	Weight	Height	Distance
212	71	205	214	73	215	152	78	198
220	71	218	186	75	249	206	72	178
176	76	224	183	69	166	211	78	199
204	77	238	202	74	232	203	69	178
152	74	211	195	73	195	183	71	182
205	69	189	185	77	243	163	73	163
173	69	182	195	76	255	160	73	169
196	76	231	198	78	258	216	74	200
202	69	183	206	68	174	205	69	179
171	72	181	205	69	170	199	68	155

$$\begin{bmatrix} \text{Dist}_{11} \\ \vdots \\ \text{Dist}_{110} \\ \vdots \\ \vdots \\ \text{Dist}_{21} \\ \text{Dist}_{31} \\ \vdots \\ \text{Dist}_{310} \end{bmatrix} = \begin{bmatrix} 1 & HT_{11} & WT_{11} & 0 & 0 & 0 & 0 & 0 & 0 \\ \vdots & \vdots & \vdots & \vdots & \vdots & \vdots & \vdots & \vdots & \vdots \\ 1 & Ht_{110} & Wt_{110} & 0 & 0 & 0 & 0 & 0 & 0 \\ 0 & 0 & 0 & 0 & Ht_{21} & Wt_{21} & 0 & 0 & 0 \\ \vdots & \vdots & \vdots & \vdots & \vdots & \vdots & \vdots & \vdots & \vdots \\ 0 & 0 & 0 & 1 & Ht_{210} & Wt_{210} & 0 & 0 & 0 \\ 0 & 0 & 0 & 0 & 0 & 0 & 1 & Ht_{31} & Wt_{31} \\ \vdots & \vdots & \vdots & \vdots & \vdots & \vdots & \vdots & \vdots & \vdots \\ 0 & 0 & 0 & 0 & 0 & 0 & 1 & Ht_{310} & Wt_{310} \end{bmatrix} \begin{bmatrix} \alpha_1 \\ \beta_{11} \\ \beta_{21} \\ \alpha_2 \\ \beta_{12} \\ \beta_{22} \\ \alpha_3 \\ \beta_{13} \\ \beta_{23} \end{bmatrix} + \varepsilon$$

First determine if the Dist means depend on the values of the two covariates. The appropriate hypothesis to be tested are

$$H_o: \beta_{11} = \beta_{12} = \beta_{13} = 0 \text{ given the } \alpha\text{'s and the } \beta_2\text{'s are in the model, vs.} \quad (4.3)$$
$$H_a: (\text{some } \beta_{1i} \neq 0)$$

and

$$H_o: \beta_{21} = \beta_{22} = \beta_{23} = 0 \text{ given the } \alpha\text{'s and the } \beta_1\text{'s are in the model, vs.} \quad (4.4)$$
$$H_a: (\text{some } \beta_{2i} \neq 0).$$

The model restricted by the null hypothesis of Equation 4.3 is

$$\text{Dist}_{ij} = \alpha_i + \beta_{2i} Wt_{ij} + \varepsilon_{ij}$$

TABLE 4.2
PROC GLM Code and Analysis of Variance Table of the Full Model for the Golf Ball Distance Data

```
proc glm data=golf; class shaft;
model dist=shaft ht*shaft wt*shaft/noint solution;
```

Source	df	SS	MS	FValue	ProbF
Model	9	1231095.962	136788.440	1312.85	0.0000
Error	21	2188.038	104.192		
Uncorrected Total	30	1233284.000			

Source	df	SS (Type III)	MS	FValue	ProbF
shaft	3	4351.213	1450.404	13.92	0.0000
ht*shaft	3	15542.402	5180.801	49.72	0.0000
wt*shaft	3	1080.395	360.132	3.46	0.0348

TABLE 4.3
Parameter Estimates of the Full Model for the Golf Ball Distance Data

Parameter	Estimate	StdErr	tValue	Probt
shaft graphite	−572.432	113.166	−5.06	0.0001
shaft steel1	−334.630	91.523	−3.66	0.0015
shaft steel2	−145.290	86.726	−1.68	0.1087
ht*shaft graphite	10.141	1.003	10.11	0.0000
ht*shaft steel1	6.451	1.109	5.82	0.0000
ht*shaft steel2	3.688	1.016	3.63	0.0016
wt*shaft graphite	0.233	0.348	0.67	0.5111
wt*shaft steel1	0.386	0.160	2.42	0.0247
wt*shaft steel2	0.306	0.152	2.02	0.0566

and the model restricted by the null hypothesis in Equation 4.4 is

$$\text{Dist}_{ij} = \alpha_i + \beta_{1i} \text{Ht}_{ij} + \varepsilon_{ij}$$

The matrix forms for the above two models are obtained by eliminating the columns of the design matrix and parameters of β corresponding to Ht and Wt, respectively. The model comparison method can be used to compute the sums of squares appropriate for testing the hypotheses in Equations 4.3 and 4.4. The PROC GLM code and analysis of variance table are in Table 4.2 and the parameter estimates are in Table 4.3.

The hypothesis in Equations 4.3 and 4.4 is tested by the lines corresponding to Ht*Shaft and Wt*Shaft, respectively, in Table 4.2. The significance levels corresponding to both of these tests are very small indicating that the Dist mean for some treatments depends on the Ht and Wt values. The estimates of the intercepts in Table 4.3 are negative, so the planes are just approximations to the unknown model in the range of the observed Ht and Wt values. The estimates of the shaft slopes for Wt are similar, while the estimates of the shaft slopes for Ht do not appear to be similar. Given that the Dist means depend on Ht and Wt, next determine if the planes are parallel in each of these directions. The parallelism hypotheses can be studied by testing

$$H_o: \beta_{11} = \beta_{12} = \beta_{13} \text{ vs. } H_a: (\text{not } H_o) \qquad (4.5)$$

and

$$H_a: \beta_{21} = \beta_{22} = \beta_{23} \text{ vs. } H_a: (\text{not } H_o). \qquad (4.6)$$

The model restricted by the null hypothesis in Equation 4.5 is

$$\text{Dist}_{ij} = \alpha_i + \beta_1 \text{Ht}_{ij} + \beta_{2i} \text{Wt}_{ij} + \varepsilon_{ij} \qquad (4.7)$$

which has a common slope in the Ht direction and unequal slopes for the levels of shaft in the Wt direction. The model can be expressed in matrix form as

$$\begin{bmatrix} \text{Dist}_{11} \\ \vdots \\ \text{Dist}_{110} \\ \text{Dist}_{21} \\ \vdots \\ \text{Dist}_{210} \\ \text{Dist}_{31} \\ \vdots \\ \text{Dist}_{310} \end{bmatrix} = \begin{bmatrix} 1 & \text{WT}_{11} & 0 & 0 & 0 & 0 & \text{Ht}_{11} \\ \vdots & \vdots & \vdots & \vdots & \vdots & \vdots & \vdots \\ 1 & \text{Wt}_{110} & 0 & 0 & 0 & 0 & \text{Ht}_{110} \\ 0 & 0 & 1 & \text{Wt}_{21} & 0 & 0 & \text{Ht}_{21} \\ \vdots & \vdots & \vdots & \vdots & \vdots & \vdots & \vdots \\ 0 & 0 & 1 & \text{Wt}_{210} & 0 & 0 & \text{Ht}_{210} \\ 0 & 0 & 0 & 0 & 1 & \text{Wt}_{31} & \text{Ht}_{31} \\ \vdots & \vdots & \vdots & \vdots & \vdots & \vdots & \vdots \\ 0 & 0 & 0 & 0 & 0 & \text{Wt}_{310} & \text{Ht}_{310} \end{bmatrix} \begin{bmatrix} \alpha_1 \\ \beta_{21} \\ \alpha_2 \\ \beta_{22} \\ \alpha_3 \\ \beta_{23} \\ \beta_1 \end{bmatrix} + \varepsilon$$

The model restricted by the null hypothesis in Equation 4.6 is

$$\text{Dist}_{ij} = \alpha_i + \beta_{1i} \text{Ht}_{ij} + \beta_2 \text{Wt}_{ij} + \varepsilon_{ij}, \qquad (4.8)$$

and the matrix form of the model can be constructed similarly to the model in Equation 4.7.

These two hypotheses can be tested using the model comparison method or they can be tested by the software by using the appropriate model. If the PROC GLM model statement contains Ht, Wt, in addition to Ht*shaft and Wt*shaft, then test

TABLE 4.4
PROC GLM Code and Analysis of Variance Table to Test the Equality of the Ht Slopes and Equality of the Wt Slopes for the Golf Ball Distance Data

```
proc glm data=golf; class shaft;
model dist=shaft ht ht*shaft wt wt*shaft/noint
   solution;
```

Source	df	SS	MS	FValue	ProbF
Model	9	1231095.96	136788.44	1312.85	0.0000
Error	21	2188.04	104.19		
Uncorrected Total	30	1233284.00			

Source	df	SS(Type III)	MS	FValue	ProbF
shaft	3	4351.21	1450.40	13.92	0.0000
ht	1	13107.88	13107.88	125.80	0.0000
ht*shaft	2	2142.75	1071.37	10.28	0.0008
wt	1	525.64	525.64	5.04	0.0356
wt*shaft	2	23.61	11.80	0.11	0.8934

statistics for Hypotheses 4.5 and 4.6 are provided by the sum of squares lines corresponding to Ht*Shaft and Wt*Shaft, respectively, as shown in Table 4.4. For this data, reject the hypothesis in 4.5 ($p = 0.0008$) and fail to reject the hypothesis in 4.6 ($p = 0.8934$). The conclusions are that the planes describing the Dist are not parallel (unequal slopes) in the Ht direction, but the planes are parallel (equal slopes) in the Wt direction.

The model of Equation 4.8 is recommend to compare the Dist means for the different types of shafts (distances between the planes). Table 4.5 contains the analysis for Model 4.8, with the analysis of variance table and parameter estimates. Since the planes are not parallel in the Ht direction, least squares means were computed at the average value of Wt (193.6 lb) and for three values of height: 68, 73, and 78 in. The adjusted means or least squares means are listed in Table 4.6 and pairwise comparisons of these means for each value of height are in Table 4.7. At Ht = 68 in., Steel1 shaft hit the ball further than either of the other two shafts ($p < 0.10$), which were not different. At Ht = 73 in., the Graphite and Steel1 shafts hit the ball further than does Steel2 shaft ($p < 0.0001$). Finally, at Ht = 78 in., all three shafts hit the ball different distances with Graphite hitting the ball the farthest and Steel2 hitting the ball the shortest. The regression planes are shown in Figure 4.1 where "o" denotes the respective adjusted means.

4.5 EXAMPLE: EFFECT OF HERBICIDES ON THE YIELD OF SOYBEANS — THREE COVARIATES

The data in Table 4.8 are the yields (in lb) of plots of soybeans where 8 herbicides were evaluated in a completely randomized design with 12 replications per treatment.

TABLE 4.5
PROC GLM Code, Analysis of Variance Table, and Parameter Estimates for Model 4.8 for the Golf Ball Data

```
proc glm data=golf; class shaft;
model dist=shaft ht*shaft wt/noint solution;
```

Source	df	SS	MS	FValue	ProbF
Model	7	1231072.35	175867.48	1828.93	0.0000
Error	23	2211.65	96.16		
Uncorrected Total	30	1233284.00			

Source	df	SS	MS	FValue	ProbF
shaft	3	7340.41	2446.80	25.45	0.0000
ht*shaft	3	16184.70	5394.90	56.10	0.0000
wt	1	1056.79	1056.79	10.99	0.0030

Parameter	Estimate	StdErr	tValue	Probt
shaft graphite	−598.142	72.661	−8.23	0.0000
shaft steel1	−319.281	81.128	−3.94	0.0007
shaft steel2	−154.622	75.572	−2.05	0.0524
ht*shaft graphite	10.220	0.932	10.97	0.0000
ht*shaft steel1	6.377	1.053	6.06	0.0000
ht*shaft steel2	3.743	0.954	3.92	0.0007
wt	0.334	0.101	3.32	0.0030

TABLE 4.6
PROC GLM Code to Compute the Adjusted Means for wt = 192.6 lb and ht = 68, 73, and 78 in for the Golf Ball Data

```
lsmeans shaft/ pdiff at (ht wt)=(68 192.6) stderr;
lsmeans shaft/ pdiff at (ht wt)=(73 192.6) stderr;
lsmeans shaft/ pdiff at (ht wt)=(78 192.6) stderr;
```

Height	shaft	LSMEAN	StdErr	LSMEAN Number
68	graphite	161.120	5.798	1
	steel1	178.642	5.559	2
	steel2	164.191	5.255	3
73	graphite	212.221	3.139	1
	steel1	210.527	3.172	2
	steel2	182.906	3.159	3
78	graphite	263.322	5.429	1
	steel1	242.412	6.680	2
	steel2	201.621	6.152	3

Three covariates were measured on each plot: the percent silt, the percent clay, and the amount of organic matter. The first step in the process is to determine if a linear regression hyper-plane describes the data for each herbicide. After plotting the data

TABLE 4.7
p-Values for Pair Wise Comparisons among the Shaft Means at Three Values of Height

Height	Row Name	_1	_2	_3
68	1		0.0385	0.6959
	2	0.0385		0.0702
	3	0.6959	0.0702	
73	1		0.7092	0.0000
	2	0.7092		0.0000
	3	0.0000	0.0000	
78	1		0.0234	0.0000
	2	0.0234		0.0001
	3	0.0000	0.0001	

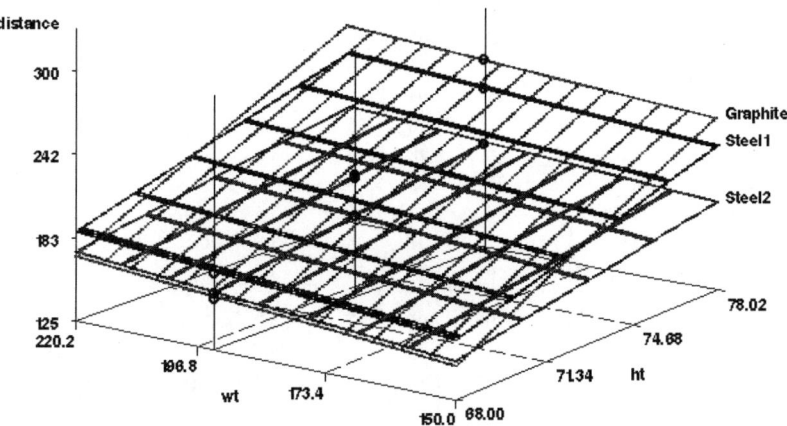

FIGURE 4.1 Graph of the estimated regression planes for each type of shaft with points of comparisons denoted by "o".

(which is not shown), the model with different slopes for each herbicide-covariate combination was fit to the data. The model with unequal slopes for each herbicide-covariate combination is

$$y_{ij} = \alpha_i + \beta_{1i} \, SILT_{ij} + \beta_{2i} \, CLAY_{ij} + \beta_{3i} \, OM_{ij} + \varepsilon_{ij} \tag{4.9}$$

$$i = 1, 2, \ldots, 8, \text{ and } j = 1, 2, \ldots, 12.$$

Figure 4.2 contains the JMP® fit model window with the model specification. Select Yield to be the dependent variable. Include Herb, Silt, Clay, OM, Herb*Silt,

TABLE 4.8
Yields of Soybeans Treated with Eight Different Herbicides from Plots with Covariates of Silt, Clay, and om

Herb	Yield	Silt	Clay	om	Herb	Yield	Silt	Clay	om
1	27.2	6	14	1.42	5	25.2	12	20	2.13
1	27.5	12	29	1.21	5	27.3	17	20	2.02
1	17.5	1	9	2.28	5	30.8	24	11	0.85
1	31.9	25	32	1.86	5	17.5	6	33	0.72
1	25	17	39	0.84	5	32.3	19	23	0.96
1	24	18	22	2.54	5	15.9	16	15	0.84
1	39.1	26	20	1.12	5	35	25	16	1.3
1	30.2	25	12	1.64	5	32	21	25	2.45
1	29.7	26	36	2.36	5	27.3	29	17	2.51
1	25	12	9	1.42	5	33.7	22	23	0.97
1	18.2	13	35	1.67	5	20	2	24	1.29
1	54.5	28	4	0.63	5	21.5	25	29	2.47
2	23.7	18	9	1.07	6	11.5	21	36	2.31
2	18.3	19	38	2.11	6	17.5	3	29	2
2	27.1	20	11	1.36	6	26.9	1	24	1.29
2	24.6	9	29	1.65	6	13	17	35	2.31
2	17.2	1	34	2.21	6	35.9	17	5	2.48
2	22.1	24	28	2.08	6	9	18	36	1.51
2	24.4	20	1	1.98	6	19.3	1	13	1.38
2	39.3	14	6	0.78	6	28.8	17	30	1.01
2	26.5	10	21	0.82	6	20.2	23	9	2.23
2	24.4	19	37	2.13	6	13.6	28	24	2.04
2	35.9	23	2	1.31	6	18.3	14	16	0.75
2	21	18	6	1.44	6	13.5	11	8	2.43
3	24.1	19	10	2.05	7	33	19	20	2.55
3	18	6	31	1.32	7	26.3	19	12	0.63
3	28.6	14	3	1.12	7	33.3	5	5	1.51
3	31.6	28	7	1.92	7	26.1	26	12	2.51
3	25.1	28	3	0.61	7	23.7	6	20	2.19
3	20.6	1	22	1.86	7	26.1	6	7	0.72
3	12.1	26	30	0.99	7	12.5	11	23	1.64
3	24.1	21	33	2.42	7	27.7	20	19	1.89
3	10.3	6	26	2.38	7	15.9	10	12	1.16
3	16.6	25	35	1.65	7	25.7	14	16	0.7
3	34.1	29	2	1.06	7	23.7	3	15	2.57
3	23.5	11	17	1.31	7	26.9	28	12	0.65
4	24.6	22	31	1.54	8	35	22	7	0.92
4	16.2	4	14	0.87	8	24.1	8	11	2.51
4	31.4	22	24	0.81	8	26.9	13	20	1.87
4	31.1	24	3	2.09	8	35.7	11	12	1.25
4	23.9	12	4	1.2	8	24	23	36	1.93
4	28.5	6	5	2.41	8	9.2	6	31	1.36

TABLE 4.8 (continued)
Yields of Soybeans Treated with Eight Different Herbicides from Plots with Covariates of Silt, Clay, and om

Herb	Yield	Silt	Clay	om	Herb	Yield	Silt	Clay	om
4	11.8	22	19	2.44	8	41	22	19	0.64
4	20	22	17	2.15	8	23.1	1	6	2.22
4	24.1	26	33	0.89	8	38.8	20	18	0.85
4	9.3	1	38	1.95	8	22.9	12	33	2.34
4	15.7	2	20	2.34	8	40.5	17	18	2.36
4	28.1	1	34	1.22	8	26.8	3	25	0.96

Note: om denotes organic matter.

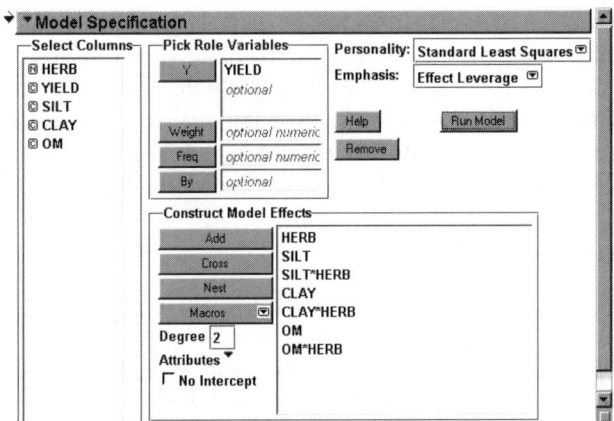

FIGURE 4.2 JMP® fit window to fit unequal slopes in each of the directions of the three covariates.

Herb*Clay, and Herb*OM as model effects and click on the "Run Model" button to fit the model. Using this model, test the hypotheses that the herbicide slopes for each of the covariates are equal, i.e., test

$$H_{01}: \beta_{11} = \beta_{12} = \ldots = \beta_{18} \text{ vs. } H_{a1}: (\text{not } H_{01}),$$

$$H_{02}: \beta_{21} = \beta_{22} = \ldots = \beta_{28} \text{ vs. } H_{a2}: (\text{not } H_{02}), \text{ and}$$

$$H_{03}: \beta_{31} = \beta_{32} = \ldots = \beta_{38} \text{ vs. } H_{a3}: (\text{not } H_{03}).$$

The analysis of variance table for the model specified in Figure 4.2 is displayed in Figure 4.3. The effect SILT*HERB tests the slopes in the SILT direction are equal,

Analysis of Variance

Source	DF	Sum of Squares	Mean Square	F Ratio
Model	31	3907.4973	126.048	3.3560
Error	64	2403.7689	37.559	Prob > F
C. Total	95	6311.2662		<.0001

Parameter Estimates

Effect Tests

Source	Nparm	DF	Sum of Squares	F Ratio	Prob > F
HERB	7	7	103.03936	0.3919	0.9038
SILT	1	1	594.10352	15.8179	0.0002
SILT*HERB	7	7	471.55978	1.7936	0.1039
CLAY	1	1	557.61049	14.8463	0.0003
CLAY*HERB	7	7	171.06552	0.6507	0.7124
OM	1	1	145.14054	3.8643	0.0537
OM*HERB	7	7	326.92867	1.2435	0.2927

FIGURE 4.3 Analysis table for the unequal slopes model in each direction of the three covariates.

the effect CLAY*HERB tests the slopes in CLAY direction are equal, and OM*HERB tests the slopes are equal in the OM direction. The results of the analysis in Figure 4.3 indicate there is not sufficient evidence to believe that the slopes are not equal in each of the covariate directions where the significance levels are 0.1039 for SILT, 0.7124 for CLAY, and 0.2927 for OM. So the next step is to fit a common slope model to the data. (A stepwise deletion process was carried out where Herb*Clay was removed first, then Herb*OM, and finally Herb*Silt. Never delete more than one term from the model at a time while simplifying the covariate part of the model.)

The common slope model is

$$y_{ij} = \alpha_i + \beta_1 \text{ SILT}_{ij} + \beta_2 \text{ CLAY}_{ij} + \beta_3 \text{ OM}_{ij} + \varepsilon_{ij} \qquad (4.10)$$

and the JMP® Fit Model window with the model specification is displayed in Figure 4.4. The results of fitting Model 4.10 to the data are in Figures 4.5 and 4.6. The estimates of the slopes are $\hat{\beta}_1 = 0.2946$, $\hat{\beta}_2 = -0.2762$, and $\hat{\beta}_3 = -2.2436$.

The analysis has established that parallel hyper-planes adequately describe the data; thus the differences between the treatments are represented by the distances between the hyper-planes. The adjusted means are in Figure 4.7 which are the herbicide hyper-planes evaluated at the average values of the covariates, 15.5833, 19.5208, and 1.6116 for SILT, CLAY, and OM, respectively. The adjusted mean for herbicide 1 is computed by using the estimates in Figure 4.5 as

$$29.1604 = 29.2354 + 4.3417 + 0.2946*15.5833 - 0.2762*19.5208 - 2.2436*1.6116.$$

Figure 4.8 is a plot of the least squares means or adjusted means with 95% confidence intervals. Figure 4.9 contains all pairwise comparisons of the herbicide means. The cells of the figure contain the difference of the means, the estimated standard error of the difference, and 95% Tukey simultaneous confidence intervals about the differences. Using the confidence intervals about the differences, the mean of herbicide 8 is significantly different from the means of herbicides 3 and 6, and the mean of herbicide 1 is significantly different from the mean of herbicide 6. There are no other differences.

Multiple Covariants on One-Way Treatment Structure

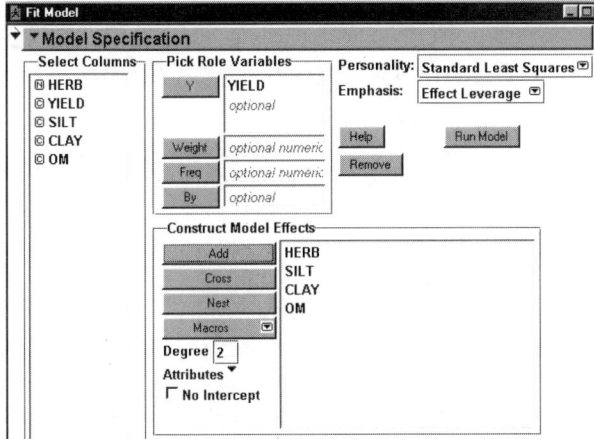

FIGURE 4.4 JMP® model specification window for common slopes in each of the directions of the three covariates.

```
Analysis of Variance
Source      DF   Sum of Squares   Mean Square   F Ratio
Model       10       2856.9924       285.699     7.0303
Error       85       3454.2739        40.639    Prob > F
C. Total    95       6311.2662                   <.0001

Parameter Estimates
Effect Tests
Source   Nparm   DF   Sum of Squares   F Ratio   Prob > F
HERB        7    7        956.62633     3.3628    0.0033
SILT        1    1        564.98238    13.9026    0.0003
CLAY        1    1        784.80406    19.3118    <.0001
OM          1    1        179.59048     4.4192    0.0385
```

FIGURE 4.5 Analysis of variance table for the common slopes model in each of the directions of the three covariates.

4.6 EXAMPLE: MODELS THAT ARE QUADRATIC FUNCTIONS OF THE COVARIATE

The yield of a process depends on the quantity of X that is present during the development time. Two additives were added to determine if they could alter the yield of the process, i.e., increase the yield. The data in Table 4.9 are the yields for three treatments where TRT = 1 is the control (without additive) TRT = 2 and 3 are the two additives and X is the covariate corresponding to each experimental unit. The design structure is a completely randomized design.

On plotting the data, one sees that the relationship between yield and X is quadratic. The analysis of covariance model used to describe this data is linear and quadratic in X with the possibility of different intercepts and slopes for the three treatments. The model used to describe the data is

$$y_{ij} = \alpha_i + \beta_{1i} X_{ij} + \beta_{2i} X_{ij}^2 + \varepsilon_{ij}, \quad i = 1, 2, 3 \text{ and } j = 1, 2, \ldots, 12. \quad (4.11)$$

Parameter Estimates

Term	Estimate	Std Error	t Ratio	Prob>|t|
Intercept	29.235454	2.483561	11.77	<.0001
HERB[1]	4.3416627	1.734624	2.50	0.0142
HERB[2]	0.0033234	1.723398	0.00	0.9985
HERB[3]	-3.562265	1.732177	-2.06	0.0428
HERB[4]	-1.910561	1.7287	-1.11	0.2722
HERB[5]	1.307541	1.74023	0.75	0.4545
HERB[6]	-4.310853	1.741336	-2.48	0.0153
HERB[7]	-0.778272	1.757427	-0.44	0.6590
SILT	0.2946062	0.079012	3.73	0.0003
CLAY	-0.276215	0.062854	-4.39	<.0001
OM	-2.243604	1.067268	-2.10	0.0385

FIGURE 4.6 Estimates of the parameters for the model with common slopes in each of the three covariates directions.

Least Squares Means Table

Level	Least Sq Mean	Std Error	Mean
1	29.160413	1.8526300	29.1500
2	24.822073	1.8421232	25.3750
3	21.256485	1.8503389	22.3917
4	22.908189	1.8470845	22.0583
5	26.126291	1.8578805	26.5417
6	20.507897	1.8589159	18.9583
7	24.040478	1.8739979	25.0750
8	29.728174	1.8501630	29.0000

FIGURE 4.7 Least squares means for each of the herbicides.

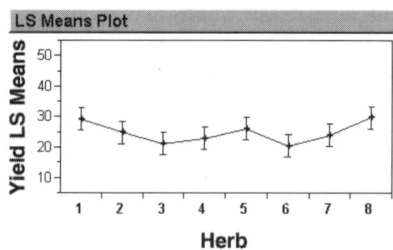

FIGURE 4.8 Plot of the herbicide least squares means.

LSMeans Differences Tukey HSD

Alpha= 0.050   Q= 3.10773

Mean[i]-Mean[i] Std Err Dif Lower CL Dif Upper CL Dif	1	2	3	4	5	6	7	8
1	0 0 0 0	4.33834 2.61267 -3.7811 12.4578	7.90393 -2.6116 -0.2122 16.0201	6.25222 2.62354 -1.901 14.4055	3.03412 2.60354 -5.057 11.1252	8.65252 2.6249 0.49501 16.81	5.11993 2.66077 -3.149 13.3889	-0.5678 2.62879 -8.7374 7.60183
2	-4.3383 2.61267 -12.458 3.78114	0 0 0 0	3.56559 2.60555 -4.5318 11.663	1.91388 2.61271 -6.2057 10.0335	-1.3042 2.61438 -9.429 6.82058	4.31418 2.62435 -3.8416 12.47	0.7816 2.62288 -7.3696 8.93281	-4.9061 2.61434 -13.031 3.21858
3	-7.9039 -2.6116 -16.02 0.21223	-3.5656 2.60555 -11.663 4.53178	0 0 0 0	-1.6517 2.62599 -9.8126 6.50918	-4.8698 2.61022 -12.982 3.24206	0.74859 2.63655 -7.4451 8.9423	-2.784 2.63413 -10.97 5.40219	-8.4717 2.62904 -16.642 -0.6013
4	-6.2522 2.62354 -14.405 1.90103	-1.9139 2.61271 -10.034 6.20574	1.6517 2.62599 -6.5092 9.81259	0 0 0 0	-3.2181 2.63143 -11.396 4.95969	2.40029 2.60998 -5.7108 10.5114	-1.1323 2.62741 -9.2976 7.03302	-6.82 2.60371 -14.912 1.27165

LSMeans Differences Tukey HSD

Mean[i]-Mean[i] Std Err Dif Lower CL Dif Upper CL Dif	1	2	3	4	5	6	7	8
5	-3.0341 2.60354 -11.125 5.05698	1.30422 2.61438 -6.8206 9.42901	4.86981 2.61022 -3.2421 12.9817	3.2181 2.63143 -4.9597 11.3959	0 0 0 0	5.61839 2.63483 -2.57 13.8068	2.08581 2.66455 -6.1949 10.3665	-3.6019 2.63645 -11.795 4.59149
6	-8.6525 2.6248 -16.81 -0.495	-4.3142 2.62435 -12.47 3.8416	-0.7486 2.63655 -8.9423 7.44512	-2.4003 2.60998 -10.511 5.71084	-5.6184 2.63483 -13.807 2.56997	0 0 0 0	-3.5326 2.65408 -11.781 4.71558	-9.2203 2.6166 -17.353 -1.088
7	-5.1199 2.66077 -13.389 3.14904	-0.7816 2.62288 -8.9328 7.36961	2.78399 2.63413 -5.4022 10.9702	1.13229 2.62741 -7.033 9.2976	-2.0858 2.66455 -10.367 6.1949	3.53258 2.65408 -4.7156 11.7807	0 0 0 0	-5.6877 2.623 -13.839 2.4639
8	0.56776 2.62879 -7.6018 8.73736	4.9061 2.61434 -3.2186 13.0308	8.47169 2.62904 0.30133 16.642	6.81999 2.60371 -1.2716 14.9116	3.60188 2.63645 -4.5915 11.7953	9.22028 2.6166 1.08795 17.3526	5.6877 2.623 -2.4639 13.8393	0 0 0 0

FIGURE 4.9 Comparisons of the herbicide least squares means using Tukey's HSD multiple comparison procedure.

At this point it has already been established that there is a relationship between Yield and X and X^2, i.e., the corresponding slopes are different than zero (analysis not shown). The next step in the analysis of covariance is to decide whether common

TABLE 4.9
Yields for Three Treatments and the Values of the Covariate

Treatment 1		Treatment 2		Treatment 3	
Yield	X	Yield	X	Yield	X
3.7	2.2	7.2	5.8	5.7	8.8
4.7	3.5	2.5	9.9	6.1	9.0
5.3	4.1	4.3	1.2	6.4	8.8
5.5	4.1	6.9	6.0	9.6	4.1
5.7	7.4	8.0	5.5	9.4	2.4
3.8	3.0	3.7	9.7	8.2	4.1
1.0	0.5	5.2	8.6	10.2	4.4
3.3	2.4	8.0	5.0	6.0	8.9
5.3	8.4	6.9	2.6	9.4	5.3
5.3	8.5	4.3	8.4	5.0	9.5
6.1	8.0	6.8	3.0	7.5	1.1
5.8	6.3	7.6	5.2	5.3	0.1

linear slopes and/or common quadratic slopes are possible simplifications of the model. The first hypotheses to be tested are

$$H_{01}: \beta_{11} = \beta_{12} = \beta_{13} \text{ vs. H: } (\text{not } H_{01}) \text{ and}$$

$$H_{02}: \beta_{21} = \beta_{22} = \beta_{23} \text{ vs. H: } (\text{not } H_{02}).$$

Table 4.10 contains the SAS® system code statements necessary to test H_{01} and H_{02} as well as the results. The significance levels for the two tests are 0.6463 for X*TRT and 0.1364 for X*X*TRT. Thus, given there are unequal slopes for the X^2 terms, then the model does not have to be unequal slopes for the X terms, or given there are unequal slopes for the X terms, then the model does not have to be unequal slopes for the X^2 terms. Since the significance level for X*TRT is larger than that for X*X*TRT, X*TRT was deleted. The resulting model has a common slope for the X term and unequal slopes for the X^2 term. Table 4.11 contains the SAS® system code for fitting the above model to provide a test of the equality of the quadratic term slopes. The results are in the lower part of Table 4.11. The significance level corresponding to X*X*TRT is 0.0000, indicating that the slopes are not equal for the three treatments in the X^2 direction.

The final model with a common slope for the X direction and unequal slopes in the X^2 direction is

$$y_{ij} = \alpha_i + \beta_1 X_{ij} + \beta_{2i} X_{ij}^2 + \varepsilon_{ij}, \quad i = 1, 2, 3 \text{ and } j = 1, 2, \ldots, 12. \quad (4.12)$$

The code in Table 4.12 fits Model 4.12 and also contains the analysis of variance table. Since the model used the NOINT option, the F statistic corresponding to TRT

TABLE 4.10
Code to Fit a Model with Unequal Slopes in Both the X and X² Directions and Provide Tests for Equality of Each of the Two Sets of Slopes

```
PROC GLM DATA=QUAD; CLASS TRT;
MODEL YIELD= TRT X X*TRT X*X X*X*TRT/SS3;
```

Source	df	SS	MS	FValue	ProbF
Model	8	140.22	17.53	72.53	0.0000
Error	27	6.53	0.24		
Corr Total	35	146.75			

Source	df	SS (Type III)	MS	FValue	ProbF
TRT	2	16.00	8.00	33.11	0.0000
X	1	47.03	47.03	194.58	0.0000
X*TRT	2	0.21	0.11	0.44	0.6463
X*X	1	50.30	50.30	208.14	0.0000
X*X*TRT	2	1.04	0.52	2.15	0.1364

TABLE 4.11
Code to Fit a Model with a Common Slope in the X Direction and Unequal Slopes in the X² Direction and Test Equality of the X² Slopes

```
PROC GLM DATA=QUAD; CLASS TRT;
MODEL YIELD= TRT X X*X X*X*TRT /SS3;
```

Source	df	SS	MS	FValue	ProbF
Model	6	140.01	23.33	100.41	0.0000
Error	29	6.74	0.23		
Corr Total	35	146.75			

Source	df	SS (Type III)	MS	FValue	ProbF
TRT	2	73.63	36.81	158.41	0.0000
X	1	49.50	49.50	212.98	0.0000
X*X	1	53.86	53.86	231.76	0.0000
X*X*TRT	2	17.67	8.83	38.01	0.0000

provides a test $H_0: \alpha_1 = \alpha_2 = \alpha_3 = 0$ vs. H_a:(not H_0). The F statistic corresponding to X provides a test of $H_0: \beta_1 = 0$ vs. H_a:(not H_0) and the F statistic corresponding to X*X*TRT provides a test of $H_0: \beta_1 = \beta_2 = \beta_3 = 0$ vs. H_a:(not H_0). The significance levels are very small, indicating there is sufficient evidence to reject the null hypotheses. The estimates of the parameters of the model are in Table 4.13.

Table 4.14 contains the code to provide estimates of the regression lines at several values of X. The first statement, LSMEANS TRT/STDERR PDIFF, provides adjusted means that are not interpretable since they are computed by evaluating the

TABLE 4.12
Code to Fit Model 4.12 with Analysis of Variance Table

```
PROC GLM DATA=QUAD; CLASS TRT;
MODEL YIELD=TRT X X*X*TRT /NOINT SOLUTION SS3;
```

Source	df	SS	MS	FValue	ProbF
Model	7	1432.41	204.63	880.49	0.0000
Error	29	6.74	0.23		
Uncor Tot	36	1439.15			

Source	df	SS (Type III)	MS	FValue	ProbF
TRT	3	96.98	32.33	139.10	0.0000
X	1	49.50	49.50	212.98	0.0000
X*X*TRT	3	91.60	30.53	131.38	0.0000

TABLE 4.13
Estimates of the Parameters of Model 4.12

Parameter	Estimate	StdErr	tValue	Probt
TRT 1	−0.0102	0.3224	−0.03	0.9751
TRT 2	2.9889	0.3885	7.69	0.0000
TRT 3	5.3598	0.3163	16.94	0.0000
X	1.8520	0.1269	14.59	0.0000
X*X*TRT 1	−0.1432	0.0133	−10.75	0.0000
X*X*TRT 2	−0.1894	0.0112	−16.88	0.0000
X*X*TRT 3	−0.2000	0.0122	−16.45	0.0000

model at the average value of X (5.4389) and the average value of X^2 (37.9478), which does not correspond to any points on the regression models since the square of the mean of X is $5.4389^2 = 29.5816$. The estimate statements in Table 4.15 are included to demonstrate the computation of the LSMEANS. The last estimate statement in Table 4.15 is used to provide the computation for Treatment 1 as used by LSMEANS TRT/STDERR PDIFF. The average value of X is 5.4388889 and the average value of X^2 is 37.9477778, while the square of the average value of X is 29.581512665. The last statement uses 37.9477778 for X^2 in the computation to provide a value of 4.630. That is the same result corresponding to LSMEAN in Table 4.14 for Treatment 1. Thus the usual LSMEAN statement provides incorrect adjusted values. The third estimate statement in Table 4.15 uses the average value of X and the square of the average value of X in the computations, providing 5.828 as the adjusted mean for Treatment 1. The last two LSMEAN statements in Table 4.14, LSMEANS TRT/PDIFF at MEANS and LSMEANS TRT/PDIFF at X=5.4388889, use the correct computations and provide the adjusted mean for Treatment 1 of 5.828. The other two LSMEAN statements are used to obtain adjusted means at a large value of X (X = 9) and a small value of X (X = 1). The first two

TABLE 4.14
Code and Results for Computing Adjusted Means

```
LSMEANS TRT/STDERR PDIFF;
LSMEANS TRT/PDIFF AT X=1;
LSMEANS TRT/PDIFF AT X=9;
LSMEANS TRT/PDIFF AT MEANS;
LSMEANS TRT/PDIFF AT X=5.4388889;
```

	TRT	LSMEAN	RowName	_1	_2	_3
LSMEAN	1	4.63	1		0.0000	0.0000
Incorrect	2	5.88	2	0.0000		0.0000
	3	7.84	3	0.0000	0.0000	
X=1	1	1.70	1		0.0000	0.0000
	2	4.65	2	0.0000		0.0000
	3	7.01	3	0.0000	0.0000	
X=9	1	5.06	1		0.0590	0.0493
	2	4.32	2	0.0590		0.0000
	3	5.83	3	0.0493	0.0000	
MEANS	1	5.83	1		0.0000	0.0000
	2	7.46	2	0.0000		0.0000
	3	9.52	3	0.0000	0.0000	
X=5.4388889	1	5.83	1		0.0000	0.0000
	2	7.46	2	0.0000		0.0000
	3	9.52	3	0.0000	0.0000	

TABLE 4.15
Estimate Statements and Results Demonstrating the Computation of the LSMEANS

```
ESTIMATE 'TRT 1 AT X=1' TRT 1 0 0 X 1 X*X*TRT 1 0 0;
ESTIMATE 'TRT 1 AT X=9' TRT 1 0 0 X 9 X*X*TRT 81 0 0;
ESTIMATE 'TRT 1 AT X=5.4388889' TRT 1 0 0 X 5.4388889
    X*X*TRT 29.581512665 0 0;
ESTIMATE 'TRT 1 LSM' TRT 1 0 0 X 5.4388889 X*X*TRT
    37.9477778 0 0;
```

Parameter	Estimate	StdErr	tValue	Probt
TRT 1 AT X=1	1.699	0.244	6.95	0.0000
TRT 1 AT X=9	5.062	0.312	16.21	0.0000
TRT 1 AT X=5.4388889	5.828	0.163	35.74	0.0000
TRT 1 LSM	4.630	0.145	31.83	0.0000

estimate statements in Table 4.15 are also used to demonstrate the computations of the adjusted means at $X = 1$ and $X = 9$. Pairwise comparisons of the treatment means for each set of adjusted means are in the columns labeled _1, _2, and _3 of Table 4.14. The graph in Figure 4.10 displays the data and estimated regression models for the three treatments with vertical lines at $X = 1$, $X = 5.4388889$ (the mean of X), and

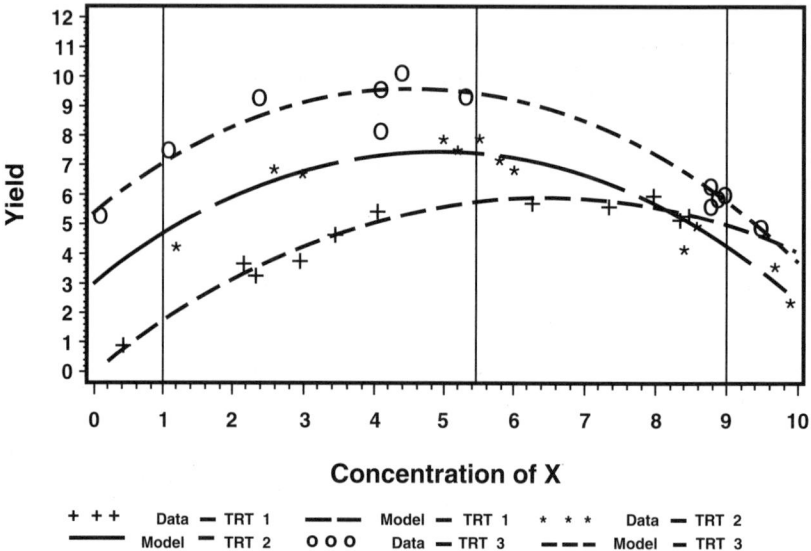

FIGURE 4.10 Graph of the quadratic regression models with the data points for Example 4.5.

X = 9. The pairwise comparison significance levels in Table 4.14 indicate the three regression lines are significantly different at X = 1 and at the mean of X, while at X = 9, Treatment 1 is not signicantly different from the other two treatments while Treatment 3 provides a significantly higher response than Treatment 2.

4.7 EXAMPLE: COMPARING RESPONSE SURFACE MODELS

A drug company had two formulation processes with which to make a certain type of pill and they wanted to determine the concentrations of two binders which produce the stronger pills. For each formulation, a treatment structure with design points from a two factor rotatable central composite design with four center points in a completely randomized design structure was used to collect information to study the relationship between strength and the concentrations of the two binders. The force (lb/in.) required to fracture the pill is the dependent measure. The data for the two formulations, coded U3Y and X2Z, the force, and the coded concentrations of the two binders (CONC1, CONC2) are in Table 4.16. The objectives of the experiment are (1) for each formulation determine an adequate model to describe the data and estimate the combination of the binders that will produce a pill with maximum strength and (2) to compare the response surfaces via analysis of covariance.

A quadratic response surface model was selected to describe the responses for each of the formulations and is expressed as

$$y_{ijkm} = \mu_i + \beta_{1i} x_{1j} + \beta_{2i} x_{2k} + \beta_{3i} x_{ij}^2 + \beta_{4i} x_{2k}^2 + \beta_{5i} x_{ij} x_{2k} + \varepsilon_{ijkm} \quad (4.13)$$

TABLE 4.16
Strength of Pills Made by Two Formulations with Concentrations of Two Binders

CONC1	CONC2	U3Y Force	X2Z Force
1.00	1.00	14.14	16.02
0.00	1.41	10.64	14.07
−1.00	1.00	1.47	7.20
−1.41	0.00	2.28	3.85
−1.00	−1.00	8.53	10.30
0.00	−1.41	6.57	12.78
1.00	−1.00	3.14	8.14
1.41	0.00	8.16	9.45
0.00	0.00	13.58	13.98
0.00	0.00	11.73	15.66
0.00	0.00	13.07	16.83
0.00	0.00	12.24	14.41

TABLE 4.17
Analysis of Variance Table to Provide the Estimate of Pure Error for the Response Surface Models Used to Compare Pill Strength

```
PROC GLM DATA=RESPSUR; CLASS FORMULA CONC1 CONC2;
MODEL FORCE = FORMULA*CONC1*CONC2;
```

Source	df	SS	MS	FValue	ProbF
Model	17	458.69	26.98	23.01	0.0004
Error	6	7.04	1.17		
Corrected Total	23	465.72			

Source	df	SS(Type III)	MS	FValue	ProbF
Formula*Conc1*Conc2	17	458.69	26.98	23.01	0.0004

where
- $i = 1, 2$ for the two formulations
- $j = 1, 2, \ldots, 5$, for the levels of x_1 (CONC1),
- $k = 1, 2, \ldots, 5$, for the levels of x_2 (CONC2),
- $m = 1$ or 4 for the replications per formulation.

Since there are four replications of the center point for each formulation, the variation between these observations within a formulation provides an estimate of pure error (or a model free estimate of the variance). The estimate of the pure error is $\hat{\sigma}^2_{PE} = 1.17$, which is the Mean Square Error from the analysis of variance in Table 4.17. To compute the estimate of σ^2_{PE} using PROC GLM of the SAS® system,

TABLE 4.18
Code and Analysis of Variance Table to Test Each Set of the Slopes are Equal to Zero

```
PROC GLM DATA=RESPSUR; CLASS FORMULA;
MODEL FORCE=FORMULA CONC1*FORMULA CONC2*FORMULA
  C11*FORMULA C22*FORMULA C12*FORMULA;
```

Source	df	SS	MS	FValue	ProbF
Model	11	456.55	41.50	54.29	0.0000
Error	12	9.17	0.76		
Corrected Total	23	465.72			

Source	df	SS (Type III)	MS	FValue	ProbF
FORMULA	1	13.16	13.16	17.21	0.0013
CONC1*FORMULA	2	56.97	28.49	37.26	0.0000
CONC2*FORMULA	2	17.20	8.60	11.25	0.0018
C11*FORMULA	2	201.94	100.97	132.08	0.0000
C22*FORMULA	2	30.97	15.49	20.26	0.0001
C12*FORMULA	2	111.68	55.84	73.04	0.0000

include Formula, Conc1, and Conc2 in the class statement and then use Formula*Conc1*Conc2 in the model statement (see Table 4.17). This process constructs a means type model where there is one mean for each combination of Formula by Conc1 by Conc2. For this example there are nine combinations for each formula. The resulting Error Sum of Squares measures the variability of those experimental units treated alike, which in this case are those with levels (0,0) of (Conc1,Conc2) for each formula. There are four center points for each formulation; thus there are six degrees of freedom for pure error with three degrees of freedom coming from each formula. Next, Model 4.13 is fit to the data to (1) determine if the quadratic response surface model adequately describes the data from each formulation, (2) to obtain estimates of the parameters of the models, and (3) to test the hypotheses that the sets of slopes are equal to zero, i.e., H_{os}: $\beta_{s1} = \beta_{s2} = 0$ vs. H_{as}: (not H_{os}), s = 1, 2, ..., 5. The SAS® system code to fit Model 4.13 is in Table 4.18 along with the results. The terms C11, C22, and C12 denote CONC1*CONC1, CONC2*CONC2, and CONC1*CONC2, respectively. The sum of squares due to lack of fit is computed by subtracting the sum of squares for pure error (see Table 4.17) from the current model error sum of squares (Table 4.18). The computations are SS Lack of Fit = 9.17 − 7.04 = 2.13, based on 6 = 12 − 6 degrees of freedom. The value of the statistic to test the lack of the ability of the model to fit the cell means is F = (2.13/6)/(7.04/6) = 0.303, indicating there is no evidence of lack of fit. The F statistics in Table 4.18 provide tests of the following hypotheses: CONC1*FORMULA tests H_0: $\beta_{11} = \beta_{12} = 0$ vs. H_a:(not H_0:) with significance level 0.0000, CONC2*FORMULA tests H_0: $\beta_{21} = \beta_{22} = 0$ vs. H_a:(not H_0:) with significance level 0.0018, C11*FORMULA tests H_0: $\beta_{31} = \beta_{32} = 0$ vs. H_a:(not H_0:) with significance level 0.0000, C22*FORMULA tests H_0: $\beta_{41} = \beta_{42} = 0$ vs. H_a:(not H_0:) with significance level 0.0001, and C12*FORMULA tests H_0: $\beta_{51} = \beta_{52} = 0$ vs. H_a:(not H_0:) with significance level 0.0000. Each

TABLE 4.19
Code and Analysis of Variance Table for Testing the Parallelism Hypothesis for Each Covariate

```
PROC GLM DATA=RESPSUR; CLASS FORMULA;
MODEL FORCE = CONC1 CONC2 C11 C22 C12
   FORMULA CONC1*FORMULA CONC2*FORMULA C11*FORMULA
   C22*FORMULA C12*FORMULA;
```

Source	df	SS	MS	FValue	ProbF
Model	11	456.55	41.50	54.29	0.0000
Error	12	9.17	0.76		
Corrected Total	23	465.72			

Source	df	SS (Type III)	MS	FValue	ProbF
CONC1	1	56.91	56.91	74.44	0.0000
CONC2	1	16.61	16.61	21.72	0.0006
C11	1	201.29	201.29	263.31	0.0000
C22	1	26.01	26.01	34.03	0.0001
C12	1	105.42	105.42	137.89	0.0000
FORMULA	1	13.16	13.16	17.21	0.0013
CONC1*FORMULA	1	0.06	0.06	0.08	0.7764
CONC2*FORMULA	1	0.60	0.60	0.78	0.3941
C11*FORMULA	1	0.65	0.65	0.85	0.3754
C22*FORMULA	1	4.96	4.96	6.49	0.0256
C12*FORMULA	1	6.27	6.27	8.20	0.0143

of these sum of squares is based on two degrees of freedom since the hypothesis corresponds to specifying two parameters are equal to zero. The results of these F statistics indicate the respective sets of slopes are not all equal to zero.

The next step in the analysis is to check for parallelism in the direction of each covariate. The SAS® system code for testing for the parallelism hypotheses for each of the covariates in the model is in Table 4.19. The model includes each of the individual terms as well as the interactions, e.g., CONC1 and CONC1*FORMULA. When both terms are included in the model, the resulting F statistics provide tests that the slopes for the two levels of formula are equal. In this case, CONC1*FORMULA tests H_0: $\beta_{11} = \beta_{12}$ vs. H_a:(not H_0:) with significance level 0.7764, CONC2*FORMULA tests H_0: $\beta_{21} = \beta_{22}$ vs. H_a:(not H_0:) with significance level 0.3941, C11*FORMULA tests H_0: $\beta_{31} = \beta_{32}$ vs. H_a:(not H_0:) with significance level 0.3754, C22*FORMULA tests H_0: $\beta_{41} = \beta_{42}$ vs. H_a:(not H_0:) with significance level 0.0256, and C12*FORMULA tests H_0: $\beta_{51} = \beta_{52}$ vs. H_a:(not H_0:) with significance level 0.0143. Each of these sum of squares is based on one degree of freedom since they correspond to specifying the difference between two parameters is equal to zero. The information in Table 4.19 indicates that (1) $\beta_{41} \neq \beta_{42}$ (C22*FORMULA) and 2) $\beta_{51} \neq \beta_{52}$ (C12*FORMULA).

The sum of squares corresponding to each specific line in Table 4.19 provides a conditional sum of squares due to that effect given the other terms in the model.

TABLE 4.20
Code and Analysis of Variance Table for the Final Model Used to Compare the Response Surfaces for the Soybean Data

```
PROC GLM DATA=RESPSUR; CLASS FORMULA;
MODEL FORCE = CONC1 CONC2 C11 FORMULA C22*FORMULA
    C12*FORMULA/SOLUTION NOINT P;
```

Source	df	SS (Type III)	MS	FValue	ProbF
Model	9	3022.87	335.87	480.56	0.0000
Error	15	10.48	0.70		
Uncorrected Total	24	3033.35			

Source	df	SS	MS	FValue	ProbF
CONC1	1	56.91	56.91	81.42	0.0000
CONC2	1	16.61	16.61	23.76	0.0002
C11	1	201.29	201.29	288.01	0.0000
FORMULA	2	1570.24	785.12	1123.34	0.0000
C22*FORMULA	2	31.95	15.98	22.86	0.0000
C12*FORMULA	2	111.68	55.84	79.90	0.0000

Thus one needs to be careful when deleting more than one term from the model at a time. If several terms are deleted at once, the fit of the resulting model needs to be checked, i.e., make sure that the MS Residual does not increase too much, etc. For this particular problem, a stepwise deletion process was used to simplify the model which resulted in

$$y_{ijkm} = \mu_i + \beta_1 x_{1j} + \beta_2 x_{2k} + \beta_3 x_{1j}^2 + \beta_{4i} x_{2k}^2 + \beta_{5i} x_{1j} x_{2k} + \varepsilon_{ijkm} \quad (4.14)$$

where
 $i = 1,2$ for the two formulations
 $j = 1,2,...,5$, for the levels of x_1 (CONC1),
 $k = 1,2,...,5$, for the levels of x_2 (CONC2),
 $m = 1$ or 4 for the replications per formulation.

The SAS® system code for fitting Model 4.14 to the pill data is in Table 4.20.

The analysis of variance table is in Table 4.20 where the mean square error is 0.70 and it is based on 15 degrees of freedom. The estimates of the parameters of the model are in Table 4.21. Table 4.22 contains SAS® system code for computing a test for lack of fit for Model 4.14. The process is to define two new variables, C1 = CONC1 and C2 = CONC2. Include C1 and C2 in the model in place of CONC1 and CONC2, but include CONC1 and CONC2 in the Class statement. Finally include CONC1*CONC2*FORMULA in the model. The Type III sum of squares for CONC1*CONC2*FORMULA (which is the same here as the Type I sum of squares since CONC1*CONC2*FORMULA is the last term in the model) is the sum of squares due to deviations of the model from the cells' means given all of the other

TABLE 4.21
Parameter Estimates for the Final Model

Parameter	Estimate	StdErr	tValue	Probt
CONC1	1.886	0.209	9.02	0.0000
CONC2	1.019	0.209	4.87	0.0002
C11	−3.966	0.234	−16.97	0.0000
FORMULA U3Y	12.835	0.374	34.33	0.0000
FORMULA X2Z	15.040	0.374	40.23	0.0000
C22*FORMULA U3Y	−2.093	0.327	−6.40	0.0000
C22*FORMULA X2Z	−0.758	0.327	−2.32	0.0350
C12*FORMULA U3Y	4.515	0.418	10.80	0.0000
C12*FORMULA X2Z	2.745	0.418	6.57	0.0000

TABLE 4.22
Code and Analysis of Variance Table to Provide a Test of Lack of Fit for the Final Model

```
PROC GLM DATA=RESPSUR; CLASS FORMULA CONC1 CONC2;
MODEL FORCE = C1 C2 C11 FORMULA C22*FORMULA C12*FORMULA
  FORMULA*CONC1*CONC2;
```

Source	df	SS	MS	FValue	ProbF
Model	17	458.69	26.98	23.01	0.0004
Error	6	7.04	1.17		
Corrected Total	23	465.72			

Source	df	SS (Type III)	MS	FValue	ProbF
C1	0	0.00			
C2	0	0.00			
C11	0	0.00			
FORMULA	1	8.41	8.41	7.17	0.0366
C22*FORMULA	0	0.00			
C12*FORMULA	0	0.00			
FORMULA*CONC1*CONC2	9	3.45	0.38	0.33	0.9356

terms are in the model. Thus, the sum of squares corresponding to CONC1*CONC2*FORMULA is the sum of squares due to the lack of the model fitting the cell means. The lack of fit sum of squares is 3.45 and the value of the F statistic is 0.33 with significance level being 0.9356, indicating there is no evidence that the model fails to fit the data.

Using the estimates of the parameters in Table 4.20, the equations for estimating the the two response surfaces are

$$\hat{y}_{U3Y} = 12.835 + 1.886\,X_1 + 1.019\,X_2 - 3.966\,X_1^2 - 2.093\,X_2^2 + 4.515\,X_1X_2$$

and

$$\hat{y}_{X2Z} = 15.040 + 1.886\ X_1 + 1.019\ X_2 - 3.966\ X_1^2 - 0.758\ X_2^2 + 2.745\ X_1X_2.$$

These response surfaces for the two formulations have common slopes in the X_1, X_2, and X_1^2 directions and different slopes in X_2^2 and X_1X_2 directions, where X_1 = CONC1 and X_2 = CONC2.

The combination of the binders at which the response surface estimates the maximum strength is determined by differentiating the model with respect to X_1 and X_2, equating the derivatives to zero, and solving the resulting equations. The two sets of equations are:

$$U3Y: \quad -2(3.966)\ X_1 + 4.515\ X_2 = -1.886$$

$$4.515\ X_1 - 2(2.093)\ X_2 = -1.019$$

and

$$X2Z: \quad -2(3.966)\ X_1 + 2.745\ X_2 = -1.886$$

$$2.745\ X_1 - 2(0.758)\ X_2 = -1.019.$$

The concentrations at which the models estimate the maximum strength occurs are (1) for U3Y (CONC1 = 0.975, CONC2 = 1.295) and for Z2X (CONC1 = 1.260, CONC2 = 2.953). The maximum for X2Z occurs outside of the range of experimentation, so one needs to be careful in assessing the usefulness of the estimate.

Table 4.23 contains the predicted values (PFORCE) for each point in the design space and the two formulations. The maximum observed response means occur at

TABLE 4.23
Predicted Values of Force for the Two Formulations at Each of the Design Points in the Experiment

		U3Y		X2Z	
CONC1	CONC2	PFORCE	STE	PFORCE	STE
1.00	1.00	14.20	0.58	15.97	0.58
0.00	1.41	10.09	0.58	14.96	0.58
−1.00	1.00	1.39	0.58	6.70	0.58
−1.41	0.00	2.24	0.52	4.44	0.52
−1.00	−1.00	8.39	0.58	10.16	0.58
0.00	−1.41	7.21	0.58	12.08	0.58
1.00	−1.00	3.13	0.58	8.44	0.58
1.41	0.00	7.57	0.52	9.78	0.52
0.00	0.00	12.84	0.37	15.04	0.37
0.00	0.00	12.84	0.37	15.04	0.37
0.00	0.00	12.84	0.37	15.04	0.37
0.00	0.00	12.84	0.37	15.04	0.37

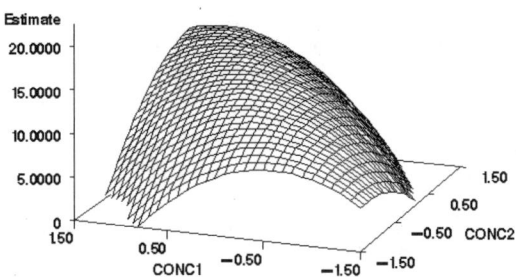

FIGURE 4.11 Predicted regression surface for formula U3Y over the range of concentrations.

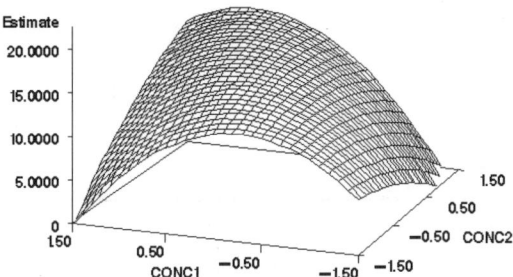

FIGURE 4.12 Predicted regression surface for formula X2Z over the range of concentrations.

(CONC1, CONC2) = (1, 1) for each of the formulations. The estimated standard errors associated with the predicted values are also included (STE). Remember these are estimated standard errors for the mean of the model and not estimated standard deviations associated with an individual measurement at a given combination of CONC1 and CONC2. The major conclusion that can be made by looking the predicted values is that X2Z produces stronger pills at each of the observed combinations of the two binders.

Figures 4.11 and 4.12 are plots of the predicted response surfaces for each of the two formulations. To make further comparisons between the two response surfaces, a set of estimate statements were generated to construct a grid over the range of (CONC1, CONC2) so that estimates of the differences of the two models could be evaluated (along with the estimated standard errors). The difference between the two models is

$$\text{Diff}_{(U3Y-X2Z)} = (\mu_1 - \mu_2) + (\beta_{41} - \beta_{42})X_2^2 + (\beta_{51} - \beta_{52})X_1 X_2$$

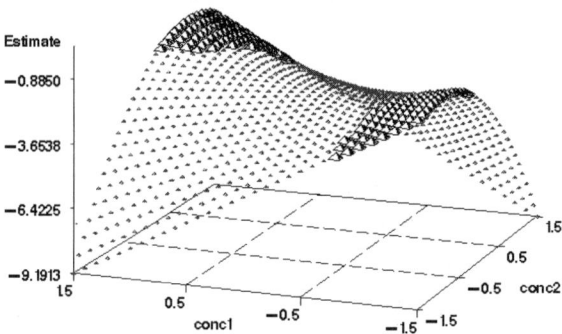

FIGURE 4.13 Surface of the difference between the two treatments' response surfaces where the dark areas are regions where the surfaces are not significantly different.

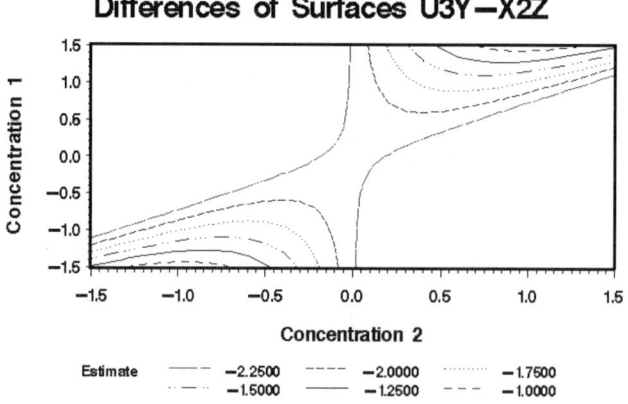

FIGURE 4.14 Contour plot for the surface of the differences between the two treatments' response surfaces.

which is a comparison that involves three linearly independent parameters (combinations of parameters) $(\mu_1 - \mu_2)$, $(B_{41} - \beta_{42})$, and $(\beta_{51} - \beta_{52})$. To compare the response surfaces over the grid, a Scheffe approach was utilized where a value D was computed as $D = \sqrt{3F_{.05,3,15}}$. If the computed difference divided by its estimated standard error at a grid point was less in absolute value than D, then it was declared that the two formulations produce similar mean responses at that point. Figure 4.13 is a graph of the difference of the two response surfaces, where the large symbols indicate grid points where the response is not significantly different from zero (points where the two formulations produce similar mean responses). Figure 4.14 is a contour plot of the differences with contours in the regions where the differences are closest to zero.

REFERENCE

Special Issue on the Analysis of Covariance, *Biometrics,* September 1957, and *Biometrics,* September 1982.

EXERCISES

EXERCISE 4.1: The average daily gain ADG during 3 to 9 months of age of 3 breeds of calves was studied. It was thought that ADG is partly an inherited trait; thus the ADG of each calf's sire (SADG) and the ADG of each calf's dam (DADG) were determined when they were growing during 3 to 9 months of age and were used as possible covariates. The data are in the following table. Use the analysis of covariance strategy to construct an appropriate model and then use the model to carry out comparisons of the three breeds.

Average Daily Gain Data for Exercise 4.1

	ADG	SADG	DADG
Breed 1	2.80	2.1	2.4
	3.36	2.8	2.2
	3.12	3.0	1.7
	2.75	2.0	2.1
	2.82	2.3	1.8
	3.08	2.8	2.0
Breed 2	2.49	1.2	2.1
	1.83	1.4	1.0
	2.54	1.8	1.7
	2.53	2.3	2.0
	2.87	1.9	2.5
	2.19	2.3	1.1
	3.03	1.2	3.0
	2.60	1.8	2.3
Breed 3	3.84	3.0	2.3
	3.93	3.1	2.0
	4.17	3.9	2.3
	3.96	3.3	2.2
	4.31	2.6	3.2
	4.45	3.0	3.2
	4.02	2.6	2.7
	4.45	3.0	3.3
	4.38	3.7	3.1

EXERCISE 4.2: Carry out an analysis of covariance for the following data set by determining the appropriate model and then making the needed treatment comparisons. Y is the response variable and X and Z are the covariates.

Data for Exercise 4.2

Treatment A			Treatment B			Treatment C		
X	Y	Z	X	Y	Z	X	Y	Z
90.5	23.3	13.4	90.1	15.7	39.9	75.0	14.3	48.3
79.4	7.0	35.1	74.2	21.8	32.6	86.6	10.3	49.1
94.7	11.8	34.1	93.6	14.9	41.1	87.7	11.3	50.9
75.4	21.2	14.0	93.0	9.7	43.7	91.5	4.0	42.9
95.9	14.8	31.4	90.9	23.3	34.1	95.2	16.4	55.8
79.6	14.5	27.1	85.7	11.4	41.1	85.9	23.6	59.8
83.7	16.6	26.5	88.2	11.1	42.5	79.8	14.8	51.1
96.1	19.5	24.1	96.9	18.5	41.3	91.2	6.1	45.6
72.0	20.3	15.4	89.6	12.4	42.3	91.8	22.3	61.0
95.6	21.8	19.6	82.2	18.1	37.0			
			83.9	22.0	33.5			
			73.1	20.7	33.1			

5 Two-Way Treatment Structure and Analysis of Covariance in a Completely Randomized Design Structure

5.1 INTRODUCTION

The main difference between the analysis of a one-way treatment structure and that of a two-way treatment structure is that the sums of squares for intercepts and slopes can be partitioned into row treatment effects, column treatment effects, and row treatment by column treatment interaction effects. One of the objectives of a good analysis is to determine if a different slope is needed for each treatment combination and, if not, to determine whether a different slope is needed for each row treatment and whether a different slope is needed for each column treatment. Once an adequate model has been determined in terms of the slope parameters, the analysis is completed by making comparisons of interest between the planes or lines at various selected values of the covariate. If there are unequal slopes for the levels of the row treatments or column treatments or both, the analysis of the intercepts provides a comparison of the regression surfaces at the value of the covariate equal to zero. Great care must be used in the interpretation of results when there are unequal slopes in two-way and higher order treatment structures. Several examples involving one covariate are used to demonstrate these concepts.

5.2 THE MODEL

The cell means model for a two-way treatment structure in a completely randomized design structure with one covariate is

$$y_{ijk} = \alpha_{ij} + \beta_{ij} x_{ijk} + \varepsilon_{ijk}$$

$$i = 1, 2, \ldots, s, \quad j = 1, 2, \ldots, t, \quad k = 1, 2, \ldots, n_{ij}$$

(5.1)

where the ijth cell consists of the ith level of the row treatment in combination with the jth level of the column treatment, α_{ij} denotes the intercept for the ijth cell, and β_{ij}

denotes the cell slope of the ijth cell. The model can also be expressed as an effects model as

$$y_{ijk} = \left(\mu + \tau_i + \rho_j + \gamma_{ij}\right) + \left(\eta + \phi_i + \theta_j + \delta_{ij}\right)\bar{X}_{ijk} + \varepsilon_{ijk}$$

where the effects intercept is $\alpha_{ij} = (\mu + \tau_i + \rho_j + \gamma_{ij})$ and the effects slope is $\beta_{ij} = (\eta + \phi_i + \theta_j + \delta_{ij})$. The general model has h covariates. When there are k covariates then in the linear model, then each cell must contain at least h + 1 observations in each cell in order to estimate the h + 1 parameter model in the model for the ijth cell. If a cell has fewer than h + 1 observations, then there are problems with the ability to estimate the parameters and the analysis will act as if the cell is missing, unless the form of the covariate part of the model can be simplified. Model 5.1 can be analyzed the same as the one-way treatment structure Model 2.1 where α_{ij} and β_{ij} denote the intercept and slope, respectively, corresponding to the ijth treatment combination. That is, this data set can be considered to have come from a one-way treatment structure with bt treatments in a completely randomized design structure. Therefore, many of the techniques used for the analysis of a one-way treatment structure (discussed in Chapters 2 through 4) can be used for two-way treatment structures. The strategy stays the same as the three steps to be used are (1) determine the simplest form of the slopes for the covariate part of the model, (2) determine the form of the intercepts, and (3) properly compare the resulting regression surfaces.

The slopes can take on one of five possible forms:

(i) Common slope denoted by $\bar{\beta}_{..}$ or η
(ii) Slope is a function of the levels of the row treatment, denoted by $\bar{\beta}_{i.}$, $\bar{\beta}_{.j}$ or $\eta + \phi_i$
(iii) Slope is a function of the levels of the column treatment $\bar{\beta}_{j}$, $\bar{\beta}_{.j}$ or $\eta + \theta_j$
(iv) Slope is an additive function of the levels of the row treatment and of the levels of the column treatment, denoted by $\beta_{Ri} + \beta_{cj}$, $\bar{\beta}_{.j} + \bar{\beta}_{i.}$, or $\eta + \theta_j + \phi_i$ where β_{Ri} is the row treatment component and β_{cj} is the column treatment components of the slope
(v) There is no pattern among the slopes or there is row treatment by column treatment interaction which is represented by β_{ij} or $\eta + \phi_i + \theta_j + Y_{ij}$

The decisions about the slopes can be made by testing the following hypotheses.

The first hypothesis to be tested is to see if for at least some of the bt treatment combinations the mean of y is a linear function of X. This can be answered by testing

$$H_0: \beta_{11} = \beta_{12} = \cdots = \beta_{1t} = \cdots = \beta_{st} = 0$$

vs.

$$H_a: \text{(at least one slope is nonzero)} \quad (5.2)$$

or

$$H_0: \mu_{Y|X} = \alpha_{ij} \text{ vs. } H_a: \mu_{Y|X} = \alpha_{ij} + \beta_{ij} X.$$

If one fails to reject the hypothesis of Equation 5.2, then fit a model with a common slope to make sure that there is not an overall relationship with the covariate. If the common slope is not significantly different from zero, then the resulting analyses follow along the lines of the analysis of a two-way treatment structure model discussed in Milliken and Johnson (1992). If the hypothesis in Equation 5.2 is rejected, then test the parallelism hypothesis, i.e.,

$$H_0: \beta_{11} = \beta_{12} = \cdots = \beta_{st} \text{ vs. } H_a: (\text{not } H_0:). \tag{5.3}$$

If there is not sufficient evidence to reject the hypothesis in Equation 5.3, then conclude that the lines are all parallel, i.e., conclude a common slope model will adequately describe the data. The common slope model is

$$y_{ijk} = \alpha_{ij} + \beta x_{ijk} + \varepsilon_{ijk}$$

$$i = 1, 2, \ldots, s, \quad j = 1, 2, \ldots, t, \quad k = 1, 2, \ldots, n_{ij}$$

In this case, the distances between the planes can be studied by comparing the intercepts. However, if Hypothesis 5.3 is rejected, a model with a different slope for each treatment combinations could be used or other hypotheses could be tested concerning the effect of the treatments on the slope parameters in an attempt to simplify the covariate part of the model. The hypotheses of interest correspond to testing hypotheses about main effects and interaction effects on the slope parameters as

(i) Average slope for the levels of the row treatments are equal or

$$H_0: \bar{\beta}_{1\cdot} = \bar{\beta}_{2\cdot} = \cdots = \bar{\beta}_{s\cdot} \text{ vs. } H_a: (\text{not } H_0) \tag{5.4}$$

(ii) Average slope for the levels of the column treatments are equal or

$$H_0: \bar{\beta}_{\cdot 1} = \bar{\beta}_{\cdot 2} = \cdots = \bar{\beta}_{\cdot t} \text{ vs. } H_a: (\text{not } H_0) \tag{5.5}$$

(iii) No interaction between the row treatment and column treatment or

$$H_0: \beta_{ij} - \beta_{i'j} - \beta_{ij'} + \beta_{i'j'} = 0 \text{ for all } i, i', j \text{ and } j' \text{ vs. } H_a: (\text{not } H_0) \tag{5.6}$$

The results of tests of these hypotheses provide a guide to constructing an adequate covariate part of the model to describe the mean of y given x. For example, if one rejects Equation 5.4 and fails to reject Equations 5.5 and 5.6, then the model can be written as

$$y_{ijk} = \alpha_{ij} + \beta_i x_{ijk} + \varepsilon_{ijk}.$$

If Equation 5.6 is rejected, then there is interaction among the slopes and the unequal slope model in Equation 5.1 should be used.

Thus, without simplifying the intercepts of the regression models, the mean of y for treatment combination ij given X can have one of the following forms:

(i) $\mu_{ij|\beta X} = \alpha_{ij} + \beta X$ (common slope)
(ii) $\mu_{ij|\beta_i X} = \alpha_{ij} + \beta_i X$ (unequal slopes for row treatments)
(iii) $\mu_{ij|\beta_j X} = \alpha_{ij} + \beta_j X$ (unequal slopes for column treatments)
(iv) $\mu_{ij|(\beta_{Ri}+\beta_{cj})X} = \alpha_{ij} + (\beta_{Ri} + \beta_{cj})X$ (additive slopes)

or

(v) $\mu_{ij|\beta_{ij}X} = \alpha_{ij} + \beta_{ij}X$ (interaction for slopes)

If there are main effects and no interaction for the intercepts, then one needs to compare the row treatment models and compare the column treatment models. For the common slope model, row mean and column mean models are

$$\bar{\mu}_{i\cdot|\beta X} = \bar{\alpha}_{i\cdot} + \beta X \quad \text{and} \quad \bar{\mu}_{\cdot j|\beta X} = \bar{\alpha}_{\cdot j} + \beta X$$

which are common slope models themselves. The row treatment comparisons are contrasts of the $\bar{\alpha}_{i\cdot}$'s as $\bar{\mu}_{1\cdot|\beta X} - \bar{\mu}_{2\cdot|\beta X} = \bar{\alpha}_{1\cdot} - \bar{\alpha}_{2\cdot}$ for any x. The column treatment comparisons are contrasts of the $\bar{\alpha}_{\cdot j}$'s as $\bar{\mu}_{\cdot 1|\beta X} - \bar{\mu}_{\cdot 2|\beta X} = \bar{\alpha}_{\cdot 1} - \bar{\alpha}_{\cdot 2}$ for any x.

If the slopes are dependent on the row treatments, then the row means models are $\bar{\mu}_{i\cdot|\beta_i X} = \bar{\alpha}_{i\cdot} + \beta_i X$, a set of nonparallel lines models, and the column means models are $\bar{\mu}_{\cdot j|\beta_i X} = \bar{\alpha}_{\cdot j} + \bar{\beta}_\cdot X$, a set of parallel lines models. The row treatment comparisons are contrasts of $\bar{\alpha}_{i\cdot} + \beta_i X$ as $\bar{\mu}_{1\cdot|\beta_i X} - \bar{\mu}_{2\cdot|\beta_i X} = \bar{\alpha}_{1\cdot} - \bar{\alpha}_{2\cdot} + (\beta_1 - \beta_2)X$, which depend on the value of X. The column treatment comparisons are contrasts of $\bar{\alpha}_{\cdot j}$ as $\bar{\mu}_{\cdot 1|\beta_i X} - \bar{\mu}_{\cdot 2|\beta_i X} = \bar{\alpha}_{\cdot 1} - \bar{\alpha}_{\cdot 2}$, which do not depend on the value of X.

If the slopes are dependent on the column treatments, comparisons are similar to the models with $\beta_i X$, except the row and column treatment comparison are reversed.

If the slopes are additive functions of the row treatments and column treatments, then the row means model is $\bar{\mu}_{i\cdot|(\beta_{Ri}+\beta_{cj})x} = \bar{\alpha}_{i\cdot} + (\beta_{Ri} + \bar{\beta}_{c\cdot})X$ and the column means model is $\bar{\mu}_{\cdot j|(\beta_{Ri}+\beta_{cj})x} = \bar{\alpha}_{\cdot j} + (\bar{\beta}_{R\cdot} + \bar{\beta}_{cj})X$, both which are sets of nonparallel lines models. Comparisons of the row treatments are contrasts of the $\bar{\alpha}_{i\cdot} + \beta_{Ri}X$ and comparisons of the column treatments are contrasts of the $\bar{\alpha}_{\cdot j} + \beta_{cj}X$. Both sets of comparisons involve unequal slopes and thus depend on the value of X.

If β_{ij} cannot be simplified (intersection exists in the slopes between the row and column treatments), then comparisons of the row means are contrasts of $\bar{\alpha}_{i\cdot} + \bar{\beta}_{i\cdot}X$ and comparisons of the column means are contrasts of $\bar{\alpha}_{\cdot j} + \bar{\beta}_{\cdot j}X$.

If the intercepts of the model exhibit interaction and cannot be simplified, the cell models must be compared. There are at least four types of comparisons of interest:

(i) Compare column treatments within same row treatment
(ii) Compare row treatments within same column treatment
(iii) Compare two cells from different rows and/or columns
(iv) Compare 2×2 table interaction

Whether the above comparisons are functions of the covariate depend on the form of the slopes. If a common slope model is adequate, then all comparisons are free of the covariate. If the slopes are functions of the row treatments, then comparisons involving different row treatments involve nonparallel lines, while comparisons within the same row treatment or 2×2 table interaction comparisons involve parallel lines. If the slopes are additive functions of the row and column treatments, then only 2×2 interaction comparisons are free of the covariate. The other cases can be derived by similar arguments using the respective models.

The basic strategy is to express the covariate part of the model as an effects structure. If the highest order interaction involving the covariate (in this case the three-way, X*row*col) is not important, then delete that term and refit the model. If all of the next level interactions with the covariate are significant, then the model is established. If not, delete the term that has the largest significance level (in this case choose between X*row and X*col) and refit the model. Continue looking at lower order interactions until one or more of those terms are significant. That term or set of terms specifies the form of the covariate part of the model. Next the intercepts are evaluated. If there is one term in the covariate part of the model that interacts with a factor in the treatment structure, then the sum of squares corresponding to the factorial effects provide tests that the respective effects are zero at the intercept or the origin or $X = 0$. The only time the test statistics from the analysis of variance table provide appropriate interpretations is when there is a common slopes model (unless comparing the models at $X = 0$ is of interest). When there is an interaction with the covariate and one or more factors in the treatment structure, the regression models need to be compared at three or more values of the covariate. It is easy to compare the models at X^*. The comparisons can be carried out by subtracting X^* from each observation from the value of the covariate and then re-running the model. The subtraction process translates the origin of the x axis to X^* and the intercepts of the model become the adjusted means or predicted values computed from the model evaluated at $X = X^*$.

5.3 USING THE SAS® SYSTEM

To use SAS® or some other model-driven computer system, construct models similar to those used in Chapters 2 and 3 except extend the models to include the two-way treatment structure in both the intercepts and the slopes. The next two sections present the code and process of specifying the models for the two-way analysis of covariance problem.

5.3.1 Using PROC GLM and PROC MIXED

Both PROC GLM and PROC MIXED are model-driven computer codes in that a model specification process is used to describe exactly the model needing to be fit to the data set. The following sets of code specify a model with factorial effects for the intercepts and the slopes of the model:

```
PROC GLM; CLASS A B;
MODEL Y = A B A*B X X*A X*B X*A*B;
```

and

```
PROC MIXED; CLASS A B;
MODEL Y = A B A*B X X*A X*B X*A*B;
```

If the model building process indicates that the X*A*B interaction is significant, then no simplification of the slopes is possible and the analysis of the intercepts would continue using the model specifications:

```
PROC GLM; CLASS A B;
MODEL Y = A B A*B X*A*B;
```

and

```
PROC MIXED; CLASS A B;
MODEL Y = A B A*B X*A*B;
```

If the process discovers that the slopes are a function of the levels of A and a function of the levels of B, then the models can be specified as:

```
PROC GLM; CLASS A B;
MODEL Y = A B A*B X*A X*B;
```

and

```
PROC MIXED; CLASS A B;
MODEL Y = A B A*B X*A X*B;
```

When there are unequal slopes, then the regression models can be evaluated as to the presence of interaction and main effects at specified values of the covariate by computing in a data step, the variable $XT = X - X^*$, where X^* is the value of X at which the comparisons are to be made. The model can be fit by:

```
PROC GLM; CLASS A B;
MODEL Y = A B A*B XT*A XT*B;
```

and

```
PROC MIXED; CLASS A B;
MODEL Y = A B A*B XT*A XT*B;
```

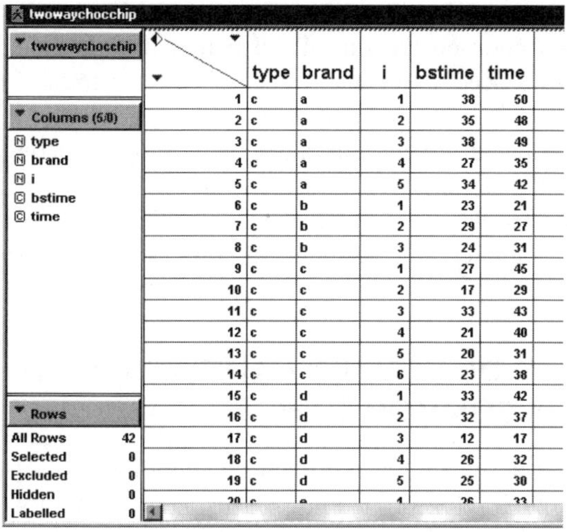

Two-Way Treatment Structure and Analysis of Covariance

FIGURE 5.1 JMP® data screen for a two-way treatment structure with one covariate.

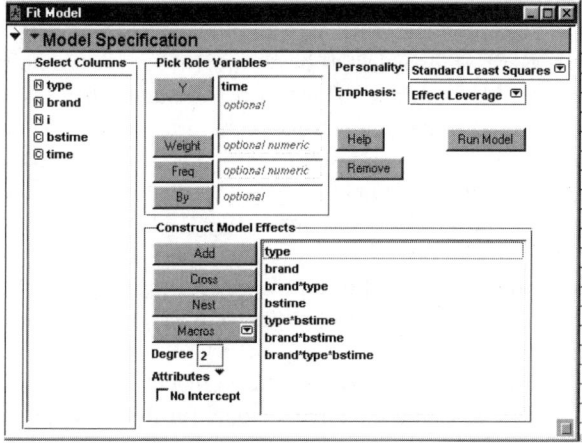

FIGURE 5.2 The fit model window with the two-way effects model for both intercepts and slopes.

5.3.2 Using JMP®

As described in Chapters 2 and 3, JMP® uses the data in a spread sheet format. Figure 5.1 is the data screen for the data in Exercise 5.1. The levels of the two factors are brands of chocolate chips (A, B, C, D, and E) and types of chocolate chips (regular and semisweet, denoted by c and s). Type, brand, and "i" or person are declared to be nominal data types and bstime and time are continuous data types. Figure 5.2 contains the fit model screen window. Time has been selected as the response variable (Y) and the model effects for the intercepts are brand, type, and brand*type and the model effects for the slopes are bstime bstime*brand, bstime*type, and bstime*brand*type. Once the model has been specified in the fit model screen, then click on the run model button to fit the model. The analysis provides the statistics needed to make decisions as to the form of the covariate part of the model. Also, if it is decided that there are unequal slopes, then the regression models can be compared at selected values of X by constructing a new covariate where the specified value of X is subtracted from the X value for each observation. Refit the necessary model by specifying the model in the model specification window.

Now several examples are presented to demonstrate the processes discussed above.

5.4 EXAMPLE: AVERAGE DAILY GAINS AND BIRTH WEIGHT — COMMON SLOPE

An animal scientist designed an experiment to study the efficacy of a growth promotion drug (at four levels) for suckling calves as measured by the average daily gain from birth to 12 weeks of age. Average daily gain is computed by taking the number of pounds the animal gains during the study divided by the number of days the study was conducted. If the compound is efficacious, the animal scientist wanted to determine if it worked similarly for male and female calves with the ultimate goal of estimating the dose producing the maximum response. Since light-weight baby calves do not necessarily gain as fast as heavy-weight baby calves, birth weight was considered as a possible covariate. Thirty-two male and thirty-two female calves were randomly assigned to the four levels of the drug with eight males and eight females assigned to each level. The resulting design is a two-way treatment structure (SEX by Drug) in a completely randomized design structure. The data in Table 5.1 are the average daily gains (ADG) and birth weights of the male and female calves which were implanted with one of four levels of a performance enhancement drug. The four levels of drug are Control (0 mg), Low (2.5 mg), Med (5.0 mg), and High (7.5 mg). After 12 weeks (84 days), the calves were weighed and their average daily gains were computed, (12 week wt – birth wt)/84.

The PROC GLM code used to fit the effects model for both the intercepts and the slopes is in Table 5.2. Using the effects model for the covariate is an attempt to determine an appropriate simple form for the covariate part of the model. The results in Table 5.2 indicate that there is little evidence to believe that the slopes are different for each sex by drug combinations as the significance level corresponding to Birth_wt*sex*drug is 0.1805. The three-way interaction was removed from the model and the reduced model was fit to the data. At this step, the Birth_wt*Sex term has the largest significance level (0.7861) and was deleted from the model (results not shown). Finally, the significance level corresponding to Birth_wt*drug in the next phase reduced model was 0.7323, so it was deleted from the model. Thus none of the interaction terms involving Birth_wt were significant, so a common slopes model was adequate to fit the data.

Next a common slope model was fit to the data to investigate the effects of SEX and Drug on the intercepts of the models. The PROC GLM code for and results of fitting the common slope model are in Table 5.3. The significance level corresponding to Birth_wt indicates there is a strong linear relationship between ADG and

TABLE 5.1
Average Daily Gains (ADG lb/day) and Birth Weights (lb) of Hereford Calves for Example 5.4

		Control (0 mg)		Low (2.5 mg)		Med (5.0 mg)		High (7.5 mg)	
Drug		Birth_wt	adg	Birth_wt	adg	Birth_wt	adg	Birth_wt	adg
Female		69	2.58	78	2.92	77	3.01	67	2.94
		70	2.60	73	2.51	85	3.13	84	2.80
		74	2.13	83	2.78	77	2.75	83	2.85
		66	2.00	79	2.91	61	2.36	73	2.44
		74	1.98	66	2.18	76	2.71	66	2.28
		78	2.31	63	2.25	74	2.79	63	2.70
		78	2.30	82	2.91	75	2.84	66	2.70
		64	2.19	75	2.42	65	2.59	83	2.85
Male		76	2.43	71	2.52	69	3.14	77	2.73
		69	2.36	81	2.54	69	2.91	79	3.17
		70	2.93	66	2.95	71	2.53	76	2.92
		72	2.58	69	2.46	68	3.09	72	2.85
		60	2.27	70	3.13	63	2.86	66	2.47
		80	3.11	76	2.72	61	2.88	84	3.28
		71	2.42	79	3.41	64	2.75	79	3.13
		77	2.66	78	3.43	73	2.91	83	2.73

TABLE 5.2
Analysis of Variance Table with Effects for Both Intercepts and Slopes

```
proc glm data=common; class Drug Sex;
model ADG=drug sex drug*sex Birth_wt Birth_wt*drug Birth_wt*sex;
```

Source	df	SS	MS	FValue	ProbF
Model	15	4.1363	0.2758	4.73	0.0000
Error	48	2.7993	0.0583		
Corrected Total	63	6.9357			

Source	df	SS (Type III)	MS	FValue	ProbF
drug	3	0.0527	0.0176	0.30	0.8245
Sex	1	0.0037	0.0037	0.06	0.8012
drug*Sex	3	0.3044	0.1015	1.74	0.1714
Birth_wt	1	0.7576	0.7576	12.99	0.0007
Birth_wt*drug	3	0.0546	0.0182	0.31	0.8163
Birth_wt*Sex	1	0.0001	0.0001	0.00	0.9690
Birth_wt*drug*Sex	3	0.2966	0.0989	1.70	0.1805

Birth_wt. The estimated slope is 0.0217 indicating for each 1 lb increase in Birth_wt, there is an estimated 0.0217 lb/day increase in ADG. The significance levels corresponding to the intercept effects are 0.0001, 0.0001, and 0.6584 respectively for Sex, Drug, and Sex*Drug. Thus there are important Sex and Drug effects, but there

TABLE 5.3
Analysis of Variance Table with Effects for the Intercepts and a Common Slope Model

```
proc glm data=common; class Drug Sex;
model ADG=sex drug drug*sex Birth_wt/solution;
```

Source	df	SS	MS	FValue	ProbF
Model	8	3.7583	0.4698	8.13	0.0000
Error	55	3.1774	0.0578		
Corrected total	63	6.9357			
Source	df	SS (Type III)	MS	FValue	ProbF
Sex	1	1.0366	1.0366	17.94	0.0001
drug	3	1.5517	0.5172	8.95	0.0001
drug*Sex	3	0.0932	0.0311	0.54	0.6584
Birth_wt	1	1.0964	1.0964	18.98	0.0001
Parameter	Estimate	StdErr	tValue	Probt	
Birth_wt	0.0217	0.0050	4.36	0.0001	

is not a significant interaction between the levels of Drug and levels of SEX. Therefore, the main effects of the treatments need to be compared. The PROC GLM code for fitting a means model to the intercepts and a common slope is in Table 5.4. The results in Table 5.4 provide the estimates of the parameters for the regression model. Each of the Sex*Drug combinations has an intercept and there is a common slope for the covariate, Birth_wt.

Since there is no interaction between the levels of sex and the levels of drug for the intercepts, a regression model can be constructed for each sex and for each level of drug. Table 5.5 contains the estimate statements used to compute the estimates of the intercepts of the regression models for each sex by averaging over the levels of drug and the estimates of the intercepts for each level of drug by averaging over the levels of sex. For example, the regression model to describe the mean of the ADG values as a function of birth weight for females and for males is $ADG_{female} = 0.997 + 0.0217$ Birth_wt and $ADG_{male} = 1.252 + 0.0217$ Birth_wt. Figure 5.3 contains a graph of the estimated regression lines for each sex. The models for each level of drug are $ADG_{Control} = 0.875 + 0.0217$ Birth_wt, $ADG_{Low} = 1.144 + 0.0217$ Birth_wt, $ADG_{Med} = 1.302 + 0.0217$ Birth_wt, and $ADG_{High} = 1.177 + 0.0217$ Birth_wt. Figure 5.4 contains a graph of the estimated regression lines for each level of drug. Tables 5.6 and 5.7 contain the adjusted means (from main effects regression lines evaluated at the average birth weight of the experiment, 73 lb) for the main effects of Drug and main effects of SEX, respectively. The vertical line at 73 in Figure 5.4 indicates the locations of the adjusted means or LSMEANS. The mean ADG for males calves is significantly larger than the mean ADG for female calves. The mean ADG for the calves on the Control is significantly less than the means of the three non-zero levels of drug.

TABLE 5.4
Analysis of Variance Table with Intercepts Expressed as Means

```
proc glm data=common; class Drug Sex;
model ADG=drug*sex Birth_wt/solution noint;
```

Source	df	SS	MS	FValue	ProbF
Model	9	471.2908	52.3656	906.44	0.0000
Error	55	3.1774	0.0578		
Uncorrected Total	64	474.4682			

Source	df	SS	MS	FValue	ProbF
drug*Sex	8	3.5166	0.4396	7.61	0.0000
Birth_wt	1	1.0964	1.0964	18.98	0.0001

Parameter	Estimate	StdErr	tValue	Probt
Control (0 mg) Female	0.7105	0.3660	1.94	0.0573
Control (0 mg) Male	1.0388	0.3672	2.83	0.0065
High (7.5 mg) Female	1.1118	0.3732	2.98	0.0043
High (7.5 mg) Male	1.2429	0.3920	3.17	0.0025
Low (2.5 mg) Female	0.9889	0.3817	2.59	0.0122
Low (2.5 mg) Male	1.2983	0.3762	3.45	0.0011
Med (5.0 mg) Female	1.1758	0.3762	3.12	0.0028
Med (5.0 mg) Male	1.4277	0.3449	4.14	0.0001
Birth_wt	0.0217	0.0050	4.36	0.0001

TABLE 5.5
Estimates of the Intercepts for the Sex Models and for the Drug Models

```
estimate 'int Female' drug*sex 1 0 1 0 1 0 1 0 /divisor=4;
estimate 'int Male' drug*sex 0 1 0 1 0 1 0 1 /divisor=4;
estimate 'int 0 mg' drug*sex 1 1 0 0 0 0 0 0 /divisor=2;
estimate 'int 2.5 mg' drug*sex 0 0 0 0 1 1 0 0 /divisor=2;
estimate 'int 5.0 mg' drug*sex 0 0 0 0 0 0 1 1/divisor=2;
estimate 'int 7.5 mg' drug*sex 0 0 1 1 0 0 0 0 /divisor=2;
```

Parameter	Estimate	StdErr	tValue	Probt
int Female	0.997	0.367	2.72	0.0088
int Male	1.252	0.363	3.45	0.0011
int 0 mg	0.875	0.362	2.42	0.0189
int 2.5 mg	1.144	0.374	3.06	0.0035
int 5.0 mg	1.302	0.355	3.66	0.0006
int 7.5 mg	1.177	0.378	3.12	0.0029

Thus far, the animal scientist has determined a common slope model using birth weight as a covariate is appropriate, that there is an effect of the drug (Table 5.7 shows the mean ADG of all non-zero levels of the Drug are significantly larger than

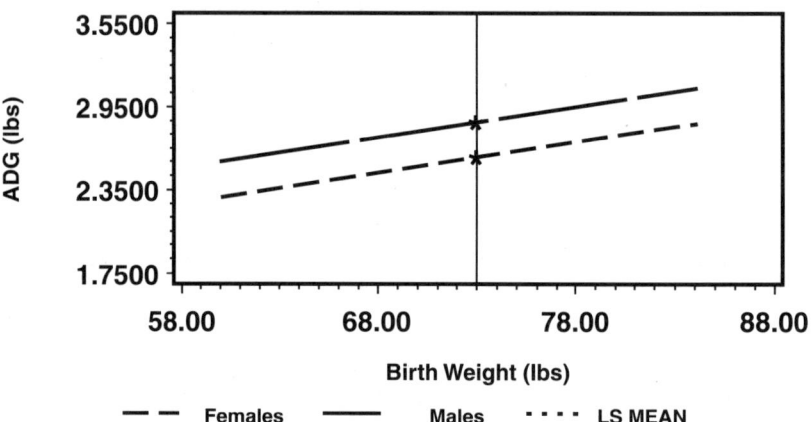

FIGURE 5.3 Plot of estimated regression models for each level of sex with least squares means.

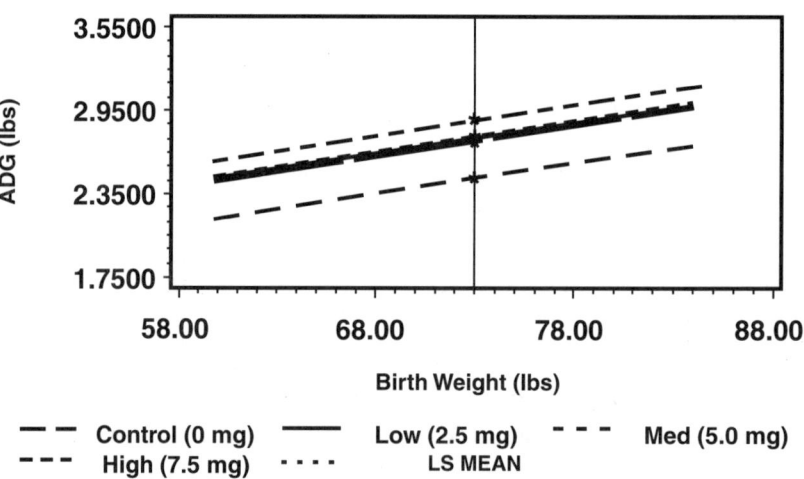

FIGURE 5.4 Plot of estimated regression models for each level of drug with least squares means.

the mean of the control. The mean of MED dose is larger than the means of all the other doses, but not significantly), that males gain significantly faster than females (see Table 5.6), and that there is no interaction between the levels of SEX and Drug.

Finally, the animal scientist wants to estimate the optimal dose, i.e., estimate that dose of the compound which should produce the maximum ADG. To accomplish this objective, the adjusted means and the corresponding standard errors were

TABLE 5.6
Least Squares Means and Comparisons for the Levels of Sex

Sex	LSMean	StdErr	ProbtDiff
Female	2.5752	0.0425	0.0001
Male	2.8304	0.0425	

TABLE 5.7
Least Squares Means and Comparisons for the Levels of Drug

Drug	LSMean	StdErr	LSMean #	_1	_2	_3	_4
Control (0 mg)	2.453	0.060	1		0.0009	0.0028	0.0000
High (7.5 mg)	2.756	0.061	2	0.0009		0.6930	0.1629
Low (2.5 mg)	2.722	0.060	3	0.0028	0.6930		0.0747
Med (5.0 mg)	2.880	0.061	4	0.0000	0.1629	0.0747	

TABLE 5.8
Results of Fitting Quadratic Regression Model to Drug Least Squares Means

```
data lsmeans; input xlsmean drug stderr @@;
wt=1/(stderr**2); drug2=drug**2;
 datalines;
2.4536 0 .06037 2.7225 2.5 .06048
2.8806 5.0 .06129 2.7562 7.5 .06102
proc reg data=lsmeans; weight wt; model
 xlsmean=drug drug2;
```

Source	df	SS	MS	FValue	ProbF
Model	2	25.9274	12.9637	32.58	0.1229
Error	1	0.3979	0.3979		
Corr Total	3	26.3253			

Variable	df	Estimate	StdErr	tValue	Probt
Intercept	1	2.4452	0.0371	65.85	0.0097
drug	1	0.1604	0.0240	6.69	0.0945
drug2	1	–0.0157	0.0031	–5.12	0.1229

obtained from Table 5.7 and used to construct a data set to which a quadratic regression model was fit using the reciprocal of the square of the standard errors as weights. The code and data are in Table 5.8. The results of fitting the quadratic regression model are in Table 5.8 and the estimated regression model is

```
ADG = 2.4452 + 0.1604 DRUG −0.0157 DRUG²
```

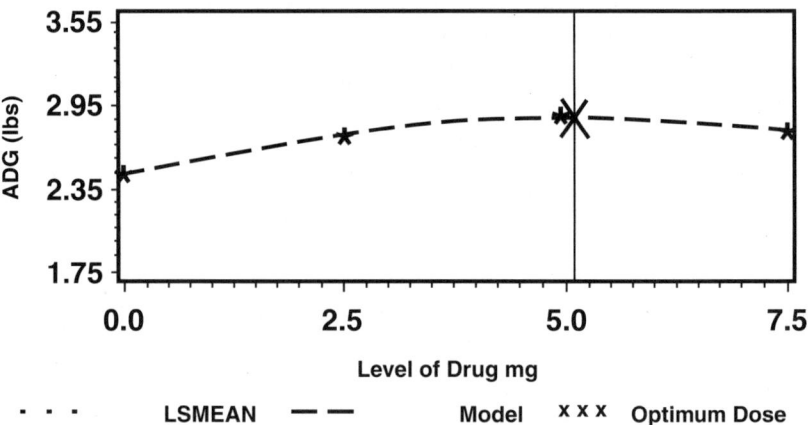

FIGURE 5.5 Plot of quadratic regression model for drug least squares means with predicted maximum.

To determine the level of drug that is estimated to yield the maximum response, differentiate the model with respect to DRUG, set the derivative equal to zero, and solve for the resulting equation for the level of drug estimated to produce the maximum response, providing

$$\text{DRUG}_{\text{Max}} = -(0.1604)/(2*(-0.0157)) = 5.11 \text{ mg}$$

This stationary point is a maximum since the coefficient of DRUG^2 is negative. Figure 5.5 is a graph of the least squares means, the estimated quadratic regression model, and the estimated level of Drug producing the maximum response from the quadratic model.

5.5 EXAMPLE: ENERGY FROM WOOD OF DIFFERENT TYPES OF TREES — SOME UNEQUAL SLOPES

A forester and a chemical engineer studied the amount of energy produced by burning wood blocks from different species of trees (four levels) in different types of stoves (three levels). It is suspected that the moisture content (%) of a block of wood affects the amount of energy (kilogram/calorie denoted by kg/cal) produced by burning a block of wood. Table 5.9 contains the data where ten blocks (uniform dimensions) of each wood type were burned in each stove type. For several runs complications occurred and the data were not usable; thus the unequal sample sizes occurred. The wood from osage orange and red oak are considered to be hard woods and the wood from white pine and black walnut are are considered to be soft woods.

TABLE 5.9
Energy Produced from Blocks of Wood with Different Moisture Contents Burned in Three Different Types of Stoves

Stove	Osage Orange Moist	Osage Orange Energy	Red Oak Moist	Red Oak Energy	White Pine Moist	White Pine Energy	Black Walnut Moist	Black Walnut Energy
Type A	8.70	7.33	16.50	5.72	8.60	2.71	15.50	1.87
	10.20	7.58	16.80	6.54	7.60	2.99	13.70	2.39
	15.60	6.91	7.20	6.09	7.40	3.29	18.70	1.14
	16.20	7.08	11.40	6.87	17.20	1.58	11.30	2.42
	17.10	6.82	9.90	6.53	9.20	2.36	7.30	3.23
	10.50	7.63			13.30	2.47	15.90	1.81
	19.00	6.68			21.00	1.12		
	13.00	7.64			17.50	1.35		
	7.20	7.52			13.30	1.88		
					7.00	2.51		
Type B	13.10	7.41	11.30	5.62	15.40	2.64	10.20	3.43
	9.40	6.66	12.60	5.37	16.60	2.68	12.80	2.79
	11.10	7.01	13.90	5.05	16.20	3.76	9.20	3.92
	17.60	5.69	12.40	5.25	8.70	4.50	13.60	2.67
	14.70	6.65	20.40	5.77	19.50	1.66	16.50	2.38
	18.40	6.24	12.10	5.43			11.80	2.49
							18.60	1.92
							14.20	2.77
Type C	14.20	6.33	12.50	4.81	15.80	1.59	8.30	4.65
	10.80	6.01	17.90	4.48	12.90	2.06	18.20	3.51
	10.40	6.62	11.20	5.92	14.80	1.66	12.60	5.13
	15.80	5.97	13.20	4.60	8.20	2.71	13.10	4.46
	10.10	5.86	16.90	4.63	16.20	1.28	9.40	4.86
	15.90	6.19	12.40	4.82	11.60	2.61	8.80	5.49
	14.30	5.30	8.70	5.50	18.30	0.34	18.00	3.25
	18.90	6.38	10.90	5.27			20.90	2.85
	20.90	5.32	13.90	4.12				

Table 5.10 contains the PROC MIXED code to fit a model with a means model for the intercepts and an effects model for the slopes. The significance level for the moist*wood*stove term is 0.0842. For this problem, the conclusion is that there is not an important three-way interaction. The next step is to remove moist*wood*stove from the model and fit a model with just the two-way interaction terms, moist*wood and moist*stove. For this model, the significance level corresponding to the moist*stove term is .3267, indicating it is not an important term (results not shown). Table 5.11 contains the results of fitting a model with unequal slopes for each level of wood and the estimates of those slopes. The slopes for the two hard woods seem to be smaller (closer to zero) than the slopes for the two soft woods. In an attempt to simplify the model, pairwise comparisons of the slopes were accomplished by using a set of estimate statements.

TABLE 5.10
Results of Fitting an Effects Model for the Slopes of the Models for Example 5.5

```
proc mixed data=common ORDER=DATA; class
 Wood Stove;
model Energy=Wood*Stove Moist Moist*Wood
 Moist*Stove Moist*Wood*Stove;
```

CovParm	Estimate
Residual	000.1431

Effect	NumDF	DenDF	FValue	ProbF
Wood*Stove	11	64	8.93	0.0000
Moist	1	64	103.50	0.0000
Moist*Wood	3	64	10.02	0.0000
Moist*Stove	2	64	1.36	0.2650
Moist*Wood*Stove	6	64	1.96	0.0842

TABLE 5.11
Results of Fitting a Model with Different Slopes for Each Type of Wood for Example 5.5

```
proc mixed data=common; class Wood Stove;
model Energy=Wood*Stove Moist*Wood /noint
 solution;
```

CovParm	Estimate
Residual	0.156

Effect	NumDF	DenDF	FValue	ProbF
Wood*Stove	12	72	139.23	0.0000
Moist*Wood	4	72	35.54	0.0000

Effect	Wood	Estimate	StdErr
Moist*Wood	Black Walnut	–0.179	0.022
Moist*Wood	Osage Orange	–0.070	0.022
Moist*Wood	Red Oak	–0.043	0.028
Moist*Wood	White Pine	–0.170	0.021

The estimate statements and the results are in Table 5.12, where blk, osag, red, and wp denote black walnut, osage orange, red oak, and white pine, respectively. The results indicate the slopes for black walnut and white pine are not unequal and the slopes for osage orange and red oak are not unequal, while all other comparisons indicate unequal slopes. Thus the last estimate statement tests the equality of the average of the soft wood slopes (black walnut and white pine) and the average of the hard wood slopes (osage orange and red oak). The results, denoted by soft-hard, indicate the slope for soft wood is significantly different from the slope of hard wood.

TABLE 5.12
Pairwise Comparisons between the Wood Slopes and between the Soft and Hard Slopes for Example 5.5

```
estimate  'blk-osag'   Moist*wood  1 -1  0  0;
estimate  'blk-red'    Moist*wood  1  0 -1  0;
estimate  'blk-wp'     Moist*wood  1  0  0 -1;
estimate  'osag-red'   Moist*wood  0  1 -1  0;
estimate  'osag-wp'    Moist*wood  0  1  0 -1;
estimate  'red-wp'     Moist*wood  0  0  1 -1;
estimate  'soft-hard'  Moist*wood  .5 -.5 -.5 .5;
```

Label	Estimate	StdErr	df	tValue	Probt
blk-osag	−0.109	0.032	72	−3.45	0.0009
blk-red	−0.136	0.036	72	−3.76	0.0003
blk-wp	−0.009	0.031	72	−0.29	0.7713
osag-red	−0.027	0.036	72	−0.74	0.4611
osag-wp	0.100	0.031	72	3.28	0.0016
red-wp	0.127	0.035	72	3.60	0.0006
soft-hard	−0.118	0.024	72	−4.99	0.0000

Using the information about the relationship between the slopes, a new model was fit using a cell means model for the intercepts and one slope for the two soft woods and another slope for the two hard woods. The PROC MIXED code and results are in Table 5.13 using the classification variable KIND with two levels, HARD or SOFT. Table 5.13 contains the estimate of the variance, 0.153, and the estimates of the 12 intercepts and 2 slopes of the model. These intercepts and slopes can be used to construct a regression model for each of the wood*stove combinations. Figures 5.6 is a graph of the regression models for the soft woods with each stove type, and Figure 5.7 is a graph of the regression models for the hard woods with each stove type.

Comparisons among the regression models are needed in order to complete the analysis. The first step is to make comparisons among the intercepts, but it is not possible to obtain wood chunks with 0% moisture content. To obtain meaningful results, comparisons were made at a low level of moisture (7%), a middle level of moisture (13.34%, the mean for the experiment), and a high level of moisture (21%). Analyses were run using new covariates where each of the above points of comparison were subtracted from the observed levels of moisture. In all three cases, there was a significant wood*stove interaction (analyses are not shown). Since there are interactions between the levels of wood and the levels of stove for the intercepts, regression models for all 12 combinations of wood and stove need to be compared. There are two types of comparison. Comparisons among the wood*stove models involving hard woods are parallel line comparisons as are comparisons among the wood*stove models involving soft wood models. Comparisons between hard wood and soft wood models involve comparing nonparallel lines, since there is one slope for the hard wood and one slope for the soft wood. It would be desirable to use the LSMEANS statement from PROC MIXED to provide these comparisons, but the

TABLE 5.13
Results of Fitting a Model with Means for the Intercepts and a Different Slope for Each Kind of Wood

```
proc mixed data=common order=data; class Wood
   Stove Kind;
model Energy=Wood*Stove Moist*Kind /noint
   solution;
```

CovParm	Estimate
Residual	0.153

Effect	NumDF	DenDF	FValue	ProbF
Wood*Stove	12	74	149.58	0.0000
Moist*Kind	2	74	72.09	0.0000

Effect	Wood	Stove	Kind	Estimate	StdErr
Wood*Stove	Osage Orange	Type A		8.019	0.261
Wood*Stove	Osage Orange	Type B		7.444	0.291
Wood*Stove	Osage Orange	Type C		6.864	0.285
Wood*Stove	Red Oak	Type A		7.084	0.277
Wood*Stove	Red Oak	Type B		6.234	0.288
Wood*Stove	Red Oak	Type C		5.682	0.262
Wood*Stove	White Pine	Type A		4.348	0.222
Wood*Stove	White Pine	Type B		5.704	0.289
Wood*Stove	White Pine	Type C		4.178	0.258
Wood*Stove	Black Walnut	Type A		4.530	0.262
Wood*Stove	Black Walnut	Type B		5.119	0.245
Wood*Stove	Black Walnut	Type C		6.650	0.248
Moist*Kind			HARD	−0.059	0.017
Moist*Kind			SOFT	−0.174	0.015

LSMEANS statement will not provide the desired results for this model. A statement that would usually provide the adjusted means is "LSMEANS wood*stove/diff at Moist=7." However, for the model with moist*kind describing the covariate part of the model and wood*stove describing the intercepts, the LSMEANS statement does not work. The adjusted mean for osage orange should be computed as

$$\hat{y}_{osage\ orange|moist=7} = \hat{\alpha}_{osage\ orange} + 7\hat{\beta}_{hard}.$$

However, what is in fact being computed by the LSMEANS statement is

$$\hat{y}_{osage\ orange|moist=7} = \hat{\alpha}_{osage\ orange} + 3.5\hat{\beta}_{hard} + 7\hat{\beta}_{soft},$$

a value that does not have any meaning. Since the LSMEANS statement is requesting adjusted means for wood*stove combinations and the model involves moist*kind, the adjusted means are computed using the means of moisture over the levels of

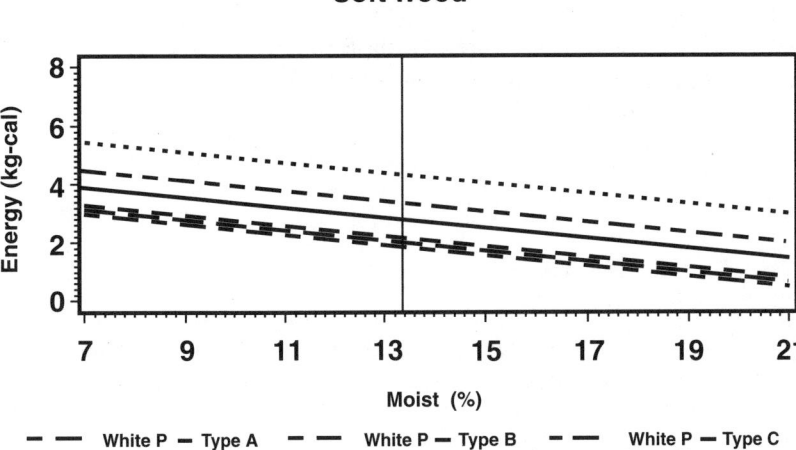

FIGURE 5.6 Graph of the estimated regression models for white pine and black walnut wood types with each stove type.

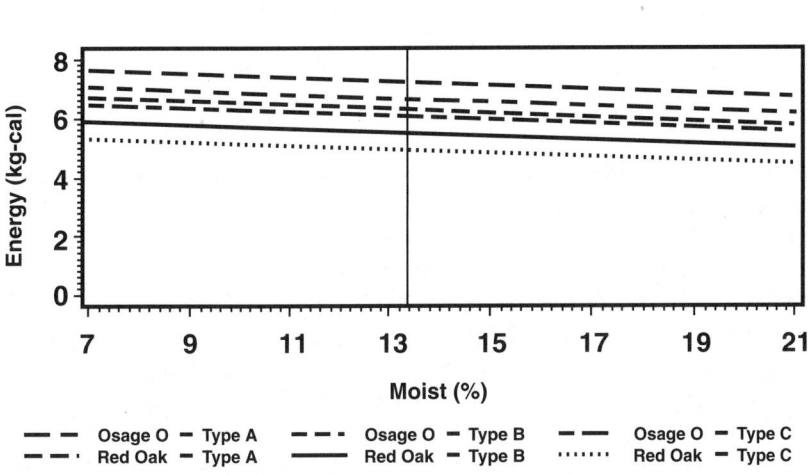

FIGURE 5.7 Graph of the estimated regression lines for osage orange and red oak wood types with each stove type.

kind. Be very careful when using the LSMEANS statement when the form of the covariate part of the model has been simplified, but is not a common slopes model. To get around this situation, estimate statements were used in order to obtain the appropriate adjusted means as well as to make comparisons among the pairs of adjusted means.

TABLE 5.14
Least Squares Means Computed Using Estimate Statements for Combinations of Wood and Stove at Moisture Contents of 7, 13.34, and 21%

```
estimate 'OO-A at 7.00' Wood*Stove 1 0 0 0 0 0 0 0 0 0 0 0
  Moist*Kind 7 0;
estimate 'OO-B at 7.00' Wood*Stove 0 1 0 0 0 0 0 0 0 0 0 0
  Moist*Kind 7 0;
estimate 'OO-C at 7.00' Wood*Stove 0 0 1 0 0 0 0 0 0 0 0 0
  Moist*Kind 7 0;
```

Wood	Stove	Moisture = 7%		Moisture = 13.34%		Moisture = 21%	
		Estimate	Stderr	Estimate	Stderr	Estimate	Stderr
Osage Orange	A	7.60	0.17	7.23	0.13	6.77	0.19
	B	7.03	0.20	6.65	0.16	6.20	0.20
	C	6.45	0.19	6.07	0.13	5.62	0.17
Red Oak	A	6.67	0.20	6.29	0.18	5.84	0.23
	B	5.82	0.20	5.44	0.16	4.99	0.20
	C	5.27	0.17	4.89	0.13	4.43	0.19
White Pine	A	3.13	0.15	2.03	0.12	0.70	0.18
	B	4.49	0.21	3.38	0.18	2.05	0.19
	C	2.96	0.18	1.86	0.15	0.53	0.18
Black Walnut	A	3.31	0.19	2.21	0.16	0.88	0.19
	B	3.90	0.17	2.80	0.14	1.47	0.18
	C	5.43	0.17	4.33	0.14	3.00	0.18

Table 5.14 contains three of the thirty-six estimate statements needed to compute the adjusted means. The label OO-A at 7 means osage orange for stove A at moisture equal to 7%. Adjusted means were computed for each of the 12 wood*stove combinations for moisture content of 7, 21, and 13.3427% (the average moisture content for the experiment). The adjusted means are in Table 5.14.

Table 5.15 contains all of the pairwise comparisons among the three levels of stove at each type of wood. These are parallel lines comparisons in that they are comparisons of stove types within a wood type. These comparisons were extracted from the LSMEANS statement where only comparisons within a wood type were retained. All stove type means are significantly different within a wood type except for stove types A and C with white pine. Stove type A has the largest mean for osage orange and red oak, stove type B has the largest mean for white pine, and stove type C has the largest mean for black walnut. Table 5.16 contains the comparisons of wood type models within each stove type that involve parallel lines comparisons.

The comparisons of the two hard woods (osage orange and red oak) and the comparisons of the two soft woods (white pine and black walnut) are parallel lines comparisons. These too are comparisons that can be extracted from the LSMEANS statement results. In this case, estimate statements for making the two types of comparisons with stove A are provided. All comparisons of means are significantly

TABLE 5.15
Pairwise Comparisons of Stove Least Squares Means for Each Level of Wood Evaluated at 13.34% Moisture

Wood	Stove	_Stove	Estimate	StdErr	df	tValue	Probt
Black Walnut	Type A	Type B	−0.588	0.211	74	−2.79	0.0067
Black Walnut	Type A	Type C	−2.119	0.211	74	−10.04	0.0000
Black Walnut	Type B	Type C	−1.531	0.195	74	−7.83	0.0000
Osage Orange	Type A	Type B	0.574	0.207	74	2.78	0.0069
Osage Orange	Type A	Type C	1.154	0.186	74	6.20	0.0000
Osage Orange	Type B	Type C	0.580	0.206	74	2.81	0.0062
Red Oak	Type A	Type B	0.850	0.238	74	3.58	0.0006
Red Oak	Type A	Type C	1.402	0.218	74	6.43	0.0000
Red Oak	Type B	Type C	0.552	0.206	74	2.68	0.0092
White Pine	Type A	Type B	−1.356	0.219	74	−6.19	0.0000
White Pine	Type A	Type C	0.170	0.194	74	0.87	0.3850
White Pine	Type B	Type C	1.525	0.230	74	6.64	0.0000

TABLE 5.16
Pairwise Comparisons of Wood Least Squares Means for Each Level of Stove Evaluated at 13.34% Moisture for Types of Wood with Equal Slopes

```
Estimate 'OO-RO stove A' wood*stove 1 0 0 -1 0 0 0 0 0 0 0 0;
estimate 'WP-BW stove A' wood*stove 0 0 0 0 0 0 1 0 0 -1 0 0;
```

Stove	Wood	_Wood	Estimate	StdErr	df	tValue	Probt
Type A	Osage Orange	Red Oak	0.935	0.218	74	4.28	0.0001
Type A	White Pine	Black Walnut	−0.182	0.203	74	−0.90	0.3727
Type B	Osage Orange	Red Oak	1.211	0.226	74	5.37	0.0000
Type B	White Pine	Black Walnut	0.585	0.225	74	2.60	0.0111
Type C	Osage Orange	Red Oak	1.183	0.186	74	6.36	0.0000
Type C	White Pine	Black Walnut	−2.471	0.202	74	−12.22	0.0000

different from zero except for the white pine and black walnut comparison with stove type A.

Finally, three of the thirty-six estimate statements are included in Table 5.17 to make comparisons among nonparallel lines models. These are comparisons that involve different kinds of wood, hard or soft. Since the lines are not parallel, these comparisons need to be made at three or more values of moisture. The nonparallel lines comparisons are osage orange to white pine and black walnut and red oak to white pine and black walnut. The comparisons were made for 7, 13.34, and 21% moisture. Table 5.17 contains the significance levels of each of the tests. All significance levels are less than 0.0001 except for the red oak to white pine comparisons for stove C at moisture levels of 7 and 13.34%.

TABLE 5.17
p-Values of Comparisons of Types of Wood (Hard and Soft) at Each Type of Stove

```
estimate 'OO-A - WP-A AT 7' WOOD*STOVE 1 0 0 0
  0 0 -1 0 0 0 0 0 Moist*KIND 7 -7;
estimate 'OO-B - WP-B AT 7' WOOD*STOVE 0 1 0 0
  0 0 0 -1 0 0 0 0 Moist*KIND 7 -7;
estimate 'OO-C - WP-C AT 7' WOOD*STOVE 0 0 1 0
  0 0 0 0 -1 0 0 0 Moist*KIND 7 -7;
```

Comparison Between Types of Wood	Stove	Moisture Content		
		7%	13.34%	21%
Osage Orange — White Pine	A	0.0000	0.0000	0.0000
	B	0.0000	0.0000	0.0000
	C	0.0000	0.0000	0.0000
Red Oak — White Pine	A	0.0000	0.0000	0.0000
	B	0.0000	0.0000	0.0000
	C	0.0000	0.0000	0.0000
Osage Orange — Black Walnut	A	0.0000	0.0000	0.0000
	B	0.0000	0.0000	0.0000
	C	0.0001	0.0000	0.0000
Red Oak — Black Walnut	A	0.0000	0.0000	0.0000
	B	0.0000	0.0000	0.0000
	C	0.4870	0.0044	0.0000

The relationships among the four wood types for each stove type are displayed in Figures 5.8 to 5.10. As in Figures 5.6 and 5.7, a vertical line denotes the mean moisture level of the experiment (13.34%) and the ends of each graph are the 7 and 21% points where the means were compared when the models have unequal slopes. The graphs for each stove easily demonstrate there are two different slopes, one for the hard woods and one for the soft woods. This example demonstrates many of the problems one faces when carrying out the analysis of covariance strategy. When the model simplification process is carried out and there are groups of treatments with common slopes within a group and unequal slopes between the groups, then the LSMEANS statement will not provide the appropriate adjusted means. In this case, estimate statements need to be used to obtain the appropriate adjusted means, although the LSMEANS results can be used to compare those combination of treatment's means involving parallel lines. You can also get around this LMEANS problem by using the model terms, "stove kind wood (kind) stove*kind stove*wood (kind) moist*kind."

5.6 MISSING TREATMENT COMBINATIONS

When there are missing treatment combinations in the analysis of covaraince model for a two-way treatment structure, one encounters problems similar to the analysis of a two-way treatment structure without covariates, as discussed in Chapters 13 and 14 of Milliken and Johnson (1992). There is still a set of observed treatment

Two-Way Treatment Structure and Analysis of Covariance

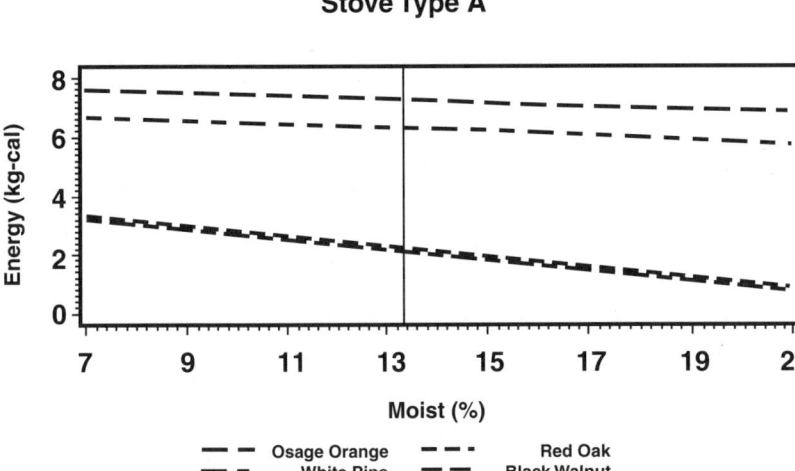

FIGURE 5.8 Graph of the regression lines for the four wood types with stove type A.

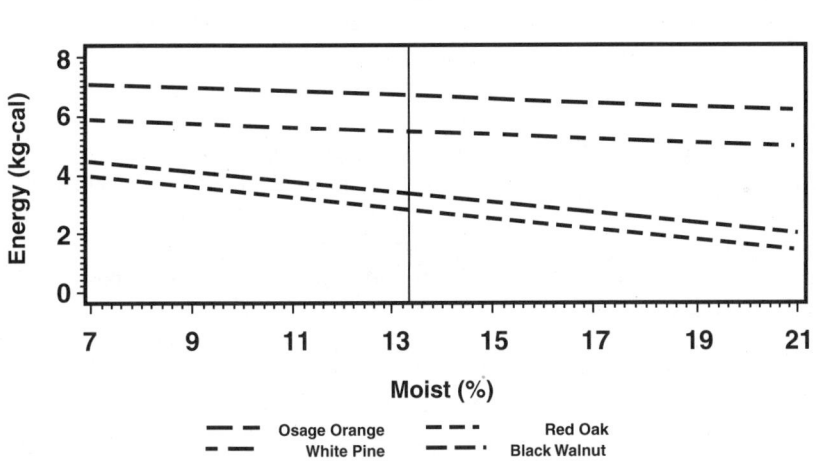

FIGURE 5.9 Graph of the regression lines for the four wood types with stove B.

combinations with a regression model for each. The observed treatment combinations could be analyzed as a one-way treatment structure as described in Chapters 2, 3, and 4, but it is also of interest to determine if there are row treatment and/or column treatment main effects and/or interaction between row and column treatments effects on the intercepts and slopes. The model in Equation 5.1 can be used to describe the data (for a linear relationship between the mean of Y and X) where some treatments are not observed. As in the analysis of the unbalanced two-way treatment structure without covariates, it is important to be able to determine which functions of the parameters are estimable and which estimable functions are associated

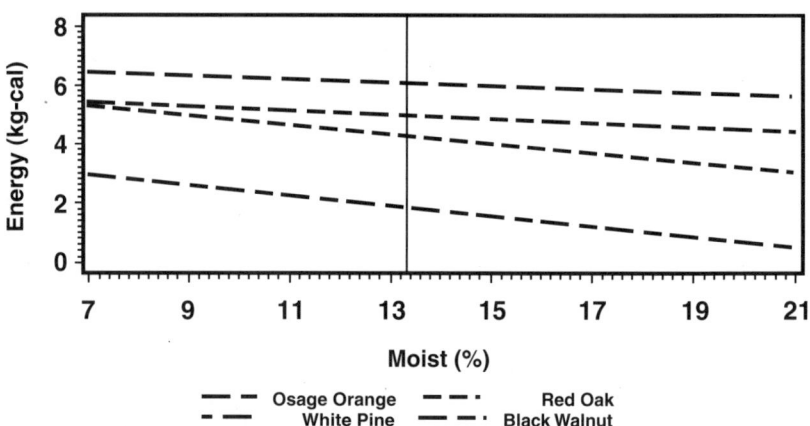

FIGURE 5.10 Graph of the regression lines of the four wood types with stove C.

with a given sum of squares. (See Milliken and Johnson, 1992, for a detailed discussion of estimability and the Type I to IV sums of squares.)

A similarity between the analysis of variance and analysis of covariance of a two-way treatment structure with missing cells is that the estimable functions for the slopes and the estimable functions for the intercepts have the same form, i.e., if a contrast of the intercepts is estimable, the same contrast of the slopes is also estimable. Thus, when the same function of the slopes and of the intercepts is estimable, then that function of the models at a given value of X is estimable.

As in the analysis of variance, the Type II and Type III estimable functions can be strange linear combinations of slopes and/or intercepts which are functions of the sample size. There are several different sets of Type IV estimable functions and it is important that one knows which Type IV estimable functions are used in each sum of squares. These statements about estimable functions depend on there being at least two observations with different values of the covariate within each cell with data.

If the structure of the treatments enables a linear combination of the intercepts to be estimable as $\sum_{(i,j) \in R} c_{ij}\alpha_{ij}$, where R is an index set containing the ordered pairs of indices of the observed treatment combinations, then the same linear combination of the slopes is estimable, as $\sum_{(i,j) \in R} c_{ij}\beta_{ij}$. Since linear combinations of estimable functions are estimable, then

$$\sum_{(i,j) \in R} c_{ij}\alpha_{ij} + X \sum_{(i,j) \in R} c_{ij}\beta_{ij} = \sum_{(i,j) \in R} c_{ij}\left(\alpha_{ij} + \beta_{ij}X\right)$$

$$= \sum_{(i,j) \in R} c_{ij}\left(\mu_{ij|\beta_{ij}X}\right)$$

is estimable, which is a linear combination of the models evaluated at X.

5.7 EXAMPLE: TWO-WAY TREATMENT STRUCTURE WITH MISSING CELLS

A mechanical engineer designed an experiment to study the characteristics of a lathe used to cut a small amount of the outside diameter from a metal rod. Two important factors which effect the roughness of the cut surface are the depth of cut and the turning speed of the surface of the uncut rod. The engineer felt the hardness of the rod being cut could influence the roughness of the surface; thus a measure of hardness was made on each rod (BHN is Brinell hardness number). The data in Table 5.18 were selected so as to have the same missing cell pattern as the example in Table 13.1 of Milliken and Johnson (1992). The missing cell structure of this data set uses the same estimable functions for the means model and the effects model described in Chapters 13 and 14, respectively, of Milliken and Johnson (1992).

Table 5.19 contains the PROC GLM code to fit a means model for both intercepts and slopes. The analysis of variance table has 14 degrees of freedom for the model

The similarities of the analysis of a two-way treatment structure with missing cells with and without a covariate (or covariates) are examined in the next example.

TABLE 5.18
Roughness Values for Rods Run on a Lathe at Various Depths of Cut and Speeds

	1 mm		2 mm		3 mm	
	BHN	Rough	BHN	Rough	BHN	Rough
100 (m/min)	254.0	3.00			238.0	4.18
	245.0	3.15			264.0	6.55
	218.0	1.77			216.0	3.14
	260.0	2.22			268.0	6.64
	242.0	2.73			263.0	6.32
	277.0	3.90				
	225.0	1.58				
150 (m/min)	269.0	3.07	256.0	4.47	256.0	6.89
	236.0	4.69	217.0	4.69	262.0	6.99
	256.0	4.06	238.0	4.49	217.0	4.27
	261.0	3.37	246.0	4.63	239.0	6.09
	211.0	4.08	245.0	4.77	218.0	5.12
	273.0	3.25	223.0	4.18	241.0	6.03
	233.0	4.25			216.0	5.46
200 (m/min)	211.0	7.76	221.0	7.94		
	263.0	3.89	275.0	4.40		
	232.0	5.93	243.0	5.76		
	253.0	5.02	237.0	6.46		
	217.0	7.49	269.0	4.72		
	240.0	5.73	226.0	7.68		
			247.0	6.11		

TABLE 5.19
PROC GLM Code to Fit the Means Model for the Intercepts and Slopes

```
PROC GLM; CLASSES TRT;
MODEL Rough=TRT BHN*TRT/E NOINT SOLUTION;
```

Source	df	SS	MS	FValue	ProbF
Model	14	1175.109	83.936	615.68	0.0000
Error	31	4.226	0.136		
Uncorrected Total	45	1179.336			

Source	df	SS	MS	FValue	ProbF
trt	7	50.341	7.192	52.75	0.0000
BHN*trt	7	39.844	5.692	41.75	0.0000

as there are 7 intercepts and 7 slopes. There are 31 degrees of freedom for estimating the error. The estimate of the error is obtained by pooling the error sums of squares across the seven observed treatment combinations. There are two basic sets of estimable functions as discussed by Milliken and Johnson (1992). First, the estimable functions of Chapter 13 based on the means are the mean of speed 2 averaged over the level of depth, the two 2 × 2 table differences measuring interaction and a comparison of the levels of speed using two contrasts, one averaging over depths 1 and 3 and the other averaging over depth 1 and 2. The linear combinations are

(a) "MU 2 DOT EX 13.1" = $\bar{\mu}_{2.} = (\mu_{21} + \mu_{22} + \mu_{23})/3$
(b) "Inter Ex 13.2" is measuring interaction by testing
$\mu_{11} - \mu_{13} - \mu_{21} + \mu_{23} = 0$ and $\mu_{21} - \mu_{22} - \mu_{31} + \mu_{32} = 0$

and

(c) "TEST IN EX 13.3" is measuring the main effect of speed by testing
$\frac{\mu_{11}+\mu_{13}}{2} = \frac{\mu_{21}+\mu_{23}}{2}$ and $\frac{\mu_{21}+\mu_{22}}{2} = \frac{\mu_{31}+\mu_{32}}{2}$.

The linear combinations of the above form can be constructed using the intercepts, the slopes, or the models evaluated at some value of BHN.

Second, the Type IV estimable hypotheses for speed (from Figure 14.2 of Milliken and Johnson, 1992) and for depth (from Figure 14.3 of Milliken and Johnson, 1992) are investigated. Again, the above contrasts are evaluated to compare intercepts, slopes, and/or models evaluated at some value of BHN. Tables 5.20 and 5.21 contain the estimate statements to evaluate the two types of estimable functions for the intercepts and the slopes. The contrasts in Table 5.20 are constructed from the intercepts as they are functions of TRT. The contrasts in Table 5.21 are constructed from the slopes as they are functions of BHN*TRT. The only difference between Tables 5.20 and 5.21 is that the contrasts in Table 5.20 use TRT and the contrasts in Table 5.21 use BHN*TRT. The results are in Table 5.22. Within each set, there is an indication of differences between the intercepts and slopes for speed

TABLE 5.20
Contrast Statements Used to Evaluate the Two Sets of Estimable Functions for the Intercepts of the Model

```
* COMPARISONS OF INTERCEPTS AS IN CHAPTER 13 OF AMD I;
  ESTIMATE 'MU BAR 2 DOT EX 13.1' TRT 0 0 .3333333 .3333333 .3333333 0 0;
  CONTRAST 'INTER EX 13.2' TRT 1 -1 -1 0 1 0 0, TRT 0 0 1 -1 0 -1 1;
  CONTRAST 'TEST IN EX 13.3' TRT 1 1 -1 0 -1 0 0, TRT 0 0 1 1 0 -1 -1;
* INTERCEPT COMPARISONS FOR CHAPTER 14 — Depth;
  CONTRAST '(M11+M21-M13-M23)/2' TRT .5 -.5 .5 0 -.5 0 0;
  CONTRAST '(M21+M31-M22-M32)/2' TRT 0 0 .5 -.5 0 .5 -.5;
  CONTRAST 'M11-M13' TRT 1 -1 0 0 0 0 0;
  CONTRAST 'M21-M22' TRT 0 0 1 -1 0 0 0;
  CONTRAST 'M21-M23' TRT 0 0 1 0 -1 0 0;
  CONTRAST 'M22-M23' TRT 0 0 0 1 -1 0 0;
  CONTRAST 'M31-M32' TRT 0 0 0 0 0 1 -1;
  CONTRAST 'SS Depth Type IV' TRT 0 0 0 1 -1 0 0, TRT .5 -.5 .5 0 -.5 0 0;
* COMPARISONS FOR CHAPTER 14 — Speed;
  CONTRAST '(M11+M13-M21-M23)/2' TRT .5 .5 -.5 0 -.5 0 0;
  CONTRAST '(M21+M22-M31-M32)/2' TRT 0 0 .5 .5 0 -.5 -.5;
  CONTRAST 'M11-M21' TRT 1 0 -1 0 0 0 0;
  CONTRAST 'M11-M31' TRT 1 0 0 0 0 -1 0;
  CONTRAST 'M21-M31' TRT 0 0 1 0 0 -1 0;
  CONTRAST 'M22-M32' TRT 0 0 0 1 0 0 -1;
  CONTRAST 'M13-M23' TRT 0 1 0 0 -1 0 0;
  CONTRAST 'SS Speed Type IV' TRT 1 0 0 0 0 -1 0, TRT 0 0 .5 .5 0 -.5 -.5;
```

TABLE 5.21
Selected Contrast Statements Used to Evaluate the Two Sets of Estimable Functions for the Slopes of the Model

```
* COMPARISONS OF SLOPES AS IN CHAPTER 13 OF AMD I;
  ESTIMATE 'Slope BAR 2 DOT like EX 13.1' BHN*TRT 0 0 .3333333
    .3333333 .3333333 0 0;
  CONTRAST 'BHN*SPEED*DEPTH-EX 13.2' BHN*TRT 1 -1 -1 0 1 0 0, BHN*TRT
    0 0 1 -1 0 -1 1;
  CONTRAST 'Slope TEST like IN EX 13.3' BHN*TRT 1 1 -1 0 -1 0 0,
    BHN*TRT 0 0 1 1 0 -1 -1;
* Slope COMPARISONS AS IN CHAPTER 14 for Depth;
  CONTRAST '(Slope 11+21-13-23)/2' BHN*TRT .5 -.5 .5 0 -.5 0 0;
  CONTRAST '(Slope 21+31-22-32)/2' BHN*TRT 0 0 .5 -.5 0 .5 -.5;
  CONTRAST 'Slope 11-13' BHN*TRT 1 -1 0 0 0 0 0;
  CONTRAST 'Slope 21-22' BHN*TRT 0 0 1 -1 0 0 0;
  CONTRAST 'Slope 21-23' BHN*TRT 0 0 1 0 -1 0 0;
  CONTRAST 'Slope 22-23' BHN*TRT 0 0 0 1 -1 0 0;
  CONTRAST 'Slope 31-32' BHN*TRT 0 0 0 0 0 1 -1;
  CONTRAST 'SS BHN*Depth Type IV' BHN*TRT 0 0 0 1 -1 0 0, BHN*TRT
    .5 -.5 .5 0 -.5 0 0;
```

TABLE 5.22
Results of the Contrast Statements for Both Types of Estimable Functions for the Intercepts and the Slopes

Source	df	Intercepts SS	Intercepts MS	Intercepts FValue	Slopes SS	Slopes MS	Slopes FValue
INTER EX 13.2	2	0.547	0.274	2.01	0.536	0.268	1.97
TEST IN EX 13.3	2	15.813	7.907	58.00	12.351	6.175	45.30
(M11+M21-M13-M23)/2	1	4.274	4.274	31.35	6.503	6.503	47.70
(M21+M31-M22-M32)/2	1	0.281	0.281	2.06	0.431	0.431	3.16
M11-M13	1	0.847	0.847	6.21	1.597	1.597	11.72
M21-M22	1	0.341	0.341	2.50	0.425	0.425	3.12
M21-M23	1	4.265	4.265	31.29	5.692	5.692	41.75
M22-M23	1	1.065	1.065	7.81	1.415	1.415	10.38
M31-M32	1	0.017	0.017	0.13	0.061	0.061	0.45
SS Depth Type IV	2	4.484	2.242	16.45	6.701	3.351	24.58
(M11+M13-M21-M23)/2	1	4.541	4.541	33.31	3.545	3.545	26.00
(M21+M22-M31-M32)/2	1	8.487	8.487	62.26	6.740	6.740	49.44
M11-M21	1	4.538	4.538	33.28	3.837	3.837	28.15
M11-M31	1	15.583	15.583	114.30	12.355	12.355	90.62
M21-M31	1	4.206	4.206	30.85	3.315	3.315	24.32
M22-M32	1	4.373	4.373	32.08	3.496	3.496	25.64
M13-M23	1	0.916	0.916	6.72	0.620	0.620	4.54
SS Speed Type IV	2	18.346	9.173	67.28	14.433	7.216	52.93

and for depth. Some of the F-values are small, but most are large. Table 5.23 contains a selected set of contrast statements to compare the models at BHN = 210. These contrast statements are identical to those in Table 5.20, except that both TRT and BHN*TRT are included. The coefficients of BHN*TRT depend on the value of BHN, unlike the contrasts in Table 5.21. The results of the contrasts for BHN = 210 and 280 are in Table 5.24. There is evidence that the depth models are different at BHN = 280 and the speed models are different at BHN = 210. Table 5.25 contains the PROC GLM code to fit an effects model for both the intercepts and the slopes. Table 5.25 contains the Type I sums of squares, while the Type II, Type III, and Type IV sums of squares are in Tables 5.26 through 5.28. There are four different sums of squares for Speed, Depth, and BHN*Speed. There are three different sums of squares for BHN and BHN*Depth. There are two different sums of squares for Speed*Depth, and there is one sum of square for BHN*Speed*Depth. When there is more than one sum of squares for a given effect, then more than one hypothesis is being tested about that effect. Chapter 14 of Milliken and Johnson (1992) presents a detailed discussion of the hypotheses being tested. The general form of the estimable function is in Table 5.29. The most important aspect of this table is that the general form of the estimable function for the intercepts is identical to the general form of the estimable function for the slopes and thus would be the general form of the estimable function for the models evaluated at a specific value of BHN. Since

Two-Way Treatment Structure and Analysis of Covariance 151

TABLE 5.23
Selected Contrasts Statements to Evaluate Both Types of Estimable Functions for Comparing the Models at BNH = 210

```
* COMPARISONS OF MODELS AS IN CHAPTER 13 OF AMD I;
  CONTRAST 'Inter of Models at BHN=210-EX 13.2' TRT 1 -1 -1 0 1 0
  0 BHN*TRT 210 -210 -210 0 210 0 0,
  TRT 0 0 1 -1 0 -1 1 BHN*TRT 0 0 210 -210 0 -210 210;
  CONTRAST 'Model TEST at BHN=210 - EX 13.3'
  TRT 1 1 -1 0 -1 0 0 BHN*TRT 210 210 -210 0 -210 0 0, TRT 0 0
  1 1 0 -1 -1 BHN*TRT 0 0 210 210 0 -210 -210;
* MODEL COMPARISONS at BHN=210 like FOR CHAPTER 14 for Depth;
  CONTRAST '(Model-210 11+21-13-23)/2' TRT .5 -.5 .5 0 -.5 0 0
  BHN*TRT 105 -105 105 0 -105 0 0;
  CONTRAST '(Model 210 21+31-22-32)/2' TRT 0 0 .5 -.5 0 .5 -.5
  BHN*TRT 0 0 105 -105 0 105 -105;
  CONTRAST 'Model 210 11-13' TRT 1 -1 0 0 0 0 0 BHN*TRT 210 -210
  0 0 0 0 0; 0 0 0 0 -210 0, TRT 0 0 .5 .5 0 -.5 -.5 BHN*TRT 0
  0 105 105 0 -105 -105;
```

TABLE 5.24
The Sums of Squares Corresponding to the Two Sets of Estimable Functions for the Models at BHN Values of 210 and 280

	df	BHN = 210			BHN = 280		
		SS	MS	FValue	SS	MS	FValue
Inter of Models -EX 13.2	2	0.502	0.251	1.84	0.341	0.170	1.25
Model TEST - EX 13.3	2	38.936	19.468	142.80	0.358	0.179	1.31
(Model- 11+21-13-23)/2	1	0.478	0.478	3.51	25.848	25.848	189.60
(Model 21+31-22-32)/2	1	0.083	0.083	0.61	1.667	1.667	12.23
Model 11-13	1	0.725	0.725	5.31	11.116	11.116	81.54
Model 21-22	1	0.012	0.012	0.09	0.950	0.950	6.97
Model 21-23	1	0.002	0.002	0.01	14.752	14.752	108.21
Model 22-23	1	0.025	0.025	0.18	3.719	3.719	27.28
Model 31-32	1	0.314	0.314	2.31	0.718	0.718	5.27
SS Depth Type IV	2	0.478	0.239	1.75	25.851	12.925	94.81
(Model 11+13-21-23)/2	1	10.154	10.154	74.48	0.009	0.009	0.07
(Model 21+22-31-32)/2	1	19.951	19.951	146.35	0.327	0.327	2.40
Model 11-21	1	7.710	7.710	56.56	0.288	0.288	2.11
Model 11-31	1	35.347	35.347	259.28	0.528	0.528	3.87
Model 21-31	1	9.389	9.389	68.87	0.099	0.099	0.73
Model 22-32	1	10.596	10.596	77.72	0.231	0.231	1.69
Model 13-23	1	3.090	3.090	22.67	0.095	0.095	0.70
SS Speed Type IV	2	43.709	21.855	160.31	0.613	0.306	2.25

TABLE 5.25
PROC GLM Code to Fit the Effects Model for Both the Intercepts and the Slope with the Type I Sums of Squares

```
proc glm data=common; class Speed Depth;
model Rough=Speed Depth Speed*Depth BHN BHN*Speed
 BHN*Depth BHN*Speed*Depth /solution ss1 ss2 ss3
 ss4 e e1 e2 e3 e4;
```

Source	df	SS	MS	FValue	ProbF
Model	13	110.088	8.468	62.12	0.0000
Error	31	4.226	0.136		
Corrected Total	44	114.314			

Source	df	SS (Type I)	MS	FValue	ProbF
Speed	2	33.648	16.824	123.41	0.0000
Depth	2	34.911	17.455	128.04	0.0000
Speed*Depth	2	1.685	0.843	6.18	0.0055
BHN	1	0.094	0.094	0.69	0.4127
BHN*Speed	2	32.375	16.188	118.74	0.0000
BHN*Depth	2	6.839	3.419	25.08	0.0000
BHN*Speed*Depth	2	0.536	0.268	1.97	0.1572

TABLE 5.26
Type II Sums of Squares

Source	df	SS (Type II)	MS	FValue	ProbF
Speed	2	20.344	10.172	74.61	0.0000
Depth	2	4.609	2.305	16.91	0.0000
Speed*Depth	2	0.547	0.274	2.01	0.1514
BHN	1	0.094	0.094	0.69	0.4127
BHN*Speed	2	15.989	7.994	58.64	0.0000
BHN*Depth	2	6.839	3.419	25.08	0.0000
BHN*Speed*Depth	2	0.536	0.268	1.97	0.1572

these two parameters have identical forms of the general estimable function, then the general form of the estimable function of the models evaluated at a specified value of BHN has the same structure.

There is no evidence that there is an interaction between the levels of speed and the levels of depth for the intercepts and covariate part of the model since the significance levels corresponding to speed*depth is 0.01514 and to BHN*speed*depth is 0.1572.

The three-way interaction term was removed and a model with unequal slopes for speeds and unequal slopes for depths and an additive effects for the intercepts

TABLE 5.27
Type III Sums of Squares

Source	df	SS (Type III)	MS	FValue	ProbF
Speed	2	20.067	10.034	73.60	0.0000
Depth	2	4.192	2.096	15.38	0.0000
Speed*Depth	2	0.547	0.274	2.01	0.1514
BHN	1	0.011	0.011	0.08	0.7778
BHN*Speed	2	15.742	7.871	57.73	0.0000
BHN*Depth	2	6.421	3.211	23.55	0.0000
BHN*Speed*Depth	2	0.536	0.268	1.97	0.1572

TABLE 5.28
Type IV Sums of Squares

Source	df	SS (Type IV)	MS	FValue	ProbF
Speed	2	18.346	9.173	67.28	0.0000
Depth	2	4.484	2.242	16.45	0.0000
Speed*Depth	2	0.547	0.274	2.01	0.1514
BHN	1	0.001	0.001	0.01	0.9208
BHN*Speed	2	14.433	7.216	52.93	0.0000
BHN*Depth	2	6.701	3.351	24.58	0.0000
BHN*Speed*Depth	2	0.536	0.268	1.97	0.1572

was used to describe the data. The effects model with additive effects for speed and depth for both the slopes and intercepts can be expressed as

$$Y_{ijk} = \left(\mu + \alpha_i + \gamma_j\right) + \left(\beta + \rho_i + \phi_j\right)X_{ij} + \varepsilon_{ijk}.$$

An assumption of no interaction is assuming there is no interaction among the speeds and depths, whether the cell was observed or not. The slope for speed 1 is computed as and the slope for $\beta_{s1} = \beta + \rho_1 + \bar{\phi}$. and the slope for depth 1 is computed as $\beta_{d1} = \beta + \bar{\rho}. + \phi_1$. The PROC GLM code in Table 5.30 fits the above model to the data and the test statistics indicate there are significant speed and depth effects for both the intercepts and the slopes. Table 5.31 contains the estimate statements needed to compte each of the slopes for the speed models and for the depth models. Table 5.32 uses the LSMEANS statement evaluated at BHN = 0 (the origin) to provide the estimates of the intercepts for each of the models.

These slope and intercept estimates were used to provide graphs of the speed models and the depth models as shown in Figures 5.11 and 5.12. The least squares means for the speed and depth models evaluated at BHN = 210, 280, and 242.6 are displayed in Table 5.33. Pairwise comparisons of the speed models at each of the three values of BHN are in Table 5.34, while Table 5.35 contains the pairwise

TABLE 5.29
General Form of the Estimable Function for the Intercepts and for the Slopes

Effect	Coefficients
Intercept	
Intercept	L1
Speed 100 (m/min)	L2
Speed 150 (m/min)	L3
Speed 200 (m/min)	L1 – L2 – L3
Depth 1 mm	L5
Depth 2 mm	L6
Depth 3 mm	L1 – L5 – L6
Speed*Depth 100 (m/min) 1 mm	L8
Speed*Depth 100 (m/min) 3 mm	L2 – L8
Speed*Depth 150 (m/min) 1 mm	L10
Speed*Depth 150 (m/min) 2 mm	–L1 + L2 + L3 + L5 + L6 – L8 – L10
Speed*Depth 150 (m/min) 3 mm	L1 – L2 – L5 – L6 + L8
Speed*Depth 200 (m/min) 1 mm	L5 – L8 – L10
Speed*Depth 200 (m/min) 2 mm	L1 – L2 – L3 – L5 + L8 + L10
Slopes	
BHN	L15
BHN*Speed 100 (m/min)	L16
BHN*Speed 150 (m/min)	L17
BHN*Speed 200 (m/min)	L15 – L16 – L17
BHN*Depth 1 mm	L19
BHN*Depth 2 mm	L20
BHN*Depth 3 mm	L15 – L19 – L20
BHN*Speed*Depth 100 (m/min) 1 mm	L22
BHN*Speed*Depth 100 (m/min) 3 mm	L16 – L22
BHN*Speed*Depth 150 (m/min) 1 mm	L24
BHN*Speed*Depth 150 (m/min) 2 mm	–L15 + L16 + L17 + L19 + L20 – L22 – L24
BHN*Speed*Depth 150 (m/min) 3 mm	L15 – L16 – L19 – L20 + L22
BHN*Speed*Depth 200 (m/min) 1 mm	L19 – L22 – L24

comparisons of the levels of depth. The three speed models are significantly different at BHN = 210 and 242.6, but are not significantly different at BHN = 280. The depth models are significantly different at BHN 242.6 and 280, but are not significantly different at BHN = 210. The graphs in Figures 5.11 and 5.12 help understand the above interpretations. The model used to describe this data set consisted of slopes that were simplified in form to be a function of the levels of speed and of the levels of depth, but not of the interaction between the levels of the two factors. Simplifying the slopes is part of the proposed analysis of covariance strategy, but the model also includes a simplified form for the intercepts. The intercepts were expressed as an additive function of the levels of speed and of the levels of depth. This is a dangerous process. The test for interaction between the levels of speed and the levels of depth for both slopes and intercepts concerns only those cells where there are data and cannot address the occurrence of interaction involving the cells with no data.

TABLE 5.30
PROC GLM Code to Fit the Reduced Model to the Roughness Data

```
proc glm data=common; class Speed Depth;
model Rough=Speed Depth BHN BHN*Speed
 BHN*Depth /ss1 ss2 ss3 ss4 e e1 e2 e3 e4;
```

Source	df	SS	MS	FValue	ProbF
Model	9	109.479	12.164	88.05	0.0000
Error	35	4.835	0.138		
Corrected Total	44	114.314			

Source	df	SS (Type III)	MS	FValue	ProbF
Speed	2	20.300	10.150	73.47	0.0000
Depth	2	4.709	2.354	17.04	0.0000
BHN	1	0.035	0.035	0.26	0.6166
BHN*Speed	2	15.922	7.961	57.63	0.0000
BHN*Depth	2	7.073	3.536	25.60	0.0000

TABLE 5.31
Estimate Statement to Provide Estimates of the Slopes for the Speed Models and the Depth Models

```
Estimate 'Slope speed 1' BHN 1 BHN*Speed 3 0 0
 BHN*Depth 1 1 1/divisor=3;
Estimate 'Slope speed 2' BHN 3 BHN*Speed 0 3 0
 BHN*Depth 1 1 1/divisor=3;
Estimate 'Slope speed 3' BHN 3 BHN*Speed 0 0 3
 BHN*Depth 1 1 1/divisor=3;
Estimate 'Slope depth 1' BHN 3 BHN*Depth 3 0 0
 BHN*Speed 1 1 1/divisor=3;
Estimate 'Slope depth 2' BHN 3 BHN*Depth 0 3 0
 BHN*Speed 1 1 1/divisor=3;
Estimate 'Slope depth 3' BHN 3 BHN*Depth 0 0 3
 BHN*Speed 1 1 1/divisor=3;
```

Label	Estimate	StdErr	df	tValue	Probt
SLOPE SPEED 1	0.0482	0.0062	35	7.75	0.0000
SLOPE SPEED 2	0.0090	0.0047	35	1.90	0.0661
SLOPE SPEED 3	−0.0525	0.0062	35	−8.52	0.0000
SLOPE DEPTH 1	−0.0197	0.0043	35	−4.64	0.0000
SLOPE DEPTH 2	−0.0080	0.0070	35	−1.14	0.2602
SLOPE DEPTH 3	0.0324	0.0061	35	5.29	0.0000

Generally, when there are missing cells, a global statement of the nonexistence of interaction cannot be made. Only a decision about the nonsignificance of interaction among the cells that have data can result. The reason the form of the intercepts was simplified is so that least squares means could be computed. When there are

TABLE 5.32
Estimates of the Intercepts for the Speed Models and the Depth Models from the LSMEANS Statement

```
lsmeans speed depth/ diff at BHN=0;
```

Speed	Depth	Estimate	StdErr
100 (m/min)		−8.164	1.534
150 (m/min)		2.672	1.140
200 (m/min)		19.269	1.481
	1 mm	8.758	1.037
	2 mm	6.650	1.681
	3 mm	−1.632	1.483

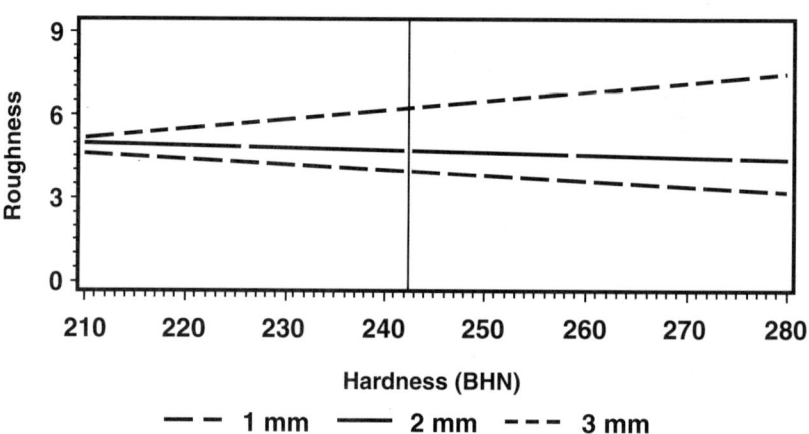

FIGURE 5.11 Graph of the depth regression models.

missing cells and the data suggest there is no interaction between the levels of the two factors, then one possible step is to simplify the form of the intercepts.

The researcher must be comfortable with the assumption across all of the cells (even those with no data), i.e., does the assumption make sense? If the no interaction assumption is reasonable, then the model can be simplified. When the model is simplified, the degrees of freedom for interaction are pooled with those of the degrees of freedom for error. Pooling degrees of freedom from interaction can have the effect of contaminating the estimate of the variance as well as increasing the number of degrees of freedom to more than were purchased when the experiment was designed. If no cells are empty, then it is not recommended to reduce the form of the intercepts.

TABLE 5.33
Least Squaes Means for the Levels of Speed and the Levels of Depth for BHN= 210, 242.6 and 280

```
lsmeans speed depth/ diff at BHN=210 e;
lsmeans speed depth/ diff at BHN=280 e;
lsmeans speed depth/ diff at means e;
```

		BHN = 210		BHN = 280		BHN = 242.6	
		Estimate	StdErr	Estimate	StdErr	Estimate	StdErr
Speed	100 (m/min)	1.963	0.252	5.339	0.244	3.535	0.121
Speed	150 (m/min)	4.559	0.166	5.188	0.207	4.852	0.086
Speed	200 (m/min)	8.251	0.217	4.579	0.274	6.541	0.118
		Estimate	StdErr	Estimate	StdErr	Estimate	StdErr
Depth	1 mm	4.616	0.164	3.236	0.179	3.974	0.085
Depth	2 mm	4.975	0.243	4.417	0.298	4.715	0.117
Depth	3 mm	5.181	0.225	7.452	0.267	6.239	0.120

TABLE 5.34
Comparisons of the Speed Models at BHN=210, 242.6, and 280

Speed	_Speed	BHN	Estimate	StdErr	tValue	Probt
100 (m/min)	150 (m/min)	210.0	−2.596	0.296	−8.78	0.0000
100 (m/min)	200 (m/min)	210.0	−6.289	0.339	−18.57	0.0000
150 (m/min)	200 (m/min)	210.0	−3.693	0.285	−12.97	0.0000
100 (m/min)	150 (m/min)	280.0	0.151	0.297	0.51	0.6153
100 (m/min)	200 (m/min)	280.0	0.760	0.409	1.86	0.0716
150 (m/min)	200 (m/min)	280.0	0.609	0.357	1.70	0.0973
100 (m/min)	150 (m/min)	242.6	−1.317	0.147	−8.97	0.0000
100 (m/min)	200 (m/min)	242.6	−3.006	0.177	−16.96	0.0000
150 (m/min)	200 (m/min)	242.6	−1.689	0.149	−11.35	0.0000

Table 5.36 contains the estimate statements needed to compute the estimates of the cell means at a BHN value of 210. Only the estimate statements for the levels of speed for depth 1 are included because the other statements can be constructed using the pattern of the statements provided. The body of the table contains the estimates of the cell means, using the additive model, evaluated at BHN = 210. The speed least squares means were computed by averaging over the levels of the depth estimated cell means. For example, the least squares mean for speed 100 (m/min) is $(1.6550 + 2.0139 + 2.2199)/3 = 1.9603$. This is identical to the least squares mean in Table 5.33. The least squares means for depth are computed by averaging over the levels of speed estimated cell means. This process of computing cell means illustrates the importance of making sure the no interaction assumption is reasonable

TABLE 5.35
Comparisons of the Depth Models at BHN = 210, 242.6, and 280

Depth	_Depth	BHN	Estimate	StdErr	tValue	Probt
1 mm	2 mm	210.0	−0.359	0.291	−1.23	0.2255
1 mm	3 mm	210.0	−0.565	0.283	−2.00	0.0537
2 mm	3 mm	210.0	−0.206	0.337	−0.61	0.5454
1 mm	2 mm	280.0	−1.181	0.366	−3.23	0.0027
1 mm	3 mm	280.0	−4.216	0.305	−13.84	0.0000
2 mm	3 mm	280.0	−3.035	0.448	−6.78	0.0000
1 mm	2 mm	242.6	−0.742	0.149	−4.98	0.0000
1 mm	3 mm	242.6	−2.265	0.145	−15.58	0.0000
2 mm	3 mm	242.6	−1.524	0.176	−8.67	0.0000

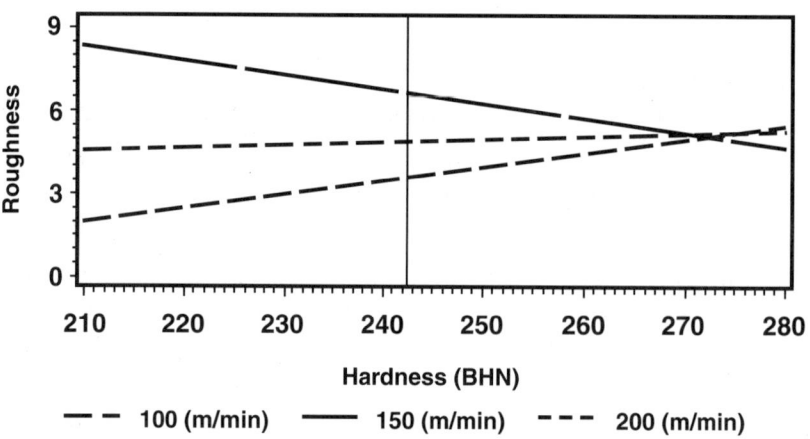

FIGURE 5.12 Graph of the speed regression models.

as the estimated cell means used in computing the depth and speed least squares means are computed using a model with additive effects for the intercepts and slopes. If the no interaction assumption is not appropriate for all of the cells in the study (including the empty cells), then the resulting depth and speed marginal or least squares means are not meaningful.

5.8 EXTENSIONS

When the treatment structure involves more than two factors and/or two or more covariates, the effects models for the treatment structure need to be fit to the

TABLE 5.36
Estimate Statements to Provide Estimates of the Cell Means with the Corresponding Least Squares Means for the Levels of Depth and Speed at BHN = 210

```
ESTIMATE 'CELL 11 AT 210' INTERCEPT 1 SPEED
1 0 0 DEPTH 1 0 0 BHN 210 BHN*SPEED 210 0
0 BHN*DEPTH 210 0 0;
ESTIMATE 'CELL 12 AT 210' INTERCEPT 1 SPEED
1 0 0 DEPTH 0 1 0 BHN 210 BHN*SPEED 210 0
0 BHN*DEPTH 0 210 0;
ESTIMATE 'CELL 13 AT 210' INTERCEPT 1 SPEED
1 0 0 DEPTH 0 0 1 BHN 210 BHN*SPEED 210 0
0 BHN*DEPTH 0 0 210;
```

		Speed			Depth LSM
		100	150	200	
Depth	1	1.6550	4.2509	7.9436	4.6165
	2	2.0139	4.6098	8.3025	4.9754
	3	2.2199	4.8157	8.5084	5.1813
Speed LSM		1.9630	4.5588	8.2515	

intercepts and the slopes corresponding to each covariate. A stepwise deletion process can be used to simplify the forms of the slopes, remembering that only one term may be eliminated at each step. Higher order interactions with the covariate should be eliminated before lower order terms: otherwise the lower order terms are pooled with the corresponding higher order term and no model reduction occurs. Once the covariate part of the model is simplified as much as possible, then compare the regression models. If the slopes for any covariate are unequal for one or more factors, comparisons of intercepts are comparisons of the regression models evaluated at the origin (all covariate set to zero). Such comparisons may not be meaningful. One can always compare models at a certain value of X^* by subtracting X^* from all values of the covariates and then fitting the model using this adjusted covariate. The intercepts of the resulting models are the adjusted means from the models at $X = X^*$.

If the slopes have been grouped in a way that does not relate to the factors defining the intercepts, such as the kind of wood factor used in Section 5.5, then the results from the LSMEANS statement may not be useful. By including the "e" option at the end of the LSMEANS statement, PROC MIXED and PROC GLM will provide those linear combinations of parameters used to evaluate the adjusted means. These linear combinations should be checked to make sure the adjusted means calculated are in fact those that need to be calculated.

If common slopes are adequate for all covariates, then the comparisons of LSMEANS or comparisons of the intercepts provide information about the distances between the lines, planes or hyper-planes. As with any modeling problem, the residuals should be plotted before a model is used to compare treatment combinations.

REFERENCE

Milliken, G. A. and Johnson, D. E. (1992) *Analysis of Messy Data, Volume I: Design Experiments*, Chapman & Hall, London.

EXERCISES

EXERCISE 5.1: The following data are from a two-way treatment structure in a completely randomized design structure. Ten types of chocolate chips were obtained from local markets consisting of two types of chocolate chips (regular and semisweet) each from five brands. Six plastic bags were prepared where one chip from a given brand and type was put into the bag and then the bag was marked for identification. In addition to the specific chocolate chip, a butterscotch chip was also included in each bag. Six bags were made up for each combination of type and brand. The bags were put into a box and mixed (randomized). Students formed a line, reached into the box, and selected one bag (another level of randomization). The students then measured the amount of time required to dissolve the butterscotch chip followed by the amount of time to dissolve the chocolate chip after putting them into their mouths. Provide an analysis of the following data set where all decisions are made at $\alpha = 0.10$. Carry out the analysis of covariance and then make any comparisons that are needed in order to develop an understanding of the relationship among the combinations of types and brands as if all persons dissolved the butterscotch chips in the same amount of time.

Data for Exercise 5.1

	Brand A		Brand B		Brand C		Brand D		Brand E	
	bstime	time	bstime	time	bstime	time	bstime	time	bstime	time
Regular	38	50	23	21	27	45	33	42	26	33
	35	48	29	27	17	29	32	37	25	37
	38	49	24	31	33	43	12	17	16	26
	27	35			21	40	26	32	21	37
	34	42			20	31	25	30		
	bstime	time	bstime	time	bstime	time	bstime	time	bstime	time
Semisweet	16	42	38	44	17	26	37	49	27	52
	32	52	39	45	19	23	39	59	24	48
	37	53	37	43	17	33	20	40	21	44
	19	38			16	22	12	40		
	30	49								

EXERCISE 5.2: Use the data set in Section 5.5, but first delete the data in cells Type A and white pine, Type B and Red Oak, and Type C and Black Walnut. Determine the estimable functions associated with a the intercepts and with the slopes. Provide a complete analysis of the data set including graphics of the resulting regression models.

EXERCISE 5.3: Carry out an analysis of covariance for the following data set, by determining the appropriate model and then making the needed treatment comparisons. Y is the response variable and X is the covariate. Note that this is a two-way treatment structure in a completely randomized design structure.

Data for Exercise 5.3

		Treatment A				
	A = 1		A = 2		A = 3	
Treatment B	X	Y	X	Y	X	Y
B = 1	26.9	30.9	19.4	24.1	37.1	29.0
	28.0	32.8	23.8	26.7	24.6	22.6
	26.8	31.4	21.4	27.5	36.9	24.1
	36.8	41.5	30.1	29.9	19.8	21.4
	28.6	34.8	32.4	32.4	24.3	21.6
	31.3	37.2	18.8	26.7	18.6	20.6
B = 2	37.1	29.0	12.7	23.3	31.1	21.3
	24.6	22.6	36.1	36.6	37.8	22.5
	36.9	24.1	17.0	28.5	18.7	18.7
	19.8	21.4	35.2	35.1	16.9	20.8
	24.3	21.6	21.4	27.0	15.0	16.8
	18.6	20.6	34.0	33.3	26.2	19.2

EXERCISE 5.4: Carry out an analysis of covariance for the following data set, by determining the appropriate model and then making the needed treatment comparisons. Y is the response variable and X is the covariate. Note that this is a two-way treatment structure in a completely randomized design structure.

Data for Exercise 5.4

	Treatment B			
	B = 1		B = 2	
Treatment A	X	Y	X	Y
A = 1	8.0	87.4	18.1	78.2
	7.2	92.3	19.8	78.2
	12.4	91.1	19.3	81.1
	5.5	94.3	18.6	79.4
	13.1	81.6	23.8	76.1
	5.4	97.9	24.0	77.2
A = 2	11.8	90.8	23.8	79.4
	8.2	103.3	18.0	84.9
	13.4	93.2	19.9	85.9
	14.0	84.0	20.2	79.3
	6.9	104.0	17.6	97.9
	11.5	94.3	23.5	80.5

6 Beta-Hat Models

6.1 INTRODUCTION

When the same form of model is used to describe data from several populations, treatments, or treatment combinations, many questions about the models can be answered by testing hypotheses and constructing confidence intervals about functions of parameters or a single parameter from each model. For example, if a simple linear regression model is fit to data from each treatment, comparisons between the slopes can answer questions about parallelism or can be used to group treatments together that have common slopes. The beta-hat model can easily be used to extract the necessary information. Beta-hat models are extremely useful in investigating the necessary form of the analysis of covariance model when there are a large number of treatments and computing resources are limited.

6.2 THE BETA-HAT MODEL AND ANALYSIS

The beta-hat model can be defined and used in a very general context, but a simplified form is described here. Assume the analysis of covariance model for the jth observation from the ith treatment is of the form

$$y_{ij} = \mu_i + \beta_{1i} x_{1ij} + \beta_{2i} x_{2ij} + \cdots + \beta_{qi} x_{qij} + \varepsilon_{ij} \tag{6.1}$$

for $i = 1, 2, \ldots, t$ and $j = 1, 2, \ldots, n_i$, where $\varepsilon_{ij} \sim \text{iidN}(0, \sigma^2)$ and the models for each treatment have their own set of parameters. The model for treatment i can be expressed in terms of matrices as

$$\mathbf{y}_i = \begin{bmatrix} y_{i1} \\ \vdots \\ y_{in_i} \end{bmatrix} = \mathbf{X}_i \boldsymbol{\beta}_i + \boldsymbol{\varepsilon}_i = \begin{bmatrix} 1 & x_{1i1} & \cdots & x_{qi1} \\ 1 & x_{1i2} & \cdots & x_{qi2} \\ \vdots & \vdots & & \vdots \\ 1 & x_{1in_i} & \cdots & x_{qin_i} \end{bmatrix} \begin{bmatrix} \mu_i \\ \beta_{1i} \\ \vdots \\ \beta_{qi} \end{bmatrix} + \begin{bmatrix} \varepsilon_{i1} \\ \vdots \\ \varepsilon_{in_i} \end{bmatrix} \tag{6.2}$$

The least squares estimates of the $\boldsymbol{\beta}_i$ are

$$\hat{\boldsymbol{\beta}}_i = (\mathbf{X}'_i \mathbf{X}_i)^{-1} \mathbf{X}'_i \mathbf{y}_i, \quad i = 1, 2, \ldots, t \tag{6.3}$$

and the sampling distributions of the $\hat{\boldsymbol{\beta}}_i$ are

$$\hat{\beta}_i \sim N\left(\beta_i, \sigma^2(X_i'X_i)^{-1}\right), \quad i = 1, 2, \ldots, t \tag{6.4}$$

where the $\hat{\beta}_i$ are pairwise independently distributed. The residual sum of squares from treatment i is denoted by $SSRES_i$ and the pooled estimate of the variance is

$$\hat{\sigma}^2 = SSRES/(N - t(q+1)) \tag{6.5}$$

where $SSRES = \sum_{i=1}^{t} SSRES_i$ and $N = \sum_{i=1}^{t} n_i$. (One should test for equality of variances before pooling.) The sufficient statistics are $\hat{\sigma}^2$ and $\hat{\beta}_i$, $i = 1, 2, \ldots, t$; thus all of the information contained in the data set and thus needed to do the analysis of covariance is contained in $\{\hat{\sigma}^2, \hat{\beta}_1, \hat{\beta}_2, \ldots, \hat{\beta}_t, (X_1'X_1)^{-1}, (X_2'X_2)^{-1}, \ldots, (X_t'X_t)^{-1}\}$.

Suppose you are interested in comparing the β_{ki} $i = 1, 2, \ldots, t$ parameters from Model 6.1 across treatments. From Equation 6.4, the sampling distribution of $\hat{\beta}_{ki}$ is $N(\beta_{ki}, \sigma^2 C_{ikk})$ where C_{ikk} is the kth diagonal element of $(X_i'X_i)^{-1}$. Let

$$\hat{\beta}_k' = [\hat{\beta}_{k1}, \hat{\beta}_{k2}, \ldots, \hat{\beta}_{kt}]$$

then the beta-hat model is

$$\hat{\beta}_k = I_t \beta_k + \varepsilon^*, \tag{6.6}$$

where I_t is a $t \times t$ identity matrix (the design matrix for this model), $\beta_k' = [\beta_{k1}, \beta_{k2}, \ldots, \beta_{kt}]$, and $\varepsilon^* \sim N(0, \sigma^2 C_k)$ where C_k is the diagonal matrix

$$C_k = \begin{bmatrix} C_{1kk} & 0 & \cdots & 0 \\ 0 & C_{2kk} & \cdots & 0 \\ \vdots & \vdots & \vdots & \vdots \\ 0 & 0 & & C_{tkk} \end{bmatrix}.$$

The model comparison method can be used with Model 6.6 to test hypotheses about β_k. Since Model 6.6 has t observations and t parameters, the residual sum of squares is zero (the residual sum of squares with no restrictions on the model).

Suppose you wish to test the hypothesis $H_0: \beta_k = H\theta$ vs. $H_a: (\beta_k$ is unspecified), where $H\theta$ has a specific structure. For example, if it is of interest to test the equality of the t slopes, the matrix H would be of the form $H = j_t$ (a $t \times 1$ vector of one's) and $\theta = \beta$, an unspecified scalar. Then

$$\hat{\beta}_k = H\theta + \varepsilon^*, \tag{6.7}$$

represents the beta-hat model under the conditions of the null hypothesis. The sum of squares due to deviations from the null hypothesis, which is the residual sum of squares for Model 6.7, is

Beta-Hat Models

$$\mathrm{SSH}_0 = \hat{\boldsymbol{\beta}}'_k \left(\mathbf{C}_k^{-1} - \mathbf{C}_k^{-1} \mathbf{H} \left(\mathbf{H}' \mathbf{C}_k^{-1} \mathbf{H} \right)^{-1} \mathbf{H} \mathbf{C}_k^{-1} \right) \hat{\boldsymbol{\beta}}_k. \tag{6.8}$$

The statistic to test the null hypothesis is

$$F_c = \frac{\mathrm{SSH}_0 / r}{\hat{\sigma}^2} \tag{6.9}$$

which is distributed as a noncentral F with r and N − t(q + 1) degrees of freedom where "r" is the number of linearly independent parameters in $H\theta$ or the rank of $\mathbf{H}$.

6.3 TESTING EQUALITY OF PARAMETERS

If data are collected from a one-way treatment structure in a completely randomized design structure with one covariate, assume a simple linear regression model can be used to carry out the analysis of covariance (as in Chapter 2). The model is

$$y_{ij} = \alpha_i + \beta_i x_{ij} + \varepsilon_{ij} \quad i = 1, 2, \ldots, r, \quad j = 1, 2, \ldots, n_i \tag{6.10}$$

To test the equality of slopes, the beta-hat model for slopes under H_a is

$$\begin{bmatrix} \hat{\beta}_1 \\ \hat{\beta}_2 \\ \vdots \\ \hat{\beta}_t \end{bmatrix} = \begin{bmatrix} 1 & 0 & 0 \\ 0 & 1 & 0 \\ \vdots & \vdots & \vdots \\ 0 & 0 & 1 \end{bmatrix} \begin{bmatrix} \beta_1 \\ \beta_2 \\ \vdots \\ \beta_t \end{bmatrix} + \varepsilon^*$$

and the beta-hat model under the equal slope hypothesis is

$$\begin{bmatrix} \hat{\beta}_1 \\ \hat{\beta}_2 \\ \vdots \\ \hat{\beta}_t \end{bmatrix} = \begin{bmatrix} 1 \\ 1 \\ \vdots \\ 1 \end{bmatrix} \beta + \varepsilon^*. \tag{6.11}$$

where β is an unspecified scaler. The SSH_0 simplifies to

$$\mathrm{SSH}_0 = \sum_{i=1}^{t} \left(\hat{\beta}_i - \hat{\beta}^* \right)^2 / C_i \tag{6.12}$$

where

$$\hat{\beta}^+ = \left(\sum_{i=1}^{t} \hat{\beta}_i/C_i\right) \bigg/ \left(\sum_{i=1}^{t} (1/C_i)\right),$$

C_i is the (2, 2) element of the 2×2 matrix $(\mathbf{X}_i' \mathbf{X}_i)^{-1}$ and

$$\mathbf{X}_i = \begin{bmatrix} 1 & x_{i1} \\ 1 & x_{i2} \\ \vdots & \vdots \\ 1 & x_{in_i} \end{bmatrix}.$$

A similar formulation can be constructed for comparing the intercepts.

The beta-hat model can be used to compare nonparallel regression models at fixed values of the covariate. For Model 6.10, let $\theta_i = \alpha_i + \beta_i X_0$ where X_0 is a fixed value of the covariate. The estimate of θ_i is $\hat{\theta}_i = \hat{\alpha}_i + \hat{\beta}_i X_0$ which has sampling distribution $\hat{\theta}_i \sim N(\theta_i, \sigma^2_{\hat{\theta}_i})$ where $\sigma^2_{\hat{\theta}_i} = \sigma^2 [1, X_0](\mathbf{X}_i' \mathbf{X}_i)^{-1}[1, X_0]' = \sigma^2 C_i$. Equation 6.12 can be used to compute the sum of squares needed to test H_0: $\theta_1 = \theta_2 = \ldots = \theta_t$.

6.4 COMPLEX TREATMENT STRUCTURES

The beta-hat model can be used to investigate the results of main effects and interactions on a particular parameter in the regression models when the treatment structure is complex. Assume data are from a two-way treatment structure in a completely randomized design structure and one covariate. Assume a simple linear regression model is adequate to describe the relationship between y and x for each treatment combination and is expressed as

$$y_{ijk} = \alpha_{ij} + \beta_{ij} x_{ijk} + \varepsilon_{ijk}$$

$i = 1, 2, \ldots, r$, $j = 1, 2, \ldots, c$ and $k = 1, \ldots, n_j$.

A beta-hat model can be used to investigate the effect the row and column treatments have on, say, β_{ij}. The beta-hat model for the slopes can be expressed as

$$\hat{\beta}_{ij} = \mu + \rho_i + \tau_j + (\rho\tau)_{ij} + \varepsilon_{ij}^*$$

for $i = 1, 2, \ldots, r$ and $j = 1, 2, \ldots, c$ where $\text{Var}(\hat{\beta}_{ij}) = \sigma^2 C_{ij}$. A weighted two-way analysis of variance can be used to investigate the structure of the slopes where the weights are $1/C_{ij}$. The pooled residual sum of squares divided by the pooled residual degrees of freedom would be used as the estimate of σ^2 and the divisor for the constructed F statistics. Next, two examples are used to demonstrate the computations for beta-hat models.

6.5 EXAMPLE: ONE-WAY TREATMENT STRUCTURE

The example in Section 3.3 evaluates exercise programs as to the effect on an individual's heart rate using the initial resting heart rate (IHR) as a covariate. Unequal slopes were determined as necessary to represent the covariate part of the model. In this section, beta-hat models are used to compare the intercepts, the slopes, and the models at predetermined values of IHR. First, contrasts were constructed to test the equality of the various sets of parameters in the context of using the analysis of variance on the complete data set.

The model to describe the heart rate is

$$y_{ij} = \alpha_i + \beta_i X_{ij} + \varepsilon_{ij}$$

$$i = 1, 2, 3, \quad j = 1, 2, \ldots, 8$$

where y_{ij} denotes the heart rate and X_{ij} denotes the initial resting heart rate. The six hypotheses of interest are

(i) Test equality of intercepts, the models at IHR = 0, as

$$H_{01}: \alpha_1 = \alpha_2 = \alpha_3 \text{ vs. } H_{a1} \text{ (not } H_0\text{)},$$

(ii) test equality of slopes as

$$H_{02}: \beta_1 = \beta_2 = \beta_3 \text{ vs. } H_{a2} \text{ (not } H_0\text{)},$$

(iii) Test equality of the models at IHR = 70.375, 55, 70, and 85 as

$$H_{03}: \alpha_1 + \beta_1 x^0 = \alpha_2 + \beta_2 X^0 = \alpha_3 + \beta_3 X^0 \text{ vs. } H_a \text{ (not } H_0\text{)}$$

where X_0 takes on a specific value of IHR.

The means model was fit to the heart rate data and the results are displayed in Table 3.16. The estimate of the variance is 28.203, based on 18 degrees of freedom. Table 3.16 also contains the estimates of the intercepts and slopes along with their estimated standard errors. Table 6.1 contains the SAS® system code to fit a beta hat model to the slopes from the heart rate data set. The first part of the table displays the input statement and data management lines to compute the weighting factor. The slopes and the estimated standard errors were obtained from Table 3.16 and the estimate of the standard deviation of the data is the square root of 28.203 or 5.31062. The value of c_i in the covariance matrix of the beta vector is (standard error/root means square)^2. PROC GLM uses $1/c_i$ as the weight for weighted least squares or $1/k^2$ for this application. The second part of the table shows the intermediate computations. The value of the SSH_0 from Equation 6.12 (or from the general

TABLE 6.1
Information to Use the Beta-Hat Model to Test the Equality of the Slopes for the Heart Rate Data of Example 3.3

```
DATA SLOPES; INPUT EPRO SLOPES STDERR;
ROOTMSE=5.31062; K=STDERR/ROOTMSE; WT=1/K**2;
DATALINES;
 1  1.4852189  0.17907182
 2  0.9379976  0.19550316
 3  0.2637550  0.17886856
PROC GLM DATA=SLOPES; WEIGHT WT; MODEL SLOPES=;
```

EPRO	SLOPES	STDERR	ROOTMSE	K	WT
1	1.49	0.18	5.31	0.0337	879.50
2	0.94	0.20	5.31	0.0368	737.88
3	0.26	0.18	5.31	0.0337	881.50
Source	**df**	**SS**	**MS**	**FValue**	**ProbF**
Error	2	658.984	329.492		

TABLE 6.2
Information to Use the Beta-Hat Model to Test the Equality of the Intercepts for the Heart Rate Data of Example 3.3

```
DATA INTERCPS; INPUT EPRO INTERCPS STDERR;
ROOTMSE=5.31062; K=STDERR/ROOTMSE; WT=1/K**2;
DATALINES;
 1   46.5383741  12.71914269
 2   97.1909199  14.12823158
 3  137.4849688  12.52814364
PROC GLM DATA=intercps; WEIGHT WT; MODEL INTERCPS=;
```

EPRO	INTERCPS	STDERR	ROOTMSE	K	WT
1	46.54	12.72	5.31	2.3950	0.1743
2	97.19	14.13	5.31	2.6604	0.1413
3	137.48	12.53	5.31	2.3591	0.1797
Source	**df**	**SS**	**MS**	**FValue**	**ProbF**
Error	2	733.916	366.958		

Equation 6.8) is 658.984 and the mean square is 329.492. The resulting F statistic is $329.492/28.203 = 11.68$, which is the same as the F value corresponding to IHR*EPRO in Table 3.17.

Table 6.2 contains the comparisons of the intercepts of the EPRO models, providing a SSH_0 equal to 733.916 with a computed value of the F statistic of 13.01. This F statistic is identical to that corresponding to EPRO in Table 3.17.

It is important to remember that when the slopes are unequal, the sum of squares corresponding to the treatments is a test of the equality of intercepts or models at the value of the covariate equal to zero. For this example, a comparison of the

TABLE 6.3
Information to Use the Beta-Hat Model to Test the Equality of the Models at IHR = 55 for the Heart Rate Data of Example 3.3

```
DATA ADJMN55; INPUT EPRO ADJMN55 STDERR;
ROOTMSE=5.31062; K=STDERR/ROOTMSE; WT=1/K**2;
DATALINES;
1 128.2254122 3.3140384
2 148.7807894 3.7535843
3 151.9914918 3.1657714
PROC GLM DATA=ADJMN55; WEIGHT WT; MODEL ADJMN55=;
```

EPRO	ADJMN55	STDERR	ROOTMSE	K	WT
1	128.23	3.31	5.31	0.6240	2.5679
2	148.78	3.75	5.31	0.7068	2.0017
3	151.99	3.17	5.31	0.5961	2.8140

Source	df	SS	MS	FValue	ProbF
Error	2	854.784	427.392		

intercepts is comparing the exercise programs when the initial heart rates are zero. Tables 6.3 through 6.6 contain the results of applying the beta-hat model approach to compare the models at initial heart rates of 55, 70, 85, and 70.375.

The error sum of squares provides the sum of squares due to deviations from the null hypothesis that all of the parameters are equal. Table 6.3 contains the information to test the equality of the regression lines at IHR = 55. The resulting F statistic is 427.392/28.203 = 15.14. This is the same F statistic obtained for EPRO in Table 3.19. The analyses in Tables 6.4 and 6.5 provide sums of squares that

TABLE 6.4
Information to Use the Beta-Hat Model to Test the Equality of the Models at IHR = 70 for the Heart Rate Data of Example 3.3

```
DATA ADJMN70; INPUT EPRO ADJMN70 STDERR;
ROOTMSE=5.31062; K=STDERR/ROOTMSE; WT=1/K**2;
DATALINES;
1 150.5036953 1.8781211
2 162.8507539 1.9042750
3 155.9478162 1.8823739
PROC GLM DATA=ADJMN70; WEIGHT WT;
MODEL ADJMN70=;
```

EPRO	ADJMN70	STDERR	ROOTMSE	K	WT
1	150.50	1.88	5.31	0.3537	7.9955
2	162.85	1.90	5.31	0.3586	7.7773
3	155.95	1.88	5.31	0.3545	7.9594

Source	df	SS	MS	FValue	ProbF
Error	2	603.219	301.610		

TABLE 6.5
Information to Use the Beta-Hat Model to Test the Equality of the Models at IHR = 85 for the Heart Rate Data of Example 3.3

```
DATA ADJMN85; INPUT EPRO ADJMN85 STDERR;
ROOTMSE=5.31062;
K=STDERR/ROOTMSE;
WT=1/K**2;
DATALINES;
1 172.7819784 3.2406558
2 176.9207183 3.2191303
3 159.9041407 3.3855335
PROC GLM DATA=ADJMN85; WEIGHT WT; MODEL ADJMN85=;
```

EPRO	ADJMN85	STDERR	ROOTMSE	K	WT
1	172.78	3.24	5.31	0.6102	2.6855
2	176.92	3.22	5.31	0.6062	2.7215
3	159.90	3.39	5.31	0.6375	2.4606

Source	df	SS	MS	FValue	ProbF
Error	2	401.660	200.830		

TABLE 6.6
Information to Use the Beta-Hat Model to Test the Equality of the Models at IHR = 70.375 for the Heart Rate Data of Example 3.3

```
DATA ADJMN70_375; INPUT EPRO ADJMN70_375 STDERR;
ROOTMSE=5.31062; K=STDERR/ROOTMSE; WT=1/K**2;
DATALINES;
1 151.060652 1.877721
2 163.202503 1.893424
3 156.046724 1.888340
PROC GLM DATA=ADJMN70_375; WEIGHT WT; MODEL ADJMN70_375=;
```

EPRO	ADJMN70_375	STDERR	ROOTMSE	K	WT
1	151.06	1.88	5.31	0.3536	7.9989
2	163.20	1.89	5.31	0.3565	7.8667
3	156.05	1.89	5.31	0.3556	7.9092

Source	df	SS	MS	FValue	ProbF
Error	2	590.351	295.175		

produce F statistics corresponding to those in Tables 3.20 and 3.21. Finally, Table 6.6 provides the information necessary to test equality of the regression models at the average value of the IHR data from the experiment, IHR = 70.375. The resulting F statistic is $295.175/28.203 = 10.47$.

The above examples show the versatility of the beta-hat model approach to deriving a test statistic. By using the beta-hat model it is obvious as to which

6.6 EXAMPLE: TWO-WAY TREATMENT STRUCTURE

The beta-hat model is very useful when the treatment structure involves more than one factor. As an example, the results from the energy from wood experiment in Section 5.5 are used to demonstrate the application to a two-way treatment structure. Table 6.7 contains the estimates of the slopes and standard errors for each of the wood by stove combinations. The beta-hat model for the slopes can be expressed as

$$\hat{\beta}_{ij} = \mu + \rho_i + T_j + (\rho T)_{ij} + \varepsilon^*_{ij}$$

for i = 1, 2, 3, 4 and j = 1, 2, 3 where $\text{Var}(\hat{\beta}_{ij}) = \sigma^2 C_{ij}$. The estimated standard errors are computed as $\sqrt{\hat{\sigma}^2 C_{ij}}$. The value of C_{ij} can be computed as C_{ij} = (standard error($\hat{\beta}_{ij}$))$^2/\hat{\sigma}^2$ and the weight for use in the analysis is $wt_{ij} = 1/C_{ij}$. The estimate of the variance from Table 5.10 is 0.1431. The data management code in Table 6.7 computes the square root of C_{ij}, called K and computes the weight as wt = $1/K^2$. The middle part of Table 6.5 contains the intermediate computations. The PROC GLM code uses a model with wood, stove, and wood*stove as terms where both wood and stove are in the class statement. The error sum of squares for this model is 0 since there are 12 observations and 12 parameters or wood*stove combinations. The bottom part of the table provides the sums of squares for the two main effects and the interaction. The values of the F statistics need be computed by dividing each of the mean squares by the estimate of the variance, 0.1431. The three F statistics are F_{wood} = 1.434/0.1413 = 10.02, F_{stove} = 0.194/0.1413 = 1.36, and $F_{wood*stove}$ = 0.281/0.1413 = 1.96. These are the identical to the values of the corresponding F statistics in Table 5.10. There 64 denominator degrees of freedom associated with each of these F statistics. The significance levels of the three tests are 0.000, 0.2650 and 0.0842, respectively, so the same conclusions can be made as were reached in Section 5.5.

Table 6.8 contains the estimates (adjusted means) of the regression models evaluated at the average value of the covariate or at the mean moisture level = 13.34%. These parameters were estimated by assuming an unequal slopes model for all of the wood by stove combinations. Thus, the estimates in Table 6.8 are estimates of the parameters $\mu_{ij|Moist=13.34} = \alpha_{ij} + \beta_{ij}$ 13.34. The software (PROC GLM in this case) evaluated the estimated standard errors for each of these adjusted means and those values are included in Table 6.8.

Once the estimates of the parameters and their standard errors have been computed, then the modeling is identical to that described for the slopes discussed above and displayed in Table 6.7. The values of the F statistics are F_{wood} = 77.456/0.1413 = 541.27, F_{stove} = 0.658/0.1413 = 4.60, and $F_{wood*stove}$ = 6.073/0.1413 = 42.44. There 64 denominator degrees of freedom associated with each of these F-statistics. The significance levels of the three tests are 0.000, 0.0136, and 0.0000, respectively.

TABLE 6.7
SAS® System Code, Data, and Results for Using the Beta-Hat Model to Evaluate the Effect of Wood, Stove and wood*stove on the Slopes

```
DATA SLOPE; input wood $ stove $ slope stderr;
ROOTMSE=SQRT(.1431); K=STDERR/ROOTMSE; WT=1/K**2;
datalines;
Black_Walnut Type_A -0.17371 0.04235
Black_Walnut Type_B -0.17832 0.04584
Black_Walnut Type_C -0.18105 0.02956
Osage_Orange Type_A -0.07641 0.03244
Osage_Orange Type_B -0.11115 0.04760
Osage_Orange Type_C -0.03840 0.03545
Red_Oak Type_A -0.01937 0.04507
Red_Oak Type_B 0.03660 0.05050
Red_Oak Type_C -0.13442 0.04629
White_Pine Type_A -0.13248 0.02516
White_Pine Type_B -0.24002 0.04740
White_Pine Type_C -0.22753 0.04594
PROC GLM DATA=SLOPE; WEIGHT WT; CLASS WOOD STOVE;
MODEL SLOPE=WOOD STOVE WOOD*STOVE;
```

Wood	Stove	Estimate	StdErr	ROOTMSE	K	WT
Black_Walnut	Type_A	−0.1737	0.0424	0.3783	0.1120	79.7845
Black_Walnut	Type_B	−0.1783	0.0458	0.3783	0.1212	68.1112
Black_Walnut	Type_C	−0.1810	0.0296	0.3783	0.0781	163.7807
Osage_Orange	Type_A	−0.0764	0.0324	0.3783	0.0858	135.9872
Osage_Orange	Type_B	−0.1112	0.0476	0.3783	0.1258	63.1680
Osage_Orange	Type_C	−0.0384	0.0354	0.3783	0.0937	113.8763
Red_Oak	Type_A	−0.0194	0.0451	0.3783	0.1191	70.4442
Red_Oak	Type_B	0.0366	0.0505	0.3783	0.1335	56.1021
Red_Oak	Type_C	−0.1344	0.0463	0.3783	0.1224	66.7726
White_Pine	Type_A	−0.1325	0.0252	0.3783	0.0665	226.1241
White_Pine	Type_B	−0.2400	0.0474	0.3783	0.1253	63.7010
White_Pine	Type_C	−0.2275	0.0459	0.3783	0.1214	67.8068

Source	df	SS	MS	FValue	ProbF
Model	11	5.910	0.537		
Error	0	0.000			
Corrected Total	11	5.910			

Source	df	SS	MS	FValue	ProbF
wood	3	4.301	1.434		
stove	2	0.388	0.194		
wood*stove	6	1.685	0.281		

TABLE 6.8
SAS® System Code, Data, and Results for Using the Beta-Hat Model to Evaluate the Effect of Wood, Stove and Wood*Stove on the Adjusted Means Computed at a Moisture Level of 13.34%

```
DATA ADJMN; length wood $12; INPUT WOOD $ STOVE $ LSMEAN STDERR;
ROOTMSE=SQRT(.1431); K=STDERR/ROOTMSE; WT=1/K**2;
DATALINES;
Black_Walnut  Type_A  2.2117  0.1553
Black_Walnut  Type_B  2.8003  0.1338
Black_Walnut  Type_C  4.3334  0.1341
Osage_Orange  Type_A  7.2216  0.1264
Osage_Orange  Type_B  6.6889  0.1581
Osage_Orange  Type_C  6.0457  0.1336
Red_Oak  Type_A  6.3310  0.1749
Red_Oak  Type_B  5.3988  0.1561
Red_Oak  Type_C  4.8688  0.1267
White_Pine  Type_A  2.0763  0.1230
White_Pine  Type_B  3.5136  0.1926
White_Pine  Type_C  1.8937  0.1459;
PROC GLM DATA=ADJMN; CLASS WOOD STOVE; WEIGHT WT;
MODEL LSMEAN=WOOD STOVE WOOD*STOVE;
```

Wood	STOVE	LSMEAN	STDERR	ROOTMSE	K	WT
Black_Walnut	Type_A	2.2117	0.1553	0.3783	0.4105	5.9333
Black_Walnut	Type_B	2.8003	0.1338	0.3783	0.3537	7.9933
Black_Walnut	Type_C	4.3334	0.1341	0.3783	0.3545	7.9576
Osage_Orange	Type_A	7.2216	0.1264	0.3783	0.3341	8.9566
Osage_Orange	Type_B	6.6889	0.1581	0.3783	0.4179	5.7250
Osage_Orange	Type_C	6.0457	0.1336	0.3783	0.3532	8.0173
Red_Oak	Type_A	6.3310	0.1749	0.3783	0.4623	4.6780
Red_Oak	Type_B	5.3988	0.1561	0.3783	0.4127	5.8726
Red_Oak	Type_C	4.8688	0.1267	0.3783	0.3349	8.9143
White_Pine	Type_A	2.0763	0.1230	0.3783	0.3252	9.4587
White_Pine	Type_B	3.5136	0.1926	0.3783	0.5091	3.8577
White_Pine	Type_C	1.8937	0.1459	0.3783	0.3857	6.7225

Source	df	SS	MS	FValue	ProbF
Model	11	293.458	26.678		
Error	0	0.000			
Corrected Total	11	293.458			

Source	df	SS	MS	FValue	ProbF
wood	3	232.369	77.456		
STOVE	2	1.316	0.658		
wood*STOVE	6	36.436	6.073		

6.7 SUMMARY

If the treatment structure is a three-way factorial with two covariates in a completely randomized design structure, the beta-hat model could be used to help study the resulting models. The model could be expressed as

$$y_{ijkm} = \alpha_{ijk} + \beta_{ijk} X_{ijkm} + \rho_{ijk} W_{ijkm} + \varepsilon_{ijkm}.$$

where $\text{Var}(\varepsilon_{ijkm}) = \sigma^2$.

Fit the model to the data from each of the treatment combinations and obtain estimates of the β_{ijk}, i.e., estimates of slopes for X_{ijkm}. The beta-hat model is then

$$\beta_{ijk} = \mu + \delta_i + \theta_j + T_k + (\delta\theta)_{ij} + (\delta T)_{ik} + (\theta T)_{jk} + (\delta\theta T)_{ijk} + \varepsilon^*_{ijk}$$

where $\text{Var}(\varepsilon^*_{ijk}) = \sigma^2 C_{ijk}$.

Thus a three-way factorial analysis of variance model could be used to evaluate the effects the main effects and interactions have on the slopes. The beta-hat model approach is a very useful tool for comparing parameters of different models, but there are limitations to the application. The major limitation is that the models must not have any parameters in common, i.e., the form of the covariate part of the model cannot be simplified. A major assumption is that the estimates of the parameters from the respective treatment groups are independently distributed, and that cannot happen if some models share a common parameter.

EXERCISES

EXERCISE 6.1: For the data of Section 3.2, use the beta-hat model to construct the test for equality of slopes. Assume the unequal slope model is needed; then use the beta-hat model to compare the 6 types of chocolate candies as if all persons were able to dissolve the butterscotch chip in 20 sec.

EXERCISE 6.2: For the average daily gain data in Section 5.4, use the beta-hat model to evaluate the effects of Sex and Diet on the slopes of the models.

EXERCISE 6.3: Use the beta-hat model to test the equality of slopes for height and for weight for the distance data in Section 4.4.

7 Variable Selection in the Analysis of Covariance Model

7.1 INTRODUCTION

Often experimenters measure several characteristics of each experimental unit as possible candidates for covariates. This process is an attempt to measure those variables that are not controlled by the experimental design in order to account for as much of the experimental unit to experimental unit variation as possible. The experimenter knows that all of the measured covariates are not necessary and wishes to operate in a model building mode and select only the important variables to be used in the final analysis of covariance for comparing the treatments. The experimenter would like to use a variable selection computer code to build the model, but the popular commercial stepwise regression packages do not allow the use of class variables (such as treatments). When the design is simple, one can use indicator variables to code the levels of the class variables and use the stepwise regression codes. When the design is complex, the use of indicator variables is quite cumbersome.

Fortunately, there is a method for building analysis of covariance models which can easily be implemented using available stepwise regression software (Yang, 1989). The method is simple to use for the equal slope covariance problem and can be used if the slopes are possibly unequal, but is more unwieldily.

This chapter describes the methodology for homogeneous slopes in Section 7.2 and a detailed example is presented in Section 7.3. The theory is presented in Section 7.4. Section 7.5 contains a discussion of how to use the process within the unequal slope framework.

7.2 PROCEDURE FOR EQUAL SLOPES

The experimental situation involves an experimental design with a treatment structure (a one-way is used here) in a design structure (a randomized block is used here) where a dependent variable, y, and several independent variables (or possible covariates), $x_1, x_2, \ldots, x_q$, which were measured on each experimental unit before the treatments were applied to the experimental units. A model for y with the treatment structure components (denoted by μ_i) and design structure components (denoted by b_j) with out the covariates is

$$y_{ij} = \mu_i + b_j + \varepsilon_{ij} \qquad (7.1)$$

where μ_i denotes the mean from the treatment structure, b_j denotes the block effect, and ε_{ij} denotes the noise or unexplained variation. The model building procedure is to help decide which of the covariates are useful in describing as much of the unexplained variation in the y values within a treatment as possible.

The model building process has two steps. The first step is to fit Model 7.1, which is a model involving all of the class variables, to the dependent variable and compute the residuals, i.e., fit

$$y_{ij} = \mu_i + b_j + \varepsilon_{ij}$$

and compute the residuals as

$$r_{ij} = y_{ij} - \hat{\mu}_i - \hat{b}_j. \tag{7.2}$$

The same process is carried out for each independent variable, i.e., fit the model

$$x_{kij} = \mu_{ki} + b_{kj} + \varepsilon_{ki}; \quad k = 1, 2, \ldots, q \tag{7.3}$$

and compute the corresponding residuals as,

$$r_{kij} = x_{kij} - \hat{\mu}_{ki} - \hat{b}_{kj}, \quad k = 1, 2, \ldots, q. \tag{7.4}$$

In general, the first step involves computing the residuals for the dependent variable and for each independent variable from a model that involves the appropriate design structure and treatment structure of the data set. The second step is to use residuals of y, denoted by r, as the dependent variable in a stepwise regression where the residuals of the x_k, denoted by r_k, $k = 1, 2, \ldots, q$ are used as the independent variables. Use model selection procedures on these residuals (as if they are the raw data) to select those variables needed in the analysis of covariance. The model involving the residuals with all of the covariates is

$$r_{ij} = \beta_1 r_{1ij} + \beta_2 r_{2ij} + \ldots + \beta_q r_{qij} + e_{ij}. \tag{7.5}$$

As is shown in Section 7.4, the estimates of the regression parameters are the same for this residual model as one would get by fitting the y's to a model involving the design and treatment structures and covariates.

The various residual sums of squares are correct and the sums of squares corresponding to testing the slope for each independent variable is zero are correct, but the degrees of freedom for the different residual sums of squares are not correct. The degrees of freedom for each computed residual sum of squares needs to be reduced by the number of parameters in Model 7.1 (where the b_j are considered to be a fixed effect), the model from which the residuals are computed. The consequence of using the incorrect degrees of freedom is that the computed F or t statistics from the stepwise regression code are larger than they should be and thus the significance

Variable Selection in the Analysis of Covariance Model

probabilities are somewhat smaller than they should be if the correct error degrees of freedom were used in computing the error mean square. An example involving a one-way treatment structure in a completely randomized design structure is used to demonstrate the computations.

7.3 EXAMPLE: ONE-WAY TREATMENT STRUCTURE WITH EQUAL SLOPES MODEL

To demonstrate the process described in Section 7.2, the data in Table 7.1 were generated from the model

$$y_{ij} = \mu_i + \beta_1 x_{1ij} + \beta_2 x_{2ij} + 0\, x_{3ij} + 0\, x_{4ij} + \beta_5 x_{5ij} + \varepsilon_{ij} \tag{7.6}$$

where i corresponds to TREAT and j indicates the observation within a treatment (this example does not have block effects). This example has been limited to five possible covariates, but the process is more useful when there are numerous possible covariates in the study.

Table 7.2 contains the PROC GLM code to fit an analysis of covariance model involving the treatment effects and the five covariates to the data set. The significance levels corresponding to X3 and X4 are 0.3292 and 0.6222, respectively, indicating

TABLE 7.1
Data for the Variable Selection Example

Treat	x1	x2	x3	x4	x5	y	Treat	x1	x2	x3	x4	x5	y
1	27.8	4.68	92.6	22.0	39	15.7	3	27.1	5.10	65.4	17.9	34	20.9
1	28.7	3.99	71.5	15.0	32	16.7	3	25.6	4.64	86.3	19.3	35	19.0
1	24.1	1.94	80.4	18.8	28	17.5	3	24.2	4.23	95.6	19.9	34	19.7
1	27.2	3.41	90.0	17.0	41	19.2	3	24.8	4.33	93.6	19.4	43	21.6
1	24.5	3.48	76.5	18.0	35	18.9	3	27.5	3.73	74.8	16.8	31	20.5
1	24.5	3.54	71.8	17.5	39	18.5	3	24.6	4.81	75.2	21.5	32	18.1
1	21.2	4.71	67.8	16.8	32	16.3	3	23.6	3.75	78.3	19.5	40	17.3
1	28.2	5.36	85.5	17.1	47	21.1	3	24.0	4.41	66.1	20.0	32	18.8
1	25.3	4.50	92.2	18.0	40	19.1	3	24.0	3.34	81.6	15.8	39	21.1
1	26.9	4.72	79.4	18.7	37	16.3	3	27.8	4.02	81.9	19.4	40	21.8
2	25.1	4.72	74.9	11.9	36	18.1	4	28.9	3.74	81.8	20.5	33	25.9
2	24.6	3.72	90.6	19.4	40	19.6	4	26.2	4.19	81.5	20.1	36	24.3
2	23.6	4.30	85.4	19.3	40	19.1	4	29.9	3.13	83.6	18.8	34	23.3
2	22.5	2.73	64.4	18.0	38	18.5	4	24.9	4.50	65.1	17.5	33	17.7
2	26.5	4.06	75.8	18.3	33	18.6	4	24.0	4.53	97.7	17.1	36	21.6
2	24.5	4.52	90.2	19.3	32	16.0	4	24.0	4.54	86.5	15.0	36	22.8
2	27.7	2.69	95.5	17.8	32	19.2	4	26.4	4.85	62.6	17.2	38	21.2
2	26.7	3.84	81.0	20.0	34	18.1	4	28.6	3.93	75.5	20.6	36	19.2
2	23.9	4.23	74.8	15.1	40	15.6	4	25.5	5.10	113.6	18.0	37	20.4
2	28.8	2.67	93.3	19.9	39	22.8	4	23.2	3.49	72.2	15.5	35	19.9

TABLE 7.2
Analysis of Covariance Using All Five of the Possible Covariates

```
proc glm data=ex_7_1; classes treat; model y=treat
  x1 x2 x3 x4 x5/solution;
```

Source	df	SS	MS	FValue	ProbF
Model	8	130.117	16.265	5.673	0.0002
Error	31	88.883	2.867		
Corr Total	39	219.000			

Source	df	SS (III)	MS	FValue	ProbF
treat	3	91.166	30.389	10.599	0.0001
x1	1	16.172	16.172	5.640	0.0239
x2	1	13.198	13.198	4.603	0.0399
x3	1	2.818	2.818	0.983	0.3292
x4	1	0.710	0.710	0.248	0.6222
x5	1	12.406	12.406	4.327	0.0459

Parameter	Estimate	StdErr	tValue	Probt
x1	0.337	0.142	2.375	0.0239
x2	−0.839	0.391	−2.145	0.0399
x3	0.027	0.027	0.991	0.3292
x4	−0.074	0.149	−0.498	0.6222
x5	0.162	0.078	2.080	0.0459

TABLE 7.3
Adjusted Means and Pairwise Comparisons Based on the Model with Five Covariates

```
lsmeans treat/pdiff stderr;
```

Treat	LSMEAN	StdErr	LSMEANNumber
1	17.720	0.542	1
2	18.312	0.555	2
3	20.280	0.560	3
4	21.688	0.551	4

LSM Num	_1	_2	_3	_4
1		0.4487	0.0028	0.0000
2	0.4487		0.0207	0.0002
3	0.0028	0.0207		0.0844
4	0.0000	0.0002	0.0844	

the two variables are possibly not needed in the model. The estimate of the variance is 2.867, which is based on 31 degrees of freedom. Table 7.3 contains the adjusted or least squares means for the four treatments as well as pairwise comparisons of the treatments. Using a Fisher's protected LSD approach, the means of Treatments 1

TABLE 7.4
PROC GLM Code to Fit the Analysis of Variance Model to the Response Variable and Each of the Possible Covariates and Compute the Residuals for Each

```
proc glm; classes treat;
model y x1 x2 x3 x4 x5=treat; * fit models 7.1 and 7.3;
output out=resids r=ry r1 r2 r3 r4 r5; * compute the residuals;
```

and 2 and of Treatments 3 and 4 are not significantly different while all other comparisons have significance levels less than 0.05.

Since at least two of the possible covariates have slopes that are not significantly different from zero, the model building process described in Section 7.2 is used to carry out variable selection for determining the adequate set of covariates for the model. The PROC GLM statement in Table 7.4 fits Model 7.1 to the response variable y and Model 7.3 to each of the possible covarites x_1, x_2, x_3, x_4, and x_5. The main product of these analyses is the computation of the sets of residuals for each of the variables. The output statement provides a file, called "resids," that contains all of the residuals, ry, r1, r2, r3, r4, and r5. The REG procedure in Tables 7.5 to 7.9 uses the computed residuals and model selection procedures to select variables for the analysis of covariance.

Five out of several available different variable selection methods (Draper and Smith, 1981; SAS, 1989; and Ott, 1988) were used to demonstrate some of the aspects of model building. The methods used were stepwise, forward, backward, adjusted R^2, and CP. There is no guarantee that these procedures will yield the same model and in most cases that involves many possible covariates, the sets of selected variables will not be identical. Tables 7.5 through 7.9 contain the results of the model building processes.

The PROC REG code and results of using the stepwise method are in Table 7.5. The stepwise variable selection method starts with no variables in the model and includes variables in a stepwise manner. At each step after including a new variable, a variable with the largest significance level is eliminated (when the significance level is greater than a pre-set value). In this case variables r1, r5, and r2 were selected.

TABLE 7.5
PROC REG Code to Use the Stepwise Variable Selection Procedure and Results

```
proc reg data=resids;
stepwise: model ry=r1 r2 r3 r4 r5/selection = stepwise;
```

Step	Entered	Var In	PartialR**2	ModelR**2	Cp	FValue	ProbF
1	r1	1	0.161	0.161	8.506	7.301	0.010
2	r5	2	0.082	0.243	6.159	4.005	0.053
3	r2	3	0.094	0.337	3.155	5.124	0.030

TABLE 7.6
PROC REG Code to Use the Forward Variable Selection Procedure and Results

```
proc reg data=resids;
forward: model ry = r1 r2 r3 r4 r5 /selection = forward;
```

Step	Entered	Var In	PartialR**2	ModelR**2	Cp	FValue	ProbF
1	r1	1	0.161	0.161	8.506	7.301	0.010
2	r5	2	0.082	0.243	6.159	4.005	0.053
3	r2	3	0.094	0.337	3.155	5.124	0.030
4	r3	4	0.017	0.354	4.272	0.902	0.349

TABLE 7.7
PROC REG Code to Use the Backward Variable Selection Procedure and Results

```
proc reg data=resids;
backward: model ry = r1 r2 r3 r4 r5 /selection = backward;
```

Step	Removed	Var In	PartialR**2	ModelR**2	Cp	FValue	ProbF
1	r4	4	0.005	0.354	4.272	0.272	0.606
2	r3	3	0.017	0.337	3.155	0.902	0.349

TABLE 7.8
PROC REG Code to Use the adjrsq Variable Selection Procedure and Results for Top Five Combinations of Variables

```
proc reg data=resids;
adjrsq: model ry = r1 r2 r3 r4 r5
  /selection = adjrsq;
```

Dependent	Var In	Adjrsq	RSquare	VarsInModel
ry	3	0.2822	0.3374	r1 r2 r5
ry	4	0.2802	0.3541	r1 r2 r3 r5
ry	5	0.2649	0.3592	r1 r2 r3 r4 r5
ry	4	0.2633	0.3389	r1 r2 r4 r5
ry	2	0.2022	0.2431	r1 r5

Thus, the analysis indicates that X1, X2, and X5 are needed as possible covariates in the analysis of the response variable.

Table 7.6 contains the PROC REG code and results of using the forward method. The forward variable selection process starts with no variables in the model and includes the next most important variable at each step. The forward variable selection method selects X1, X2, X3, and X5, although the significance level for X3 in

TABLE 7.9
PROC REG Code to Use the CP Variable Selection Procedure and Results for Top Five Combinations of Variables

```
proc reg data=resids;
cp: model ry = r1 r2 r3 r4 r5 /selection = cp;
```

Dependent	Var In	Cp	RSquare	VarsInModel
ry	3	3.155	0.337	r1 r2 r5
ry	4	4.272	0.354	r1 r2 r3 r5
ry	4	5.078	0.339	r1 r2 r4 r5
ry	5	6.000	0.359	r1 r2 r3 r4 r5
ry	2	6.159	0.243	r1 r5

the final model is 0.349. This indicates that X3 is most likely not needed in the model.

The backward variable selection PROC REG code and results are in Table 7.7. The backward variable selection method starts with all covariates in the model and eliminates the least important variable at each step (that variable with the largest significance level). The backward method eliminated variables r3 and r4, indicating that r1, r2, and r5 are remaining in the model.

Table 7.8 contains the PROC REG code to use the method "adjrsq" to select variables for the model. The process is to fit models that include all possible combinations of the variables and compute the adjusted R^2 for each model. The selected model consists of that set of variables with the largest adjusted R^2. With five variables, this process fits $2^5 - 1 = 31$ models. The results of the five sets of variables with the largest adjusted R^2 are included in Table 7.8. The set of variables with the largest adjusted R^2 consists of r1, r2, and r5.

Finally, Table 7.9 contains the PROC REG code and results of using the CP method of variable selection. As for the "adjrsq" method, the CP method fits models with all possible combinations of variables and selects that model where CP approaches "p," the number of parameters in the model including the intercept. That combination of variables with the CP value closest to "p" is r1, r2, r3, and r5 with CP = 4.272. When fitting a model with all four of these variables, the significance level corresponding to X3 is 0.3705, indicating that given the other variables are in the model, variable X3 is not needed. Just as for the adjusted R^2 method, the CP method fits all possible combinations of the variables, which can become an unmanageable number when the number of possible covariates becomes large.

Using the approach of not including any variables with large significance levels in the model, all of the procedures indicate that the needed variables are X_1, X_2, and X_5. Remember that the degrees of freedom associated with the residual sum of squares for any of the above models is larger than they are supposed to be since the regression code does not take into account the fact that the data being analyzed are residuals. In this case, four degrees of freedom for residual were used to estimate the means for the treatments for use in computing the residuals. Thus, all of the

TABLE 7.10
PROC GLM Code to Fit the Final Model with Three Covariates, Sums of Squares, and Estimates of the Slopes for Each of the Covariates

```
proc glm data=ex_7_1; classes treat;
model y=treat x1 x2 x5/solution;
lsmeans treat/pdiff stderr;
```

Source	df	SS	MS	FValue	ProbF
Model	6	127.097	21.183	7.606	0.0000
Error	33	91.903	2.785		
Corr Total	39	219.000			

Source	df	SS (III)	MS	FValue	ProbF
treat	3	91.799	30.600	10.988	0.0000
x1	1	17.678	17.678	6.348	0.0168
x2	1	13.081	13.081	4.697	0.0375
x5	1	17.419	17.419	6.255	0.0175

Parameter	Estimate	StdErr	tValue	Probt
x1	0.340	0.135	2.519	0.0168
x2	−0.833	0.384	−2.167	0.0375
x5	0.184	0.073	2.501	0.0175

degrees of freedom associated with a residual sum of squares are inflated by four. This means that the computed t-statistics, adjusted R^2 values, and CP values are not correct. The values of these statistics could be recomputed before decisions are made concerning the variables to be included in the model, but, the results without recomputation from the model building procedures provide really good approximations and provide adequate means for making decisions.

Table 7.10 contains the PROC GLM code to fit the final model with X_1, X_2, and X_5 as covariates. The mean square error has a value of 2.785 as compared to the mean square error for the model with all five covariates (see Table 7.2) which has a value of 2.867. When covariates are included in the model that are not needed, the degrees of freedom for error are reduced more than the error sum of squares are reduced, thus increasing the value of the estimate of the variance. The significance levels corresponding to the statistics for testing the individual slopes of the covariates are equal to zero are 0.0168, 0.0375, and 0.0175 for X1, X2, and X5, respectively. The significance level corresponding to source Treat is 0.0000, indicating the intercepts are not equal or that distances between the various parallel hyper-planes are not zero. Table 7.11 contains the adjusted means, predicted values on the hyper-planes at the average values of X1, X2, and X5 which are 25.68, 4.05, and 36.20, respectively. Using a Fisher's protected LSD method to make pairwise comparisons of the treatment means indicates that Treatments 1 and 2 are not significantly different while all other comparisons have significance levels less than 0.05. There is one additional comparison, 3 vs. 4, that is significant for the model with three covariates than for the model with five covariates.

TABLE 7.11
Adjusted Means and Pairwise Comparisons Using the Final Model with Three Covariates

```
lsmeans treat/pdiff stderr;
```

treat	LSMean	StdErr	Probt	LSM Num
1	17.710	0.532	0.0000	1
2	18.366	0.544	0.0000	2
3	20.190	0.535	0.0000	3
4	21.734	0.541	0.0000	4

LSM Num	_1	_2	_3	_4
1		0.3925	0.0025	0.0000
2	0.3925		0.0242	0.0001
3	0.0025	0.0242		0.0492
4	0.0000	0.0001	0.0492	

TABLE 7.12
PROC GLM Code to Fit the Residual Model with Three Covariates to Provide, Sums of Squares, and Estimates of the Slopes for Each of the Covariates

```
proc glm data=resids;
model ry=r1 r2 r5/solution;
```

Source	df	SS	MS	FValue	ProbF
Model	3	46.799	15.600	6.111	0.0018
Error	36	91.903	2.553		
Corrected Total	39	138.702			

Source	df	SS(III)	MS	FValue	ProbF
r1	1	17.678	17.678	6.925	0.0124
r2	1	13.081	13.081	5.124	0.0297
r5	1	17.419	17.419	6.823	0.0130

Parameter	Estimate	StdErr	tValue	Probt
Intercept	0.000	0.253	0.000	1.0000
r1	0.340	0.129	2.631	0.0124
r2	−0.833	0.368	−2.264	0.0297
r5	0.184	0.070	2.612	0.0130

For comparison purposes, the residuals of y were regressed on the residuals of X_1, X_2, and X_5 and the results are in Table 7.12. The error sum of squares is 91.903, the same as in Table 7.10. The mean square error is 2.553 = 91.903/36 instead of 2.785 = 91.903/33 since the degrees of freedom for error from the residual model are 36 instead of the 33 as in Table 7.10. The estimates of the slopes are identical for both models (as shown by the theory), but the estimated standard errors from the residual model are smaller than those from the final model. Again this is the result of using 36 degrees of freedom for error rather than using 33 degrees of freedom. The standard errors from Table 7.12 can be recomputed as

$$\text{stderr}_{\text{slope}} = \text{stderr}_{\text{slope from residual model}} \sqrt{36/33}$$

$$0.135 = 0.129\sqrt{36/33}$$

Even though the variable selection procedure is not exact, the results are adequate enough that effective models can be selected for carrying out analysis of covariance.

7.4 SOME THEORY

The analysis of covariance model can be expressed in general matrix notation as

$$\mathbf{y} = \mathbf{M}\boldsymbol{\mu} + \mathbf{X}\boldsymbol{\beta} + \boldsymbol{\varepsilon} \qquad (7.7)$$

where $\mathbf{y}$ is $n \times 1$ vector of the dependent variable, $\mathbf{M}$ is the design matrix, $\boldsymbol{\mu}$ is the associated parameters corresponding to the treatment and design structures (all considered as fixed effects for this purpose), $\mathbf{X}$ is the matrix of possible covariates, $\boldsymbol{\beta}$ is the vector of slopes corresponding to each of the covariates, and $\boldsymbol{\varepsilon}$ is the error distributed $N(\mathbf{0}, \sigma^2 \mathbf{I}_n)$. The estimates of the slopes can be obtained by using a stepwise process where the first step is to fit the $\mathbf{M}\boldsymbol{\mu}$ part of the model, computing the residuals, and then the second step is to fit the $\mathbf{X}\boldsymbol{\beta}$ part of the model, i.e., first fit

$$\mathbf{y} = \mathbf{M}\boldsymbol{\mu} + \boldsymbol{\varepsilon} \qquad (7.8)$$

and compute the residuals as

$$\mathbf{r} = (\mathbf{I} - \mathbf{M}\,\mathbf{M}^-)\mathbf{y}$$

where $\mathbf{M}^-$ denotes a generalized inverse of $\mathbf{M}$ (Graybill, 1976).

A model for these residuals is a model that is free of the $\mathbf{M}\boldsymbol{\mu}$ effects since the model for $\mathbf{r}$ is

$$\mathbf{r} = (\mathbf{I} - \mathbf{M}\,\mathbf{M}^-)\mathbf{X}\boldsymbol{\beta} + \boldsymbol{\varepsilon}^+$$

where $\boldsymbol{\varepsilon}^+ \sim N(\mathbf{0}, \sigma^2(\mathbf{I} - \mathbf{M}\,\mathbf{M}^-))$. Next the BLUE of $\boldsymbol{\beta}$ (assuming $\boldsymbol{\beta}$ is estimable) is

$$\begin{aligned}\hat{\boldsymbol{\beta}} &= \left[\mathbf{X}'(\mathbf{I}-\mathbf{M}\mathbf{M}^-)(\mathbf{I}-\mathbf{M}\mathbf{M}^-)\mathbf{X}\right]^{-1}\mathbf{X}'(\mathbf{I}-\mathbf{M}\mathbf{M}^-)(\mathbf{I}-\mathbf{M}\mathbf{M}^-)\mathbf{y} \\ &= \left[\mathbf{X}'(\mathbf{I}-\mathbf{M}\mathbf{M}^-)\mathbf{X}\right]^{-1}\mathbf{X}'(\mathbf{I}-\mathbf{M}\mathbf{M}^-)\mathbf{y}.\end{aligned} \qquad (7.9)$$

Variable Selection in the Analysis of Covariance Model

The estimate of β is a function of $\mathbf{r} = (\mathbf{I} - \mathbf{M}\,\mathbf{M}^-)\mathbf{y}$, the residuals of $\mathbf{y}$ from Model 7.8, and is a function of $(\mathbf{I} - \mathbf{M}\,\mathbf{M}^-)\mathbf{X}$, but each column of $(\mathbf{I} - \mathbf{M}\,\mathbf{M}^-)\mathbf{X}$ is as a set of residuals computed from fitting the model $\mathbf{x}_k = \mathbf{M}\boldsymbol{\mu}_k + \boldsymbol{\varepsilon}_k$ where $\mathbf{x}_k$ is the kth column of $\mathbf{X}$. Thus, computing the residuals of $\mathbf{y}$ and of each candidate covariate from a model with the design matrix of the treatment and design structures and then performing a variable selection procedure using those residuals provides the appropriate estimates of the slopes. Since the covariance matrix of the residuals of $\mathbf{y}$, $\mathbf{r}$, is not positive definite [it is of rank n-Rank $(\mathbf{M})$], the error degrees of freedom from using variable selection method on the residuals is inflated by Rank $(\mathbf{M})$. The correct degrees of freedom could be used in the final steps of the variable selection procedure to compute the appropriate significance levels. The overall effect of the inflated error degrees of freedom depends on the sample size and the Rank $(\mathbf{M})$. For example if n = 100, R $(\mathbf{M})$ = 30, and q = 10 (number of candidate covariates), there is not much difference between t percentage points with 60 and 90 degrees of freedom. On the other hand if n = 50, R(M) = 30, and q = 10, there is a big difference between t percentage points with 10 and 40 degrees of freedom.

7.5 WHEN SLOPES ARE POSSIBLY UNEQUAL

When slopes are unequal, the procedure in Section 7.1 may not determine the appropriate covariates, particularly when some treatments have positive slopes and others have negative slopes. To extend the procedure to handle unequal slopes for each covariate, an independent variable needs to be constructed for each level of the treatment (or levels of treatment combinations) which has the value of the covariate corresponding to observations of that treatment and has the value zero for observations not belonging to that treatment. In effect, the following model needs to be constructed

$$y_{ij} = \alpha_i + \beta_{i1}X_{ij1} + \beta_{i2}X_{ij2} + \ldots + \beta_{ik}X_{ijk} + \varepsilon_{ij}.$$

For "t" treatments and two covariates, construct the matrix model

$$\begin{bmatrix} y_{11} \\ y_{12} \\ \vdots \\ y_{1n} \\ y_{21} \\ y_{22} \\ \vdots \\ y_{2n} \\ \vdots \\ y_{t1} \\ y_{t2} \\ \vdots \\ y_{tn} \end{bmatrix} = \begin{bmatrix} 1 & 0 & & 0 & X_{111} & 0 & & 0 & X_{112} & 0 & & 0 \\ 1 & 0 & & 0 & X_{121} & 0 & & 0 & X_{122} & 0 & & 0 \\ \vdots & \vdots & \ldots & \vdots & \vdots & \vdots & \ldots & \vdots & \vdots & \vdots & \ldots & \vdots \\ 1 & 0 & & 0 & X_{1n1} & 0 & & 0 & X_{1n2} & 0 & & 0 \\ 0 & 1 & & 0 & 0 & X_{211} & & 0 & 0 & X_{212} & & 0 \\ 0 & 1 & & 0 & 0 & X_{221} & & 0 & 0 & X_{222} & & 0 \\ \vdots & \vdots & \ldots & \vdots & \vdots & \vdots & \ldots & \vdots & \vdots & \vdots & \ldots & \vdots \\ 0 & 1 & & 0 & 0 & X_{2n1} & & 0 & 0 & X_{2n2} & & 0 \\ \vdots & \vdots & \ldots & \vdots & \vdots & \vdots & \ldots & \vdots & \vdots & \vdots & \ldots & \vdots \\ 0 & 0 & & 1 & 0 & 0 & & X_{t11} & 0 & 0 & & X_{t12} \\ 0 & 0 & & 1 & 0 & 0 & & X_{t21} & 0 & 0 & & X_{t22} \\ \vdots & \vdots & \ldots & \vdots & \vdots & \vdots & \ldots & \vdots & \vdots & \vdots & \ldots & \vdots \\ 0 & 0 & & 1 & 0 & 0 & & X_{tn1} & 0 & 0 & & X_{tn2} \end{bmatrix} \begin{bmatrix} \alpha_1 \\ \alpha_2 \\ \vdots \\ \alpha_t \\ \beta_{11} \\ \beta_{21} \\ \vdots \\ \beta_{t1} \\ \beta_{12} \\ \beta_{22} \\ \vdots \\ \beta_{t2} \end{bmatrix} + \boldsymbol{\varepsilon}$$

$$y = [D, \ x_{11}, \ x_{21}, \ \ldots, \ x_{t1}, x_{12}, x_{22} \ \ldots, \ x_{t2}]\beta + \varepsilon$$

where **D** denotes the part of the design matrix with ones and zeros and

$$x'_{is} = (0, 0, \ldots, 0, x_{i1s}, x_{i2s}, \ldots, x_{ins}, 0, \ldots 0).$$

Next fit the models

$$y = D\alpha + \varepsilon$$
$$x_{is} = D\alpha_{is} + \varepsilon_{is} \quad i = 1, 2, \ldots, t, \quad s = 1, 2, \ldots, k$$

and compute the residuals, denoted by

$$r, \text{ and } r_{ik} \quad i = 1, 2, \ldots, t, \quad s = 1, 2, \ldots, k.$$

Finally, the variable selection procedure can be applied to the resulting sets of residuals as in Section 7.1.

REFERENCES

Draper, N. R. and Smith, H. (1981). *Applied Regression Analysis, Second Edition,* New York: John Wiley & Sons.

Graybill, F. A. (1976). *Theory and Application of the Linear Model,* Pacific Grove, CA. Wadsworth and Brooks/Cole.

Ott, Lyman (1988). *An Introduction to Statistical Methods and Data Analysis,* Boston: PWS-Kent.

SAS Institute Inc. (1989). *SAS/STAT® User's Guide, Version 6, Fourth Edition, Volume 2,* Cary, NC.

Yang, S. S. (1989). Personal communication.

EXERCISES

EXERCISE 7.1: Carry out an analysis of covariance for the following data set by determining the appropriate model and then making the needed treatment comparisons. Y is the response variable and X, Z, and W are the covariates. Use a regression model building strategy.

EXERCISE 7.2: Use the data in Section 4.6 with the variable selection procedures to select variables to be included in the model. The discussion in Section 4.6 indicates there are some unequal slopes, so the method in Section 7.5 will need to be utilized.

EXERCISE 7.3: Use the data in Section 4.4 with the variable selection process to determine if the models can be improved by including the square of height, the square of weight, and the cross-product of height and weight in addition to height and weight as possible covariates. Make the necessary treatment comparisons using the final model.

Data for Exercise 7.1

TRT	X	Z	W	Y	TRT	X	Z	W	Y
A	2.9	4.9	2.2	11.9	C	5.2	4.2	5.6	13.3
A	7.3	4.2	3.2	17.5	C	7.4	4.7	5.4	15.0
A	4.5	4.2	1.9	21.5	C	5.6	4.2	9.3	15.4
A	4.0	4.8	9.2	18.1	C	5.1	4.5	3.8	12.3
A	2.8	4.6	6.6	9.5	C	2.4	4.3	5.4	13.6
A	6.2	4.3	5.5	16.8	C	4.2	4.9	6.9	18.4
A	5.5	4.3	1.0	14.0	C	8.6	4.3	6.3	13.0
A	3.1	5.0	3.7	16.3	C	6.2	4.6	0.2	12.6
A	3.0	4.7	0.4	13.4	C	6.9	4.0	9.1	17.9
A	3.8	4.3	7.7	15.6	C	7.8	4.2	1.4	18.1
A	5.9	4.7	2.7	20.8	C	6.2	4.1	6.8	16.7
A	2.1	4.7	2.4	13.3	C	3.0	4.6	0.5	21.1
A	3.5	4.7	6.1	13.9	C	2.4	4.5	5.6	15.4
A	6.9	4.7	7.1	15.7	C	8.6	4.3	4.5	13.2
A	4.5	4.7	9.4	16.1	C	6.0	4.7	4.3	14.5
B	8.1	4.6	8.3	11.4	D	3.7	4.2	8.3	19.6
B	2.8	4.1	5.9	13.2	D	2.9	4.5	1.0	20.5
B	6.2	4.9	5.5	16.5	D	4.6	4.8	3.4	12.8
B	3.3	4.8	0.9	6.9	D	2.0	4.2	1.0	23.5
B	4.1	4.8	0.6	8.9	D	7.4	4.1	6.6	17.9
B	5.9	4.9	7.3	12.1	D	5.3	4.1	9.7	11.4
B	5.1	4.1	7.6	8.4	D	4.0	4.4	1.4	21.6
B	8.1	4.1	3.2	14.1	D	5.7	4.5	8.8	24.6
B	8.8	4.6	9.9	12.9	D	7.5	4.7	4.6	17.0
B	7.0	4.1	9.4	10.4	D	7.2	4.0	7.7	18.4
B	5.7	4.8	5.5	12.2	D	2.2	4.7	4.0	16.3
B	2.0	4.7	9.4	15.0	D	6.5	4.7	8.1	15.0
B	5.7	4.7	4.3	10.3	D	7.3	4.9	4.9	16.9
B	5.8	4.5	8.0	12.8	D	8.8	4.2	8.1	12.7
B	3.9	4.1	7.4	12.5	D	8.4	4.9	3.3	18.4

8 Comparing Models for Several Treatments

8.1 INTRODUCTION

Once an adequate covariance model has been selected to describe the relationship between the dependent variable and the covariates, it often is of interest to see if the models differ from one treatment to the next or from treatment combination to treatment combination. If one is concerned about the experiment-wise error rate in an analysis involving many tests of hypotheses, this procedure can provide that protection if it is used as a first step in comparing the treatments' models. Suppose the selected analysis of covariance model is

$$y_{ij} = \alpha_i + \beta_{i1}x_{1ij} + \beta_{i2}x_{2ij} + \cdots + \beta_{iq}x_{qij} + \varepsilon_{ij} \qquad (8.1)$$

for $i = 1, 2, \ldots, t$ and $j = 1, 2, \ldots, n_i$. The equal model hypothesis is

$$H_0: \begin{bmatrix} \alpha_1 \\ \beta_{11} \\ \beta_{12} \\ \vdots \\ \beta_{1q} \end{bmatrix} = \begin{bmatrix} \alpha_2 \\ \beta_{21} \\ \beta_{22} \\ \vdots \\ \beta_{2q} \end{bmatrix} = \cdots = \begin{bmatrix} \alpha_t \\ \beta_{t1} \\ \beta_{t2} \\ \vdots \\ \beta_{tq} \end{bmatrix} \text{ vs. } H_a: (\text{not } H_0).$$

This type of hypothesis can be tested by constructing a set of contrast statements in either PROC GLM or PROC MIXED or the model comparison method can be used to compute the value of the test statistic. The methodology described in this chapter is an application of the model comparison method that can easily be used to test the equality of models in many different settings. Schaff et al. (1988) and Hinds and Milliken (1987) used the method to compare nonlinear models. Section 8.2 describes the methodology to develop the statistics to test the equal model hypothesis for a one-way treatment structure, and methodology for the two-way treatment structure is discussed in Section 8.3. For two-way and higher order treatment structures, this process generates Type II sums of squares (Milliken and Johnson, 1992). Three examples are used to demonstrate the methods.

8.2 TESTING EQUALITY OF MODELS FOR A ONE-WAY TREATMENT STRUCTURE

The model comparison method (Milliken and Johnson, 1992) is used to construct the test of the equal models hypotheses. The first step is to use a linear regression procedure to fit Model 8.1 to the data from each treatment. Let SSRES (T_i) denote the sum of squares residual resulting from fitting Model 8.1 to the data from the ith treatment, $i = 1, 2, \ldots, t$. The sum of squares residuals for the complete model is obtained by pooling or adding the sum of squares residuals from each of the treatments, as

$$\text{SSRES(POOLED)} = \text{SSRES}(T_1) + \text{SSRES}(T_2) + \cdots + \text{SSRE}(T_t).$$

The number of degrees of freedom (df) associated with SSRES(POOLED) is the sum of the numbers of degrees of freedom of the respective SSRES(T_i). Since the ith treatment has n_i observations and there are $q + 1 = p$ parameters in the model, the number of degrees of freedom associated with SSRES(T_i) is df(SSRES(T_i)) = $n_i - (q + 1) = n_i - p$. Thus the number of degrees of freedom associated with SSRES(POOLED) is

$$\sum_{i=1}^{t}(n_i - p) = n. - tp$$

where

$$n. = \sum_{i=1}^{t} n_i$$

The hypothesis to be tested is that there are no differences among the models describing the data for the treatments or the models are identical or that one common model will describe the data from all treatments. The common model or model under the conditions of H_0 is

$$y_{ij} = \alpha + \beta_1 x_{1ij} + \beta_2 x_{2ij} + \cdots + \beta_q x_{qij} + \varepsilon_{ij}.$$

The next step in the analysis is to use a linear regression procedure to fit one model to the data combined across all treatments. Let SSRES(COMBINED) denote the sum of squares residual resulting from fitting one model to the combined data set. The common model has $q + 1 = p$ parameters thus df(SSRES(POOLED)) = $n. - p$. Using the model comparison method, the sum of squares due to deviations from the equal models hypothesis is

$$\text{SSH}_0 = \text{SSRES(COMBINED)} - \text{SSRES(POOLED)}$$

Comparing Models for Several Treatments

TABLE 8.1
Analysis of Variance Table to Summarize the Testing of Equality Analysis of Covariance Models

Source of Variation	df	SS	F
TREATMENTS	$(t-1)p$	SSH_0	F_c
ERROR	$n_. - tp$	SSRES(POOLED)	

Note: t = number of treatments; p = number of parameters in the model; $n_.$ = total number of observations.

and the degrees of freedom associated with SSH_0 is

$$df(SSH_0) = df(SSRES(COMBINED)) - df(SSRES(POOLED)) = (t-1)p$$

where t is the number of treatments and p is the number of parameters in the model used to describe the response for each of the treatments.

The statistic to test the equal model hypothesis is

$$F_c = \frac{SSH_0/df(SSH_0)}{SSRES(POOLED)/df(SSRES(POOLED))}.$$

The decision is to reject the equal models hypothesis if $F_c > F_{\alpha,\,(t-1)p,\,n_.-tp}$. If the test statistics indicate there are significant differences among the treatments' models, the treatments can be compared pairwise by constructing confidence bands about the difference of two models over a range of x values or by using the methods of analysis described in Chapter 2. The results of the analysis to test for equality of the models are summarized in Table 8.1.

8.3 COMPARING MODELS FOR A TWO-WAY TREATMENT STRUCTURE

When the treatments in the experiment consist of treatment combinations formed by combining the levels of two factors, a two-way analysis for comparing models is an extension of the one-way analysis process where there are tests for main effects and interaction. For demonstration purposes, suppose there are a set of row treatments, $R_1, R_2, \ldots, R_r$, and a set of column treatments, $C_1, C_2, \ldots, C_c$. Each of the r row treatments is combined with each of the c column treatments, generating rc treatment combinations. The experiment consists of randomly assigning n_{ij} experimental units to each treatment combination where q covariates were measured. Assume the adequate analysis of covariance model is

$$y_{ijk} = \alpha_{ij} + \beta_{ij1}x_{1ijk} + \beta_{ij2}x_{2ijk} + \cdots + \beta_{ijq}x_{qijk} + \varepsilon_{ij} \tag{8.2}$$

The first step in the analysis could be to consider this experiment as a one-way treatment structure and test the equality of the rc models using the results of Section 8.2. If the models are determined to be different, then the main effects and interaction hypotheses can be investigated. The process to compute the necessary sums of squares to compare regression models in a two-way treatment structure involves using a linear regression procedure to fit Model 8.2 and obtain the sum of squares residuals and corresponding df for each of the following combinations of the data. Fit model to

1. All of the data and compute SSRES(COMBINED)
2. The data from each row treatment, combined over the column treatments and compute SSRES(R_i), i = 1, 2, ..., r
3. The data from each column treatment combined over the row treatments and compute SSRES(C_j), j = 1, 2, ..., c
4. The data from each row by column treatment combination and compute SSRES(R_i, C_j), i = 1, 2, ..., r, j = 1, 2, ..., c.

Next, add or pool the above sums of squares in various ways as

1. SSRES(ROW) = SSRES(R_1) + $\cdots$ + SSRES(R_r)
2. SSRES(COL) = SSRES(C_1) + $\cdots$ + SSRES(C_c)
3. SSRES(POOLED) = SSRES(R_1, C_1) + $\cdots$ + SSRES(R_r, C_c)

The numbers of degrees of freedom associated with the above sums of squares are obtained by adding the respective numbers of degrees of freedom of those sums of squares involved in the sum. The Type II sums of squares for the analysis of covariance table are

SSROW = SSRES(COMBINED) − SSRES(ROW)
SSCOLUMN = SSRES(COMBINED) − SSRES(COL)
SSROW*COLUMN = SSRES(ROW) + SSRES(COL) − SSRES(COMBINED) − SSRES(POOLED)

The numbers of degrees of freedom associated with each of the above sums of squares are df(SSROW) = (r − 1)p, df(SSCOLUMN) = (c − 1)p, df(SSROW*COLUMN) = (r − 1)(c − 1)p, and SSRES(POOLED) = $n_{..}$ − rcp where $n_{..}$ is the total number of observations in the study.

The statistic to test the equality of row treatment models averaged over column treatments is

$$F_{c(R)} = \frac{SSROW/((r-1)p)}{SSRES(POOLED)/(n_{..}-rcp)}.$$

where r is the number of levels of the row treatment and p is the number of parameters in the model used to describe the response.

TABLE 8.2
Analysis of Variance Table for the Two-Way Nonlinear Analysis of Covariance

Source of Variation	df	SS	F
ROW	$(r-1)p$	SSROW	$F_{c(T)}$
COLUMN	$(c-1)p$	COLUMN	$F_{c(C)}$
ROW*COLUMN	$(r-1)(c-1)p$	SSROW*COLUMN	$F_{c(R \times C)}$
ERROR	$n_{..} - rcp$	SSRES(POOLED)	

Note: r = number of row treatments; c = number of column treatments; p = number of parameters in the model; $n_{..}$ = total number of observations.

The statistic to test the equality of the column treatment models averaged over the row treatments is

$$F_{c(C)} = \frac{\text{SSCOLUMN}/(c-1)p}{\text{SSRES(POOLED)}/(n_{..} - rcp)}.$$

The statistic to test for row by column interaction effects on the models is

$$F_{c(R \times C)} = \frac{\text{SSROW} * \text{COLUMN}/(r-1)(c-1)p}{\text{SSRES(POOLED)}/(n_{..} - rcp)}.$$

The results of this two-way analysis of covariance are summarized in Table 8.2.

The process can be extended to more than two factors. Three examples are used to demonstrate the process for one-way and two-way treatment structures.

8.4 EXAMPLE: ONE-WAY TREATMENT STRUCTURE WITH ONE COVARIATE

The exercise program data from Section 3.3 are used to demonstrate the procedure to compare models for a one-way treatment structure with one covariate. The model used to describe the heart rate (HR) for the jth person assigned to the ith exercise program is

$$HR_{ij} = \alpha_i + \beta_i IHR_{ij} + \varepsilon_{ij}, \quad i = 1, 2, 3, \quad j = 1, 2, \ldots, 8$$

where IHR_{ij} is the heart rate of the jth person assigned to the ith exercise program and IHR_{ij} is the initial resting heart rate. The models are identical if the intercepts are all equal and if the slopes are all equal. The identical models hypothesis can be stated in terms of the parameters of the model as

$$H_0: \begin{bmatrix} \alpha_1 \\ \beta_1 \end{bmatrix} = \begin{bmatrix} \alpha_2 \\ \beta_2 \end{bmatrix} = \begin{bmatrix} \alpha_3 \\ \beta_3 \end{bmatrix} \text{ vs. } H_a: (\text{not } H_0).$$

TABLE 8.3
SAS System Code to Perform Preliminary Computations to Compare the Exercise Program Models

```
PROC REG; MODEL HR = IHR;  *Fit one model to all data;
PROC SORT; by EPRO;
PROC REG; BY EPRO; MODEL HR = IHR;*Fit model to each level of EPRO;
```

TABLE 8.4
Residual Sums of Squares and Degrees of Freedom for Each Level of EPRO and Combined with Computations of the F Statistic

SSRES(COMBINED) = 1767.82 df = 22

EPRO	SSRES(EPRO$_i$)	df	
1	316.81	6	
2	102.66	6	
3	88.18	6	
SSRES(POOLED)	507.65	18	$\sigma^2 = 28.20$

$SSH_0 = 1767.82 - 507.65 = 1260.17$

$$F_c = \frac{1260.17/4}{28.20} = 11.17$$

A contrast statement involving a series of comparisons could be used to test this hypothesis in the analysis of variance framework of Chapter 2, but it is much simpler to use the methodology described in Section 8.2. The method is easily applied by (1) fitting one simple linear regression model to all data and (2) then fitting a simple linear regression model to the data for each EPRO. The only information needed from each of these analyses are the sums of squares residuals and the corresponding numbers of degrees of freedom.

Table 8.3 contains the necessary PROC REG code to fit one model to all of the data and to fit a model to the data from each of the exercise programs. Table 8.4 displays the respective sums of squares and their respective numbers of degrees of freedom. The sum of squares residuals from fitting one model is SSRES(COMBINED) = 1767.82 with 22 degrees of freedom. The sums of squares residuals for the three exercise programs are 316.81, 102.66, and 88.18, each with 6 degrees of freedom. The pooled sum of squares residuals is 507.65 with 18 pooled degrees of freedom. The sum of squares due to deviations from the null hypothesis is $SSH_0 =$ 1767.82 − 507.65 = 1260.17 which is based on 22 − 18 = 4 = (# parameters)(# treatments − 1) degrees of freedom. The resulting F statistic is 11.17 which is based on 4 and 18 degrees of freedom. The significance level is less than 0.0001, so the

TABLE 8.5
Analysis of Variance Table Which Summarizes the Comparison of the EPRO Models

Source	df	Sum of Squares	F
EPRO	4	1260.17343	11.17
ERROR	18	507.64821	

researcher would conclude there is sufficient evidence to say the exercise programs' models are not identical. The results of the testing for the equality of the three exercise program models can be summarized in an analysis of variance table. Table 8.5 is a display of the analysis of variance, but there is one difference from the usual analysis of variance table in that the degrees of freedom corresponding to EPRO is 4 instead of the expected 2. There are 4 degrees of freedom associated with EPRO because the models with 2 parameters are being compared so the degrees of freedom are (# parameters)(#treatments − 1). The usual analysis of variance model has one parameter for each treatment group, so the number of degree of freedom for treatments is the usual 1*(#treatments − 1). At this point, the analysis followed in Section 3.3 would be used to determine if the form of the models could be simplified and then make comparisons among the regression models

8.5 EXAMPLE: ONE-WAY TREATMENT STRUCTURE WITH THREE COVARIATES

The data in Section 4.4 are used to compare the equality of eight herbicide models where three covariates were measured. The response variable is the yield in pounds of soybeans per plot. The model to describe yield as a linear function of silt, clay, and organic matter for the jth plot of the ith herbicide is

$$\text{Yield}_{ij} = \alpha_i + \beta_{i1}\text{SILT}_{ij} + \beta_{i2}\text{CLAY}_{ij} + \beta_{i3}\text{OM}_{ij} + \varepsilon_{ij}, \quad i = 1, 2, \ldots, 8, \ j = 1, 2, \ldots, 12.$$

The herbicide models are identical if the eight intercepts are equal, the eight slopes corresponding to silt are equal, the eight slopes for clay are equal, and the eight slopes for organic matter are equal. A set of contrasts can be constructed to provide the statistic to test the equal models hypothesis, but the method described in Section 8.2 is much easier to implement. The identical model hypothesis can be stated in terms of the parameters as

$$H_0: \begin{bmatrix} \alpha_1 \\ \beta_{11} \\ \beta_{12} \\ \beta_{13} \end{bmatrix} = \begin{bmatrix} \alpha_2 \\ \beta_{21} \\ \beta_{22} \\ \beta_{23} \end{bmatrix} = \cdots = \begin{bmatrix} \alpha_8 \\ \beta_{81} \\ \beta_{82} \\ \beta_{83} \end{bmatrix} \text{ vs. } H_a: (\text{not } H_0).$$

TABLE 8.6
SAS® System Code to Perform Preliminary Computations to Compare Herbicide Models

```
PROC REG; MODEL YIELD = SILT CLAY OM;*Fit one model to all data;
PROC SORT; BY HERB; *Fit model to each of level of Herbicide;
PROC REG; BY HERB; MODEL YIELD = SILT CLAY OM;
```

TABLE 8.7
Residual Sums of Squares and Degrees of Freedom for each Level of Herb and Combined Data with Computations of the F Statistic

SSRES(COMBINED) = 4410.90 d.f. = 92

Herbicide	SSRes(Herb$_i$)	df	
1	164.18	8	
2	214.69	8	
3	198.08	8	
4	409.78	8	
5	291.53	8	
6	501.81	8	
7	266.43	8	
8	357.26	8	
SSRes(POOLED)	2403.76	64	$\sigma^2 = 37.56$

SSH$_0$ = 4410.90 − 2403.76 = 2007.14

$$F_c = \frac{2007.14/28}{37.56} = 1.91$$

The PROC REG code required to fit the necessary models is in Table 8.6. The first part of the code fits one model to the combined data set. The SSRES(COMBINED) = 4410.90 and is based on 92 degrees of freedom as displayed in Table 8.7. The code in the second part of Table 8.6 fits the model to the data for each of the herbicides and the results are summarized in Table 8.7. Each of the sum of squares residuals is based on 8 degrees of freedom. The pooled sum of squares residuals is SSRES(POOLED) = 2403.76 and is based on 64 degrees of freedom. The sum of squares due to deviations from the equal model hypothesis is 4410.90 − 2403.76 = 2007.14 and is based on 92 − 64 = 28 degrees of freedom. Since each model consists of 4 parameters, the degrees of freedom associated with the identical herbicide models hypothesis is 4(8 − 1) = 28. The corresponding F statistic is equal to 1.91 with significance level 0.0169, which is generally sufficient evidence to conclude that the models are not all identical. These results are summarized in the analysis of variance table displayed in Table 8.8. The analysis is completed by using methods to

TABLE 8.8
Analysis of Variance Table to Summarize the Comparison of the Herbicide Models

Source	df	Sum of Squares	F
HERB	28	2007.13131	1.91
ERROR	64	2403.76892	

simplify the models and then comparing the resulting regression models as described in Chapter 4.

8.6 EXAMPLE: TWO-WAY TREATMENT STRUCTURE WITH ONE COVARIATE

The experiment of Section 5.5 investigated the effect of moisture content on the amount of energy from burning a block of wood from four different species of trees in three types of stoves and is used to demonstrate the procedure for testing the equality of models in a two-way treatment structure. The cell means model expressing a linear relationship between the amount of energy and the moisture content is

$$y_{ijk} = \alpha_{ij} + \beta_{ij} M_{ijk} + \varepsilon_{ijk}, \quad i=1, 2, 3, 4, \quad j=1, 2, 3, \quad k=1, 2, \ldots, n_{ij}.$$

The first step in the process could be to test for the equality of all 12 of the models in the treatment structure, i.e., test the equality of the 12 intercepts and the equality of the 12 slopes as is related in the following hypothesis:

$$H_{0_S}: \begin{bmatrix} \alpha_{11} \\ \beta_{11} \end{bmatrix} = \begin{bmatrix} \alpha_{12} \\ \beta_{12} \end{bmatrix} = \cdots = \begin{bmatrix} \alpha_{43} \\ \beta_{43} \end{bmatrix} \text{ vs. } H_a: (\text{not } H_0)$$

Testing this hypothesis is identical to testing the equality of models from a one-way treatment structure as described in Section 8.2. Table 8.9 contains the PROC REG code needed to fit the model to the various data sets. The first part of Table 8.9 contains the code to fit the model to the combined data set and the last part of Table 8.9 contains the code to fit the model to the data for each of the wood*stove combinations. The SSRES(COMBINED) = 326.64870 and is based on 86 degrees of freedom as displayed in part (1) of Table 8.10. The SSRES(POOLED) = 9.159410 and is based on 64 degrees of freedom as obtained from Table 8.11. The sum of squares due to deviations from the equal models hypothesis is 326.64870 − 9.15941 = 317.48929 and is based on 86 − 64 = 22 degrees of freedom. The correspond F statistic is F = 100.84 with significance level much less than 0.0001. Thus there is sufficient evidence to conclude that the wood*stove models are not all identical.

The next step in the analysis is to investigate the effect of the factorial structure on the regression models. This is accomplished by testing the equality of the models

TABLE 8.9
SAS® System Code to Perform Preliminary Computations to Compare WOOD*STOVE models.

```
*Fit one model to all data;
PROC REG; MODEL ENERGY = MOISTURE;

*Fit a model to each level of wood;
PROC SORT; BY WOOD;
PROC REG; BY WOOD; MODEL ENERGY = MOISTURE;

*Fit a model to each level of stove type;
PROC SORT; BY STOVE;
PROC REG; BY STOVE; MODEL ENERGY = MOISTURE;

*Fit a model to each combination of WOOD*STOVE;
PROC SORT; BY STOVE WOOD;
PROC REG; BY STOVE WOOD; MODEL ENERGY = MOISTURE;
```

TABLE 8.10
Residual Sums of Squares and Degrees of Freedom for One Model for All Data, One Model for Each Wood and One Model for Each Stove to Compute SSRES(COMBINED), SSRES(WOOD), and SSRES(STOVE)

(1) One model SSRES(COMBINED) = 326.64870 d.f. = 86

(2) **WOOD**	SSRes(WOOD$_i$)	df
Black Walnut	19.05508	20
Osage Orange	8.70268	22
Red Oak	9.66393	18
White Pine	10.71815	20
SSRes(WOOD)	48.13984	80
(3) **STOVE**	SSRes(STOVE$_j$)	df
Type A	173.27729	28
Type B	70.59879	23
Type C	82.32894	31
SSRES(STOVE)	326.20502	82

for the levels of wood, the equality of models for the levels of stove, and testing for wood by stove interaction effects on the models. By summing the model over the levels of stove for each level of wood, a wood level model can be obtained as

$$\bar{y}_{i \cdot k} = \bar{\alpha}_{i \cdot} + \bar{\beta}_{i \cdot} \bar{M}_{i \cdot k} + \bar{\varepsilon}_{i \cdot k}$$

TABLE 8.11
Residual Sums of Squares and Degrees of Freedom for Various Wood*Stove Combinations Used to Compute SSRES(POOLED)

STOVE	WOOD	SSRes(W_{ij}, S_j)	df
Type A	Black Walnut	0.10304	4
Type A	Osage Orange	0.34459	7
Type A	Red Oak	0.77695	3
Type A	White Pine	0.66661	8
Type B	Black Walnut	0.54988	6
Type B	Osage Orange	1.00689	4
Type B	Red Oak	0.25559	4
Type B	White Pine	1.17348	3
Type C	Black Walnut	1.02172	6
Type C	Osage Orange	1.47882	7
Type C	Red Oak	1.29265	7
Type C	White Pine	0.48916	5
SSRES(POOLED)		9.15941	64

To compute the sum of squares due to test for equality of wood models, fit a model to the data from each level of wood. The code in the second part of Table 8.9 is used to fit a model to each level of wood. The sums of squares residuals for each level of wood are in part (2) of Table 8.10 where SSRES(WOOD) = 48.13984 and is based on 80 degrees of freedom. The sum of squares due to deviations from the equal models for the levels of wood hypothesis is SS_WOOD = 326.64870 − 48.13984 = 278.50886 and is based on 86 − 80 = 6 degrees of freedom. The value of the F statistic is 324.34 as displayed in Table 8.12. The significance level is very

TABLE 8.12
Sums of Squares, Degrees of Freedom and F Statistics for Wood, Stove and Wood*Stove

$\hat{\sigma}^2 = 0.143116$

SSWOOD = 326.64870 − 48.13984 = 317.48929 df = 6

$F_{c_w} = \dfrac{52.91488}{0.143116} = 324.34$

SS STOVE = 326.64870 − 326.20502 = 0.44368 df = 4

$F_{c_s} = \dfrac{0.11092}{0.143116} = 0.78$

SS WOOD × STOVE = 48.13984 + 326.20502 − 326.64820 − 9.159410
       = 38.53675, d.f. = 80 + 82 − 86 − 64 = 12

$F_{c_{ws}} = \dfrac{3.21140}{0.143116} = 22.4391$

small, providing sufficient evidence to conclude the levels of wood models are not all identical. Recall this testing a Type II hypothesis (Milliken and Johnson (1992)).

Next investigate the effect of the levels of stove on the regression models. This is accomplished by testing the equality of the models for the levels of stove. By summing the model over the levels of wood for each level of stove, a stove level model can be obtained as

$$\bar{y}_{.jk} = \bar{\alpha}_{.j} + \bar{\beta}_{.j}\bar{M}_{.jk} + \bar{\varepsilon}_{.jk}$$

To compute the sum of squares due to test for equality of stove models, fit a model to the data from each level of stove. The code in the third part of Table 8.9 is used to fit a model to each level of stove. The sums of squares residuals for each level of stove are in part (3) of Table 8.10 where SSRES(STOVE) = 326.20502 and is based on 82 degrees of freedom. The sum of squares due to deviations from the equal models for the levels of stove hypothesis is SS_STOVE = 326.64870 − 326.20502 = 0.44368 and is based on 86 − 82 = 4 degrees of freedom. The value of the F statistic is 0.78 as displayed in Table 8.12. The significance level is very large, so there is no evidence the levels of stove models are not all identical.

The final step is to investigate the possibility of wood by stove interaction effect on the regression models. The no interaction effect on the regression models hypothesis is equivalent to the hypothesis that all 2 × 2 table differences (Milliken and Johnson (1992)) of the vectors of parameters are equal to zero. This no interaction hypothesis can be stated as

$$H_{0_{WS}}: \begin{bmatrix}\alpha_{ij}\\ \beta_{ij}\end{bmatrix} - \begin{bmatrix}\alpha_{is}\\ \beta_{is}\end{bmatrix} - \begin{bmatrix}\alpha_{rj}\\ \beta_{rj}\end{bmatrix} + \begin{bmatrix}\alpha_{rs}\\ \beta_{rs}\end{bmatrix} = \mathbf{0} \text{ for all } i, j, s, r \text{ vs. } H_a: \left(\text{not } H_{0_{WS}}\right)$$

The sum of squares due to deviations from the above hypothesis is SS_WOOD × STOVE = 48.13984 + 326.20502 − 326.64870 − 9.159410 − 38.53675 and is based on 80 + 82 − 86 − 64 = 12 degrees of freedom. The value of the F statistic is 22.44 as shown in Table 8.12. The significance level is <0.0001, indicating that there is sufficient evidence to conclude there is an interaction effect on the wood by stove models. Table 8.13 contains the analysis of variance table summarizing the above factorial effects on the models. Remember the test statistics are comparing the models not just an individual parameter of the models. So the numbers of degrees of freedom for the main effects and interaction in Table 8.13 are the usual numbers of degrees of freedom times the number of parameters in a treatment combination model. Remember, these are Type II sums of squares.

8.7 DISCUSSION

A very useful method is presented to test identical model hypotheses. The method could be used to perform prior tests of equality of models before examining the individual parameters in order to control the Type I error rate. That is, one would

TABLE 8.13
Analysis of Variance Table to Summarize the Factorial Effects of Wood and Stove on the Regression Models

Source	df	SS(II)	F
WOOD	6	317.48929	324.35
STOVE	4	0.44368	0.78
WOOD*STOVE	12	38.53675	22.44
ERROR	64	9.15941	

only continue with the analysis of covariance process of simplifying the form of the model if there were adequate evidence to believe the models are not identical. The process is easily implemented with a multiple regression computer package, i.e., these comparisons can be accomplished without a general linear models computer package.

REFERENCES

Hinds, M. A. and Milliken, G. A. (1987). Statistical methods to use nonlinear models to compare silage treatments, *Biometrical Journal* 29(6), 825–834.

Milliken, G. A. and Johnson, D. E. (1992). *Analysis of Messy Data, Volume I: Design Experiments*, London: Chapman & Hall.

Schaff, D. A., Milliken, G. A., and Clayberg, C. D. (1988). A method for analyzing nonlinear models when the data are from a split-plot or repeated measures design, *Biometrical Journal* 30(2), 139–146.

EXERCISES

EXERCISE 8.1: Compare the equality of the models for the data in Section 3.4.

EXERCISE 8.2: Compare the equality of the models for the data in Section 4.5.

EXERCISE 8.3: Compare the equality of the models for the data in Section 5.4.

EXERCISE 8.4: Compare the equality of the models for the data in Section 5.7.

EXERCISE 8.5: For the data in Section 5.5, use contrast statements to obtain tests for equal wood models, equal stove models, and the no interaction hypothesis and compare the results to those in Section 8.6. The contrast statements will provide tests of Type III hypotheses.

9 Two Treatments in a Randomized Complete Block Design Structure

9.1 INTRODUCTION

The introduction of blocking into the analysis of covariance models presents another dimension in the analysis. That dimension involves obtaining information about the slopes of the lines from the block means or totals, a process called the recovery of interblock information, as well as obtaining information about the slopes from the within block comparisons. The recovery of interblock information about treatment effects is used in the analysis of incomplete block designs, but there is no interblock information about treatment effects in complete block designs. This chapter develops the general methodology for analyzing analysis of covariance models when the data are collected in complete blocks. The next step is to consider the analysis of covariance in incomplete block designs which include split-plot and repeated measures designs (discussed in later chapters). A simple experiment involving two treatments in six blocks is used throughout this chapter to demonstrate the various concepts. The last section gives an example with equal slopes in 20 blocks.

9.2 COMPLETE BLOCK DESIGNS

An experiment was conducted to evaluate the effect of two herbicides on soybean yield. The experimental design consists of a one-way treatment structure in a randomized complete block design structure. The herbicides were preemergence herbicides and were incorporated into the soil. The activity of the herbicides was thought to be influenced by the amount of organic matter in the soil, so the organic matter content of each plot was determined and was used as a possible covariate. For purposes of discussion, it is assumed that organic matter affects the yield of the plots linearly. As in any modeling process, this assumption must be substantiated before the analysis can continue. Since the herbicides were of differing chemical compositions, there was the possibility of unequal slopes.

A model that can be used to describe the yield of the soybeans grown on a plot in the jth block treated by the ith herbicide is

$$y_{ij} = \alpha_i + \beta_i x_{ij} + b_j + \varepsilon_{ij}, \quad i = 1, 2, \quad j = 1, 2, \ldots, 6, \tag{9.1}$$

where y_{ij} denotes the observed yield per plot in bushels per acre,
   α_i denotes the mean response of the ith herbicide when the value of the covariate (organic matter) is zero,
   x_{ij} denotes the value of the covariate (organic matter) measured on the experimental unit in the jth block receiving the ith herbicide,
   β_i is the slope of the regression line for the ith herbicide,
   b_j is the random block effect associated with the jth block where the block effects are assumed to be distributed iid N $(0, \sigma_b^2)$, and
   ε_{ij} denotes the random error where the errors are assumed to be distributed iid N $(0, \sigma_\varepsilon^2)$.

Model 9.1 is a mixed model in that its parameters include two components of variance (Milliken and Johnson, 1992, and Littel et al., 1996). To help understand how the mixed models analysis operates, the within block analysis, the between block analysis, and the combined within and between block analysis are discussed. Before the development of mixed models software, the analysis of Model 9.1 could be carried out in two parts: the **Within Block Analysis** (that is done by most computer codes) and the **Between Block Analysis** (which is not done by most computer codes).

9.3 WITHIN BLOCK ANALYSIS

The within block analysis provides estimates of the slopes which are based on the within block information, i.e., the estimates of the slopes are based on contrasts of the observations computed within each block. Within block information is free of block effects and the variance of a within block estimate is a scalar multiple of σ_ε^2. To demonstrate this idea, consider the data in Table 9.1 which represent the yield of soybeans in bushels per acre where the treatments are two herbicides, the covariate is the percent organic matter, and there are six blocks.

The within block analysis of Model 9.1 can be carried out by taking contrasts within the blocks and analyzing the corresponding models. Since there are only two treatments per block, the only contrast within each block is the difference $d_j = y_{1j} - y_{2j}$. The within block model is constructed by taking the difference of the models, i.e., the model for d_j is

$$\begin{aligned} d_j &= \alpha_1 + \beta_1 x_{1j} + b_j + \varepsilon_{1j} - \left(\alpha_2 + \beta_2 x_{2j} + b_j + \varepsilon_{2j}\right) \\ &= \alpha_1 + \beta_1 x_{1j} + \varepsilon_{1j} - \alpha_2 - \beta_2 x_{2j} - \varepsilon_{2j} \\ &= \alpha_1 - \alpha_2 + \beta_1 x_{1j} - \beta_2 x_{2j} + \varepsilon_{1j} - \varepsilon_{2j} \end{aligned}$$

The difference model is free of the block effects and variance of each difference is $2\sigma_\varepsilon^2$. By letting $\alpha_d = \alpha_1 - \alpha_2$ and $e_j = \varepsilon_{1j} - \varepsilon_{2j}$, the model for the differences (called

TABLE 9.1
Yield of Soybeans (in bu/acre) for Herbicide Treatments with Percent Organic Matter as a Covariate in RCB Design Structure

	Herbicide 1		Herbicide 2	
Block	Yield	OM	Yield	OM
1	26.6	0.91	30.2	1.02
2	31.1	1.22	29.2	0.89
3	34.7	1.43	32.1	1.39
4	34.4	1.45	31.9	1.47
5	32.1	1.33	30.2	1.27
6	28.5	1.10	31.0	1.12

a within block model) can be expressed as the two independent variable multiple regression model

$$d_j = \alpha_d + \beta_1 x_{1j} - \beta_2 x_{2j} + e_j.$$

By fitting the multiple regression model to the data in Table 9.1, one obtains estimates of the estimable functions of the parameters of the original model, i.e., $\alpha_1 - \alpha_2$, β_1, β_2, and σ_ε^2. The matrix form of the model for the differences computed from the data in Table 9.1 is

$$\begin{bmatrix} -3.60 \\ 1.90 \\ 2.60 \\ 2.50 \\ 2.00 \\ -2.50 \end{bmatrix} = \begin{bmatrix} 1 & 0.91 & -1.02 \\ 1 & 1.22 & -0.89 \\ 1 & 1.43 & -1.39 \\ 1 & 1.45 & -1.47 \\ 1 & 1.33 & -1.27 \\ 1 & 1.10 & -1.12 \end{bmatrix} \begin{bmatrix} \alpha_d \\ \beta_1 \\ \beta_2 \end{bmatrix} + \mathbf{e}.$$

or $\mathbf{d} = \mathbf{Z}\,\boldsymbol{\eta} + \mathbf{e}$, where $\boldsymbol{\eta}' = [\alpha_d, \beta_1, \beta_2]$.

The least squares estimates of the parameters of the model for the differences are

$$\begin{bmatrix} \hat{\alpha}_d \\ \hat{\beta}_1 \\ \hat{\beta}_2 \end{bmatrix} = (\mathbf{Z}'\mathbf{Z})^{-1}\mathbf{Z}'\mathbf{d} = \begin{bmatrix} -13.82302281 \\ 16.95662402 \\ 5.64513210 \end{bmatrix}. \qquad (9.2)$$

The residual sum of squares for the difference model is 1.44, which is based on 3 degrees of freedom. The residual mean square, $1.44/3 = 0.48$, is an estimate of $2\hat{\sigma}_\varepsilon^2$, the variance of the d_j. The estimated covariance matrix of the estimated parameter vector is

$$2\hat{\sigma}_\varepsilon^2(\mathbf{Z'Z})^{-1} = \begin{bmatrix} 3.633765 & -2.0436818 & 0.8542064 \\ -2.0436818 & 5.1684106 & 3.6579537 \\ 0.8542064 & 3.65479537 & 4.5168137 \end{bmatrix} = \hat{\Sigma}_w. \quad (9.3)$$

The estimated covariance matrix corresponding to the within block analysis estimates of the two slopes is the lower 2×2 partition of $\hat{\Sigma}_w$, or

$$\hat{\Sigma}_{\beta_w} = \begin{bmatrix} 5.1684106 & 3.6579537 \\ 3.6579537 & 4.168137 \end{bmatrix}.$$

The estimates of the standard errors of the parameter estimates are obtained by taking the square root of the diagonal elements of the covariance matrix. These standard errors can be used to construct confidence intervals about the respective parameters. The above estimates of the parameters are what would be obtained from a computer code that performs the within block analysis, but there is additional information about the slopes and treatment effects from the between block analysis.

9.4 BETWEEN BLOCK ANALYSIS

The between block analysis utilizes the model of the block totals and provides information about the parameters contained in those block totals. Let $t_j = y_{1j} + y_{2j}$ denote the total of the two observations in the jth block. The block total model is constructed by taking the totals of the corresponding models, i.e.,

$$t_j = \alpha_1 + \beta_1 x_{1j} + \varepsilon_{1j} + \alpha_2 + \beta_2 x_{2j} + b_j + \varepsilon_{2j}$$
$$= \alpha_1 + \alpha_2 + \beta_1 x_{1j} + \beta_2 x_{2j} + 2b_j + \varepsilon_{1j} + \varepsilon_{2j}$$

which can be expressed as: $t_j = \alpha_t + \beta_1 x_{1j} + \beta_2 x_{2j} + r_j$, where $\alpha_t = \alpha_1 + \alpha_2$ and $r_j = 2b_j + \varepsilon_{1j} + \varepsilon_{2j}$. The variance of a block total is $2(\sigma_\varepsilon^2 + 2\sigma_b^2)$. A multiple regression program can be used to fit the above model to the block totals and obtain the between block estimates of α_t, β_1, β_2, and $2(\sigma_\varepsilon^2 + 2\sigma_b^2)$.

The matrix form of the block total model for the data in Table 9.1 is

$$\begin{bmatrix} 56.8 \\ 60.3 \\ 66.8 \\ 66.3 \\ 62.4 \\ 59.5 \end{bmatrix} = \begin{bmatrix} 1 & 0.91 & 1.02 \\ 1 & 1.22 & 0.89 \\ 1 & 1.43 & 1.39 \\ 1 & 1.45 & 1.47 \\ 1 & 1.33 & 1.27 \\ 1 & 1.10 & 1.12 \end{bmatrix} \begin{bmatrix} \alpha_t \\ \beta_1 \\ \beta_2 \end{bmatrix} + \mathbf{r}.$$

or $t = M\tau + e$, where $\tau' = [\alpha_t, \beta_1, \beta_2]$.

The between block estimates are

$$\begin{bmatrix} \hat{\alpha}_t \\ \hat{\beta}_1 \\ \hat{\beta}_2 \end{bmatrix} = (M'M)^{-1}M't = \begin{bmatrix} 38.48854914 \\ 13.83912245 \\ 5.32201593 \end{bmatrix} \quad (9.4)$$

The residual sum of squares from the between block model is 3.25 and is based on 3 degrees of freedom. The between block residual mean square is 3.25/3, which is an estimated $2(\sigma_\varepsilon^2 + 2\sigma_b^2)$. The estimated covariance matrix of the between block estimates is

$$(3.25/3)(M'M)^{-1} = \begin{bmatrix} 8.218374 & -4.622627 & -1.932139 \\ -4.622627 & 11.690547 & -8.274010 \\ -1.932139 & -8.274010 & 10.216685 \end{bmatrix} = \hat{\Sigma}_b. \quad (9.5)$$

The estimated covariance matrix corresponding to the between block analysis estimates of the two slopes is the lower 2×2 partition of $\hat{\Sigma}_b$, or

$$\hat{\Sigma}_{\beta_b} = \begin{bmatrix} 11.690547 & -8.274010 \\ -8.274010 & 10.211685 \end{bmatrix}.$$

9.5 COMBINING WITHIN BLOCK AND BETWEEN BLOCK INFORMATION

The vector of parameters for Model 9.1 is $\theta = [\alpha_1, \alpha_2, \beta_1, \beta_2]'$. The within block model provides estimates of $\theta_1 = [\alpha_1 - \alpha_2, \beta_1, \beta_2]'$ and the between block model provides estimates of $\theta_2 = [\alpha_1 + \alpha_2, \beta_1, \beta_2]$. θ_1 and θ_2 are linear transforms of θ expressed as $\theta_1 = H_1\theta$ and $\theta_2 = H_2\theta$ where

$$H_1 = \begin{bmatrix} 1 & -1 & 0 & 0 \\ 0 & 0 & 1 & 0 \\ 0 & 0 & 0 & 1 \end{bmatrix} \text{ and } H_2 = \begin{bmatrix} 1 & 1 & 0 & 0 \\ 0 & 0 & 1 & 0 \\ 0 & 0 & 0 & 1 \end{bmatrix}.$$

The estimators can be expressed as beta-hat models (Chapter 6)

$$\hat{\theta}_1 = H_1\theta + e_1 \text{ where } e_1 \sim N(0, \Sigma_w)$$

and

$$\hat{\theta}_2 = H_2\theta + e_2 \text{ where } e_2 \sim N(0, \Sigma_b)$$

The two estimators are independently distributed; thus the joint model is

$$\begin{bmatrix} \hat{\theta}_1 \\ \hat{\theta}_2 \end{bmatrix} = \begin{bmatrix} \mathbf{H}_1 \\ \mathbf{H}_2 \end{bmatrix} \theta + \begin{bmatrix} \mathbf{e}_1 \\ \mathbf{e}_2 \end{bmatrix} \text{ where } \begin{bmatrix} \mathbf{e}_1 \\ \mathbf{e}_2 \end{bmatrix} \sim N\left[\begin{pmatrix} 0 \\ 0 \end{pmatrix}, \begin{pmatrix} \Sigma_w & 0 \\ 0 & \Sigma_b \end{pmatrix} \right].$$

If the variances are known, then the BLUE (Best Linear Unbiased Estimator) of θ is

$$\hat{\theta}_B = \left[\mathbf{H}_1' \Sigma_w^{-1} \mathbf{H}_1 + \mathbf{H}_2' \Sigma_b^{-1} \mathbf{H}_2 \right]^{-1} \left[\mathbf{H}_1' \Sigma_w^{-1} \hat{\theta}_1 + \mathbf{H}_2' \Sigma_b^{-1} \hat{\theta}_2 \right]$$

with sampling distribution

$$\hat{\theta}_B \sim N\left[\theta, \left(\mathbf{H}_1' \Sigma_w^{-1} \mathbf{H}_1 + \mathbf{H}_2' \Sigma_b^{-1} \mathbf{H}_2 \right)^{-1} \right]$$

or $\hat{\theta}_B \sim N(\theta, \Sigma_\theta)$.

The variance components are most likely unknown; thus the weighted least squares combined estimator of θ is

$$\hat{\theta}_B = \left[\mathbf{H}_1' \hat{\Sigma}_w^{-1} \mathbf{H}_1 + \mathbf{H}_2' \hat{\Sigma}_b^{-1} \mathbf{H}_2 \right]^{-1} \left[\mathbf{H}_1' \hat{\Sigma}_w^{-1} \hat{\theta}_1 + \mathbf{H}_2' \hat{\Sigma}_b^{-1} \hat{\theta}_2 \right]$$

with approximate estimated covariance matrix

$$\widehat{\mathrm{Var}(\hat{\theta})} = \left[\mathbf{H}_1' \hat{\Sigma}_w^{-1} \mathbf{H}_1 + \mathbf{H}_2' \hat{\Sigma}_b^{-1} \mathbf{H}_2 \right]^{-1} = \hat{\Sigma}_\theta.$$

$\hat{\theta}$ is the mixed models estimate of the parameters obtained when the above estimates of the variance components are used in place of the actual variance components (Littell et al., 1996). The combined estimate of θ for the data in Table 9.1 is

$$\hat{\theta} = \begin{bmatrix} \hat{\alpha}_1 \\ \hat{\alpha}_2 \\ \hat{\beta}_1 \\ \hat{\beta}_2 \end{bmatrix} = \begin{bmatrix} 11.94177 \\ 25.41262 \\ 15.55772 \\ 4.48663 \end{bmatrix}$$

with estimated covariance matrix

$$\hat{\Sigma}_\theta = \begin{bmatrix} 2.6379 & 0.7031 & -2.0748 & -0.5681 \\ 0.7031 & 2.1475 & -0.5467 & -1.7450 \\ -2.0748 & -0.5467 & 1.6733 & 0.4581 \\ -0.5681 & -1.7450 & 0.4581 & 1.4623 \end{bmatrix}$$

where $\hat{\theta}_1$, $\hat{\Sigma}_w$, $\hat{\theta}_2$, and $\hat{\Sigma}_b$ are from equations 9.2, 9.3, 9.4, and 9.5, respectively.

The combined estimate of the slopes should be used when the variance of the combined estimate is smaller than the variance of the within block estimates of the slopes (Ash, 1982). When the number of blocks is small, the between block information may not be useful. There needs to be one more block than treatments before a between block estimate can be computed and there needs to be more blocks in order to obtain a within block residual mean square with adequate degrees of freedom.

The within block model was used to obtain estimates of $\alpha_1 - \alpha_2$, β_1, and β_2 and the between block model was used to obtain estimates of $\alpha_1 + \alpha_2$, β_1, and β_2. Neither the within block model nor the between block model provides estimates of α_1 and α_2, but the combined estimator does provide estimates of all of the parameters, α_1, α_2, β_1, and β_2, where $\alpha_1 = (\alpha_t + \alpha_d)/2$ and $\alpha_2 = (\alpha_t - \alpha_d)/2$. For further discussion on intra-block models (within block), interblock models (between block), and the process of combining estimators, see John (1971) and Fergen (1997).

Once the estimate of the parameters of θ have been obtained and the covariance matrix has been estimated, it is generally of interest to estimate linear combinations of θ, such as $\mathbf{a}'\theta$. Some choices for $\mathbf{a}$ are to (1) provide estimates of the regression models evaluated at $X = X_0$ by letting $\mathbf{a}_1' = (1, 0, X_0, 0)$ and $\mathbf{a}_2' = (0, 1, 0, X_0)$ or (2) to provide estimates of the differences of the regression models evaluated at $X = X_0$ by letting $\mathbf{a}' = (1, -1, X_0, -X_0)$. The approximate sampling distribution of a linear combination of $\hat{\theta}$ is $\mathbf{a}'\hat{\theta} \sim N(\mathbf{a}'\theta, \mathbf{a}'\Sigma_\theta \mathbf{a})$.

9.6 DETERMINING THE FORM OF THE MODEL

Thus far the discussion about the model in Equation 9.1 assumes that the slopes are unequal. The next step in the analysis, after determining that straight lines are adequate to describe the data for each treatment, is to test the equality of slopes. Generally there is sufficient within block information to test the equal slopes hypothesis without considering the between block information, although, if there were many blocks, a test based on the combined estimate could be quite a bit more powerful. The model comparison method can be used to construct a statistic based on the within block information to test the equal slope hypothesis

$$H_0: \beta_1 = \beta_2 = \beta_0 \text{ vs. } H_a: (\text{not } H_0:)$$

where β_0 is unspecified.

The model under the conditions of the null hypothesis is

$$y_{ij} = \alpha_i + \beta_0 x_{ij} + b_j + \varepsilon_{ij}, \quad i = 1, 2, \quad j = 1, 2, \ldots, a. \tag{9.6}$$

Let RSS(H_0) denote the residual sum of squares for the model under the conditions of H_0: which is based on $(2-1)(a-1) - 1 = df_{RSS(H_0)}$ degrees of freedom, where "a" is the number of blocks in the experiment. Let RSS denote the residual sum of squares for the unrestricted Model 9.1, which is based on $(2-1)(a-1) - 2 = df_{RSS}$ degrees of freedom. The sum of squares due to deviations from the equal slope hypothesis is: $SSH_0 = RSS(H_0) - SSR$, which is based on $df_{SSH_0} = df_{RSS(H_0)} - df_{RSS} = 1$ degree of freedom. The test statistic is

$$F_c = \frac{SSH_0/df_{SSH_0}}{RSS/df_{RSS}},$$

which is distributed as F with df_{SSH_0} and df_{RSS} degrees of freedom. This statistic uses only the within block information available about the slopes.

When there are sufficient numbers of blocks, a statistic to test the equal slope hypothesis based on the combined estimate can be computed.

The estimates of the slopes, $\hat{\beta}_c$, from combining the within and between block information is

$$\hat{\beta}_c = H_\beta \hat{\theta} \text{ where } H_\beta = \begin{bmatrix} 0 & 0 & 1 & 0 \\ 0 & 0 & 0 & 1 \end{bmatrix}$$

which has sampling distribution

$$\hat{\beta}_c \sim N(\beta, \Sigma_c) \text{ where}$$

$$\Sigma_c = H_\beta \Sigma_\theta H_\beta'.$$

Under the equal slopes hypothesis, the beta-hat model for $\hat{\beta}_c$ is $\hat{\beta}_c = j\beta_0 + e_\beta$ where j is a 2×1 vector of ones, β_0 is an unknown parameter, and $e_\beta \sim N(0, \Sigma_c)$. The weighted residual mean square from the beta-hat model is the statistic to test the equal slope hypothesis and is computed as

$$u_c = \hat{\beta}_c' \left[\hat{\Sigma}_c^{-1} - \{\hat{\Sigma}_c^{-1} j j' \hat{\Sigma}_c^{-1}\} / \{j' \hat{\Sigma}_c^{-1} j\} \right] \hat{\beta}_c / (2-1),$$

where u_c is compared to the percentage points of the F distribution with 1 and df_{RSS} degrees of freedom (the large sample distribution of u_c is $X^2_{(1)}$, but the F distribution is used to approximate the small sample size distribution). If the null hypothesis is rejected, then there is enough evidence to believe that the slopes are unequal. If one fails to reject the null hypothesis, then an equal slopes models would be appropriate. For the herbicide example, the combined estimate of the slopes is

$$\hat{\beta}_c = \begin{bmatrix} \hat{\beta}_{1c} \\ \hat{\beta}_{2c} \end{bmatrix} = \begin{bmatrix} 15.557715 \\ 4.486631 \end{bmatrix}$$

with estimated covariance matrix

$$\hat{\Sigma}_c = \begin{bmatrix} 1.673255 & 0.458145 \\ 0.458145 & 1.462303 \end{bmatrix}$$

Using $\hat{\boldsymbol{\beta}}_c$ and $\hat{\boldsymbol{\Sigma}}_c$ the value of u_c becomes 55.229407 with approximate sampling distribution $F_{(1,3)}$. The corresponding significance level is .0050, indicating the slopes are not equal and a nonparallel lines model is needed to describe the data.

The estimated regression model for herbicide 1 evaluated at $X = X_0$ is

$$\hat{\mu}_{1|X_0} = \hat{\alpha}_{1c} + \hat{\beta}_{1c} X_0$$

and for herbicide 2 is

$$\hat{\mu}_{2|X_0} = \hat{\alpha}_{2c} + \hat{\beta}_{2c} X_0.$$

The variance of the estimate of the mean response evaluated at $X = X_0$ for herbicide 1 is

$$\mathrm{Var}(\hat{\mu}_{1|X_0}) = [1 \ 0 \ X_0 \ 0] \hat{\boldsymbol{\Sigma}}_\theta [1 \ 0 \ X_0 \ 0]'$$

and for herbicide 2 is

$$\mathrm{Var}(\hat{\mu}_{2|X_0}) = [0 \ 1 \ 0 \ X_0] \hat{\boldsymbol{\Sigma}}_\theta [0 \ 1 \ 0 \ X_0]'$$

The variance of the estimate of the difference between the mean responses of the two herbicides at $X = X_0$ is

$$\mathrm{Var}(\hat{\mu}_{1|X_0} - \hat{\mu}_{2|X_0}) = [1 \ -1 \ X_0 \ -X_0] \hat{\boldsymbol{\Sigma}}_\theta [1 \ -1 \ X_0 \ -X_0]'.$$

9.7 COMMON SLOPE MODEL

For the common slope model, there are two estimators for the slope, one from the within block information and one from the between block information (as is the case for the unequal slope model). The two treatment example is used to demonstrate the computation of the two types of estimators. If one assumes that the common slope model is appropriate to describe the herbicide data, then the data in Table 9.1 can be described by

$$y_{ij} = \alpha_i + \beta x_{ij} + b_j + \varepsilon_{ij}, \quad i = 1, 2, \quad j = 1, 2, \ldots, 6.$$

As in the unequal slope case, the method of carrying out the within block analysis is constructed by taking contrasts between the observations within the blocks and analyzing the resulting model, as in Section 9.3. Within each block, compute the contrast $d_j = y_{1j} - y_{2j}$. The model for the within block contrast is constructed by taking the same contrast of the models as

$$d_j = \alpha_1 + \beta x_{1j} + b_j + \varepsilon_{1j} - \left(\alpha_2 + \beta x_{2j} + b_j + \varepsilon_{2j}\right)$$

$$= \alpha_1 + \beta x_{1j} + \varepsilon_{1j} - \alpha_2 - \beta x_{2j} - \varepsilon_{2j}$$

$$= \alpha_1 - \alpha_2 + \beta \left[x_{1j} - x_{2j}\right] + \varepsilon_{1j} - \varepsilon_{2j}$$

The model for the differences is free of the block effects and has variance $2\sigma_\varepsilon^2$. By letting $\alpha_d = \alpha_1 - \alpha_2$ and $e_j = \varepsilon_{1j} - \varepsilon_{2j}$, then the model for the differences can be expressed as the simple linear regression model

$$d_j = \alpha_d + \beta z_j + e_j,$$

where the independent variable z_j is the difference between the two covariates, i.e., $z_j = x_{1j} - x_{2j}$. On fitting the simple linear regression model to the data in Table 9.1, one obtains estimates of the estimable functions of the parameters of the original model, i.e., $\alpha_1 - \alpha_2$, β, and σ_ε^2. The matrix form of the model for the differences is

$$\begin{bmatrix} -3.60 \\ 1.90 \\ 2.60 \\ 2.50 \\ 2.00 \\ -2.50 \end{bmatrix} = \begin{bmatrix} 1 & -0.11 \\ 1 & 0.33 \\ 1 & 0.04 \\ 1 & -0.02 \\ 1 & 0.06 \\ 1 & -0.02 \end{bmatrix} \begin{bmatrix} \alpha_d \\ \beta \end{bmatrix} + \mathbf{e} \text{ or } \mathbf{d} = \mathbf{Z}_0 \mathbf{\eta}_0 + \mathbf{e}$$

The within block estimates of the parameters are

$$\hat{\mathbf{\eta}} = \begin{bmatrix} \hat{\alpha}_d \\ \hat{\beta}_w \end{bmatrix} = \left(\mathbf{Z}_0'\mathbf{Z}_0\right)^{-1}\mathbf{Z}_0'\mathbf{d} = \begin{bmatrix} .0119 \\ 9.7455 \end{bmatrix}.$$

The residual sum of squares from the within block model is 27.313 based on 4 degrees of freedom. The estimate of $2\sigma_\varepsilon^2$ is $27.313/4 = 6.828$. The estimate of the covariance matrix of the within block estimates is

$$2\hat{\sigma}_\varepsilon^2\left(\mathbf{Z}_0'\mathbf{Z}_0\right)^{-1} = 6.828 \begin{bmatrix} .1859 & -.4099 \\ -.4099 & 8.7771 \end{bmatrix} = \hat{\Sigma}_w$$

The between block analysis utilizes the block totals and provides information about the parameters contained in those block totals. Let $t_j = y_{1j} + y_{2j}$ denote the total of the two observations in the jth block. The block total model is constructed by taking the totals of the corresponding models, as

Two Treatments in a Randomized Complete Block Design Structure

$$t_j = \alpha_1 + \beta\, x_{1j} + b_j + \varepsilon_{1j} + \alpha_2 + \beta\, x_{2j} + b_j + \varepsilon_{2j}$$

$$= \alpha_1 + \alpha_2 + \beta\left[x_{1j} + x_{2j}\right] + 2b_j + \varepsilon_{1j} + \varepsilon_{2j}$$

which can be expressed as the following simple linear regression model

$$t_j = \alpha_t + \beta w_j + r_j,$$

where $\alpha_t = \alpha_1 + \alpha_2$, $w_j = x_{1j} + x_{2j}$, and $r_j = 2b_j + \varepsilon_{1j} + \varepsilon_{2j}$. The variance of these block totals is $2(\sigma_\varepsilon^2 + 2\sigma_b^2)$. A linear regression program can be used to fit the above model to the block totals to obtain the between block estimates of α_t, β, and $2(\sigma_\varepsilon^2 + 2\sigma_b^2)$.

The matrix form for the block total model is

$$\begin{bmatrix} 56.8 \\ 60.3 \\ 66.8 \\ 66.3 \\ 62.3 \\ 59.5 \end{bmatrix} = \begin{bmatrix} 1 & 1.93 \\ 1 & 2.11 \\ 1 & 2.82 \\ 1 & 2.92 \\ 1 & 2.60 \\ 1 & 2.22 \end{bmatrix} \begin{bmatrix} \alpha_t \\ \beta_b \end{bmatrix} + \mathbf{r} \text{ or } \mathbf{t} = \mathbf{M}_1\, \mathbf{u} + \mathbf{r}$$

The between block estimates of the parameters are

$$\hat{\mathbf{u}} = \begin{bmatrix} \hat{\alpha}_t \\ \hat{\beta}_b \end{bmatrix} = \begin{bmatrix} 39.08 \\ 9.42 \end{bmatrix}.$$

The residual sum of squares from the between block model is 5.29, which is based on 4 degrees of freedom. The between block residual mean square is $5.29/4 = 1.324$, which is an estimate of $2(\sigma_\varepsilon^2 + 2\sigma_b^2)$. The estimated covariance matrix of the between block estimate is

$$\widehat{\left(2(\sigma_\varepsilon^2 + 2\sigma_b^2)\right)}(\mathbf{M}_1'\mathbf{M}_1)^{-1} = 1.324 \begin{bmatrix} 7.4093 & -2.9764 \\ -2.9764 & 1.2232 \end{bmatrix} = \hat{\Sigma}_b.$$

The between and within block information can be combined as

$$\hat{\boldsymbol{\theta}} = \left[\mathbf{H}_1'\hat{\boldsymbol{\Sigma}}_w\mathbf{H}_1 + \mathbf{H}_2'\hat{\boldsymbol{\Sigma}}_b\mathbf{H}_2\right]^{-1}\left[\mathbf{H}_1'\hat{\boldsymbol{\Sigma}}_w\hat{\boldsymbol{\theta}}_1 + \mathbf{H}_2'\hat{\boldsymbol{\Sigma}}_b\hat{\boldsymbol{\theta}}_2\right]$$

where the beta-hat models for $\hat{\boldsymbol{\theta}}_1$ and $\hat{\boldsymbol{\theta}}_2$ are

$$\hat{\boldsymbol{\theta}}_1 = \mathbf{H}_1\boldsymbol{\theta} + \mathbf{e}_1, \quad \hat{\boldsymbol{\theta}}_2 = \mathbf{H}_2\boldsymbol{\theta} + \mathbf{e}_2$$

$$\mathbf{e}_1 \sim N(\mathbf{0}, \boldsymbol{\Sigma}_w), \quad \mathbf{e}_2 \sim N(\mathbf{0}, \boldsymbol{\Sigma}_b)$$

$$\mathbf{H}_1 = \begin{bmatrix} 1 & -1 & 0 \\ 0 & 0 & 1 \end{bmatrix}, \quad \mathbf{H}_2 = \begin{bmatrix} 1 & 1 & 0 \\ 0 & 0 & 1 \end{bmatrix}$$

$$\boldsymbol{\theta} = (\alpha_1, \alpha_2, \beta)'.$$

The combined estimator of $\boldsymbol{\theta}$ is

$$\hat{\boldsymbol{\theta}} = \begin{bmatrix} 19.5429 \\ 19.5162 \\ 9.4286 \end{bmatrix}$$

The estimate of the approximate covariance matrix of is $\hat{\boldsymbol{\theta}}$

$$\hat{\boldsymbol{\Sigma}}_\theta = \left[\mathbf{H}_1' \hat{\boldsymbol{\Sigma}}_w \mathbf{H}_1 + \mathbf{H}_2' \hat{\boldsymbol{\Sigma}}_b \mathbf{H}_2\right]^{-1}$$

$$= \begin{bmatrix} 2.7645 & 2.1039 & -1.9553 \\ 2.1039 & 2.5853 & -1.8817 \\ -1.9553 & -1.8817 & 1.5769 \end{bmatrix}.$$

The estimated regression model for herbicide 1 is

$$\hat{\mu}_{1|x_0} = \hat{\alpha}_{1c} + \hat{\beta}_c x_0$$

and the estimated regression model for herbicide 2 is

$$\hat{\mu}_{2|x_0} = \hat{\alpha}_{2c} + \hat{\beta}_c x_0.$$

The variances of and are $\hat{\mu}_{1|x_0}$ and $\hat{\mu}_{2|x_0}$ are $\text{Var}(\hat{\mu}_{1|x_0}) = [1 \; 0 \; X_0]\hat{\boldsymbol{\Sigma}}_{\hat{\theta}}[1 \; 0 \; X_0]'$ and $\text{Var}(\hat{\mu}_{2|x_0}) = [0 \; 1 \; X_0]\hat{\boldsymbol{\Sigma}}_{\hat{\theta}}[0 \; 1 \; X_0]'$, respectively.

9.8 COMPARING THE TREATMENTS

As discussed in Chapter 2, the method used to compare the treatments depends on the form of the model: either use the equal slopes models procedure or use the unequal slopes models procedure.

9.8.1 Equal Slopes Models

When the equal slopes models are adequate to describe the data, the comparison between treatments is not effected by the magnitude of the covariate. The models describing the treatment means as a function of x_0 are two parallel lines. Thus, the distance between the two lines is the same for all values of x_0. The estimate of the distance between the two lines at $x = x_0$ is

$$\hat{\mu}_{1|x_0} - \hat{\mu}_{2|x_0} = \hat{\alpha}_1 + \hat{\beta} x_0 - \left[\hat{\alpha}_2 + \hat{\beta} x_0\right] = \hat{\alpha}_1 - \hat{\alpha}_2,$$

where $\hat{\mu}_{1|x_0}$ and $\hat{\mu}_{2|x_0}$ denote the estimated mean values of the models evaluated at $x = x_0$ or adjusted means for Treatments 1 and 2, respectively. One could use the within block information about $\hat{\alpha}_1 - \hat{\alpha}_2$ and the corresponding adjusted means and standard errors for the difference of two adjusted means to make comparisons between the lines. The estimate of $\alpha_1 - \alpha_2$ is obtained from the within block information; thus the comparisons of adjusted means are based only on the within block information, but the adjusted means depend on the estimate of the slope. The combined estimator should provide better estimates of the adjusted means, $\hat{\mu}_{1|x_0}$ and $\hat{\mu}_{2|x_0}$. If the number of blocks is large, there should be an advantage to using the combined estimators.

9.8.2 Unequal Slopes Model

When the unequal slopes model is used to describe the data, the comparison between treatments is affected by the magnitude of the covariate. The models describing the treatment means as a function of x are no longer a series of parallel lines. Thus, the distance between any two of the lines depends on the values of x. The estimate of the distance between the two lines at $x = x_0$ is

$$\hat{\mu}_{1|x_0} - \hat{\mu}_{2|x_0} = \hat{\alpha}_1 + \hat{\beta}_1 x_0 - \left[\hat{\alpha}_2 + \hat{\beta}_2 X_0\right] = \hat{\alpha}_1 - \hat{\alpha}_2 + \left[\hat{\beta}_1 - \hat{\beta}_2\right] X_0$$

where $\hat{\mu}_{1|x_0}$ and $\hat{\mu}_{2|x_0}$ denote the estimated mean values of the models evaluated at $x = x_0$ or adjusted means for Treatments 1 and 2, respectively. For the data in Table 9.1, the estimated difference between the two treatments' adjusted means at OM = 0.9 using the combined information is $(11.9418 - 25.4126) + (15.5577 - 4.4866)0.9 = -3.5069$ with an estimated standard error of 0.5547.

9.9 CONFIDENCE INTERVALS ABOUT DIFFERENCES OF TWO REGRESSION LINES

Confidence intervals about the differences of the regression models can be constructed from either the within block information or from the combined within block between block information. Both the within block model and the combined information provide estimates of $\alpha_1 - \alpha_2$, β_1, and β_2, the parameters needed to estimate the differences between regression models.

9.9.1 WITHIN BLOCK ANALYSIS

When there is insufficient data to do the combined analysis, a confidence interval about $E[\hat{\mu}_{1|x_0} - \hat{\mu}_{2|x_0}]$ can be constructed from the within block information. The estimate of the variance of the within block estimate of $\hat{\mu}_{1|x_0} - \hat{\mu}_{2|x_0}$ is

$$\hat{\sigma}^2_{\mu 12_w} = [1 \ x_0 - x_0] \hat{\Sigma}_w [1 \ x_0 - x_0]'.$$

A $(1 - \alpha)100\%$ confidence interval about $E[\hat{\mu}_{1|x_0} - \hat{\mu}_{2|x_0}]$, the difference between the two regression lines at x = x_0, is

$$\hat{\mu}_{1|x_0} = \hat{\mu}_{2|x_0} \pm \{t_{\alpha/2, m}\} \hat{\sigma}_{\mu 12_w}$$

where m is the degrees of freedom for the within block sum of squares residuals. The variance for comparing the two treatments at OM = 0.9 is $(.580)^2$ and the 95% confidence interval is $-3.69 \pm \{t_{.025,\ 3}\}(.580)$.

9.9.2 COMBINED WITHIN BLOCK AND BETWEEN BLOCK ANALYSIS

When there is sufficient information to combine the estimates of the slopes, a confidence interval can be constructed using the combined information. The difference between the two models at x = x_0 is a linear combination of θ as

$$\mu_{1|x_0} - \mu_{2|x_0} = (\alpha_1 - \alpha_2) + (\beta_1 - \beta_2)X_0$$

$$= [1 \ -1 \ X_0 \ -X_0]\theta = \mathbf{a}'_{x_0}\theta$$

The estimate of the difference between the two regression lines or models or adjusted means at x = x_0 is

$$\hat{\mu}_{1|x_0} - \hat{\mu}_{2|x_0} = \mathbf{a}'_{x_0}\hat{\theta}_c$$

with estimated variance

$$\text{Var}(\hat{\mu}_{1|x_0} - \hat{\mu}_{2|x_0}) = \mathbf{a}'_{x_0}\hat{\Sigma}_\theta \mathbf{a}_{x_0}.$$

An approximate $(1 - \alpha)100\%$ confidence interval about $\hat{\mu}_{1|x_0} - \hat{\mu}_{2|x_0}$ is

$$\hat{\mu}_{1|x_0} - \hat{\mu}_{2|x_0} \pm \{t_{\alpha/2, m}\} \sqrt{\text{Var}(\hat{\mu}_{1|x_0} - \hat{\mu}_{2|x_0})}$$

where m is selected as appropriate number of degrees of freedom. One could use m = df_w, which will in general be too liberal, i.e., df_w is too large. There are several approximations to the number of degrees of freedom than can be used to enable the above confidence interval to have better coverage properties. Two approximations used by PROC MIXED are Satterthwaite and Kenward-Roger (SAS Institute Inc., 2000).

An approximation to an appropriate value for $t_{\alpha/2,m}$ can also be obtained from a weighted average of $t_{\alpha/2,df_w}$ and $t_{\alpha/2,df_b}$ similar to that which was done for split-plot type designs (Milliken and Johnson, 1992). The weighted average is obtained by computing $\hat{\Sigma}_\theta^*$ as

$$\hat{\Sigma}_\theta^* = \left[\left\{ t_{\alpha/2,df_w}^2 \right\} \mathbf{H}_1 \hat{\Sigma}_w \mathbf{H}_1 + \left\{ t_{\alpha/2,df_b}^2 \right\} \mathbf{H}_2' \hat{\Sigma}_b \mathbf{H}_2 \right]^{-1}$$

and then computing $d^* = \mathbf{a}_{X_0}' \hat{\Sigma}_\theta^* \mathbf{a}_{X_0}$. The approximate t value is

$$t_{\alpha/2,m} = d^* \big/ \mathbf{a}_{X_0}' \hat{\Sigma}_\theta \mathbf{a}_{X_0}$$

The value of the number of degrees of freedom, m, can be determined by comparing the value of $t_{\alpha/2,m}$ to the tabulated t values for $\alpha/2$. For the examples in this chapter, $df_b = df_w$, thus $t_{\alpha/2,m} = t_{\alpha/2,df_w} = t_{\alpha/2,df_b}$ or $m = df_w = df_b$.

If simultaneous confidence intervals are desired as in the construction of a confidence band about the difference of the two models for the values of X_0 in an interval, the value of $t_{\alpha/2,m}$ can be replaced by $(2 F_{\alpha/2,2,m})^{1/2}$ using the approximate degrees of freedom from above or replace $t_{\alpha/2,df_w}$ and $t_{\alpha/2,df_b}$ by $(2 F_{\alpha/2,2,df_w})^{1/2}$ and $(2 F_{\alpha/2,2,df_b})^{1/2}$, respectively, to compute an approximate percentage point for $(2 F_{\alpha/2,2,m})^{1/2}$.

9.10 COMPUTATIONS FOR MODEL 9.1 USING THE SAS® SYSTEM

The computations described in the previous sections can be accomplished by using PROC GLM to fit the within and between block models and then using either PROC IML or PROC MIXED to finish computing the results. Table 9.2 contains the herbicide data set where there is code to read the input data set and perform the preliminary computations to be used in the modeling process. The variables y1 and om1 are the yield and organic matter for herbicide 1 and y2 and om2 are the yield and organic matter for herbicide 2. Dif_y and sum_y are the difference and sum of the two yields within each block and negom2 is the negative of the organic matter for herbicide 2.

The within block model is fit using the PROC GLM code in Table 9.3. The difference of the two yields is the dependent measure and om1 and negom2 are the independent variables. The mean square error from the third part of Table 9.3 provides an estimate of $2\sigma_\epsilon^2$. The second part of Table 9.3 contains the elements of the inverse of $\mathbf{Z}'\mathbf{Z}$ which are used to compute $\hat{\Sigma}_w$. The type III sums of squares are provided for om1 and negom2. Finally, the within block model parameter estimates are in the lower part of Table 9.3.

The PROC GLM code in Table 9.4 fits the between block model where the dependent variable is sum_y and the independent variables are om1 and om2. The

TABLE 9.2
Herbicide Data Set and Computation of Preliminary Variables to be Used by the Models

```
data raw; input block y1 om1 y2 om2;
dif_y=y1-y2; sum_y=y1+y2;
negom2=-om2;
```

block	y1	om1	y2	om2	dif_y	sum_y	negom2
1	26.6	0.91	30.2	1.02	−3.6	56.8	−1.02
2	31.1	1.22	29.2	0.89	1.9	60.3	−0.89
3	34.7	1.43	32.1	1.39	2.6	66.8	−1.39
4	34.4	1.45	31.9	1.47	2.5	66.3	−1.47
5	32.1	1.33	30.2	1.27	1.9	62.3	−1.27
6	28.5	1.10	31.0	1.12	−2.5	59.5	−1.12

TABLE 9.3
PROC GLM Code to Fit the Within Block Model to the Herbicide Data Set with Results

```
PROC GLM data=raw;
MODEL DIF_Y=OM1 NEGOM2/SOLUTION INVERSE;
*--model for differences;
```

Parameter	Intercept	om1	negom2		
Intercept	7.5830	−4.2652	1.7828		
om1	−4.2652	10.7867	7.6343		
om2	1.7828	7.6343	9.4268		

Source	df	SS	MS	FValue	ProbF
Model	2	36.6959	18.3479	38.29	0.0073
Error	3	1.4374	0.4791		
Corr Total	5	38.1333			

Source	df	SS(III)	MS	FValue	ProbF
om1	1	26.6556	26.6556	55.63	0.0050
negom2	1	3.3805	3.3805	7.06	0.0766

Parameter	Estimate	StdErr	tValue	Probt
Intercept	−13.8230	1.9061	−7.25	0.0054
om1	16.9566	2.2734	7.46	0.0050
negom2	5.6451	2.1253	2.66	0.0766

second part of the table contains the inverse of $\mathbf{M'M}$ and the third part contains the estimate of 2 $(\sigma_\varepsilon^2 + 2\sigma_b^2)$, which when multiplied together provide $\hat{\Sigma}_b$. The type III sums of squares for om1 and om2 are in the fourth part of Table 9.4 and the between block model parameter estimates are in the bottom part of the table. The final step is to combine the within and between block estimates. This can be accomplished in

TABLE 9.4
PROC GLM Code to Fit the Between Block Model to the Herbicide Data with Results

```
PROC GLM data=raw;
MODEL SUM_Y=OM1 OM2/SOLUTION INVERSE; *--model for block sums;
```

Parameter	Intercept	om1	om2
Intercept	7.583005543	−4.26524796	−1.78276375
om1	−4.26524796	10.78674209	−7.63433986
om2	−1.78276375	−7.63433986	9.426825569

Source	df	SS	MS	FValue	ProbF
Model	2	74.5486	37.2743	34.39	0.0085
Error	3	3.2514	1.0838		
Corr Total	5	77.8000			

Source	df	SS(III)	MS	FValue	ProbF
om1	1	17.7553	17.7553	16.38	0.0272
om2	1	3.0046	3.0046	2.77	0.1945

Parameter	Estimate	StdErr	tValue	Probt
Intercept	38.4885	2.8668	13.43	0.0009
om1	13.8391	3.4191	4.05	0.0272
om2	5.3220	3.1964	1.67	0.1945

at least two ways. The first is to use a matrix language such as Proc IML to complete the computations. Table 9.14 contains the necessary PROC IML code to finish the computations. The other method is to use PROC MIXED and specify the estimates of the variance components obtained from the between and within block analyses. As will be discussed in Chapter 13, the mixed models estimator is a combined estimator and hence is used here to obtain the combined within and between block estimates. Table 9.5 contains the PROC MIXED code necessary to fit the model. The fixed effects are herb and om*herb and the random effect is block. The parms statement contains the estimates of the two variance components, $\hat{\sigma}_\varepsilon^2 = 0.239573625$ and $\hat{\sigma}_b^2 = 0.151160305$. The option "hold=1,2" causes PROC MIXED to use those values as the estimates of the variance components for the rest of the analysis.

The fourth part of Table 9.5 contains the combined estimates of the models parameters as described in Section 9.5. The bottom part of Table 9.5 contains the test of the equal slopes hypothesis, which was generated because of the presence of the contrast statement. The combined estimates of the model's parameters depend on the values of the estimates of the variance components. PROC MIXED has several methods of estimating the variance components and when the data set involves unbalance or covariates or both, each method can have a unique set of estimates of the variances (see Chapter 13). Table 9.6 contains the PROC MIXED code to fit the model where the type III mean squares and expected mean squares are used to estimate the variance components. The estimates of the variance components are $\hat{\sigma}_\varepsilon^2 = 0.2396$ and $\hat{\sigma}_b^2 = 0.0820$, i.e., the block variance component is a little smaller

TABLE 9.5
PROC MIXED Code to Provide the Combined Estimates of the Models Parameters Using the Estimates of the Variance Components from the Within and Between Block Analyses

```
proc mixed data=long;
class herb block;
model y=herb om*herb/solution noint;
contrast 'b1=b2' om*herb 1 -1;
random block;
parms (.151160305) (.239573625)/hold=1,2;
```

CovParm	Estimate
block	0.1512
Residual	0.2396

Effect	NumDF	DenDF	FValue	ProbF
herb	2	3	153.08	0.0010
om*herb	2	3	72.35	0.0029

Effect	herb	Estimate	StdErr	df	tValue	Probt
herb	1	11.9418	1.6242	3	7.35	0.0052
herb	2	25.4126	1.4654	3	17.34	0.0004
om*herb	1	15.5577	1.2935	3	12.03	0.0012
om*herb	2	4.4866	1.2093	3	3.71	0.0340

Label	NumDF	DenDF	FValue	ProbF
b1=b2	1	3	55.23	0.0050

than the estimate obtained in Section 9.4. The estimate of the parameters are somewhat different from those in Table 9.5, but the estimated standard errors are smaller in Table 9.6 than in Table 9.5. They are smaller because the estimate of the block variance component is smaller.

Table 9.7 contains the PROC MIXED code to fit the model where the REML method is are used to estimate the variance components. The estimates of the variance components are $\hat{\sigma}_\varepsilon^2 = 0.128$ and $\hat{\sigma}_b^2 = 0.1065$, i.e., the block variance component is a little larger and the within block variance component is a little smaller than the estimates obtained in Table 9.6. The estimates of the parameters are somewhat different from those in Table 9.6, but the estimated standard errors are smaller in Table 9.7 than in Table 9.6. They are smaller because the estimate of the within block variance component is smaller.

The LSMEANS statements in Table 9.7 were included to provide estimates of the regression models for a collection of values of OM as well as the differences of the models. The "cl" option was included to provide confidence intervals about the models as well as about the differences of the models. The results are not shown, but Figure 9.1 contains the graph of the models with nonsimultaneous 95% confidence bands. Figure 9.2 is a graph of the difference of the two models with a nonsimultaneous confidence band. The PROC IML code in Table 9.15 can be used to estimate the regression models and construct the confidence bands.

TABLE 9.6
PROC MIXED Code to Use Type3 Estimates of the Variance Components in the Combining of Within and Between Block Information

```
proc mixed data=long method=type3;
class herb block;
model y=herb om*herb/solution noint;
contrast 'b1=b2' om*herb 1 -1;
random block;
```

CovParm	Estimate
block	0.0820
Residual	0.2396

Effect	NumDF	DenDF	FValue	ProbF
herb	2	3	186.08	0.0007
om*herb	2	3	84.09	0.0023

Effect	herb	Estimate	StdErr	df	tValue	Probt
herb	1	12.0791	1.5048	3	8.03	0.0040
herb	2	25.4797	1.3575	3	18.77	0.0003
om*herb	1	15.4470	1.1991	3	12.88	0.0010
om*herb	2	4.4304	1.1209	3	3.95	0.0289

Label	NumDF	DenDF	FValue	ProbF
b1=b2	1	3	55.79	0.0050

9.11 EXAMPLE: EFFECT OF DRUGS ON HEART RATE

A physician conducted an experiment to determine which of two drugs was more effective in lowering a person's heart rate. A person was to walk on a treadmill at 2.5 mi/hr for 10 min to stabilize the person's heart rate. At the 10-min mark, one of two drugs was injected into the bloodstream and the person continued to walk for 5 more minutes. After 5 min, the person's heart rate was determined. Since all people did not have the same stabilized heart rate before the injection, the stabilized or initial heart rate was used as a covariate. (If the physician would use change from base line or initial heart rate as the measurement for analysis, the physician would be assuming the lines of the two drugs are parallel with slope equal to one. In this case the physician wanted the model to provide information about the form of the slopes, as discussed in Section 3.5). A second restriction was placed on the operation of the experiment. Two identical treadmills were used where two people were run through the process together. Thus 40 people were randomly split into 20 pairs. A pair was selected at random and the two persons were randomly assigned to a treadmill. Then the two drugs were randomly assigned to the two persons within each pair, where the pairs form the blocks.

The data from 40 human females, age 18 to 24, with body fat less than 24% are in Table 9.8 where HR1 and IHR1 are the heart rates and initial heart rates for those

TABLE 9.7
PROC MIXED Code to Use REML Estimates of the Variance Components to Provide Combined Within and Between Block Estimates of the Models Parameters with Results

```
proc mixed data=long method=reml;
class herb block;
model y=herb om*herb/solution noint;
contrast 'b1=b2' om*herb 1 -1;
random block;
lsmeans herb/diff at om=0.9 cl; lsmeans herb/diff at om=1.0 cl;
lsmeans herb/diff at om=1.1 cl; lsmeans herb/diff at om=1.2 cl;
lsmeans herb/diff at om=1.3 cl; lsmeans herb/diff at om=1.4 cl;
lsmeans herb/diff at om=1.5 cl;
```

CovParm	Estimate
block	0.1065
Residual	0.2128

Effect	NumDF	DenDF	FValue	ProbF
herb	2	3	186.47	0.0007
om*herb	2	3	86.47	0.0022

Effect	herb	Estimate	StdErr	df	tValue	Probt
herb	1	12.0000	1.4829	3	8.09	0.0039
herb	2	25.4433	1.3379	3	19.02	0.0003
om*herb	1	15.5107	1.1813	3	13.13	0.0010
om*herb	2	4.4609	1.1044	3	4.04	0.0273

Label	NumDF	DenDF	FValue	ProbF
b1=b2	1	3	62.41	0.0042

assigned drug 1 and HR2 and IHR2 are the heart rates and initial heart rates for those assigned to drug 2.

The variable dif_HR is the difference of the heart rates for drug 1 and drug 2 and the variable dif_IHR is the same difference for the initial heart rates. The variables sum_HR and sum_IHR are the sums of the respective values within each block and negIHR2 is the negative of IHR2. The main conclusion from the analysis of this data set will be that there is not sufficient evidence to conclude the slopes are unequal. This part of the analysis is accomplished by using PROC MIXED. Table 9.9 contains the PROC MIXED code to fit the unequal slopes models to the data set. The contrast statement is used to extract the statistic to test the equal slopes hypothesis. The last part of Table 9.9 contains the equal slopes test statistic, which has a significance level of 0.8643, indicating there is not sufficient evidence to conclude the slopes are unequal.

With that step out of the way, the next step is to fit a common slopes model. The PROC GLM code in Table 9.10 fits the within block common slope model of Section 9.7. The estimate of the within block variance component is $\hat{\sigma}_\varepsilon^2 = 12.5364/2 = 6.2682$. The second part of Table 9.10 is the inverse of $\mathbf{Z}_0'\mathbf{Z}_0$, which is used to

Two Treatments in a Randomized Complete Block Design Structure 223

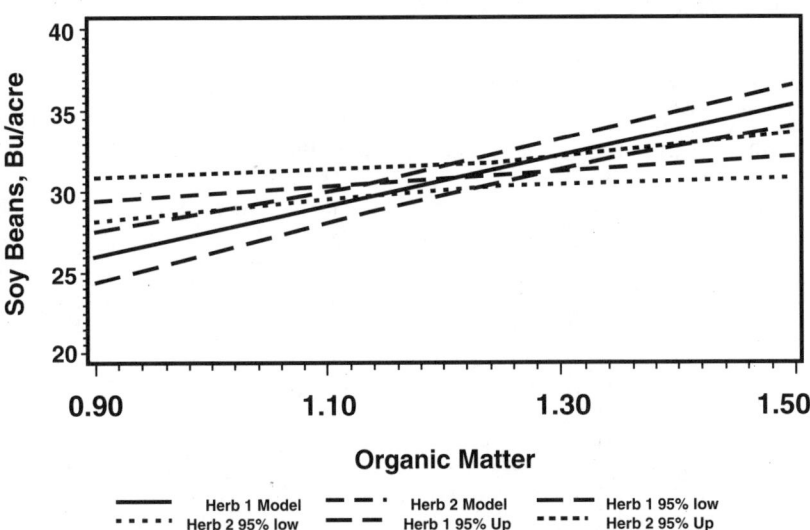

FIGURE 9.1 Plot of estimated herbicide models with nonsimultaneous 95% confidence bands.

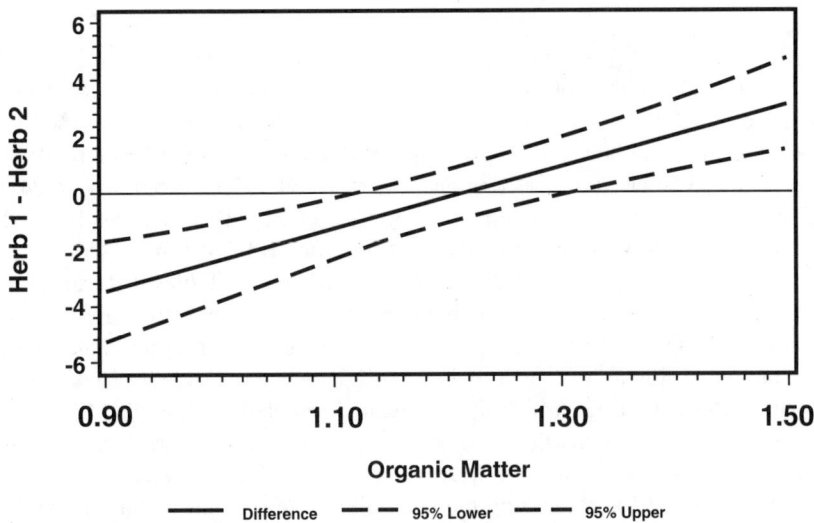

FIGURE 9.2 Graph of the difference of the two herbicide models with nonsimultaneous 95% confidence band.

TABLE 9.8
Data for Effect of Drugs of Heart Rate

```
data heart; input pair HR1 IHR1 HR2 IHR2;
dif_HR=HR1-HR2; sum_HR=HR1+HR2;dif_IHR=IHR1-IHR2; Sum_IHR=IHR1+IHR2;
negIHR2=-IHR2;
```

pair	HR1	IHR1	HR2	IHR2	dif_HR	sum_HR	dif_IHR	Sum_IHR	negIHR2
1	60	85	56	78	4	116	7	163	−78
2	45	67	69	94	−24	114	−27	161	−94
3	73	87	60	68	13	133	19	155	−68
4	63	83	60	76	3	123	7	159	−76
5	66	87	76	89	−10	142	−2	176	−89
6	51	67	69	83	−18	120	−16	150	−83
7	77	99	64	80	13	141	19	179	−80
8	80	105	57	67	23	137	38	172	−67
9	69	87	86	104	−17	155	−17	191	−104
10	53	67	71	81	−18	124	−14	148	−81
11	58	68	62	72	−4	120	−4	140	−72
12	55	68	79	90	−24	134	−22	158	−90
13	52	77	69	87	−17	121	−10	164	−87
14	86	104	78	90	8	164	14	194	−90
15	60	77	65	84	−5	125	−7	161	−84
16	59	68	94	105	−35	153	−37	173	−105
17	53	75	63	74	−10	116	1	149	−74
18	74	97	62	85	12	136	12	182	−85
19	64	83	76	97	−12	140	−14	180	−97
20	66	92	82	103	−16	148	−11	195	−103

compute $\hat{\Sigma}_w$. The within block estimate of the common slope is 0.8245 and the estimate of the difference of the intercepts is −4.0617. Table 9.11 contains the PROC GLM code to fit the between block common slope model to the block totals. The estimate of the block variance component is $\hat{\sigma}_b^2 = (69.5030/2 − \hat{\sigma}_\varepsilon^2)/2 = 14.24166$. The second part of the table contains the inverse of $\mathbf{M}_1' \mathbf{M}_1$, which can be used to compute $\hat{\Sigma}_b$. The between block estimate of the slope is 0.7465 and the estimate of the sum of the intercepts is 8.0531. The PROC IML code in Tables 9.16 can be used to compute the combined within and between block parameter estimates. Table 9.12 contains the PROC MIXED code to carry out the combined analysis of the data set. The combined estimate of the slope is 0.8148. Part five of Table 9.12 contains the estimated adjusted means, which have estimated standard error of 1.0149. The last part of Table 9.12 contains the estimated difference of the two regression models (they are parallel) with estimate difference of −4.0925 and estimated standard error of 0.8031. The test for comparing the two adjusted means has a significance level of 0.0001, indicating the two drugs are different.

Finally, Table 9.13 contains the PROC MIXED code to use REML to estimate the variance components. The estimates are $\hat{\sigma}_\varepsilon^2 = 6.2414$ and $\hat{\sigma}_b^2 = 13.7478$. The

TABLE 9.9
PROC MIXED Code to Fit the Unequal Slopes Model to the Drug Data Set with Results

```
proc mixed data=long method=reml;
class drug pair;
model hr=drug ihr*drug/solution noint;
contrast 'b1=b2' ihr*drug 1 -1;
random pair;
```

CovParm	Estimate
pair	13.8550
Residual	6.4962

Effect	NumDF	DenDF	FValue	ProbF
drug	2	17	0.38	0.6904
ihr*drug	2	17	182.02	0.0000

Effect	drug	Estimate	StdErr	df	tValue	Probt
drug	1	−4.2987	5.0154	17	−0.86	0.4033
drug	2	1.1552	5.7988	17	0.20	0.8445
ihr*drug	1	0.8217	0.0598	17	13.74	0.0000
ihr*drug	2	0.8054	0.0669	17	12.04	0.0000

Label	NumDF	DenDF	FValue	ProbF
b1=b2	1	17	0.03	0.8643

TABLE 9.10
PROC GLM Code to Fit the Within Block Common Slopes Model to the Heart Rate Data with Results

```
PROC GLM data=heart;
MODEL DIF_hr=dif_IHR/SOLUTION INVERSE;*--model
  for differences;
```

Parameter	Intercept	dif_IHR
Intercept	0.0516	0.0005
dif_IHR	0.0005	0.0002

Source	df	SS	MS	FValue
Model	1	4220.5451	4220.5451	336.66
Error	18	225.6549	12.5364	
Corr Total	19	4446.2000		

Source	df	SS (III)	MS	FValue	ProbF
dif_IHR	1	4220.5451	4220.5451	336.66	0.0000

Parameter	Estimate	StdErr	tValue	Probt
Intercept	−4.0617	0.8047	−5.05	0.0001
dif_IHR	0.8245	0.0449	18.35	0.0000

TABLE 9.11
PROC GLM Code to Fit the Between Common Slopes Model to the Heart Rate Data with Results

```
PROC GLM data=heart;
MODEL SUM_hr=sum_ihr/SOLUTION INVERSE;*--model for pair sums;
```

Parameter	Intercept	Sum_IHR			
Intercept	5.8312	−0.0345			
Sum_IHR	−0.0345	0.0002			

Source	df	SS	MS	FValue	ProbF
Model	1	2704.7453	2704.7453	38.92	0.0000
Error	18	1251.0547	69.5030		
Corr Total	19	3955.8000			

Source	df	SS (III)	MS	FValue	ProbF
Sum_IHR	1	2704.7453	2704.7453	38.92	0.0000

Parameter	Estimate	StdErr	tValue	Probt
Intercept	8.0531	20.1317	0.40	0.6938
Sum_IHR	0.7465	0.1197	6.24	0.0000

combined estimate of the slope is 0.8146. The fifth part of Table 9.13 contains the estimated adjusted means, which are very similar to those in Table 9.12, but the estimates of the standard deviations are a little smaller when computed using the REML estimates of the variance components. Again the combined estimates of the model's parameters depend on the values of the variance components, which depend on the method of variance component estimation.

9.12 SUMMARY

This chapter presents a very interesting problem that occurs when blocking or some other random effects are used in the model where there are covariates to be considered. Information about the slopes of the model can come from within the intersection of all of the random effects as well as from between the levels of each of the random effects. This is an important problem for split-plot and repeated measures types of designs. These topics are to be discussed in the following chapters. Finally, the combining of within and between block estimates occurs when the mixed models analysis is used.

TABLE 9.12
PROC MIXED Code to Carry Out the Combined Within and Between Block Analysis for the Common Slopes Model

```
proc mixed data=long method=reml;
class drug pair;
model hr=drug ihr/solution noint;
random pair;
lsmeans drug/diff;
parms (14.24166) (6.26819)/hold=1,2;
```

CovParm	Estimate
pair	014.2417
Residual	006.2682

Effect	NumDF	DenDF	FValue	ProbF
drug	2	18	13.84	0.0002
ihr	1	18	375.21	0.0000

Effect	drug	Estimate	StdErr	df	tValue	Probt
drug	1	−3.7381	3.6010	18	−1.04	0.3130
drug	2	0.3544	3.7304	18	0.10	0.9254
ihr		0.8148	0.0421	18	19.37	0.0000

Effect	drug	Estimate	StdErr	df	tValue	Probt
drug	1	64.5037	1.0149	18	63.56	0.0000
drug	2	68.5963	1.0149	18	67.59	0.0000

Effect	drug	_drug	Estimate	StdErr	df	tValue	Probt
drug	1	2	−4.0925	0.8031	18	−5.10	0.0001

TABLE 9.13
PROC MIXED Code to Use REML As a Method to Estimate the Variance Components for Use in the Combined Within and Between Block Analysis with Results

```
proc mixed data=long method=reml;
class drug pair;
model hr=drug ihr/solution noint;
random pair;
lsmeans drug/diff;
```

CovPar m	Estimate
pair	13.7478
Residu al	6.2414

Effect	NumDF	DenDF	FValue	ProbF
drug	2	18	13.90	0.0002
ihr	1	18	377.81	0.0000

Effect	drug	Estimate	StdErr	df	tValue	Probt
drug	1	−3.7204	3.5851	18	−1.04	0.3131
drug	2	0.3728	3.7140	18	0.10	0.9212
ihr		0.8146	0.0419	18	19.44	0.0000

Effect	drug	Estimate	StdErr	df	tValue	Probt
drug	1	64.5034	1.0020	18	64.38	0.0000
drug	2	68.5966	1.0020	18	68.46	0.0000

Effect	drug	_drug	Estimate	StdErr	df	tValue	Probt
drug	1	2	−4.0932	0.8013	18	−5.11	0.0001

Two Treatments in a Randomized Complete Block Design Structure

TABLE 9.14
PROC IML Code to Combine Within and Between Block Information and Finish the Computations for the Herbicide Example

```
proc iml;
 bw={-13.82302281, 16.95662402, 5.645132210};
 sbw=.479144725*{7.5830055428 -4.265247955 1.7827637526,
 -4.265247955
10.78674209 7.63433986,1.7827637526 7.63433986 9.42682557};
 bb={38.48854914, 13.83912245, 5.32201593};
 sbb=1.08378847*{7.5830055428 -4.265247955 -1.7827637526, -4.265247955
10.78674209 -7.63433986,-1.7827637526 -7.63433986 9.42682557};
 print bw sbw bb sbb;
 hw={1 -1 0 0 , 0 0 1 0 , 0 0 0 1};
 hb={1 1 0 0 , 0 0 1 0 , 0 0 0 1};
 isbw=inv(sbw); isbb=inv(sbb);
 bc=inv(hw`*isbw*hw+hb`*isbb*hb)*(hw`*isbw*bw+hb`*isbb*bb);
 sbc=inv(hw`*isbw*hw+hb`*isbb*hb);
 dia=diag(sbc); dia=dia*{1 1 1 1}`;
 dia=sqrt(dia); si=inv(sbc);
 print bc dia; print bc sbc si;
 * compute test of b1=b2;
 j={1, 1};
 h={0 0 1 0, 0 0 0 1};
 betas=h*bc; vbetas=h*sbc*h`;
 iv=inv(vbetas); uc=betas`*(iv-iv*j*inv(j`*iv*j)*j`*iv)*betas;
 print uc;
```

TABLE 9.15
PROC IML Code to Construct Confidence Band about Difference of the Herbicide Models

```
x={1 -1 .9 -.9, 1 -1 1 -1, 1 -1 1.1 -1.1, 1 -1 1.2 -1.2,
 1 -1 1.3 -1.3, 1 -1 1.4 -1.4, 1 -1 1.5 -1.5, 1 -1 1.6 -1.6};
dif=x*bc; vardif=x*sbc*x`;
dia=diag(vardif);
dia=dia*{1 1 1 1 1 1 1 1}`;
dia=sqrt(dia);
*dif is difference and dia is standard error;
print dif dia x;
*compute adjusted means at x=0.9 and x=1.6;
x={1 0 .9 0, 0 1 0 .9, 1 0 1.6 0 , 0 1 0 1.6};
dif=x*bc; vardif=x*sbc*x`;
dia=diag(vardif); dia=dia*{1 1 1 1 }`;
dia=sqrt(dia);
*dif is adjusted mean and dia is standard error;
print dif dia x;
```

TABLE 9.16
PROC IML Code to Combine Within and Between Pair Estimates of the Slopes and Test the Equality of Slopes Using the Combined Estimates for the Drug Data Set

```
Proc IML;
bw={-7.336075 , .842023 , .8030000};
sbw=13.1665*{5.9293706 -.031022 0.0390255 , -.031022 .0003302
 -.000045 , 0.0390255 -.000045 .0004133 };
bb={7.136951325, .713955852, .788653490};
sbb=73.08841*{5.9293706 -.031022 -0.0390255 , -.031022 .0003302
 .0000455 , -0.0390255 .0000455 .0004133665};
print bw sbw bb sbb;
hw={1 -1 0 0 , 0 0 1 0 , 0 0 0 1}; hb={1 1 0 0 , 0 0 1 0 , 0 0 0 1};
isbw=inv(sbw); isbb=inv(sbb);
bc=inv(hw`*isbw*hw+hb`*isbb*hb)*(hw`*isbw*bw+hb`*isbb*bb);
sbc=inv(hw`*isbw*hw+hb`*isbb*hb);
dia=diag(sbc); dia=dia*{1 1 1 1}`; dia=sqrt(dia);
si=inv(sbc); print bc dia;
*test b1=b2;
j={1, 1}; h={0 0 1 0, 0 0 0 1};
betas=h*bc; vbetas=h*sbc*h`;
iv=inv(vbetas); uc=betas`*(iv-iv*j*inv(j`*iv*j)*j`*iv)*betas;
print uc;
```

TABLE 9.17
PROC IML Code to Combine Between and Within Block Information for Common Slope Model for the Drug Data Set

```
Proc IML; bw={-4.061747085, .824454036};
sbw=12.53638*{0.0516491658 0.0005153 , 0.0005153 0.0001610513};
bb={8.053121780, 0.746548527};
sbb=69.50304*{5.8312178034 -.034514733, -.034514733 .0002060581};
print bw sbw bb sbb;
hw={1 -1 0 , 0 0 1}; hb={1 1 0 , 0 0 1};
isbw=inv(sbw); isbb=inv(sbb);
bc=inv(hw`*isbw*hw+hb`*isbb*hb)*(hw`*isbw*bw+hb`*isbb*bb);
sbc=inv(hw`*isbw*hw+hb`*isbb*hb);
dia=diag(sbc); dia=dia*{1 1 1}`;
dia=sqrt(dia); si=inv(sbc);
print bc dia; print bc sbc si;
*compare the intercepts-like comparing LSMEANS for parallel lines model;
x={1 -1 0}; dif=x*bc; vardif=x*sbc*x`;
dia=diag(vardif); dia=dia*{1 }`; dia=sqrt(dia);
print dif dia x;
Compute adjusted means for x=70, 100 and 83.75;
x={1 0 70, 0 1 70, 1 0 100 , 0 1 100, 1 0 83.75, 0 1 83.75};
dif=x*bc; vardif=x*sbc*x`; dia=diag(vardif); dia=dia*{1 1 1 1 1 1}`;
dia=sqrt(dia); print dif dia x; run;
```

REFERENCES

Ash, Katherine (1982). Use of Inter-Block Information in Analysis of Covariance and Some Repeated Measures Covariance Models, Ph.D. dissertation, Department of Statistics, Kansas State University, Manhattan.
Fergen, Brian (1997). On the Mixed-Models Estimator, Ph.D. dissertation, Department of Statistics, Kansas State University, Manhattan.
John, P. W. M. (1971). *Statistical Design and Analysis of Experiments*, New York: MacMillan.
Littell, R., Milliken, G. A., Stroup, W., Wolfinger, R. (1996). *SAS System for Mixed Models*, Cary NC: SAS Institute Inc.
Milliken, G. A. and Johnson, D. E. (1992). *Analysis of Messy Data, Volume I: Design Experiments*, London: Chapman & Hall.

EXERCISES

EXERCISE 9.1: A boy and a girl were selected from each of 16 sixth grade classes to determine if there were gender differences on the ability to perform on an engineering and mathematics aptitude test. Each student's IQ was obtained as a possible covariate. Provide the estimates of the between block model parameters and the within block model parameters. Determine if the slopes are similar or different. Provide a combined within and between block analysis and make the necessary comparisons between the regression line for the boys and the regression line for the girls. School is the blocking factor.

Data for Exercise 9.1

School	Male_score	Male_IQ	Female_score	Female_IQ
1	83	120	75	113
2	76	115	55	96
3	81	126	62	102
4	85	123	59	96
5	70	95	68	105
6	65	93	78	117
7	65	94	74	115
8	82	124	76	115
9	79	119	88	127
10	81	116	59	95
11	67	89	86	125
12	72	102	81	120
13	83	125	63	103
14	79	111	86	124
15	63	88	79	119
16	84	127	71	109

EXERCISE 9.2: A chemical engineer designed a study to evaluate the ability of a wafer cleaning machine to remove particles from the surface of a silicon wafer used in the manufacture of semi-conductors. The specific study was to see if the machine cleaned two types of wafers differently after each had been exposed to a processing step that leaves particles on the surface. The particles must be removed before continuing with the development. The cleaning machine had two slots, so two wafers could be cleaned at the same time or during one run. The process was to randomly select a wafer of each type from a lot of wafers and then randomly assign the two wafers to the two wafer slots in the machine. The number of particles on the surface of each wafer was determined before the cleaning process started and was to be used as a possible covariate. After the cleaning process, the number of particles remaining on the surface was determined. Use the following data to obtain the within and between block or run estimates of the parameters of the models and then obtain the combined estimates. Determine if there is evidence that the two slopes are unequal. Complete the analysis by comparing the resulting regression models for the two wafer types.

Data set for Exercise 9.2

Run	A_pre_part	A_rem_part	B_pre_part	B_rem_part
1	214	56	236	77
2	159	36	219	63
3	179	43	255	73
4	160	44	245	80
5	188	51	154	53
6	225	56	148	39
7	218	52	153	42
8	221	56	248	75
9	259	63	234	68
10	155	45	222	73
11	252	62	137	48
12	235	58	179	58

EXERCISE 9.3: Use change from baseline (pre_part − rem_part) in Exercise 9.2 and use the randomized complete block analysis of variance to compare the wafer types. Were the conclusions similar?

EXERCISE 9.4: Use the change from baseline (IHR − HR) in Section 9.11 and use the randomized complete block analysis of variance to compare the two drugs. Are the conclusions similar?

10 More Than Two Treatments in a Blocked Design Structure

10.1 INTRODUCTION

Two treatments in a randomized complete block (RCB) design structure were discussed in Chapter 9. In addition to the usual problems associated with the analysis of covariance, when blocks are present, information about the parameters comes from within block comparisons and between block comparisons. A process of combining the between block and within block information within the context of analysis of covariance becomes necessary. If the blocks are complete, the within block information provides estimates of the slopes and of contrasts of the intercepts while the between block information provides estimates of the slopes and the average intercept. The two pieces of information can be combined to obtain estimates of the parameters of the models.

If an incomplete block design structure is used, then the within block information consists of estimates of slopes and estimates of contrasts of intercepts (if the design is connected). The between block information provides estimates of the intercepts and of the slopes where there are enough blocks to estimate all of the parameters. Again these two pieces of information can be combined.

This chapter discusses the various aspects of the analysis for a RCB structure and an incomplete block design structure. Two examples will be used to demonstrate the ideas.

10.2 RCB DESIGN STRUCTURE — WITHIN AND BETWEEN BLOCK INFORMATION

A model to represent a one-way treatment structure with one covariate in a RCB design structure is

$$y_{ij} = \alpha_i + \beta_i X_{ij} + b_j + \varepsilon_{ij} \tag{10.1}$$

$$i = 1, 2, ..., t, \quad j = 1, 2, ..., b$$

where y_{ij} denotes the response from treatment i in block j, α_i is the intercept, and β_i is the slope of the regression model for treatment i, $b_j \sim$ iid $N(0, \sigma_b^2)$, and $\varepsilon_{ij} \sim$

iid $N(0, \sigma_\varepsilon^2)$. Let C_i, $i = 1, 2, \ldots, t$ be a set of constants such that $\sum_{i=1}^{t} C_i = 0$ (a contrast). Then any within block contrast $\sum_{i=1}^{t} C_i y_{ij}$ is free of block effects as shown by the model

$$\sum_{i=1}^{t} C_i y_{ij} = \sum_{i=1}^{5} C_i \alpha_i + \sum_{i=1}^{t} C_i \beta_i X_{ij} + \sum_{i=1}^{t} C_i \varepsilon_{ij}$$

Unless all X_{ij} are equal within the same block (see Chapter 11), this model provides estimates of contrasts of the α_i and estimates of the β_i, or the within block analysis provides estimates of

$$\alpha_1 - \bar{\alpha}., \alpha_2 - \bar{\alpha}., \ldots, \alpha_t - \bar{\alpha}., \beta_1, \beta_2, \ldots, \beta_t.$$

The between block model (or block totals) is

$$y_{.j} = \alpha. + \sum_{i=1}^{t} \beta_i X_{ij} + t b_j + \sum_{i=1}^{t} \varepsilon_{ij}$$

$$= \alpha. + \sum_{i=1}^{t} \beta_i X_{ij} + e_{j}.$$

(10.2)

where $e_j = t b_j + \sum_{i=1}^{t} \varepsilon_{ij} \sim$ iid $N(0, t^2 \sigma_b^2 + t \sigma_\varepsilon^2)$. The between block model is a multiple regression model with intercept $\alpha.$, slopes $\beta_1, \beta_2, \ldots, \beta_t$, and variance $\sigma_{BLK}^2 = t^2 \sigma_b^2 + t \sigma_\varepsilon^2$. Thus the between block analysis provides estimates of $\bar{\alpha}., \beta_1, \beta_2, \ldots, \beta_t$, where $\bar{\alpha}. = \alpha./t$.

By combining $(\alpha_i - \bar{\alpha}.)$ and $\bar{\alpha}.$, the intercept for the ith model can be estimated as

$$\hat{\alpha}_i = \hat{\bar{\alpha}}. + \widehat{(\alpha_i - \bar{\alpha}.)} \quad i = 1, 2, \ldots, t.$$

The general process of combining within and between block information is discussed in Section 10.4.

10.3 INCOMPLETE BLOCK DESIGN STRUCTURE — WITHIN AND BETWEEN BLOCK INFORMATION

An incomplete block design structure is one where the blocks do not contain all treatments, i.e., the block size is less than the number of treatments. If the treatments are arranged in blocks such that the design is connected (Milliken and Johnson, 1992), then the within block information provides estimates of all contrasts of the intercepts and estimates of the individual slopes. Thus the connected incomplete

More Than Two Treatments in a Blocked Design Structure

block design provides the same within block information as the RCB, i.e., estimates of $\alpha_1 - \bar{\alpha}., \alpha_2 - \bar{\alpha}., \ldots, \alpha_t - \bar{\alpha}., \beta_1, \beta_2, \ldots, \beta_t$.

The between block information from the incomplete block design is different than that of the RCB. The reason is that not all treatments occur in each block; thus the block total models are different. Let $R_s = \{\text{indices of treatments in block s}\}$. Then the block total model for block s is

$$\sum_{i \in R_s} y_{is} = \sum_{i \in R_s} \alpha_i + \sum_{i \in R_s} \beta_i X_{is} + r_s b_s + \sum_{i \in R_s} \varepsilon_{is} \qquad (10.3)$$

where r_s is number of treatments in block s. The variance of a block total is

$$\mathrm{Var}\left(\sum_{i \in R_s} y_{is}\right) = r_s\left(\sigma_\varepsilon^2 + r_s \sigma_b^2\right).$$

For simplicity, assume the r_s are all equal to r (equal block sizes); then

$$\mathrm{Var}\left(\sum_{i \in R_s} y_{is}\right) = r\left(\sigma_\varepsilon^2 + r\sigma_b^2\right). \qquad (10.4)$$

If there is a sufficient number of blocks (more than 2t), then between block information provides estimates of $\alpha_1, \alpha_2, \ldots, \alpha_t, \beta_1, \beta_2, \ldots, \beta_t$. Table 10.1 contains an incomplete block design with three treatments in eight blocks. Let T_j denote the total for block j; then the between block or block total model is

$$\begin{bmatrix} T_1 \\ T_2 \\ T_3 \\ T_4 \\ T_5 \\ T_6 \\ T_7 \\ T_8 \end{bmatrix} = \begin{bmatrix} 1 & 1 & 0 & X_{11} & X_{21} & 0 \\ 0 & 1 & 1 & 0 & X_{22} & X_{32} \\ 1 & 0 & 1 & X_{13} & 0 & X_{33} \\ 0 & 1 & 1 & 0 & X_{24} & X_{34} \\ 1 & 1 & 0 & X_{15} & X_{25} & 0 \\ 1 & 1 & 0 & X_{16} & X_{26} & 0 \\ 0 & 1 & 1 & 0 & X_{27} & X_{37} \\ 0 & 1 & 1 & 0 & X_{28} & X_{38} \end{bmatrix} \begin{bmatrix} \alpha_1 \\ \alpha_2 \\ \alpha_3 \\ \beta_1 \\ \beta_2 \\ \beta_3 \end{bmatrix} + \mathbf{e}$$

The model is full rank, thus all parameters are estimable and therefore the between block model provides estimates of $\alpha_1, \alpha_2, \alpha_3, \beta_1, \beta_2,$ and β_3. The between block information needs to be combined with the within block information to obtain better estimators. There are some incomplete block designs that do not allow all intercepts to be estimated from the between block model. Those types of designs are not considered here.

TABLE 10.1
An Incomplete Block Arrangement with Three Treatments in Eight Blocks

Block	1	2	3	4	5	6	7	8
Treatments	1	2	3	3	2	1	2	3
	2	3	1	2	1	2	3	2

10.4 COMBINING BETWEEN BLOCK AND WITHIN BLOCK INFORMATION

When combining between and within block information, the functions of the parameters need to be consistent for both models. For example, the RCB provides within block estimates of $\alpha_i - \bar{\alpha}.$ and β_j and between block estimates of $\bar{\alpha}.$ and β_j. The goal is to obtain a combined estimate of the vector of parameters $\theta = (\bar{\alpha}., \alpha_1 - \bar{\alpha}., \alpha_2 - \bar{\alpha}., \ldots, \alpha_t - \bar{\alpha}., \beta_1, \beta_2, \ldots, \beta_b)'$.

An additional complication occurs when the solution to the normal equations does not yield estimates of the desired functions of the parameters. That is the case for PROC GLM of the SAS® system where the within block information provides estimates of

$$\alpha_1 - \alpha_t, \alpha_2 - \alpha_t, \ldots, \alpha_t - \alpha_t = 0, \beta_1, \beta_2, \ldots, \beta_t.$$

Let $\alpha_i^* = (\alpha_i - \alpha_t)$, then

$$\alpha_i^* - \bar{\alpha}.^* = \alpha_i - \bar{\alpha}..$$

Thus, before continuing with the combining process, the estimates of estimable functions of the α_i's need to be transformed into estimates of $\alpha_i - \bar{\alpha}.$.

For three treatments in a RCB design structure, the within block estimates are

$$\hat{\theta}_1 = \left(\widehat{\alpha_1 - \bar{\alpha}.}, \widehat{\alpha_2 - \bar{\alpha}.}, \widehat{\alpha_3 - \bar{\alpha}.}, \hat{\beta}_1, \hat{\beta}_2, \hat{\beta}_3\right)' \qquad (10.5)$$

and the between block estimates are

$$\hat{\theta}_2 = \left(\hat{\bar{\alpha}}., \hat{\beta}_1, \hat{\beta}_2, \hat{\beta}_3\right)'. \qquad (10.6)$$

The within block estimates can be expressed as the beta-hat model

$$\hat{\boldsymbol{\theta}}_1 = \mathbf{H}_1\boldsymbol{\theta} + \boldsymbol{\varepsilon}_1 \text{ where } \mathbf{H}_1 = \begin{bmatrix} 0 & 1 & 0 & 0 & 0 & 0 & 0 \\ 0 & 0 & 1 & 0 & 0 & 0 & 0 \\ 0 & 0 & 0 & 1 & 0 & 0 & 0 \\ 0 & 0 & 0 & 0 & 1 & 0 & 0 \\ 0 & 0 & 0 & 0 & 0 & 1 & 0 \\ 0 & 0 & 0 & 0 & 0 & 0 & 1 \end{bmatrix}, \quad (10.7)$$

$\boldsymbol{\varepsilon}_1 \sim N(\mathbf{0}, \boldsymbol{\Sigma}_1)$, $\boldsymbol{\Sigma}_1 = \sigma_\varepsilon^2 \mathbf{G}_1$,

$\mathbf{G}_1$ is a matrix of constants from the respective partition of the inverse of $\mathbf{X}'\mathbf{X}$, and $\mathbf{X}$ is the design matrix including, treatment effects, block effects, and covariates.

The between block information from the RCB can be expressed as the beta-hat model

$$\hat{\boldsymbol{\theta}}_2 = \mathbf{H}_2\boldsymbol{\theta} + \boldsymbol{\varepsilon}_2$$

$$\text{where } \mathbf{H} = \begin{bmatrix} 1 & 0 & 0 & 0 & 0 & 0 & 0 \\ 0 & 0 & 0 & 0 & 1 & 0 & 0 \\ 0 & 0 & 0 & 0 & 0 & 1 & 0 \\ 0 & 0 & 0 & 0 & 0 & 0 & 1 \end{bmatrix}, \quad (10.8)$$

$\boldsymbol{\varepsilon}_2 \sim N(0, \boldsymbol{\Sigma}_2)$, $\boldsymbol{\Sigma}_2 = \sigma_B^2 \mathbf{G}_2$, $\sigma_B^2 = r\,(\sigma_\varepsilon^2 + r\sigma_b^2)$, and $\mathbf{G}_2$ is the inverse of $\mathbf{WW}'$ where $\mathbf{W}$ is the between block design matrix.

The between block estimates and the within block estimates are distributed independently (under normality and the independence of the b's and the ε's); thus the joint within/between beta-hat model is

$$\begin{bmatrix} \hat{\boldsymbol{\theta}}_1 \\ \hat{\boldsymbol{\theta}}_2 \end{bmatrix} = \begin{bmatrix} \mathbf{H}_1 \\ \mathbf{H}_2 \end{bmatrix} \boldsymbol{\theta} + \begin{bmatrix} \boldsymbol{\varepsilon}_1 \\ \boldsymbol{\varepsilon}_2 \end{bmatrix}$$

where

$$\begin{bmatrix} \boldsymbol{\varepsilon}_1 \\ \boldsymbol{\varepsilon}_2 \end{bmatrix} \sim N\left[\begin{pmatrix} \mathbf{0} \\ \mathbf{0} \end{pmatrix}, \begin{pmatrix} \boldsymbol{\Sigma}_1 & \mathbf{0} \\ \mathbf{0} & \boldsymbol{\Sigma}_2 \end{pmatrix} \right]$$

The best linear unbiased estimator (assuming σ_ε^2 and σ_b^2 are known) of $\boldsymbol{\theta}$ is

$$\hat{\boldsymbol{\theta}} = \left(\mathbf{H}_1'\boldsymbol{\Sigma}_1^{-}\mathbf{H}_1 + \mathbf{H}_2'\boldsymbol{\Sigma}_2^{-}\mathbf{H}_2 \right)^{-1} \left(\mathbf{H}_1'\boldsymbol{\Sigma}_1^{-}\hat{\boldsymbol{\theta}}_1 + \mathbf{H}_2'\boldsymbol{\Sigma}_2^{-}\hat{\boldsymbol{\theta}}_2 \right).$$

The weighted least squares estimate of θ or combined between/within block estimate of θ (assuming σ_ε^2 and σ_b^2 are unknown) is

$$\hat{\theta}_c = \left(\mathbf{H}_1' \hat{\boldsymbol{\Sigma}}_1^- \mathbf{H}_1 + \mathbf{H}_2' \hat{\boldsymbol{\Sigma}}_2^- \mathbf{H}_2\right)^{-1} \left(\mathbf{H}_1' \hat{\boldsymbol{\Sigma}}_1^- \hat{\boldsymbol{\theta}}_1 + \mathbf{H}_2' \hat{\boldsymbol{\Sigma}}_2^- \hat{\boldsymbol{\theta}}_2\right) \quad (10.9)$$

where $\hat{\boldsymbol{\Sigma}}_1 = \hat{\sigma}_\varepsilon^2 \mathbf{G}_1$ and $\hat{\boldsymbol{\Sigma}}_2 = r(\hat{\sigma}_\varepsilon^2 + r\hat{\sigma}_b^2)\mathbf{G}_2$, $\hat{\sigma}_\varepsilon^2$ is the within block residual mean square and $r(\hat{\sigma}_\varepsilon^2 + r\hat{\sigma}_b^2)$ is the between block residual mean square. (If the block sizes are not equal, then the function of σ_ε^2 and σ_b^2 cannot be factored out of Σ_2 as well as the block totals will have unequal variances. (See Chapter 13 when you have unequal block sizes.) The estimated approximate variance of $\hat{\theta}_c$ is

$$\text{Var}(\hat{\theta}_c) = \left(\mathbf{H}_1' \hat{\boldsymbol{\Sigma}}_1^- \mathbf{H}_1 + \mathbf{H}_2' \hat{\boldsymbol{\Sigma}}_2^- \mathbf{H}_2\right)^- \quad (10.10)$$

and the variance of an estimable linear combination of θ, say $\mathbf{a}'\hat{\theta}_c$, is $\text{Var} = (\mathbf{a}'\hat{\theta}_c) = \mathbf{a}'\text{Var}(\hat{\theta}_c)\mathbf{a}$. The number of degrees of freedom associated with this estimated variance needs to be approximated. The Satterthwaite or Kenward and Roger approximation or a weighted average of t-values can be used. An approximate $(1-\alpha)100\%$ LSD value can be computed (using a weighted t-value similar to Chapter 24 of Milliken and Johnson, 1992) by replacing $\hat{\Sigma}_1$ by $(t_{\alpha/2,df_1})\hat{\Sigma}_1$ and $\hat{\Sigma}_2$ by $(t_{\alpha/2,df_2})\hat{\Sigma}_2$ in the $\text{Var}(\hat{\theta}_c)$ where df_1 is the degrees of freedom of the residual mean square from the within block analysis and df_2 is the degrees of freedom of the residual mean square from the within block analysis. The resulting value of $\text{Var}(\hat{\theta}_c)$ is the approximate LSD value. The approximate t-value used in this LSD computation is

$$t_{\alpha/2}^* = \frac{\mathbf{a}'\left(\mathbf{H}_1'\left(t_{\alpha/2,df_1}\right)\hat{\boldsymbol{\Sigma}}_1^- \mathbf{H}_1 + \mathbf{H}_2'\left(t_{\alpha/2,df_2}\right)\hat{\boldsymbol{\Sigma}}_2^- \mathbf{H}_2\right)^- \mathbf{a}}{\mathbf{a}'\text{Var}(\hat{\theta}_c)\mathbf{a}}. \quad (10.11)$$

Approximate degrees of freedom can be computed by matching $t_{\alpha/2}^*$ to the $t_{\alpha/2,df}$ values in the t-table. A Satterthwaite approximation and a Kenward-Roger (Kenward and Roger, 1997) approximation to the degrees of freedom are available as options in PROC MIXED. The statistic to test the parallelism hypothesis can be computed by constructing a beta-hat model for the slopes. Let $\hat{\boldsymbol{\beta}}_c = \mathbf{W}\hat{\theta}_c$ where $\mathbf{W} = [\mathbf{0}, \mathbf{0}, \mathbf{I}]$; then the asymptotic sampling distribution of $\hat{\boldsymbol{\beta}}_c$ is $N(\boldsymbol{\beta}, \boldsymbol{\Sigma}_{\hat{\beta}})$ where $\boldsymbol{\Sigma}_{\hat{\beta}} = \mathbf{W}\,\text{Var}(\hat{\theta}_c)\mathbf{W}'$. The beta-hat model under the equal slope hypothesis is

$$\hat{\boldsymbol{\beta}}_c = \mathbf{j}\beta + \varepsilon^*$$

where $\mathbf{j}$ is a $t \times 1$ vector of ones. The residual mean square from the beta-hat model is the test statistic, i.e.,

$$\mathbf{u}_c = \hat{\boldsymbol{\beta}}_c' \left(\hat{\boldsymbol{\Sigma}}_{\hat{\beta}}^{-1} - \hat{\boldsymbol{\Sigma}}_{\hat{\beta}}^{-1}\mathbf{j}\,\mathbf{j}'\hat{\boldsymbol{\Sigma}}_{\hat{\beta}}^{-1} / \left(\mathbf{j}'\hat{\boldsymbol{\Sigma}}_{\hat{\beta}}^{-1}\mathbf{j}\right)\right)\hat{\boldsymbol{\beta}}_c / (t-1) \quad (10.12)$$

which has an approximate small sample size distribution of $F(t-1, df_1)$.

Another small sample approximation can be obtained by recomputing u_c where $\hat{\Sigma}_1$ and $\hat{\Sigma}_2$ are replaced with $(F_{\alpha,(t-1),df_1})\hat{\Sigma}_1$ and $(F_{\alpha,(t-1),df_2})\hat{\Sigma}_2$, respectively. Denote this value of u_c by u_c^*. An approximate F value is

$$F^*_{(\alpha,t-1,df_c)} = u_c/u_c^*. \qquad (10.13)$$

The approximate degrees of freedom df_c are determined by matching $F^*_{(\alpha,t-1,df_c)}$ to an F-table with $t-1$ degrees of freedom in the numerator and significance level α. The approximate small sampling distribution for u_c is $F_{(t-1,df_c)}$.

For the three-treatment incomplete block design structure in Table 10.1, the within block estimates and corresponding beta-hat model are the same as above for the RCB (Equations 10.5 through 10.7). The between block estimates are

$$\hat{\theta}_2 = \left(\hat{\alpha}_1, \hat{\alpha}_2, \hat{\alpha}_3, \hat{\beta}_1, \hat{\beta}_2, \hat{\beta}_3\right)'.$$

Next transform the intercepts to $\hat{\theta}_3$ where

$$\hat{\theta}_3 = T\hat{\theta}_2 \text{ where}$$

$$\hat{\theta}_3 = \left(\widehat{\alpha.}, \widehat{\alpha_1 - \alpha.}, \widehat{\alpha_2 - \alpha.}, \widehat{\alpha_3 - \alpha.}, \hat{\beta}_1, \hat{\beta}_2, \hat{\beta}_3\right)'$$

and

$$T = \begin{bmatrix} 1/3 & 1/3 & 1/3 & 0 & 0 & 0 \\ 2/3 & -1/3 & -1/3 & 0 & 0 & 0 \\ -1/3 & 2/3 & -1/3 & 0 & 0 & 0 \\ -1/3 & -1/3 & 2/3 & 0 & 0 & 0 \\ 0 & 0 & 0 & 1 & 0 & 0 \\ 0 & 0 & 0 & 0 & 1 & 0 \\ 0 & 0 & 0 & 0 & 0 & 1 \end{bmatrix}$$

The beta-hat model for $\hat{\theta}_3$ is

$$\hat{\theta}_3 = H_3\theta + \varepsilon_3 \qquad (10.14)$$

where $H_3 = I_7$ and $\text{Var}(\varepsilon_3) = \Sigma_3 = T\Sigma_2 T'$.

The combined estimate of θ can be computed from Equation 10.9 where $\hat{\theta}_2$, H_2 and $\hat{\Sigma}_2$ are replaced by $\hat{\theta}_3$, H_3, and $\hat{\Sigma}_3$, respectively. Two examples are used to demonstrate the above ideas.

TABLE 10.2
Data for the Example in Section 10.5 Involving Five Treatments in a RCB Design Structure Where Y is the Response Variable and X is the Covariate

	Treatment 1		Treatment 2		Treatment 3		Treatment 4		Treatment 5	
BLOCK	Y1	X1	Y2	X2	Y3	X3	Y4	X4	Y5	X5
1	55.1	14.0	59.6	11.4	63.5	14.4	66.1	12.2	80.5	21.1
2	50.7	15.4	59.1	20.6	65.7	20.3	59.9	11.9	78.6	24.3
3	58.4	13.5	65.5	21.2	66.1	15.6	77.8	21.6	76.4	16.1
4	60.5	19.0	57.9	9.0	72.7	20.8	73.5	17.0	78.7	17.5
5	60.6	12.0	65.6	16.9	70.6	16.5	91.1	31.4	82.2	19.0
6	60.4	24.1	64.8	21.3	62.4	14.2	69.7	16.2	67.0	12.5
7	62.1	28.9	64.7	23.6	66.0	17.8	76.2	22.9	86.8	26.9
8	59.5	18.7	64.0	16.0	76.1	24.8	78.3	19.1	81.2	19.8
9	68.1	25.6	64.9	13.2	80.5	27.2	79.8	19.4	96.4	28.6
10	66.7	22.7	75.9	28.5	82.8	28.5	84.0	22.4	79.0	16.7
11	68.7	25.3	68.5	14.8	75.3	19.2	87.8	26.1	76.1	14.9
12	64.9	22.6	63.6	11.1	75.5	18.4	80.2	19.8	88.5	22.4

10.5 EXAMPLE: FIVE TREATMENTS IN RCB DESIGN STRUCTURE

The data in Table 10.2 are yields (Y) (kilograms per hectare) of winter wheat where the treatments are herbicides applied during the spring and the covariate is the depth of adequate moisture (centimeters) measured on each plot. The design structure is a RCB.

Table 10.3 contains the SAS® system code used to extract the within block information from the data set. The estimate of σ_ε^2 is 0.9965 and the parameter estimates provide estimates of the slopes (denoted by x*trt 1, ..., x*trt 5) and estimates of $\widehat{\alpha_i - \alpha_5}$ i = 1, 2, ..., 5, (denoted by trt 1, ..., trt 5). Estimate statements have been included in Table 10.4 to provide estimates of $\widehat{\alpha_i - \bar{\alpha}.}$, quantities that are needed in the combined estimator process. Table 10.5 contains the PROC GLM code to provide the statistic to test the equal slopes hypothesis using the within block information. The value of the F statistic is 32.23 with a significance level less than 0.0001, indicating there is sufficient within block information to conclude the slopes are not all equal. Using the expected mean square for BLOCKS and ERROR, the estimate of the block variance component is $\hat{\sigma}_b^2 = (59.8718 - 0.99650/4.5455 = 12.9524$. The block totals for Y are listed in Table 10.6 and are denoted by SY, where SX1, ..., SX5 are the sums of the covariates within each block. The PROC GLM code to fit the between block model is in Table 10.7. Also included are estimate statements to provide estimates of $\bar{\alpha}$ and the five slopes. The contrast statement provides a between block test of the equal slopes hypothesis. The resulting F statistic is 5.50 with significance level 0.0330; again there is sufficient between block information to conclude the slopes are not all equal. The between block analysis of

TABLE 10.3
PROC GLM Code to Fit the Within Block Model with Parameter Estimates

```
PROC GLM DATA=LONG10; CLASSES TRT BLOCK;
MODEL Y=BLOCK TRT X*TRT/SOLUTION INVERSE;
ESTIMATE "A1-Abar" TRT  .8 -.2 -.2 -.2 -.2;
ESTIMATE "A2-Abar" TRT -.2  .8 -.2 -.2 -.2;
ESTIMATE "A3-Abar" TRT -.2 -.2  .8 -.2 -.2;
ESTIMATE "A4-Abar" TRT -.2 -.2 -.2  .8 -.2;
ESTIMATE "A5-Abar" TRT -.2 -.2 -.2 -.2  .8;
```

Source	df	SS	MS	FValue	ProbF
Model	20	5724.9054	286.2453	287.25	0.0000
Error	39	38.8640	0.9965		
Corr Total	59	5763.7693			

Source	df	SS(III)	MS	FValue	ProbF
BLOCK	11	658.5902	59.8718	60.08	0.0000
trt	4	6.5423	1.6356	1.64	0.1834
x*trt	5	1040.1444	208.0289	208.76	0.0000

Parameter	Estimate	Biased	StdErr	tValue	Probt
trt 1	−1.3249	1	1.8508	−0.72	0.4783
trt 2	0.5019	1	1.7831	0.28	0.7798
trt 3	0.9823	1	1.9378	0.51	0.6151
trt 4	3.0808	1	1.9275	1.60	0.1180
trt 5	0.0000	1			
x*trt 1	0.5100	0	0.0613	8.32	0.0000
x*trt 2	0.6734	0	0.0579	11.63	0.0000
x*trt 3	0.9135	0	0.0708	12.90	0.0000
x*trt 4	1.0798	0	0.0611	17.67	0.0000
x*trt 5	1.4309	0	0.0693	20.66	0.0000

TABLE 10.4
PROC GLM Code for the Within Block Model Parameter Estimates from Estimate Statements

```
ESTIMATE "A1-Abar" TRT  .8 -.2 -.2 -.2 -.2;
ESTIMATE "A2-Abar" TRT -.2  .8 -.2 -.2 -.2;
ESTIMATE "A3-Abar" TRT -.2 -.2  .8 -.2 -.2;
ESTIMATE "A4-Abar" TRT -.2 -.2 -.2  .8 -.2;
ESTIMATE "A5-Abar" TRT -.2 -.2 -.2 -.2  .8;
```

Parameter	Estimate	StdErr	tValue	Probt
A1-Abar	−1.9730	1.0980	−1.80	0.0801
A2-Abar	−0.1461	0.9878	−0.15	0.8832
A3-Abar	0.3343	1.2054	0.28	0.7830
A4-Abar	2.4328	1.1397	2.13	0.0391
A5-Abar	−0.6480	1.2343	0.53	−0.6026

TABLE 10.5
PROC GLM Code to Fit Model to Provide Expected Mean Squares and Within Block Test of the Equal Slopes Hypothesis

```
PROC GLM DATA=LONG10; CLASSES BLOCK TRT;
MODEL Y=BLOCK TRT X X*TRT;
RANDOM BLOCK;
```

Source	df	SS	MS	FValue	ProbF
Model	20	5724.9054	286.2453	287.25	0.0000
Error	39	38.8640	0.9965		
Corrected Total	59	5763.7693			

Source	df	SS(I)	MS	FValue	ProbF
BLOCK	11	1424.9533	129.5412	129.99	0.0000
trt	4	3259.8077	814.9519	817.80	0.0000
x	1	911.6756	911.6756	914.87	0.0000
x*trt	4	128.4688	32.1172	32.23	0.0000

Source	df	SS(III)	MS	FValue	ProbF
BLOCK	11	658.5902	59.8718	60.08	0.0000
trt	4	6.5423	1.6356	1.64	0.1834
x	1	909.9734	909.9734	913.16	0.0000
x*trt	4	128.4688	32.1172	32.23	0.0000

Source	Expected Mean Square (III)
BLOCK	Var(Error) + 4.5455 Var(BLOCK)

TABLE 10.6
Data for the Between Block Analysis

BLOCK	SY	SX1	SX2	SX3	SX4	SX5
1	324.8	14.0	11.4	14.4	12.2	21.1
2	314.0	15.4	20.6	20.3	11.9	24.3
3	344.2	13.5	21.2	15.6	21.6	16.1
4	343.3	19.0	9.0	20.8	17.0	17.5
5	370.1	12.0	16.9	16.5	31.4	19.0
6	324.3	24.1	21.3	14.2	16.2	12.5
7	355.8	28.9	23.6	17.8	22.9	26.9
8	359.1	18.7	16.0	24.8	19.1	19.8
9	389.7	25.6	13.2	27.2	19.4	28.6
10	388.4	22.7	28.5	28.5	22.4	16.7
11	376.4	25.3	14.8	19.2	26.1	14.9
12	372.7	22.6	11.1	18.4	19.8	22.4

TABLE 10.7
PROC GLM Code to Fit the Between Block Model, Test Equality of Slopes and Estimate the Parameters of the Model

```
PROC GLM data=new; MODEL SY = SX1 SX2 SX3 SX4
  SX5/INVERSE SOLUTION;
CONTRAST 'SLOPES =' SX1 1 SX2 -1, SX1 1 SX3 -1, SX1
  1 SX4 -1, SX1 1 SX5 -1;
ESTIMATE "Abar" Intercept 1;
ESTIMATE "B1" SX1 1; ESTIMATE "B2" SX2 1; ESTIMATE
  "B3" SX3 1;
ESTIMATE "B4" SX4 1; ESTIMATE "B5" SX5 1;
```

Source	df	SS	MS	FValue	ProbF
Model	5	6331.0082	1266.2016	9.57	0.0080
Error	6	793.7584	132.2931		
Corrected Total	11	7124.7667			
SLOPES =	4	2910.1908	727.5477	5.50	0.0330

Parameter	Estimate	StdErr	tValue	Probt
Abar	228.8619	24.9251	9.18	0.0001
B1	0.9776	0.6784	1.44	0.1996
B2	−0.9333	0.6271	−1.49	0.1872
B3	2.8539	0.8005	3.57	0.0119
B4	3.1500	0.6504	4.84	0.0029
B5	0.1646	0.7776	0.21	0.8394

variance table is a partition of the Type I block sum of squares from the within block analyses. The block totals are based on five observations; thus the corrected total sum of squares in Table 10.7 is five times larger than the Type I SS BLOCKS in Table 10.5. The MSERROR from Table 10.7 provides the estimate of $5(\hat{\sigma}_\varepsilon^2 + 5\hat{\sigma}_b^2) = 132.2931$ which yields $\hat{\sigma}_b^2 = 5.0924$. Using the Type III expected mean squares in Table 10.5, the method of moments estimate of σ_b^2 is $\hat{\sigma}_b^2 = 12.9524$. The reason for this discrepancy is that the between block covariate variation is contaminating the between block analysis (which uses Type 1 analysis) but is removed from the within block analysis (which uses the Type 3 analysis).

Table 10.8 contains the within block estimates of the slopes, the between block estimates of the slopes, their estimated covariance matrices, and the combined estimates of the slopes with the corresponding estimated covariance matrix. The covariance matrices for the within and between block estimates are from the respective partitions of the inverse of the $\mathbf{X'X}$ matrix from each model. The statistic to test the parallelism hypothesis based on the combined information using Equation 10.12 is 32.519. Using the approximation in Equation 10.13, the value of $F^*_{.05,4,dfc} = 2.623$, which provides the approximate number of denominator degrees of freedom as 37. Again, the test for equal slopes using the combined estimates of the slopes indicates there is sufficient evidence to conclude the slopes are not all equal.

TABLE 10.8
Between and Within Block Estimates of Slopes, Estimated Covariance Matrices and the Resulting Combined Estimate of the Slopes and Covariance Matrix

Within Block Estimate of Slopes and Covariance Matrix

Slope	Estimate	Covariance Matrix				
β_1	0.5100	0.003773	0.000156	0.000376	0.000093	0.000227
β_2	0.6734	0.000156	0.003363	0.000160	0.000175	−0.000138
β_3	0.9135	0.000376	0.000160	0.005030	0.000058	0.000371
β_4	1.0798	0.000093	0.000175	0.000058	0.003749	−0.000175
β_5	1.4309	0.000227	−0.000138	0.000371	−0.000175	0.004812

Between Block Estimate of Slopes and Covariance Matrix

Slope	Estimate	Covariance Matrix				
β_1	0.9776	0.003479	−0.000447	−0.001007	−0.000291	−0.000626
β_2	−0.9333	−0.000447	0.002972	−0.000566	−0.000407	0.000717
β_3	2.8539	−0.001007	−0.000566	0.004844	−0.000200	−0.001300
β_4	3.1500	−0.000291	−0.000407	−0.000200	0.003198	0.000653
β_5	0.1646	−0.000626	0.000717	−0.001300	0.000653	0.004571

Combined Estimates of the Slopes With Standard Errors (STD), t-Test of Each Slope Equal to Zero (T), and Significance Level (P)

Slope	Estimate	STD	T	P	LSD(.05)
β_1	0.518	0.060993	8.485	0.0000	0.123574
β_2	0.668	0.057599	11.596	0.0000	0.116688
β_3	0.928	0.070419	13.184	0.0000	0.142675
β_4	1.099	0.060825	18.075	0.0000	0.123220
β_5	1.425	0.068891	20.685	0.0000	0.139574

Test of the Equal Slopes Hypothesis Using the Combined Information
UC = 32.519 sign. level < .0001

Next the within block solution vector in Table 10.3 is transformed from the set-to-zero restrictions to the sum-to-zero restrictions. The required matrix is in Table 10.9. The $\mathbf{H}_1$ design matrix for the beta-hat model in Equation 10.7 is in Table 10.9. The set-to-zero estimates (of the intercepts) are in Table 10.10 and the sum-to-zero estimates (of the intercepts) are in Table 10.11.

Table 10.12 contains the between block information where the intercept is $5\bar{\alpha}$. The first step is to transform to $\bar{\alpha}$. by multiplying by the transformation matrix in Table 10.12 with the rescaled between block estimate of the parameters. The resulting combined estimate of θ is in Table 10.13, which was computed by Equation 10.9. The LSD(.05) values were computed using the method described following Equation 10.10.

This computation of the combined estimates of the parameters can be accomplished using PROC MIXED. Table 10.14 contains the PROC MIXED code to fit the desired model where the fixed effects are TRT and X*TRT and the random effect is BLOCK. The "PARMS (5.09) (.99651)/ HOLD = 1,2" statement was used to

TABLE 10.9
Within Block Transformation Matrices

Transformation Matrix to Convert Set-to-Zero Estimates
of Intercepts to Sum-to-Zero Estimates

0.0	0.0	0.0	0.0	0.0	0.0	0.0	0.0	0.0	0.0
0.8	−0.2	−0.2	−0.2	−0.2	0.0	0.0	0.0	0.0	0.0
−0.2	0.8	−0.2	−0.2	−0.2	0.0	0.0	0.0	0.0	0.0
−0.2	−0.2	0.8	−0.2	−0.2	0.0	0.0	0.0	0.0	0.0
−0.2	−0.2	−0.2	0.8	−0.2	0.0	0.0	0.0	0.0	0.0
−0.2	−0.2	−0.2	−0.2	0.8	0.0	0.0	0.0	0.0	0.0
0.0	0.0	0.0	0.0	0.0	1.0	0.0	0.0	0.0	0.0
0.0	0.0	0.0	0.0	0.0	0.0	1.0	0.0	0.0	0.0
0.0	0.0	0.0	0.0	0.0	0.0	0.0	1.0	0.0	0.0
0.0	0.0	0.0	0.0	0.0	0.0	0.0	0.0	1.0	0.0
0.0	0.0	0.0	0.0	0.0	0.0	0.0	0.0	0.0	1.0

Design Matrix of the Within Block Estimates
Used to Combine the Estimates

0.0	0.0	0.0	0.0	0.0	0.0	0.0	0.0	0.0	0.0	0.0
0.0	1.0	0.0	0.0	0.0	0.0	0.0	0.0	0.0	0.0	0.0
0.0	0.0	1.0	0.0	0.0	0.0	0.0	0.0	0.0	0.0	0.0
0.0	0.0	0.0	1.0	0.0	0.0	0.0	0.0	0.0	0.0	0.0
0.0	0.0	0.0	0.0	1.0	0.0	0.0	0.0	0.0	0.0	0.0
0.0	0.0	0.0	0.0	0.0	1.0	0.0	0.0	0.0	0.0	0.0
0.0	0.0	0.0	0.0	0.0	0.0	1.0	0.0	0.0	0.0	0.0
0.0	0.0	0.0	0.0	0.0	0.0	0.0	1.0	0.0	0.0	0.0
0.0	0.0	0.0	0.0	0.0	0.0	0.0	0.0	1.0	0.0	0.0
0.0	0.0	0.0	0.0	0.0	0.0	0.0	0.0	0.0	1.0	0.0
0.0	0.0	0.0	0.0	0.0	0.0	0.0	0.0	0.0	0.0	1.0

TABLE 10.10
Within Block Estimates of Regression Model Parameters for Set-to-Zero Restrictions on the Intercepts

Parameters and Estimates

$\alpha_1 - \alpha_5$	$\alpha_2 - \alpha_5$	$\alpha_3 - \alpha_5$	$\alpha_4 - \alpha_5$	$\alpha_5 - \alpha_5$	β_1	β_2	β_3	β_4	β_5
−1.3249	0.5019	0.9823	3.0808	0.0000	0.5100	0.6734	0.9135	1.0798	1.4309

cause PROC MIXED to use those specified values as the estimates of the variance components. The estimates of the slopes are directly comparable to those in Table 10.12 and the estimates of the intercepts in Table 10.14 can be computed by combining the average intercept with each of the estimates of the intercepts deviating from the average. A contrast statement is included in Table 10.14 to provide a test of the equal slopes hypothesis. The F statistic has a value of 32.52 with 4 and

TABLE 10.11
Within Block Estimates of Regression Model Parameters with Sum-to-Zero Restrictions on the Intercepts

			Parameters and Estimates							
$\bar{\alpha}$	$\alpha_1 - \bar{\alpha}$	$\alpha_2 - \bar{\alpha}$	$\alpha_3 - \bar{\alpha}$	$\alpha_4 - \bar{\alpha}$	$\alpha_5 - \bar{\alpha}$	β_1	β_2	β_3	β_4	β_5
0.	−1.973	−0.146	0.334	2.432	−0.648	0.510	0.673	0.913	1.07	1.43

TABLE 10.12
Between Block Estimates of Regression Model Parameters and Estimated Covariance Matrices

		Parameters			
$5\bar{\alpha}$	β_1	β_2	β_3	β_4	β_5
		Parameter Estimates			
228.862	0.9776	−0.9333	2.8539	3.1500	0.1646
		Covariance Matrix			
4.696085	−0.024078	−0.037388	−0.035901	−0.060125	−0.078422
−0.024078	0.003479	−0.000447	−0.001007	−0.000291	−0.000626
−0.037388	−0.000447	0.002972	−0.000566	−0.000407	0.000717
−0.035901	−0.001007	−0.000566	0.004844	−0.000200	−0.001300
−0.060125	−0.000291	−0.000407	−0.000200	0.003198	0.000653
−0.078422	−0.000626	0.000717	−0.001300	0.000653	0.004571
		Transformation matrix to rescale $\bar{\alpha}$			
0.2	0	0	0	0	0
0	1	0	0	0	0
0	0	1	0	0	0
0	0	0	1	0	0
0	0	0	0	1	0
0	0	0	0	0	1
	Rescaled between block estimate of parameters and covariance matrix.				
$\bar{\alpha}$	β_1	β_2	β_3	β_4	β_5
45.7724	0.9776	−0.9333	2.8539	3.1500	0.1646

39 degrees of freedom. The least squares means from the analysis in Table 10.14 are in Table 10.15. The least squares means are predicted values evaluated at X = 20.

Table 10.16 contains the PROC MIXED code that uses REML estimates of the variance components to carry out the analysis. The estimates of the block and residual variance components are 13.6077 and 0.9967, respectively. The estimates of the slopes and intercepts in Table 10.16 are quite similar to those in Table 10.14, the difference being that different sets of estimates of variances components were used in the computations. A contrast statement is included in Table 10.16 to provide a test of the equal slopes hypothesis. The F statistic has a value of 32.28 with 4 and 39.2 degrees of freedom. The number of denominator degrees of freedom was

TABLE 10.13
Combined Estimates of the Parameters with Standard Errors (STD), t-Test of Each Parameter Equal to Zero (T), and Significance Level (P) for Example 9.1

Parameter	Estimate	Std Err	t-value	Sign Level	LSD(.05)
$\bar{\alpha}$	52.879	0.889635	59.439	0.0000	2.020255
$\alpha_1 - \bar{\alpha}$	−1.999	1.094679	−1.826	0.0755	2.216329
$\alpha_2 - \bar{\alpha}$	0.075	0.983428	0.076	0.9399	1.991971
$\alpha_3 - \bar{\alpha}$	0.165	1.201541	0.137	0.8916	2.432836
$\alpha_4 - \bar{\alpha}$	2.166	1.133910	1.910	0.0635	2.297224
$\alpha_5 - \bar{\alpha}$	−0.406	1.228441	−0.331	0.7428	2.488505
β_1	0.518	0.060993	8.485	0.0000	0.123574
β_2	0.668	0.057599	11.596	0.0000	0.116688
β_3	0.928	0.070419	13.184	0.0000	0.142675
β_4	1.099	0.060825	18.075	0.0000	0.123220
β_5	1.425	0.068891	20.685	0.0000	0.139574

obtained using the KR (Kenward-Roger) option. This test statistic is very similar to that obtained in Table 10.8. Finally, Table 10.17 contains the least squares means for the five treatments evaluated at X = 15, 20, and 25, respectively. For this data set, all pairwise comparisons between pairs of least squares means evaluated at each of the values of X are significantly different.

10.6 EXAMPLE: BALANCED INCOMPLETE BLOCK DESIGN STRUCTURE WITH FOUR TREATMENTS

The data in Table 10.18 consist of 4 treatments in blocks of size 3 arranged in 16 blocks to form a BIB design structure where Y is the response variable and X is the covariate. The results in the previous sections could be used to obtain the within block and the between block estimates of the parameters of the models and then those estimates could be combined. That part of the process is left to the reader as an exercise. The objective of the analysis in this section is to use PROC MIXED to provide estimates of the model's parameters and then demonstrate the process of computing the least squares means for the incomplete block design structure. The PROC MIXED code in Table 10.19 contains the terms necessary to fit the desired model. The fixed effects are TRT and X*TRT and the random effect is BLOCK. The estimates of the block and residual variance components are 15.5409 and 35.9386, respectively. The estimates of the model parameters are included in the lower part of Table 10.19. The intercept is the estimate of the intercept for treatment 4 and the estimates corresponding to the levels of TRT are the deviations of the intercepts from that of treatment 4. The least squares means evaluated at X = 100 are in Table 10.20. Since the model has unequal slopes, the least squares means need to be evaluated at (at least) three values of X (only one was chosen for demonstration purposes).

TABLE 10.14
PROC MIXED Code to Use the Method of Moments Estimates of the Variance Components by Employing the Hold Option

```
PROC MIXED CL COVTEST DATA=LONG10;;
CLASS BLOCK TRT;
MODEL Y=TRT X*TRT/NOINT SOLUTION DDFM=KR;
RANDOM BLOCK;
LSMEANS TRT/DIFF AT X=20;
PARMS (5.09) (.99651)/HOLD=1,2;
CONTRAST 'EQUAL SLOPES' X*TRT 1 -1 0 0 0, X*TRT 1 0 -1 0 0,
  X*TRT 1 0 0 -1 0, x*TRT 1 0 0 0 -1;
```

CovParm	Estimate
BLOCK	5.090000
Residual	0.996510

Effect	NumDF	DenDF	FValue	ProbF
trt	5	39	730.39	
x*trt	5	39	213.81	

Label	NumDF	DenDF	FValue	ProbF
EQUAL SLOPES	4	39	32.52	0.0000

Effect	trt	Estimate	StdErr	df	tValue	Probt
trt	1	50.8797	1.4204	1	35.82	0.0178
trt	2	52.9537	1.2248	1	43.23	0.0147
trt	3	53.0436	1.5662	1	33.87	0.0188
trt	4	55.0445	1.4096	1	39.05	0.0163
trt	5	52.4730	1.5500	1	33.85	0.0188
x*trt	1	0.5176	0.0610	1	8.49	0.0747
x*trt	2	0.6679	0.0576	1	11.60	0.0548
x*trt	3	0.9284	0.0704	1	13.18	0.0482
x*trt	4	1.0994	0.0608	1	18.08	0.0352
x*trt	5	1.4250	0.0689	1	20.69	0.0308

TABLE 10.15
Least Squares Means Computed Using the Method of Moments Estimates of the Variance Components from Table 10.14

Effect	trt	x	Estimate	StdErr	df	tValue	Probt
trt	1	20.00	61.2307	0.7122	1	85.97	0.0074
trt	2	20.00	66.3117	0.7290	1	90.97	0.0070
trt	3	20.00	71.6113	0.7123	1	100.53	0.0063
trt	4	20.00	77.0333	0.7122	1	108.16	0.0059
trt	5	20.00	80.9738	0.7122	1	113.70	0.0056

TABLE 10.16
PROC MIXED Code to Use REML to Estimate the Variance Components and Complete the Analysis

```
Proc Mixed CL COVTEST DATA=LONG10;
CLASS BLOCK TRT;
MODEL Y=TRT X*TRT/NOINT SOLUTION DDFM=KR;
CONTRAST '=SLOPES' X*TRT 1 -1 0 0 0, X*TRT 1 0 -1 0 0,
   X*TRT 1 0 0 -1 0, X*TRT 1 0 0 0 -1;
```

CovParm	Estimate	StdErr	ZValue	ProbZ	Alpha	Lower	Upper
BLOCK	13.6077	5.9019	2.31	0.0106	0.05	6.7626	40.1423
Residual	0.9967	0.2258	4.41	0.0000	0.05	0.6688	1.6435

Effect	NumDF	DenDF	FValue	ProbF
trt	5	40.89	374.54	0.0000
x*trt	5	39.28	210.16	0.0000

Label	NumDF	DenDF	FValue	ProbF
=slopes	4	39.2	32.28	0.0000

Effect	trt	Estimate	StdErr	df	tValue	Probt
trt	1	50.9726	1.6554	40.28	30.79	0.0000
trt	2	52.8946	1.4893	32.64	35.52	0.0000
trt	3	53.2248	1.7832	44.5	29.85	0.0000
trt	4	55.2857	1.6458	39.92	33.59	0.0000
trt	5	52.4010	1.7687	44.11	29.63	0.0000
x*trt	1	0.5129	0.0613	39.3	8.37	0.0000
x*trt	2	0.6713	0.0578	39.28	11.61	0.0000
x*trt	3	0.9192	0.0707	39.31	13.00	0.0000
x*trt	4	1.0874	0.0611	39.28	17.81	0.0000
x*trt	5	1.4286	0.0692	39.3	20.65	0.0000

The estimate statements in Table 10.21 are used to provide predicted values for treatment 1 evaluated at X = 100 for each of the 16 blocks, even though treatment 1 did not occur in all of the blocks. Only the estimate statements for treatment one are included in the table, but predicted values for all four treatments were obtained. The lower part of Table 10.21 lists the predicted values of each of the treatments at each of the blocks. The last line in Table 10.21 contains the means of the predicted values in the column. The column means correspond exactly to the least squares means in Table 10.20. Thus, when an incomplete block design is used, predicted values for every treatment are obtained for every block, whether or not that treatment actually occurred in the block, and the least squares means are the means of those predicted values. So, the least squares means are the means of the predicted responses as if all treatments had actually occurred in all of the blocks. Because this is in fact what is being estimated by the computational process, it is very important that the factors used to form blocks do not interact with the treatments in the treatment structure (Milliken and Johnson, 1992).

TABLE 10.17
Least Squares Means at X=15, 20 and 25 from the REML Estimates of the Variance Components

```
LSMEANS TRT/DIFF AT X=15; LSMEANS TRT/DIFF AT X=20;
LSMEANS TRT/DIFF AT X=25;
```

Effect	trt	x	Estimate	StdErr	df	tValue	Probt
trt	1	15.00	58.6667	1.1474	14	51.13	0.0000
trt	2	15.00	62.9643	1.1112	13	56.66	0.0000
trt	3	15.00	67.0133	1.1544	15	58.05	0.0000
trt	4	15.00	71.5964	1.1447	14	62.55	0.0000
trt	5	15.00	73.8306	1.1558	15	63.88	0.0000
trt	1	20.00	61.2314	1.1032	12	55.50	0.0000
trt	2	20.00	66.3209	1.1142	13	59.52	0.0000
trt	3	20.00	71.6095	1.1033	12	64.91	0.0000
trt	4	20.00	77.0333	1.1032	12	69.83	0.0000
trt	5	20.00	80.9738	1.1032	12	73.40	0.0000
trt	1	25.00	63.7961	1.1425	14	55.84	0.0000
trt	2	25.00	69.6775	1.1897	16	58.57	0.0000
trt	3	25.00	76.2057	1.1627	15	65.54	0.0000
trt	4	25.00	82.4702	1.1447	14	72.05	0.0000
trt	5	25.00	88.1170	1.1565	15	76.19	0.0000

TABLE 10.18
Data Set for Four Treatments in a Balanced Incomplete Block Design Structure Where Y is the Response and X is the Possible Covariate

BLOCK	TRT	Y	X	TRT	Y	X	TRT	Y	X
1	1	118	127	2	134	129	3	105	105
2	1	101	98	2	109	110	4	121	107
3	1	111	90	3	124	128	4	143	123
4	2	103	81	3	116	102	4	121	106
5	1	123	116	2	99	90	3	108	91
6	1	121	96	2	119	121	4	128	108
7	1	101	92	3	110	112	4	117	102
8	2	122	123	3	114	119	4	124	110
9	1	108	110	2	101	91	3	124	127
10	1	102	104	2	113	123	4	111	105
11	1	108	94	3	128	109	4	142	122
12	2	119	109	3	117	116	4	135	120
13	1	128	105	2	108	83	3	101	79
14	1	113	92	2	117	93	4	118	102
15	1	87	83	3	108	95	4	120	104
16	2	134	121	3	119	125	4	146	123

TABLE 10.19
PROC MIXED Code to Provide Analysis of BIB Design Structure with Four Treatments for the Unequal Slopes Model

```
proc mixed data=bib10 cl covtest; class block trt;
model y=trt x*trt/solution; random block/solution;
```

CovParm	Estimate	StdErr	ZValue	ProbZ	Alpha	Lower	Upper
BLOCK	15.5409	11.6085	1.34	0.0903	0.05	5.3497	153.8921
Residual	35.9386	10.2345	3.51	0.0002	0.05	22.0401	68.8343

Effect	NumDF	DenDF	FValue	ProbF
TRT	3	25	1.71	0.1904
X*TRT	4	25	16.27	0.0000

Effect	TRT	Estimate	StdErr	df	tValue	Probt
Intercept		0.2895	26.5828	15	0.01	0.9915
TRT	1	52.8686	30.9324	25	1.71	0.0998
TRT	2	51.5408	28.7706	25	1.79	0.0853
TRT	3	65.8460	29.1279	25	2.26	0.0327
TRT	4	0.0000				
X*TRT	1	0.5676	0.1627	25	3.49	0.0018
X*TRT	2	0.5942	0.1151	25	5.16	0.0000
X*TRT	3	0.4412	0.1325	25	3.33	0.0027
X*TRT	4	1.1433	0.2389	25	4.79	0.0001

TABLE 10.20
Least Squares Means Computed at X = 100 for the BIB Design Structure

```
lsmeans trt/at x=100;
```

TRT	X	Estimate	StdErr	df	tValue	Probt
1	100.00	109.9168	2.0331	25	54.06	0.0000
2	100.00	111.2536	2.1537	25	51.66	0.0000
3	100.00	110.2540	2.3539	25	46.84	0.0000
4	100.00	114.6167	3.3181	25	34.54	0.0000

10.7 EXAMPLE: BALANCED INCOMPLETE BLOCK DESIGN STRUCTURE WITH FOUR TREATMENTS USING JMP®

The incomplete block data in Table 10.18 is analyzed using JMP® and the data are displayed in the data screen of Figure 10.1, where Y and X are continuous variables and TRT and BLOCK are nominal variables. Figure 10.2 is the fit model screen where the response variable is Y and the model effects are BLOCK, TRT X, and X*TRT. Specify the BLOCK effects to be RANDOM by selecting BLOCK and use the Attributes button to select RANDOM. The "no center" option was selected from

TABLE 10.21
Estimate Statements Used to Provide Predicted Values for Treatment 1 for Each of the 16 Blocks

```
estimate '11' intercept 1 trt 1 0 0 0 x*trt 100|block 1;
estimate '12' intercept 1 trt 1 0 0 0 x*trt 100|block 0 1;
estimate '13' intercept 1 trt 1 0 0 0 x*trt 100|block 0 0 1;
estimate '14' intercept 1 trt 1 0 0 0 x*trt 100|block 0 0 0 1;
estimate '15' intercept 1 trt 1 0 0 0 x*trt 100|block 0 0 0 0 1;
estimate '16' intercept 1 trt 1 0 0 0 x*trt 100|block 0 0 0 0 0 1;
estimate '17' intercept 1 trt 1 0 0 0 x*trt 100|block 0 0 0 0 0 0 1;
estimate '18' intercept 1 trt 1 0 0 0 x*trt 100|block 0 0 0 0 0 0 0 1;
estimate '19' intercept 1 trt 1 0 0 0 x*trt 100|block 0 0 0 0 0 0 0 0 1;
estimate '110' intercept 1 trt 1 0 0 0 x*trt 100|block 0 0 0 0 0 0 0 0 0 1;
estimate '111' intercept 1 trt 1 0 0 0 x*trt 100|block 0 0 0 0 0 0 0 0 0 0 1;
estimate '112' intercept 1 trt 1 0 0 0 x*trt 100|block 0 0 0 0 0 0 0 0 0 0 0 1;
estimate '113' intercept 1 trt 1 0 0 0 x*trt 100|block 0 0 0 0 0 0 0 0 0 0 0 0 1;
estimate '114' intercept 1 trt 1 0 0 0 x*trt 100|block 0 0 0 0 0 0 0 0 0 0 0 0 0 1;
estimate '115' intercept 1 trt 1 0 0 0 x*trt 100|block 0 0 0 0 0 0 0 0 0 0 0 0 0 0 1;
estimate '116' intercept 1 trt 1 0 0 0 x*trt 100|block 0 0 0 0 0 0 0 0 0 0 0 0 0 0 0 1;
```

Treatment 1		Treatment 2		Treatment 3		Treatment 4	
Block	Estimate	Block	Estimate	Block	Estimate	Block	Estimate
1	108.1873	1	109.5241	1	108.5245	1	112.8872
2	106.6044	2	107.9412	2	106.9416	2	111.3043
3	111.8443	3	113.1811	3	112.1815	3	116.5442
4	111.3143	4	112.6511	4	111.6515	4	116.0142
5	109.8052	5	111.1420	5	110.1424	5	114.5051
6	112.3372	6	113.6740	6	112.6743	6	117.0370
7	108.0669	7	109.4037	7	108.4041	7	112.7668
8	108.1085	8	109.4453	8	108.4456	8	112.8083
9	107.9095	9	109.2463	9	108.2466	9	112.6093
10	103.9985	10	105.3353	10	104.3357	10	108.6984
11	113.2101	11	114.5469	11	113.5472	11	117.9099
12	109.8421	12	111.1789	12	110.1793	12	114.5420
13	114.0777	13	115.4145	13	114.4148	13	118.7775
14	113.4230	14	114.7598	14	113.7602	14	118.1229
15	107.5628	15	108.8996	15	107.9000	15	112.2627
16	112.3774	16	113.7142	16	112.7146	16	117.0773
Means	109.9168		111.2536		110.2540		114.6167

the model specification menu to enable the fitted model to match the models fit by PROC MIXED. Click on the Run Model button to carry out the analysis. The parameter estimates are in Figure 10.3 where the ones of interest correspond to Intercept, TRT[1], TRT[2], TRT[3], X, TRT[1]*X, TRT[2]*X, and TRT[3]*X. The

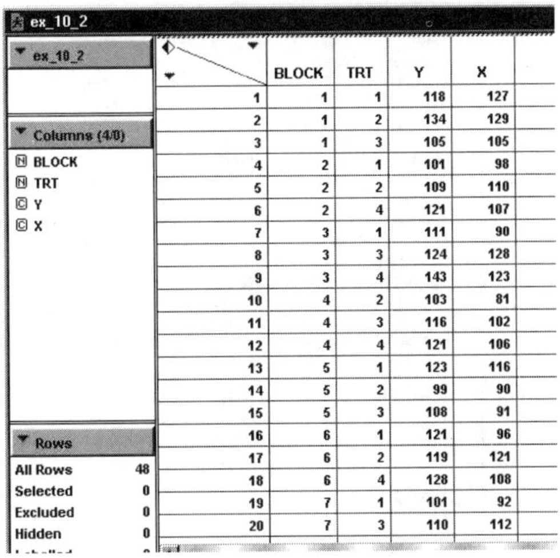

More Than Two Treatments in a Blocked Design Structure

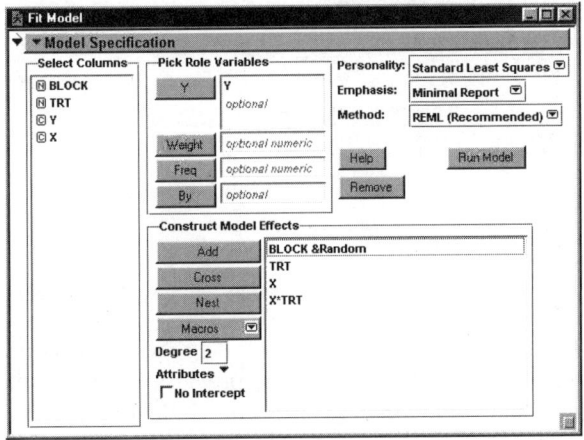

FIGURE 10.1 JMP® data screen for the incomplete block example.

FIGURE 10.2 Fit model table where block is declared as a random effect and no center is selected from model specification.

estimates of the slopes and intercepts can be constructed as described in Chapter 3. The estimates of the variance components and tests for the fixed effects are in Figure 10.4. The estimates of the variance components are the same as obtained by PROC MIXED, displayed in Table 10.19, as is the test for equal intercepts, TRT. Since the model included X and TRT*X, a test of the slopes equal to zero hypothesis is not obtained, but a test of the equal slopes hypothesis is provided through the TRT*X term. The significance level is 0.0897. The custom test screen needs to be used to provide the test of the slopes equal to zero hypothesis. Such a screen is in Figure 10.5 where there are four columns used to construct the estimates of the slopes. The value line provides the estimates of the slopes, which are identical to

Parameter Estimates

Term	Estimate	Std Error	t Ratio	Prob>|t|
Intercept	42.853381	9.675366	4.43	0.0002
BLOCK[1]	-1.729817	2.890387	-0.60	0.5549
BLOCK[2]	-3.312883	2.729791	-1.21	0.2362
BLOCK[3]	1.927825	2.846532	0.68	0.5045
BLOCK[4]	1.3977081	2.793086	0.50	0.6212
BLOCK[5]	-0.111577	2.835865	-0.04	0.9689
BLOCK[6]	2.4206136	2.747132	0.88	0.3866
BLOCK[7]	-1.850165	2.764868	-0.67	0.5095
BLOCK[8]	-1.808644	2.760442	-0.66	0.5183
BLOCK[9]	-2.007633	2.785866	-0.72	0.4778
BLOCK[10]	-5.919202	2.760411	-2.14	0.0419
BLOCK[11]	3.2937371	2.774256	1.19	0.2463
BLOCK[12]	-0.074671	2.76242	-0.03	0.9786
BLOCK[13]	4.161457	2.890951	1.44	0.1624
BLOCK[14]	3.5066368	2.782437	1.26	0.2192
BLOCK[15]	-2.35431	2.817835	-0.84	0.4113
BLOCK[16]	2.4609252	2.839888	0.87	0.3944
TRT[1]	10.303955	15.06679	0.68	0.5003
TRT[2]	8.9756126	12.30855	0.73	0.4726
TRT[3]	23.281578	13.27283	1.75	0.0917
X	0.6865694	0.088819	7.73	<.0001
TRT[1]*X	-0.118974	0.14546	-0.82	0.4211
TRT[2]*X	-0.092324	0.113877	-0.81	0.4252
TRT[3]*X	-0.24538	0.121348	-2.02	0.0540

FIGURE 10.3 Parameter estimates from JMP®.

REML Variance Component Estimates

Random Effect	Var Ratio	Var Component	Std Error	95% Lower	95% Upper	Pct of Total
BLOCK&Random	0.4325617	15.544708	13.819544	4.6539914	317.87939	30.195
Residual		35.936398				69.805
Total		51.481106				100.000

-2 LogLikelihood = 313.76572

Effect Tests

Source	Nparm	DF	DFDen	Sum of Squares	F Ratio	Prob > F	
BLOCK&Random	16	15	25	607.7211	1.1274	0.3835	Shrunk
TRT	3	3	25	184.4179	1.7106	0.1904	
X	1	1	25	2147.2797	59.7522	<.0001	
TRT*X	3	3	25	260.9839	2.4208	0.0897	

Tests on Random effects refer to shrunken predictors rather than traditional estimates.

FIGURE 10.4 Estimates of the variance components and tests for the effects.

those from PROC MIXED; thus JMP® is providing the combined estimates of the model parameters (Table 10.19). The F Ratio provides the test of the slopes equal to zero hypothesis and is the same as obtained by PROC MIXED in Table 10.19. Finally, the custom test screen is used to provide estimates of the models for each of the treatments evaluated at X = 100, as displayed in Figure 10.6. The results of the custom test to provide the least squares means are in Figure 10.7. These are the same least squares means as obtained by PROC MIXED and have the same interpretation (Table 10.20). The least squares means are the mean of the predicted values at X = 100 from all of the blocks, whether or not a treatment occurred in all blocks.

10.8 SUMMARY

When the design structure involves blocking, information about slopes and intercepts can be extracted from within block comparisons and between block comparisons. The type of information available depends on the type of blocking. If the designs are connected, the same within block information is available from all designs, but

More Than Two Treatments in a Blocked Design Structure

Custom Test

Parameter				
BLOCK[9]	0	0	0	0
BLOCK[10]	0	0	0	0
BLOCK[11]	0	0	0	0
BLOCK[12]	0	0	0	0
BLOCK[13]	0	0	0	0
BLOCK[14]	0	0	0	0
BLOCK[15]	0	0	0	0
BLOCK[16]	0	0	0	0
TRT[1]	0	0	0	0
TRT[2]	0	0	0	0
TRT[3]	0	0	0	0
X	1	1	1	1
X*TRT[1]	1	0	0	-1
X*TRT[2]	0	1	0	-1
X*TRT[3]	0	0	1	-1
Value	0.567595098	0.5942458094	0.4411893339	1.1432474537
Std Err	0.1627350897	0.1151460305	0.132467314	0.2388526944
T-Ratio	3.4878470225	5.160801523	3.3305524263	4.7864122126
Prob>|t|	0.0018201113	0.0000245981	0.0026936637	0.0000648081
SS	437.16904711	957.12544537	398.6273538	823.29360874

Sum of Squares	2338.1322656
Numerator DF	4
F Ratio	16.265766578
Prob > F	0.0000011059

FIGURE 10.5 Custom tests screen used to provide estimates of the slopes and a test of the slopes equal to zero hypothesis.

Custom Test

Parameter				
Intercept	1	1	1	1
BLOCK[1]	0	0	0	0
BLOCK[2]	0	0	0	0
BLOCK[3]	0	0	0	0
BLOCK[4]	0	0	0	0
BLOCK[5]	0	0	0	0
BLOCK[6]	0	0	0	0
BLOCK[7]	0	0	0	0
BLOCK[8]	0	0	0	0
BLOCK[9]	0	0	0	0
BLOCK[10]	0	0	0	0
BLOCK[11]	0	0	0	0
BLOCK[12]	0	0	0	0
BLOCK[13]	0	0	0	0
BLOCK[14]	0	0	0	0
BLOCK[15]	0	0	0	0
BLOCK[16]	0	0	0	0
TRT[1]	1	0	0	-1
TRT[2]	0	1	0	-1
TRT[3]	0	0	1	-1
X	100	100	100	100
TRT[1]*X	100	0	0	-100
TRT[2]*X	0	100	0	-100
TRT[3]*X	0	0	100	-100

FIGURE 10.6 Coefficients used to provide least squares means evaluated at X = 100.

the between block information depends on the type of blocking used. RCB designs have different information than incomplete block designs and these differences are evident when the within block and between block estimates are combined. The Type I SS BLOCKS from the within block analysis contains variation due to covariates; thus biased estimates of variance components occur from the within block comparisons. Most of the difficulties occurring because of blocking are avoided when a mixed model approach is used to provide the analysis. The details of the mixed model approach are discussed in Chapter 13.

The final topic discussed was that of using an incomplete block design structure. The example in Section 10.6 indicates the interpretation of the least squares means

Parameter				
BLOCK[10]	0	0	0	0
BLOCK[11]	0	0	0	0
BLOCK[12]	0	0	0	0
BLOCK[13]	0	0	0	0
BLOCK[14]	0	0	0	0
BLOCK[15]	0	0	0	0
BLOCK[16]	0	0	0	0
TRT[1]	1	0	0	-1
TRT[2]	0	1	0	-1
TRT[3]	0	0	1	-1
X	100	100	100	100
TRT[1]*X	100	0	0	-100
TRT[2]*X	0	100	0	-100
TRT[3]*X	0	0	100	-100
Value	109.916846	111.25357472	110.2538931	114.61698055
Std Err	2.0331411469	2.1536686021	2.3539314203	3.3180767868

FIGURE 10.7 Least squares means evaluated at X = 100 for each of the treatments.

is the means of the predicted values of the treatments from all blocks, whether the treatments were observed in all blocks or not. This is another reason that the choice of the blocking factor is very important. The choice of the blocking factor must not be able to interact with the treatment or else the least squares means are likely not meaningful.

REFERENCES

Kenward, M. G. and Roger, J. H. (1997). Small sample inference for fixed effects from restricted maximum likelihood, *Biometrics* 54:983.
Milliken, G. A. and Johnson, D. E. (1992). *Analysis of Messy Data, Volume I: Design Experiments*, London, Chapman & Hall.

EXERCISES

EXERCISE 10.1: For the data set in Section 10.6, obtain the within block and the between block estimates of the model's parameters. The process is to show that there is between block information about the intercepts when an incomplete block design structure is used while it does not exist when a complete block design structure is used for the data set in Section 10.5.

EXERCISE 10.2: For the data in Section 10.6, use a common slope model and obtain the within block and between block estimates of the parameters.

EXERCISE 10.3: The data in the following table consist of eight incomplete blocks where Treatments 1 and 2 occur in blocks 1 through 4 and Treatments 3 and 4 occur in blocks 5 through 8. Use a common slopes model to describe the data for each treatment. Obtain the within block estimates and the between block estimates of the parameters. Since the design is not connected, care must be taken in constructing the parameters for each of the models. Use a mixed models code such as PROC MIXED to obtain predicted values for each of the treatments at each of the blocks evaluated at X = 40. Show that the means of these predicted values are the same as the least squares means evaluated at X = 40.

Data for Exercise 10.3

block	treat	x	y	treat	x	y
1	1	40.7	62.1	2	37.5	63.1
2	1	41.9	61.1	2	40.2	62.9
3	1	41.2	62.6	2	39.0	60.6
4	1	38.3	57.7	2	40.0	63.3
5	3	40.7	69.3	4	41.0	70.6
6	3	39.3	70.2	4	39.0	71.7
7	3	39.5	66.0	4	39.1	63.5
8	3	39.0	64.0	4	40.8	68.5

11 Covariate Measured on the Block in RCB and Incomplete Block Design Structures

11.1 INTRODUCTION

Measuring the value of the covariate on the block, i.e., setting up blocks where all experimental units within a block have the same value of the covariate, is often used as a method of constructing blocks of experimental units. Animals are grouped by age, weight, or stage of life. Students are grouped by class, age, or by IQ. The grouping of the experimental units by the value of some covariate forms more homogeneous groups on which to compare the treatments than by not blocking. However, one must be much more concerned that there is no interaction between the levels of the treatments and the levels of the factor used to construct the blocks. An assumption of the RCB (randomized complete block), or as a matter of fact any blocked design structure, is that there is no interaction between the factors in the treatment structure and the factors in the design structure (Milliken and Johnson, 1992). It is not unusual to see researchers construct blocks by using a factor such as initial age or weight or current thickness. However, some thought should be taken into account about the possibility of interaction with the levels of the treatments. The usual approach to the analysis of such data sets is to remove the block to block variation by the analysis of variance and not consider doing an analysis of covariance, i.e., ignore the fact that a covariate was measured. That strategy is appropriate if the slopes of the treatments' regression lines are equal, but if the slopes of the lines are unequal, a model with a covariate is required in order to extract the necessary information from the data. A block total or mean model must be used to make decisions about the slopes of the model before the analysis can continue.

There are some changes in the form of the analysis when the covariate is measured on the block rather than being measured on the experimental units within each block. The data in Table 11.1 are from a blocked experiment with two treatments where the covariate is measured on the block, i.e., the covariate has the same value for each of the two experimental units within the block. The within block analysis does not provide information about the magnitudes of the slopes. The sum of squares due to the slopes being zero after the block effects have been removed does not test the hypothesis that the slopes are zero, in fact the sum of squares is equal to zero providing no test. If the slopes are assumed to be equal, then the block effects

TABLE 11.1
Data for Example in Section 11.7

BLK	X	y1	y2	y_sum	y_dif
1	23.2	60.4	76.0	136.4	−15.6
2	26.9	59.9	76.3	136.2	−16.4
3	29.4	64.4	77.8	142.2	−13.4
4	22.7	63.5	75.6	139.1	−12.1
5	30.6	80.6	94.6	175.2	−14.0
6	36.9	75.9	96.1	172.0	−20.2
7	17.6	53.7	62.3	116.0	−8.6
8	28.5	66.3	81.6	147.9	−15.3

Note: Involves two treatments in a one-way treatment structure in a RCB design structure where y1 is the response for Treatment 1 and y2 is the response for Treatment 2; x is the covariate; BLK is the block; y_sum and y_dif are the sum and difference, respectively, of the observations within each block.

removes all information about the slopes. However, the information about the slopes can be extracted by constructing and analyzing the proper models. The proper models are the within block model and the between block model. The above described example with two treatments is analyzed in detail in Sections 11.2 to 11.7. Two additional examples involving more than two treatments, one in a RCB design structure and the other in a balanced incomplete block design structure, are included to demonstrate additional complications in the analyses.

11.2 THE WITHIN BLOCK MODEL

The basic model to describe the data in Table 11.1 is

$$y_{ij} = \alpha_i + \beta_i x_j + b_j + \varepsilon_{ij}, \quad i = 1, 2, \; j = 1, 2, \ldots, 6 \tag{11.1}$$

where $b_j \sim$ iid $N(0, \sigma_b^2)$, $\varepsilon_{ij} \sim$ iid $N(0, \sigma_\varepsilon^2)$ and the value of the covariate is the same for both experimental units in each block denoted by x_j (only one subscript on X).

The within block model is constructed by subtracting the values of the two observations within each block of Model 11.1, say the observation for Treatment 1 minus the observation for Treatment 2, yielding the model:

$$\begin{aligned} y_{1j} - y_{2j} &= \alpha_1 - \alpha_2 + (\beta_1 - \beta_2) x_j + \varepsilon_{1j} - \varepsilon_{2j} \\ &= \alpha_d + \beta_d x_j + \varepsilon_{dj}, \text{ where } \varepsilon_{dj} \sim N(0, 2\sigma_\varepsilon^2). \end{aligned} \tag{11.2}$$

This is a simple linear regression model where $y_{1j} - y_{2j}$ is the dependent variable and x_j is the independent variable with intercept $\alpha_1 - \alpha_2$ and slope $\beta_1 - \beta_2$. Fitting

Covariate Measured on the Block in RCB

the model to the data provides estimates of $\alpha_1 - \alpha_2$, $\beta_1 - \beta_2$, and σ_ε^2. If $\beta_1 = \beta_2$, the appropriate analysis for comparing the treatments is a paired t-test or a RCB analysis of variance with two treatments. The within block analysis provides statistics to test H_0: $\alpha_1 - \alpha_2$ and H_0: $\beta_1 - \beta_2$, but it does not enable the individual parameters (α_1, α_2, β_1, and β_2) to be estimated.

11.3 THE BETWEEN BLOCK MODEL

The between block or block sum or block total model is:

$$y_{1j} + y_{2j} = \alpha_1 + \alpha_2 + (\beta_1 + \beta_2)x_j + 2b_j + \varepsilon_{1j} + \varepsilon_{2j}$$
$$= \alpha_s + \beta_s x_j + e_{sj}^*, \text{ where } e_{sj}^* \sim N(0, 2(\sigma_\varepsilon^2 + 2\sigma_b^2)). \tag{11.3}$$

Model 11.3 is a simple linear regression model where where $y_{1j} + y_{2j}$ is the dependent variable and x_j is the independent variable with intercept $\alpha_1 + \alpha_2$ and slope $\beta_1 + \beta_2$. Fitting this model to the data provides estimates of $\alpha_1 + \alpha_2$, $\beta_1 + \beta_2$ and $\sigma_\varepsilon^2 + 2\sigma_b^2 = \sigma^2 e^*$.

11.4 COMBINING WITHIN BLOCK AND BETWEEN BLOCK INFORMATION

By combining the estimates from the within block model and the between block model, estimates of all of the parameters can be obtained. The estimates are

$$\hat{\alpha}_1 = (\hat{\alpha}_s + \hat{\alpha}_d)/2, \quad \hat{\alpha}_2 = (\hat{\alpha}_s - \hat{\alpha}_d)/2,$$
$$\hat{\beta}_1 = (\hat{\beta}_s + \hat{\beta}_d)/2, \quad \hat{\beta}_2 = (\hat{\beta}_s - \hat{\beta}_d)/2 \tag{11.4}$$
$$\hat{\sigma}_b^2 = (\hat{\sigma}_{e*}^2 - \hat{\sigma}_\varepsilon^2)/2.$$

The estimates of α_1, α_2, β_1, and β_2 involve both between block information and within block information. Assuming the data are normally distributed, the estimators from within blocks are independently distributed of the estimators from between blocks.

As in Chapters 9 and 10, the beta-hat model can be used to combine the within block and the between block information to obtain the estimators of the model's parameters and the corresponding covariance matrix. Let $\boldsymbol{\theta}_1 = \{(\alpha_1 - \alpha_2), (\beta_1 - \beta_2)\}'$, $\boldsymbol{\theta}_2 = (\alpha_1 + \alpha_2, \beta_1 + \beta_2)'$, $\text{Var}(\hat{\boldsymbol{\theta}}_1) = \Sigma_w$, $\text{Var}(\hat{\boldsymbol{\theta}}_2) = \Sigma_b$ and $\boldsymbol{\theta} = (\alpha_1, \alpha_2, \beta_1, \beta_2)'$. The beta-hat models relating $\hat{\boldsymbol{\theta}}_1$ and $\hat{\boldsymbol{\theta}}_2$ to $\boldsymbol{\theta}$ are

$$\hat{\boldsymbol{\theta}}_1 = \mathbf{H}_1 \boldsymbol{\theta} + \mathbf{e}_1, \quad \mathbf{e}_1 \sim N(\mathbf{0}, \Sigma_w) \text{ and}$$
$$\hat{\boldsymbol{\theta}}_2 = \mathbf{H}_2 \boldsymbol{\theta} + \mathbf{e}_2, \quad \mathbf{e}_2 \sim N(\mathbf{0}, \Sigma_b) \tag{11.5}$$

where

$$\mathbf{H}_1 = \begin{bmatrix} 1 & -1 & 0 & 0 \\ 0 & 0 & 1 & -1 \end{bmatrix} \text{ and } \mathbf{H}_2 = \begin{bmatrix} 1 & 1 & 0 & 0 \\ 0 & 0 & 1 & 1 \end{bmatrix}.$$

The combined estimator of θ is

$$\hat{\boldsymbol{\theta}}_c = \left[\mathbf{H}_1' \hat{\boldsymbol{\Sigma}}_w^{-1} \mathbf{H}_1 + \mathbf{H}_2' \hat{\boldsymbol{\Sigma}}_b^{-1} \mathbf{H}_2 \right]^{-1} \left[\mathbf{H}_1' \hat{\boldsymbol{\Sigma}}_w^{-1} \hat{\boldsymbol{\theta}}_1 + \mathbf{H}_2' \hat{\boldsymbol{\Sigma}}_b^{-1} \hat{\boldsymbol{\theta}}_2 \right]$$

with estimated approximate covariance matrix

$$\hat{\boldsymbol{\Sigma}}_\theta = \left[\mathbf{H}_1' \hat{\boldsymbol{\Sigma}}_w^{-1} \mathbf{H}_1 + \mathbf{H}_2' \hat{\boldsymbol{\Sigma}}_b^{-1} \mathbf{H}_2 \right]^{-1}. \tag{11.6}$$

To demonstrate the above described process, the data in Table 11.1 are used where the within block analysis is in Table 11.5 and the between block analysis is in Table 11.6. From Table 11.5, the within block information is

$$\hat{\boldsymbol{\theta}}_1 = \begin{bmatrix} -1.585320 \\ -0.476912 \end{bmatrix} \hat{\boldsymbol{\Sigma}}_2 = (2.05389)^2 \begin{bmatrix} 3.153408 & -0.112267 \\ -0.112267 & 0.00416 \end{bmatrix}$$

and from Table 11.6 the between block information is

$$\hat{\boldsymbol{\theta}}_b = \begin{bmatrix} 66.442387 \\ 2.935407 \end{bmatrix} \hat{\boldsymbol{\Sigma}}_b = (10.09140)^2 \begin{bmatrix} 3.153408 & -0.112267 \\ -0.112267 & 0.00416 \end{bmatrix}.$$

The combined estimator of θ and the approximate covariance matrix are

$$\hat{\boldsymbol{\theta}}_c = \begin{bmatrix} 32.4285 \\ 34.0138 \\ 1.2292 \\ 1.7062 \end{bmatrix} \hat{\boldsymbol{\Sigma}}_\theta = \begin{bmatrix} 83.608516 & 76.957248 & -2.976619 & -2.739822 \\ 76.957248 & 83.608516 & -2.739822 & -2.976619 \\ -2.976619 & -2.739822 & 0.110347 & 0.101569 \\ -2.739822 & 2.976619 & 0.101569 & 0.113047 \end{bmatrix}$$

A matrix manipulation software such as PROC IML described in Chapter 9 can be used to carry out the above computations. PROC MIXED was used to provide the computations of the combined estimator, which are displayed in the fourth part of Table 11.7. To test equality of slopes, test H_0: $\beta_1 - \beta_2 = 0$ vs. H_a: (not H_a). Let $\mathbf{a} = [0 \ 0 \ 1 \ -1]'$, then

$$\hat{\beta}_1 - \hat{\beta}_2 = \mathbf{a}' \hat{\boldsymbol{\theta}}_c = -0.4769$$

with variance $\text{Var}(\hat{\beta}_1 - \hat{\beta}_2) = \mathbf{a}'\hat{\Sigma}_\theta \mathbf{a} = (.132502)^2$. The test statistic is $t_c = -0.4769/.132502 = -3.5993$, and when compared to a t-distribution with 6 d.f., the significance level is 0.0114. The contrast statement was used in Table 11.7 to provide the test for equal slopes, labeled as b1 = b2.

11.5 COMMON SLOPE MODEL

When the slopes are equal, Model 11.1 becomes

$$y_{1j} = \alpha_i + \beta X_j + b_j + \varepsilon_{ij}. \tag{11.7}$$

The within block model is

$$y_{1j} - y_{2j} = \alpha_1 - \alpha_2 + \varepsilon_{1j} - \varepsilon_{2j}, \tag{11.8}$$

a model with intercept $\alpha_1 - \alpha_2$ and does not contain any information about the covariate. The between block model is

$$y_{1j} + y_{2j} = (\alpha_1 + \alpha_2) + 2\beta X_j + 2b_j + \varepsilon_{ij}, \tag{11.9}$$

a simple linear regression model with intercept $\alpha_1 + \alpha_2$ and slope 2β. All information about the covariate is contained in the between block model.

In order to compute adjusted means or LSMEANS, the estimate of $\boldsymbol{\theta} = [\alpha_1, \alpha_2, \beta]'$ needs to be computed. Let $\boldsymbol{\theta}_1 = (\alpha_1 - \alpha_2)$, $\boldsymbol{\theta}_2 = (\alpha_1 + \alpha_2, \beta)$, $\text{Var}(\hat{\boldsymbol{\theta}}_1) = \Sigma_w$, $\text{Var}(\hat{\boldsymbol{\theta}}_2) = \Sigma_b$, then the beta-hat models relating $\boldsymbol{\theta}_1$ and $\boldsymbol{\theta}_2$ to $\boldsymbol{\theta}$ are

$$\boldsymbol{\theta}_1 = \mathbf{H}_1 \boldsymbol{\theta} + \mathbf{e}_1 \quad \mathbf{e}_1 \sim N(\mathbf{0}, \Sigma_2)$$

$$\boldsymbol{\theta}_2 = \mathbf{H}_2 \boldsymbol{\theta} + \mathbf{e}_2 \quad \mathbf{e}_2 \sim N(\mathbf{0}, \Sigma_b)$$

where $\mathbf{H}_1 = [1 \; -1 \; 0]$ and $\mathbf{H}_2 = \begin{bmatrix} 1 & 1 & 0 \\ 0 & 0 & 1 \end{bmatrix}$.

The weight least squares estimate of $\boldsymbol{\theta}$ is

$$\hat{\boldsymbol{\theta}}_c = \left[\mathbf{H}'_1 \hat{\Sigma}_2 \mathbf{H}_1 + \mathbf{H}'_2 \hat{\Sigma}_b \mathbf{H}_2\right]^{-1} \left[\mathbf{H}'_1 \hat{\Sigma}_w \hat{\boldsymbol{\theta}}_1 + \mathbf{H}'_2 \hat{\Sigma}_b \hat{\boldsymbol{\theta}}_2\right]$$

with estimated approximate covariance matrix

$$\hat{\Sigma}_\theta = \left[\mathbf{H}'_1 \hat{\Sigma}_w \mathbf{H}_1 + \mathbf{H}'_2 \hat{\Sigma}_b \mathbf{H}_2\right]^{-1}.$$

11.6 ADJUSTED MEANS AND COMPARING TREATMENTS

11.6.1 COMMON SLOPE MODEL

For the common slope model, the adjusted means at $X = X_0$ are $\hat{\mu}_{i|X_0} = \hat{\alpha}_i + \hat{\beta}X_0$ or $\hat{\mu}_{1|X_0} = [1 \; 0 \; X_0]\hat{\theta}_c$ and $\hat{\mu}_{2|X_0} = [0 \; 1 \; X0]\hat{\theta}_c$ with variances

$$\text{Var}(\hat{\mu}_{1|X_0}) = (1 \quad 0 \quad X_0)\hat{\Sigma}_\theta(1 \quad 0 \quad X_0)'$$

and

$$\text{Var}(\hat{\mu}_{2|X_0}) = (0 \quad 1 \quad X_0)\hat{\Sigma}_\theta(0 \quad 1 \quad X_0)'.$$

The difference between two adjusted means at the same value of X is $\hat{\mu}_{1|X_0} - \hat{\mu}_{2|X_0} = \hat{\alpha}_1 - \hat{\alpha}_2$ which is independent of the covariate. The difference between the treatment means is estimated from the within block model and its variance depends only on the within block information. In order to compute the adjusted means and their standard errors, one must combine the between block and within block information.

11.6.2 NON-PARALLEL LINES MODEL

For the unequal slope model, the adjusted means at $X = X_0$ are

$$\hat{\mu}_{1|X_0} = \hat{\alpha}_1 + \hat{\beta}_1 X_0 = \begin{bmatrix} 1 & 0 & X_0 & 0 \end{bmatrix}\hat{\theta}_C$$

and

$$\hat{\mu}_{2|X_0} = \hat{\alpha}_2 + \hat{\beta}_2 X_0 = \begin{bmatrix} 0 & 1 & 0 & X_0 \end{bmatrix}\hat{\theta}_C$$

with variances

$$\text{Var}(\hat{\mu}_{1|X_0}) = (1 \quad 0 \quad X_0 \quad 0)\hat{\Sigma}_\theta(1 \quad 0 \quad X_0 \quad 0)'$$

and

$$\text{Var}(\hat{\mu}_{2|X_0}) = (0 \quad 1 \quad 0 \quad X_0)\hat{\Sigma}_\theta(0 \quad 1 \quad 0 \quad X_0)'.$$

The difference between two adjusted means at the same value of X_o is

$$\hat{\mu}_{1|X_0} - \hat{\mu}_{2|X_0} = (\hat{\alpha}_1 - \hat{\alpha}_2) + (\hat{\beta}_1 - \hat{\beta}_2)X_0 = (1 \quad -1 \quad X_0 \quad -X_0)\hat{\theta}_c = \mathbf{a}'_{X_0}\hat{\theta}_c$$

with variance $\text{Var}(\hat{\mu}_{1|X_0} - \hat{\mu}_{2|X_0}) = \mathbf{a}'_{X_0}\hat{\Sigma}_\theta \mathbf{a}_{X_0}.$

TABLE 11.2
PROC GLM Code and Analysis for Model with X and X*TREAT for the Two Treatment Data Set

```
proc glm data=III117;
class blk treat;
model yield=blk treat x x*treat/solution;
random blk/test;
```

Source	df	SS	MS	FValue	ProbF
Model	9	2203.2222	244.8025	116.06	0.0000
Error	6	12.6553	2.1092		
Corrected Total	15	2215.8775			

Source	df	SS (I)	MS	FValue	ProbF
BLK	7	1340.6875	191.5268	90.80	0.0000
TREAT	1	835.2100	835.2100	395.98	0.0000
X	0	0.0000			
X*TREAT	1	27.3247	27.3247	12.95	0.0114

Source	df	SS (III)	MS	FValue	ProbF
BLK	6	305.5088	50.9181	24.14	0.0006
TREAT	1	0.3985	0.3985	0.19	0.6790
X	0	0.0000			
X*TREAT	1	27.3247	27.3247	12.95	0.0114

11.7 EXAMPLE: TWO TREATMENTS

The data in Table 11.1 are from an experiment with two treatments in a randomization complete block design structure with the covariate measured on the block. The variables y_sum and y_diff represent the sum of the two yields and the difference of the two yields (TRT 1 − TRT 2), respectively, for each block. Table 11.2 contains the PROC GLM code to fit the unequal slopes model to the data. The model statement involves both X and X*TREAT. However, the sum of squares due to X is zero. This is occurring because both the Type I and Type III sums of squares are adjusted for the block effect. Adjusting for the block effect removes the average slope effect since the covariate is measure on the block. Table 11.3 contains the PROC GLM code to fit the uequal slopes model where X*TREAT is included without the X term. As discussed in previous chapters, the X*TREAT term is used to test the slopes equal to zero hypothesis. However, the number degrees of freedom associated with the X*TREAT term in either the Type I or Type III sums of squares is one instead of the predicted two. The sum of squares X*TREAT is adjusted for the other terms in the model, including the block effects. Since the covariate is measured on the block, when the block effect is removed, the average slope is also removed, thus leaving one degree of freedom for the X*TREAT sum of squares. In this case, the X*TREAT sums of squares provides a test of the equal slopes hypothesis. The least squares solution in the last part of Table 11.3 indicates that only one linear combination of the slopes is estimable. Also of interest are the numbers of degrees of

TABLE 11.3
PROC GLM Code and Results for Fitting the Unequal Slope Model that Extracts Only Within Block Information for the Two Treatment Experiment

```
proc glm data=III117;
class blk treat;
model yield=blk treat x*treat/solution;
random blk/test;
```

Source	df	SS	MS	FValue	ProbF
Model	9	2197.7203	244.1911	166.81	0.0000
Error	5	7.3197	1.4639		
Corrected Total	14	2205.0400			

Source	df	SS (I)	MS	FValue	ProbF
BLK	7	1451.530	207.361	141.65	0.0000
TREAT	1	714.286	714.286	487.92	0.0000
X*TREAT	1	31.905	31.905	21.79	0.0055

Source	df	SS (III)	MS	FValue	ProbF
BLK	6	303.764	50.627	34.58	0.0006
TREAT	1	0.023	0.023	0.02	0.9058
X*TREAT	1	31.905	31.905	21.79	0.0055

Parameter	Estimate	StdErr	tValue	Probt
Intercept	81.3559	0.9164	88.78	0.0000
BLK 1	−8.9729	1.6132	−5.56	0.0026
BLK 2	−6.2770	1.2134	−5.17	0.0035
BLK 3	−2.6098	1.2110	−2.16	0.0837
BLK 4	−5.9478	1.2545	−4.74	0.0051
BLK 5	14.2104	1.2159	11.69	0.0001
BLK 6	14.2917	1.3017	10.98	0.0001
BLK 7	−18.8589	1.3610	−13.86	0.0000
BLK 8	0.0000			
TREAT 1	0.3997	3.2115	0.12	0.9058
TREAT 2	0.0000			
X*TREAT 1	−0.5337	0.1143	−4.67	0.0055
X*TREAT 2	0.0000			

freedom associated with the two BLK sum of squares. The number of degrees of freedom associated with the Type I sum of squares for BLK is 7 = (8 − 1) since for this model the block term is first in the model statement. The number of degrees of freedom associated with the Type III sum of squares for BLK is 6 since this sum of squares is adjusted for the other terms in the model, including X*TREAT. In this case one of the degrees of freedom for BLK is totally tied up with (or confounded) one of the degrees of freedom for X*TREAT. The least squares means are in Table 11.4, where PROC GLM was able to provide the adjusted means at the means of the X values, but the other adjusted means were declared to be nonestimable

TABLE 11.4
Least Squares Means Statements and Results for the Two Treatment Experiment (If specify value of x, the least squares means are not estimable.)

```
lsmeans treat/pdiff at means stderr;
lsmeans treat/pdiff at x=15 stderr;
lsmeans treat/pdiff at x=20 stderr;
lsmeans treat/pdiff at x=25 stderr;
lsmeans treat/pdiff at x=30 stderr;
lsmeans treat/pdiff at x=35 stderr;
lsmeans treat/pdiff at x=40 stderr;
```

TREAT	LSMean	StdErr	Probt	ProbtDiff
1	65.5875	0.5135	0.0000	0.0000
2	80.0375	0.5135	0.0000	

TABLE 11.5
PROC REG Code and Within Block Analysis Based on the Block Differences, Including the Inverse of X′X

```
proc reg data=wide;
model y_dif=x/xpx i;
```

Variable	Intercept	X
Intercept	3.15340755	–0.11226719
X	–0.11226719	0.00416190

Source	df	SS	MS	FValue	ProbF
Model	1	54.6493	54.6493	12.95	0.0114
Error	6	25.3107	4.2184		
Corrected Total	7	79.9600			

Variable	df	Estimate	StdErr	tValue	Probt
Intercept	1	–1.5853	3.6473	–0.43	0.6790
X	1	–0.4769	0.1325	–3.60	0.0114

when using X = 15, or 20, or 25, or 30, or 35, or 40. Basically, the least squares means are not estimable from the within block information, but PROC GLM has been set up to provide the adjusted means at the mean of the X values. These least squares means have an estimated standard error of 0.5135, which does not take the block variance component into account and are too small (See Table 11.8 for appropriate estimated standard errors.) The information for computing the combined estimates in Section 11.3 can be extracted from two tables. Table 11.5 contains the PROC REG code to fit the within block model and Table 11.6 contains the PROC REG code to fit the between block model. Both analyses provide estimates of the models' parameters and the inverse of the $X'X$ matrices (which are the same for

TABLE 11.6
PROC REG Code and Between Block Analysis Based on the Block Sums, Including the Inverse of X'X

```
proc reg data=wide;
  model y_sum=x/xpx i;
```

Variable	Intercept	X
Intercept	3.15340755	–0.11226719
X	–0.11226719	0.00416190

Source	df	SS	MS	FValue	ProbF
Model	1	2070.3575	2070.3575	20.33	0.0041
Error	6	611.0175	101.8363		
Corrected Total	7	2681.3750			

Variable	df	Estimate	StdErr	tValue	Probt
Intercept	1	66.4424	17.9201	3.71	0.0100
X	1	2.9354	0.6510	4.51	0.0041

this example). The error sum of squares from Table 11.5 (estimates $2\sigma_\varepsilon^2$) and Table 11.6 (estimates $2(\sigma_\varepsilon^2 + 2\sigma_\delta^2)$) correspond to twice the error sum of squares and Type III block sum of squares in Table 11.2.

Table 11.7 contains the PROC MIXED code to fit the unequal slopes model. This is the ideal way to analyze the data since the mixed models approach carries out the combined within-block and between-block analysis. The DDFM = KR option was used to provide approximate numbers of degrees of freedom when needed. The fourth part of Table 11.7 contains the combined estimates of the intercepts and slopes. The contrast statements are used to test the equality of the intercepts (when meaningful) and the equality of the slopes. The significance level corresponding to the equal slope hypothesis is 0.0114, indicating there is sufficient evidence to conclude the slopes are not equal. Adjusted means at X = 26.98, 15, 20, 25, 30, 35, and 40 are provided in Table 11.8. The least squares means at X = 26.98 are the same as provided by PROC GLM in Table 11.4. However, the mixed model estimated standard errors are larger and more appropriate than those provided by PROC GLM. The differences of the adjusted means are in the lower part of Table 11.8. Treatment 2 mean is significantly larger than the mean for Treatment 1 for each value of X.

A common thought is that if the blocks are formed by using the value of a covariate, then the analysis of covariance is not necessary since the blocking effect will account for the covariate effect. To evaluate that thought, an analysis of variance was carried out. Table 11.9 contains the PROC MIXED code to provide the RCB analysis of variance of the one-way treatment structure. The first difference is in the estimate of the BLK variance component: it is 24.4045 when the covariate is included and 92.9077 when the covariate is not included. When the covariate is measured on the block, including the covariate in the analysis reduces the estimate of the block variance component, i.e., it provides the block variance component after removing the effect of the covariate. The major effect of this inflated variance component is

TABLE 11.7
PROC MIXED Code and Analysis of the Data in Table 11.1 Providing the Combined Within-Between Block Analysis

```
proc mixed cl covtest data=III117;
class blk treat;
model yield=treat x*treat/solution noint;
contrast 'a1=a2' treat 1 -1;
contrast 'b1=b2' x*treat 1 -1;
random blk;
```

CovParm	Estimate	StdErr	ZValue	ProbZ	Alpha	Lower	Upper
BLK	24.4045	14.7114	1.66	0.0486	0.05	9.8367	130.4912
Residual	2.1092	1.2178	1.73	0.0416	0.05	0.8758	10.2278

Effect	NumDF	DenDF	FValue	ProbF
TREAT	2	7.3	6.39	0.0248
X*TREAT	2	7.3	15.27	0.0025

Effect	TREAT	Estimate	StdErr	df	tValue	Probt
TREAT	1	32.4285	9.1438	6.5	3.55	0.0121
TREAT	2	34.0138	9.1438	6.5	3.72	0.0099
X*TREAT	1	1.2292	0.3322	6.5	3.70	0.0101
X*TREAT	2	1.7062	0.3322	6.5	5.14	0.0021

Label	NumDF	DenDF	FValue	ProbF
a1=a2	1	6	0.19	0.6790
b1=b2	1	6	12.95	0.0114

on the estimated standard error of the least squares means. The estimated standard error for a least squares mean from the analysis of variance is 3.511, while it is 1.8205 for the analysis of covariance (at $X = 26.98$). The difference between the two least squares means is the same for both analyses, but again the estimated standard errors are smaller when the covariate is used in the analysis.

11.8 EXAMPLE: FOUR TREATMENTS IN RCB

Four rations were evaluated to determine how they effected the way female calves gain weight. The calves were grouped by initial weight and blocks of four calves were formed. The rations were randomly assigned to one calf within each block yielding a one-way treatment structure in a RCB design structure. The average daily gain was computed for each of the calves in the study by ADG = (final weight − initial weight)/number of days on study. The average pen weight was the only information available about the calves weight prior to the experiment, thus the average pen weight was used as the covariate, i.e., all calves in a pen have the same value of the covariate. The data are in Table 11.10 where $ration_i$ denotes ADG of the calves assigned to that ration and wt denotes the mean weight of the pen in pounds divided by 1000.

TABLE 11.8
Least Squares Means and Comparisons at x = 26.98 (Mean), 15, 20, 25, 30, 35, and 40

```
lsmeans treat/diff at means;
lsmeans treat/diff at x=15;
lsmeans treat/diff at x=20;
lsmeans treat/diff at x=25;
lsmeans treat/diff at x=30;
lsmeans treat/diff at x=35;
lsmeans treat/diff at x=40;
```

TREAT	X	Estimate	StdErr	df	tValue	Probt
1	26.98	65.5875	1.8205	6	36.03	0.0000
2	26.98	80.0375	1.8205	6	43.96	0.0000
1	15	50.8673	4.3747	6	11.63	0.0000
2	15	59.6062	4.3747	6	13.63	0.0000
1	20	57.0135	2.9466	6	19.35	0.0000
2	20	68.1370	2.9466	6	23.12	0.0000
1	25	63.1597	1.9351	6	32.64	0.0000
2	25	76.6678	1.9351	6	39.62	0.0000
1	30	69.3060	2.0794	6	33.33	0.0000
2	30	85.1986	2.0794	6	40.97	0.0000
1	35	75.4522	3.2281	6	23.37	0.0000
2	35	93.7294	3.2281	6	29.04	0.0000
1	40	81.5985	4.6941	6	17.38	0.0000
2	40	102.2602	4.6941	6	21.78	0.0000

TREAT	_TREAT	X	Estimate	StdErr	df	tValue	Probt
1	2	26.98	−14.4500	0.7262	6	−19.90	0.0000
1	2	15	−8.7390	1.7450	6	−5.01	0.0024
1	2	20	−11.1235	1.1754	6	−9.46	0.0001
1	2	25	−13.5081	0.7719	6	−17.50	0.0000
1	2	30	−15.8927	0.8294	6	−19.16	0.0000
1	2	35	−18.2772	1.2876	6	−14.19	0.0000
1	2	40	−20.6618	1.8724	6	−11.04	0.0000

The model assuming there is a linear relationship between ADG and WT is

$$\text{ADG}_{ij} = \alpha_i + \beta_i \text{WT}_j + b_j + \varepsilon_{ij}. \qquad (11.10)$$

The within block analysis provides estimates of the contrasts of the α_i's and contrasts of the β_i's. Table 11.11 contains the PROC GLM code to extract the within block information. The results are in Tables 11.11 and 11.12 which includes that partition of $(\mathbf{X \cdot X})^-$ and parameter estimates corresponding to the α's and β's. The solutions for the $\hat{\alpha}_i$'s and $\hat{\beta}_i$'s satisfy the set-to-zero restrictions. The sum of squares corresponding to wt*Ration tests the equal slopes hypothesis (df = 3) instead of testing

TABLE 11.9
PROC MIXED Code and Analysis of Variance without the Covariate

```
proc mixed cl covtest data=III117;
class blk treat;
model yield=treat/solution;
lsmeans treat/diff;
random blk;
```

CovParm	Estimate	StdErr	ZValue	ProbZ	Alpha	Lower	Upper
BLK	92.9077	51.2104	1.81	0.0348	0.05	39.8015	408.7899
Residual	5.7114	3.0529	1.87	0.0307	0.05	2.4968	23.6586

Effect	NumDF	DenDF	FValue	ProbF
TREAT	1	7	146.23	0.0000

Effect	TREAT	Estimate	StdErr	df	tValue	Probt
TREAT	1	65.5875	3.5110	7	18.68	0.0000
TREAT	2	80.0375	3.5110	7	22.80	0.0000

Effect	TREAT	_TREAT	Estimate	StdErr	df	tValue	Probt
TREAT	1	2	−14.5500	1.1949	7	−12.09	0.0000

TABLE 11.10
Data for Example in Section 11.8

blk	wt	ration1	ration2	ration3	ration4
1	0.51	2.19	2.44	3.02	2.66
2	0.52	2.11	2.47	3.01	2.96
3	0.54	2.06	2.28	2.82	2.36
4	0.55	1.70	1.91	2.60	2.35
5	0.57	2.00	1.99	2.42	2.22
6	0.59	2.59	2.59	3.37	2.90
7	0.61	2.06	2.34	3.16	2.59
8	0.62	1.91	1.99	2.62	2.29
9	0.64	1.98	2.06	2.96	2.39
10	0.65	1.73	1.72	2.65	2.26
11	0.67	1.80	1.86	2.81	2.10
12	0.69	2.37	2.12	2.99	2.41

Note: Rationi corresponds to the average daily gain for the ith ration and wt is initial weight in pounds divided by 1000.

the slopes equal to zero hypothesis (df = 4). The significance level corresponding to the equal slopes hypothesis is 0.0052, indicating there is sufficient evidence to conclude the slopes are not equal. The least squares means were requested at the

TABLE 11.11
PROC GLM Code and Within Block Analysis
for Data of Section 11.8

```
proc glm data=III118;
class blk ration;
model adg=blk ration wt*ration/solution i;
random blk/test;
```

Source	df	SS	MS	FValue	ProbF
Model	17	7.7508	0.4559	39.89	0.0000
Error	30	0.3429	0.0114		
Corrected Total	47	8.0936			

Source	df	SS (I)	MS	FValue	ProbF
blk	11	2.6066	0.2370	20.73	0.0000
Ration	3	4.9657	1.6552	144.84	0.0000
wt*Ration	3	0.1785	0.0595	5.21	0.0052

Source	df	SS (III)	MS	FValue	ProbF
blk	10	2.2988	0.2299	20.12	0.0000
Ration	3	0.1585	0.0528	4.62	0.0090
wt*Ration	3	0.1785	0.0595	5.21	0.0052

mean of the wt values and at wt = 0.510 and 0.680. Only the adjusted means at the mean of the wt values (0.596) are estimable and are provided in Table 11.13.

The block sum or total model is

$$\text{ADG}_{\cdot j} = \alpha_{\cdot} + \beta_{\cdot} \text{WT}_j + e_j \tag{11.11}$$

where $e_j = 4b_j + \varepsilon_{\cdot j} \sim N(0, 4(\sigma_\varepsilon^2 + 4\sigma_b^2))$, which is a simple linear regression model. The block total data are in Table 11.14 (note that model depends on WT not 4*WT) and the PROC REG code to fit the blocked total model is in Table 11.15. This analysis provides estimates of $\alpha_{\cdot}$, $\beta_{\cdot}$ and $4(\sigma_\varepsilon^2 + 4\sigma_b^2)$ as $\hat{\alpha}_{\cdot} = 12.7662$, $\hat{\beta}_{\cdot} = -5.4521$ and $4(\widehat{\sigma_\varepsilon^2 + 4\sigma_b^2}) = 0.9195$. The $\mathbf{X}'\mathbf{X}^{-1}$ matrix is also included in Table 11.15. Using the between block and the within block information, combined estimators of the slopes and intercepts can be computed using the process outlined in Chapter 10. The next step is to combine the within block information and between block information. The vector of parameters to be estimate is

$$\boldsymbol{\theta} = \left(\overline{\alpha}_{\cdot}, \overline{\beta}_{\cdot}, \alpha_1 - \overline{\alpha}_{\cdot}, \alpha_2 - \overline{\alpha}_{\cdot}, \alpha_3 - \overline{\alpha}_{\cdot}, \alpha_4 - \overline{\alpha}_{\cdot}, \beta_1 - \overline{\beta}_{\cdot}, \beta_2 - \overline{\beta}_{\cdot}, \beta_3 - \overline{\beta}_{\cdot}, \beta_4 - \overline{\beta}_{\cdot}\right)' \tag{11.12}$$

Let

$$\boldsymbol{\theta}_1^* = \left(\alpha_1 - \alpha_4, \alpha_2 - \alpha_4, \alpha_3 - \alpha_4, \alpha_4 - \alpha_4, \beta_1 - \beta_4, \beta_2 - \beta_4, \beta_3 - \beta_4, \beta_4 - \beta_4\right)' \tag{11.13}$$

TABLE 11.12
Partition of Inverse of X'X Corresponding to the Intercepts and Slopes and Within Block Estimates of the Slopes and Intercepts for the Data of Section 11.8

Parameter	Ration_1	Ration_2	Ration_3	Ration_4	wt_Ration_1	wt_Ration_2	wt_Ration_3	wt_Ration_4
Ration 1	17.332451	8.666225	8.666225	0.000000	−28.793599	−14.396799	−14.396799	0.000000
Ration 2	8.666225	17.332451	8.666225	0.000000	−14.396799	−28.793599	−14.396799	0.000000
Ration 3	8.666225	8.666225	17.332451	0.000000	−14.396799	−14.396799	−28.793599	0.000000
Ration 4	0.000000	0.000000	0.000000	0.000000	0.000000	0.000000	0.000000	0.000000
wt*Ration 1	−28.793599	−14.396799	−14.396799	0.000000	48.297901	24.148951	24.148951	0.000000
wt*Ration 2	−14.396799	−28.793599	−14.396799	0.000000	24.148951	48.297901	24.148951	0.000000
wt*Ration 3	−14.396799	−14.396799	−28.793599	0.000000	24.148951	24.148951	48.297901	0.000000
wt*Ration 4	0.000000	0.000000	0.000000	0.000000	0.000000	0.000000	0.000000	0.000000

Parameter	Estimate	StdErr	tValue	Probt
Ration 1	−1.4789	0.4451	−3.32	0.0024
Ration 2	−0.1669	0.4451	−0.37	0.7103
Ration 3	−0.8355	0.4451	−1.88	0.0702
Ration 4	0.0000			
wt*Ration 1	1.7832	0.7429	2.40	0.0228
wt*Ration 2	−0.2400	0.7429	−0.32	0.7489
wt*Ration 3	2.0919	0.7429	2.82	0.0085
wt*Ration 4	0.0000			

TABLE 11.13
Least Squares Means and Comparisons at the Mean Value of wt (0.596) as Others are Not Estimable from the Within Block Information

```
lsmeans ration/pdiff at means stderr;
lsmeans ration/pdiff at wt=.510 stderr;
lsmeans ration/pdiff at wt=.680 stderr;
```

Ration	LSMean	StdErr	Probt	LSMean#
1	2.0417	0.0309	0.0000	1
2	2.1475	0.0309	0.0000	2
3	2.8692	0.0309	0.0000	3
4	2.4575	0.0309	0.0000	4

RowName	_1	_2	_3	_4
1		0.0215	0.0000	0.0000
2	0.0215		0.0000	0.0000
3	0.0000	0.0000		0.0000
4	0.0000	0.0000	0.0000	

TABLE 11.14
Sums of the ADG Values within Each of the Blocks, Information for the Between Block Analysis

BLK	sADG	mWT
1	10.31	0.505
2	10.55	0.520
3	9.52	0.539
4	8.56	0.551
5	8.63	0.569
6	11.45	0.585
7	10.15	0.606
8	8.81	0.619
9	9.39	0.640
10	8.36	0.654
11	8.57	0.674
12	9.89	0.692

denote the parameter estimated by the set to zero restriction solution. Convert θ_1^* to a sum-to-zero restriction θ_1 by $\theta_1 = \mathbf{T}_1 \theta_1^*$ where

$$\mathbf{T}_1 = \begin{bmatrix} \mathbf{I}_4 - \frac{1}{4}\mathbf{J}_4 & 0 \\ 0 & \mathbf{I}_4 - \frac{1}{4}\mathbf{J}_4 \end{bmatrix}.$$

TABLE 11.15
PROC REG Code and Between Block Analysis with the Inverse of X′X and the Parameter Estimates

```
proc reg data=blktotal;
model sadg=mwt/xpx i;
```

Variable	Intercept	mWT
Intercept	8.666225278	−14.3967995
mWT	−14.3967995	24.14895073

Source	df	SS	MS	FValue	ProbF
Model	1	1.2309	1.2309	1.34	0.2742
Error	10	9.1954	0.9195		
Corrected Total	11	10.4263			

Variable	df	Estimate	StdErr	tValue	Probt
Intercept	1	12.7662	2.8229	4.52	0.0011
mWT	1	−5.4521	4.7123	−1.16	0.2742

and $\boldsymbol{\theta}_1 = \left(\alpha_1 - \bar{\alpha}., \alpha_2 - \bar{\alpha}., \alpha_3 - \bar{\alpha}., \alpha_4 - \bar{\alpha}., \beta_1 - \bar{\beta}., \beta_2 - \bar{\beta}., \beta_3 - \bar{\beta}., \beta_4 - \bar{\beta}.\right)'$,

The beta-hat model relating $\hat{\boldsymbol{\theta}}_1$ to $\boldsymbol{\theta}$ is

$$\hat{\boldsymbol{\theta}}_1 = \mathbf{H}_1 \boldsymbol{\theta} + \mathbf{e}_1 \tag{11.14}$$

where

$$\mathbf{H}_1 = \begin{bmatrix} 0 & 0 & 1 & 0 & 0 & 0 & 0 & 0 & 0 & 0 \\ 0 & 0 & 0 & 1 & 0 & 0 & 0 & 0 & 0 & 0 \\ 0 & 0 & 0 & 0 & 1 & 0 & 0 & 0 & 0 & 0 \\ 0 & 0 & 0 & 0 & 0 & 1 & 0 & 0 & 0 & 0 \\ 0 & 0 & 0 & 0 & 0 & 0 & 1 & 0 & 0 & 0 \\ 0 & 0 & 0 & 0 & 0 & 0 & 0 & 1 & 0 & 0 \\ 0 & 0 & 0 & 0 & 0 & 0 & 0 & 0 & 1 & 0 \\ 0 & 0 & 0 & 0 & 0 & 0 & 0 & 0 & 0 & 1 \end{bmatrix} \tag{11.15}$$

with $\widehat{\text{Var}(\mathbf{e}_1)} = \hat{\Sigma}_2$.

The between block information estimates $\boldsymbol{\theta}_2^* = (\alpha., \beta.)'$ which can be transformed to $\boldsymbol{\theta}_2 = (\bar{\alpha}., \bar{\beta}.)'$ by $\boldsymbol{\theta}_2 = \mathbf{T}_2$ where $\mathbf{T}_2 = (1/4)\mathbf{I}_2$.

The beta-hat model relating $\hat{\boldsymbol{\theta}}_2$ to $\boldsymbol{\theta}$ is $\hat{\boldsymbol{\theta}}_2 = \mathbf{H}_2 \boldsymbol{\theta} + \mathbf{e}_2$ where

$$\mathbf{H}_2 = \begin{bmatrix} 1 & 0 & 0 & 0 & 0 & 0 & 0 & 0 & 0 & 0 \\ 0 & 1 & 0 & 0 & 0 & 0 & 0 & 0 & 0 & 0 \end{bmatrix}$$

and $\widehat{\text{Var}(\mathbf{e}_2)} = \hat{\Sigma}_b$.

The two beta-hat models are combined as $\begin{bmatrix}\hat{\theta}_1\\\hat{\theta}_2\end{bmatrix}=\begin{bmatrix}H_1\\H_2\end{bmatrix}\theta+\begin{pmatrix}e_1\\e_2\end{pmatrix}$ and the results in Section 11.3 are applied to provide the combined estimate of θ and its corresponding estimated covariance matrix. In this chapter the combining process is accomplished by using the mixed models approach. The intercepts and slopes are computed as $\begin{bmatrix}\hat{\alpha}\\\hat{\beta}\end{bmatrix}=H_0\hat{\theta}$ where

$$H = \begin{bmatrix} 1 & 0 & 1 & 0 & 0 & 0 & 0 & 0 & 0 & 0 \\ 1 & 0 & 0 & 1 & 0 & 0 & 0 & 0 & 0 & 0 \\ 1 & 0 & 0 & 0 & 1 & 0 & 0 & 0 & 0 & 0 \\ 1 & 0 & 0 & 0 & 0 & 1 & 0 & 0 & 0 & 0 \\ 0 & 1 & 0 & 0 & 0 & 0 & 1 & 0 & 0 & 0 \\ 0 & 1 & 0 & 0 & 0 & 0 & 0 & 1 & 0 & 0 \\ 0 & 1 & 0 & 0 & 0 & 0 & 0 & 0 & 1 & 0 \\ 0 & 1 & 0 & 0 & 0 & 0 & 0 & 0 & 0 & 1 \end{bmatrix}$$ (11.16)

The adjusted mean for, say, ration 1 at $X = X_0$ is $\hat{\mu}_{1|X_0} = a'_{X_0}\hat{\theta}$ where $a'_{X_0} = (1\ X_0\ 1\ 0\ 0\ 0\ X_0\ 0\ 0\ 0)$ and differences between, say rations 1 and 2 at $X = X_0$ is $\hat{\mu}_{1|X_0} - \hat{\mu}_{2|X_0} = b'_{X_0}\hat{\theta}$ where $b'_{X_0} = (0\ 0\ 1\ -1\ 0\ 0\ X_0\ -X_0\ 0\ 0)$.

In this chapter the combining process is accomplished by using the mixed models approach. The PROC MIXED code in Table 11.16 fits the unequal slopes model to the data set with contrast statements to provide tests of the equal slopes and of the equal intercepts hypotheses. Since the results in Table 11.16 are based on combined estimators, the Kenward-Roger adjustment to the denominator of freedom was used (Kenward and Roger, 1997). The lower partition of Table 11.16 contains the combined estimates of the intercepts and slopes of the four regression models. The F-statistic corresponding to wt*Ration tests the slopes equal to zero hypothesis and the contrast with label "b1 = b2 = b3 = b4" tests the equal slopes hypotheses. There is sufficient information to reject both hypotheses as the significance levels are 0.0095 and 0.0052, respectively. The least squares means evaluated at wt = 0.596, 0.510, and 0.680 are included in Table 11.17 where the lower part of the table contains the pairwise comparisons of the ration means within each level of wt. All adjusted means within a value of wt are significantly different ($p \leq 0.05$) except for rations 1 and 2 at wt = 0.680 which has a significance level of 0.4083. Finally, Table 11.18 contains the PROC MIXED code to carry out an analysis of variance on the data set. The estimate of the blk variance component is 0.0553 for the analysis of variance and is 0.0546 when the covariate is taken into account. The least squares means in the fourth partition of Table 11.18 are identical to the least squares means in Table 11.17 evaluated at wt = 0.596. The analysis of covariance has smaller estimated standard errors for the least squares means and for the comparison of the least squares means. The analysis of covariance is providing additional information

TABLE 11.16
PROC MIXED Code and Combined Within-Between Block Analysis with Parameter Estimates for the Unequal Slopes Model

```
proc mixed cl covtest data=III118;
class blk ration;
model adg=ration wt*ration/solution noint ddfm=kr;
contrast 'a1=a2=a3=a4' ration 1 -1, ration 1 0 -1, ration 1 0 0 -1;
contrast 'b1=b2=b3=b4' wt*ration 1 -1, wt*ration 1 0 -1,
  wt*ration 1 0 0 -1;
random blk;
```

CovParm	Estimate	StdErr	ZValue	ProbZ	Alpha	Lower	Upper
blk	0.0546	0.0257	2.12	0.0168	0.05	0.0259	0.1817
Residual	0.0114	0.0030	3.87	0.0001	0.05	0.0073	0.0204

Effect	NumDF	DenDF	FValue	ProbF
Ration	4	28.6	8.30	0.0001
wt*Ration	4	28.6	4.10	0.0095

Label	NumDF	DenDF	FValue	ProbF
a1=a2=a3=a4	3	30	4.62	0.0090
b1=b2=b3=b4	3	30	5.21	0.0052

Effect	Ration	Estimate	StdErr	df	tValue	Probt
Ration	1	2.3330	0.7565	13.1	3.08	0.0086
Ration	2	3.6450	0.7565	13.1	4.82	0.0003
Ration	3	2.9764	0.7565	13.1	3.93	0.0017
Ration	4	3.8119	0.7565	13.1	5.04	0.0002
wt*Ration	1	−0.4886	1.2629	13.1	−0.39	0.7050
wt*Ration	2	−2.5119	1.2629	13.1	−1.99	0.0680
wt*Ration	3	−0.1799	1.2629	13.1	−0.14	0.8889
wt*Ration	4	−2.2718	1.2629	13.1	−1.80	0.0951

about the relationship among the four ration means than is obtained by ignoring the covariate because it was used as a blocking factor.

Figure 11.1 contains the model specification screen for analyzing the data using JMP®. ADG was selected as the response variable. The model effects are blk (specified as being random), ration, wt and wt*ration. Do not use the center model option from the model specification menu (deactivate it), then click on run model. Figure 11.2 contains the estimates of the variance components and the test of the effects in the model. These are identical to those of PROC MIXED in Table 11.16. The custom test menu can be used to provide least squares means at various values of wt as was done in Chapter 10, but those results are not included here.

11.9 EXAMPLE: FOUR TREATMENTS IN BIB

The data in Table 11.19 are surface finish measurements (Finish) of rods turned or cut on a lathe. Sixteen rods were available for the study involving four types of

TABLE 11.17
Least Squares Means and Comparisons at wt = 0.596, 0.510, and 0.0680

```
lsmeans  ration/diff  at  means;
lsmeans  ration/diff  at  wt=.510;
lsmeans  ration/diff  at  wt=.680;
```

Effect	Ration	wt	Estimate	StdErr	df	tValue	Probt
Ration	1	0.596	2.0417	0.0742	13.1	27.52	0.0000
Ration	2	0.596	2.1475	0.0742	13.1	28.95	0.0000
Ration	3	0.596	2.8692	0.0742	13.1	38.68	0.0000
Ration	4	0.596	2.4575	0.0742	13.1	33.13	0.0000
Ration	1	0.510	2.0838	0.1317	13.1	15.82	0.0000
Ration	2	0.510	2.3639	0.1317	13.1	17.95	0.0000
Ration	3	0.510	2.8847	0.1317	13.1	21.90	0.0000
Ration	4	0.510	2.6533	0.1317	13.1	20.15	0.0000
Ration	1	0.680	2.0007	0.1293	13.1	15.48	0.0000
Ration	2	0.680	1.9369	0.1293	13.1	14.98	0.0000
Ration	3	0.680	2.8541	0.1293	13.1	22.08	0.0000
Ration	4	0.680	2.2670	0.1293	13.1	17.54	0.0000

Effect	Ration	_Ration	wt	Estimate	StdErr	df	tValue	Probt
Ration	1	2	0.596	−0.1058	0.0436	30	−2.42	0.0215
Ration	1	3	0.596	−0.8275	0.0436	30	−18.96	0.0000
Ration	1	4	0.596	−0.4158	0.0436	30	−9.53	0.0000
Ration	2	3	0.596	−0.7217	0.0436	30	−16.54	0.0000
Ration	2	4	0.596	−0.3100	0.0436	30	−7.10	0.0000
Ration	3	4	0.596	0.4117	0.0436	30	9.43	0.0000
Ration	1	2	0.510	−0.2802	0.0775	30	−3.62	0.0011
Ration	1	3	0.510	−0.8009	0.0775	30	−10.34	0.0000
Ration	1	4	0.510	−0.5695	0.0775	30	−7.35	0.0000
Ration	2	3	0.510	−0.5207	0.0775	30	−6.72	0.0000
Ration	2	4	0.510	−0.2893	0.0775	30	−3.73	0.0008
Ration	3	4	0.510	0.2314	0.0775	30	2.99	0.0056
Ration	1	2	0.680	0.0638	0.0761	30	0.84	0.4083
Ration	1	3	0.680	−0.8534	0.0761	30	−11.22	0.0000
Ration	1	4	0.680	−0.2663	0.0761	30	−3.50	0.0015
Ration	2	3	0.680	−0.9172	0.0761	30	−12.06	0.0000
Ration	2	4	0.680	−0.3301	0.0761	30	−4.34	0.0001
Ration	3	4	0.680	0.5870	0.0761	30	7.72	0.0000

cutting tools. Three pieces could be turned or cut from each rod. The rods varied in hardness; thus the rods were considered as blocks and the measure of hardness (Hard) of a given rod was used as a covariate. Since there were four tool types and blocks of size three, a balanced incomplete block (BIB) design structure was used as shown in Table 11.19.

The model using a linear relationship between surface finish and hardness is

$$\text{Finish}_{ij} = \alpha_i + \beta_i \text{Hard}_j + b_j + \varepsilon_{ij}. \tag{11.17}$$

TABLE 11.18
PROC MIXED Code and Analysis of Variance with Comparisons of the Four Rations without the Covariate Information

```
proc mixed cl covtest data=III118;
class blk ration;
model adg=ration/solution ddfm=kr;
lsmeans ration/diff;
random blk;
```

CovParm	Estimate	StdErr	ZValue	ProbZ	Alpha	Lower	Upper
blk	0.0553	0.0253	2.19	0.0144	0.05	0.0266	0.1759
Residual	0.0158	0.0039	4.06	0.0000	0.05	0.0103	0.0274

Effect	NumDF	DenDF	FValue	ProbF
Ration	3	33	104.78	0.0000

Effect	Ration	Estimate	StdErr	df	tValue	Probt
Ration	1	2.0417	0.0770	15.6	26.53	0.0000
Ration	2	2.1475	0.0770	15.6	27.90	0.0000
Ration	3	2.8692	0.0770	15.6	37.28	0.0000
Ration	4	2.4575	0.0770	15.6	31.93	0.0000

Effect	Ration	_Ration	Estimate	StdErr	df	tValue	Probt
Ration	1	2	–0.1058	0.0513	33	–2.06	0.0471
Ration	1	3	–0.8275	0.0513	33	–16.13	0.0000
Ration	1	4	–0.4158	0.0513	33	–8.10	0.0000
Ration	2	3	–0.7217	0.0513	33	–14.06	0.0000
Ration	2	4	–0.3100	0.0513	33	–6.04	0.0000
Ration	3	4	0.4117	0.0513	33	8.02	0.0000

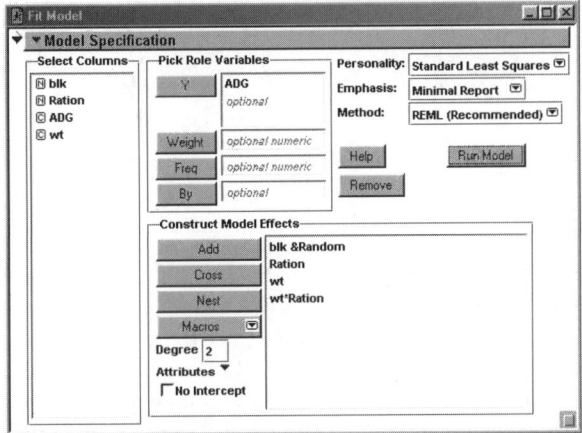

FIGURE 11.1 Fit model screen for the BIB where the center model specification is not used and blk is declared as a random effect.

Response ADG

Parameter Estimates

Term	Estimate	Std Error	t Ratio	Prob>|t|
Ration[2]	0.4534274	0.272546	1.66	0.1069
Ration[3]	-0.215157	0.272546	-0.79	0.4363
wt	-1.363037	1.178076	-1.16	0.2567
wt*Ration[1]	0.8744234	0.45496	1.92	0.0645
wt*Ration[2]	-1.148816	0.45496	-2.53	0.0173
wt*Ration[3]	1.1831678	0.45496	2.60	0.0145

REML Variance Component Estimates

Random Effect	Var Ratio	Var Component	Std Error	95% Lower	95% Upper	Pct of Total
blk&Random	4.7787617	0.0546138	0.0296779	0.0236176	0.2335883	82.695
Residual		0.0114284				17.305
Total		0.0660423				100.000

-2 LogLikelihood = -32.58838

Effect Tests

Source	Nparm	DF	DFDen	Sum of Squares	F Ratio	Prob > F	
blk&Random	12	11	29	2.1845533	17.3773	<.0001	Shrunk
Ration	3	3	29	0.1585319	4.6239	0.0092	
wt	1	1	29	0.0152988	1.3387	0.2567	
wt*Ration	3	3	29	0.1784820	5.2058	0.0053	

FIGURE 11.2 Estimates of the variance components and tests for the effects in the model using JMP®.

TABLE 11.19
Data for Balanced Incomplete Block Design Structure with Four Types of Tools

BLOCK	HARD	TOOL1	TOOL2	TOOL3	TOOL4
1	23	63	73	64	
2	44	78	72		85
3	40	66		69	86
4	27		73	75	97
5	26	71	77	88	
6	33	85	87		103
7	39	87		72	101
8	50		88	87	94
9	20	66	67	76	
10	44	97	91		97
11	28	82		84	90
12	26		72	76	91
13	27	87	85	86	
14	38	87	90		107
15	44	84		72	96
16	48		104	103	128

Table 11.20 contains the PROC GLM code to extract the within block analysis. The same problems as observed in the previous two examples occur in this analysis. The sum of squares corresponding to Hard is zero since the average value of the covariate is completely confounded with the blocks. Only least squares means at the average hardness value are estimable by PROC GLM and they are displayed in fourth part of Table 11.20 and the p-values for pairwise differences are in the lower part of Table 11.20. The best approach is to use mixed models software to carry out the analysis because it will provide the combined within and between block analysis.

TABLE 11.20
PROC GLM Code and Within Block Analysis for the BIB Example
(The LSMEANS for HARD = 25 and 45 are nonestimable.)

```
PROC GLM data=bibhard; CLASS BLOCK Tool;
MODEL Finish=BLOCK Tool Hard Hard*Tool/SOLUTION;
LSMEANS Tool/STDERR PDIFF AT MEAN;
LSMEANS Tool/STDERR PDIFF AT HARD=25;
LSMEANS Tool/STDERR PDIFF AT HARD=45;
RANDOM BLOCK/TEST;
```

Source	df	SS	MS	FValue	ProbF
Model	21	7517.9156	357.9960	14.55	0.0000
Error	26	639.8969	24.6114		
Corrected Total	47	8157.8125			

Source	df	SS (I)	MS	FValue	ProbF
BLOCK	15	5389.1458	359.2764	14.60	0.0000
Tool	3	1893.4375	631.1458	25.64	0.0000
HARD	0	0.0000			
HARD*Tool	3	235.3322	78.4441	3.19	0.0403

Source	df	SS (III)	MS	FValue	ProbF
BLOCK	14	3350.9528	239.3538	9.73	0.0000
Tool	3	354.8133	118.2711	4.81	0.0086
HARD	0	0.0000			
HARD*Tool	3	235.3322	78.4441	3.19	0.0403

Tool	LSMean	StdErr	Probt	LSMean#
1	81.6313	1.5133	0.0000	1
2	80.3581	1.5100	0.0000	2
3	79.9083	1.5285	0.0000	3
4	97.0543	1.7047	0.0000	4

RowName	_1	_2	_3	_4
1		0.5636	0.4403	0.0000
2	0.5636		0.8388	0.0000
3	0.4403	0.8388		0.0000
4	0.0000	0.0000	0.0000	

Table 11.21 contains the PROC MIXED code to provide the combined analysis. The NOINT option was not used in this model so the estimates of the intercepts satisfy the set-to-zero restriction. The test of the slopes equal to zero hypothesis is provided by the F value corresponding to HARD*Tool, which has a significance level of 0.0253. The contrast "b1 = b2 = b3 = b4" provides the F value to test the equal slopes hypothesis which has a significance level of 0.0565. Two of the slopes are not significantly different from zero, HARD*Tool 3 and 4. A simpler model possibly could be obtained, but it was not attempted here. Least squares means were computed at Hard = 34.8125 (mean), 25, and 45 and those results are in Table 11.22. The least squares means are graphed in Figure 11.3. The means for Tool 4 are significantly

TABLE 11.21
PROC MIXED Code and Combined Within and Between Block Analysis with Parameter Estimates for BIB

```
PROC MIXED CL COVTEST data=bibhard; CLASS BLOCK Tool;
MODEL Finish=Tool hard Hard*Tool/SOLUTION DDFM=KR;
CONTRAST 'b1=b2=b3=b4' HARD*TOOL 1 -1, HARD*TOOL 1 0 -1, HARD*TOOL
    1 0 0 -1;
RANDOM BLOCK/solution;
```

CovParm	Estimate	StdErr	ZValue	ProbZ	Alpha	Lower	Upper
BLOCK	75.5758	31.9598	2.36	0.0090	0.05	38.1043	215.4836
Residual	24.6198	6.8304	3.60	0.0002	0.05	15.2668	46.2483

Effect	NumDF	DenDF	FValue	ProbF
Tool	3	26.8	4.52	0.0109
HARD*Tool	4	29.5	3.25	0.0253

Effect	Tool	Estimate	StdErr	df	tValue	Probt
Intercept		84.3294	11.5223	31.1	7.32	0.0000
Tool	1	−36.1543	10.6914	27.8	−3.38	0.0022
Tool	2	−24.4205	9.5039	27.1	−2.57	0.0160
Tool	3	−14.0808	9.4157	27.0	−1.50	0.1464
Tool	4	0.0000				
HARD*Tool	1	0.9557	0.2966	26.3	3.22	0.0034
HARD*Tool	2	0.5939	0.2744	21.1	2.16	0.0421
HARD*Tool	3	0.2779	0.2757	21.4	1.01	0.3248
HARD*Tool	4	0.3630	0.3059	28.3	1.19	0.2452

Label	NumDF	DenDF	FValue	ProbF
b1=b2=b3=b4	3	26.8	2.84	0.0565

larger than the means of the other three tools for all hardness values. For Hard = 34.8125 and 25, the adjusted means for Tools 1, 2, and 3 are not significantly different. For Hard = 45, the mean of Tool 1 is significantly larger than the mean for Tool 3. The mean of Tool 1 is not significantly different from the mean of Tool 2 and the means of Tools 2 and 3 are not significantly different. From the graph in Figure 11.3, it can be concluded that Tool 4 provides the largest mean finish value for any of the hardness values in the range of the data.

11.10 SUMMARY

When blocking is used in a design, information about the parameters can be extracted from more than one source. The usual analysis extracts only the within block information and ignores the between block information. The process described in this chapter extracts both the within block and between block information and then combines it when possible. The discussion here considered equal block size designs and could be extended to unequal block size design structures. For unequal block size designs, the process would be to group blocks of the same size together and

TABLE 11.22
Least Squares Means for the Four Levels of Tool Evaluated at the Mean Hardness (34.81) and at 25 and 45

```
LSMEANS Tool/DIFF at MEAN;
LSMEANS Tool/DIFF at HARD=25;
LSMEANS Tool/DIFF at HARD=45;
```

Effect	Tool	HARD	Estimate	StdErr	df	tValue	Probt
Tool	1	34.81	81.4451	2.6459	23.1	30.78	0.0000
Tool	2	34.81	80.5840	2.6440	23.1	30.48	0.0000
Tool	3	34.81	79.9222	2.6543	23.3	30.11	0.0000
Tool	4	34.81	96.9651	2.7565	25.9	35.18	0.0000
Tool	1	25.00	72.0674	3.8324	23.1	18.80	0.0000
Tool	2	25.00	74.7564	3.7019	20.7	20.19	0.0000
Tool	3	25.00	77.1955	3.6655	20.1	21.06	0.0000
Tool	4	25.00	93.4035	4.4639	32.8	20.92	0.0000
Tool	1	45.00	91.1812	4.1165	26.6	22.15	0.0000
Tool	2	45.00	86.6344	3.9199	23.3	22.10	0.0000
Tool	3	45.00	82.7531	3.9873	24.4	20.75	0.0000
Tool	4	45.00	100.6628	3.7233	19.8	27.04	0.0000

Tool	_Tool	HARD	Estimate	StdErr	df	tValue	Probt
1	2	34.81	0.8610	2.1684	26.7	0.40	0.6945
1	3	34.81	1.5229	2.1891	26.8	0.70	0.4926
1	4	34.81	−15.5200	2.2904	26.9	−6.78	0.0000
2	3	34.81	0.6618	2.1806	26.7	0.30	0.7639
2	4	34.81	−16.3811	2.3043	27.0	−7.11	0.0000
3	4	34.81	−17.0429	2.3036	26.9	−7.40	0.0000
1	2	25.00	−2.6890	2.9267	26.4	−0.92	0.3665
1	3	25.00	−5.1282	2.8751	26.4	−1.78	0.0860
1	4	25.00	−21.3361	4.0252	27.7	−5.30	0.0000
2	3	25.00	−2.4391	2.7050	26.2	−0.90	0.3754
2	4	25.00	−18.6471	3.7731	27.3	−4.94	0.0000
3	4	25.00	−16.2080	3.7058	27.2	−4.37	0.0002
1	2	45.00	4.5468	3.5010	27.1	1.30	0.2050
1	3	45.00	8.4281	3.6082	27.2	2.34	0.0271
1	4	45.00	−9.4816	3.2099	26.7	−2.95	0.0065
2	3	45.00	3.8813	3.3224	26.8	1.17	0.2530
2	4	45.00	−14.0284	2.9209	26.3	−4.80	0.0001
3	4	45.00	−17.9097	3.0118	26.4	−5.95	0.0000

obtain between block information from each group. Then the estimators from each block size analysis and the within block analysis would be combined. This process is tedious when there are several different block sizes, but the mixed models approach can provide the appropriate analysis. It is important to provide a combined analysis

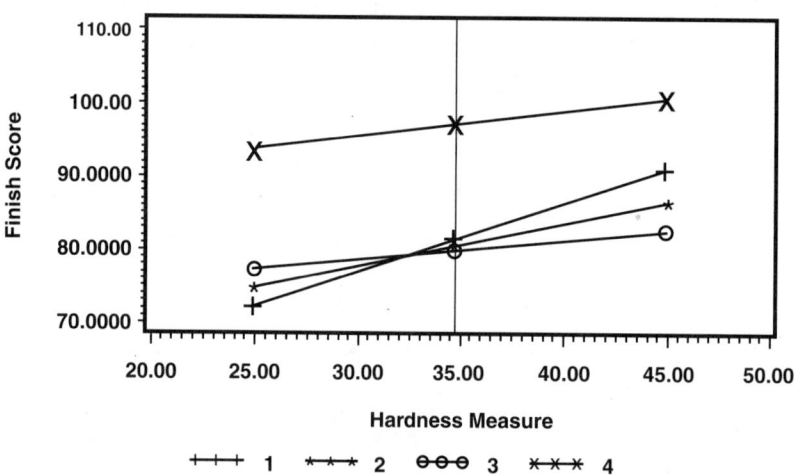

FIGURE 11.3 Predicted finish scores for the four tools over a range of hardness measures. The vertical line is the mean hardness measure, 34.8125.

since experiments involving split-plot and repeated measures designs are composed of many small blocks and much of the covariate information is contained in the between block comparisons.

Contrary to common thought and method of analysis, even when the covariate is used to form blocks, including the covariate in the analysis can improve the comparisons of the treatments when there is a relationship between the mean of the treatments and the values of the covariate.

REFERENCES

Kenward, M. G. and Roger, J. H. (1997). Small sample inference for fixed effects from restricted maximum likelihood, *Biometrics* 54:983.

Milliken, G. A. and Johnson, D. E. (1992). *Analysis of Messy Data, Volume I: Design Experiments*, Chapman & Hall, London.

EXERCISES

EXERCISE 11.1: Determine the proper form for the analysis of covariance model to describe the following data. Construct the appropriate analysis of variance table. Provide estimates of the intercepts and slopes for each treatment. Plot the regression lines. Compare the regression lines at $x = .40$, $x = .60$, and $x = .80$. Determine the LSMEANS, estimate their standard errors, and construct 95% confidence intervals about each.

Covariate Measured on the Block in RCB

BLK	TRT	Y	X	BLK	TRT	Y	X
1	1	84.6	0.834	6	1	90.4	0.579
1	2	86.8	0.834	6	2	91.6	0.579
1	3	90.8	0.834	6	3	91.4	0.579
1	4	90.3	0.834	6	4	95.5	0.579
1	5	90.2	0.834	6	5	97.3	0.579
2	1	86.8	0.163	7	1	82.2	0.603
2	2	86.7	0.163	7	2	93.3	0.603
2	3	91.2	0.163	7	3	88.0	0.603
2	4	85.1	0.163	7	4	94.8	0.603
2	5	92.1	0.163	7	5	94.1	0.603
3	1	84.6	0.356	8	1	94.0	0.649
3	2	87.8	0.356	8	2	89.0	0.649
3	3	93.3	0.356	8	3	95.4	0.649
3	4	89.0	0.356	8	4	92.3	0.649
3	5	97.6	0.356	8	5	96.7	0.649
4	1	89.0	0.593	9	1	71.9	0.174
4	2	86.9	0.593	9	2	81.3	0.174
4	3	89.6	0.593	9	3	74.5	0.174
4	4	85.8	0.593	9	4	84.7	0.174
4	5	87.7	0.593	9	5	86.8	0.174
5	1	80.8	0.395	10	1	84.4	0.824
5	2	76.6	0.395	10	2	86.9	0.824
5	3	81.2	0.395	10	3	87.3	0.824
5	4	81.0	0.395	10	4	89.9	0.824
5	5	83.2	0.395	10	5	99.2	0.824

EXERCISE 11.2: Determine the proper form for the analysis of covariance model to describe the following data. Construct the appropriate analysis of variance table. Provide estimates of the intercepts and slopes for each treatment. Plot the regression lines. Compare the regression lines at x = 6, x = 12, and x = 18. Determine the LSMEANS, estimate their standard errors, and construct 95% confidence intervals about each.

BLK	TRT	Y	X	BLK	TRT	Y	X	BLK	TRT	Y	X
1	1	15.4	17.1	5	1	9.2	9.9	9	2	7.7	6.3
1	2	20.7	17.1	5	3	15.3	9.9	9	3	8.7	6.3
2	1	13.4	6.1	6	2	19.3	12.9	10	1	15.1	16.9
2	3	13.5	6.1	6	3	23.3	12.9	10	2	19.8	16.9
3	2	17.3	9.3	7	1	16.7	13.3	11	1	10.3	13.6
3	3	20.9	9.3	7	2	17.7	13.3	11	3	15.2	13.6
4	1	15.4	13.1	8	1	20.5	14.1	12	2	10.0	7.4
4	2	19.9	13.1	8	3	28.5	14.1	12	3	10.2	7.4

12 Random Effects Models with Covariates

12.1 INTRODUCTION

A treatment is called a random effect when the levels of the treatment are a random sample of levels from a population of possible levels (Milliken and Johnson, 1992). Assume that there is one covariate and that a simple linear model describes the relationship between the mean of the response and the covariate for each treatment. Since the levels of the treatments are randomly selected from a population of possible levels, the mean of the response given the value of the covariate or the model for a selected treatment is also a random variable. More specifically, the coefficients of the model corresponding to the randomly selected treatment are random variables. These models are referred to as random coefficient models. Three examples are used to demonstrate the use of these models.

12.2 THE MODEL

To describe the model, assume the relationship between the response variable, y, and the value of a single covariate, x, is the simple linear regression model. The model for the jth observation from the ith randomly selected level of the population of treatments (called a random effect) is

$$y_{ij} = a_i + b_i x_{ij} + \varepsilon_{ij}$$

$$i = 1, 2, \ldots, t, \quad j = 1, 2, \ldots, n_i$$

(12.1)

where the intercept and slope, a_i and b_i, are random variables with joint distribution

$$\begin{pmatrix} a_i \\ b_i \end{pmatrix} \sim \text{iid } N\left[\begin{pmatrix} \alpha \\ \beta \end{pmatrix}, \Sigma_m \right]$$

where

$$\Sigma_m = \begin{bmatrix} \sigma_a^2 & \sigma_{ab} \\ \sigma_{ab} & \sigma_b^2 \end{bmatrix}.$$

To carry out an experiment, involving the random effect, a random sample of size $N = \sum_{i=1}^{t} n_i$ experimental units is selected from a population of experimental units. Measure the values of the covariates on each experimental unit. The values of the covariate are the same no matter which level of the treatment is to be assigned to an experimental unit. In fact, the values of the covariate do not depend on whether the treatment structure is a set of fixed effects, random effects, or mixed effects. Thus the analysis for a random effects treatment structure is similar to the analysis of a fixed effects treatment structure in that the models are compared for the given values of the covariate. The analysis of a random effects treatment structure also involves estimating the variances and covariances of Σ_m and the experimental unit variance, σ_ε^2.

The model in Equation 12.1 can also be expressed as

$$y_{ij} = \alpha + \beta x_{ij} + e_{ij} \qquad (12.2)$$
$$i = 1, 2, \ldots, t, \quad j = 1, 2, \ldots, n_i$$

where $e \sim N\{0, \mathbf{X}(\Sigma_m \otimes \mathbf{I}_t)\mathbf{X}' + \sigma_\varepsilon^2 \mathbf{I}_N\}$,

$$\mathbf{X} = \begin{pmatrix} \mathbf{X}_1 & 0 & \cdots & 0 \\ 0 & \mathbf{X}_2 & \cdots & 0 \\ \vdots & \vdots & \ddots & \vdots \\ 0 & 0 & \cdots & \mathbf{X}_t \end{pmatrix}, \quad \mathbf{X}_i = \begin{bmatrix} 1 & x_{i1} \\ 1 & x_{i2} \\ \vdots & \vdots \\ 1 & x_{in_i} \end{bmatrix}, \text{ and } \Sigma_m \otimes \mathbf{I}_t = \begin{bmatrix} \Sigma_m & 0 & \cdots & 0 \\ 0 & \Sigma_m & \cdots & 0 \\ \vdots & \vdots & \ddots & \vdots \\ 0 & 0 & \cdots & \Sigma_m \end{bmatrix}.$$

(See Graybill (1969) for a discussion of direct products like $\mathbf{A} \otimes \mathbf{B}$.) The model of 12.1 can be expressed as

$$y_{ij} = (\alpha + a_i^*) + (\beta + b_i^*) x_{ij} + \varepsilon_{ij}$$

$$\begin{pmatrix} a_i^* \\ b_i^* \end{pmatrix} \sim \text{iid } N\left[\begin{pmatrix} 0 \\ 0 \end{pmatrix}, \Sigma_m\right]$$

where

$$\Sigma_m = \begin{bmatrix} \sigma_a^2 & \sigma_{ab} \\ \sigma_{ab} & \sigma_b^2 \end{bmatrix}.$$

Models 12.1 and 12.2 can be extended to multiple covariates as

$$y_{ij} = a_i + b_{1i} x_{1ij} + b_2 x_{2ij} + \ldots + b_{ki} x_{kij} + \varepsilon_{ij} \qquad (12.3)$$
$$i = 1, 2, \ldots, t, \quad j = 1, 2, \ldots, n_i$$

where $\varepsilon_{ij} \sim$ iid $N(0, \sigma_\varepsilon^2)$ and

$$\begin{pmatrix} a_i \\ b_{1i} \\ \vdots \\ b_{ki} \end{pmatrix} \sim \text{iid } N \begin{bmatrix} \begin{pmatrix} \alpha \\ \beta_1 \\ \beta_2 \\ \vdots \\ \beta_k \end{pmatrix}, \Sigma_m \end{bmatrix}$$

where

$$\Sigma_m = \begin{bmatrix} \sigma_a^2 & \sigma_{ab_1} & \sigma_{ab_2} & \cdots & \sigma_{ab_k} \\ \sigma_{ab_1} & \sigma_{b_1}^2 & \sigma_{b_1 b_2} & \cdots & \sigma_{b_1 b_k} \\ \sigma_{ab_2} & \sigma_{b_1 b_2} & \sigma_{b_2}^2 & \cdots & \sigma_{b_2 b_k} \\ \vdots & \vdots & \vdots & & \vdots \\ \sigma_{ab_k} & \sigma_{b_1 b_k} & \sigma_{b_2 b_k} & \cdots & \sigma_{b_k}^2 \end{bmatrix}$$

or

$$y_{ij} = \alpha + \beta_1 x_{1ij} + \beta_2 x_{2ij} + \ldots + \beta_k x_{kij} + e_{ij} \tag{12.4}$$

where $e \sim N\{0, \mathbf{X}(\Sigma_m \otimes \mathbf{I}_t)\mathbf{X}' + \sigma_\varepsilon^2 \mathbf{I}_N]$

$$\mathbf{X} = \begin{pmatrix} \mathbf{X}_1 & 0 & \cdots & 0 \\ 0 & \mathbf{X}_2 & \cdots & 0 \\ \vdots & \vdots & \ddots & \vdots \\ 0 & 0 & \cdots & \mathbf{X}_t \end{pmatrix}, \text{ and } \mathbf{X}_i = \begin{bmatrix} 1 & x_{1i1} & \cdots & x_{ki1} \\ 1 & x_{1i2} & \cdots & x_{ki2} \\ \vdots & \vdots & & \vdots \\ 1 & x_{1i n_i} & \cdots & x_{ik n_i} \end{bmatrix}$$

The model for a two-way random effects treatment structure with one covariate in a linear form can be expressed similar to the model in Equation 12.1 as

$$\begin{aligned} y_{ijk} &= (\alpha + a_i + c_j + f_{ij}) + (\beta + b_i + d_j + g_{ij}) x_{ijk} + \varepsilon_{ijk} \\ &= (\alpha + \beta x_{ijk}) + (a_i + b_i x_{ijk}) + (c_j + d_j x_{ijk}) + (f_{ij} + g_{ij} x_{ijk}) + \varepsilon_{ijk} \end{aligned} \tag{12.5}$$

$i = 1, 2, \ldots, r, \ j = 1, 2, \ldots, s, \ k = 1, 2, \ldots, n_{ij}$

where

$$\begin{bmatrix} a_i \\ b_i \end{bmatrix} \sim \text{iid } N[\mathbf{0}, \Sigma_R], \quad \begin{bmatrix} c_j \\ d_j \end{bmatrix} \sim \text{iid } N[\mathbf{0}, \Sigma_S], \quad \begin{bmatrix} f_{ij} \\ g_{ij} \end{bmatrix} \sim \text{iid } N[\mathbf{0}, \Sigma_{RS}],$$

$$\Sigma_R = \begin{bmatrix} \sigma_a^2 & \sigma_{ab} \\ \sigma_{ab} & \sigma_b^2 \end{bmatrix}, \quad \Sigma_S = \begin{bmatrix} \sigma_c^2 & \sigma_{cd} \\ \sigma_{cd} & \sigma_d^2 \end{bmatrix}, \quad \text{and } \Sigma_{RS} = \begin{bmatrix} \sigma_f^2 & \sigma_{fg} \\ \sigma_{fg} & \sigma_g^2 \end{bmatrix}.$$

The two-way random effects treatment structure can also be expressed in a form similar to the model in Equation 12.2 as

$$y_{ijk} = \alpha + \beta x_{ijk} + e_{ijk}$$

$$i = 1, 2, \ldots, r, \quad j = 1, 2, \ldots, s, \quad k = 1, 2, \ldots, n_{ij}$$

where $e \sim N[\mathbf{0}, \Sigma_e]$,

$$\Sigma_e = X_R (\Sigma_R \otimes I_r) X_R' + X_S (\Sigma_S \otimes I_s) X_S' + X_{RS} (\Sigma_{RS} \otimes I_{RS}) X_{RS}' + \sigma_\varepsilon^2 I_N,$$

$$N = \sum_{i=1}^{r} \sum_{j=1}^{s} n_{ij},$$

and

$$X_R = \begin{bmatrix} X_{11} & 0 & \cdots & 0 \\ X_{12} & 0 & \cdots & 0 \\ \vdots & \vdots & & \vdots \\ X_{1s} & 0 & \cdots & 0 \\ 0 & X_{21} & \cdots & 0 \\ 0 & X_{22} & \cdots & 0 \\ \vdots & \vdots & & \vdots \\ 0 & X_{2s} & \cdots & 0 \\ 0 & 0 & \cdots & 0 \\ \vdots & \vdots & & \vdots \\ 0 & 0 & \cdots & X_{r1} \\ 0 & 0 & \cdots & X_{r2} \\ \vdots & \vdots & & \vdots \\ 0 & 0 & \cdots & X_{rs} \end{bmatrix},$$

Random Effects Models with Covariates

$$\mathbf{X}_S = \begin{bmatrix} \mathbf{X}_{11} & 0 & \cdots & 0 \\ 0 & \mathbf{X}_{12} & \cdots & 0 \\ \vdots & \vdots & & \vdots \\ 0 & 0 & \cdots & \mathbf{X}_{1s} \\ \mathbf{X}_{21} & 0 & \cdots & 0 \\ 0 & \mathbf{X}_{22} & \cdots & 0 \\ \vdots & \vdots & & \vdots \\ 0 & 0 & \cdots & \mathbf{X}_{2s} \\ \mathbf{X}_{31} & 0 & \cdots & 0 \\ \vdots & \vdots & & \vdots \\ \mathbf{X}_{r1} & 0 & \cdots & 0 \\ 0 & \mathbf{X}_{r2} & \cdots & 0 \\ \vdots & \vdots & & \vdots \\ 0 & 0 & \cdots & \mathbf{X}_{rs} \end{bmatrix},$$

$$\mathbf{X}_{RS} = \begin{bmatrix} \mathbf{X}_{11} & 0 & 0 & \cdots & 0 & 0 \\ 0 & \mathbf{X}_{12} & 0 & \cdots & 0 & 0 \\ \vdots & \vdots & \vdots & & \vdots & \vdots \\ 0 & 0 & 0 & \cdots & \mathbf{X}_{r(s-1)} & 0 \\ 0 & 0 & 0 & \cdots & 0 & \mathbf{X}_{rs} \end{bmatrix}.$$

and

$$\mathbf{X}_{ij} = \begin{bmatrix} 1 & X_{ij1} \\ 1 & X_{ij2} \\ \vdots & \vdots \\ 1 & X_{ijn_{ij}} \end{bmatrix}.$$

The two-way model can be extended to k covariates where the Σ_m, m = R, S, RS become $(k + 1) \times (k + 1)$ matrices of parameters and

$$\mathbf{X}_{ij} = \begin{bmatrix} 1 & X_{1ij1} & X_{2ij1} & \cdots & X_{kij1} \\ 1 & X_{1ij2} & X_{2ij2} & \cdots & X_{kij2} \\ \vdots & \vdots & & & \\ 1 & X_{1ijn_{ij}} & X_{2ijn_{ij}} & \cdots & X_{kijn_{ij}} \end{bmatrix}.$$

The one-way random effects treatment structure with covariates model is an extension of the model without covariates (Milliken and Johnson, 1992, page 221) where columns of ones are replaced with X_i matrices. The two-way random effects treatment structure with covariates is an extension of the model without covariates (Milliken and Johnson, 1992, page 224) where columns of ones are replaced with X_{ij} matrices. Thus extensions to other random effects treatment structures can be accomplished by expressing the model without covariates in matrix form and then replacing the columns of ones for each cell with the corresponding X matrix which includes the necessary or possible covariates.

A set of simultaneous prediction intervals about the population of regression models evaluated at specific values of the covariates can be constructed by using a Scheffé (Milliken and Johnson, 1992) approach. A set of simultaneous prediction intervals can be used to form a prediction band for the population of regression models. To demonstrate the process, simultaneous prediction intervals are constructed about the one-way random coefficient model of Equation 12.1. The variance of an observation with covariate x_0 from a randomly selected level of the population of treatments is

$$\text{Var}(y_{ij}|X=x_0) = \text{Var}(\alpha + a_i^* + (\beta + b_i^*)x_0 + \varepsilon_{ij})$$

$$= \text{Var}(a_i^* + b_i^* x_0 + \varepsilon_{ij})$$

$$= \sigma_a^2 + 2x_0 \sigma_{ab} + x_0^2 \sigma_b^2 + \sigma_\varepsilon^2$$

The estimate of the standard error of prediction is

$$SEP(y_{ij|x_0}) = \sqrt{\hat{\sigma}_a^2 + 2x_0 \hat{\sigma}_{ab} + x_0^2 \hat{\sigma}_b^2 + \hat{\sigma}_\varepsilon^2}$$

The upper and lower $(1-\alpha)$ 100% prediction limits about the model at x_0 is

$$\hat{\alpha} + \hat{\beta} x_0 \pm \sqrt{2 F_{\alpha,2,df}} \; SEP(y_{ij|x_0})$$

where "df" is the approximate number of degrees of freedom associated with SEP $(y_{ij|x_0})$. By constructing a series of these prediction intervals over a range of x_0, a prediction band about the population of regression models can be constructed. This process can be extended to models with more than two parameters.

12.3 ESTIMATION OF THE VARIANCE COMPONENTS

Maximum likelihood methods seem to be the most appropriate methods of estimating parameters of the random effects models with covariates. Using the assumed form of the normal distribution for the specified model, the appropriate likelihood function can be constructed and then an iterative method can be used to obtain maximum likelihood (ML) or residual maximum likelihood estimates (REML) of the parameters

Random Effects Models with Covariates

(see Chapter 19 of Milliken and Johnson, 1992). With existing software, the maximum likelihood methods are the most accessible and the resulting estimates have the usual large sample properties. Most of the examples in the remainder of the book use the REML estimates of the variance and/or covariance components (Littell et al., 1996).

Method of moments estimates can be computed for models that involve covariates, but such estimators are not available with existing commercial software. In an attempt to motivate the process of estimating parameters for the random coefficient model, the following discussion provides the method of moments estimators for a very simple one-way random effects treatment structure in a completely randomized design structure with one covariate. Assume there are t treatments in a random effects treatment structure. Assume that the values of the covariate are the same for each of the treatments, i.e., the set of experimental units assigned to treatment one has identical covariate values as the set of experimental units assigned to each of the other treatments. This assumption is similar to having the covariate assigned to a block as described in Chapter 11, but the block is not considered as part of the model. The model can be expressed as

$$y_{ij} = (\alpha + a_i) + (\beta + b_i)X_j + \varepsilon_{ij}$$

$$i = 1, 2, \ldots, t, \quad j = 1, 2, \ldots, n$$

Further assume the values of the X_j's within each treatment are scaled such that $\sum_{j=1}^{n} X_j = 0 = \mathbf{j}_n'\mathbf{x}$, where $\mathbf{j}_n$ use a $n \times 1$ vector of ones and $\mathbf{x}' = (x_1, x_2, \ldots, x_n)$. Let $\mathbf{X} = [\mathbf{j}_n, \mathbf{x}]$, then the model can be expressed as

$$\mathbf{y} = (\mathbf{X} \otimes \mathbf{I}_t)\boldsymbol{\beta} + \mathbf{e}$$

where $\boldsymbol{\beta}' = (\alpha, \beta)$ and $\text{Var}(\mathbf{e}) = \mathbf{X}\boldsymbol{\Sigma}\mathbf{X}' \otimes \mathbf{I}_t + \sigma_\varepsilon^2 \mathbf{I}_n \otimes \mathbf{I}_t$.

The predicted values for the elements of **a** and **b** are

$$\hat{\mathbf{a}} = \left[\frac{1}{n}\mathbf{j}' \otimes \left(\mathbf{I}_t - \frac{1}{t}\mathbf{J}_t\right)\right]\mathbf{y} \text{ and}$$

$$\hat{\mathbf{b}} = \left[\mathbf{x}^- \otimes \left(\mathbf{I}_t - \frac{1}{t}\mathbf{J}_t\right)\right]\mathbf{y}$$

where $\mathbf{x}^- = \mathbf{x}'/\mathbf{x}'\mathbf{x}$ (see Graybill, 1969) for discussion of generalized inverses). If the model consists of all fixed effects, the covariance matrices of $\hat{\mathbf{a}}$ and $\hat{\mathbf{b}}$ would be

$$\text{Cov}(\hat{\mathbf{a}}) = \sigma_\varepsilon^2 \mathbf{V}_a \text{ where } \mathbf{V}_a = \frac{1}{n}\left(\mathbf{I}_t - \frac{1}{t}\mathbf{J}_t\right) \text{ and}$$

$$\text{Cov}(\hat{\mathbf{b}}) = \sigma_\delta^2 \mathbf{V}_b = \frac{1}{\mathbf{x}'\mathbf{x}}\left(\mathbf{I}_t - \frac{1}{t}\mathbf{J}_t\right).$$

The sum of squares due to "a" is

$$SSA = \hat{a}'V_a^- V_a^- \hat{a} = n\, y'\left[\frac{1}{n}J_n \otimes \left(I_t - \frac{1}{t}J_t\right)\right]y$$

$$= n^2\, \hat{a}'\hat{a}.$$

The sum of squares due to "b" is

$$SSB = \hat{b}'V_b^- V_b^- \hat{b} = (x'x)\, y'\left[x\,x^- \otimes \left(I_t - \frac{1}{t}J_t\right)\right]y$$

$$= (x'x)^2\, \hat{b}'\hat{b}.$$

The sum of cross products of **a** and **b** is

$$SCPAB = \hat{b}'V_b^- V_a^- \hat{a}$$

$$= n\,(x'x)y'\left[\left(x^-\left(\frac{1}{n}j_n'\right)\right) \otimes \left(I_t - \frac{1}{t}J_5\right)\right]y$$

$$= n\,(x'x)\,\hat{b}'\hat{a}.$$

Finally, the sum of squares error for the model is

$$SSERROR = y'\left((I_n - XX^-) \otimes I_t\right)y.$$

The expectations of the sums of squares and cross products using the random effects model assumptions (Chapter 18, Milliken and Johnson, 1992) are

$$E(SSA) = n(t-1)\,\sigma_\varepsilon^2 + (t-1)\,n^2\,\sigma_a^2$$

$$E(SSB) = (x'x)(t-1)\,\sigma_\varepsilon^2 + (t-1)(x'x)^2\,\sigma_b^2$$

$$E(SCPAB) = (t-1)\,n\,x'x\,\sigma_{ab} \text{ and}$$

$$E(SSERROR) = t\,(n-2)\,\sigma_\varepsilon^2.$$

By equating the observed sums of squares to their expected values and then solving, the method of moments solution to the system of equations is

$$\hat{\sigma}_{\varepsilon o}^2 = SSERROR/(t(n-2)),$$

$$\hat{\sigma}_{ao}^2 = \frac{SSA - n(t-1)\,\hat{\sigma}_{\varepsilon o}^2}{(t-1)\,n},$$

Random Effects Models with Covariates

$$\hat{\sigma}_{bo}^2 = \frac{SSB - (\mathbf{x'x})(t-1)\hat{\sigma}_{\varepsilon o}^2}{(t-1)n(\mathbf{x'x})}.$$

and

$$\hat{\sigma}_{abo} = \frac{SCPAB}{(t-1)n(\mathbf{x'x})}.$$

The solution to the method of moments equations can provide negative values for some variance (which must be positive); thus, the method of moments estimates of the variance and covariance components are usually taken as

$$\hat{\sigma}_{\varepsilon}^2 = \hat{\sigma}_{\varepsilon o}^2$$

$$\hat{\sigma}_a^2 = \begin{bmatrix} \hat{\sigma}_{ao}^2 & \text{if } \hat{\sigma}_{ao}^2 > 0 \\ 0 & \text{otherwise} \end{bmatrix}$$

$$\hat{\sigma}_b^2 = \begin{bmatrix} \hat{\sigma}_{bo}^2 & \text{if } \hat{\sigma}_{bo}^2 > 0 \\ 0 & \text{otherwise} \end{bmatrix}$$

$$\hat{\sigma}_{ab}^2 = \hat{\sigma}_{abo}^2.$$

This process can be extended to unbalanced models using $SSA = \hat{\mathbf{a}}'\mathbf{V}_a^-\mathbf{V}_a^-\hat{\mathbf{a}}$, $SSB = \hat{\mathbf{b}}'\mathbf{V}_b^-\mathbf{V}_b^-\hat{\mathbf{b}}$, $SCPAB = \hat{\mathbf{b}}'\mathbf{V}_b^-\mathbf{V}_a^-\hat{\mathbf{a}}$, and $SSERROR = \mathbf{y}'(\mathbf{I} - \mathbf{X}\mathbf{X}^-)\mathbf{y}$. The expectations of these sums of squares can be evaluated using Hartley's method of synthesis (Chapter 18 of Milliken and Johnson, 1992).

If the model has two covariates, the model can be described as

$$y_{ij} = (\alpha + a_i) + (\beta + b_i)x_{ij} + (\gamma + c_i)z_{ij} + \varepsilon_{ij}$$

where

$$\begin{bmatrix} a_i \\ b_i \\ c_i \end{bmatrix} \sim \text{iid } N[0, \Sigma] \text{ and } \Sigma = \begin{bmatrix} \sigma_a^2 & \sigma_{ab} & \sigma_{ac} \\ \sigma_{ab} & \sigma_b^2 & \sigma_{bc} \\ \sigma_{ac} & \sigma_{bc} & \sigma_c^2 \end{bmatrix}.$$

The sums of squares and cross products can be computed as

$$SSA = \hat{\mathbf{a}}'\mathbf{V}_a^-\mathbf{V}_a^-\mathbf{a}, \quad SSB = \hat{\mathbf{b}}'\mathbf{V}_b^-\mathbf{V}_b^-\hat{\mathbf{b}},$$

$$SSC = \hat{\mathbf{c}}'\mathbf{V}_c^-\mathbf{V}_c^-\hat{\mathbf{c}}, \quad SCPAB = \hat{\mathbf{b}}\mathbf{V}_b^-\mathbf{V}_a^-\hat{\mathbf{a}},$$

$$\text{SCPAC} = \hat{\mathbf{c}}\mathbf{V}_c^-\mathbf{V}_a\hat{\mathbf{a}}, \quad \text{SCPBC} = \hat{\mathbf{c}}\mathbf{V}_d^-\mathbf{V}_b\hat{\mathbf{b}}, \text{ and}$$

$$\text{SSERROR} = \mathbf{y}'(\mathbf{i} - \mathbf{X}^-)\mathbf{y}.$$

The expectations of these sums of squares can be evaluated via Hartley's method of synthesis and then the method of moments estimates can be computed.

For the two-way random effects treatment structure of Equation 12.5, the variance components can be estimated by the method of moments by computing the sums of squares and sum of cross products of the predicted values at each level and then using Hartley's method of synthesis to evaluate their expectations. The necessary sums of squares and cross products are

$$\text{SSA} = \hat{\mathbf{a}}'\mathbf{V}_a^-\mathbf{V}_a\hat{\mathbf{a}}, \quad \text{SSB} = \hat{\mathbf{b}}'\mathbf{V}_b^-\mathbf{V}_b\hat{\mathbf{b}}, \quad \text{SCPAB} = \hat{\mathbf{b}}\mathbf{V}_b^-\mathbf{V}_a\hat{\mathbf{a}},$$

$$\text{SSC} = \hat{\mathbf{c}}'\mathbf{V}_c^-\mathbf{V}_c\hat{\mathbf{c}}, \quad \text{SSD} = \hat{\mathbf{d}}'\mathbf{V}_d^-\mathbf{V}_d\hat{\mathbf{d}}, \quad \text{SSCPCD} = \hat{\mathbf{c}}'\mathbf{V}_c^-\mathbf{V}_d\hat{\mathbf{d}},$$

$$\text{SSF} = \hat{\mathbf{f}}\mathbf{V}_f^-\mathbf{V}_f\hat{\mathbf{f}}, \quad \text{SSG} = \hat{\mathbf{g}}\mathbf{V}_g^-\mathbf{V}_g\hat{\mathbf{g}}, \quad \text{SSCPFG} = \hat{\mathbf{f}}\mathbf{V}_f^-\mathbf{V}_g\hat{\mathbf{g}}, \text{ and}$$

$$\text{SSERROR} = \mathbf{y}'(\mathbf{I} - \mathbf{X}\mathbf{X}^-)\mathbf{y}.$$

For a balanced model, where experimental units within each cell have the same set of covariates as every other cell, the model can be expressed as

$$y_{ijk} = (\alpha + a_i + c_j + f_{ij}) + (\beta + b_i + d_j + g_{ij})X_k + \varepsilon_{ij} \qquad (12.6)$$

where it is assumed that $\sum_{k=1}^{N} X_k = 0$, then

$$\hat{\mathbf{a}} = \frac{1}{n}\mathbf{j}'_n \otimes \frac{1}{s}\mathbf{j}'_s \otimes \left(\mathbf{I}_r - \frac{1}{r}\mathbf{J}_r\right), \quad \mathbf{V}_a = \frac{1}{sn}\left(\mathbf{I}_r - \frac{1}{r}\mathbf{J}_r\right)$$

$$\hat{\mathbf{b}} = \mathbf{x}^- \otimes \frac{1}{s}\mathbf{j}'_s \otimes \left(\mathbf{I}_r - \frac{1}{r}\mathbf{J}_r\right), \quad \mathbf{V}_b = \frac{1}{s\mathbf{x}'\mathbf{x}}\left(\mathbf{I}_r - \frac{1}{r}\mathbf{J}_r\right)$$

$$\hat{\mathbf{c}} = \frac{1}{n}\mathbf{j}'_n \otimes \left(\mathbf{I}_s - \frac{1}{s}\mathbf{J}_s\right) \otimes \frac{1}{r}\mathbf{j}'_r, \quad \mathbf{V}_c = \frac{1}{nr}\left(\mathbf{I}_s - \frac{1}{s}\mathbf{J}_s\right)$$

$$\hat{\mathbf{d}} = \mathbf{x}^- \otimes \left(\mathbf{I}_s - \frac{1}{s}\mathbf{J}_s\right) \otimes \frac{1}{r}\mathbf{j}'_r, \quad \mathbf{V}_d = \frac{1}{s\mathbf{x}'\mathbf{x}}\left(\mathbf{I}_s - \frac{1}{s}\mathbf{J}_s\right)$$

$$\hat{\mathbf{f}} = \frac{1}{n}\mathbf{j}'_n \otimes \left(\mathbf{I}_s - \frac{1}{s}\mathbf{J}_s\right) \otimes \left(\mathbf{I}_r - \frac{1}{r}\mathbf{J}_r\right), \quad \mathbf{V}_f = \frac{1}{n}\left(\mathbf{I}_s - \frac{1}{s}\mathbf{J}_s\right) \otimes \left(\mathbf{I}_r - \frac{1}{r}\mathbf{J}_r\right)$$

$$\hat{\mathbf{g}} = \mathbf{x}^- \otimes \left(\mathbf{I}_s - \frac{1}{s}\mathbf{J}_s\right) \otimes \left(\mathbf{I}_r - \frac{1}{r}\mathbf{J}_r\right), \quad \mathbf{V}_g = \frac{1}{\mathbf{x}'\mathbf{x}}\left(\mathbf{I}_s - \frac{1}{s}\mathbf{J}_s\right) \otimes \left(\mathbf{I}_r - \frac{1}{r}\mathbf{J}_r\right)$$

Random Effects Models with Covariates

TABLE 12.1
Expectations of Sums of Squares and Cross Products for Model 12.6

SOURCE	E(SS or SCP)
Intercepts	
A	$ns(r-1)\left[\sigma_e^2 + n\sigma_f^2 + sn\sigma_a^2\right]$
C	$nr(s-1)\left[\sigma_e^2 + n\sigma_f^2 + rn\sigma_c^2\right]$
F or AxC	$n(r-1)(s-1)\left[\sigma_e^2 + n\sigma_f^2\right]$
Slopes	
B	$\mathbf{x'x}s(r-1)\left[\sigma_e^2 + \mathbf{x'x}\,\sigma_g^2 + s\,\mathbf{x'x}\,\sigma_b^2\right]$
D	$\mathbf{x'x}r(n-1)\left[\sigma_e^2 + \mathbf{x'x}\,\sigma_g^2 + r\,\mathbf{x'x}\,\sigma_d^2\right]$
G or BxD	$\mathbf{x'x}(r-1)(s-1)\left[\sigma_e^2 + \mathbf{x'x}\,\sigma_g^2\right]$
Cross Products	
AB	$s\mathbf{x'x}(r-1)\left[\sigma_{fg} + s\,\sigma_{ab}\right]$
CD	$r\mathbf{x'x}(s-1)\left[\sigma_{fg} + r\,\sigma_{cd}\right]$
FG	$\mathbf{x'x}(r-1)(s-1)\,\sigma_{fg}$

The expectations of the sums of squares and cross products for Model 12.6 are in Table 12.1.

Type I sums of squares or the Henderson Method III method can be used to obtain sums of squares, their expected values, and the resulting estimates of the variance components, but methods using different types of sums of squares cannot provide estimates of the covariance between the intercepts and slopes. The individual vectors of predictors are still needed to compute the sums of cross products in order to estimate the covariance components. As the experiment becomes more complex, these methods are more difficult to implement and the iterative methods used to obtain REML or ML estimates are much easier to implement with current software.

12.4 CHANGING LOCATION OF THE COVARIATE CHANGES THE ESTIMATES OF THE VARIANCE COMPONENTS

A change in the location of the covariate results in a change of the intercepts of the model and thus a change in the variance components associated with the intercepts. In previous chapters, one could construct a test of the equality of the adjusted means at a given value of the covariate, say at $X = X_0$, by using $X - X_0$ as the covariate. When the random coefficient model is used, the variance of the random intercepts changes as the location of the covariate is changed. As in the fixed effects treatment structure case, comparing the models at $x = 0$ may not be meaningful. The same is true for a random effects treatment structure in that the variance components computed from the models at $x = 0$ may not reflect the reality of what the variance components are for the models evaluated at $x = x^*$ where x^* is some value of x in the range of the data. The one-way treatment structure in a completely randomized design structure random effects model can be expressed as $y_{ij} = \alpha + a_i + \beta x_{ij} + b_i x_{ij} + \varepsilon_{ij}$. Next add and subtract $\beta x_0 + b_i x_0$ to the model to get $y_{ij} = (\alpha + \beta x_0) + a_i + b_i x_0) + \beta (x_{ij} - x_0) + b_i (x_{ij} - x_0) + \varepsilon_{ij}$ which can be expressed as $y_{ij} = (\alpha^*) + (a_i^*) + \beta (x_{ij} - x_0) + b_i (x_{ij} - x_0) + \varepsilon_{ij}$ where $\alpha^* = (\alpha + \beta x_0)$ and $a_i^* = a_i + b x_0 (a_i + b_i x_0)$. The distribution of the new random part of the intercept a_i^* is $a_i^* \sim$ iid $N(0, \sigma_{a^*}^2)$ where $\sigma_{a^*}^2 = \sigma_a^2 + 2 x_0 \sigma_{ab} + x_0^2 \sigma_b^2$. The covariance between a_i^* and b_i is $\sigma_{a^*b} = \sigma_{ab} + x_0 \sigma_b^2$. Thus when the location of the covariate is changed, changes in the estimates of the variance components of the models evaluated at the new value of x are predictable. This phenomenon is demonstrated in the examples of Sections 12.6 and 12.7.

Prediction intervals are obtainable by using the above information. The question to be answered is "If a city is randomly selected and has X percent unemployment, what is a 95% prediction interval about the amount that would be spent on vocational training?" This question can be answered by constructing a 95% (or whatever level of confidence) prediction interval for the distribution at a selected value of X. The variance of an observation from the distribution with selected value of X is $\text{Var}(y_{|x_0}) = \text{var}(\alpha + \beta x_0 + a_i + b_i x_0 + \varepsilon_{ij}) = \sigma_a^2 + 2x_0\sigma_{ab} + x_0^2\sigma_b^2 + \sigma_\varepsilon^2$, so the estimated standard error of a prediction is $\text{StdErrP}(y_{|x_0}) = \sqrt{\hat{\sigma}_a^2 + 2x_0\hat{\sigma}_{ab} + x_0^2\hat{\sigma}_b^2 + \hat{\sigma}_\varepsilon^2}$. Since the regression model has two parameters, α and β, a set of simultaneous prediction intervals can be constructed to form a confidence band by using the Scheffé method where the critical point is $\sqrt{2 F_{\alpha, 2, df}}$. The upper and lower prediction limits y at $x = x_0$ are $\hat{\alpha} + \hat{\beta} x_0 \pm \sqrt{2 F_{\alpha, 2, df}} \text{StdErrP}(y_{|x_0})$.

The previous two sections discuss the construction of the analysis of covariance model for the random effects treatment structure. The construction of the final model follows the same philosophy as used in Chapter 2 in that the final analysis of covariance model should be as simple as possible. The model building process starts with the full model and then deletes parameters that are not significantly greater than zero for variance components and not significantly different from zero for covariance components. For each covariate, start the process of determining if any of the covariance components are not significantly different from zero. If all covariances involving the covariate are zero, then determine if the variance component associated with the slopes for the covariate is greater than zero. If the variance component is not significantly greater than zero, then remove the covariate from the model. Obviously, as discussed in the previous chapters, the relationship between the response and the covariate must be modeled appropriately. Is the linear relationship adequate or does the data require a more complex model? The final sections present a series of examples to demonstrate the points discussed in the first four sections.

TABLE 12.2
Data for Section 12.5, a Balanced Random Effects Treatment Structure with One Covariate Where Yield is in Pounds Per Plot and N2 is the Coded Nitrogen Level

Variety 1		Variety 2		Variety 3		Variety 4		Variety 5		Variety 6		Variety 7	
N2	Yield	N2	Yield	N2	Yield	N2	Yield	N2	Yield	N2	Yield	N2	Yield
−3	4.0	−3	6.1	−3	5.5	−3	3.9	−3	5.8	−3	6.2	−3	1.0
−1	9.7	−1	8.5	−1	9.3	−1	10.0	−1	8.2	−1	9.6	−1	3.4
1	14.2	1	13.2	1	15.6	1	13.4	1	11.7	1	14.4	1	7.2
3	18.5	3	16.0	3	21.2	3	14.6	3	14.0	3	18.3	3	10.0

TABLE 12.3
PROC MIXED Code to the Random Coefficient Model with Uncorrelated Slopes and Intercepts and Results

```
proc mixed data=ex_12_5 cl covtest ic;
class variety;
model yield=N2/solution ddfm=kr;
random variety N2*variety/solution;
```

CovParm	Estimate	StdErr	ZValue	ProbZ	Alpha	Lower	Upper
variety	5.8518	3.4794	1.68	0.0463	0.05	2.3813	30.3058
N2*variety	0.1785	0.1238	1.44	0.0747	0.05	0.0650	1.3886
Residual	0.6944	0.2624	2.65	0.0041	0.05	0.3722	1.7270

Neg2LogLike	Parameters	AIC	AICC	HQIC	BIC	CAIC
104.75	3	110.75	111.84	108.74	110.59	113.59

Effect	Estimate	StdErr	DF	tValue	Probt
Intercept	10.4821	0.9278	6	11.30	0.0000
N2	1.9379	0.1745	6	11.10	0.0000

Effect	NumDF	DenDF	FValue	ProbF
N2	1	6	123.26	0.0000

12.5 EXAMPLE: BALANCED ONE-WAY TREATMENT STRUCTURE

The data in Table 12.2 are plot yields of seven randomly selected varieties of corn with four levels of nitrogen coded as −3, −1, 1, and 3. Within each variety there was one plot with each level of nitrogen which is considered as the covariate. This is a balanced model since there are equal numbers of observations per variety and since each variety has experimental units with all values of the covariate.

Table 12.3 contains the PROC MIXED code to fit the model in Equation 12.1 where the covariance between the slopes and intercepts is specified to be zero. Using the statement

```
random variety n2*variety/solution;
```

specifies that there are variance components for the intercepts (variety) and for the slopes (n2*variety), but does not specify a value for the covariance between the slopes and intercepts, so the covariance sets to zero. The REML estimates of the variance components are $\hat{\sigma}_a^2 = 5.818$, $\hat{\sigma}_b^2 = 0.1785$, and $\hat{\sigma}_\varepsilon^2 = 0.6944$. The mixed model estimates of the parameters of the model are $\hat{\alpha} = 10.4821$ and $\hat{\beta} = 1.9379$.

Table 12.4 contains the PROC MIXED code to fit the model in Equation 12.1 where the random statement allows for the slopes and intercepts to be correlated. The statement

```
random Int N2/subject=variety type=un solution;
```

specifies that the intercepts (Int) and slopes (N2) within each level of variety (subject = variety) has an unstructured correlation structure (type = un). When the random statement involves a "/", the interpretation of the terms on the left side of the "/" includes the term or terms on the right side of the "/". For this statement, Int means Variety and n2 means n2*variety. The REML estimates of the parameters are $\hat{\sigma}_a^2 = 5.8518$ from UN(1,1), $\hat{\sigma}_b^2 = 0.1785$ from UN(2,2), $\hat{\sigma}_{ab} = 0.7829$ from UN(2,1), and $\hat{\sigma}_\varepsilon^2 = 0.6944$. The mixed models estimates of the mean of the intercepts and slopes are $\hat{\alpha} = 10.4821$, $\hat{\beta} = 1.9379$, so the estimate of the population mean yield at a given level of nitrogen is $\hat{\mu}_{y|N2} = 10.4821 + 1.9379N2$. Since the model is balanced, all parameters in Table 12.4 have the same estimates in Table 12.3, except $\sigma_{ab} = 0$. The solutions and standard errors for the fixed effects are identical, but the solutions and standard errors for the random effects are different. The information criteria indicate that the model with the non-zero covariance fits the data better than the one with the covariance set to zero (the smaller the information criteria the better the covariance structure). Table 12.5 contains the solution for the individual intercepts and slopes for each of the varieties in the study based on the model with the non-zero covariance between the slopes and intercepts. These two sets of solutions satisfy the sum-to-zero restriction, i.e., the intercepts sum-to-zero and the slopes sum-to-zero. These are the estimated best linear unbiased predictors of the intercepts and slopes (Littel et al., 1996). The solution for the random effects based on the model with a zero covariance between the slopes and intercepts are somewhat different than the ones in Table 12.5 (results not shown here). The main difference is the estimated standard errors for the independent slope and intercept model are smaller than for the correlated slope and intercept model. The predicted values of the slope and intercept for the ith variety are obtained by $\hat{a}_{i(p)} = \hat{\alpha} + \hat{a}_i$ and $\hat{b}_{i(p)} = \hat{\beta} + \hat{b}_i$, where $\hat{\alpha}$ and $\hat{\beta}$ are the estimates of the population intercept and slope and $\hat{a}_i$ and $\hat{b}_i$ are the predicted values of the intercepts and slopes obtained from the solution for the random effects. Estimate statements in Table 12.6 are used to obtain the predicted slopes and intercepts for each of the varieties. The statements

```
estimate 'slope 1' N2 1 | N2 1/subject 1;
estimate 'slope 2' N2 1 | N2 1/subject 0 1;
```

TABLE 12.4
PROC MIXED Code to Fit the Random Coefficient Model with the Correlated Slopes and Intercepts Covariance Structure and Results

```
Proc mixed data=ex_12_5 cl covtest ic;
class variety;
model yield=N2/solution ddfm=kr;
random Int N2/subject=variety type=un solution;
```

CovParm	Subject	Estimate	StdErr	ZValue	ProbZ	Alpha	Lower	Upper
UN(1,1)	Variety	5.8518	3.4794	1.68	0.0463	0.05	2.3813	30.3058
UN(2,1)	Variety	0.7829	0.5624	1.39	0.1639	0.05	−0.3194	1.8852
UN(2,2)	Variety	0.1785	0.1238	1.44	0.0747	0.05	0.0650	1.3886
Residual		0.6944	0.2624	2.65	0.0041	0.05	0.3722	1.7270

Neg2LogLike	Parameters	AIC	AICC	HQIC	BIC	CAIC
100.86	4	108.86	110.76	106.19	108.64	112.64

Effect	Estimate	StdErr	DF	tValue	Probt
Intercept	10.4821	0.9278	6	11.30	0.0000
N2	1.9379	0.1745	6	11.10	0.0000

Effect	NumDF	DenDF	FValue	ProbF
N2	1	6	123.26	0.0000

TABLE 12.5
Solution for the Random Effect Slopes and Intercepts for Each Variety

Effect	Variety	Estimate	StdErrPred	df	tValue	Probt
Intercept	1	1.1497	0.9935	7.6	1.16	0.2823
N2	1	0.3635	0.2406	9.6	1.51	0.1632
Intercept	2	0.3980	0.9935	7.6	0.40	0.6997
N2	2	−0.1311	0.2406	9.6	−0.54	0.5983
Intercept	3	2.4327	0.9935	7.6	2.45	0.0416
N2	3	0.6020	0.2406	9.6	2.50	0.0323
Intercept	4	−0.0397	0.9935	7.6	−0.04	0.9692
N2	4	−0.1125	0.2406	9.6	−0.47	0.6507
Intercept	5	−0.6342	0.9935	7.6	−0.64	0.5420
N2	5	−0.3895	0.2406	9.6	−1.62	0.1379
Intercept	6	1.5760	0.9935	7.6	1.59	0.1533
N2	6	0.1471	0.2406	9.6	0.61	0.5552
Intercept	7	−4.8826	0.9935	7.6	−4.91	0.0014
N2	7	−0.4796	0.2406	9.6	−1.99	0.0755

provide predicted values for the slopes of varieties 1 and 2 (similar statements were used for the other varieties). The fixed effects are on the left hand side of "|" and the random effects are on the right hand side. On the left hand side using N2 1

TABLE 12.6
Estimate Statements Used to Obtain Predicted Values for Each of the Varieties At N2 = –3 and N2 = 3 and the Predicted Slopes and Intercepts

```
estimate 'Var 1 at N2=-3' intercept 1 N2 -3|Int 1 N2 -3/subject 1;
estimate 'Var 2 at N2=-3' intercept 1 N2 -3|Int 1 N2 -3/subject 0 1;
estimate 'Var 1 at N2= 3' intercept 1 N2  3|Int 1 N2  3/subject 1;
estimate 'Var 2 at N2= 3' intercept 1 N2  3|Int 1 N2  3/subject 0 1;
estimate 'slope 1' N2 1 | N2 1/subject 1;
estimate 'slope 2' N2 1 | N2 1/subject 0 1;
estimate 'Int 1' Int 1 | Int 1/subject 1;
estimate 'Int 2' Int 1 | Int 1/subject 0 1;
```

	Slopes		Intercepts		Predicted values at N2 = –3		Predicted values at N2 = –3	
Variety	Value	Stderr	Value	Stderr	Value	Stderr	Value	Stderr
1	2.3014	0.1933	11.6319	0.4194	4.7278	0.7229	18.5360	0.7084
2	1.8067	0.1933	10.8802	0.4194	5.4600	0.7229	16.3004	0.7084
3	2.5399	0.1933	12.9149	0.4194	5.2952	0.7229	20.5345	0.7084
4	1.8254	0.1933	10.4425	0.4194	4.9662	0.7229	15.9187	0.7084
5	1.5483	0.1933	9.8479	0.4194	5.2029	0.7229	14.4929	0.7084
6	2.0850	0.1933	12.0582	0.4194	5.8032	0.7229	18.3131	0.7084
7	1.4583	0.1933	5.5995	0.4194	1.2246	0.7229	9.9744	0.7084

selects the value of $\hat{\beta}$ and N2 1/subject 1 selects the value of $\hat{b}_1$. Since the random statement includes "/subject = variety", then the "/subject 1" on the estimate statement specifies to select the first level of variety. Using "/subject 0 0 0 1" would select the fourth level of variety. The statements

```
estimate 'Int 1' Int 1 | Int 1/subject 1;
estimate 'Int 2' Int 1 | Int 1/subject 0 1;
```

provide predicted values for the intercepts of varieties 1 and 2 (similar statements were used for the other varieties). Using Int 1 selects the value of $\hat{\alpha}$ and Int 1/subject 1 selects the value of $\hat{a}_1$.

The predicted values of the slopes and intercepts and the corresponding estimated standard errors for each of the varieties are displayed in Table 12.6. The variance of these predicted slopes and of these predicted intercepts are proportional to the estimates of the variance components σ_a^2 and σ_b^2. The estimate statements

```
estimate 'Var 1 at N2=-3' intercept 1 N2 -3|Int 1
  N2 -3/subject 1;
estimate 'Var 2 at N2=-3' intercept 1 N2 -3|Int 1
  N2 -3/subject 0 1;
estimate 'Var 1 at N2= 3' intercept 1 N2 3|Int 1
  N2 3/subject 1;
estimate 'Var 2 at N2= 3' intercept 1 N2 3|Int 1
  N2 3/subject 0 1;
```

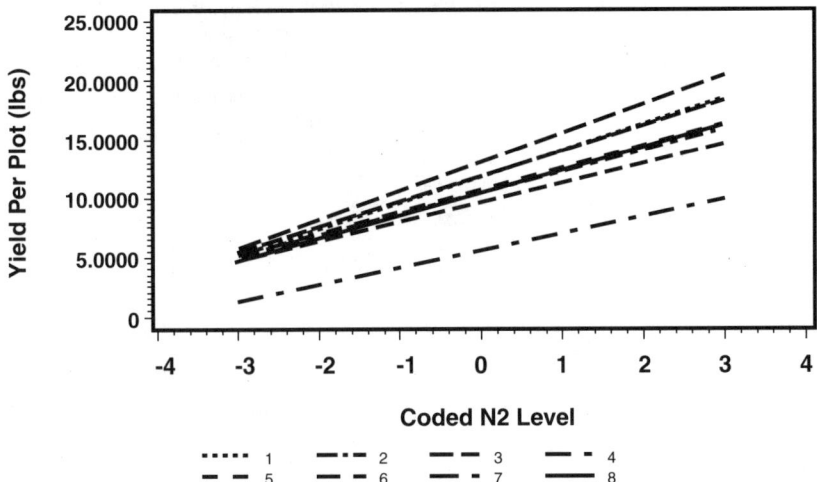

FIGURE 12.1 Graph of the predicted models for each variety where variety 8 corresponds to the mean model.

provide predicted values for varieties 1 and 2 at N2 = –3 and N2 = 3. These are predicted values from the regression equations using the intercepts and slopes for each variety. The set of statements can be extended to include all of the varieties. The predicted values at N2 = –3 and 3 for all of the varieties are included in Table 12.6. Those predicted values were used to construct a graph of the set of regression lines for these varieties which are displayed in Figure 12.1. The methods of moments estimates of the variance components process described in Section 12.3 are summarized in Table 12.7. Since the model is balanced, the method of moments estimates of the variance components are identical to the REML estimates displayed in Table 12.4.

TABLE 12.7
Sums of Squares, the Cross Product and Their Expectations for Computing the Methods of Moments Estimates of the Variance and Covariance Components for Example 12.5

Sum of Squares	Expected Sum of Squares
SS(INTERCEPTS) = 578.434286	$24\,\sigma_\varepsilon^2 + 96\,\sigma_I^2$
SS(SLOPES) = 511.817143	$120\,\sigma_\varepsilon^2 + 2400\,\sigma_S^2$
SSCP = 375.788571	$480\,\sigma_{IS}^2$
SS(ERROR) = 9.721000	$14\,\sigma_\varepsilon^2$
$\sigma_\varepsilon^2 = 0.6943571$	$\sigma_I^2 = 5.8517679$
$\sigma_S^2 = 0.1785393$	$\sigma_{IS}^2 = 0.7828929$

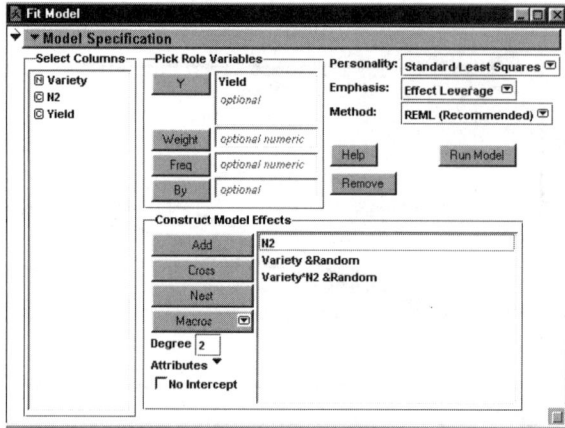

FIGURE 12.2 Fit model screen to fit the random coefficient model with independent slopes and intercepts.

REML Variance Component Estimates

Random Effect	Var Ratio	Var Component	Std Error	95% Lower	95% Upper	Pct of Total
Variety&Random	8.4276051	5.8517678	4.1578914	2.0928899	49.063778	87.019
Variety*N2&Random	0.2571289	0.1785393	0.1471614	0.0568613	2.5825108	2.655
Residual		0.6943571				10.326
Total		6.7246642				100.000

-2 LogLikelihood = 104.74837

Effect Tests

Source	Nparm	DF	DFDen	Sum of Squares	F Ratio	Prob > F	
N2	1	1	6	85.58958	123.2645	<.0001	
Variety&Random	7	6	6	140.44243	33.7104	0.0002	Shrunk
Variety*N2&Random	7	6	14	21.42471	5.1426	0.0055	Shrunk

FIGURE 12.3 REML estimates of the variance components and tests for the fixed effects and the random effects.

The JMP® software can fit the independent slope and intercept model to the data set, but it cannot fit the model with non-zero covariance. Figure 12.2 contains the fit model screen where yield is specified to be the response variable and the model contains N2, Variety, and N2*Variety, where the last two terms are specified as being random by using the attributes menu. Figure 12.3 contains the REML estimates of the variance components as well as test for the fixed and random effects. The indication is that the variance components are significantly different from zero, or they are important components in describing the variability in the model. The custom test menu can be used to provide predicted values for each variety's slope, intercept, or model evaluated at some value for N2. Those results were not included in this example. PROC MIXED can perform similar tests for the independent slope and intercept model by specifying method = type1, or type2, or type3 (an example is included in Section 12.7).

12.6 EXAMPLE: UNBALANCED ONE-WAY TREATMENT STRUCTURE

The data in Table 12.8 are from randomly selected cities and then randomly selected school districts with each city where the response variable is the amount of income

TABLE 12.8
Data for Example in Section 12.6 with One-Way Random Effects Treatment Structure (Cities) Where the Response Variable is the Amount Spent on Vocational Training and the Covariate is the Percent Unemployment

City 1		City 2		City 3		City 4		City 5	
Spent	Unemp	Spent	Unemp	Spent	Unemp	Spent	Unemp	Spent	Unemp
27.9	4.9	29.6	8.1	17.5	3.1	28.0	9.6	22.0	6.7
43.0	9.4	12.7	0.9	21.2	6.4	16.0	3.3	22.6	6.9
26.0	4.2	19.9	4.6	19.0	4.0	16.8	4.2	12.8	2.2
34.4	6.6	12.3	1.9	14.5	2.0	26.5	8.6	11.8	1.4
43.4	9.5	23.8	5.9	25.2	8.4	23.1	6.2	23.5	8.4
42.4	8.8	28.8	8.0	15.0	2.4	18.0	3.1	26.0	9.3
27.9	4.6	21.0	4.4	24.6	8.7	17.8	3.9	24.5	8.0
25.9	4.6	13.9	2.1			11.6	0.5	18.8	5.7
23.3	4.2	13.2	1.3			24.2	7.4	11.2	1.8
34.3	6.8								

City 6		City 7		City 8		City 9		City 10	
Spent	Unemp	Spent	Unemp	Spent	Unemp	Spent	Unemp	Spent	Unemp
28.0	6.3	37.6	9.9	10.6	0.0	10.9	2.9	8.6	0.8
18.2	2.6	34.7	8.8	21.4	3.4	12.0	2.8	9.1	0.7
29.2	6.7	38.6	9.9	32.1	7.0	14.1	3.6	13.6	3.4
25.0	5.4	33.8	8.2	28.3	5.8	15.4	5.2	13.7	3.1
22.6	3.8	10.1	0.1	38.7	9.2	11.2	1.4	13.4	3.6
34.9	8.4	29.8	7.0	19.8	2.7	17.9	5.6	15.3	4.4
39.3	9.8	15.1	1.8	40.0	9.7	26.2	10.0	23.1	7.8
37.3	9.1	35.0	8.5	36.3	8.2	14.2	3.6	22.2	7.4
30.5	6.8			29.0	5.9	21.4	7.2		

to be spent on vocational training (y) and the possible covariate is the level of unemployment (x). The PROC MIXED code to fit Model 12.1 is in Table 12.9 where the estimates of the parameters are $\hat{\sigma}_a^2 = 2.5001$ from UN(1,1), $\hat{\sigma}_b^2 = 0.4187$ from UN(2,2), $\hat{\sigma}_{ab} = 0.3368$ from UN(2,1), and $\hat{\sigma}_\varepsilon^2 = 0.6832$. The mixed models estimates of the mean of the intercepts and slopes are $\hat{\alpha} = 9.6299$, $\hat{\beta} = 2.3956$, so the estimate of the population mean spent income on vocational training at a given level of unemployment is $\hat{\mu}_{y|EMP} = 9.6299 + 2.3956$ UNEMP. The solution for the random effects that satisfy the sum-to-zero restriction within the intercepts and within the slopes are in Table 12.10. The estimate statements in Table 12.11 provide predicted values (estimated BLUPS) for each of the selected cities evaluated at 10% unemployment. Table 12.11 contains the predicted values for each of the cities evaluated at 0, 8, and 10%. Tables 12.12, 12.13, and 12.14 contain the PROC MIXED code and results for fitting Model 12.1 where the values of the covariate have been altered by subtracting 2, then mean (5.4195402) and 8 from the percent unemployment. The estimate of the variance component for the intercepts depends on the amount subtracted from the covariate as described in Section 12.4. When X_0 is subtracted

TABLE 12.9
PROC MIXED Code to Fit the Random Coefficients Model to the Vocational Training Spending Data

```
PROC MIXED CL COVTEST DATA=CITY IC;
CLASS CITY;
MODEL Y=X/SOLUTION DDFM=KR;
RANDOM INT X/TYPE=UN SUB=CITY SOLUTION;
```

CovParm	Subject	Estimate	StdErr	ZValue	ProbZ	Alpha	Lower	Upper
UN(1,1)	CITY	2.5001	1.3572	1.84	0.0327	0.05	1.0819	10.6709
UN(2,1)	CITY	0.3368	0.3910	0.86	0.3890	0.05	−0.4295	1.1031
UN(2,2)	CITY	0.4187	0.2035	2.06	0.0198	0.05	0.1944	1.4648
Residual		0.6832	0.1180	5.79	0.0000	0.05	0.5003	0.9892

Neg2LogLike	Parameters	AIC	AICC	HQIC	BIC	CAIC
284.40	4	292.40	292.90	291.07	293.61	297.61

Effect	Estimate	StdErr	DF	tValue	Probt
Intercept	9.6299	0.5401	9.2	17.83	0.0000
X	2.3956	0.2074	8.9	11.55	0.0000

TABLE 12.10
Predicted Intercepts and Slopes for Each of the Cities in the Study

Effect	CITY	Estimate	StdErrPred	df	tValue	Probt
Intercept	1	0.9717	0.8266	23	1.18	0.2516
X	1	1.1022	0.2280	13	4.83	0.0003
Intercept	2	−0.2659	0.6818	19	−0.39	0.7009
X	2	0.0511	0.2249	12	0.23	0.8239
Intercept	3	1.7366	0.7532	22	2.31	0.0308
X	3	−0.7753	0.2282	13	−3.40	0.0049
Intercept	4	0.8222	0.7136	21	1.15	0.2624
X	4	−0.5249	0.2225	12	−2.36	0.0366
Intercept	5	−1.0013	0.7233	21	−1.38	0.1807
X	5	−0.4962	0.2215	11	−2.24	0.0457
Intercept	6	0.7862	0.8307	24	0.95	0.3535
X	6	0.5100	0.2271	13	2.25	0.0435
Intercept	7	0.3338	0.7395	22	0.45	0.6561
X	7	0.4642	0.2180	11	2.13	0.0570
Intercept	8	1.3563	0.7176	21	1.89	0.0727
X	8	0.6307	0.2201	11	2.87	0.0151
Intercept	9	−2.3763	0.7147	21	−3.33	0.0033
X	9	−0.5371	0.2260	12	−2.38	0.0344
Intercept	10	−2.3633	0.6952	20	−3.40	0.0029
X	10	−0.4249	0.2290	13	−1.85	0.0865

TABLE 12.11
Estimate Statements Used to Provide Predicted Values for Each City at 10% Unemployment with Predicted Values at 0, 8, and 10% Unemployment

```
estimate 'city  1 at 10' intercept 1 x 10 |int 1 x 10/subject 1;
estimate 'city  2 at 10' intercept 1 x 10 |int 1 x 10/subject 0 1;
estimate 'city  3 at 10' intercept 1 x 10 |int 1 x 10/subject 0 0 1;
estimate 'city  4 at 10' intercept 1 x 10 |int 1 x 10/subject 0 0 0 1;
estimate 'city  5 at 10' intercept 1 x 10 |int 1 x 10/subject 0 0 0 0 1;
estimate 'city  6 at 10' intercept 1 x 10 |int 1 x 10/subject 0 0 0 0 0 1;
estimate 'city  7 at 10' intercept 1 x 10 |int 1 x 10/subject 0 0 0 0 0 0 1;
estimate 'city  8 at 10' intercept 1 x 10 |int 1 x 10/subject 0 0 0 0 0 0 0 1;
estimate 'city  9 at 10' intercept 1 x 10 |int 1 x 10/subject 0 0 0 0 0 0 0 0 1;
estimate 'city 10 at 10' intercept 1 x 10 |int 1 x 10/subject 0 0 0 0 0 0 0 0 0 1;
```

	At 0%		At 10%		At 8%	
City	Estimate	StdErr	Estimate	StdErr	Estimate	StdErr
1	10.6016	0.7186	45.5793	0.4894	38.5837	0.3255
2	9.3641	0.4799	33.8311	0.6548	28.9377	0.4814
3	11.3666	0.6041	27.5696	0.6448	24.3290	0.4650
4	10.4522	0.5370	29.1593	0.5329	25.4178	0.3858
5	8.6286	0.5538	27.6221	0.4907	23.8234	0.3561
6	10.4162	0.7246	39.4721	0.4746	33.6609	0.3241
7	9.9638	0.5807	38.5619	0.3932	32.8423	0.3091
8	10.9862	0.5438	41.2490	0.4620	35.1964	0.3401
9	7.2536	0.5392	25.8383	0.6225	22.1214	0.4469
10	7.2667	0.5048	26.9738	0.7496	23.0324	0.5517

TABLE 12.12
PROC MIXED Code to Fit the Random Coefficients Model Using X2 = X − 2 as the Covariate

```
PROC MIXED CL COVTEST DATA=CITY IC;
CLASS CITY;
MODEL Y=X2 /SOLUTION DDFM=KR;
RANDOM INT X2/TYPE=UN SUB=CITY SOLUTION;
```

CovParm	Subject	Estimate	StdErr	ZValue	ProbZ	Alpha	Lower	Upper
UN(1,1)	CITY	5.5225	2.7060	2.04	0.0206	0.05	2.5516	19.5882
UN(2,1)	CITY	1.1742	0.6471	1.81	0.0696	0.05	−0.0942	2.4425
UN(2,2)	CITY	0.4187	0.2034	2.06	0.0198	0.05	0.1944	1.4648
Residual		0.6832	0.1180	5.79	0.0000	0.05	0.5003	0.9892

Neg2LogLike	Parameters	AIC	AICC	HQIC	BIC	CAIC
284.40	4	292.40	292.90	291.07	293.61	297.61

Effect	Estimate	StdErr	DF	tValue	Probt
Intercept	14.4211	0.7576	9.0	19.04	0.0000
X2	2.3956	0.2074	8.9	11.55	0.0000

TABLE 12.13
PROC MIXED Code to Fit the Random Coefficients Model Using XMN = X − 5.4195402, the Mean of the X Values or Percent Unemployment Data

```
PROC MIXED CL COVTEST DATA=CITY IC;
CLASS CITY;
MODEL Y=XMN /SOLUTION DDFM=KR;
RANDOM INT XMN/TYPE=UN SUB=CITY SOLUTION;
```

CovParm	Subject	Estimate	StdErr	ZValue	ProbZ	Alpha	Lower	Upper
UN(1,1)	CITY	18.4486	8.7381	2.11	0.0174	0.05	8.7033	61.9455
UN(2,1)	CITY	2.6059	1.2808	2.03	0.0419	0.05	0.0956	5.1162
UN(2,2)	CITY	0.4187	0.2034	2.06	0.0198	0.05	0.1944	1.4648
Residual		0.6832	0.1180	5.79	0.0000	0.05	0.5003	0.9892

Neg2LogLike	Parameters	AIC	AICC	HQIC	BIC	CAIC
284.40	4	292.40	292.90	291.07	293.61	297.61

Effect	Estimate	StdErr	DF	tValue	Probt
Intercept	22.6128	1.3615	9.0	16.61	0.0000
XMN	2.3956	0.2074	8.9	11.55	0.0000

TABLE 12.14
PROC MIXED Code to Fit the Random Coefficients Model Using X8 = X − 8 as the Covariate

```
PROC MIXED CL COVTEST DATA=CITY IC;
CLASS CITY;
MODEL Y=X8 /SOLUTION DDFM=KR;
RANDOM INT X8/TYPE=UN SUB=CITY SOLUTION;
```

CovParm	Subject	Estimate	StdErr	ZValue	ProbZ	Alpha	Lower	Upper
UN(1,1)	CITY	34.6853	16.4247	2.11	0.0174	0.05	16.3654	116.4205
UN(2,1)	CITY	3.6863	1.7897	2.06	0.0394	0.05	0.1785	7.1941
UN(2,2)	CITY	0.4187	0.2034	2.06	0.0198	0.05	0.1944	1.4648
Residual		0.6832	0.1180	5.79	0.0000	0.05	0.5003	0.9892

Neg2LogLike	Parameters	AIC	AICC	HQIC	BIC	CAIC
284.40	4	292.40	292.90	291.07	293.61	297.61

Effect	Estimate	StdErr	DF	tValue	Probt
Intercept	28.7945	1.8670	9.0	15.42	0.0000
X8	2.3956	0.2074	8.9	11.55	0.0000

from X, the resulting variance component is related to the variance of the models evaluated at X_0. Table 12.15 contains the estimates of the variance components for the intercepts evaluated at X = 0, 2, 5.4195402, and 8 obtained from Tables 12.9, 12.12, 12.13, and 12.14 as well as the sample variance of the predicted values at each of those values of X. The variance at X = 0 is the variance of the intercept

TABLE 12.15
Estimates of the Intercept Variance Components and the Variance of the Predicted Values at Four Values of X

Amount Subtracted from X	Estimate of the Intercept Variance Component	Variance of the Predicted Values
0	2.5001	2.1678
2	5.5225	5.3460
5.4195402	18.4486	18.3628
8	34.6853	34.5209

predictions from Table 12.10. The variance at X = 8 is the variance of the predicted values from Table 12.11. The sample variances of the predicted values are not quite equal to the REML estimates of the variance components since the sample variances do not take into account the number of other parameters being estimated in the models. However, the estimates of the variance components of the models evaluated at a given value of X are predictable using the results from Section 12.4. In this case, using the estimates of the variance components from Table 12.9, the estimate of the variance component for the intercepts at X = 8 is computed as

$$\hat{\sigma}_{a*}^2 = \hat{\sigma}_a^2 + 2 \times 8 \times \hat{\sigma}_{ab} + 8^2 \times \hat{\sigma}_b^2$$

$$= 2.5001 + 2 \times 8 \times 0.3368 + 8^2 \times 0.4187$$

$$= 34.6853$$

where a* denotes the intercepts of the models with the covariate taking on the values of X − 8. The REML estimates of variance components for the slopes [UN(2,2)] and residual are the same in all four of the analyses: Tables 12.9, 12.12, 12.13, and 12.14.

Using the results in Section 12.2, a prediction band can be constructed about the population of regression models. Table 12.16 contains the estimates of the amount to be spent on vocational training from the estimated population regression model at 0, 1, ...,10% unemployment. The estimated standard error of prediction at each of the values of x and the simultaneous lower and upper 95% prediction intervals are also included in Table 12.16. The Scheffé percentage point used was $\sqrt{2 \, F_{0.05, \, 2, \, 70}} = 2.50$. Figure 12.4 is a graph of the ten predicted regression lines, the mean and the upper and lower prediction limits over the range of 0 to 10% unemployment. The graph clearly shows that as the percent of unemployment increases, the width of the prediction interval also increases. The next section applies the analysis to a two-way random effects treatment structure.

12.7 EXAMPLE: TWO-WAY TREATMENT STRUCTURE

Three genetic lines of Hereford cattle were randomly selected from a population of Hereford cattle producers, and three genetic lines of Black Angus cattle were randomly

TABLE 12.16
The Mean, Estimated Standard Error of Prediction, and Lower and Upper Prediction Intervals for Selected Values of X (% Unemployment)

x	mean	se	low	up
0	9.63	1.78	5.17	14.09
1	12.03	2.07	6.85	17.20
2	14.42	2.49	8.19	20.65
3	16.82	3.00	9.32	24.31
4	19.21	3.55	10.34	28.08
5	21.61	4.13	11.29	31.93
6	24.00	4.72	12.19	35.81
7	26.40	5.33	13.07	39.73
8	28.79	5.95	13.92	43.67
9	31.19	6.57	14.76	47.62
10	33.59	7.20	15.59	51.58

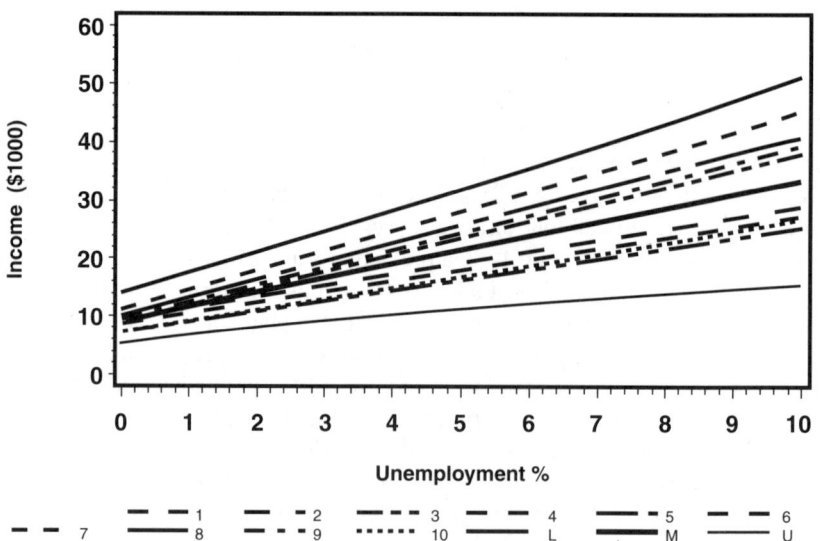

FIGURE 12.4 Graph of the predicted models for each city with the lower (L) and upper (U) prediction band and the mean (M) of the regression model.

selected from a population of Black Angus cattle producers. The set of three genetic lines from each population is not a large enough sample to provide reliable estimates of the variance components in a genetic study, but the smaller number was used for

Random Effects Models with Covariates

demonstration purposes. Cows (20) from each line of the Hereford cattle were crossed with bulls from each line of the Black Angus, and 20 cows from each line of Black Angus cattle were crossed with bulls from each line of Hereford. Thus, there were 180 cows in the study where each cow gave birth to 1 calf. Five pastures (denoted as Rep) were used for the study where 36 cows with calves were put into each pasture where there were 4 cows and calves from each of the 9 crosses. The response variable was the weight of the calf at weaning time and the birth weight was considered as a possible covariate. Table 12.17 contains the data where Rep represents the pasture and (y_{ij}, x_{ij}) represent the weaning weight and the birth weight, respectively, for the ij^{th} cross.

The two-way random coefficient model of Equation 12.5 was to be fit to the data set. The PROC MIXED code in Table 12.18 fits Model 12.5 where all of the covariances between the slopes and intercepts for each set are zero, "a" denotes Hereford, "b" denotes Black Angus, y denotes the weaning weight and x denotes the birth weight. In this analysis, method = type3 was used to provide method of moments estimates of the variance components using type III sums of squares. This option causes PROC MIXED to compute sums of squares, mean squares, expected mean squares, appropriate error terms (Milliken and Johnson, 1992), approximate error degrees of freedom, and F-statistics to test a hypothesis about each of the effects in the model. Using the parameters in the description of Model 12.5, source "a" tests $H_0: \sigma_a^2 = 0$ vs Ha: $\sigma_a^2 > 0$, source "x*a" tests $H_0: \sigma_b^2 = 0$ vs. $H_a: \sigma_b^2 > 0$, source "b" tests $H_0: \sigma_c^2 = 0$ vs. $H_a: \sigma_c^2 > 0$, source "x*b" tests $H_0: \sigma_d^2 = 0$ vs. $H_a: \sigma_d^2 > 0$, source "a*b" tests $H_0: \sigma_f^2 = 0$ vs. $H_a: \sigma_f^2 > 0$, and source "x*a*b" tests $H_0: \sigma_g^2 = 0$ vs. $H_a: \sigma_g^2 > 0$. The results indicate that all of the variance components are important except σ_a^2 (source a or Hereford) and σ_c^2 (source b or Black Angus) with significance levels of 0.1565 and 0.2140, respectively. The method of moments estimates of the variance components are in Table 12.19 as well as the estimates of the population intercept (230.2420) and slope (0.6281).

Based on the results of Section 12.4, the interpretations of the variance components σ_a^2, σ_c^2, and σ_f^2 are that they represent the variation of the intercepts of the regression models or the variation of the predicted models evaluated at X = 0 or at a birth weight of 0, a place that is not meaningful to compare the regression models. The researcher wanted to evaluate the variance components of the intercepts at a birth weight of 75 pounds so a new covariate was computed as x75 = x − 75. The PROC MIXED code in Table 12.20 fits the independent slopes and intercepts model using x75 as the covariate. The sums of squares and tests for all terms involving x75 provides identical values as those same terms in Table 12.18. The sums of squares and tests for a, b, and a*b have changed as they are evaluating the variance components of the new intercepts or the models evaluated at x = 75. At x = 75, the significance levels corresponding to sources a, b, and a*b are 0.0569, 0.1039, and 0.000, respectively. The method of moments estimates of the variance components are in Table 12.21. The estimates of the three variance components corresponding to the models at x = 75 are $\hat{\sigma}_a^2 = 217.3864$, $\hat{\sigma}_c^2 = 129.4336$, and $\hat{\sigma}_f^2 = 121.1095$. These variance components are much larger than those obtained for the models evaluated at x = 0. The next step is to fit the correlated slopes and intercepts model to the data. Table 12.22 contains the PROC MIXED code to fit the correlated slopes and

TABLE 12.17
Weights of Calves at Weaning (y_{ij}) and Birth Weights (x_{ij}) of Crosses of the ith Hereford Genetic Line with the jth Black Angus Genetic Line

Rep	y11	x11	y12	x12	y13	x13	y21	x21	y22	x22	y23	x23	y31	x31	y32	x32	y33	x33
1	283	94	262	92	261	58	304	86	280	100	283	94	334	86	282	58	278	77
1	258	53	248	53	258	52	293	72	259	53	266	59	336	88	281	57	280	82
1	285	96	252	64	272	85	311	96	271	80	270	66	344	94	288	65	278	79
1	270	71	261	88	263	63	303	88	279	96	264	55	302	57	297	77	275	73
2	273	63	264	71	285	95	293	61	285	89	294	97	345	88	308	80	295	94
2	285	83	270	87	286	98	287	55	287	96	277	64	321	67	287	54	276	58
2	279	71	273	97	267	52	286	54	279	77	286	83	331	76	315	88	272	50
2	295	98	268	83	278	78	288	56	273	63	273	57	319	66	319	94	273	53
3	243	63	237	83	244	71	261	61	241	61	264	99	320	93	266	65	252	70
3	244	65	241	92	255	98	283	88	250	80	245	60	300	76	283	86	266	96
3	258	90	241	94	236	53	281	85	254	89	247	64	303	78	285	89	263	90
3	247	70	240	91	239	60	283	88	237	55	254	79	312	86	288	92	246	59
4	283	88	263	88	273	83	281	56	266	63	282	86	329	80	294	71	269	56
4	260	53	252	58	267	69	287	64	274	81	283	89	321	73	292	67	278	74
4	270	69	264	92	266	64	310	91	261	53	277	77	337	86	291	67	288	92
4	259	50	261	84	274	85	290	67	277	87	284	91	337	87	291	68	291	98
5	252	51	254	87	259	68	309	100	254	57	266	71	318	78	298	86	268	71
5	270	82	257	91	264	79	306	96	270	91	275	89	324	83	301	90	258	54
5	269	80	247	64	269	91	287	73	260	67	272	82	343	99	294	80	282	97
5	280	99	249	72	257	64	299	87	262	72	278	94	313	73	286	71	269	72

TABLE 12.18
PROC MIXED Code to Fit the Independent Slope and Intercept Model Using Type III Sums of Squares with Expected Mean Squares and Error Terms

```
proc mixed scoring=2 data=twoway method=type3;
class rep a b;
model y=x/solution ddfm=kr;
random rep a x*a b x*b a*b x*a*b;
```

Source	df	SS	MS	EMS
x	1	12837.5138	12837.5138	Var(Residual) + 3616.3 Var(x*a*b) + 10849 Var(x*b) + 10849 Var(x*a) + Q(x)
rep	4	18310.2743	4577.5686	Var(Residual) + 34.044 Var(rep)
a	2	25.8699	12.9349	Var(Residual) + 0.6016 Var(a*b) + 1.8047 Var(a)
x*a	2	686.0917	343.0459	Var(Residual) + 3696.7 Var(x*a*b) + 11090 Var(x*a)
b	2	19.7484	9.8742	Var(Residual) + 0.5877 Var(a*b) + 1.763 Var(b)
x*b	2	811.8033	405.9017	Var(Residual) + 3601.7 Var(x*a*b) + 10805 Var(x*b)
a*b	4	17.2849	4.3212	Var(Residual) + 0.6183 Var(a*b)
x*a*b	4	212.1297	53.0324	Var(Residual) + 3762.7 Var(x*a*b)
Residual	158	41.8099	0.2646	Var(Residual)

Source	ErrorDF	FValue	ProbF	ErrorTerm
x	3.4	18.55	0.0176	0.9783 MS(x*a) + 1.0041 MS(x*b) – 0.9611 MS(x*a*b) – 0.0212 MS(Residual)
rep	158.0	17298.68	0.0000	MS(Residual)
a	4.0	3.07	0.1552	0.9729 MS(a*b) + 0.0271 MS(Residual)
x*a	4.0	6.58	0.0543	0.9825 MS(x*a*b) + 0.0175 MS(Residual)
b	4.0	2.40	0.2063	0.9505 MS(a*b) + 0.0495 MS(Residual)
x*b	4.0	7.99	0.0400	0.9572 MS(x*a*b) + 0.0428 MS(Residual)
a*b	158.0	16.33	0.0000	MS(Residual)
x*a*b	158.0	200.41	0.0000	MS(Residual)

intercepts model where four random statements are needed to complete the model specification. The statement "`random rep;`" specifies that the replications or pastures are a random effect. The statement "`random int x75/type = un subject = a solution;`" specifies covariance structure for the intercepts and slopes for the main effect of a or Hereford genetic lines. The statement "`random int x75/type = un subject = b solution;`" specifies covariance structure for the intercepts and slopes for the main effect of b or Black Angus genetic lines. The statement "`random int x75/type = un subject = a*b solution;`" specifies covariance structure for the intercepts and slopes for the interaction between the levels of a and the levels of b or a cross between a Hereford line and a Black Angus line. PROC MIXED would not fit this model without including some initial values for the variance components. The estimates of the variance components obtained from the uncorrelated model in Table 12.21 were used as starting values where "0" was inserted in place of the covariances of the slopes and intercepts for the main effect of a, for the main effect of b, and for the interaction of a and b.

TABLE 12.19
Method of Moments Estimates of the Variance Components and Resulting Estimates and Tests of the Fixed Effects

CovParm	Estimate				
rep	134.4528				
a	4.8338				
x*a	0.0262				
b	3.2637				
x*b	0.0329				
a*b	6.5611				
x*a*b	0.0140				
Residual	0.2646				

Effect	Estimate	StdErr	df	tValue	Probt
Intercept	230.2420	5.5108	5.0	41.78	0.0000
x	0.6281	0.1458	3.9	4.31	0.0131

Effect	NumDF	DenDF	FValue	ProbF
x	1	3.9	18.55	0.0131

These starting values are included in the "Parameters" statement in Table 12.22. Table 12.22 contains the REML estimates of the variance components, which are similar to those obtained from the method of moments, but not identical since covariances of the slopes and intercepts were also estimated. The REML estimates of the three variance components from the correlated slopes and intercepts structure corresponding to the models evaluated at x = 75 are $\hat{\sigma}_a^2 = 225.0213$ (from U(1,1) for subject = a), $\hat{\sigma}_c^2 = 132.7903$ (from UN(1,1) for subject = b), and $\hat{\sigma}_f^2 = 119.1411$ (from UN(1,1) for subject = a*b). These are the important variance components to be used in any genetic analysis of this data set (no genetic analysis is included here). Table 12.23 contains the predicted values for the slopes (x75) and the models at X = 75 (Intercept) where each set satisfies the sum to zero restriction within each type of effect. These predicted values can be used to obtain the predicted slope and intercept for each level of a (Hereford), for each level of b (Black Angus), and for each combination of levels of a and b (crosses of Hereford and Black Angus). The predicted intercept for the first Hereford genetic line is $\hat{a}_{p\,a_1} = \hat{\alpha} + \hat{a}_{a_1} = 277.3511 - 11.9311 = 265.4200$ and the predicted slope is $\hat{b}_{p\,a_1} = \hat{\beta} + \hat{b}_{a_1} = 0.6275 - 0.1312 = 0.4963$. The predicted model for the first Hereford genetic is $\hat{wt}_{p\,a_1} = 265.42 + .4963$ (birth wt − 75). Figure 12.5 is a graph of the three predicted models for the three Hereford genetic lines. Similar models can be constructed for the three Black Angus genetic lines, and those predicted models are displayed in Figure 12.6. The predicted slope and intercept for a cross of the first Hereford genetic line and the first Black Angus genetic line is $\hat{b}_{p\,ab_{11}} = \hat{\beta} + \hat{b}_{a_1} + \hat{b}_{b_1} + \hat{b}_{ab_1} = 0.6275 - 0.1312 + 0.1862 - 0.0793 = 0.6032$ and $\hat{a}_{p\,ab_{11}} = \hat{\alpha} + \hat{a}_{a_1} + \hat{a}_{b_1} + \hat{a}_{ab_1} = 277.3511 - 11.9311 + 11.3392 - 8.2703 = 268.4889$. Thus, the model for the (1,1) combination is $\hat{wt}_{p\,ab_{11}} = 268.4889 + 0.6032$ (birth wt − 75). The nine predicted regression lines for the Hereford and Black Angus crosses are displayed in Figure 12.7.

TABLE 12.20
PROC MIXED Code to Fit the Independent Slope and Intercept Model with x75 = x − 75 as the Covariate Using Type III Sums of Squares with Expected Mean Squares and Error Terms

```
proc mixed scoring=2 data=twoway method=type3;
class rep a b;
model y=x75/solution ddfm=kr;
random rep a x75*a b x75*b a*b x75*a*b;
```

Source	df	SS	MS	EMS
x75	1	12837.5138	12837.5138	Var(Residual) + 3616.3 Var(x75*a*b) + 10849 Var(x75*b) + 10849 Var(x75*a) + Q(x75)
rep	4	18310.2743	4577.5686	Var(Residual) + 34.044 Var(rep)
a	2	28862.0375	14431.0187	Var(Residual) + 18.662 Var(a*b) + 55.986 Var(a)
x75*a	2	686.0917	343.0459	Var(Residual) + 3696.7 Var(x75*a*b) + 11090 Var(x75*a)
b	2	19003.6174	9501.8087	Var(Residual) + 18.652 Var(a*b) + 55.956 Var(b)
x75*b	2	811.8033	405.9017	Var(Residual) + 3601.7 Var(x75*a*b) + 10805 Var(x75*b)
a*b	4	9067.4377	2266.8594	Var(Residual) + 18.715 Var(a*b)
x75*a*b	4	212.1297	53.0324	Var(Residual) + 3762.7 Var(x75*a*b)
Residual	158	41.8099	0.2646	Var(Residual)

Source	ErrorDF	FValue	ProbF	ErrorTerm
x75	3.4	18.55	0.0176	0.9783 MS(x75*a) + 1.0041 MS(x75*b) − 0.9611 MS(x75*a*b) − 0.0212 MS(Residual)
rep	158.0	17298.68	0.0000	MS(Residual)
a	4.0	6.38	0.0569	0.9972 MS(a*b) + 0.0028 MS(Residual)
x75*a	4.0	6.58	0.0543	0.9825 MS(x75*a*b) + 0.0175 MS(Residual)
b	4.0	4.21	0.1039	0.9966 MS(a*b) + 0.0034 MS(Residual)
x75*b	4.0	7.99	0.0400	0.9572 MS(x75*a*b) + 0.0428 MS(Residual)
a*b	158.0	8566.49	0.0000	MS(Residual)
x75*a*b	158.0	200.41	0.0000	MS(Residual)

12.8 SUMMARY

Random effects treatment structures with covariates provides the usual random coefficient regression model. The strategy of determining the simplest covariate part of the model is applied here as one would for a fixed effects treatment structure, except we are looking for variance and covariance components which help describe the data instead of regression models. Predictions of the coefficients for a family of regression models can be obtained and for some cases it might be important to obtain the BLUP of the model for a specific level of the random effect and to compare BLUPs for two or more specific levels of the random effect. It is also important to evaluate the intercept variance components at values of the covariates that are important to the study. The default is to provide estimates of the variance components for the intercepts of the models, which may not be meaningful.

TABLE 12.21
Method of Moments Estimates of the Variance Components and Resulting Estimates and Tests of the Fixed Effects with x75 = x − 75 as the Covariate

CovParm	Estimate
rep	134.4528
a	217.3864
x75*a	0.0262
b	129.4336
x75*b	0.0329
a*b	121.1095
x75*a*b	0.0140
Residual	0.2646

Effect	Estimate	StdErr	df	tValue	Probt
Intercept	277.3469	12.4882	4.6114479	22.21	0.0000
x75	0.6279	0.1458	3.91563788	4.31	0.0132

Effect	NumDF	DenDF	FValue	ProbF
x75	1	3.9	18.54	0.0132

TABLE 12.22
PROC MIXED Code to Fit the Correlated Slopes and Intercepts Model to Obtain REML Estimates of the Variance Components Using Starting Values from the Independent Slopes and Intercept Model

```
proc mixed cl covtest scoring=2 data=twoway; class rep a b;
model y=x75/solution ddfm=kr;
random rep;
random int x75/type=un subject=a solution;
random int x75/type=un subject=b solution;
random int x75/type=un subject=a*b solution;
parms 134.5 217.4 0 0.0262 129.4 0 0.0329 121.11 0 0.0140 0.2646;
```

CovParm	Subject	Estimate	StdErr	ZValue	ProbZ	Alpha	Lower	Upper
rep		134.5898	95.1750	1.41	0.0787	0.05	48.3103	1111.5577
UN(1,1)	a	225.0213	266.2253	0.85	0.1990	0.05	52.6302	31896.9335
UN(2,1)	a	2.6013	3.0448	0.85	0.3929	0.05	−3.3664	8.5691
UN(2,2)	a	0.0302	0.0350	0.86	0.1936	0.05	0.0072	3.5032
UN(1,1)	b	132.7903	174.7797	0.76	0.2237	0.05	28.2152	55751.2871
UN(2,1)	b	2.1664	2.6132	0.83	0.4071	0.05	−2.9554	7.2883
UN(2,2)	b	0.0347	0.0394	0.88	0.1893	0.05	0.0084	3.4530
UN(1,1)	a*b	119.1411	84.2553	1.41	0.0787	0.05	42.7632	984.1469
UN(2,1)	a*b	1.2668	0.9001	1.41	0.1593	0.05	−0.4975	3.0310
UN(2,2)	a*b	0.0137	0.0097	1.41	0.0797	0.05	0.0049	0.1147
Residual		0.2644	0.0297	8.89	0.0000	0.05	0.2146	0.3340

Effect	Estimate	StdErr	df	tValue	Probt
Intercept	277.3511	12.6265	4.40353	21.97	0.0000
x75	0.6275	0.1522	3.4939	4.12	0.0192

Effect	NumDF	DenDF	FValue	ProbF
x75	1	3.5	16.99	0.0192

TABLE 12.23
Predicted Values for the Random Effects for the Model Using x75 = x − 75 as the Covariate (intercepts are at x = 75)

Effect	rep	a	b	Estimate	StdErrPred	df	tValue	Probt
rep	1			3.3838	5.1890	6.3	0.65	0.5375
rep	2			12.1698	5.1890	6.3	2.35	0.0557
rep	3			−18.4913	5.1890	6.3	−3.56	0.0110
rep	4			5.5019	5.1890	6.3	1.06	0.3282
rep	5			−2.5642	5.1890	6.3	−0.49	0.6380
Intercept		1		−11.9311	9.7610	27.6	−1.22	0.2319
x75		1		−0.1312	0.1120	26.4	−1.17	0.2519
Intercept		2		−3.4067	9.7606	27.6	−0.35	0.7297
x75		2		−0.0511	0.1120	26.4	−0.46	0.6520
Intercept		3		15.3379	9.7624	27.6	1.57	0.1275
x75		3		0.1823	0.1121	26.5	1.63	0.1156
Intercept			1	11.3392	4.3461	2.3	2.61	0.1046
x75			1	0.1862	0.0963	8.5	1.93	0.0868
Intercept			2	2.9709	4.3409	2.3	0.68	0.5564
x75			2	0.0173	0.0962	8.5	0.18	0.8616
Intercept			3	−14.3101	4.2957	2.2	−3.33	0.0704
x75			3	−0.2035	0.0959	8.4	−2.12	0.0651
Intercept		1	3	−8.2703	2.8119	178.0	−2.94	0.0037
x75		1	3	−0.0793	0.0388	178.0	−2.04	0.0425
Intercept		2	1	−15.5268	2.9098	178.0	−5.34	0.0000
x75		2	1	−0.1557	0.0390	178.0	−3.99	0.0001
Intercept		2	2	13.2504	2.7889	178.0	4.75	0.0000
x75		2	2	0.1303	0.0384	178.0	3.40	0.0008
Intercept		2	3	5.8315	2.8476	178.0	2.05	0.0420
x75		2	3	0.0510	0.0390	178.0	1.31	0.1920
Intercept		3	1	−11.2035	2.8291	178.0	−3.96	0.0001
x75		3	1	−0.1292	0.0389	178.0	−3.32	0.0011
Intercept		3	2	10.9771	2.8258	178.0	3.88	0.0001
x75		3	2	0.1246	0.0385	178.0	3.24	0.0014
Intercept		3	3	13.8910	2.9458	178.0	4.72	0.0000
x75		3	3	0.1496	0.0387	178.0	3.87	0.0002
Intercept		1	1	−3.5521	2.8408	178.0	−1.25	0.2128
x75		1	1	−0.0276	0.0384	178.0	−0.72	0.4740
Intercept		1	2	−5.3973	2.7446	178.0	−1.97	0.0508
x75		1	2	−0.0639	0.0379	178.0	−1.68	0.0940

Random Effects Models with Covariates

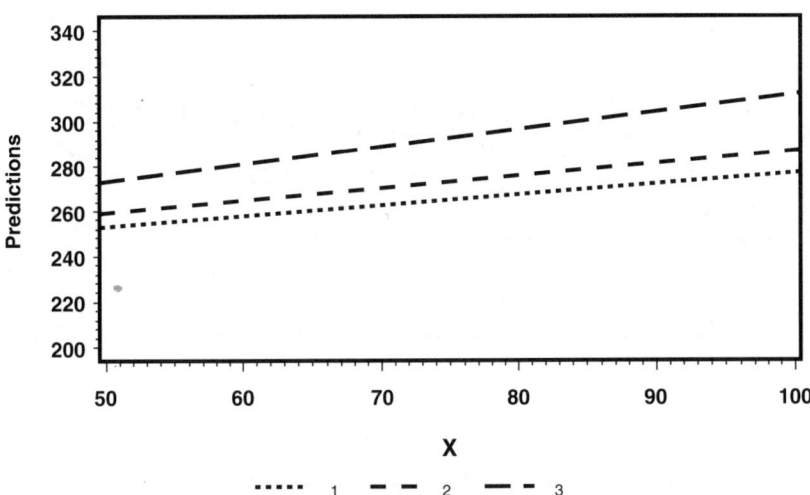

FIGURE 12.5 Graph of the models for each Hereford genetic line.

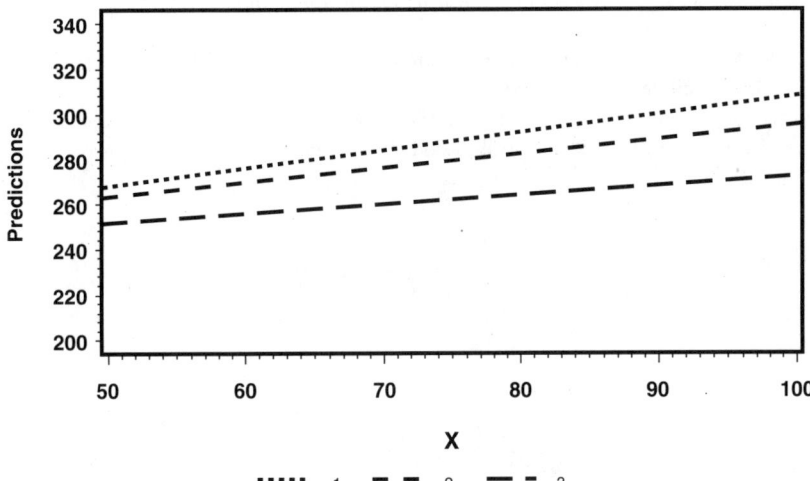

FIGURE 12.6 Graph of the models for Black Angus genetic line.

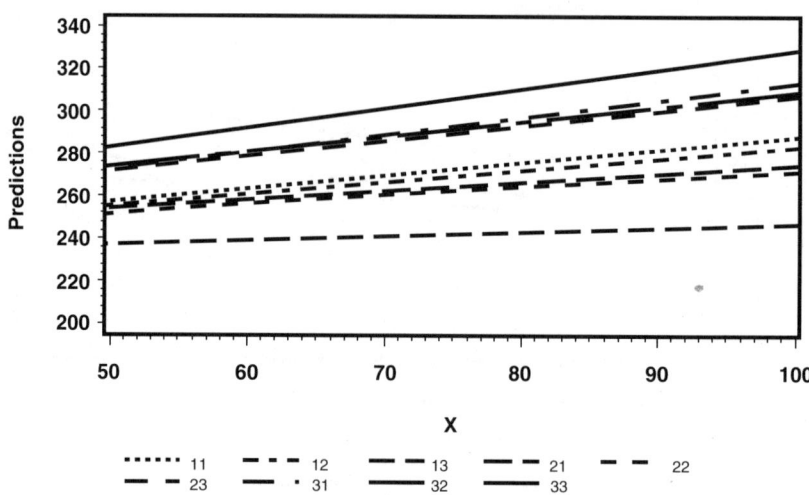

FIGURE 12.7 Graph of the Hereford by Black Angus cross models.

REFERENCES

Graybill, F. A. (1969) *Introduction to Matrices with Applications in Statistics*, Wadsworth, Belmont, CA.

Littell, R., Milliken, G., Stroup, W., and Wolfinger, R. (1996) *SAS System for Mixed Models*, SAS Institute Inc., Cary, NC.

Milliken, G. A. and Johnson, D. E. (1992) *Analysis of Messy Data, Volume I: Design Experiments*, Chapman & Hall, London.

EXERCISES

EXERCISE 12.1: Determine the proper form for the analysis of covariance model to describe the following data. The treatments are a random sample from a population of treatments. Construct the appropriate analysis of variance table. Provide estimates of the intercepts and slopes for each treatment. Plot the regression lines. Provide estimates of the variance components of the intercepts at $X = 2, 4, 6$, and 8.

Data for Exercise 12.1

TRT	SUB	Y	X	TRT	SUB	Y	X	TRT	SUB	Y	X
1	1	134.2	8.2	4	1	155.3	2.7	7	1	152.0	1.2
1	2	151.4	5.9	4	2	151.1	3.5	7	2	149.5	3.6
1	3	158.7	0.1	4	3	142.5	5.3	7	3	151.6	4.2
1	4	155.7	2.3	4	4	143.9	2.1	7	4	140.1	8.8
1	5	138.9	7.5	4	5	153.8	1.3	7	5	155.7	9.3
1	6	111.3	9.9	4	6	146.3	4.7	7	6	155.2	7.0
1	7	155.2	0.7	4	7	149.0	3.2	7	7	146.3	5.8
1	8	148.0	7.6	4	8	115.4	9.3	7	8	155.1	6.3
1	9	147.2	6.8	4	9	143.0	5.1	7	9	156.8	6.1
2	1	122.5	4.6	5	1	153.3	1.7	8	1	110.3	5.2
2	2	116.6	7.0	5	2	138.6	5.3	8	2	97.6	5.8
2	3	105.3	9.1	5	3	154.7	4.8	8	3	107.2	5.2
2	4	126.8	4.5	5	4	110.2	8.2	8	4	126.9	4.5
2	5	132.9	3.0	5	5	157.7	0.6	8	5	114.1	4.8
2	6	124.5	4.4	5	6	164.6	3.5	8	6	97.1	6.6
2	7	140.7	2.2	5	7	154.6	0.5	8	7	125.9	2.9
2	8	118.2	7.0	5	8	103.6	8.7	8	8	138.8	1.3
2	9	132.8	5.0	5	9	82.4	9.8	8	9	128.6	4.1
3	1	81.5	8.0	6	1	128.8	6.6	9	1	135.8	7.4
3	2	136.7	4.5	6	2	146.3	3.1	9	2	179.0	0.8
3	3	141.2	4.6	6	3	120.7	7.6	9	3	135.1	7.2
3	4	161.3	2.5	6	4	110.7	8.6	9	4	163.1	4.1
3	5	45.3	9.4	6	5	120.8	7.8	9	5	168.5	1.0
3	6	135.6	4.7	6	6	138.2	4.5	9	6	172.7	3.6
3	7	173.7	0.9	6	7	144.8	0.5	9	7	149.6	6.1
3	8	47.3	9.3	6	8	138.9	4.8	9	8	162.9	4.5
3	9	95.5	7.3	6	9	103.4	9.6	9	9	177.9	1.3
10	1	161.3	1.5								
10	2	113.6	5.7								
10	3	41.0	9.1								
10	4	78.1	7.3								
10	5	120.9	5.0								
10	6	128.8	4.5								
10	7	47.9	8.4								
10	8	144.7	3.3								
10	9	95.5	6.8								

EXERCISE 12.2: A banking corporation wanted to study the variability in credit scores of customers who applied for loans to finance a new automobile. The corporation randomly selected 20 branches from the population of 237 branches. The data in the following two tables consists of the branch, customer number (cust), the amount of the loan (in $1000), the applicants' income (in $1000), the age of the applicant, the amount of average credit card balance the past 12 months, and the credit score. Construct a random coefficient model for this one-way random effects treatment structure involving four possible covariates. Provide an estimate of the branch variance component at the mean of the four covariates.

Part 1 of the Data for Exercise 12.2

Branch	Cust	Loan	Age	Income	Ccard	Score	Branch	Cust	Loan	Age	Income	Ccard	Score
1	1	24	43	21	1736	678	7	1	24	43	33	2180	658
1	2	21	37	21	2766	679	7	2	15	30	21	299	669
1	3	9	42	24	714	674	7	3	15	31	72	1785	651
1	4	18	31	18	1597	680	7	4	18	29	21	2953	669
1	5	27	25	27	1924	682	7	5	27	41	18	2143	664
1	6	24	32	45	321	675	7	6	30	40	30	2175	661
							7	7	27	41	39	984	658
2	1	18	26	18	2973	724	7	8	12	32	30	49	664
2	2	21	34	57	2250	734	7	9	21	44	123	303	628
2	3	18	30	27	763	731	7	10	21	44	27	1502	658
2	4	21	32	18	2007	735							
2	5	24	35	24	1523	739	8	1	18	41	39	868	630
2	6	18	38	27	2175	744	8	2	18	41	24	1694	635
							8	3	24	30	21	1864	653
3	1	27	32	30	2234	669	8	4	18	43	21	2010	634
3	2	21	36	18	452	670	8	5	15	38	27	2806	636
3	3	21	29	21	1465	674	8	6	15	30	30	1938	647
3	4	21	47	24	2645	662	8	7	24	42	27	2304	634
3	5	9	33	18	507	670	8	8	24	31	27	357	649
3	6	24	29	27	2903	671	8	9	21	35	39	2736	639
4	1	27	40	18	2407	642	9	1	24	36	21	421	697
4	2	18	40	18	1533	641	9	2	24	35	24	45	696
4	3	27	45	18	1724	636	9	3	12	35	21	571	695
4	4	18	34	126	370	609	9	4	18	34	18	630	696
4	5	24	29	39	2929	648	9	5	12	28	21	2356	694
4	6	18	35	21	384	646	9	6	18	29	24	1338	694
4	7	21	43	21	1140	636							
4	8	21	34	21	2238	649	10	1	21	36	18	2451	661
							10	2	24	31	21	2739	665
5	1	24	39	21	1399	699	10	3	24	37	30	244	656
5	2	21	30	36	315	692	10	4	12	39	33	253	652
5	3	24	30	18	701	698	10	5	21	35	27	476	658
5	4	21	41	18	1084	700	10	6	18	28	21	2563	666
5	5	21	29	21	1570	696							
5	6	24	28	21	1368	697	11	1	27	38	21	2292	591
							11	2	24	35	24	2438	597
6	1	15	44	48	521	718	11	3	18	44	18	1507	571
6	2	21	32	24	2119	713	11	4	21	39	24	272	584
6	3	9	41	21	3	719	11	5	27	41	24	150	580
6	4	12	31	21	487	711	11	6	21	34	24	1975	598
6	5	12	37	66	153	710							
6	6	27	36	21	1318	717							
6	7	24	38	27	999	718							

Part 2 of the Data for Exercise 12.2

Branch	Cust	Loan	Age	Income	Ccard	Score	Branch	Cust	Loan	Age	Income	Ccard	Score
12	1	24	40	21	1194	695	17	1	18	41	24	2196	554
12	2	21	32	66	1406	679	17	2	24	40	21	2524	566
12	3	15	32	18	2840	693	17	3	24	40	24	1096	561
12	4	27	38	21	286	695	17	4	27	27	18	767	609
12	5	21	34	24	1251	692	17	5	24	41	21	1165	560
12	6	27	39	30	777	692	17	6	21	36	21	863	577
							17	7	21	37	27	1415	569
13	1	24	35	30	1443	712	17	8	24	30	30	2423	593
13	2	18	36	21	1389	714	17	9	21	36	24	1884	573
13	3	21	34	18	2066	713	17	10	15	34	30	1488	574
13	4	15	32	42	1370	707							
13	5	30	35	24	1762	714	18	1	21	31	57	1123	677
13	6	21	31	27	1131	710	18	2	24	34	21	1054	695
13	7	33	47	21	2703	722	18	3	15	41	27	2918	690
13	8	21	36	24	1673	713	18	4	24	38	42	2739	687
13	9	27	33	21	17	713	18	5	27	31	24	1449	695
							18	6	24	28	36	1381	688
14	1	18	33	30	1218	718	18	7	21	30	21	2302	695
14	2	30	33	24	1005	721	18	8	18	29	18	1507	695
14	3	24	38	24	2450	726	18	9	30	38	24	1461	695
14	4	12	36	21	2976	722							
14	5	21	33	21	2174	721	19	1	12	30	18	105	699
14	6	21	35	24	1617	722	19	2	24	26	30	2586	696
							19	3	21	43	87	2144	688
15	1	24	39	21	1270	678	19	4	18	46	24	2184	704
15	2	27	31	21	1712	681	19	5	24	33	24	1832	700
15	3	27	39	27	2968	677	19	6	27	33	54	1693	694
15	4	15	33	36	1553	674	19	7	21	37	48	1636	695
15	5	21	41	21	2750	678	19	8	15	35	21	1784	700
15	6	27	26	24	1854	683							
15	7	24	34	21	1338	680	20	1	21	18	21	22	707
15	8	15	43	30	2427	673	20	2	21	26	27	403	714
							20	3	24	35	21	2149	723
16	1	24	37	36	1442	597	20	4	21	36	21	2402	724
16	2	30	33	24	1815	613	20	5	24	38	24	2101	726
16	3	21	35	18	1272	609	20	6	18	36	18	141	725
16	4	18	29	18	850	622							
16	5	21	37	63	2485	580							
16	6	21	26	27	2803	626							
16	7	21	41	27	1533	589							
16	8	24	29	24	717	620							
16	9	18	32	48	2940	598							

13 Mixed Models

13.1 INTRODUCTION

The discussions in Chapters 9 through 12 have been building towards the general method of analyzing mixed models. The mixed model occurs when the model has more than one parameter in the distribution of the error term. There are three types of mixed models. The first type of mixed model occurs when there is a fixed effects treatment structure and some type of blocking in the design structure which requires a model with more than one random effect and thus more than one variance component. The second type of mixed model occurs when the treatment structure consists of fixed effects and random effects. The third type of mixed model occurs when the treatment structure consists of fixed effects and there are unequal variances; thus there are more than one variance component in the model. If all of the factors in the treatment structure are random effects, no matter the type of design structure, the model can be analyzed using the methods of Chapter 12.

From Chapter 4 of Milliken and Johnson (1992), a basic assumption is that no interaction exists between the components of the treatment structure and the components of the design structure. An additional assumption is that the slopes of the regression lines are not affected by the components of the design structure. As presented in Chapters 9 and 10, the slopes were not expressed as functions of the blocks. Thus, the slopes corresponding to covariates are considered to depend only on the levels of the factors in the treatment structure and not on blocks or any sizes of experimental units or any factors used to construct the design structure.

This chapter starts with a short discussion of the matrix form of the mixed model and of the solution to the mixed model equations. Then the analysis of models with fixed effects treatment structures and multiple error terms is discussed, followed by the presentation of the analysis of a mixed effects treatment structure.

13.2 THE MATRIX FORM OF THE MIXED MODEL

The mixed model can be expressed as

$$\mathbf{y} = \mathbf{x}\boldsymbol{\beta} + \mathbf{Z}\mathbf{u} + \boldsymbol{\varepsilon} \tag{13.1}$$

where $\mathbf{X}\boldsymbol{\beta}$ represents the fixed effects part of the model, $\mathbf{X}$ is the fixed effects design matrix, $\boldsymbol{\beta}$ is the vector of parameters corresponding to the fixed effects, $\mathbf{Z}\mathbf{u} + \boldsymbol{\varepsilon}$ is the random effects part of the model, $\mathbf{Z}$ is the random effects design matrix, $\mathbf{u}$ is the vector of random effects with $E(\mathbf{u}) = \mathbf{0}$, $\boldsymbol{\varepsilon}$ is the vector of errors with $E(\boldsymbol{\varepsilon}) = \mathbf{0}$, and

$$\text{Var}\begin{bmatrix} \mathbf{u} \\ \boldsymbol{\varepsilon} \end{bmatrix} = \begin{bmatrix} \mathbf{G} & \mathbf{0} \\ \mathbf{0} & \mathbf{R} \end{bmatrix}.$$

The random effects part of the model consists of the random effects of the treatment structure (if there are any), the random effects from the design structure, and the residual errors.

The model can also be expressed as

$$\mathbf{y} = \mathbf{x}\boldsymbol{\beta} + \mathbf{e}$$

where $E(\mathbf{e}) = \mathbf{0}$ and $\text{Var}(\mathbf{e}) = \mathbf{Z}\mathbf{G}\mathbf{Z}' + \mathbf{R} = \boldsymbol{\Sigma}$. Henderson (1984) provided the mixed model equations

$$\begin{bmatrix} \mathbf{X}'\mathbf{R}^{-1}\mathbf{X} & \mathbf{X}'\mathbf{R}^{-1}\mathbf{Z} \\ \mathbf{Z}'\mathbf{R}^{-1}\mathbf{X} & \mathbf{Z}'\mathbf{R}^{-1}\mathbf{Z} + \mathbf{G}^{-1} \end{bmatrix} \begin{bmatrix} \hat{\boldsymbol{\beta}} \\ \tilde{\mathbf{u}} \end{bmatrix} = \begin{bmatrix} \mathbf{X}'\mathbf{R}^{-1}\mathbf{y} \\ \mathbf{Z}'\mathbf{R}^{-1}\mathbf{y} \end{bmatrix}$$

and showed the solution for $\hat{\boldsymbol{\beta}}$ is

$$\hat{\boldsymbol{\beta}} = \left(\mathbf{X}'\boldsymbol{\Sigma}^{-1}\mathbf{X}\right)^{-1}\mathbf{X}'\boldsymbol{\Sigma}^{-1}\mathbf{y}.$$

When the parameters of $\boldsymbol{\Sigma}$ are known then $\mathbf{a}'$ is the BLUE (best linear unbiased estimator) of an estimable function $\mathbf{a}'\hat{\boldsymbol{\beta}}$ and the solution for $\tilde{\mathbf{u}}$ is the BLUP (best linear unbiased predictor) of $\tilde{\mathbf{u}}$. Generally the parameters of $\boldsymbol{\Sigma}$ are unknown and need to be estimated, thus the solutions to the mixed model equations for $\hat{\boldsymbol{\beta}} = (\mathbf{X}'\hat{\boldsymbol{\Sigma}}^{-1}\mathbf{X})^{-1}\mathbf{X}'\hat{\boldsymbol{\Sigma}}^{-1}\mathbf{y}$ provides weighted least squares estimates of the estimable functions and $\tilde{\mathbf{u}}$ is the estimated BLUP of the random effect. A very important aspect of the solution of the mixed model equations is that $\hat{\boldsymbol{\beta}} = (\mathbf{X}'\hat{\boldsymbol{\Sigma}}^{-1}\mathbf{X})^{-1}\mathbf{X}'\hat{\boldsymbol{\Sigma}}^{-1}\mathbf{y}$ is a combined estimator as described in Chapters 9, 10, and 11. The solution of the mixed model equations combines the information from all parts of the model into a common estimator, meaning the solution is based on all available information about estimable functions of the fixed effects parameters. The value of $\tilde{\mathbf{u}}$ computed using $\hat{\boldsymbol{\Sigma}}$ is called the estimated best linear unbiased predictor.

To demonstrate the various terms of the mixed model, a two-way mixed effects treatment structure with one covariate in a completely randomized design structure is described. Assume factor S has s fixed levels and factor T has t randomly selected levels and there are n_{ij}, $i = 1, 2, \ldots, s$, $j = 1, 2, \ldots, t$ observations per treatment combination. Since the levels of T are a random effect, then the interaction between the levels of T and the levels of S is also a random effect (Milliken and Johnson, 1992).

A model to describe the data is

$$y_{ijk} = \left(\alpha_i + a_j + c_{ij}\right) + \left(\beta_i + b_j + d_{ij}\right)X_{ijk} + \varepsilon_{ijk} \tag{13.2}$$

$$i = 1, 2, \ldots, s, \quad j = 1, 2, \ldots, t, \quad k = 1, 2, \ldots, n_{ij}$$

where

$$\begin{bmatrix} \mathbf{a}_j \\ \mathbf{b}_j \end{bmatrix} \sim \text{iidN}[\mathbf{0}, \Sigma_{ab}],$$

$$\begin{bmatrix} \mathbf{c}_{ij} \\ \mathbf{d}_{ij} \end{bmatrix} \sim \text{iidN}[\mathbf{0}, \Sigma_{cd}],$$

$$\varepsilon_{ij} \sim \text{iidN}(\mathbf{0}, \sigma_\varepsilon^2).$$

$$\Sigma_{ab} = \begin{bmatrix} \sigma_a^2 & \sigma_{ab} \\ \sigma_{ab} & \sigma_b^2 \end{bmatrix} \text{ and } \Sigma_{cd} = \begin{bmatrix} \sigma_c^2 & \sigma_{cd} \\ \sigma_{cd} & \sigma_d^2 \end{bmatrix}.$$

The part of the model $\alpha_i + \beta_i X_{ijk}$ corresponds to the levels of the fixed effect factor, S, $a_j + b_j X_{ijk}$ corresponds to the levels of the random effect factor, T, and $c_{ij} + d_{ij} X_{ijk}$ corresponds to the interaction between the levels of S and the levels of T.

$$\text{Let } \mathbf{X}_{ij} = \begin{bmatrix} 1 & X_{ij1} \\ 1 & X_{ij2} \\ \vdots & \vdots \\ 1 & X_{ijn_{ij}} \end{bmatrix}$$

then the matrix form of the model can be expressed as $\mathbf{y} = \mathbf{X}\beta + \mathbf{e}$ where

$$\mathbf{X} = \begin{bmatrix} \mathbf{X}_{11} & 0 & \cdots & 0 \\ \mathbf{X}_{12} & 0 & \cdots & 0 \\ \vdots & \vdots & & \\ \mathbf{X}_{1t} & 0 & \cdots & 0 \\ 0 & \mathbf{X}_{21} & \cdots & 0 \\ 0 & \mathbf{X}_{22} & \cdots & 0 \\ \vdots & \vdots & & \vdots \\ 0 & \mathbf{X}_{2t} & \cdots & 0 \\ \vdots & \vdots & & \vdots \\ 0 & 0 & \cdots & \mathbf{X}_{st1} \\ 0 & 0 & \cdots & \mathbf{X}_{st2} \\ \vdots & \vdots & & \vdots \\ 0 & 0 & \cdots & \mathbf{X}_{st} \end{bmatrix}$$

$$\boldsymbol{\beta} = (\alpha_1, \beta_1, \alpha_2, \beta_2, \ldots, \alpha_s, \beta_s)',$$

$$\mathbf{e} \sim N[\mathbf{0}, \boldsymbol{\Sigma}]$$

$$\boldsymbol{\Sigma} = \mathbf{X}_1[\boldsymbol{\Sigma}_{ab} \otimes \mathbf{I}_t]\mathbf{X}_1' + \mathbf{X}_2[\boldsymbol{\Sigma}_{cd} \otimes \mathbf{I}_{st}]\mathbf{X}_2' + \sigma_\varepsilon^2 \mathbf{I}_N$$

$$\mathbf{X}_1 = \begin{bmatrix} \mathbf{X}_{11} & 0 & \cdots & 0 \\ 0 & \mathbf{X}_{12} & \cdots & 0 \\ \vdots & \vdots & & \vdots \\ 0 & 0 & \cdots & \mathbf{X}_{1t} \\ \mathbf{X}_{21} & 0 & \cdots & 0 \\ 0 & \mathbf{X}_{22} & \cdots & 0 \\ \vdots & \vdots & & \vdots \\ 0 & 0 & \cdots & \mathbf{X}_{2t} \\ \vdots & \vdots & & \vdots \\ \mathbf{X}_{s1} & 0 & \cdots & 0 \\ 0 & \mathbf{X}_{s2} & \cdots & 0 \\ \vdots & \vdots & \ddots & \vdots \\ 0 & 0 & \cdots & \mathbf{X}_{st} \end{bmatrix}$$

and

$$\mathbf{X}_2 = \begin{bmatrix} \mathbf{X}_{11} & 0 & \cdots & 0 \\ 0 & \mathbf{X}_{12} & \cdots & 0 \\ \vdots & \vdots & \ddots & \vdots \\ 0 & 0 & \cdots & \mathbf{X}_{st} \end{bmatrix}$$

The fixed effects part of the model is $\mathbf{X}\boldsymbol{\beta}$ which consists of the design matrix and the vector of intercepts and slopes corresponding to the levels of the fixed effect factor, S. The random effects part of the model is

$$\mathbf{Z}\mathbf{u} + \boldsymbol{\varepsilon} = [\mathbf{X}_1, \mathbf{X}_2] \begin{bmatrix} \mathbf{a} \\ \mathbf{c} \end{bmatrix} + \boldsymbol{\varepsilon}$$

where
$\mathbf{a} = (a_1, b_1, a_2, b_2, \ldots, a_t, b_t)'$,
$\mathbf{c} = (c_{11}, d_{11}, c_{12}, d_{12}, \ldots, c_{st}, d_{st})'$,

$$G = \begin{bmatrix} \Sigma_{ab} \otimes I_t & 0 \\ 0 & \Sigma_{cd} \otimes I_{st} \end{bmatrix} \text{ and } R = \sigma_\varepsilon^2 I_N.$$

The solution of the mixed model equations provides estimators of estimable functions of β and estimated BLUPs of u. However, before the mixed model equations can be solved, the estimates of the parameters of Σ must be obtained. As in Chapter 12, several methods are available and each has its own properties. Generally, maximum likelihood estimators or residual maximum likelihood estimators are available in commercial computer codes such as PROC MIXED and JMP®. Most of the examples discussed that involve the mixed model analysis are based on REML estimates of the parameters of Σ. As in Chapter 12, the method of moments can be used for some simple examples.

13.3 FIXED EFFECTS TREATMENT STRUCTURE

Most applications of the mixed model involve a fixed effects treatment structure with more than one error term. Examples of such experiments involve blocking such as the experiments discussed in Chapters 9, 10, and 11. Experiments involving more than one size of experimental unit, such as split-plot types or repeated measures types, most always involve a fixed effects treatment structure with more than one error term. Chapters 15 and 16 discuss the analysis specific to experiments with more than one size of experimental unit.

An important feature of blocked design structures is that there is information about the model's parameters from more than one source, say, within block and between block, and a combined analysis must be conducted in order to construct an appropriate analysis. The examples in Chapters 9 to 11 all involved design structures with equal block sizes. When there are several different block sizes, information must be extracted from each block size and then combined which can become quite a tedious process.

This is when the mixed model analysis comes to the rescue. The estimate of an estimable function of the fixed effects parameters from the mixed model equations is a combined estimator. Since the mixed model solution is a combined estimator, approximate small sample confidence intervals and test statistics can be obtained in fashions similar to those described in Chapters 9 to 11. The next section presents some general comments about estimation of fixed effects parameters and small sample size approximations for confidence intervals.

13.4 ESTIMATION OF FIXED EFFECTS AND SOME SMALL SAMPLE SIZE APPROXIMATIONS

The estimate of an estimable function of β is a function of the estimated variance and covariance components, i.e., the estimate of $a'\beta$ is $a'\hat{\beta}$ where

$$\hat{\beta} = \left(X'\hat{\Sigma}^{-1}X\right)^{-} X'\hat{\Sigma}^{-1}y$$

(or some other solution to the weighted least squares equations). The approximate estimated variance of $\mathbf{a'\beta}$ is

$$\widehat{\mathrm{Var}}(\mathbf{a'\hat{\beta}}) = \mathbf{a'}\left(\mathbf{X'\hat{\Sigma}^{-1}X}\right)^{-}\mathbf{a}.$$

This approximate estimated variance is an underestimate of the variance and should be altered to decrease the bias as well as provide more appropriate approximate numbers of degrees of freedom (Kenward, M. G. and Roger, J. H., 1997). PROC MIXED incorporates these adjustments when the denominator degrees of freedom option DDFM=KenwardRoger=KR is used, a highly recommended option.

An approximate $(1-\alpha)100\%$ asymptotic confidence interval about $\mathbf{a'\beta}$ is $\mathbf{a'\hat{\beta}} \pm Z_{\alpha/2}\sqrt{\widehat{\mathrm{Var}}(\mathbf{a'\hat{\beta}})}$, but if sample sizes are small, a small sample approximation is desirable. One process is to base the small sample approximation on the method of moments equations and estimators. Suppose there are three variance components and the method of moments equations are

$$\mathrm{MS}_1 = \hat{\sigma}_1^2 + c_2\hat{\sigma}_2^2 + c_3\hat{\sigma}_3^2$$

$$\mathrm{MS}_2 = \hat{\sigma}_1^2 + d_2\hat{\sigma}_2^2$$

$$\mathrm{MS}_3 = \hat{\sigma}_1^2$$

where $\hat{\sigma}_1^2$, $\hat{\sigma}_2^2$, and $\hat{\sigma}_3^2$ denote the method of moments estimators. The values of $\hat{\sigma}_1^2$, $\hat{\sigma}_2^2$, and $\hat{\sigma}_3^2$ are used to compute $\mathbf{a'\hat{\beta}}$ and $\widehat{\mathrm{Var}}(\mathbf{a'\hat{\beta}})$.

Next construct a new set of equations where MS_i is replaced with $[t_{\alpha/2,df_i}]^2\mathrm{MS}_i$ and df_i is the degrees of freedom associated with MS_i as

$$\left[t_{\alpha/2,df_1}\right]^2 \mathrm{MS}_1 = \hat{\sigma}_{10}^2 + c_2\hat{\sigma}_{20}^2 + c_3\hat{\sigma}_{30}^2$$

$$\left[t_{\alpha/2,df_2}\right]^2 \mathrm{MS}_2 = \hat{\sigma}_{10}^2 + d_2\hat{\sigma}_{20}^2$$

$$\left[t_{\alpha/2,df_3}\right]^2 \mathrm{MS}_3 = \hat{\sigma}_{10}^2$$

The values $\hat{\sigma}_{10}^2$, $\hat{\sigma}_{20}^2$, and $\hat{\sigma}_{30}^2$ are used to evaluate $\widehat{\mathrm{Var}}(\mathbf{a'\hat{\beta}})_0$, the variance based on the mean squares times the square of the t-values. The value $\sqrt{\widehat{\mathrm{Var}}(\mathbf{a'\hat{\beta}})_0}$ is an approximate LSD(α) value and the $(1-\alpha)100\%$ approximate confidence interval about $\mathbf{a'\beta}$ is

$$\mathbf{a'\hat{\beta}} \pm \sqrt{\widehat{\mathrm{Var}}(\mathbf{a'\hat{\beta}})_0}.$$

Approximate degrees of freedom can be determined by computing $t^*_{\alpha/2} = \sqrt{\widehat{\mathrm{Var}}(\mathbf{a'\hat{\beta}})_0 / \widehat{\mathrm{Var}}(\mathbf{a'\hat{\beta}})}$ and then use the t-table to match degrees of freedom to $t^*_{\alpha/2}$.

Mixed Models

For REML estimates of the variance components, the Satterwaite and Kenward-Roger methods can be used to provide approximate numbers of degrees of freedom associated with a computed estimated standard error. PROC MIXED implements both of these approximations.

13.5 FIXED TREATMENTS AND LOCATIONS RANDOM

One example of a mixed-effects treatment structure is where a set of treatments are evaluated at several randomly selected locations. The resulting treatment structure is a two-way with the levels of treatment crossed with the levels of location. Further, assume there are r blocks of size t in the design structure at each location. A model to describe data involving a linear relationship with one covariate is

$$y_{ijk} = (\alpha_i + a_j + c_{ij}) + (\beta_i + b_j + d_{ij})X_{ijk} + r_{jk} + \varepsilon_{ijk}$$

$$= (\alpha_i + \beta_i X_{ijk}) + (a_j + b_j X_{ijk}) + (c_{ij} + d_{ijk} X_{ijk}) + r_{ij} + \varepsilon_{ijk}$$

where

$$\begin{bmatrix} a_j \\ b_j \end{bmatrix} \sim \text{iidN}[0, \Sigma_{ab}],$$

$$\begin{bmatrix} c_{ij} \\ d_{ij} \end{bmatrix} \sim \text{iidN}[0, \Sigma_{cd}],$$

$r_{jk} \sim$ iid $N(0, \sigma_r^2)$, $\varepsilon_{ijk} \sim$ iidN$(0, \sigma_\varepsilon^2)$, and the random variables are independently distributed.

The analysis of covariance is a strategy for making decisions about the covariate part of the model. The first step is to test individual hypotheses

$$H_0: \sigma_b^2 = 0 \text{ vs. } H_a: \sigma_b^2 > 0, H_0: \sigma_d^2 = 0 \; H_a: \sigma_d^2 > 0,$$

$$H_0: \sigma_{ab} = 0 \text{ vs. } H_a: \sigma_{ab} \neq 0 \text{ and } H_0: \sigma_{cd} = 0 \text{ vs. } H_a: \sigma_{cd} \neq 0$$

and determine which components are meaningful. The variance cannot be zero if the associated covariance is not zero. After decisions have been made about the variances and covariances associated with the covariate part of the model, estimate those parameters and the other variances and covariances not involved with the covariate part of the model. At this point, the fixed effects analysis is started using the process described in Chapter 2, i.e., test hypotheses such as

$$H_{01}: \beta_1 = \beta_2 = \ldots = \beta_t = 0 \text{ vs. } H_{a1}: (\text{not } H_{01}:) \text{ and}$$

$$H_{02}: \beta_1 = \beta_2 = \ldots = \beta_t. \text{ vs. } H_{a1}: (\text{not } H_{02}:).$$

Once the fixed effects covariate part of the model is selected, comparisons between the fixed effects regression models need to be made.

Since the above model contains three random effects, locations, blocks (location), and error, information about the treatment effects and the slopes for the covariate comes from more than one source. The solution to the mixed model equations combines all sources of information. This chapter is concluded by analyzing data sets from two problems to demonstrate the procedures just described.

13.6 EXAMPLE: TWO-WAY MIXED EFFECTS TREATMENT STRUCTURE IN A CRD

The data in Table 13.1 are from an experiment to evaluate the effects of four levels of fertilizer (fixed effect) on the yield (pounds per plot) of six varieties randomly selected from a population of hard red winter wheat varieties grown in Kansas. The

TABLE 13.1
Data for Example in Section 13.6 with Levels of Variety as a Random Effect and Levels of Fertility as a Fixed Effect

Variety	Fertility 1		Fertility 2		Fertility 3		Fertility 4	
	Yield	Weeds	Yield	Weeds	Yield	Weeds	Yield	Weeds
1.0	72.8	26	44.3	45	45.6	46	64.7	35
1.0	49.5	41	55.1	39	63.0	34	41.4	48
1.0	50.4	42	44.6	45	46.2	44	76.8	27
1.0	73.2	27	77.1	26	42.7	46	45.9	46
2.0	53.9	41	71.1	32	64.0	39	54.3	47
2.0	49.7	45	73.2	29	54.1	46	79.6	29
2.0	69.3	32	82.9	23	64.1	39	61.5	42
2.0	40.6	49	73.1	30	72.5	32	89.5	23
3.0	57.6	40	84.8	24	93.3	21	98.7	20
3.0	71.1	31	86.2	24	77.8	31	63.9	45
3.0	55.1	41	83.9	26	69.0	37	79.0	35
3.0	77.0	27	74.8	31	86.0	26	71.8	41
4.0	48.6	47	61.7	42	76.7	35	96.8	26
4.0	64.2	37	85.3	25	77.8	33	85.6	33
4.0	44.4	50	83.0	27	85.3	30	97.1	25
4.0	51.2	45	76.4	32	75.9	36	82.6	35
5.0	49.1	45	78.5	32	98.9	22	94.0	30
5.0	80.2	25	78.3	32	99.0	23	72.8	46
5.0	76.5	27	61.7	46	59.8	50	98.4	28
5.0	78.6	27	90.2	24	72.7	40	70.0	50
6.0	61.9	42	90.9	27	97.1	29	82.6	48
6.0	60.2	42	75.1	39	90.6	34	108.1	27
6.0	89.5	23	65.1	49	98.7	28	101.4	32
6.0	70.3	38	100.8	21	97.3	28	99.9	33

design structure is a completely randomized design with four replications per fertilizer-variety combination. Weeds are a competitor of the wheat plants and since there was not to be any weed control (or any other chemicals used) used on the plots, a sample of the top one inch of the soil of each plot was obtained (100 g) and the number of weed seeds were counted. The number of weed seeds per sample was to be considered as a possible covariate. Table 13.2 contains the PROC MIXED code to fit the independent slopes and intercepts model where the fixed effects are Fert, Weed_s, and Weed_s*Fert and the random effects are Variety, Weed_s*Variety, Variety*Fert and Weed_s*Varity*Fert. The significance levels associated with the random effects involving the covariate are 0.0099 and 0.1275 for Weed_s*Variety and Weed_s*Variety*Fert, respectively. This is an indication that possible Weed_s*Variety*Fert is not needed in the model. To continue the investigation, REML solutions were obtained with and without Weed_s*Variety*Fert and the AIC were obtained to determine the need for the term. Table 13.3 contains the PROC MIXED code to fit the full model and Table 13.4 contains the PROC MIXED code to fit the reduced model. In both cases, the independent slopes and intercepts form of the models were used, i.e., σ_{ab} and σ_{cd} of Model 13.2 were set to zero. The AIC for the full model is 417.83 and for the reduced model is 422.61, indicating that Weed_s*Variety*Fert is a useful term in the model. The next step was to fit the correlated slopes and intercepts model to the data. To specify the correlated slopes and intercepts model or specify Σ_{ab} and Σ_{cd}, the following random statements need to be used:

```
Random int weed_s/type=un subject = variety;  for Σab and
Random int weed_s/type=un subject=variety*fert;  for Σcd.
```

Unfortunately, that model with these two statements would not converge, so a simpler covariance matrix was used. The model that did converge involved setting $\sigma_{ab} = 0$, which was accomplished by using the random statement:

```
Random int weed_s/type=un(1) subject = variety;
```

which sets the off diagonal elements of Σ_{ab} equal to zero. Table 13.5 contains the PROC MIXED code using the above random statements to fit Model 13.2 with $\sigma_{ab} = 0$. The elements U(1,1), U(2,1), and U(2,2) corresponding to variety in the subject column provide the elements of Σ_{ab} and the elements U(1,1), U(2,1) and U(2,2) corresponding to variety*fert in the subject column provide the elements of Σ_{cd}. Element U(2,1) for variety is zero which is the result of using type = un(1) in the random statement. The bottom part of Table 13.5 contains the tests for the fixed effects in the model. The significance level associated with the weed_s*fert term in the model is 0.0868 which is a test of the hypothesis that the fertility slopes are equal. One could conclude there is not sufficient evidence (since the significance level is greater than 0.05) to say the slopes are unequal and delete the weed_s*fert term from the model. In which case an equal slopes model would be used to compare the levels of fertilizer. However, on the other hand, one could say there is reasonable evidence (since the significance level is close to 0.05) to conclude the slopes are likely to be unequal and use the unequal slopes model to compare the levels of fertilizer. The LSMEAN statements in Table 13.6 provide estimates of the regression models at

TABLE 13.2
PROC MIXED Code to Fit the Independent Intecepts and Slopes Model with Type III Sums of Squares to Test Hypotheses about the Variance Components

```
proc MIXED cl covtest cl method=type3 data=ex_13_5;
class variety fert;
MODEL yield=fert weed_s weed_s*fert/ddfm=kr;
RANDOM variety weed_s*variety variety*fert weed_s*variety*fert;
```

Source	df	SS	MS	EMS
fert	3	54.40389	18.13463	Var(Residual) + 0.1001 Var(variety*fert) + Q(fert)
weed_s	1	4866.61369	4866.61369	Var(Residual) + 95.402 Var(weed_s*variety*fert) + 381.61 Var(weed_s*variety) + Q(weed_s,weed_s*fert)
weed_s*fert	3	10.77039	3.59013	Var(Residual) + 117.5 Var(weed_s*variety*fert) + Q(weed_s*fert)
variety	5	44.87971	8.97594	Var(Residual) + 0.1047 Var(variety*fert) + 0.4188 Var(variety)
weed_s*variety	5	34.91409	6.98282	Var(Residual) + 123.46 Var(weed_s*variety*fert) + 493.85 Var(weed_s*variety)
variety*fert	15	29.52654	1.96844	Var(Residual) + 0.1342 Var(variety*fert)
weed_s*variety*fert	15	29.45335	1.96356	Var(Residual) + 171.48 Var(weed_s*variety*fert)
Residual	48	61.07080	1.27231	Var(Residual)

Source	ErrorDF	FValue	ProbF	ErrorTerm
fert	22.0	10.12	0.0002	0.7458 MS(variety*fert) + 0.2542 MS(Residual)
weed_s	5.5	856.05	0.0000	0.7727 MS(weed_s*variety) + 0.2273 MS(Residual)
weed_s*fert	24.6	2.06	0.1321	0.6852 MS(weed_s*variety*fert) + 0.3148 MS(Residual)
variety	20.7	4.94	0.0039	0.7805 MS(variety*fert) + 0.2195 MS(Residual)
weed_s*variety	23.1	3.95	0.0099	0.72 MS(weed_s*variety*fert) + 0.28 MS(Residual)
variety*fert	48	1.55	0.1262	MS(Residual)
weed_s*variety*fert	48	1.54	0.1275	MS(Residual)

Source	Estimate	StdErr	ZValue	ProbZ
variety	17.09539	10.37430	1.65	0.0994
weed_s*variety	0.01056	0.00724	1.46	0.1446
variety*fert	5.18887	2.95962	1.75	0.0796
weed_s*variety*fert	0.00403	0.00223	1.81	0.0708
Residual	1.27231	0.23428	5.43	0.0000

TABLE 13.3
PROC MIXED Code to Fit the Independent Slopes and Intercepts Model

```
proc mixed cl covtest ic data=ex_13_5;
class variety fert;
model yield=fert weed_s weed_s*fert/ddfm=kr;
random variety weed_s*variety variety*fert weed_s*variety*fert;
```

CovParm	Estimate	StdErr	ZValue	ProbZ	Alpha	Lower	Upper
variety	22.31045	16.53183	1.35	0.0886	0.05	7.72770	214.86710
weed_s*variety	0.01170	0.00914	1.28	0.1003	0.05	0.00389	0.13646
variety*fert	6.50305	3.94944	1.65	0.0498	0.05	2.60757	35.39015
weed_s*variety*fert	0.00496	0.00293	1.69	0.0455	0.05	0.00202	0.02538
Residual	1.19896	0.22239	5.39	0.0000	0.05	0.85950	1.78936

Neg2LogLike	Parameters	AIC	AICC	HQIC	BIC	CAIC
407.83	5	417.83	418.56	413.66	416.79	421.79

Effect	NumDF	DenDF	FValue	ProbF
fert	3	26.0	9.76	0.0002
weed_s	1	5.3	845.95	0.0000
weed_s*fert	3	26.5	2.08	0.1264

TABLE 13.4
PROC MIXED Code to Fit the Reduced Independent Slopes and Intercepts Model

```
proc mixed cl covtest ic data=ex_13_5;
class variety fert;
model yield=fert weed_s weed_s*fert/ddfm=kr;
random variety weed_s*variety variety*fert;
```

CovParm	Estimate	StdErr	ZValue	ProbZ	Alpha	Lower	Upper
variety	19.96009	16.22121	1.23	0.1093	0.05	6.42950	272.16194
weed_s*variety	0.01321	0.00937	1.41	0.0794	0.05	0.00473	0.11026
variety*fert	13.15558	4.94595	2.66	0.0039	0.05	7.07120	32.52760
Residual	1.41400	0.24846	5.69	0.0000	0.05	1.03013	2.06208

Neg2LogLike	Parameters	AIC	AICC	HQIC	BIC	CAIC
414.61	4	422.61	423.09	419.27	421.77	425.77

Effect	NumDF	DenDF	FValue	ProbF
fert	3	37.3	6.26	0.0015
weed_s	1	5.3	834.73	0.0000
weed_s*fert	3	67.5	3.78	0.0144

weed_s = 0, 25, 34.79 (the mean of weed_s), and 45 seeds per plot using the unequal slopes model. For this study it is reasonable to estimate and compare the levels of fert at weed_s equal to zero since with the use of chemical the researcher could eliminate the weeds from the plots. The regression lines for the levels of fert are

TABLE 13.5
PROC MIXED Code to Fit the Final Model to the Data in Section 13.5

```
proc mixed cl covtest data=ex_13_5;
class variety fert;
model yield=fert weed_s weed_s*fert/solution;
random int weed_s/type=un(1) subject=variety solution;
random int weed_s/type=un subject=variety*fert solution;
```

CovParm	Subject	Estimate	StdErr	ZValue	ProbZ	Alpha	Lower	Upper
UN(1,1)	variety	21.62078	15.68449	1.38	0.0840	0.05	7.61169	193.86792
UN(2,1)	variety	0.00000						
UN(2,2)	variety	0.01156	0.00875	1.32	0.0931	0.05	0.00394	0.11970
UN(1,1)	variety*fert	3.83690	4.95720	0.77	0.2195	0.05	0.82909	1308.83111
UN(2,1)	variety*fert	0.07919	0.10718	0.74	0.4600	0.05	−0.13088	0.28926
UN(2,2)	variety*fert	0.00311	0.00355	0.88	0.1907	0.05	0.00075	0.32554
Residual		1.25670	0.25640	4.90	0.0000	0.05	0.87408	1.96094

Effect	NumDF	DenDF	FValue	ProbF
fert	3	15.0	12.59	0.0002
weed_s	1	5.0	893.71	0.0000
weed_s*fert	3	15.0	2.65	0.0868

TABLE 13.6
Least Squares Means for the Levels of Fertilizer at Four Levels of Weed Infestation

```
LSMEANS fert/diff at means adjust=simulate;
LSMEANS fert/diff at weed_s=45 adjust=simulate;
LSMEANS fert/diff at weed_s=25 adjust=simulate;
LSMEANS fert/diff at weed_s=0 adjust=simulate;
```

	Weeds = 34.79		Weeds = 45		Weeds = 25		Weeds = 0	
fert	Estimate	StdErr	Estimate	StdErr	Estimate	StdErr	Estimate	StdErr
1	65.749	2.861	50.050	3.242	80.808	2.580	119.254	2.414
2	71.224	2.863	56.790	3.268	85.070	2.557	120.420	2.353
3	75.371	2.860	60.301	3.261	89.826	2.575	126.732	2.500
4	80.903	2.860	66.623	3.244	94.600	2.562	129.571	2.297

displayed in Figure 13.1. Table 13.7 contains the pairwise comparisons between the levels of fert at each value of weed_s where the "adjust=simulate" option was selected to control the error rates for multiple comparisons (Westfall et al., 1999). The adjusted significance levels are in the column titled "Adjp." At the 0.05 level, the means for Fert levels 1 and 2 are never significantly different, the mean of Fert 1 is always significantly less than the means of Fert 3 and Fert 4, the mean of Fert 2 is significantly less than the mean of Fert 3 only at weed_s = 0, the mean of Fert 2 is always less than the mean of Fert 4, and the means of Fert 3 and 4 are never different.

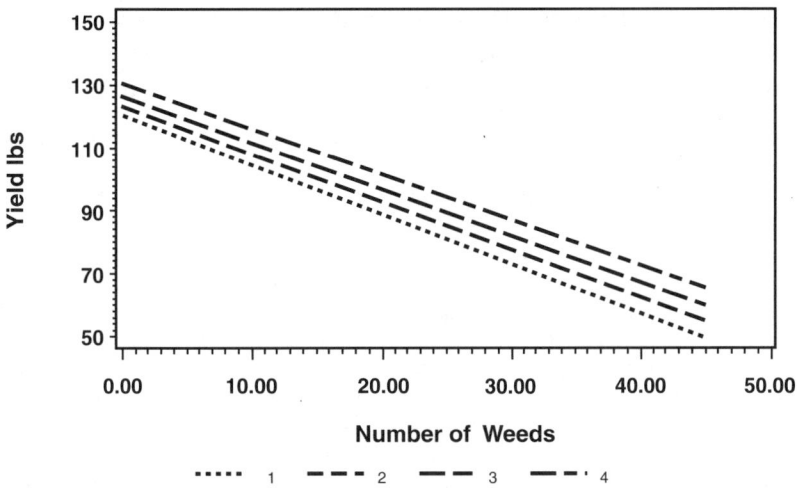

FIGURE 13.1 Graph of the fertility level models with unequal slopes.

The final step taken in this analysis was to provide predicted values for the slopes and intercepts for each variety grown at each level fertilizer. The predicted values were constructed by using the estimates of the slopes and intercepts for each fertilizer level using the fixed effects solution and the predicted values of the slopes and intercepts for each variety and for each variety*fert combination from the random effects part of the solution. The predicted values are $\tilde{a}_{pij} = \hat{\alpha}_i + \tilde{a}_j + \tilde{c}_{ij}$ and $\tilde{b}_{pij} = \hat{\beta}_i + \tilde{b}_j + \tilde{d}_{ij}$, where $\tilde{a}_j$ are the predicted intercepts correspond to the jth variety, $\tilde{c}_{ij}$ are the predicted intercepts corresponding to the ij^{th} combination of fert and variety, $\tilde{b}_j$ are the predicted slopes correspond to the jth variety, and $\tilde{d}_{ij}$ are the predicted slopes corresponding to the ij^{th} combination of fert and variety. The predicted intercepts are in Table 13.8 and the predicted slopes are in Table 13.9. Figures 13.2 through 13.5 are plots of the predicted regression models for each of the fertilizer levels. Prediction bands could be constructed for the population of variety models at each level of fertilizer using the method in Chapter 12. The construction of the prediction bands is left as an exercise.

13.7 EXAMPLE: TREATMENTS ARE FIXED AND LOCATIONS ARE RANDOM WITH A RCB AT EACH LOCATION

The data in Table 13.10 is from an experiment to evaluate the effect of three levels of a compound (denoted by Drug) designed to promote faster growth of calves. A diet with none of the drug was included as a control and used as a fourth level of drug (0 mg/kg). The experiment involved six blocks of four pens of eight calves at each of seven locations where the levels of the drug were randomly assigned to the four pens within each block. The response variable was the average daily gain (ADG)

TABLE 13.7
Pairwise Comparisons of the Fertilizer Means Using the Simulate Multiple Comparison Adjustment Method at Weed Seed Densities of 0, 25, 34.79, and 45 per Sample for the Data in Section 13.5

Effect	fert	_fert	weed_s	Estimate	StdErr	df	tValue	Probt	Adjp
fert	1	2	34.79	−5.475	2.125	15	−2.58	0.0210	0.0893
fert	1	3	34.79	−9.621	2.120	15	−4.54	0.0004	0.0023
fert	1	4	34.79	−15.153	2.121	15	−7.14	0.0000	0.0000
fert	2	3	34.79	−4.147	2.121	15	−1.95	0.0695	0.2512
fert	2	4	34.79	−9.679	2.121	15	−4.56	0.0004	0.0022
fert	3	4	34.79	−5.532	2.117	15	−2.61	0.0196	0.0833
fert	1	2	45	−6.739	2.485	15	−2.71	0.0161	0.0673
fert	1	3	45	−10.250	2.478	15	−4.14	0.0009	0.0047
fert	1	4	45	−16.573	2.455	15	−6.75	0.0000	0.0001
fert	2	3	45	−3.511	2.511	15	−1.40	0.1823	0.5244
fert	2	4	45	−9.833	2.489	15	−3.95	0.0013	0.0062
fert	3	4	45	−6.322	2.474	15	−2.56	0.0219	0.0908
fert	1	2	25	−4.262	1.891	15	−2.25	0.0396	0.1534
fert	1	3	25	−9.018	1.909	15	−4.72	0.0003	0.0018
fert	1	4	25	−13.792	1.899	15	−7.26	0.0000	0.0000
fert	2	3	25	−4.756	1.884	15	−2.52	0.0234	0.0973
fert	2	4	25	−9.530	1.868	15	−5.10	0.0001	0.0009
fert	3	4	25	−4.774	1.892	15	−2.52	0.0234	0.0973
fert	1	2	0	−1.166	2.022	15	−0.58	0.5728	0.9378
fert	1	3	0	−7.478	2.168	15	−3.45	0.0036	0.0187
fert	1	4	0	−10.317	1.963	15	−5.25	0.0001	0.0009
fert	2	3	0	−6.312	2.143	15	−2.94	0.0100	0.0463
fert	2	4	0	−9.151	1.911	15	−4.79	0.0002	0.0017
fert	3	4	0	−2.839	2.052	15	−1.38	0.1867	0.5319

TABLE 13.8
Predicted Intercepts for Each Variety at Each Level of Fert for the Data in Section 13.5

Fertilizer	variety					
	1	2	3	4	5	6
1	116.84	116.38	118.17	121.08	119.32	123.75
2	116.53	115.73	119.28	121.68	122.58	126.72
3	120.81	121.76	124.56	128.55	130.03	134.69
4	121.99	122.96	127.68	131.93	133.64	139.23

of the calves within each pen (grams/day). All of the calves did not weigh the same nor were they of the same age at the start of the experiment. Eight calves were randomly assigned to each pen. The average age in days and the average weight in

TABLE 13.9
Predicted Slopes for Each Variety at Each Level of Fert for the Data in Section 13.5

Fertilizer	Variety					
	1	2	3	4	5	6
1	−1.63	−1.51	−1.53	−1.54	−1.55	−1.47
2	−1.58	−1.43	−1.41	−1.43	−1.35	−1.29
3	−1.68	−1.49	−1.50	−1.48	−1.41	−1.31
4	−1.66	−1.46	−1.40	−1.39	−1.30	−1.18

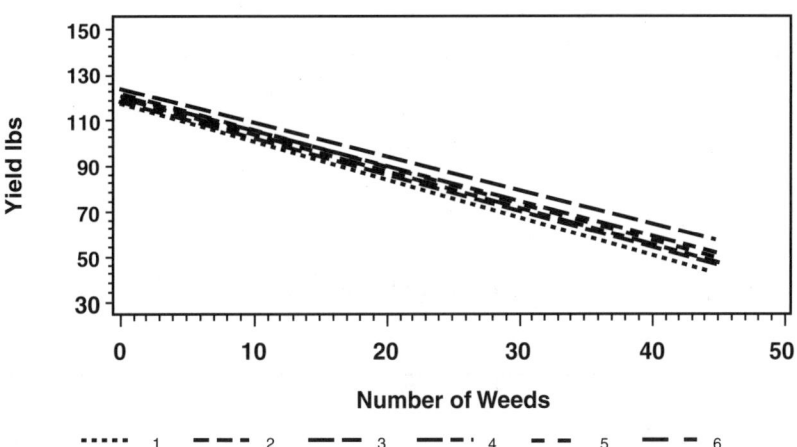

FIGURE 13.2 Graph of the variety models evaluated at fertilizer level 1.

kilograms of the calves in each pen at the start of the experiment were determined so they could be used as possible covariates. The model used to describe the ADG as a linear function of age and weight was

$$y_{ijk} = (\alpha_i + a_j + d_{ij}) + (\beta_i + b_i + f_{ij})AGE_{ijk} + (\gamma_i + c_j + g_{ij})WT_{ijk} + r_{jk} + \varepsilon_{ijk}$$

$$= (\alpha_i + \beta_i AGE_{ijk} + \gamma_i WT) + (a_j + b_j AGE_{ijk} + c_j WT_{ijk}) +$$

$$(d_{ij} + f_{ij} AGE_{ijk} + g_{ij} WT_{ijk}) + r_{jk} + \varepsilon_{ijk}$$

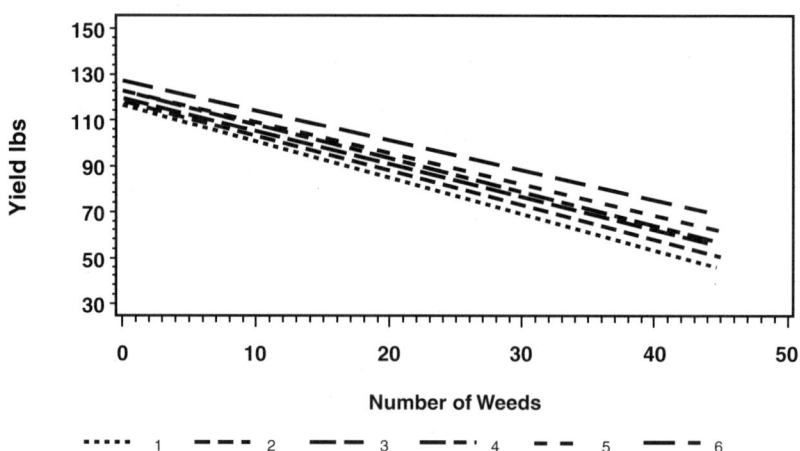

FIGURE 13.3 Graph of the variety models evaluated at fertilizer level 2.

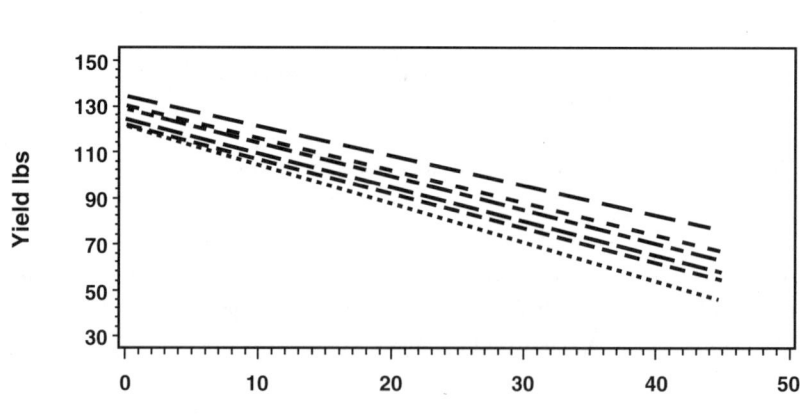

FIGURE 13.4 Graph of the variety models evaluated at fertilizer level 3.

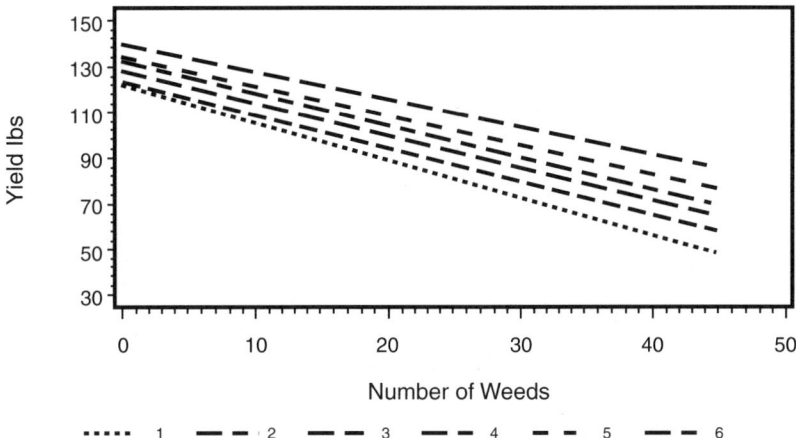

FIGURE 13.5 Graph of the variety models evaluated at fertilizer level 4.

where

$$\begin{bmatrix} a_j \\ b_j \\ c_j \end{bmatrix} \sim \text{iidN}\left[\mathbf{0}, \Sigma_{abc}\right] \text{ where } \Sigma_{abc} = \begin{bmatrix} \sigma_a^2 & \sigma_{ab} & \sigma_{ac} \\ \sigma_{ab} & \sigma_b^2 & \sigma_{bc} \\ \sigma_{ac} & \sigma_{bc} & \sigma_c^2 \end{bmatrix},$$

$$\begin{bmatrix} d_{ij} \\ f_{ij} \\ g_{ij} \end{bmatrix} \sim \text{iidN}\left[\mathbf{0}, \Sigma_{dfg}\right] \text{ where } \Sigma_{dfg} = \begin{bmatrix} \sigma_d^2 & \sigma_{df} & \sigma_{dg} \\ \sigma_{df} & \sigma_f^2 & \sigma_{fg} \\ \sigma_{dg} & \sigma_{fg} & \sigma_g^2 \end{bmatrix},$$

$r_{jk} \sim \text{iidN}(0, \sigma_r^2)$, $\varepsilon_{ijk} \sim \text{iidN}(0, \sigma_\varepsilon^2)$, and all of random variables are independent.
The first step in the analysis was to fit the above model using

```
random int age wt/type=un subject=loc;
```

to specify Σ_{abc} and

```
random int age wt/type=un subject=loc*dose;
```

to specify Σ_{dfg}.

Unfortunately, PROC MIXED was not able to fit the above model to the data set, so an alternative process was used. The next step in the analysis was to fit a model with zero covariances in Σ_{abc} and Σ_{dfg} and provide tests of individual hypotheses

TABLE 13.10
Average Daily Gain Data in g/day with Age and Weight as Possible Covariates for Four Doses in Seven Blocks at Each of Seven Locations

LOC	BLOCK	ADG0	AGE0	WT0	ADG10	AGE10	WT10	ADG20	AGE20	WT20	ADG30	AGE30	WT30
A	1	836	146	188	659	137	166	896	145	191	652	134	173
A	2	897	149	186	671	129	167	669	131	172	789	142	173
A	3	720	136	173	573	129	159	717	144	168	417	132	150
A	4	755	130	172	884	149	178	715	138	179	652	137	161
A	5	536	132	156	596	132	157	783	144	177	541	129	158
A	6	467	124	149	567	123	154	784	140	180	741	146	182
B	1	543	122	145	777	123	171	872	132	168	651	121	157
B	2	770	135	182	794	129	171	833	144	169	650	128	157
B	3	730	126	174	779	133	165	821	125	172	861	142	168
B	4	769	138	170	1003	148	194	912	138	181	839	134	166
B	5	662	140	166	600	129	152	734	121	153	947	141	173
B	6	591	122	159	661	140	163	810	123	164	908	149	176
C	1	858	131	167	977	143	178	1159	147	196	612	121	145
C	2	714	134	157	1151	147	195	870	129	160	574	127	147
C	3	692	125	159	1073	147	193	871	132	168	732	121	161
C	4	922	133	181	667	125	152	880	138	175	713	127	154
C	5	513	124	141	1026	137	183	1043	145	191	852	130	166
C	6	981	141	173	1008	136	177	906	135	168	1173	149	192

D	1	653	125	153	740	121	167	587	130	154	601	126	160
D	2	638	135	153	652	129	162	755	138	165	623	125	156
D	3	772	127	169	885	138	188	385	121	139	625	134	157
D	4	723	130	165	808	149	174	616	125	153	582	129	164
D	5	806	149	177	809	144	170	634	122	158	848	141	187
D	6	819	139	179	740	130	169	684	141	164	713	147	171
E	1	851	143	172	631	126	150	990	144	191	898	144	176
E	2	761	133	176	899	135	171	866	130	176	893	130	179
E	3	755	138	172	887	128	171	1133	147	193	699	123	159
E	4	472	121	140	782	141	162	1034	146	189	791	132	169
E	5	683	142	169	946	147	195	709	126	151	924	137	184
E	6	588	136	160	797	127	162	863	148	170	1057	144	193
F	1	715	139	171	777	135	176	765	131	173	638	123	168
F	2	653	132	156	734	131	168	646	128	149	854	144	174
F	3	823	125	174	982	148	190	789	132	167	624	136	158
F	4	738	126	171	820	139	164	837	138	179	649	125	155
F	5	680	127	159	740	130	153	841	147	174	680	126	165
F	6	797	134	171	777	145	168	732	134	165	778	129	169
G	1	718	144	167	770	130	159	714	128	150	889	127	172
G	2	984	150	191	1009	148	188	1072	149	187	1006	148	192
G	3	627	127	149	834	125	167	713	126	153	510	122	140
G	4	653	123	158	1024	141	179	853	130	174	709	123	159
G	5	687	132	160	904	136	176	724	130	147	943	132	177
G	6	830	121	167	825	130	165	742	128	151	907	133	176

about the variances of Σ_{abc} and Σ_{dfg}, or the diagonal elements. The PROC MIXED code in Table 13.11 was used to fit the independent slopes and intercepts model to provide the type III sums of squares. Block(loc) was used to specify the distribution of r_{jk}, loc, age*loc and wt*loc were used to specify the diagonal elements of Σ_{abc}, and loc*dose, age*loc*dose, and wt*loc*dose were used to specify the diagonal elements of Σ_{dfg}. In the lower part of Table 13.11, from the column ProbF, the significance level corresponding to wt*dose*loc is 0.1952, indicating there is not enough evidence to believe that σ_g^2 is important in the description of the data set or σ_g^2 is negligible as compared to the other sources of variation in the data set. The method of moments solution for the variance components and tests for the fixed effects are in Table 13.12. Using a step-wise deletion process, the terms corresponding to age*dose*loc, age*loc, and wt*loc were determined to be negligible as compared to the other sources of variation in the data set and hence were deleted from the model (analyses not shown). The next model fit to the data set was $y_{ijk} = (\alpha_i + a_j + d_{ij}) + \beta_i AGE_{ijk} + \gamma_i WT_{ijk} + r_{jk} + \varepsilon_{ijk}$ using the PROC MIXED code in Table 13.13. The results in Table 13.13 provide the REML estimates of the variance components and tests for the fixed effects. The F value corresponding to AGE*DOSE tests $H_0: \beta_1 = \beta_2 = \beta_3 = \beta_4$ vs. H_a: (not H_0:), which has significance level 0.3811, indicating there is not sufficient evidence to conclude the age slopes for each level of dose are unequal. The term AGE*DOSE was deleted from the model and the reduced model was fit using the code in Table 13.14. The significance level corresponding to WT*DOSE is 0.0416 indicating there is sufficient evidence to conclude that $H_0: \gamma_1 = \gamma_2 = \gamma_3 = \gamma_4$ is not reasonable, so unequal wt slopes for the levels of dose is a reasonable model.

The model with unequal dose slopes for wt and equal dose slopes for age, $y_{ijk} (\alpha_i + a_j + d_{ij}) + \beta AGE_{ijk} + \gamma_i WT_{ijk} + r_{jk} + \varepsilon_{ijk}$, was used to compare the dose effects. The estimate statements in Table 13.15 were used to investigate the form of the response curve over the four equally spaced levels of dose, 0, 10, 20, and 30 mg/kg. The first set of estimate statements evaluated the relation between the intercepts or at wt = 0, not a reasonable place to compare the models. The next three sets of three estimates evaluate the linear, quadratic, and cubic effects at three values of wt, 140, 165, and 190 kg. There are strong quadratic effects at 140 and 165 kg and a marginal linear effect at 190 kg. Table 13.16 contains the LSMEANS statements to provide estimates of the dose response curves at wt = 140, 165, and 190 kg. The average value of AGE (134.14 days) was used in computing the adjusted means. Figure 13.6 is a graph of the response curves constructed from the estimated means in Table 13.16. Table 13.17 contains the pairwise comparisons between the dose means within each value of wt using the simulate method of controlling the error rates within each value of wt. The results indicate that at 140 kg the mean of dose 20 is larger than the means of doses 0 and 30, at 165 kg the mean of dose 0 is less than the means of doses 10 and 20, and there are no differences at 190 kg.

The conclusions from the above analyses are that random effects corresponding to the two covariates were negligible as compared to the other variation in the model and there were equal slopes in the age direction, but unequal slopes in the wt direction. Before the levels of dose were compared the simplest form of the covariate part of the model was determined.

TABLE 13.11
ROC MIXED Code to Fit the Independent Slopes and Intercepts Model to Obtain the Type III Sums of Squares and Corresponding Tests

```
PROC MIXED cl covtest method=type3 data=ex136;
CLASS DOSE LOC BLOCK;
MODEL ADG=DOSE AGE WT AGE*DOSE WT*DOSE/ddfm=kr;
RANDOM LOC BLOCK(LOC) LOC*DOSE AGE*LOC WT*LOC AGE*LOC*DOSE WT*LOC*DOSE;
```

Source	df	SS	MS	EMS
DOSE	3	11526.6287	3842.2096	Var(Residual) + 0.0073 Var(DOSE*LOC) + Q(DOSE)
AGE	1	1193.4043	1193.4043	Var(Residual) + 38.809 Var(AGE*DOSE*LOC) + 155.24 Var(AGE*LOC) + Q(AGE,AGE*DOSE)
WT	1	240173.1462	240173.1462	Var(Residual) + 102.27 Var(WT*DOSE*LOC) + 409.08 Var(WT*LOC) + Q(WT,WT*DOSE)
AGE*DOSE	3	1029.6457	343.2152	Var(Residual) + 53.72 Var(AGE*DOSE*LOC) + Q(AGE*DOSE)
WT*DOSE	3	8692.5244	2897.5081	Var(Residual) + 105.44 Var(WT*DOSE*LOC) + Q(WT*DOSE)
LOC	6	11214.7150	1869.1192	Var(Residual) + 0.0073 Var(DOSE*LOC) + 0.0049 Var(BLOCK(LOC)) + 0.0292 Var(LOC)
BLOCK(LOC)	35	83154.6831	2375.8481	Var(Residual) + 2.4 Var(BLOCK(LOC))
DOSE*LOC	18	60191.9854	3343.9992	Var(Residual) + 0.0106 Var(DOSE*LOC)
AGE*LOC	6	13638.2266	2273.0378	Var(Residual) + 83.228 Var(AGE*DOSE*LOC) + 332.91 Var(AGE*LOC)
WT*LOC	6	9679.1807	1613.1968	Var(Residual) + 136.21 Var(WT*DOSE*LOC) + 544.83 Var(WT*LOC)
AGE*DOSE*LOC	18	44379.8742	2465.5486	Var(Residual) + 91.326 Var(AGE*DOSE*LOC)
WT*DOSE*LOC	18	39439.6110	2191.0895	Var(Residual) + 191.58 Var(WT*DOSE*LOC)
Residual	49	78991.3411	1612.0682	Var(Residual)

Source	ErrorDF	FValue	ProbF	ErrorTerm
DOSE	26.3	1.37	0.2735	0.6877 MS(DOSE*LOC) + 0.3123 MS(Residual)
AGE	18.2	0.62	0.4406	0.4663 MS(AGE*LOC) + 0.5337 MS(Residual)
WT	10.5	148.91	0.0000	0.7508 MS(WT*LOC) + 0.2492 MS(Residual)
AGE*DOSE	35.5	0.16	0.9210	0.5882 MS(AGE*DOSE*LOC) + 0.4118 MS(Residual)
WT*DOSE	40.7	1.50	0.2288	0.5504 MS(WT*DOSE*LOC) + 0.4496 MS(Residual)
LOC	26.2	0.67	0.6780	0.002 MS(BLOCK(LOC)) + 0.6899 MS(DOSE*LOC) + 0.3081 MS(Residual)
BLOCK(LOC)	49.0	1.47	0.1041	MS(Residual)
DOSE*LOC	49.0	2.07	0.0223	MS(Residual)
AGE*LOC	20.3	0.95	0.4815	0.9113 MS(AGE*DOSE*LOC) + 0.0887 MS(Residual)
WT*LOC	29.4	0.80	0.5798	0.711 MS(WT*DOSE*LOC) + 0.289 MS(Residual)
AGE*DOSE*LOC	49.0	1.53	0.1201	MS(Residual)
WT*DOSE*LOC	49.0	1.36	0.1952	MS(Residual)

TABLE 13.12
Method of Moments Solution for the Variance Components and Tests for the Fixed Effects in the Model

CovParm	Estimate	StdErr	ZValue	ProbZ	Alpha	Lower	Upper
LOC	−32176.4130	35179.5109	−0.91	0.3604	0.05	−101126.9874	36774.1614
BLOCK(LOC)	318.2416	241.4323	1.32	0.1875	0.05	−154.9569	791.4402
DOSE*LOC	163713.5246	22940.9219	7.14	0.0000	0.05	118750.1439	208676.9054
AGE*LOC	−0.3509	0.0492	−7.14	0.0000	0.05	−0.4473	−0.2545
WT*LOC	−0.7535	0.9779	−0.77	0.4410	0.05	−2.6701	1.1631
AGE*DOSE*LOC	9.3454	1.3096	7.14	0.0000	0.05	6.7787	11.9121
WT*DOSE*LOC	3.0223	0.4235	7.14	0.0000	0.05	2.1923	3.8524
Residual	1612.0682	225.8966	7.14	0.0000	0.05	1124.8736	2503.2974

Effect	NumDF	DenDF	FValue	ProbF
DOSE	3	120.0	0.90	0.4425
AGE	1	145.1	4.28	0.0403
WT	1	4.0	292.27	0.0001
AGE*DOSE	3	131.8	0.46	0.7074
WT*DOSE	3	139.3	1.33	0.2676

TABLE 13.13
PROC MIXED Code to Fit a Model with Unequal Dose Slopes for Both Weight and Age and Provide Tests for Equality of Slopes Both Covariates

```
PROC MIXED cl covtest data=ex136;
CLASS DOSE LOC BLOCK;
MODEL ADG=DOSE AGE WT AGE*DOSE WT*DOSE/ddfm=kr;
RANDOM LOC BLOCK(LOC) LOC*DOSE;
```

CovParm	Estimate	StdErr	ZValue	ProbZ	Alpha	Lower	Upper
LOC	3763.6072	2366.8954	1.59	0.0559	0.05	1472.3659	22315.6202
BLOCK(LOC)	204.5725	174.8996	1.17	0.1211	0.05	63.2482	3475.1285
DOSE*LOC	857.4266	394.6405	2.17	0.0149	0.05	411.5402	2754.8013
Residual	1866.2822	264.9196	7.04	0.0000	0.05	1439.1607	2517.4807

Effect	NumDF	DenDF	FValue	ProbF
DOSE	3	137.4	2.58	0.0562
AGE	1	135.2	7.57	0.0068
WT	1	139.6	387.14	0.0000
AGE*DOSE	3	136.9	1.03	0.3811
WT*DOSE	3	139.3	1.63	0.1845

TABLE 13.14
PROC MIXED Code to Fit the Final Model as Well as Test the Equality of the Dose Slopes for Weight

```
PROC MIXED cl covtest data=ex136;
CLASS DOSE LOC BLOCK;
MODEL ADG=DOSE AGE WT WT*DOSE/ddfm=kr;
RANDOM LOC BLOCK(LOC) LOC*DOSE;
```

CovParm	Estimate	StdErr	ZValue	ProbZ	Alpha	Lower	Upper
LOC	3705.1510	2327.5840	1.59	0.0557	0.05	1450.6282	21908.2228
BLOCK(LOC)	198.2868	174.5112	1.14	0.1279	0.05	59.8631	3859.6929
DOSE*LOC	826.5531	383.9756	2.15	0.0157	0.05	394.5256	2693.1898
Residual	1881.7969	264.4138	7.12	0.0000	0.05	1454.7029	2530.1426

Effect	NumDF	DenDF	FValue	ProbF
DOSE	3	143.6	3.29	0.0224
AGE	1	138.3	6.83	0.0100
WT	1	142.1	395.95	0.0000
WT*DOSE	3	140.8	2.81	0.0416

TABLE 13.15
Estimate Statements to Investigate the Form of the Dose Response Curve at Four Values of Weight

```
estimate 'linear' dose -3 -1 1 3;
estimate 'quad' dose 1 -1 -1 1;
estimate 'cubic' dose 1 -3 3 -1;
estimate 'linear at 140' dose -3 -1 1 3 wt*dose -420 -140 140 420;
estimate 'quad at 140' dose 1 -1 -1 1 wt*dose 140 -140 -140 140;
estimate 'cubic at 140' dose 1 -3 3 -1 wt*dose 140 -420 420 -140;
estimate 'linear at 165' dose -3 -1 1 3 wt*dose -495 -165 165 495;
estimate 'quad at 165' dose 1 -1 -1 1 wt*dose 165 -165 -165 165;
estimate 'cubic at 165' dose 1 -3 3 -1 wt*dose 165 -495 495 -165;
estimate 'linear at 190' dose -3 -1 1 3 wt*dose -570 -190 190 570;
estimate 'quad at 190' dose 1 -1 -1 1 wt*dose 190 -190 -190 190;
estimate 'cubic at 190' dose 1 -3 3 -1 wt*dose 190 -570 570 -190;
```

Label	Estimate	StdErr	df	tValue	Probt
Linear	−535.38	443.58	136.3	−1.21	0.2295
Quad	−562.48	203.14	145.8	−2.77	0.0064
Cubic	400.52	462.18	149.4	0.87	0.3876
Linear at 140	−21.28	91.20	81.6	−0.23	0.8160
Quad at 140	−161.95	42.46	86.3	−3.81	0.0003
Cubic at 140	63.43	98.43	89.8	0.64	0.5210
Linear at 165	70.52	57.40	18.2	1.23	0.2348
Quad at 165	−90.42	25.92	18.9	−3.49	0.0025
Cubic at 165	3.23	58.53	19.6	0.06	0.9566
Linear at 190	162.32	83.34	65.1	1.95	0.0558
Quad at 190	−18.90	36.53	59.9	−0.52	0.6068
Cubic at 190	−56.97	79.38	53.1	−0.72	0.4761

TABLE 13.16
Adjusted Dose Means Evaluated at Three Values of Weight and Age = 134.14 with Estimated Standard Errors

```
lsmeans dose/diff at wt=140 adjust=simulate;
lsmeans dose/diff at wt=165 adjust=simulate;
lsmeans dose/diff at wt=190 adjust=simulate;
```

Dose	Age	Weight	Estimate	StdErr	df	tValue	Probt
0	134.14	140	500.3	31.7	17	15.79	0.0000
10	134.14	140	566.4	33.7	22	16.82	0.0000
20	134.14	140	583.3	32.6	19	17.91	0.0000
30	134.14	140	487.5	32.3	19	15.07	0.0000
0	134.14	165	715.6	26.4	8.7	27.08	0.0000
10	134.14	165	767.3	26.6	9	28.79	0.0000
20	134.14	165	775.3	26.5	8.8	29.23	0.0000
30	134.14	165	736.5	26.5	8.8	27.83	0.0000
0	134.14	190	931.0	30.6	15	30.46	0.0000
10	134.14	190	968.1	29.8	14	32.49	0.0000
20	134.14	190	967.2	29.9	14	32.33	0.0000
30	134.14	190	985.4	30.5	15	32.36	0.0000

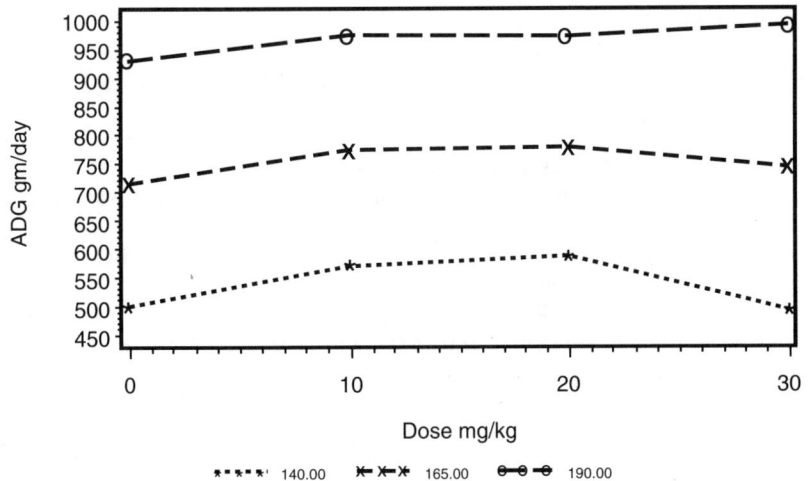

FIGURE 13.6 ADG curves vs. dose for three values of initial weight.

TABLE 13.17
Comparisons of the Dose Adjusted Means within Each Value of Weight Using the Simulate Adjustment for Multiple Comparisons (Adjp)

Dose	_Dose	Age	Weight	Estimate	StdErr	df	tValue	Probt	Adjp
0	10	134.14	140	−66.2	30.2	92	−2.19	0.0312	0.1329
0	20	134.14	140	−83.1	29.3	79	−2.83	0.0058	0.0278
0	30	134.14	140	12.7	28.6	80	0.45	0.6572	0.9689
10	20	134.14	140	−16.9	31.4	91	−0.54	0.5914	0.9470
10	30	134.14	140	78.9	30.7	93	2.57	0.0116	0.0545
20	30	134.14	140	95.8	29.8	80	3.21	0.0019	0.0098
0	10	134.14	165	−51.6	18.4	19	−2.81	0.0112	0.0284
0	20	134.14	165	−59.6	18.2	18	−3.27	0.0041	0.0077
0	30	134.14	165	−20.8	18.1	18	−1.15	0.2649	0.6521
10	20	134.14	165	−8.0	18.6	20	−0.43	0.6702	0.9718
10	30	134.14	165	30.8	18.4	19	1.67	0.1109	0.3356
20	30	134.14	165	38.8	18.3	19	2.12	0.0473	0.1503
0	10	134.14	190	−37.1	25.8	62	−1.44	0.1554	0.4905
0	20	134.14	190	−36.2	26.0	60	−1.39	0.1690	0.5213
0	30	134.14	190	−54.4	26.5	67	−2.05	0.0442	0.1727
10	20	134.14	190	0.9	24.9	52	0.03	0.9727	1.0000
10	30	134.14	190	−17.3	25.5	59	−0.68	0.5001	0.9114
20	30	134.14	190	−18.2	25.8	57	−0.70	0.4842	0.9018

REFERENCES

Henderson, C. R. (1984) *Applications of Linear Models in Animal Breeding*, University of Guelph, Canada.

Kenward, M. G. and Roger, J. H. (1997) Small sample inference for fixed effects from restricted maximum likelihood, *Biometrics* 54:983.

R. Littell, G. A. Milliken, W. Stroup, R. Wolfinger (1996) *SAS System for Mixed Models*. SAS Institute Inc., Cary, NC.

Milliken, G. A. and Johnson, D. E. (1992) *Analysis of Messy Data, Volume I: Design Experiments*, Chapman & Hall, London.

Westfall, P. H., Tobias, R. D., Rom, D., Wolfinger, R. D., and Hochberg, Y. (1999) *Multiple Comparisons and Multiple Tests Using the SAS® System*, SAS Institute Inc., Cary, NC.

EXERCISES

EXERCISE 13.1: Construct 95% prediction bands about the population of variety regression models for each level of fertilizer using the data in Table 13.7.

EXERCISE 13.2: Use the data in Section 12.6 where cities 1, 2, and 3 are classified as large cities and cities 4, 5, and 6 are classified as small cities. Fit the random coefficient model where city size is a fixed effect and city within a city size is a random effect (this is like a hierarchical linear model (Littell et al., 1996).

EXERCISE 13.3: The data set in the following table for a two-way mixed effects treatment structure where ROWs are the fixed effects and COLs are the random effects. Determine the proper form for the analysis of covariance model to describe the following data. Construct the appropriate analysis of variance table. Provide estimates of the intercepts and slopes for each treatment. Make the appropriate comparisons between the regression models.

Data for EXERCISE 13.3

ROW	COL	SUB	Y	X	ROW	COL	SUB	Y	X	ROW	COL	SUB	Y	X
1	1	1	91.3	38.8	1	3	1	54.7	38.4	1	5	1	48.4	41.2
1	1	2	97.2	40.4	1	3	2	47.3	36.9	1	5	2	30.3	33.8
1	1	3	113.7	46.5	1	3	3	70.7	49.7	1	5	3	33.1	37.6
1	1	4	107.0	43.0	1	3	4	63.2	42.6	1	5	4	22.6	25.6
1	1	5	82.3	35.3	1	3	5	55.8	42.4	1	5	5	35.2	30.5
2	1	1	146.6	40.7	2	3	1	123.5	44.3	2	5	1	125.4	45.9
2	1	2	96.5	28.5	2	3	2	111.3	40.1	2	5	2	129.7	48.6
2	1	3	168.8	47.1	2	3	3	127.5	41.5	2	5	3	77.6	32.1
2	1	4	115.9	29.4	2	3	4	76.1	25.9	2	5	4	99.4	40.4
2	1	5	96.0	25.6	2	3	5	61.1	21.0	2	5	5	82.7	32.2
3	1	1	230.1	41.1	3	3	1	108.8	21.1	3	5	1	155.0	36.4
3	1	2	212.1	38.6	3	3	2	138.1	29.3	3	5	2	193.7	45.5
3	1	3	162.7	28.7	3	3	3	231.6	47.2	3	5	3	169.1	40.3
3	1	4	186.0	33.9	3	3	4	202.8	41.6	3	5	4	122.7	31.7
3	1	5	136.6	24.2	3	3	5	177.3	37.0	3	5	5	145.3	33.8
4	1	1	268.0	27.5	4	3	1	298.0	37.1	4	5	1	173.7	29.8
4	1	2	317.2	32.8	4	3	2	209.4	26.8	4	5	2	338.6	49.9
4	1	3	431.4	44.8	4	3	3	392.5	47.9	4	5	3	154.6	27.4
4	1	4	241.8	25.1	4	3	4	165.1	21.3	4	5	4	198.9	33.0
4	1	5	224.9	22.4	4	3	5	230.2	29.1	4	5	5	263.8	41.1
1	2	1	94.5	25.7	1	4	1	109.9	47.9	1	6	1	148.8	39.6
1	2	2	157.9	46.8	1	4	2	56.0	25.2	1	6	2	127.6	32.3
1	2	3	68.3	20.7	1	4	3	116.0	49.6	1	6	3	100.5	27.5
1	2	4	85.2	23.8	1	4	4	110.8	47.2	1	6	4	87.9	24.3
1	2	5	114.3	33.9	1	4	5	88.1	37.4	1	6	5	102.0	27.9
2	2	1	215.8	49.4	2	4	1	83.1	25.1	2	6	1	205.7	47.4
2	2	2	188.7	43.5	2	4	2	171.7	48.2	2	6	2	159.6	35.9
2	2	3	196.6	45.6	2	4	3	151.2	43.0	2	6	3	213.9	47.1
2	2	4	179.2	41.9	2	4	4	109.3	31.4	2	6	4	134.3	29.6
2	2	5	130.2	29.1	2	4	5	111.6	31.4	2	6	5	123.1	26.4
3	2	1	198.1	31.3	3	4	1	127.1	25.9	3	6	1	211.6	34.3
3	2	2	299.8	47.0	3	4	2	166.5	30.3	3	6	2	197.2	33.0
3	2	3	282.0	44.7	3	4	3	268.2	48.2	3	6	3	130.6	21.2
3	2	4	133.5	20.2	3	4	4	115.4	22.2	3	6	4	234.0	38.5
3	2	5	276.7	44.0	3	4	5	206.3	37.0	3	6	5	132.4	21.9
4	2	1	400.0	37.5	4	4	1	366.4	39.8	4	6	1	167.4	20.1
4	2	2	477.3	45.3	4	4	2	373.3	41.6	4	6	2	189.0	22.3
4	2	3	435.1	41.6	4	4	3	270.3	30.5	4	6	3	331.0	35.2
4	2	4	289.1	27.9	4	4	4	400.1	44.2	4	6	4	225.5	25.6
4	2	5	283.1	26.2	4	4	5	360.4	39.9	4	6	5	296.4	31.7

14 Analysis of Covariance Models with Heterogeneous Errors

14.1 INTRODUCTION

All of the models discussed to this point have been such that the treatments can have an effect on the mean (or model) of the response, but not on the variance of the response. A more general model allows for the variances to also be affected by the application of the treatments as well as by the factors defining the design structure. The heterogeneous variance model could involve unequal variances for each treatment group or unequal variances for groups of treatments. An appropriate analysis includes the estimation of the different variances and then uses the estimated variances to estimate the parameters of the regression models and to compare the regression models. Some discussions of the unequal variance problem involve transforming the response variable in an attempt to stabilize the variances. Such transformations could change the relationship between the mean of the transformed response and the covariates. The process used in this chapter is to identify the source of the unequal variances and then to use software that allows for unequal variance models to be fit to the data set, i.e., no transformations are to be used. This chapter begins with a description of the model with unequal variances, followed by procedures to test the equal variance hypothesis. The analysis of covariance strategy is described for the case involving unequal variances, including the estimation of the parameters of the model, testing the equal slopes and slopes equal to zero hypotheses, and making comparisons between models. Three examples are presented to demonstrate the required analysis of covariance process. The process consists of (1) selecting an adequate form of the regression models, (2) selecting the simplest form of the variance structure, (3) simplifying the form of the regression model, and (4) comparing the resulting regression models.

14.2 THE UNEQUAL VARIANCE MODEL

The model for the unequal variance analysis of covariance problem for a one-way treatment structure in a completely randomized design structure with one covariate in a linear relationship with the mean of y for each treatment is

$$y_{ij} = \alpha_i + \beta_i X_{ij} + \varepsilon_{ij}$$

$$i = 1, 2, \ldots, t \quad j = 1, 2, \ldots, n_i$$

(14.1)

where $\varepsilon_{ij} \sim N(0, \sigma_i^2)$ and the ε_{ij} i = 1, 2, ..., t, j = 1, 2, ..., n_i, are independent random variables. The model in Equation 14.1 has possibly t different slopes, t different intercepts, and t different variances. The appropriate analysis provides estimates of the models parameters which is described in Section 14.3. Since this model involves more than one variance, the model is a mixed model and the appropriate analysis is a mixed models analysis.

The model can be extended to include any of the situations previously described such as two-way and higher-way treatment structures with or without controls and any type of design structure. The strategy is to use the simplest possible structure for the variances, i.e., only use the unequal variance model when there is sufficient evidence to believe that the variances are not equal. It may be possible for some experiments with unequal variances to group the treatments into groups where within a group the variances are equal and between groups the variances are unequal. Any simplification of the unequal variance portion of the model will simplify the analysis, in particular, exact distributions exist for comparisons between models within an equal variance group whereas the distributions are approximate (rely on asymptotic results) for comparisons between models with unequal variances. The next section presents a discussion of some methods to test for the equality of variances between treatment groups.

14.3 TESTS FOR HOMOGENEITY OF VARIANCES

Before continuing with the analysis using unequal variances, you need to be sure you need to use the unequal variance model. The usual tests for unequal variances described in Chapter 2 of Milliken and Johnson (1992) can be applied to this model. Four procedures for testing the equality of variances are described: Levene's test, Hartley's Max-F test, Bartlett's test for homogeneity of variances, and the likelihood ratio test. The discussion considers testing the equality of "t" variances, i.e., H_0: $\sigma_1^2 = \sigma_2^2 = ... = \sigma_t^2$ vs. H_a: (some variances are not equal), where each estimated variance is based on v_i degrees of freedom. One should fit the full covariate part of the model to the data for each treatment when computing the estimate of the variance for that treatment group. That is, do not reduce the form of the covariate part of the model before computing the estimate of the variance. If a single covariate is being used and a linear relationship is determined to be adequate to describe the data from each treatment, use the unequal slopes model to describe the data when computing the estimates of the variances for use in the following tests.

14.3.1 LEVENE'S TEST FOR EQUAL VARIANCES

Levene's test is described first because it is the easiest procedure to automate or carry out the computations. Levene's test is constructed by fitting the individual models to the separate data sets (in this case, fitting a simple linear regression model to the data from each treatment). Compute the residuals and carry out an analysis of variance on the absolute value of the residuals. If the F test corresponding to treatments is significant at a predetermined value (such as 0.01), then conclude the variances are unequal. If the treatment structure involves more than a one-way, a factorial analysis can be used to determine if the unequal variances are due solely to just the levels of one factor, etc. Unlike the use of Levene's test in models without covariates (such as in Milliken and Johnson, 1992) where one computes the absolute value of the residuals and then analyzes the absolute value of the residuals using the model used to compute the residuals, in analysis of covariance only the model with the treatment structure variables is used, i.e., do not include the covariates in the analysis of the absolute value of the residuals. If one fails to reject the equal variance hypothesis, the analysis would continue as described in the previous chapters (see Chapter 2). If one rejects the equal variance hypothesis, then the unequal variance analysis using the mixed model approach is required.

The Levene's test for unequal variances can be extended to models with factorial treatment structures and models with more than one covariate. There are other generalizations of Levene's test, such as using the squared residuals instead of using the absolute values of the residuals (SAS Institute Inc., 1999).

14.3.2 Hartley's F-Max Test for Equal Variances

Hartley's F-Max test is a general-purpose statistic used to compare variances which are based on the same number of degrees of freedom. The test statistic is

$$F_{max} = \frac{\max\left(\hat{\sigma}_i^2\right)}{\min\left(\hat{\sigma}_i^2\right)}$$

where $\hat{\sigma}_i^2$ $i = 1, 2, ..., t$ are the estimates of the variances from the t individual models.

The value of F_{max} is compared to percentage points which are reproduced in Table A.1 of Milliken and Johnson (1992). A discussion of using Hartley's F-Max test when the degrees of freedom associated with the variances are not equal is also included in Milliken and Johnson (1992).

14.3.3 Bartlett's Test for Equal Variances

One of the first tests for testing the equality of treatment variances is the one proposed by Bartlett. Bartlett's test is applicable for situations where the variances are based on equal or unequal numbers of degrees of freedom. The statistic is

$$U = \frac{1}{C}\left[v \log_e\left(\hat{\sigma}^2\right) - \sum_{i=1}^{t} v_i \log_e\left(\hat{\sigma}_i^2\right)\right]$$

where $v_i = df_i$ (the number of degrees of freedom associated with $\hat{\sigma}_i^2$)

$$v = \sum_{i=1}^{t} v_i, \quad \hat{\sigma}^2 = \sum_{i=1}^{t} \frac{v_i \hat{\sigma}_i^2}{v}, \text{ and}$$

$$C = 1 + \frac{1}{3(t-1)} \left[\sum_{i=1}^{t} \frac{1}{v_i} - \frac{1}{v} \right].$$

The hypothesis of homogeneous variances is rejected if $U > \chi^2_{\alpha,(t-1)}$.

14.3.4 Likelihood Ratio Test for Equal Variances

Some computer programs compute the value of the likelihood function for the model fitted to the data set (SAS Institute Inc., 1999). The $-2 \log_e$ (likelihood) values can be compared from two models with different covariance structures when the fixed effects parts of the models are identical. Let $\Omega(H_a)$ denote the value of $-2 \log_e$ (likelihood) for the model under the conditions of the alternative hypothesis or the full model and let $\Omega(H_o)$ denote the value of $-2 \log_e$ (likelihood) for the model under the conditions of the null hypothesis or the reduced model. In our case, the full model consists of different variances for each of the "t" treatment groups and the reduced model consists of equal variances for all of the "t" treatment groups (or one variance). The test statistic is $\Omega = \Omega(H_o) - \Omega(H_a)$ and the equal variance hypothesis is rejected if $\Omega > \chi^2_{\alpha,(t-1)}$. There are situations where some of the treatment groups can be combined into sets where equal variances are assumed for treatments within the sets and unequal variances between sets. The likelihood method can be carried out by computing and comparing various information criteria. PROC MIXED provides several information criteria where the model with the smallest value of, say AIC, is the best fitting model (SAS Institute Inc., 1999).

14.4 ESTIMATING THE PARAMETERS OF THE REGRESSION MODEL

Estimates of the parameters for Model 14.1 can be computed using least squares methods or maximum likelihood methods. For situations where the model for each treatment has a unique set of parameters (different intercepts and different slopes), the least squares and maximum likelihood estimates are identical. For models with common parameters across treatments (such as a common slope model), some type of weighting process, such as weighted least squares or maximum likelihood method, is required (weighting by the inverse of the estimated variances).

14.4.1 Least Squares Estimation

For Model 14.1 where all of the parameters of the regression models are unique, the estimates of the parameters for the ith treatment can be obtained by applying the method of least squares to the data from each treatment. For example to estimate the parameters for the first treatment's model, $y_{1j} = \alpha_1 + \beta_1 X_{1j} + \varepsilon_{1j}$ for $j = 1, 2, \ldots, n_i$, and the $\varepsilon_{1j} \sim$ iid $N(0, \sigma_1^2)$, select the values for α_1 and β_1 which minimize $SS(1) = \sum_{j=1}^{n_1}(y_{1j} - \alpha_1 - \beta_1 X_{1j})^2$ and the estimate of the variance for Treatment 1 is $\hat{\sigma}_1^2 = \sum_{j=1}^{n_1}(y_{1j} - \hat{\alpha}_1 - \hat{\beta}_1 X_{1j})^2 (n_i - 2)$. The estimates of the slope and intercept are as in Chapter 2, i.e.,

$$\hat{\beta}_1 = \frac{\sum_{j=1}^{n_1}(y_{1j}-\bar{y}_{1.})(x_{1j}-\bar{x}_{1.})}{\sum_{j=1}^{n_1}(x_{1j}-\bar{x}_{1.})^2}$$

$$\hat{\alpha}_1 = \bar{y}_{1.} - \hat{\beta}_1 \bar{x}_{1.}$$

The estimated standard errors for functions of the parameters for the data from Treatment 1 are multiples of $\hat{\sigma}_1$. The estimated standard errors for the estimates of the model's intercept and slope are

$$\hat{\sigma}_{\hat{\beta}_1} = \sqrt{\frac{\hat{\sigma}_1^2}{\sum_{j=1}^{n_1}(x_{1j}-\bar{x}_{1.})}}$$

$$\hat{\sigma}_{\hat{\alpha}_1} = \sqrt{\frac{\hat{\sigma}_1^2 \bar{x}_{1.}^2}{\sum_{j=1}^{n_1}(x_{1j}-\bar{x}_{1.})^2}}.$$

14.4.2 MAXIMUM LIKELIHOOD METHODS

The maximum likelihood methods can be used for the models with unique parameters or for models with common parameters. Many models will fall into this category and the methods used to analyze mixed models are appropriate. Mixed models analyses as described in Chapters 12 and 13 and in Chapters 19 and 22 of Milliken and Johnson (1992) can be used to estimate the parameters of these models and provide tests of hypotheses and estimated standard errors for the construction of confidence intervals. The residual maximum likelihood estimates of the parameters for Model 14.1 are the same as the least squares estimates based on the within treatment information. Maximum likelihood estimates and least squares estimates are not necessarily the same when treatments share common parameters or if there are additional variance components in the model. Appropriate computational algorithms are iterative in nature and must be capable of maximizing the likelihood function over the parameter space, i.e., provide solutions of the variances that are non-negative. Before one continues the analysis, a decision about the form of the covariate part of the model is in order, the topic of the next section.

14.5 DETERMINING THE FORM OF THE MODEL

Once it has been decided that the model with unequal variances is required to describe the data from the various treatments, the next step in the analysis is to determine the appropriate form of the covariate part of the model. The structure of the variances of the model cannot be treated independently of the regression part of the model. An adequate form of the regression model for each treatment must be determined before one computes the estimates of the variances. Questions, such as "Is the relationship between the response and the covariate linear?" or "Is the relationship between the response and all of the possible covariates adequately described by a plane for each treatment?," must be answered before decisions about the variances can be made. Thus in order to be at this point in the analysis, an adequate form of the covariate part of the model must have been determined followed by determining an adequate form of the variance structure of the model. The next step is to attempt to simplify the covariate part of the model. Assume that a simple linear regression model is adequate to describe the relationship between the response and a single covariate, i.e., assume that Model 14.1 is adequate. Referring to Chapter 2 for the strategies of determining the form of the covariate part of the model, the first step, after determining that the simple linear regression model describes the data for all treatments, is the test to see if the covariate is needed (are all slopes equal to zero), followed by the test for parallelism (are all slopes equal or will a common slope model be sufficient).

The all slopes equal zero hypothesis is

$$H_0: \beta_1 = \beta_2 = \ldots = \beta_t = 0 \text{ vs. } H_a: (\text{at least one } \beta_i \neq 0).$$

The beta hat model of Chapter 6 can be used to compute a test of the slopes equal to zero hypothesis. The beta hat model corresponding to the zero slopes hypothesis is $\hat{\beta}_i = e_i, i = 1, 2, \ldots, t$, where the e_i, $i = 1, 2, \ldots, t$ are independent random variables distributed as $N(0, \sigma^2_{\hat{\beta}_i})$. The means of the models are zero; thus the sum of squares due to deviations from the null hypothesis (the residual sum of squares from the beta hat model) is

$$SSH_o(\beta_1 = \beta_2 = \ldots = \beta_t = 0) = \sum_{i=1}^{t} \frac{\hat{\beta}_i^2}{\hat{\sigma}^2_{\hat{\beta}_i}}$$

which is asymptotically distributed as a chi square with t degrees of freedom. A small sample size distribution analogous to the Behrens-Fisher problem is that the $MSH_o(\beta_1 = \beta_2 = \ldots = \beta_t = 0) = SSH_o(\beta_1 = \beta_2 = \ldots = \beta_t = 0)/t$ is approximately distributed as $F_{(t, v)}$ where v is the minimum number of degrees of freedom of the estimates of the individual variances, $v = \min(n_i - 2, i = 1, 2, \ldots, t)$, where each model has two parameters. A second procedure is to compare $MSH_o(\beta_1 = \beta_2 = \ldots = \beta_t = 0)$ to $F_{(t, v)}$ where v is computed via a Satterthwaite or Kenward-Roger approximation.

If the all slopes equal to zero hypothesis is rejected the next step is to test the parallelism hypothesis, i.e., $H_o: \beta_1 = \beta_2 = \ldots = \beta_t = \beta$ vs. $H_a:$ (at least one inequality) where the value of β is unspecified. Again the beta hat model is useful as the model under the conditions of the equal slope hypothesis is $\hat{\beta}_i = \beta + e_i, i = 1, 2, \ldots, t$. The weighted residual sums of squares from the equal slope model is the sum of squares due to deviations from the equal slope hypothesis which is computed as

$$SSH_o(\beta_1 = \beta_2 = \ldots = \beta_t) = \sum_{i=1}^{t} \frac{(\hat{\beta}_i - \beta^*)^2}{\hat{\sigma}_{\hat{\beta}_i}^2}$$

where β^* is the weighted least squares estimate of β computed as

$$\beta^* = \frac{\sum_{i=1}^{t}\left(\hat{\beta}_i / \hat{\sigma}_{\hat{\beta}_i}^2\right)}{\sum_{i=1}^{t} 1/\hat{\sigma}_{\hat{\beta}_i}^2}.$$

The sum of squares $SSH_o(\beta_1 = \beta_2 = \ldots = \beta_t)$ is asymptotically distributed as a chi square random variable with $t - 1$ degrees of freedom. A small sample size distribution analogous to the Behrens-Fisher problem is that $MSH_o(\beta_1 = \beta_2 = \ldots = \beta_t) = SSH_o(\beta_1 = \beta_2 = \ldots = \beta_t)/(t - 1)$ is approximately distributed as $F_{(t-1,\,v)}$ where v is the minimum number of degrees of freedom of the estimates of the individual variances, $v = \min(n_i - 2,\ i = 1, 2, \ldots, t)$, where each model has two parameters. A second procedure is to compare $MSH_o(\beta_1 = \beta_2 = \ldots = \beta_t)$ to $F_{\alpha,(t-1,\,v)}$ where v is computed via a Satterthwaite or Kenward-Roger approximation.

14.6 COMPARING THE MODELS

The method of comparing the models depends on the form of the slopes in the model. If the equal slopes hypothesis is rejected, then the unequal slopes models are used to describe the data and the models must be compared to at least three different values of X. If the equal slopes hypothesis is not rejected and a plot of the residuals show that the equal slopes model is adequate to describe the data, then the models can be compared by comparing the intercepts or by comparing adjusted means computed at a single value of X. The nonparallel lines models are discussed first, followed by a discussion of the parallel lines models.

14.6.1 COMPARING THE NONPARALLEL LINES MODELS

When the lines are not parallel, several hypotheses may be of interest. One might wish to compare the slopes or compare the intercepts or compare the models at specified values of x. The slopes were compared in the Section 14.4 using the beta hat model described in Chapter 6.

The beta hat model of Chapter 6 can be used to compare the intercepts (which are the models evaluated at $X = 0$) or the models evaluated at predetermined values of X. The equal intercepts hypothesis is $H_0: \alpha_1 = \alpha_2 = \ldots = \alpha_t = \alpha$ vs. H_a: (at least one inequality) where the value of α is unspecified. (One must be very careful when testing this hypothesis as comparing the regression models at $X = 0$ may be foolish as the models may not be interpretable at $x = 0$.) Again the beta hat model (used to describe the α_i's) is useful as the model under the conditions of the equal intercept hypothesis is, $\hat{\alpha}_i = \alpha + e_i$ $i = 1, 2, \ldots, t$, where the e_i are independently distributed as $N(0, \sigma_{\hat{\alpha}_i}^2)$. The weighted residual sums of squares from the equal intercept model is the sum of squares due to deviations from the equal intercept hypothesis, which is computed as

$$SSH_o(\alpha_1 = \alpha_2 = \ldots = \alpha_t) = \sum_{i=1}^{t} \frac{(\hat{\alpha}_i - \alpha^*)^2}{\hat{\sigma}_{\hat{\alpha}_i}^2}$$

where α^* is the weighted least squares estimate of α computed as

$$\alpha^* = \frac{\sum_{i=1}^{t} (\hat{\alpha}_i / \hat{\sigma}_{\hat{\alpha}_i}^2)}{\sum_{i=1}^{t} 1/\hat{\sigma}_{\hat{\alpha}_i}^2}.$$

The sum of squares $SSHo(\alpha_1 = \alpha_2 = \ldots = \alpha_t)$ is asymptotically distributed as a chi square random variable with $t - 1$ degrees of freedom. A small sample size distribution analogous to the Behrens-Fisher problem is that the $MSH_o(\alpha_1 = \alpha_2 = \ldots = \alpha_t) = SSH_o(\alpha_1 = \alpha_2 = \ldots = \alpha_t)/(t - 1)$ is approximately distributed as $F_{(t-1, \nu)}$ where ν is the minimum number of degrees of freedom of the estimates of the individual variances, $\nu = \min(n_i - 2, i = 1, 2, \ldots, t)$, where each model has two parameters. A second procedure is to compare $MSH_o(\alpha_1 = \alpha_2 = \ldots = \alpha_t)$ to $F_{\alpha, (t-1, \nu)}$ where ν is computed via a Satterthwaite or Kenward-Roger approximation.

Finally, and likely most often needed, the models can be compared at specific values of the covariate, X_0. The models at $x = X_0$ are $\mu_{y_i|x=X_0} = \alpha_i + \beta_i X_0$. The hypothesis that the means at $x = X_0$ are equal is

$$H_0: \mu_{y_1|x=X_0} = \mu_{y_2|x=X_0} \ldots = \mu_{y_t|x=X_0} = \mu_{y|x=X_0} \text{ vs.}$$

$$H_a: \text{(at least one inequality)}$$

where the value of $\mu_{y|x=X_0}$ is unspecified. Again the beta hat model is useful as the model under the conditions of the equal means at $x = X_0$ hypothesis is $\hat{\mu}_{y_i|x=X_0} = \mu_{y|x=X_0} + e_i$ $i = 1, 2, \ldots, t$, where $e_i \sim N(0, \sigma_{\hat{\mu}_{y_i|x=X_0}}^2)$. The weighted residual sums of squares from the equal means at $x = X_0$ model is the sum of squares due to deviations from the equal slope hypothesis which is computed as

$$SSH_o(\mu_{y_1|x=X_0} = \ldots = \mu_{y_t|x=X_0}) = \sum_{i=1}^{t} \frac{\left(\hat{\mu}_{y_i|x=X_0} - \hat{\mu}_{y|x=X_0}^*\right)^2}{\hat{\sigma}_{\hat{\mu}_{y_i|x=X_0}}^2}$$

where $\hat{\mu}^*_{y|x=X_0}$ is the weighted least squares estimate of $\mu_{y|x=X_0}$ computed as

$$\hat{\mu}^*_{y|x=X_0} = \sum_{i=1}^{t} \frac{\left(\hat{\mu}^*_{y_i|x=X_0} / \hat{\sigma}^2_{\hat{\mu}_{y_i|x=X_0}}\right)}{\sum_{i=1}^{t} 1/\hat{\sigma}^2_{\hat{\mu}_{y_i|x=X_0}}}.$$

The equal means at $x = X_0$ sum of squares is asymptotically distributed as a chi square random variable with $t - 1$ degrees of freedom. A small sample size distribution analogous to the Behrens-Fisher problem is that the $MSH_o(\hat{\mu}_{y_1|x=X_0} = \cdots = \hat{\mu}_{y_t|x=X_0}) = SSH_o(\hat{\mu}_{y_1|x=X_0} = \cdots = \hat{\mu}_{y_t|x=X_0})/(t - 1)$ is approximately distributed as $F_{(t-1, v)}$ where v is the minimum number of degrees of freedom of the estimates of the individual variances, $v = \min(n_i - 2, i = 1, 2, \ldots, t)$ where each model has two parameters. A second procedure is to compare $MSH_o(\hat{\mu}_{y_1|x=X_0} = \cdots = \hat{\mu}_{y_t|x=X_0})$ to $F_{\alpha, (t-1, v)}$ where v is computed via a Satterthwaite or Kenward-Roger approximation.

14.6.2 Comparing the Parallel Lines Models

If the parallel lines model adequately describes the data for each treatment (you should plot the residuals for each treatment to verify), then fit the model

$$y_{ij} = \alpha_i + \beta X_{ij} + \varepsilon_{ij}$$

$$i = 1, 2, \ldots, t, \quad j = 1, 2, \ldots, n_i$$

where $\varepsilon_{ij} \sim$ independent $N(0, \sigma_i^2)$

The models can be compared by comparing the distances between pairs of models. Since the models are parallel, the distance between any pair of models is the same for any value of X. Thus the models can be compared at $X = 0$ (i.e., compare the intercepts) or compare the models at $X = X_0$ or compare adjusted means at $X = X_0$. The adjusted means at $X = X_0$ are computed as $\hat{\mu}_{y_i|x=X_0} = \hat{\alpha}_i + \hat{\beta}X_0$ and the difference between adjusted means for treatments one and two is $\hat{\mu}_{y_1|x=X_0} - \hat{\mu}_{y_2|x=X_0} = \hat{\alpha}_1 - \hat{\alpha}_2$, a quantity that depends only on the intercepts (it does not depend on the value of X_0). The usual adjusted means are values of the regression lines evaluated at $X_0 = \bar{X}$. Most analyses involve the computation of the adjusted means at $X_0 = \bar{X}$ and comparisons among those adjusted means. Other comparisons may be made depending on the relationships between the levels of the treatment. For example, if the levels of the treatments are quantitative, you can evaluate the relationship between the adjusted means and the levels of the treatment by testing for linear, quadratic, cubic, etc., trends by appropriate choices of the coefficients. For example if there are four equally spaced levels of treatment, you can test for the linear effect by testing H_o: $-3\hat{\mu}_{y_1|x=X_0} - 1\hat{\mu}_{y_2|x=X_0} + 1\hat{\mu}_{y_3|x=X_0} + 3\hat{\mu}_{y_4|x=X_0} = 0$ vs. H_a: (not H_o).

14.7 COMPUTATIONAL ISSUES

The computations needed for the above procedures can be accomplished by software that can carry out a weighted least squares procedure. To accomplish the weighted least squares, you first have to estimate the variances for each treatment group and then use the inverse of the variances as weights. A mixed models methodology estimates the treatments' variances and the fixed effects, all in one process. In addition, the mixed models software provides (or should) appropriate estimates of the standard errors for each parameter or combination of parameters and Satterthwaite or Kenward-Roger approximations to the numbers of degrees of freedom for the estimates of the respective standard errors. The next three sections contain examples and the major computations are carried out using PROC MIXED (SAS Institute Inc., 1999).

14.8 EXAMPLE: ONE-WAY TREATMENT STRUCTURE WITH UNEQUAL VARIANCES

The data in Table 14.1 are the scores on an exam denoted by Post and the scores on a pretest denoted by Pre. Five different study methods (denoted by Method) were evaluated to determine if they could help learning as measured by the exam. Table 14.2 contains the estimated variances for each of the study methods where a simple linear regression model was fit to the data from each of the methods and the mean square residual was computed as the displayed variance. The variances do not look equal as they range from 0.2266 to 17.4375. The PROC GLM code in Table 14.3 is used to fit Model 14.1 to the data and compute the residuals (stored in data set prebymethod). The data step computes the absolute value of the residuals (absr) and then the second PROC GLM code is used to fit a one-way analysis of variance model to the absolute values of the residuals. The middle part of Table 14.3 contains the analysis of variance where the F value corresponding to method provides the statistic for Levene's test for equality of variances. The significance level is 0.0095, indicating there is sufficient evidence to conclude the variances are not equal. The LSMEANS

TABLE 14.1
Exam Scores (Post) and Pretest Scores (Pre) for Five Study Methods

Study Method 1		Study Method 2		Study Method 3		Study Method 4		Study Method 5	
Post1	Pre1	Post2	Pre2	Post3	Pre3	Post4	Pre4	Post5	Pre5
59	23	60	26	72	37	63	11	70	22
64	31	57	11	62	30	79	29	73	27
60	31	59	20	70	25	72	17	72	29
53	15	60	21	68	38	78	32	72	34
53	17	61	30	66	20	64	13	73	32
61	11	59	17	61	16	65	18	72	32
59	20			70	36	64	22	70	11
62	15			60	27				

TABLE 14.2
Variances for Each of the Study Methods Using Unequal Slopes Model with Pre as the Covariate

```
proc sort data=oneway; by method;
proc reg; model post=pre; by method;
```

Method	df	MS
1	6	15.6484
2	4	0.2266
3	6	16.5795
4	5	17.4375
5	5	0.7613

TABLE 14.3
PROC GLM Code to Carry Out Levene's Test of Equality of Variances

```
proc glm data = oneway; class method;
model post=method pre*method/ solution;
output out=prebymethod r=r;

data prebymethod; set prebymethod; absr=abs(r);

proc glm data = prebymethod;
class method;
model absr=method; lsmeans method/pdiff;
```

Source	df	SS	MS	FValue	ProbF
Model	4	48.1557	12.0389	4.04	0.0095
Error	31	92.4285	2.9816		
Corrected Total	35	140.5842			

Source	df	SS	MS	FValue	ProbF
Method	4	48.1557	12.0389	4.04	0.0095

Method	LSMean
1	2.8598
2	0.3488
3	3.0895
4	2.6136
5	0.5984

RowName	_1	_2	_3	_4	_5
1		0.0113	0.7920	0.7848	0.0167
2	0.0113		0.0062	0.0249	0.7967
3	0.7920	0.0062		0.5982	0.0090
4	0.7848	0.0249	0.5982		0.0367
5	0.0167	0.7967	0.0090	0.0367	

TABLE 14.4
PROC GLM Code to Provide Tests for Equality of Variances between Groups (Code) and within Groups of Study Methods

```
proc glm data = prebymethod;
class code method;
model absr=code method(code);
lsmean code/pdiff;
```

Source	df	SS	MS	FValue	ProbF
Model	4	48.1557	12.0389	4.04	0.0095
Error	31	92.4285	2.9816		
Corrected Total	35	140.5842			

Source	df	SS(III)	MS	FValue	ProbF
Code	1	46.8294	46.8294	15.71	0.0004
Method(code)	3	1.0468	0.3489	0.12	0.9494

Effect	Code	Method	LSMean	ProbtDiff
Code	1		2.8543	0.0004
Code	2		0.4736	

statement was used to provide the means of the absolute values of the residuals for each of the study methods in an attempt to possibly group the methods as to the magnitude of the variances. The bottom part of Table 14.3 contains the significance levels for the pairwise comparisons of the study method means of the absolute values of residuals. The approximate multiple comparison process indicates that methods 2 and 5 have smaller variances than methods 1, 3, and 4. A new variable, code, was constructed where code=1 for methods 1, 3, and 4 and code=2 for methods 2 and 5. The PROC GLM code in Table 14.4 provides another analysis of the absolute values of the residuals where the effects in the model are code and method(code). The Fvalue corresponding to source code provides a Levene's type test of the hypothesis of the equality of the two code variances. The significance level is 0.0004, indicating there is sufficient evidence to conclude that the two variances are not equal. The Fvalue corresponding to the source method(code) provides a test of the hypothesis that the three variances for methods 1, 3, and 4 are equal and the two variances for methods 2 and 5 are equal. The significance level is 0.9494, indicating there is not sufficient evidence to conclude the variances within each value of code are unequal. Thus, the simplest model for the variances involves two variances one for each level of code. Table 14.5 contains the PROC GLM code used to compute the mean square residual for each code. The variance for code =1 is 16.5033 which is based on 17 degrees of freedom and for code=2 is 0.5237 which is based on 9 degrees of freedom. These estimates of the variances can be computed by pooling the respective variances from Table 14.2.

Another method that can be used to determine the adequate form of the variance part of the model is to use some information criteria (Littell et al., 1996) that is

TABLE 14.5
Variances for Each Group of Study Methods (Code)

```
proc sort data=oneway; by code;
proc glm data=oneway; class method;
model post=method pre*method; by code;
```

Code	df	MS
1	17	16.5033
2	9	0.5237

TABLE 14.6
PROC MIXED Code to Fit Equal Variance Model with Unequal Slopes to the Study Method Data

```
proc mixed data = oneway cl covtest ic;
class method;
model post=method pre pre*method/ddfm=satterth solution;
```

CovParm	Estimate	StdErr	ZValue	ProbZ	Alpha	Lower	Upper
Residual	10.9719	3.0430	3.61	0.0002	0.05	6.8046	20.6061

Neg2LogLike	Parameters	AIC	AICC	HQIC	BIC	CAIC
175.3	1	177.3	177.5	177.6	178.5	179.5

Effect	NumDF	DenDF	FValue	ProbF
method	4	26	1.77	0.1642
pre	1	26	15.92	0.0005
pre*method	4	26	2.04	0.1184

available in PROC MIXED. For the demonstration here, the value of AIC was used where the smaller the AIC value, the better the variance structure. Table 14.6 contains the PROC MIXED code to fit Model 14.1 where the variance structure consists of assuming all variances are equal. The value of AIC is 177.3. Table 14.7 contains the PROC MIXED code to fit a variance model with unequal variances for each of the methods. The statement "repeated/group=method;" specifies that the residuals have a different variance for each level of method. This model does not involve repeated measures, but the repeated statement is being used. In this case just think of the repeated statement as the residual statement since it is the variances of the residual part of the model that are being specified. The value of AIC is 163.3. The estimates of the variances in Table 14.7 are the same as those in Table 14.2. Finally, the PROC MIXED code in Table 14.8 fits the model with different variances for each level of code. The value of AIC is 158.8. Based on the assumption that the variance structure with the smaller AIC is more adequate, the structure with different variances for each level of code would be selected. When the variance structures of

TABLE 14.7
PROC MIXED Code to Fit the Unequal Variance Model for Each Study Method with Unequal Slopes

```
proc mixed data = oneway cl covtest ic;
class method;
model post=method pre pre*method/ddfm=satterth solution;
repeated/group=method;
```

CovParm	Group	Estimate	StdErr	ZValue	ProbZ	Alpha	Lower	Upper
Residual	Method 1	15.6484	9.0346	1.73	0.0416	0.05	6.4979	75.8805
Residual	Method 2	0.2266	0.1602	1.41	0.0786	0.05	0.0814	1.8713
Residual	Method 3	16.5795	9.5722	1.73	0.0416	0.05	6.8845	80.3958
Residual	Method 4	17.4375	11.0285	1.58	0.0569	0.05	6.7943	104.8923
Residual	Method 5	0.7613	0.4815	1.58	0.0569	0.05	0.2966	4.5796

Neg2LogLike	Parameters	AIC	AICC	HQIC	BIC	CAIC
153.3	5	163.3	166.3	166.1	171.3	176.3

Effect	NumDF	DenDF	FValue	ProbF
method	4	8.6	22.12	0.0001
pre	1	17.1	20.15	0.0003
pre*method	4	8.8	2.32	0.1371

TABLE 14.8
PROC MIXED Code to Fit the Unequal Variance Model for Each Group of Study Methods with Unequal Study Method Slopes

```
proc mixed data = oneway cl covtest ic;
class method;
model post=method pre pre*method/ddfm=satterth solution;
repeated/group=code;
```

CovParm	Group	Estimate	StdErr	ZValue	ProbZ	Alpha	Lower	Upper
Residual	Group 1	16.5033	5.6606	2.92	0.0018	0.05	9.2927	37.0899
Residual	Group 2	0.5237	0.2469	2.12	0.0169	0.05	0.2478	1.7454

Neg2LogLike	Parameters	AIC	AICC	HQIC	BIC	CAIC
154.8	2	158.8	159.4	159.9	162.0	164.0

Effect	NumDF	DenDF	FValue	ProbF
method	4	15.2	21.75	0.0000
pre	1	18.0	20.11	0.0003
pre*method	4	14.6	2.51	0.0870

models become more and more complicated, a Levene's type test statistic will not necessarily exist and so an information criteria can be used to select an adequate covariance structure.

For this model, one could use Bartlett's test and/or Hartley's test (modified for unequal sample sizes) to test the equal variance hypothesis. The Levene's type test is easy to compute when your software has the ability to compute the residuals, store them, and then analyze the absolute value of the residuals. PROC GLM has the ability to provide several different tests of homogeneity of variance when the model is a one-way treatment structure in a CRD design structure. Some of those methods use other functions of the residuals which can be easily adapted for the analysis of covariance model.

Based on the above selected covariance structure, the next step is to investigate the form of the covariate part of the model. (Note: There is the covariate part of the model that has to do with the covariates and the covariance part of the model that has to do with the variances and covariances of the data.) Using the fixed effects analysis from Table 14.8, the significance level corresponding to pre*method is 0.0870. The conclusion is that there is not sufficient evidence to conclude the slopes are unequal, so an equal slopes model was selected to continue the analysis (a plot of the residuals should be carried out for each method before this type of conclusion is reached). The PROC MIXED code in Table 14.9 is used to fit a common slope model with a different variance for each level of code. (Note that code is included

TABLE 14.9
PROC MIXED Code to Fit the Unequal Variance Model for Each Group of Study Methods with Equal Study Method Slopes

```
proc mixed data = oneway cl covtest ic;
class method code;
model post=method pre/ddfm=satterth solution;
repeated/group=code;
```

CovParm	Group	Estimate	StdErr	ZValue	ProbZ	Alpha	Lower	Upper
Residual	Group 1	21.0000	6.7071	3.13	0.0009	0.05	12.2351	44.1750
Residual	Group 2	0.5555	0.2521	2.20	0.0138	0.05	0.2689	1.7473

Neg2LogLike	Parameters	AIC	AICC	HQIC	BIC	CAIC
156.4	2	160.4	160.9	161.6	163.6	165.6

Effect	NumDF	DenDF	FValue	ProbF
method	4	16.9	166.97	0.0000
pre	1	10.8	30.02	0.0002

Effect	Method	Estimate	StdErr	df	tValue	Probt
Intercept		67.3975	0.8367	10.66	80.55	0.0000
method	1	−11.8149	1.6551	21.31	−7.14	0.0000
method	2	−11.4306	0.4495	9.865	−25.43	0.0000
method	3	−5.8980	1.6455	20.86	−3.58	0.0018
method	4	−1.3898	1.7650	21.11	−0.79	0.4398
method	5	0.0000				
pre		0.1616	0.0295	10.78	5.48	0.0002

in the class statement, but it does not have to be in the model statement.) Since the covariate part of the model has been simplified, the estimates of the two variances are a little larger. The estimate of the slope is 0.1616, indicating that the post exam score increased 0.1616 points for each additional point on the pre exam. The F-value corresponding to source method provides a test of the equal intercepts hypothesis, but since the lines are parallel, it is also a test of the equal models evaluated at some value of pre hypothesis. In this case, the significance level is 0.000, indicating there is sufficient evidence to conclude that the models are not equal. The LSMEAN statements in Table 14.10 provide estimated values for the post scores at three values

TABLE 14.10
LSMEAN Statements to Provide Adjusted Means at Three Values of Pre and Pairwise Comparison of the Means at Pre = 23.5, the Mean of the Pre Scores

```
lsmeans method/pdiff at means;
lsmeans method/pdiff at pre=10;
lsmeans method/pdiff at pre=40;
```

Method	Pre	Estimate	StdErr	df	tValue	Probt
1	23.5	59.3800	1.6228	19.7	36.59	0.0000
2	23.5	59.7642	0.3143	9.8	190.17	0.0000
3	23.5	65.2968	1.6272	19.9	40.13	0.0000
4	23.5	69.8051	1.7346	19.7	40.24	0.0000
5	23.5	71.1949	0.2972	9.8	239.54	0.0000
1	10.0	57.1985	1.6488	21.0	34.69	0.0000
2	10.0	57.5828	0.4412	10.3	130.52	0.0000
3	10.0	63.1153	1.7108	23.7	36.89	0.0000
4	10.0	67.6236	1.7584	20.8	38.46	0.0000
5	10.0	69.0134	0.5677	10.5	121.56	0.0000
1	40.0	62.0462	1.7205	24.1	36.06	0.0000
2	40.0	62.4305	0.6419	10.5	97.25	0.0000
3	40.0	67.9631	1.6546	21.2	41.08	0.0000
4	40.0	72.4714	1.8270	23.6	39.67	0.0000
5	40.0	73.8612	0.4826	10.4	153.06	0.0000

Method	_Method	Pre	Estimate	StdErr	df	tValue	Probt
1	2	23.5	−0.3843	1.6486	21.0	−0.23	0.8179
1	3	23.5	−5.9169	2.3042	20.0	−2.57	0.0183
1	4	23.5	−10.4251	2.3717	19.6	−4.40	0.0003
1	5	23.5	−11.8149	1.6551	21.3	−7.14	0.0000
2	3	23.5	−5.5326	1.6644	21.7	−3.32	0.0031
2	4	23.5	−10.0409	1.7586	20.8	−5.71	0.0000
2	5	23.5	−11.4306	0.4495	9.9	−25.43	0.0000
3	4	23.5	−4.5083	2.3844	20.0	−1.89	0.0732
3	5	23.5	−5.8980	1.6455	20.9	−3.58	0.0018
4	5	23.5	−1.3898	1.7650	21.1	−0.79	0.4398

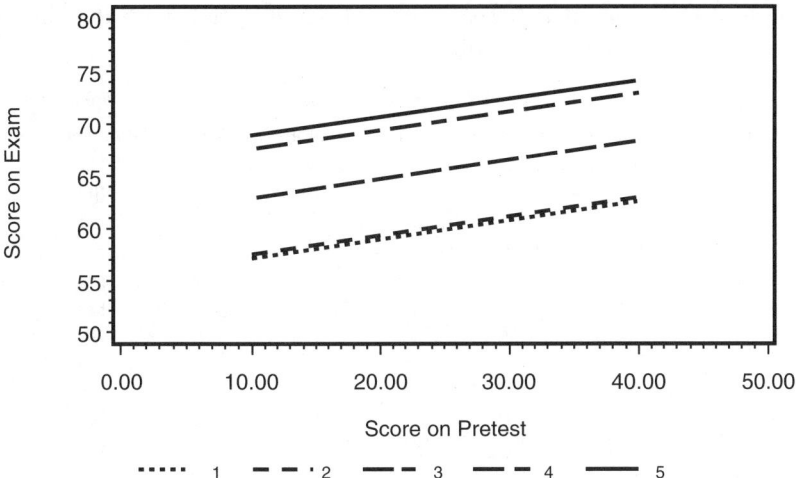

FIGURE 14.1 Plot of the regression lines for each of the five study methods using the parallel lines model.

of the pre scores, 23.4 (mean), 10, and 40. The adjusted means at 23.4 can be used to report the results and the adjusted means at 10 and 40 can be used to provide a graph of the estimated regression lines as displayed in Figure 14.1. The bottom part of Table 14.10 provides LSD type pairwise comparisons between the study method means where (1,2), (3,4), and (4,5) methods are not significantly different and all other comparisons are significantly different ($p < 0.05$).

The analysis of this model is a preview of the process to be used in later chapters. The process is to first determine an adequate set of regression models that describes the mean of the dependent variable as a function of the covariates. Then use that covariate model to study the relationship among the variances of the treatment combinations. Once an adequate covariance structure is selected, go back to the regression or covariate part of the model and simplify it as much as possible using strategies described in previous chapters.

14.9 EXAMPLE: TWO-WAY TREATMENT STRUCTURE WITH UNEQUAL VARIANCES

The data in Table 14.11 are from an experiment designed to evaluate the effect of the speed in rpm of a circular bar of steel and the feed rate into a cutting tool on the roughness of the surface finish of the final turned product, where the depth of cut was set to 0.02 in. The treatment structure is a two-way with four levels of speed (rpm) and four levels of feed rate feed. The design structure is a completely randomized design with eight replications per treatment combination. There is variation

TABLE 14.11
Cutting Tool Data Ran at Different Feed Rates and Speeds with Roughness the Dependent Variable and Hardness the Possible Covariate

Feed Rate	Speed 100 rpm		Speed 200 rpm		Speed 400 rpm		Speed 800 rpm	
	Roughness	Hardness	Roughness	Hardness	Roughness	Hardness	Roughness	Hardness
0.01	50	61	65	59	84	64	111	50
0.01	53	65	55	44	104	70	142	52
0.01	56	57	59	52	73	50	147	47
0.01	41	43	55	46	89	55	135	44
0.01	44	46	62	56	89	58	134	62
0.01	43	51	63	59	87	55	148	53
0.01	42	41	67	66	84	58	162	68
0.01	48	53	59	48	83	56	139	65
0.02	64	54	81	65	108	41	192	68
0.02	61	58	81	53	118	67	152	49
0.02	64	60	70	43	136	69	190	46
0.02	62	57	61	48	109	69	166	46
0.02	65	60	67	43	101	61	167	65
0.02	58	50	71	44	104	62	152	45
0.02	72	67	79	69	93	44	139	64
0.02	63	63	69	40	104	62	188	60
0.04	97	62	103	61	123	41	197	59
0.04	79	48	105	53	137	41	216	46
0.04	86	55	121	69	153	64	190	55
0.04	90	54	101	50	137	58	187	58
0.04	92	59	107	61	111	50	212	63
0.04	74	42	103	57	131	46	211	47
0.04	92	57	102	57	137	42	207	40
0.04	82	44	108	63	155	68	220	64
0.08	141	66	158	56	192	69	279	55
0.08	142	61	154	49	195	57	293	52
0.08	132	49	166	59	180	52	266	52
0.08	119	41	159	55	187	56	303	69
0.08	147	69	164	65	210	62	281	61
0.08	119	42	171	68	174	44	256	41
0.08	145	57	162	64	204	63	289	58
0.08	147	63	163	68	198	48	287	47

in the hardness values of the bar stock used in the experiment; thus the hardness of each piece of bar was measured to be used as a possible covariate. A model to describe the linear relationship between roughness and hardness (plot the data to see if this is an adequate assumption) is

Analysis of Covariance Models with Heterogeneous Errors

$$y_{ijk} + \alpha_{ij} + \beta_{ij} x_{ijk} + \varepsilon_{ijk},$$

$$i = 1, 2, 3, 4, \quad j = 1, 2, 3, 4, \quad k = 1, 2, \ldots, 8, \qquad (14.2)$$

$$\varepsilon_{ijk} \sim N\left(0, \sigma_{ij}^2\right)$$

where y_{ijk} denotes the roughness measure, x_{ijk} denotes the hardness values, and α_{ij} and β_{ij} denote the intercept and slope of the simple linear regression model for i^{th} level of speed and the jth level of feed rate. The cell variances are displayed in Table 14.12 and since all of the cells are based on eight observations Hartley's Max-F test can be used to test for the equality of variances. The Max-Fvalue is 457.98/1.21 = 378.50. The .01 percentage point from the Max-F distribution for 16 treatments and seven degrees of freedom per variance is 30.5. Thus there is sufficient evidence to conclude that the variances are not equal. The first PROC GLM code in Table 14.12 is used to fit Model 14.2 to the data and compute the residuals, and the second PROC GLM code is used to compute the value of Levene's statistic. The results are in the middle of Table 14.12. Levene's approach has been extended to the two-way treatment structure by fitting a factorial treatment structure to the absolute value of the residuals instead of just a one-way treatment structure. The results show that there are unequal variances (use the F-value for Model) and that the levels of rpm are contributing the most to the unequal variances while the levels of feed and the interaction between the levels of feed and the levels of rpm contribute less. The LSMEANS statement was used to provide multiple comparisons among the rpm means of the absolute values of the residuals using an LSD type of approach. The lower part of Table 14.12 are the significance levels of the pairwise comparisons between the rpm means. The means of 1 and 2 are not different, but all other comparisons are significantly different. The means of the absolute values of the residuals for the levels of feed are also presented, but they are not significantly different. The results in Table 14.12 indicate the variances are unequal for the levels of speed (denoted by rpm) and are not unequal for the other parts of the treatment structure. One possible simplification is to group rpm levels of 100 and 200 and fit a three variance model, but that was not considered here. The PROC MIXED code in Table 14.13 fits Model 14.2 with unequal variances for all combinations of the levels of feed and the levels of rpm as specified by

`"repeated/group=feed*rpm;"`

The results in Table 14.13 consist of the list of estimated variances for each cell (as also shown in Table 14.12) and the information criteria. For this model, the value of AIC is 780.6. Model 14.2 with unequal variances for each level of rpm is

$$y_{ijk} + \alpha_{ij} + \beta_{ij} x_{ijk} + \varepsilon_{ijk},$$

$$i = 1, 2, 3, 4, \quad j = 1, 2, 3, 4, \quad k = 1, 2, \ldots, 8, \qquad (14.3)$$

$$\varepsilon_{ijk} \sim N\left(0, \sigma_i^2\right)$$

TABLE 14.12
Treatment Combination Variances and PROC GLM Code to Compute the Residuals and to Compute Levene's Type Tests of Equality of Variances for the Cutting Tool Data Set

```
proc glm data=res;
class feed rpm; model rough=feed|rpm rpm*hard*feed;
output out=res r=r;

data res; set res; absr=abs(r);

proc glm data=res;; class feed rpm; model
 absr=feed|rpm;
lsmeans feed|rpm/pdiff;
```

Variances	Feed=.01	Feed=.02	Feed=.04	Feed=.08
rpm=100	9.76	6.29	7.47	21.61
rpm=200	1.21	29.49	16.06	10.69
rpm=400	30.87	135.55	140.85	108.85
rpm=800	211.48	457.98	173.04	127.78

Source	df	SS	MS	FValue	ProbF
Model	15	2182.3828	145.4922	6.77	0.0000
Error	112	2406.1654	21.4836		
Corrected Total	127	4588.5482			

Source	df	SS(III)	MS	FValue	ProbF
Feed	3	152.8896	50.9632	2.37	0.0742
rpm	3	1783.3713	594.4571	27.67	0.0000
Feed*rpm	9	246.1220	27.3469	1.27	0.2594

Effect	Feed	LSMean
Feed	0.01	4.4171
Feed	0.02	7.4871
Feed	0.04	5.7339
Feed	0.08	5.6678

Effect	Rpm	LSMean
rpm	100	2.3131
rpm	200	2.4260
rpm	400	7.2930
rpm	800	11.2737

RowName	_1	_2	_3	_4
1		0.9225	0.0000	0.0000
2	0.9225		0.0001	0.0000
3	0.0000	0.0001		0.0008
4	0.0000	0.0000	0.0008	

which is fit by the PROC MIXED code in Table 14.14. The results consist of the four estimated variances for the levels of rpm and the associated information criteria. The value of the AIC is 777.7, indicating that the four variance model is fitting the

TABLE 14.13
PROC MIXED Code to Fit a Model with Unequal Variances for Each Feed*rpm Combination with the Results Including the Information Criteria

```
proc mixed data=twoway cl covtest ic;
class feed rpm;
model rough=feed|rpm hard*feed*rpm/ddfm=satterth;
repeated/group=rpm*feed;
```

CovParm	Group	Estimate	StdErr	ZValue	Lower	Upper
Residual	Feed*rpm 0.01 100	9.76	5.63	1.73	4.05	47.33
Residual	Feed*rpm 0.01 200	1.21	0.70	1.73	0.50	5.86
Residual	Feed*rpm 0.01 400	30.87	17.82	1.73	12.82	149.67
Residual	Feed*rpm 0.01 800	211.48	122.10	1.73	87.81	1025.47
Residual	Feed*rpm 0.02 100	6.29	3.63	1.73	2.61	30.49
Residual	Feed*rpm 0.02 200	29.49	17.03	1.73	12.25	143.01
Residual	Feed*rpm 0.02 400	135.55	78.26	1.73	56.29	657.31
Residual	Feed*rpm 0.02 800	457.98	264.42	1.73	190.17	2220.80
Residual	Feed*rpm 0.04 100	7.47	4.31	1.73	3.10	36.24
Residual	Feed*rpm 0.04 200	16.06	9.27	1.73	6.67	77.87
Residual	Feed*rpm 0.04 400	140.85	81.32	1.73	58.49	682.98
Residual	Feed*rpm 0.04 800	173.04	99.91	1.73	71.85	839.10
Residual	Feed*rpm 0.08 100	21.61	12.48	1.73	8.98	104.81
Residual	Feed*rpm 0.08 200	10.69	6.17	1.73	4.44	51.84
Residual	Feed*rpm 0.08 400	108.85	62.84	1.73	45.20	527.82
Residual	Feed*rpm 0.08 800	127.78	73.77	1.73	53.06	619.61

Neg2LogLike	Parameters	AIC	AICC	HQIC	BIC	CAIC
748.6	16	780.6	787.5	799.2	826.3	842.3

Effect	NumDF	DenDF	FValue	ProbF
Feed	3	24.6	15.66	0.0000
rpm	3	28.7	20.68	0.0000
Feed*rpm	9	18.9	1.18	0.3592
Hard*Feed*rpm	16	6	17.67	0.0010

variance structure of the data as well as the sixteen variance structure. Also included in Table 14.14 is the statistic to test the slopes equal to zero hypothesis which is the F statistic corresponding to Hard*Feed*rpm. The resulting significance level is 0.0000, indicating there is sufficient evidence to believe that the slopes are not all equal to zero.

The PROC MIXED code in Table 14.15 fits the model

$$y_{ijk} = \left(\mu + \tau_i + \gamma_j + \delta_{ij}\right) + \left(\lambda + \rho_i + \phi_j + \theta_{ij}\right) x_{ijk} + \varepsilon_{ijk},$$

$$i = 1, 2, 3, 4, \quad j = 1, 2, 3, 4, \quad k = 1, 2, \ldots, 8, \quad (14.4)$$

$$\varepsilon_{ijk} \sim N\left(0, \sigma_i^2\right),$$

TABLE 14.14
PROC MIXED Code to Fit a Model with Unequal Variances for Each Level of rpm for the Cutting Tool Study

```
proc mixed data=twoway cl covtest ic;
class feed rpm;
model rough=feed|rpm hard*rpm*feed/ddfm=satterth;
repeated/group=rpm;
```

CovParm	Group	Estimate	StdErr	ZValue	Lower	Upper
Residual	rpm 100	11.28	3.26	3.46	6.88	21.84
Residual	rpm 200	14.36	4.15	3.46	8.76	27.80
Residual	rpm 400	104.03	30.03	3.46	63.43	201.33
Residual	rpm 800	242.57	70.02	3.46	147.89	469.45

Neg2LogLike	Parameters	AIC	AICC	HQIC	BIC	CAIC
769.7	4	777.7	778.2	782.4	789.1	793.1

Effect	NumDF	DenDF	FValue	ProbF
Feed	3	52.6	10.77	0.0000
rpm	3	34.3	18.58	0.0000
Feed*rpm	9	35.9	0.85	0.5795
Hard*Feed*rpm	16	24	12.11	0.0000

TABLE 14.15
PROC MIXED Code to Fit a Model with Unequal Variances for Each Level of rpm and a Factorial Structure for the Covariate Part of the Model

```
proc mixed data=twoway cl covtest ic;
class feed rpm;
model rough=feed|rpm hard hard*feed hard*rpm hard*feed*rpm/ddfm=satterth;
repeated/group=rpm;
```

CovParm	Group	Estimate	StdErr	ZValue	Lower	Upper
Residual	rpm 100	11.28	3.26	3.46	6.88	21.84
Residual	rpm 200	14.36	4.15	3.46	8.76	27.80
Residual	rpm 400	104.03	30.03	3.46	63.43	201.33
Residual	rpm 800	242.57	70.02	3.46	147.89	469.45

Effect	NumDF	DenDF	FValue	ProbF
Feed	3	52.6	10.77	0.0000
rpm	3	34.3	18.58	0.0000
Feed*rpm	9	35.9	0.85	0.5795
Hard	1	52.4	45.04	0.0000
Hard*Feed	3	51.5	0.71	0.5485
Hard*rpm	3	34.0	0.79	0.5082
Hard*Feed*rpm	9	35.5	0.54	0.8349

TABLE 14.16
PROC MIXED Code to Fit a Model with Unequal Variances for Each Level of rpm and a Factorial Structure for the Covariate Part of the Model without the Hard*Feed*rpm Term

```
proc mixed data=twoway cl covtest ic;
class feed rpm;
model rough=feed|rpm hard hard*feed
  hard*rpm/ddfm=satterth;
repeated/group=rpm;
```

CovParm	Group	Estimate	StdErr	ZValue	Lower	Upper
Residual	rpm 100	10.91	3.07	3.56	6.73	20.70
Residual	rpm 200	13.95	3.88	3.60	8.64	26.24
Residual	rpm 400	96.17	26.22	3.67	60.07	178.40
Residual	rpm 800	237.74	64.89	3.66	148.43	441.38

Effect	NumDF	DenDF	FValue	ProbF
Feed	3	67.6	40.30	0.0000
rpm	3	37.7	20.62	0.0000
Feed*rpm	9	37.3	8.18	0.0000
Hard	1	53.6	48.36	0.0000
Hard*Feed	3	60.2	2.72	0.0522
Hard*rpm	3	37.4	0.80	0.5010

in which unequal variances for the levels of rpm and the slopes and intercepts of Model 14.2 are expressed as factorial effects. The significance level corresponding to Hard*Feed*rpm is 0.8349, indicating that there is not a significant interaction effect on the slopes. A reduced model is

$$y_{ijk} = \left(\mu + \tau_i + \gamma_j + \delta_{ij}\right) + \left(\lambda + \rho_i + \phi_j\right) x_{ijk} + \varepsilon_{ijk},$$

$$i = 1, 2, 3, 4, \quad j = 1, 2, 3, 4, \quad k = 1, 2, \ldots, 8, \tag{14.5}$$

$$\varepsilon_{ijk} \sim N\left(0, \sigma_i^2\right),$$

where $\alpha_{ij} = \mu + \tau_i + \gamma_j + \delta_{ij}$ and $\beta_{ij} = \lambda + \rho_i + \phi_j$.

Table 14.16 contains the PROC MIXED code to fit the reduced model obtained excluding the Hard*Feed*rpm term and the results indicate that the slopes are not unequal levels of rpm since the significance level corresponding to Hard*rpm is 0.5010. A reduced model without the Hard*rpm and Hard*Feed*rpm terms is

$$y_{ijk} = \left(\mu + \tau_i + \gamma_j + \delta_{ij}\right) + \left(\lambda + \phi_j\right) x_{ijk} + \varepsilon_{ijk},$$

$$i = 1, 2, 3, 4, \quad j = 1, 2, 3, 4, \quad k = 1, 2, \ldots, 8, \tag{14.6}$$

$$\varepsilon_{ijk} \sim N\left(0, \sigma_i^2\right),$$

TABLE 14.17
PROC MIXED Code to Fit a Model with Unequal Variances for Each Level of rpm and a Factorial Structure for the Covariate Part of the Model without the rpm and Hard*feed*rpm Term

```
proc mixed data=twoway cl covtest ic;
class feed rpm;
model rough=feed|rpm hard hard*feed/ddfm=satterth;
repeated/group=rpm;
```

CovParm	Group	Estimate	StdErr	ZValue	Lower	Upper
Residual	rpm 100	10.92	3.06	3.57	6.75	20.67
Residual	rpm 200	14.27	3.95	3.62	8.86	26.74
Residual	rpm 400	93.88	25.15	3.73	59.06	172.01
Residual	rpm 800	232.83	62.46	3.73	146.41	427.00

Effect	NumDF	DenDF	FValue	ProbF
Feed	3	67.9	39.89	0.0000
rpm	3	38.5	767.09	0.0000
Feed*rpm	9	38.7	8.18	0.0000
Hard	1	64.4	174.68	0.0000
Hard*Feed	3	60.4	3.91	0.0129

where $\alpha_{ij} = \mu + \tau_i + \gamma_j + \delta_{ij}$ and $\beta_j = \lambda + \phi_j$. This reduced model is fit by the PROC MIXED code in Table 14.17. The model could also be expressed as

$$y_{ijk} = (\mu + \tau_i + \gamma_j + \delta_{ij}) + (\beta_j) x_{ijk} + \varepsilon_{ijk},$$

$$i = 1, 2, 3, 4, \quad j = 1, 2, 3, 4, \quad k = 1, 2, \ldots, 8, \quad (14.7)$$

$$\varepsilon_{ijk} \sim N(0, \sigma_i^2),$$

The results indicate that there is sufficient information to conclude that the slopes for the levels of Feed are unequal, since the significance level corresponding to Hard*Feed is 0.0129. There is an indication that there is an interaction effect on the intercepts; thus the LSMEANS statements in Table 14.18 were used to provide adjusted means or predicted values from the regression models at three values of hardness, i.e., at Hard = 55.45 (the mean), 40, and 70. The estimated values from Hard = 40 and 70 were used to provide plots of the predicted regression models and they are displayed in Figures 14.2 through 14.9. Multiple comparisons using the simulate adjustment were carried out for making comparisons of the levels of feed within each level of rpm and comparing the levels of rpm within each level of feed. The means of all levels of rpm with each level of feed were significantly different (not displayed here). Table 14.19 contains all of the comparisons of the level of feed within each level of rpm computed at each of the three values of Hard. The columns Adjp(40), Adjp(mn), and Adjp(70) are the adjusted significance levels using the

TABLE 14.18
Adjusted Means for the Feed*rpm Combinations Computed at Three Values of Hardness

```
lsmeans feed*rpm/diff at means adjust=simulate;
lsmeans feed*rpm/diff at hard=40 adjust=simulate;
lsmeans feed*rpm/diff at hard=70 adjust=simulate;
```

		Hard=55.45		Hard=40		Hard=70	
Feed	rpm	Estimate	StdErr	Estimate	StdErr	Estimate	StdErr
0.01	100	49.05	1.23	40.12	1.79	57.45	2.31
0.01	200	61.61	1.35	52.68	2.03	70.01	2.25
0.01	400	85.01	3.44	76.08	3.98	93.41	3.67
0.01	800	139.94	5.39	131.01	5.65	148.34	5.64
0.02	100	61.96	1.22	53.85	2.31	69.60	1.69
0.02	200	74.91	1.43	66.80	1.75	82.55	2.46
0.02	400	107.07	3.45	98.95	4.00	114.70	3.61
0.02	800	168.29	5.39	160.18	5.64	175.93	5.62
0.04	100	89.08	1.22	74.98	1.99	102.35	2.50
0.04	200	103.13	1.40	89.03	2.75	116.40	1.95
0.04	400	139.33	3.47	125.24	3.71	152.61	4.18
0.04	800	206.33	5.40	192.23	5.68	219.60	5.77
0.08	100	135.99	1.17	121.64	1.95	149.50	1.80
0.08	200	157.44	1.42	143.09	2.41	170.95	1.63
0.08	400	191.64	3.43	177.29	3.78	205.15	3.67
0.08	800	282.75	5.40	268.40	5.57	296.26	5.61

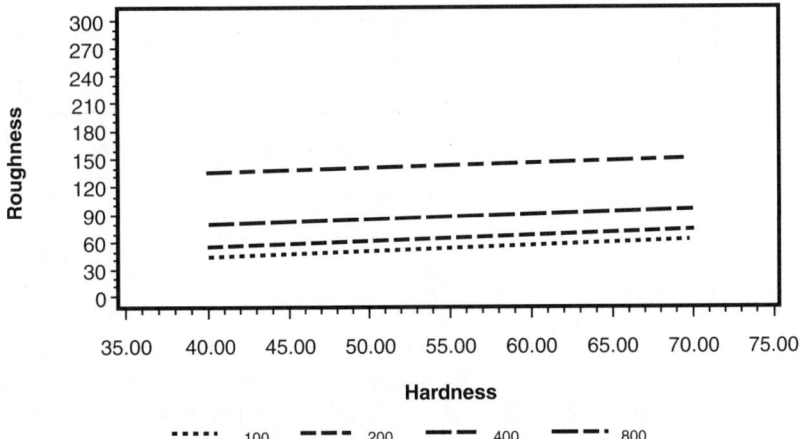

FIGURE 14.2 RPM models at Feed = 0.10.

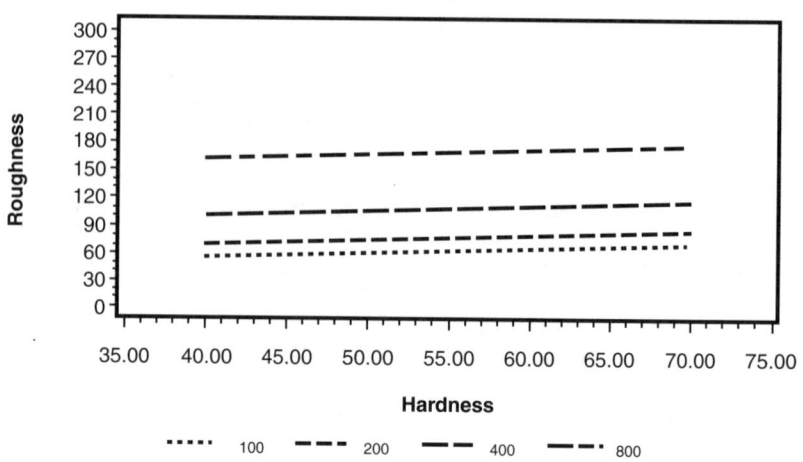

FIGURE 14.3 RPM models at Feed = 0.02.

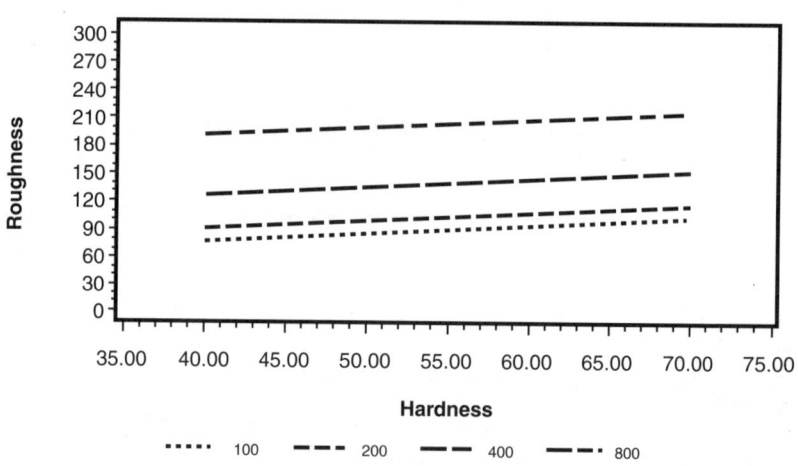

FIGURE 14.4 RPM models at Feed = 0.04.

simulate method (Westfall, 1996) for comparing the feed means within each level of rpm at Hard = 40, 55.45, and 70, respectively. All feed means within each level of rpm are significantly different (0.05) except for 0.01 and 0.02 at rpm = 800 and Hard = 70, which has a significance level of 0.0586. The final part of the analysis was to investigate how the factorial effects change the relationships between the

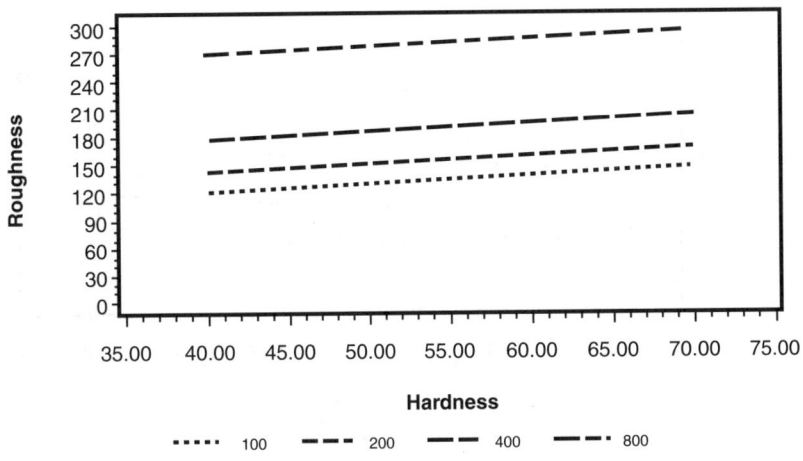

FIGURE 14.5 RPM models at Feed = 0.08.

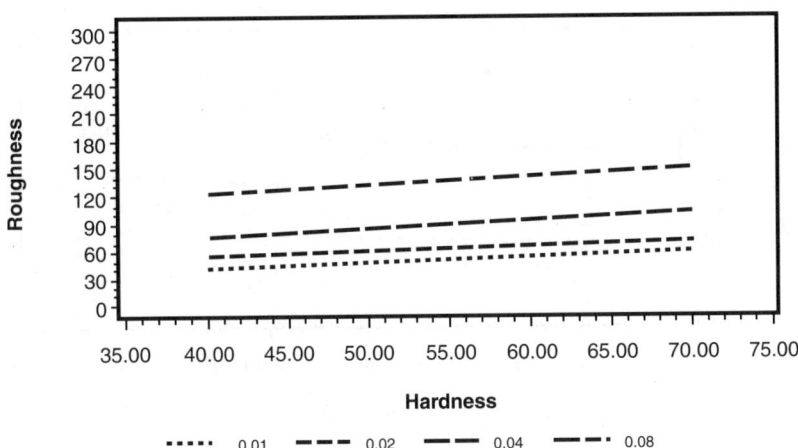

FIGURE 14.6 Feed models at rpm = 100.

models evaluated at different values of hardness. Table 14.20 contains the PROC MIXED code to fit Model 14.7 to the data using three different adjusted values of the covariate, Hard40 = Hard − 40, Hardmn = Hard − 55.45, and Hard70 = Hard − 70. The factorial effects in the analyses in Table 14.20 are measuring the effect of the levels of Feed, levels of rpm, and their interaction on the models evaluated at

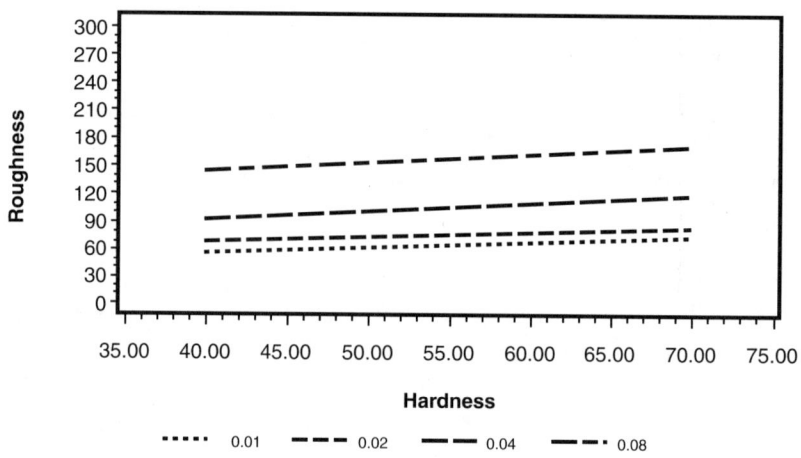

FIGURE 14.7 Feed models at rpm = 200.

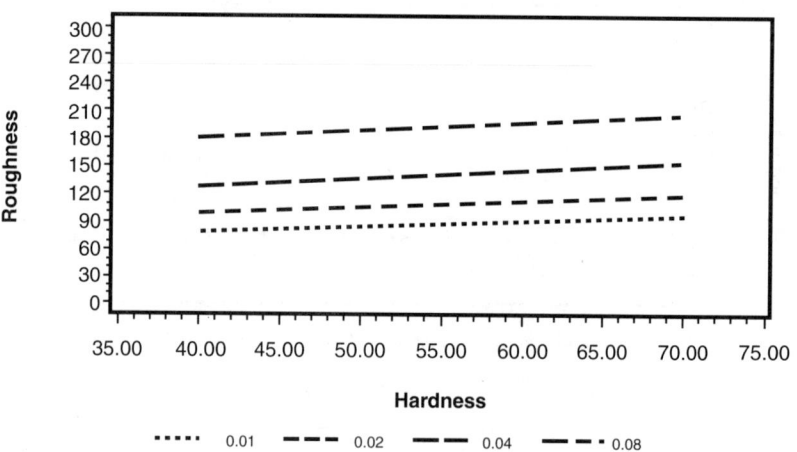

FIGURE 14.8 Feed models at rpm = 400.

the three levels of Hard. All of the factorial effects are highly significant, indicating that comparisons among the cell means is a reasonable approach to evaluating the effect of the levels of feed and the levels of rpm.

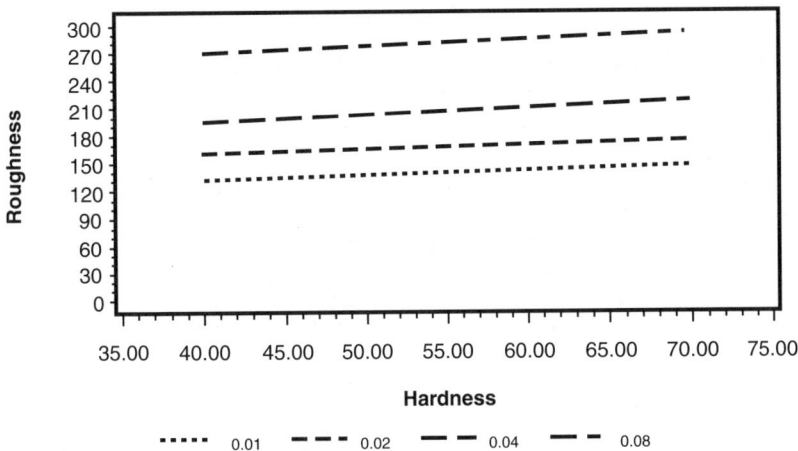

FIGURE 14.9 Feed models at rpm = 800.

Since the levels of feed and the levels of rpm are continuous, a response surface type model could be fit to the data using the unequal variance structure. That analysis is left as an exercise.

As in the previous section, the first step in the analysis was (1) to make sure there was a linear relationship between the roughness measurements and the hardness measurements (not shown), (2) determine the simplest adequate form of the covariance structure, and (3) determine the simplest adequate form of the covariate part of the model. After the above three steps, the regression lines need to be compared as in this case, using pairwise comparisons within three levels of the covariate.

14.10 EXAMPLE: TREATMENTS IN MULTI-LOCATION TRIAL

The data in Table 14.21 are the results of an experiment to evaluate the relationship between the level of a drug and the ability of a beef animal to gain weight as measured by the average daily gain (ADG). The experiment consisted of randomly forming 16 pens of 5 Hereford cross steer calves at each of 6 locations. At each location, the levels of DOSE were randomly assigned to four of the pens. Thus the design structure at each location is a completely randomized design with four replications per level of dose. The treatment structure is a two-way with the levels of location crossed with the levels of dose. The average initial weight of the calves in each pen was measured and used as a possible covariate. The following model can be used to describe the data which involves a mixed effects treatment structure, i.e., the levels of location are a random effect and the levels of dose are a fixed effect:

TABLE 14.19
Simulate Adjusted Significance Levels for Comparing the Levels of Feed within Each Level of rpm for the Three Levels of Hardness

Effect	rpm	Feed	_Feed	Adjp(40)	Adjp(mn)	Adjp(70)
Feed*rpm	100	0.01	0.02	0.0028	0.0000	0.0079
Feed*rpm	100	0.01	0.04	0.0000	0.0000	0.0000
Feed*rpm	100	0.01	0.08	0.0000	0.0000	0.0000
Feed*rpm	100	0.02	0.04	0.0000	0.0000	0.0000
Feed*rpm	100	0.02	0.08	0.0000	0.0000	0.0000
Feed*rpm	100	0.04	0.08	0.0000	0.0000	0.0000
Feed*rpm	200	0.01	0.02	0.0005	0.0000	0.0285
Feed*rpm	200	0.01	0.04	0.0000	0.0000	0.0000
Feed*rpm	200	0.01	0.08	0.0000	0.0000	0.0000
Feed*rpm	200	0.02	0.04	0.0000	0.0000	0.0000
Feed*rpm	200	0.02	0.08	0.0000	0.0000	0.0000
Feed*rpm	200	0.04	0.08	0.0000	0.0000	0.0000
Feed*rpm	400	0.01	0.02	0.0170	0.0051	0.0107
Feed*rpm	400	0.01	0.04	0.0000	0.0000	0.0000
Feed*rpm	400	0.01	0.08	0.0000	0.0000	0.0000
Feed*rpm	400	0.02	0.04	0.0021	0.0000	0.0000
Feed*rpm	400	0.02	0.08	0.0000	0.0000	0.0000
Feed*rpm	400	0.04	0.08	0.0000	0.0000	0.0000
Feed*rpm	800	0.01	0.02	0.0413	0.0344	0.0586
Feed*rpm	800	0.01	0.04	0.0000	0.0000	0.0000
Feed*rpm	800	0.01	0.08	0.0000	0.0000	0.0000
Feed*rpm	800	0.02	0.04	0.0190	0.0017	0.0002
Feed*rpm	800	0.02	0.08	0.0000	0.0000	0.0000
Feed*rpm	800	0.04	0.08	0.0000	0.0000	0.0000

$$ADG_{ijk} = \delta_i + \beta_i\, WT_{ijk} + L_j + b_j\, WT_{ijk} + LT_{ij} + c_{ij}\, WT_{ijk} + \varepsilon_{ijk} \tag{14.8}$$

$$i = 1, 2, 3, 4, \quad j = 1, 2, 3, 4, \quad k = 1, 2, \ldots, 8,$$

$$\begin{bmatrix} L_j \\ b_j \end{bmatrix} \sim \text{iid } N\left(\begin{bmatrix} 0 \\ 0 \end{bmatrix}, \begin{bmatrix} \sigma_L^2 & \sigma_{Lb} \\ \sigma_{Lb} & \sigma_b^2 \end{bmatrix}\right), \begin{bmatrix} LT_{ij} \\ c_{ij} \end{bmatrix} \sim \text{iid } N\left(\begin{bmatrix} 0 \\ 0 \end{bmatrix}, \begin{bmatrix} \sigma_{LT}^2 & \sigma_{LTc} \\ \sigma_{LTc} & \sigma_c^2 \end{bmatrix}\right), \varepsilon_{ijk} \sim N(0, \sigma^2)$$

The following random statements are needed with PROC MIXED to specify the above covariance structure:

```
random int wt/type=un subject=loc;
random int wt/type=un subject=loc*dose;
```

However, PROC MIXED could not fit the model to the data set as it was exceeding the likelihood evaluation limits and was not converging. A simpler model could be fit to the data set that involves setting $\sigma_{Lb} = 0$, $\sigma_b^2 = 0$, $\sigma_{LTc} = 0$, and $\sigma_c^2 = 0$. The

TABLE 14.20
Sets of PROC MIXED Code to Fit Models Using Adjusted Values of the Covariate Using the Unequal Variance Model with Different Variances for Each Level of rpm

```
proc mixed data=twoway cl covtest ic;
class feed rpm;
model rough=feed|rpm hard40*feed/ddfm=satterth;
repeated/group=rpm;
proc mixed data=twoway cl covtest ic;
class feed rpm;
model rough=feed|rpm hard70*feed/ddfm=satterth;
repeated/group=rpm;
proc mixed data=twoway cl covtest ic;
class feed rpm;
model rough=feed|rpm hardmn*feed/ddfm=satterth;
repeated/group=rpm;
```

CovParm	Group	Estimate	StdErr	ZValue	Lower	Upper
Residual	rpm 100	10.92	3.06	3.57	6.75	20.67
Residual	rpm 200	14.27	3.95	3.62	8.86	26.74
Residual	rpm 400	93.88	25.15	3.73	59.06	172.01
Residual	rpm 800	232.83	62.46	3.73	146.41	427.00

Effect	NumDF	DenDF	FValue	ProbF
Feed	3	100.3	361.62	0.0000
rpm	3	38.5	767.09	0.0000
Feed*rpm	9	38.7	8.18	0.0000
hard40*Feed	4	54.5	48.20	0.0000

Effect	NumDF	DenDF	FValue	ProbF
Feed	3	54.7	808.45	0.0000
rpm	3	38.5	767.09	0.0000
Feed*rpm	9	38.7	8.18	0.0000
hardmn*Feed	4	54.5	48.20	0.0000

Effect	NumDF	DenDF	FValue	ProbF
Feed	3	98.63	518.29	0.0000
rpm	3	38.52	767.09	0.0000
Feed*rpm	9	38.68	8.18	0.0000
hard70*Feed	4	54.51	48.20	0.0000

PROC MIXED code in Table 14.22 fits Model 14.8 with the four variances and covariances set to zero to the data set. Table 14.22 contains the estimates of the variance components and the information criteria as well as tests for the fixed effects. At this point the within location residual variances have been assumed to be equal. Table 14.23 contains the PROC GLM code to fit a one-way treatment structure in a CRD design structure with one covariate with unequal slopes to the data from each location. The table contains the mean square error from each location's model which are each based on 8 degrees of freedom. Since the data from each location is

TABLE 14.21
Data from a Multilocation Experiment with Four Levels of Dose Where Average Daily Gain is the Response Variable and Initial Weight is the Possible Covariate

	Dose = 0 mg		Dose = 1.2 mg		Dose = 2.4 mg		Dose = 3.6 mg	
loc	adg1	wt1	adg2	wt2	adg3	wt3	adg4	wt4
1	1.21	415	1.27	422	1.20	407	1.36	420
1	1.47	444	1.43	437	1.42	429	1.22	402
1	1.19	413	1.31	422	1.37	425	1.15	402
1	1.11	410	1.27	420	1.14	404	1.20	407
2	1.18	412	1.40	425	1.33	413	1.41	411
2	1.48	442	1.23	412	1.38	417	1.56	421
2	1.49	446	1.25	417	1.68	437	1.45	425
2	1.29	409	1.20	407	1.23	407	1.65	435
3	1.40	425	1.55	432	1.74	433	1.51	408
3	1.51	439	1.31	423	1.49	429	1.70	417
3	1.51	442	1.43	421	1.56	415	1.59	416
3	1.56	432	1.76	433	1.45	402	1.74	443
4	1.82	436	1.71	438	1.71	428	1.80	426
4	1.53	445	1.74	448	1.46	419	1.63	401
4	1.07	406	1.41	405	1.56	411	1.85	447
4	1.35	426	1.33	409	1.62	402	2.33	440
5	1.20	429	1.92	450	2.16	446	1.56	422
5	1.46	449	1.67	433	1.70	440	1.81	447
5	1.61	449	1.74	448	1.63	418	1.47	413
5	1.72	449	1.63	435	1.39	410	1.70	421
6	1.39	411	1.51	427	1.81	411	2.53	448
6	1.84	428	1.78	422	1.88	444	2.21	428
6	1.68	420	1.86	448	1.78	428	1.96	416
6	1.75	442	1.73	430	1.74	411	2.15	447

independent of the data from any other location, Hartley's Max-F test can be used to test the equality of the within location variances. The value of the test statistic is Max-F = 0.03754/0.00079 = 47.52. The 0.01 percentage point for the Max-F distribution comparing six variances each based on eight degrees of freedom is 30; thus there is sufficient evidence to conclude the variances are not equal. The unequal slope model with unequal within location variances is

$$ADG_{ijk} = \delta_i + \beta_i \, WT_{ijk} + L_j + LT_{ij} + \varepsilon_{ijk},$$

$$i = 1, 2, 3, 4, \quad j = 1, 2, \ldots, 6, \quad k = 1, 2, 3, 4, \tag{14.9}$$

$$L_j \sim \text{iid } N(0, \sigma_L^2), \, LT_{ij} \sim \text{iid } N(0, \sigma_{LT}^2), \, \varepsilon_{ijk} \sim N(0, \sigma_j^2).$$

TABLE 14.22
PROC MIXED Code to Fit the Model with Four Variance and Covariance Parameters Set to Zero

```
proc mixed cl covtest ic data=ex148;
class loc dose;
model adg=dose wt wt*dose/ddfm=kr;
random loc dose*loc;
```

CovParm	Estimate	StdErr	ZValue	ProbZ	Alpha	Lower	Upper
loc	0.02380	0.01612	1.48	0.0699	0.05	0.00882	0.17254
loc*dose	0.00281	0.00267	1.05	0.1467	0.05	0.00079	0.08253
Residual	0.01424	0.00248	5.73	0.0000	0.05	0.01040	0.02070

Neg2LogLike	Parameters	AIC	AICC	HQIC	BIC	CAIC
−53.2	3	−47.2	−46.9	−49.7	−47.8	−44.8

Effect	NumDF	DenDF	FValue	ProbF
dose	3	82.3	0.08	0.9701
wt	1	78.1	106.34	0.0000
wt*dose	3	82.2	0.12	0.9466

TABLE 14.23
PROC GLM Code to Compute the within Location Variances Based on a One-Way Analysis of Covariance Model with Unequal Slopes at Each Location

```
proc sort data=ex148; by loc;
proc glm data=ex148; by loc;
class dose;
model adg=dose wt*dose;
```

			Location			
1	2	3	4	5	6
0.00079	0.00288	0.00979	0.03754	0.01734	0.02234

and the PROC MIXED code in Table 14.24 fits this model to the ADG data. The results consist of the estimates of the variance components and the information criteria where the value of AIC is −75.4. The AIC value for the unequal variance model is considerably less than the value of AIC for the equal variance model in Table 14.22, indicating that the unequal variance model is more adequate than the model with equal variances. The statistic corresponding to wt*dose tests the equal slopes hypothesis, H_0: $\beta_1 = \beta_2 = \beta_3 = \beta_4$ vs. H_a: (not H_o:) with significance level 0.4332, indicating there is not sufficient evidence to conclude that the slopes are not equal, so a common slope model is adequate to describe the data. The common slope model with unequal within location variances is

TABLE 14.24
PROC MIXED Code to Fit the Unequal Slopes Model with Unequal Variances within Each of the Locations

```
proc mixed cl covtest ic data=ex148;
class loc dose;
model adg=dose wt wt*dose/ddfm=kr;
random loc dose*loc;
repeated/group=loc;
```

CovParm	Group	Estimate	StdErr	ZValue	Lower	Upper
loc		0.0243	0.0163	1.49	0.0091	0.1702
loc*dose		0.0017	0.0020	0.84	0.0004	0.2450
Residual	loc 1	0.0008	0.0004	2.04	0.0004	0.0027
Residual	loc 2	0.0029	0.0012	2.33	0.0014	0.0083
Residual	loc 3	0.0107	0.0041	2.61	0.0057	0.0271
Residual	loc 4	0.0310	0.0121	2.55	0.0163	0.0802
Residual	loc 5	0.0182	0.0076	2.40	0.0093	0.0508
Residual	loc 6	0.0258	0.0118	2.18	0.0124	0.0824

Neg2LogLike	Parameters	AIC	AICC	HQIC	BIC	CAIC
−91.4	8	−75.4	−73.6	−82.1	−77.1	−69.1

Effect	NumDF	DenDF	FValue	ProbF
dose	3	26.4	0.63	0.6006
wt	1	26.1	227.95	0.0000
wt*dose	3	26.3	0.94	0.4332

$$ADG_{ijk} = \delta_i + \beta \, WT_{ijk} + L_j + LT_{ij} + \varepsilon_{ijk},$$

$$i = 1, 2, 3, 4, \quad j = 1, 2, \ldots, 6, \quad k = 1, 2, 3, 4, \tag{14.10}$$

$$L_i \sim \text{iid } N(0, \sigma_L^2), \, LT_{ij} \sim \text{iid } N(0, \sigma_{LT}^2), \, \varepsilon_{ijk} \sim N(0, \sigma_j^2).$$

The PROC MIXED code to fit Model 14.10 to the ADG data is in Table 14.25 where the random statement specifies the variance components σ_L^2 and σ_{LT}^2 and the repeated statement specifies the residual variances, $\sigma_1^2, \sigma_2^2, \ldots, \sigma_6^2$. The estimates of the variance components, estimates of the intercepts and common slope and the tests for the fixed effects are in Table 14.25. The significance level for wt indicates there is sufficient evidence to conclude the common slope is not zero and should be included in the model. The significance level for dose indicates the intercepts are significantly different, which also means the models are different at any fixed value of the covariate, wt, since the models are a set of parallel lines. The estimate of the common slope is 0.0100 lb/day per lb of initial weight. Table 14.26 contains the adjusted means for the levels of dose evaluated at the mean of the initial pen weights across all of the locations, which was 425.33 lb. The estimated standard errors of the

TABLE 14.25
PROC MIXED Code to Fit the Common Slope Model with Unequal within Location Variances

```
proc mixed cl covtest ic data=ex148;
class loc dose;
model adg=dose wt/solution ddfm=kr;
random loc dose*loc;
repeated/group=loc;
```

CovParm	Group	Estimate	StdErr	ZValue	Lower	Upper
loc		0.0248	0.0167	1.49	0.0092	0.1757
loc*dose		0.0028	0.0022	1.28	0.0009	0.0326
Residual	loc 1	0.0006	0.0002	2.38	0.0003	0.0017
Residual	loc 2	0.0036	0.0014	2.53	0.0019	0.0094
Residual	loc 3	0.0103	0.0039	2.62	0.0055	0.0258
Residual	loc 4	0.0283	0.0109	2.58	0.0150	0.0722
Residual	loc 5	0.0184	0.0076	2.41	0.0094	0.0511
Residual	loc 6	0.0222	0.0099	2.24	0.0108	0.0682

Effect	Dose	Estimate	StdErr	df	tValue	Probt
Intercept		−2.5795	0.2400	26.1	−10.75	0.0000
dose	1	−0.2501	0.0448	8.5	−5.58	0.0004
dose	2	−0.1780	0.0447	8.4	−3.98	0.0037
dose	3	−0.0762	0.0446	8.3	−1.71	0.1247
dose	4	0.0000				
wt		0.0100	0.0005	22.1	18.47	0.0000

Effect	NumDF	DenDF	FValue	ProbF
dose	3	8.4	12.04	0.0021
wt	1	22.1	340.98	0.0000

adjusted means are all equal to 0.072 lb, where all of the variance components are used to provide the estimated standard errors (something a non-mixed models approach cannot do). The lower part of Table 14.26 contains the pairwise comparisons of the dose means using the simulate adjustment for multiple comparisons. The results indicate that dose levels 2.4 and 3.6 mg have means significantly greater than the mean of dose 0 mg. The mean of dose 3.6 mg is significantly larger than the mean of dose 1.2 mg. The estimate statements in Table 14.27 provide information about the trend the ADG means as a function of the levels of dose. The orthogonal polynomial coefficients are selected for four equally spaced levels. The results of the estimate statements indicate there is a significant linear trend ($p = 0.0002$), but there is no evidence of curvature since the quadratic and cubic effects are not significantly different from zero. All of the tests and comparisons of the treatments are carried out using KR approximate numbers of degrees of freedom in the denominator which are less than the loc*dose interaction degrees of freedom. If the type III sums of squares and an equal variance model were used to describe the data, then

TABLE 14.26
LSMEAN Statement to Provide Adjusted Means Evaluated at the Mean Initial Pen Weight Using the Simulate Multiple Comparison Method

```
lsmeans dose/diff at means adjust=simulate;
```

Effect	Dose	wt	Estimate	StdErr	df	tValue	Probt
dose	1	425.33	1.428	0.072	6.5	19.90	0.0000
dose	2	425.33	1.500	0.072	6.5	20.92	0.0000
dose	3	425.33	1.601	0.072	6.5	22.33	0.0000
dose	4	425.33	1.678	0.072	6.5	23.40	0.0000

Effect	Dose	_dose	Estimate	StdErr	df	tValue	Adjp
dose	1	2	−0.072	0.045	8.3	−1.62	0.4310
dose	1	3	−0.174	0.045	8.4	−3.87	0.0171
dose	1	4	−0.250	0.045	8.5	−5.58	0.0025
dose	2	3	−0.102	0.045	8.3	−2.27	0.1802
dose	2	4	−0.178	0.045	8.4	−3.98	0.0148
dose	3	4	−0.076	0.045	8.3	−1.71	0.3854

TABLE 14.27
Estimate Statements to Evaluate the Linear, Quadratic, and Cubic Trends across the Four Equally Spaced Dose Levels

```
estimate 'linear' dose -3 -1 1 3;
estimate 'quad' dose 1 -1 -1 1;
estimate 'cubic' dose 1 -3 3 -1;
```

Label	Estimate	StdErr	DF	tValue	Probt
linear	0.8520	0.1420	8.5	6.00	0.0002
quad	0.0041	0.0631	8.3	0.07	0.9495
cubic	0.0551	0.1413	8.3	0.39	0.7064

the LOC*DOSE interaction would be the error term selected to make comparisons among the adjusted means.

This example demonstrates the process of selecting a model to fit a mixed effects treatment structure. First the form of the residual variances was selected, followed by selecting the form of the variances and covariances associated with the covariate part of the model. Finally the form of the fixed effects of the covariate part of the model was simplified and then comparisons among the levels of the fixed effects in the treatment structure were made using this final model.

14.11 SUMMARY

The problem of unequal variances in the analysis of covariance model is discussed where several strategies are developed. First, the simplest form of the variance structure should be determined to improve the power of the resulting tests involving the fixed effects. Second, once the form of the variances is determined, then the usual strategies of analysis of covariance described in previous chapters are applied to the fixed effects of the model. The examples demonstrate various aspects of the developed methods. Larholt and Sampson (1995) discuss the estimation of the point of intersection of two regression lines when the variances are not equal. They provide a confidence interval about the point of intersection and a simultaneous confidence region about the difference between the two regression lines. Using the process described in this and the previous chapter, the results of Larholt and Sampson (1995) can be extended to models involving random effects in either the design structure or the treatment structure.

REFERENCES

Larholt, K. M. and Sampson, A. R. (1995) Effects of Heteroscedasticity Upon Certain Analyses When the Regression Lines Are Not Parallel, *Biometrics* 51, 731-737.

Littell, R., Milliken, G. A., Stroup, W., and Wolfinger, R. (1996), *SAS System for Mixed Models*. SAS Institute Inc., Cary, NC.

Milliken, G. A. and Johnson, D. E. (1992) *Analysis of Messy Data, Volume I: Design Experiments*, Chapman & Hall, London.

SAS Institute Inc. (1999) *SAS/STAT® Users Guide, Version 8,* SAS Institute Inc., Cary, NC.

Westfall, P. H., Tobias, R. D., Rom D., Wolfinger, R. D., and Hochberg, Y. (1999) *Multiple Comparisons and Multiple Tests Using the SAS® System*, SAS Institute Inc., Cary, NC.

EXERCISES

EXERCISE 14.1: Use Levene's procedure to test for equality of variances for the data in Problem 3.1. Use the unequal variance model to compare the treatments (even though the variances may not be unequal).

EXERCISE 14.2: Use Levene's procedure to test for equality of variances for the data in Problem 5.1. Use the unequal variance model to compare the treatments (even though the variances may not be unequal).

EXERCISE 14.3: Use the following data set to carry out the analysis of covariance strategy of finding the simplest covariance part of the model and then the simplest form of the covariate part of the model before comparing the treatments.

Data for Exercise 14.3

Treatment 1		Treatment 2		Treatment 3		Treatment 4		Treatment 5	
Y	X	Y	X	X	Y	X	Y	X	Y
59	23	60	26	72	37	63	11	70	22
64	31	57	11	62	30	79	29	73	27
60	31	63	20	70	25	72	17	72	29
53	15	60	21	68	38	78	32	72	34
53	17	61	30	66	20	64	13	73	32
61	11	59	17	61	16	65	18	72	32
59	20	43	9	70	36			70	11
62	15			60	27				

EXERCISE 14.4: Build a response surface model for the roughness data in Section 14.7. Use an unequal variance model with the variance structure selected in Section 14.7.

15 Analysis of Covariance for Split-Plot and Strip-Plot Design Structures

15.1 INTRODUCTION

In the previous chapters, discussions were centered around specifying the model, specifying the variances and covariances for the covariate part of the model, and modeling the variances of the residual part of the model. This chapter involves the split-plot and strip-plot type of design structures and the issues are similar to those in the previous chapters, except the treatment structures considered are factorial arrangements with at least two factors. In Chapter 16, the repeated measures design structures and the additional concerns about modeling the correlation structure of the repeated measures are discussed. Repeated measures and split-plot design structures are similar in structure except for the repeated measures design where the levels of one or more of the factors cannot be randomly assigned to the experimental units, thus the name repeated measures (Milliken and Johnson, 1992). This chapter starts with a description of the split-plot design structure and a discussion of the different types of models that provide estimates of the fixed effects parameters.

Repeated measures or split-plot designs involve more than one size of experimental unit (Milliken and Johnson, 1992, Chapter 5) and the covariate can be measured on any of the different sizes of experimental units. The experimental unit on which the covariate is measured must be considered when constructing the covariance model in order to understand the sources of the information used in an appropriate analysis. This chapter presents an introduction to the analysis of covariance by describing three different cases for split-plot designs with two sizes of experimental units where whole-plot is used to describe the large size experimental unit and sub-plot is used to describe the small size experimental unit. Specifically, the situations discussed are a model with the covariate measured on the whole plot or larger sized experimental unit, a model with the covariate measured on the subplot or smaller sized experimental unit, and a model with two covariates where one is measured on the whole plot and the other is measured on the subplot. There are two levels of analysis: the whole plot analysis and the sub-plot analysis. The discussion in the next section explores the two types of analyses on the covariate part of the model detailing the cases for different assumptions on the form of the slopes. The chapter ends with detailed analyses of four examples.

A word of caution is that you must be sure the values of the covariates measured on smaller sized experimental units are not affected by the applications of the treatments on the larger sized experimental units. This problem exists when a repeated measures design is used and the smaller sized experimental unit is a time interval. It is very easy to incorrectly select variables as possible covariates that have been affected by the application of treatments to the larger sized experimental units.

15.2 SOME CONCEPTS

The analysis of covariance models for split-plot design structures take on the relevant characteristics pertaining to the analysis covariance models which were discussed in the previous chapters and the problems encountered when analyzing split-plot designs that are discussed in Chapters 5 and 24 of Milliken and Johnson (1992). In particular, the concept of size of experimental unit must be used to specify an appropriate model and is used in understanding the analysis. For example, if an experimental design involves three sizes of experimental units, it is possible for a covariate to be measured on any of the three sizes of experimental units. Thus, in the terminology of a split-split plot experiment, the covariate could be measured on a whole plot (largest experimental unit), or a covariate could be measured on a subplot (middle size experimental unit), or a covariate could be measured on a sub-sub-plot (smallest size experimental unit).

As described in Milliken and Johnson (1992), there is an analysis for each size of experimental unit since the model can be partitioned into groups of terms corresponding to each size of experimental unit. The situations discussed in this chapter involve two different sizes of experimental units, but the discussion can be extended to designs involving more than two sizes of experimental units. Kempthorne (1952) discusses the split-plot model where two covariates with corresponding slopes were constructed with one covariate and slope for the whole plot part of the model and one covariate and slope for the sub-plot part of the model. The models described here will have just one slope for each covariate since there is no distinction between the whole plot information and the sub-plot information about the slope parameter.

The general thought is the covariate will have an affect on the analysis of the size of experimental unit on which the covariate is measured as well as on larger sizes of experimental units. However, because of the possibility of heterogeneous slopes, the covariate can have an affect on the analysis of all sizes of experimental units. Even when the slopes are homogeneous, the covariate will have an affect on all sizes of experimental units larger than the one on which the covariate is measured.

When constructing an appropriate covariance model, the covariate term is included with the segment of the model corresponding to the size of the experimental unit on which the covariate is measured. Expressing the model in this manner enables the analyst to determine the number of degrees of freedom removed by estimating the covariate parameters from each of the error terms. The following three situations illustrate these concepts.

15.3 COVARIATE MEASURED ON THE WHOLE PLOT OR LARGE SIZE OF EXPERIMENTAL UNIT

A split-plot design involves a factorial arrangement treatment structure and an incomplete block design structure. What distinguishes the split-plot and strip-plot

Analysis of Covariance for Split-Plot and Strip-Plot Design Structures

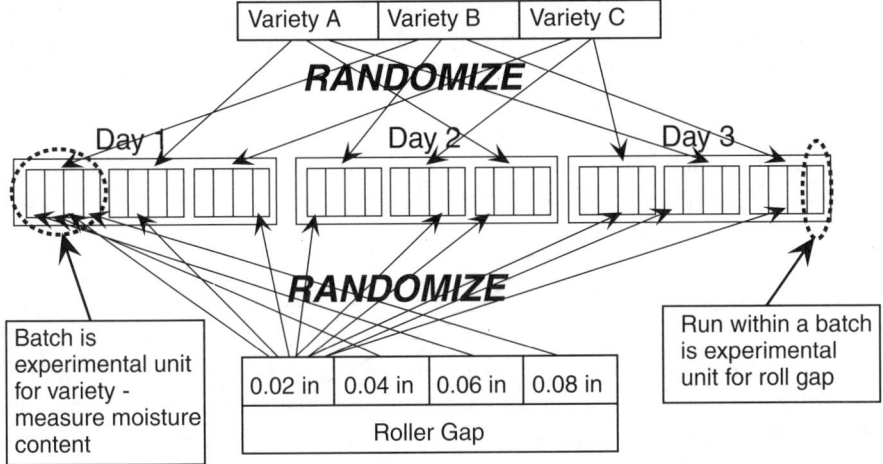

Randomization Scheme for Flour Milling Experiment

FIGURE 15.1 Graphical representation of randomization scheme for flour milling experiment.

design from usual incomplete block designs is that the main effects of one or more of the factors in the treatment structure are confounded with blocks. A flour miller wanted to study the effect that four types of roller gaps have on a flour mill during the first break (first grinding step of many) as to the amount of flour produced from three varieties of wheat. The process is to take a batch of wheat from one variety and run that batch of wheat with each of the roll gaps. The order of the roll gaps was randomized for each batch of wheat from the selected variety and the order in which the varieties were run was randomized. Twelve runs could be accomplished during one day, so the process was repeated on three different days; thus day is a blocking factor. The moisture content of the wheat being milled can have an affect on the flour production at each break, so the moisture content of each of the batches of wheat from a given variety was measured before starting the milling test. Figure 15.1 is a display of a portion of the randomization process. During each day, three batches of wheat were milled — one from each variety. The order of the three batches was randomly assigned to three varieties within each day. The batch is the large experimental unit or the whole plot, i.e., batch is the experimental unit for the levels of variety, and the moisture content of each batch was measured as a possible covariate. The day forms a block of three batches; thus the batch design structure is a randomized complete block. Next, within each batch, four runs were made where the levels of roll gap were randomly assigned to the run order. The run within a batch is the small size experimental unit or subplot, i.e., the experimental unit for the levels of roll gap. Since each level of roll gap is observed on a run within each batch, the batch is a block with 4 of the possible 12 treatment combinations. The design structure for the run within a variety is a randomized complete block design with blocks of size four. This experiment involves a two-way treatment structure in a incomplete block design structure since there are four runs within a batch (or block). By the way the levels of the treatment combinations are assigned to the runs within each batch, two sizes of experimental units are generated. The batch is the experimental unit to which the varieties were randomly assigned and the run within a batch is the experimental unit to which the roll gaps were randomly assigned. Since this is an incomplete block design structure, there is information about the fixed effects from within the blocks (batch) as well as from between the blocks as described in Chapter 10. The model without a covariate that describes flour yield (percent of total run by weight) is given by

$$FL_{ijk} = \mu_{ik} + d_j + a_{ij} + \varepsilon_{ijk} \quad i = 1, 2, 3, \quad j = 1, 2, 3, \quad k = 1, 2, 3, 4 \quad (15.1)$$

where μ_{ik} denotes the expected response of the ith variety milled with the kth roll gap, d_j denotes the random effect of the jth day or block which are assumed to be distributed iid N(0, σ_d^2), a_{ij} denotes the random effect of batch j assigned to variety i, which are assumed to be distributed iid N(0, σ_a^2), and ε_{ijk} denotes the random effect of the kth run of the roll gap of the jth batch of the ith variety, which are assumed to be distributed iid N(0, σ_ε^2). By expressing μ_{ik} as

$$\mu_{ik} = \mu + \sqrt{}_i + \rho_k + \tau_k + (\upsilon\rho)_{ik},$$

the flour yield model can be displayed in terms of the two sizes of experimental units as

$$\begin{aligned} FL_{ijk} = \mu + v_i + d_j + a_{ij} \} &\quad \text{batch part of model} \\ + \rho_k + (\upsilon\rho)_{ik} + \varepsilon_{ijk} \} &\quad \text{run part of model} \end{aligned} \quad (15.2)$$

where i = 1, 2, 3, j = 1, 2, 3, and k = 1, 2, 3, 4. Since the covariate, moisture content, was measured on each batch, the covariate part of the model is to be included with the batch part of Model 15.2. The analysis of covariance model which allows for different slope parameters for each variety and roll gap combination is given by

$$\begin{aligned} FL_{ijk} = \mu + v_i + \beta_{ik} M_{ij} + d_j + a_{ij} \} &\quad \text{batch part of model} \\ + \rho_k + (\upsilon\rho)_{ik} + \varepsilon_{ijk} \} &\quad \text{run part of model} \end{aligned} \quad (15.3)$$

where i = 1, 2, 3, j = 1, 2, 3, and k = 1, 2, 3, 4, M_{ij} denotes the moisture content of batch j from variety i, and β_{ik} denotes the slope parameter of the regression line for variety i and roll gap k. This model is a little different than the usual split-plot models in that the large size of experimental unit part of the model includes a term with a subplot subscript k. This model is similar to the randomized complete block analysis of covariance model where the covariate is the same for all experimental units within a block (see Chapter 11). The difference here is that one set of blocks (whole plots or batches) is assigned to variety 1 and another set is assigned to variety 2, etc.; thus there can be one set of slopes for the roll gaps at variety A, another set of slopes for the roll gaps for variety B, and still another set of slopes for the roll gaps at variety C. By expressing the slopes of this model as a two-way factorial effects model, as was done for the two-way treatment structures in Chapter 5, Model 15.3 can be written as

$$FL_{ijk} = \mu + v_i + \theta M_{ij} + \phi_i M_{ij} + d_j + a_{ij} \} \quad \text{batch part of model}$$

$$+ \rho_k + (v\rho)_{ik} + \gamma_k M_{ij} + \delta_{ik} M_{ij} + \varepsilon_{ijk} \} \quad \text{run part of model}$$

(15.4)

Since this is an incomplete block design structure, there is information about the fixed effects from within the blocks (batches) as well as from between the blocks as described in Chapter 10, where i = 1, 2, 3, j = 1, 2, 3, and k = 1, 2, 3, 4, and $\beta_{ik} = \theta + \phi_i + \gamma_k + \delta_{ik}$. The batch part of Model 15.4 now depends only on subscripts i and j. The next example illustrates the case when a covariate is measured on the small size of the experimental unit, the time interval.

15.4 COVARIATE IS MEASURED ON THE SMALL SIZE OF EXPERIMENTAL UNIT

This split-plot example evolved when a baker wanted to study the effect of baking temperature on the diameter (mm) of baked cookies made from prepacked refrigerator cookies. Three cookie types were selected for the experiment: sugar, peanut butter, and chocolate chip. The recommended baking temperature is 350°F, so the baker selected three level of temperature for the study: 300, 350, and 400°F. Four replications of each of the treatment combinations were obtained. The process is to cut a slice of the cookie dough from a package of each of the cookie types, place all three cookies on a baking tray, and place the tray in the oven where the oven was set to one of the three temperatures. The diameter of the cookie after baking depends on the thickness of the slice of cookie dough. The directions on the wrappers are to make 1/4-inch or 6.25-mm slices, but it is not easy to cut the slices to exactly 6.25 mm. Thus the thickness of the slices of cookie dough was measured as a possible covariate. The oven to which the temperature was assigned is the experimental unit for the levels of temperature or the whole plot. The slice of cookie dough is the experimental unit for the levels of cookie type or sub-plot. For this example, it is assumed that position on the baking tray is not important since there were just three cookies being baked at the same time (position on the tray could be an important source of variability and should be investigated). The process for conducting the experiment was to randomly select one of the temperatures, set the oven to that temperature, and randomly assign one cookie of each cookie type to the three positions on the baking tray when the tray is placed into the oven. The total time required to conduct the experiment was such that no blocking on the levels of temperature was required; thus the whole plot or oven design structure is a completely randomized

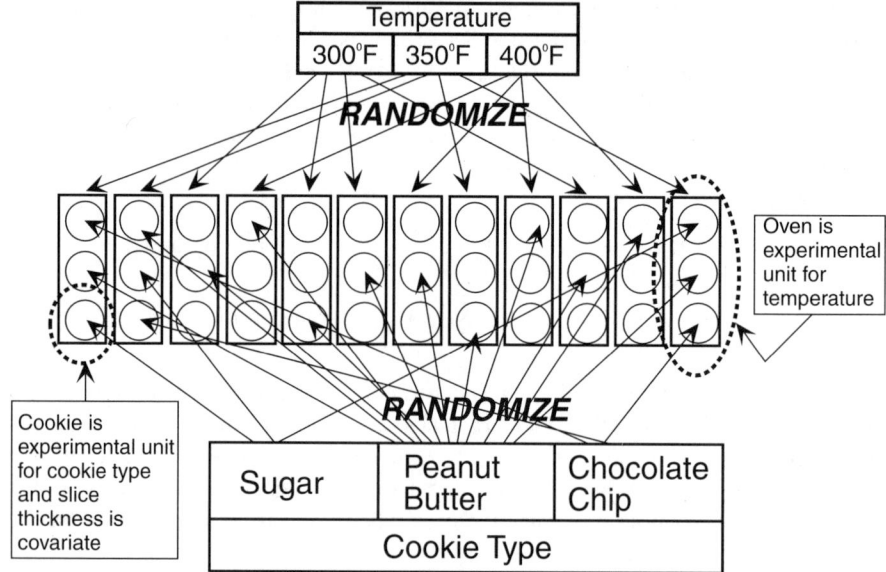

FIGURE 15.2 Graphical representation of the randomization scheme for the baking cookie experiment.

design. The randomization scheme for assigning temperatures to ovens (runs) and cookies to positions within the oven is shown in Figure 15.2. The randomization of the cookie types to the oven positions has been shown for only three ovens and for all ovens for peanut butter; thus the drawing is not complete because inserting all of the lines would clutter the picture. The oven is a blocking factor for the levels of cookie type; thus the cookie design structure is a randomized complete block within each level of temperature. Each oven is an incomplete block since it consists of three of the possible nine treatment combinations. Since this is an incomplete block design structure, there is information about the fixed effects from within the blocks (ovens) as well as from between the blocks as described in Chapter 10. The thickness of each slice of cookie dough is a covariate that is measured on the small sized experimental unit or the sub-plot. An appropriate analysis of covariance model is

$$D_{ijk} = \mu + \rho_i + a_{ij} \} \text{ oven part of model}$$
$$+ \tau_k + (\rho\tau)_{ik} + \gamma_{ik} T_{ijk} + \varepsilon_{ijk} \} \text{ cookie part of the model} \quad (15.5)$$

where $i = 1, 2, 3$, $j = 1, 2, 3, 4$, and $k = 1, 2, 3$, T_{ijk} is the thickness (mm) of the jth cookie dough slice of cookie dough type k assigned to temperature i, and γ_{ik} is the slope for the regression line for the covariate thickness corresponding to temperature i and cookie dough type k. By expressing the slopes of this model as factorial effects,

Analysis of Covariance for Split-Plot and Strip-Plot Design Structures

as was done for the two-way treatment structures in Chapter 5, Model 15.5 can be written as

$$D_{ijk} = \mu + \tau_i + \theta T_{ijk} + \phi_i T_{ijk} d_j + a_{ij} \} \quad \text{oven part of model}$$

$$+ \rho_k + (\tau\rho)_{ik} + \gamma_k T_{ijk} + \delta_{ik} T_{ijk} + \varepsilon_{ijk} \} \quad \text{cookie part of the model} \quad (15.6)$$

where $i = 1, 2, 3$, $j = 1, 2, 3, 4$, and $k = 1, 2, 3$, and $\beta_{ik} = \theta + \phi_i + \gamma_k + \delta_{ik}$. When the covariate is measured on the sub-plot, the whole plot part of the model involves subscripts associated with the sub-plot experimental units. Since this is an incomplete block design structure, there is information about the fixed effects from within the blocks (ovens) as well as from between the blocks as described in Chapter 10.

A common slope model can be expressed as

$$y_{ijk} = \mu_{ik} + \beta X_{ijk} + a_{ij} + \varepsilon_{ijk}$$

$$i = 1, 2, 3, \quad j = 1, 2, \ldots, 5, \quad k = 1, 2, 3 \quad (15.7)$$

where μ_{ik} is the intercept and β is the common slope of the regression lines. The within-oven contrasts provide information about the common slope as well as does the between-oven part of the model. The contrasts of cookie types within the jth oven assigned the ith temperature are

$$\sum_{k=1}^{r} c_k y_{ijk} = \sum_{k=1}^{r} c_k \mu_{ik} + \beta \sum_{k=1}^{r} c_k x_{ijk} + \sum_{k=1}^{r} c_k \varepsilon_{ijk},$$

which is a function of β. The oven model is $\bar{y}_{ij.} = \bar{\mu}_{i.} + \beta \bar{x}_{ij.} + e_{ij}^*$, a common slope model for a one-way treatment structure where $e_{ij}^* = a_{ij} + \bar{\varepsilon}_{ij.} \sim$ iid $N(0, \sigma_a^2 + \sigma_\varepsilon^2/3)$. This oven model provides estimates of $\bar{\mu}_{i.}$ and β and information for tests of hypotheses about the $\bar{\mu}_{i.}$, including $H_0: \bar{\mu}_{1i.} = \bar{\mu}_{2i.} = \bar{\mu}_{3i.}$ vs. H_a: (not H_0:). The variances associated with these oven comparisons are functions of $\sigma_a^2 + \sigma_\varepsilon^2/3$.

The contrasts of the observations made on the cookies within an oven provide information about $(\mu_{ik} - \bar{\mu}_{i.})$ and β. The variances of these within oven comparisons are functions of σ_ε^2 only. These parameters provide information about the cookie types, $(\bar{\mu}_{.k} - \bar{\mu}_{..})$, and interaction comparisons between the levels of temperature and the cookie types $(\mu_{ik} - \bar{\mu}_{i.} - \bar{\mu}_{.k} + \bar{\mu}_{..})$. Unlike the model in Section 15.7, both models provide an estimate of the common slope β, and the information from both sources should be combined to provide a better estimate of the common slope. In order to provide estimates of the individual intercepts (μ_{ik}), information from both models should be combined. The estimates of the intercepts are obtained from $\mu_{ik} = [\bar{\mu}_{i.}] + (\mu_{ik} - \bar{\mu}_{i.})$. The quantities in [ ] are obtained from the oven or large size experimental unit model and the quantities in ( ) are obtained from the cookie or small size experimental units within the large size experimental units. The variance of a combined estimate is the sum of the variances of the two parts, e.g., $\text{Var}(\hat{\mu}_{ik}) = \text{Var}(\text{estimate of } \bar{\mu}_{i.}) + \text{Var}(\text{estimate of } (\mu_{ik} - \bar{\mu}_{i.}))$. Kempthorne (1952) used the model $y_{ijk} = \mu_{ik} + \beta_b \bar{x}_{ij.} + \beta_w (x_{ijk} - \bar{x}_{ij.}) + a_{ij} + \varepsilon_{ijk}$ where $\bar{x}_{ij.}$ is the whole plot covariate and $(x_{ijk} - \bar{x}_{ij.})$ is the sub-plot covariate. He did not combine β_b and β_w into a single combined estimate of β.

The next example combines the examples in Sections 15.3 and 15.4 by including two covariates where one covariate is measured on the larger experimental unit and one measured on the smaller experimental unit.

15.5 COVARIATE IS MEASURED ON THE LARGE SIZE OF EXPERIMENTAL UNIT AND A COVARIATE IS MEASURED ON THE SMALL SIZE OF EXPERIMENTAL UNIT

A school district set up a study to evaluate three methods of teaching mathematics: method I was lecture and recitation, method II was lecture and computer activities, and method III was lecture, recitation, and computer activities. There was a concern that the method of teaching could have different effects on male and female students in the class. Thus, sex of student was also considered to be part of the treatment structure with the resulting treatment structure being a two-way with levels of teaching method by the two sexes. The school district randomly selected nine classes of seventh graders (with at least ten female and ten male students) and selected nine teachers to carry out the experiment. Three teachers were trained to deliver each of the teaching methods. The nine teachers were randomly assigned to a class (Figure 15.3). There was concern that there was variability among the students as

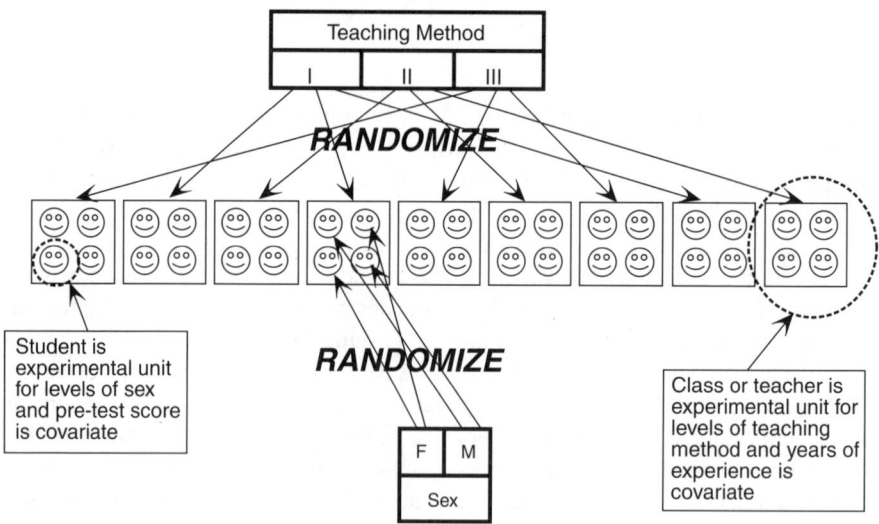

Randomization Scheme for Teaching Method Study

FIGURE 15.3 Graphical representation of the randomization for the teaching method study.

to their knowledge of mathematics at the beginning of the study; thus, a pre-test was administered to the students before starting the study to provide a gauge of initial knowledge. There was variability in the years of teaching experience among the nine teachers and there was a concern that the years of experience might have an influence on the ability to deliver the teaching method. Thus the number of years of experience was selected to be a possible covariate associated with the teacher (or class) and the pre-test score was selected as a possible covariate associated with the student. An appropriate analysis covariance model is given by

$$Sc_{ijkm} = \mu + \rho_i + \beta_{ik} Yrs_{ij} + a_{ij} \} \quad \text{teacher or class part of model}$$
$$+ \tau_k + (\rho\tau)_{ik} + \gamma_{ik} Pre_{ijkm} + \varepsilon_{ijkm} \} \quad \text{student part of the model} \quad (15.8)$$

where Sc_{ijkm} and Pre_{ijkm} are the test scores and the pre-test scores for the mth student of the kth sex in the jth class assigned to the ith method, Yrs_{ij} is the number of years of experience of the jth teacher assigned to the ith teaching method, β_{ik} is the slope for the number of years of experience, and γ_{ik} is the slope for the pre-test scores for the ith method and k^{th} sex. The slopes can be expressed as factorial effects as $\beta_{ik} = \theta + \phi_i + \lambda_k + \omega_{ik}$ and $\gamma_{ik} = \delta + \kappa_i + \zeta_k + \upsilon_{ik}$; then Model 15.8 becomes

$$Sc_{ijkm} = \mu + \rho_i + \theta\, Yrs_{ij} + \phi_i Yrs_{ij} + \delta\, Pre_{ijkm} + \kappa_i\, Pre_{ijkm} + a_{ij} \} \quad \text{teacher}$$
$$+ \tau_k + (\rho\tau)_{ik} + \lambda_k Yrs_{ij} + \omega_{ik} Yrs_{ij} + \zeta_k Pre_{ijkm} + \upsilon_{ik} Pre_{ijkm} + \varepsilon_{ijkm} \} \quad \text{student} \quad (15.9)$$

The teacher part of the model involves the covariate measured on the student and the student part of the model involves the covariate measured on the teacher. This relationship can change if the forms of the slopes can be simplified — a topic discussed in the next section.

The principles described in the above three examples can easily be used to construct analysis of covariance models for other more complex repeated measures or split-plot experimental designs. The analyst must be careful because the analysis of such models using existing statistical packages is not straightforward. In fact, blindly using computer packages will usually produce incorrect results. If you have access to a statistical package that carries out a mixed models analysis using the mixed models equations as described in Chapter 13, you will be more successful in carrying out an appropriate analysis than by using a regression based statistical package. The next section explores these models by looking at between whole plot comparisons and within whole plot comparisons and the affects on the covariate part of the model.

15.6 GENERAL REPRESENTATION OF THE COVARIATE PART OF THE MODEL

The covariance model for a two-way treatment structure in a split-plot design where the whole plot design structure is a completely randomized design with a single covariate and equal numbers of observations per treatment combination can be expressed as

$$y_{ijk} = \mu_{ik} + \beta_{ik} x_{ijk} + a_{ij} + \varepsilon_{ijk}, \quad i=1, 2, \ldots, a, \quad j=i, 2, \ldots, n, \quad k=1, 2, \ldots, t$$
$$a_{ij} \sim iid\, N(0, \sigma_a^2) \text{ and } \varepsilon_{ijk} \sim iid\, N(0, \sigma_\varepsilon^2)$$
(15.10)

where a_{ij} represents the whole plot error and ε_{ijk} represents the sub-plot error. The model has an intercept and a slope for each of the treatment combinations. The whole plots are blocks of size k sub-plots; thus there are two sources of information that can be extracted: between blocks or whole plots information and within blocks or whole plots information. The between whole plot or between large size experimental unit information is obtained from the whole plot or block totals or means as

$$\bar{y}_{ij\cdot} = \sum_{k=1}^{t} y_{ijk}/k \qquad (15.11)$$

and within a large size experimental unit or a whole plot or a block information is contained in contrasts among the small size experimental units within each large size experimental unit as

$$\sum_{k=1}^{t} c_k y_{ijk} \text{ where } \sum_{k=1}^{t} c_k = 0. \qquad (15.12)$$

A model can be constructed for the whole plot means providing the large size experimental unit model as

$$\bar{y}_{ij} = \bar{\mu}_{i\cdot} + \frac{1}{t}\sum_{k=1}^{t} \beta_{ik} x_{ijk} + e_{ij}^* \text{ where } e_{ij}^* = a_{ij} + \bar{\varepsilon}_{ij\cdot} \text{ with } Var(e_{ij}^*) = \frac{(\sigma_\varepsilon^2 + t\sigma_a^2)}{t} \qquad (15.13)$$

A contrast of the small size experimental units (sub-plots) within the ij^{th} large size experimental unit provides the model

$$\sum_{k=1}^{t} c_k y_{ijk} = \sum_{k=1}^{t} c_k \beta_{ik} x_{ikj} + \sum_{k=1}^{t} c_k \varepsilon_{ijk} \qquad (15.14)$$

which has variance

$$Var\left(\sum_{k=1}^{t} c_k y_{ijk}\right) = \sigma_\varepsilon^2 \sum_{k=1}^{t} c_k^2.$$

The expression in Equation 15.14 is free of the large size experimental unit effects, a_{ij}.

Analysis of Covariance for Split-Plot and Strip-Plot Design Structures

Models 15.13 and 15.14 show that information about the covariate can be obtained from both models. There are conditions that provide simplifications of the covariate relationship that need to be explored in order to provide a better understanding of the appropriate approach to the analysis of the model.

There are two possible cases for the value of the covariate as described in Sections 15.3 and 15.4: (1) the covariate can be measured on the large size experimental units, i.e., the covariate can be expressed as $x_{ijk} = x_{ij}$, or (2) the covariate can be measured on the small size of experimental unit, represented by x_{ijk}.

Since the treatment structure involves a two-way (see Chapter 5), there are five possible forms for which the slope can enter into the model: (1) common slope, $\beta_{ik} = \beta$, (2) slope is a function of the levels of the whole plot treatment as $\beta_{ik} = \beta_i$, (3) slope is a function of the levels of the sub-plot treatment as $\beta_{ik} = \beta_k$, (4) slope is an additive function of the levels of both treatments as $\beta_{ik} = \beta_{Bi} + \beta_{Wk}$, where β_{Bi} denotes that part of the slope which is a function of the whole plot treatment (from between block comparisons) and β_{Wk} denotes that part of the slope which is a function of the sub-plot treatment (within block comparisons), or (5) slope is a function of the levels of both treatments as β_{ik}.

To help determine the proper form of the models, the covariate part of the model can be expressed as

$$\beta_{ik} x_{ijk} = \bar{\beta}_{i.} \bar{x}_{ij.} + (\beta_{ik} - \bar{\beta}_{i.})(x_{ijk} - \bar{x}_{ij.}) + (\beta_{ik} - \bar{\beta}_{i.})\bar{x}_{ij.} + \bar{\beta}_{i.}(x_{ijk} - \bar{x}_{ij.}) \quad (15.15)$$

The next few subsections describe the affects the above ten (five forms for the slopes by two types of covariates) cases have on the form of the covariance model, and thus how the model should be used to obtain an appropriate analysis.

15.6.1 COVARIATE MEASURED ON LARGE SIZE OF EXPERIMENTAL UNIT

When the covariate is measured on the large size of experimental unit, the $x_{ijk} = x_{ij}$, $k = 1, 2, \ldots, t$ and thus $\bar{x}_{ij} = x_{ij}$. The covariate part of the model in Equation 15.15 becomes

$$\beta_{ik} x_{ijk} = \bar{\beta}_{i.} x_{ij} + (\beta_{ik} - \bar{\beta}_{i.}) x_{ij} \quad (15.16)$$

Thus, the large size experimental unit model becomes

$$\bar{y}_{ij.} = \bar{\mu}_{i.} + \bar{\beta}_{i.} x_{ij} + e^*_{ij} \text{ since } \sum_{k=1}^{t}(\beta_{ik} - \bar{\beta}_{i.}) = 0 \quad (15.17)$$

Model 15.17 is an analysis of covariance model for a one-way treatment structure in a completely randomized design structure with unequal slopes as described in Chapter 2. Hence, the between large size experimental analysis provides estimates of $\bar{\beta}_{i.}$ and tests of hypotheses about the $\bar{\beta}_{i.}$ including H_0: $\bar{\beta}_{1.} = \bar{\beta}_{2.} = \ldots = \bar{\beta}_{a.}$ vs.

H_a: (not H_0). Contrasts of the small size experimental unit observations within a large size experimental unit are

$$\sum_{k=1}^{t} c_k \bar{y}_{ijk} = \sum_{k=1}^{t} c_k (\mu_{ik} - \bar{\mu}_{i.}) + \sum_{k=1}^{t} c_k (\beta_{ik} - \bar{\beta}_{i.}) x_{ij} + \sum_{k=1}^{t} c_k \varepsilon_{ijk}, \qquad (15.18)$$

thus the within large size experimental unit analysis provides estimates of the $\beta_{ik} - \bar{\beta}_{i.}$ which includes the small size experimental unit treatment slope main effects, $\bar{\beta}_{.k} - \bar{\beta}_{..}$ and the interaction effects of the slopes as, $\beta_{ik} - \bar{\beta}_{ik} - \bar{\beta}_{i.} - \bar{\beta}_{.k} + \bar{\beta}_{..}$. The within large size experimental unit analysis provides tests for parallelism of the regression lines for the sub-plot treatments and of the no interaction between the levels of the whole plot and the levels of the sub-plot treatments hypothesis for the slopes.

If the model has a common slope as $\beta_{ik} = \beta$, then Model 15.17 becomes

$$\bar{y}_{ij.} = \bar{\mu}_{i.} + \beta x_{ij} + e_{ij}^* \qquad (15.19)$$

a common slope analysis of covariance model for a one-way treatment structure in a completely randomized design structure. The within large size of experimental unit comparison in Equation 15.18 becomes

$$\sum_{k=1}^{t} c_k \bar{y}_{ijk} = \sum_{k=1}^{t} c_k (\mu_{ik} - \bar{\mu}_{i.}) + \sum_{k=1}^{t} c_k \varepsilon_{ijk}, \qquad (15.20)$$

which does not depend on the values of the covariate. Thus when the model has a whole plot covariate and a common slope, the within whole plot analysis or the between subplot analysis does not depend on the values of covariate.

If the model has slopes that are different for each level of the large size of experimental unit treatment as, $\beta_{ik} = \beta_i$, then Model 15.16 becomes

$$\bar{y}_{ij.} = \bar{\mu}_{i.} + \beta_i x_{ij} + e_{ij}^* \qquad (15.21)$$

an unequal slope analysis of covariance model for a one-way treatment structure in a completely randomized design structure. The within large size of experimental unit comparisons in Equation 15.18 with $\beta_{ik} = \beta_i$ are the same as those in Equation 15.12. Again, the within large sample size experimental unit analysis is not effected by the values of the covariate and all of the covariate information is contained in the whole plot model.

If the slopes are different for each level of the small size of experimental unit treatment, then Model 15.17 becomes

$$\bar{y}_{ij.} = \bar{\mu}_{i.} + \bar{\beta}_. x_{ij} + e_{ij}^*, \qquad (15.22)$$

a common slope analysis of covariance model for a one-way treatment structure in a completely randomized design structure. The within large size of experimental unit comparisons are

$$\sum_{k=1}^{t} c_k y_{ijk} = \sum_{k=1}^{t} c_k (\mu_{ik} - \bar{\mu}_{i.}) + x_{ij} \sum_{i=1}^{t} c_k (\beta_k - \bar{\beta}.) + \sum_{k=1}^{t} c_k \varepsilon_{ijk}. \quad (15.23)$$

The small size of experimental unit analysis provides information about $\beta_k - \bar{\beta}.$ and a test of the hypothesis H_0: $\beta_1 = \beta_2 = \ldots = \beta_t$ vs. H_a: (not H_o:). Model 15.22 provides information about $\bar{\beta}.$ and a test to the hypothesis, H_o: $\bar{\beta}. = 0$ vs. H_a: $\bar{\beta}. \neq 0$. The estimate of the slope for the kth level of the sub-plot treatment, β_k, is obtained by combining the estimates of $\beta_k - \bar{\beta}.$ from the subplot analysis with the estimate of $\bar{\beta}.$ from the whole plot analysis, i.e., $\beta_k = \bar{\beta}. + (\beta_k - \bar{\beta}.)$.

When the slope is an additive function of the levels of both treatments as $\beta_{ik} = \beta_{Bi} + \beta_{Wk}$, Model 15.17 becomes

$$\bar{y}_{ij.} = \bar{\mu}_{i.} + (\beta_{Bi} + \bar{\beta}_{W.}) x_{ij} + e_{ij}^* \quad (15.24)$$

an unequal slope analysis of covariance model for a one way treatment structure in a completely randomized design structure where the slope for the ith level of the whole plot treatment is $\beta_{Bi} + \bar{\beta}_{W.}$. The information from the model in Equation 15.24 can be used to estimate $\beta_{Bi} + \bar{\beta}_{W.}$ and test the hypotheses H_{01}: $\beta_{B1} = \beta_{B2} = \ldots = \beta_{Ba}$ vs. H_{a1}: (not H_{01}:). The within whole plot comparisons are

$$\sum_{k=1}^{t} c_k y_{ijk} = \sum_{k=1}^{t} c_k \mu_{ik} + x_{ij} \sum_{k=1}^{t} c_k \beta_{Wk} + \sum_{k=1}^{t} c_k \varepsilon_{ijk}. \quad (15.25)$$

The within whole plot analysis provides estimates of $\beta_{Wk} - \bar{\beta}_{W.}$ and a test of the hypothesis H_0: $\beta_{W1} = \beta_{W2} = \ldots = \beta_{Wt}$ vs. H_a: (not H_{01}:). The estimates of the slopes $\beta_{Bi} + \beta_{Wk}$ are obtained by combining the estimates of $\beta_{Bi} + \bar{\beta}_{W.}$ from the whole plot analysis with the estimates of $\beta_{Wk} - \bar{\beta}_{W.}$ from the sub-plot analysis.

Finally, if the slopes depend on the levels of both treatments, the whole plot model is as in Equation 15.13 and the subplot comparisons are as in Equation 15.14. As in the analysis of the randomized complete block, information about the slopes is obtained from the whole plot model and from the subplot analysis and these estimates should be combined in order to obtain the best estimates of the slopes of the model. This combining process can be accomplished by a mixed models analysis that uses the mixed models equations to obtain the estimates of the fixed effects.

15.6.2 Covariate Measured on the Small Size of Experimental Units

When the covariate is measured on the small size of experimental unit, the covariate part of the model can be simplified by making assumptions on the form of the slope.

If the common slope model, $\beta_{ik} = \beta$, is appropriate, the large size experimental unit model of Equation 15.13 is

$$\bar{y}_{ij\cdot} = \bar{\mu}_{i\cdot} + \beta \bar{x}_{ij\cdot} + e^*_{ij}, \qquad (15.26)$$

a common slope analysis of covariance model for an one-way treatment structure in a completely randomized design structure, and the comparisons of small size experimental units within a large size of experimental unit are

$$\sum_{k=1}^{t} c_k y_{ijk} = \sum_{k=1}^{t} c_k \mu_{ik} + \beta \sum_{k=1}^{t} c_k \left(x_{ijk} - \bar{x}_{ij\cdot} \right) + \sum_{k=1}^{t} c_k \varepsilon_{ijk}. \qquad (15.27)$$

For this case, two estimates of the common slope are obtained: one from Model 15.25 and one from the small size of experimental unit comparisons using Equation 15.27. If there are enough large size experimental units or whole plots, the two estimators could be combined into a more efficient estimator, as would be accomplished by using a mixed models analysis. Kempthorne (1952) kept the two slopes separate, i.e., not combined, and carried out the analysis with two slopes.

If the slope is a function of the levels of the large size of experimental unit treatment, $\beta_{ik} = \beta_i$, the large size of experimental unit model is

$$\bar{y}_{ij\cdot} = \bar{\mu}_{i\cdot} + \beta_i \bar{x}_{ij\cdot} + e^*_{ij}, \qquad (15.28)$$

an unequal slope analysis of covariance model for a one-way treatment structure in a completely randomized design structure, and the comparisons of the small size of experimental units within a large size of experimental unit are

$$\sum_{k=1}^{t} c_k y_{ijk} = \sum_{k=1}^{t} c_k \mu_{ik} + \beta_i \sum_{k=1}^{t} c_k \left(x_{ijk} - \bar{x}_{ij\cdot} \right) + \sum_{k=1}^{t} c_k \varepsilon_{ijk}. \qquad (15.29)$$

Again, estimates of β_i, i = 1, 2, ..., a, can be obtained from Model 15.28 and as well as from the small experimental unit comparisons. These two sets of estimates should be combined to provide more efficient estimators of the slopes, as would be done by using a mixed models analysis. There also exists tests of the equal slope hypothesis, $H_0: \beta_1 = \beta_2 \ldots = \beta_a$ vs. $H_a:$ (not $H_0:$) from each part of the analysis and it is possible to construct one test based on the combined estimates of the slopes as was discussed in Chapter 10. The mixed models analysis provides a test of the above hypothesis based on the combined estimates of the slopes.

If the slope is a function of the levels of the small size of experimental unit or sub-plot treatment, then the large size of experimental unit model is

$$\bar{y}_{ij\cdot} = \bar{\mu}_{i\cdot} + \frac{1}{k}\sum_{k=1}^{t} \beta_k x_{ijk} + e^*_{ij} \qquad (15.30)$$

which is a multiple covariate model (k covariates) for a one-way treatment structure in a completely randomized design structure. The multiple covariates are the values of the covariate for each of the levels of the sub-plot treatment within each of the whole plots. The within large size of experimental unit comparisons are

$$\sum_{k=1}^{t} c_k y_{ijk} = \sum_{k=1}^{t} c_k \mu_{ik} + \sum_{k=1}^{t} c_k (\beta_k - \bar{\beta}.)(x_{ijk} - \bar{x}_{ij.}) + \sum_{k=1}^{t} c_k \varepsilon_{ijk}. \quad (15.31)$$

The small size of experimental unit analysis provides estimates of contrasts of the β_k and a test of the equal slopes hypothesis. Estimates of the β_k, $k = 1, 2, \ldots, t$, can be obtained from Model 15.29, though the information from the large size experimental unit model can be combined with the information from the small size experimental unit comparisons, as would be done when using a mixed models analysis.

If the slopes are additive functions of the levels of the two treatments as $\beta_{ik} = \beta_{Bi} + \beta_{Wk}$, the large size experimental unit model is

$$\bar{y}_{ij.} = \bar{\mu}_{i.} + \beta_{Bi}\bar{x}_{ij.} + \sum_{k=1}^{t} \beta_{Wk}\, x_{ijk}/t + e_{ij}^*, \quad (15.32)$$

a multiple covariate model with covariates $\bar{x}_{ij.}$, x_{ij1}, x_{ij2}, …, x_{ijt} for a one way treatment structure in a completely randomized design structure where slopes corresponding to $\bar{x}_{ij.}$ are unequal for each level of the whole plot treatment.

The within large size of experimental unit comparisons are

$$\sum_{k=1}^{t} c_k y_{ijk} = \sum_{k=1}^{t} c_k \mu_{ik} + \sum_{k=1}^{t} c_k \beta_{Wk}(x_{ijk} - \bar{x}_{ij.})$$
$$+ \beta_{Bi} \sum_{k=1}^{t} c_k (x_{ijk} - \bar{x}_{ij.}) + \sum_{k=1}^{t} c_k \varepsilon_{ijk}. \quad (15.33)$$

If there are no simplifying assumptions about the slopes, the large size of experimental unit model is the one in Equation 15.13 and the within large size of experimental unit comparisons are

$$\sum_{k=1}^{t} c_k y_{ijk} = \sum_{k=1}^{t} c_k \mu_{ik} + \sum_{k=1}^{t} c_k (\beta_{ik} - \bar{\beta}_{i.}) x_{ijk} + \sum_{k=1}^{t} c_k \varepsilon_{ijk}, \quad (15.34)$$

15.6.3 Summary of General Representation

This section describes how to dissect the analysis of covariance model in order to determine the sources of information you have for estimating the models parameters and for testing hypotheses about the models parameters or linear combinations of the models parameters. An understanding of the between large size experimental unit model and the information it provides is important in being able to access the degree on confidence you have in the resulting estimates and/or comparisons. An understanding of the within large size experimental unit comparisons enables one to see which parameters or linear combinations of the parameters are estimable from the comparisons with the most precision. A reduction in the degrees of freedom for error corresponding to a specific size of experimental unit can occur when information about slopes is extracted from the analysis of that size of experimental unit. When a single parameter or function of the parameters has estimates available from both models, that information should be combined into a single estimate. The combining process is accomplished by using a mixed models analysis. An understanding of the sources of information helps interpret the approximate denominator degrees of freedom associated with the various tests and parameter estimates provided by the mixed models analysis.

15.7 EXAMPLE: FLOUR MILLING EXPERIMENT — COVARIATE MEASURED ON THE WHOLE PLOT

The data in Table 15.1 are from a study to evaluate the effect of roll gap and variety on the amount of flour produced (percent of total wheat ground) during a run of a pilot flour mill, as described in Section 15.3. Three batches of wheat from each of the three varieties were used in the study where a single batch of wheat was used for all four roll gap settings. Figure 15.1 is a graphical representation of the randomization process for this experiment. The large rectangles correspond to the 3 days

TABLE 15.1
Data for Amount of Flour Milled (Percent) during the First Break of a Flour Mill Operation from Three Varieties of Wheat Using Four Roll Gaps

			Roll Gap			
Day	Variety	Moist	0.02 in	0.04 in	0.06 in	0.08 in
1	A	12.4	18.3	14.6	12.2	9.0
1	B	12.8	18.8	14.9	11.7	8.3
1	C	12.1	18.7	15.5	12.7	9.2
2	A	14.4	19.1	15.4	12.4	8.7
2	B	12.4	17.5	14.4	11.3	8.3
2	C	13.2	18.9	15.4	12.5	8.2
3	A	13.1	18.2	15.0	12.1	8.4
3	B	14.0	20.4	16.2	12.9	9.1
3	C	13.4	19.7	16.9	13.4	9.5

Note: The percent moisture content of the batch of wheat is the possible covariate.

TABLE 15.2
Analysis of Variance Table without Covariate Information for the Flour Milling Experiment

Source	df	EMS
Day	2	$\sigma_\varepsilon^2 + 4\sigma_a^2 + 12\sigma_{day}^2$
Variety	2	$\sigma_\varepsilon^2 + 4\sigma_a^2 + \phi^2(\text{variety})$
Error(batch) = Day*Variety	4	$\sigma_\varepsilon^2 + 4\sigma_a^2$
Roll Gap	3	$\sigma_\varepsilon^2 + \phi^2(\text{rollgap})$
Variety*Roll Gap	4	$\sigma_\varepsilon^2 + \phi^2(\text{variety} * \text{rollgap})$
Error(run)	18	σ_ε^2

on which the experiment was conducted. Within each day there are three rectangles of four columns where the rectangles represent a batch of wheat. The first process is to randomly assign the varieties to the batches or batch run order within each day. Thus the batch is the experimental unit for the levels of variety. The batch design is a one-way treatment structure (levels of varieties) in a randomized complete block design structure (days are the blocks). The moisture level of each batch was measured as a possible covariate. The second step in the randomization is to randomly assign the levels of roll gap to the four columns (run order) within each batch. This step in the randomization implies the four roll gap settings were observed in a random order for each batch of wheat from a particular variety. The run is the experimental unit for the levels of roll gap. The run design is a one-way treatment structure in a randomized complete block design structure where the batches form the blocks. The analysis of variance table corresponding to fitting the split-plot model with a two-way treatment structure without the covariate (the model in Equation 15.2) information is in Table 15.2. There are two error terms, one for each size of experimental unit. The batch error term measures the variability of batches treated alike within a day, is based on four degrees of freedom, and is computed as the day*variety interaction. The run error term measures the variability of the runs treated alike within a batch and is based on 18 degrees of freedom. The run error term is computed as day*roll gap(variety). The analysis of covariance process starts by investigating the type of relationship between the flour yield and the moisture content of the grain. One method of looking at the data is to plot the (yield, moisture) pairs for each combination of the variety and roll gap, for each variety, and for each roll gap. For this data set, there are just a few observations (three) for each treatment combination, so those plots do not provide much information. The plots for each roll gap indicate a slight linear relationship, but do not indicate the linear model is not adequate. Thus, a model that is linear in the covariate was selected for further study. This design consists of nine blocks of size four in three groups; thus the between block comparisons contain plenty of information about the parameters. A mixed models analysis is needed to extract the pieces of information and combine them into the

TABLE 15.3
PROC MIXED Code and Analysis for a Model with Factorial Effects for Both the Intercepts and the Slopes of the Model

```
proc mixed covtest cl data=mill;
class day variety rollgap;
model yield=variety|rollgap moist moist*rollgap moist*variety
  moist*variety*rollgap/ddfm=kr;
random day day*variety;
```

CovParm	Estimate	StdErr	ZValue	ProbZ	Alpha	Lower	Upper
day	0.0000						
day*variety	0.1887	0.1661	1.14	0.1279	0.05	0.0570	3.6741
Residual	0.0582	0.0274	2.12	0.0169	0.05	0.0275	0.1940

Effect	NumDF	DenDF	FValue	ProbF
variety	2	3.0	1.67	0.3258
rollgap	3	9.0	0.23	0.8703
variety*rollgap	6	9.0	0.62	0.7114
moist	1	3.0	6.00	0.0917
moist*variety	2	3.0	1.66	0.3268
moist*rollgap	3	9.0	8.58	0.0053
moist*variet*rollgap	6	9.0	0.73	0.6408

parameter estimates. Table 15.3 contains the PROC MIXED code and results from fitting Model 15.4 to the flour mill data. The first phase of the analysis of covariance process is to simplify the form of the covariate part of the model. So, the model reduction process begins with using the results in Table 15.3 where moist*variety*rollgap is deleted since it has the largest significance level. The next step deletes moist*variety, however, the analysis is not shown. The final form of the model is

$$FL_{ijk} = \mu + v_i + \theta M_{ij} + d_j + a_{ij} \} \quad \text{batch part of model}$$
$$+ \rho_k + (v\rho)_{ik} + \gamma_k M_{ij} + \varepsilon_{ijk} \} \quad \text{run part of model} \tag{15.35}$$

and is fit to the data using the PROC MIXED code in Table 15.4. Model 15.35 has unequal slopes for each level of roll gap and the slopes do not depend on the levels of variety. The moist*rollgap term provides a test of the equal roll gap slopes hypothesis, i.e., H_0: $\gamma_1 = \gamma_2 = \gamma_3 = \gamma_4$ vs. H_a: (not H_o:). The significance level associated with this hypothesis is 0.0009, indicating there is sufficient evidence to conclude that the slopes are not equal. The denominator degrees of freedom associated with the batch comparisons, variety and moisture, are 3.2 and 4.1, respectively, indicating that the batch error term has essentially been used as a gauge. The denominator degrees of freedom associated with rollgap, variety*rollgap, and moist*rollgap are 15, indicating that the run error term has essentially been used as a divisor. The three degrees of freedom for moist*rollgap have been removed from Error(run) and the single degree of freedom for moist has been removed from Error(batch). The

TABLE 15.4
PROC MIXED Code and Analysis to Fit the Final Model with Factorial Effects for Both the Intercepts and the Slopes of the Model Where the Slopes Part Has Been Reduced

```
proc mixed covtest cl data=mill;
class day variety rollgap;
model yield=variety|rollgap moist moist*rollgap/ddfm=kr;
random day day*variety;
```

CovParm	Estimate	StdErr	ZValue	ProbZ	Alpha	Lower	Upper
day	0.1405	0.1817	0.77	0.2198	0.05	0.0303	48.6661
day*variety	0.0943	0.0850	1.11	0.1336	0.05	0.0279	2.0664
Residual	0.0518	0.0189	2.74	0.0031	0.05	0.0283	0.1241

Effect	NumDF	DenDF	FValue	ProbF
variety	2	3.2	5.04	0.1018
rollgap	3	15.0	0.17	0.9127
variety*rollgap	6	15.0	3.45	0.0241
moist	1	4.1	6.97	0.0560
moist*rollgap	3	15.0	9.64	0.0009

covariate causes the model to be unbalanced; thus the error terms for variety and moist are like combinations of Error(batch) and Error(run) with the most weight being given to Error(batch). The degrees of freedom are larger than three because of this combination of error terms. Using the ddfm=kr approximation provides appropriate denominator degrees of freedom in most cases, but the analyst must be careful to not allow the analysis to use degrees of freedom that are not warranted. The remaining analysis can be accomplished by fitting Model 15.35 to the data set, but the estimates of the slopes and intercepts satisfy the set-to-zero restriction (Milliken and Johnson, 1992). The estimates of the intercepts and slopes can be obtained by fitting a means model to both the intercepts and the slopes. The full rank means model is

$$FL_{ijk} = \mu_{ij} + \gamma_k M_{ij} + d_j + a_{ij} + \varepsilon_{ijk} \tag{15.36}$$

where μ_{ik} if the intercept for variety i and roll gap k and γ_k is the slope for moisture for roll gap k. The PROC MIXED code in Table 15.5 fits Model 15.36 to the flour data. The second part of Table 15.5 contains the REML estimates of the variance components for the model. The third section of Table 15.5 contains the estimates of the intercepts (variety*rollgap) and slopes (moist*rollgap). The slopes decrease as the roller gap increases. The analysis of variance table in the bottom part of Table 15.5 provides tests that the intercepts are all equal to zero ($p = 0.03868$) and that the slopes are all equal to zero ($p = 0.0017$). The intercepts themselves are not of interest since it is not reasonable to have a batch of grain with zero moisture content. Table 15.6 contains the PROC MIXED code to provide adjusted means for the variety by roll gap combinations evaluated at 12, 14, and 16% moisture content.

TABLE 15.5
PROC MIXED Code and Analysis to Fit the Means Model with a Two-Way Means Representation for the Intercepts and a One-Way Means Representation for the Slopes

```
proc mixed covtest cl data=mill;
class day variety rollgap;
model yield=variety*rollgap moist*rollgap/noint solution ddfm=kr;
random day day*variety;
```

CovParm	Estimate	StdErr	ZValue	ProbZ	Alpha	Lower	Upper
day	0.1405	0.1817	0.77	0.2198	0.05	0.0303	48.6661
day*variety	0.0943	0.0850	1.11	0.1336	0.05	0.0279	2.0664
Residual	0.0518	0.0189	2.74	0.0031	0.05	0.0283	0.1241

Effect	variety	rollgap	Estimate	StdErr	df	tValue	Probt
variety*rollgap	A	0.02	5.8747	3.3062	6.2	1.78	0.1243
variety*rollgap	A	0.04	4.7550	3.3062	6.2	1.44	0.1988
variety*rollgap	A	0.06	5.2933	3.3062	6.2	1.60	0.1589
variety*rollgap	A	0.08	6.3569	3.3062	6.2	1.92	0.1013
variety*rollgap	B	0.02	6.4635	3.2487	6.2	1.99	0.0922
variety*rollgap	B	0.04	5.1014	3.2487	6.2	1.57	0.1658
variety*rollgap	B	0.06	5.1484	3.2487	6.2	1.58	0.1625
variety*rollgap	B	0.08	6.2647	3.2487	6.2	1.93	0.1005
variety*rollgap	C	0.02	6.8221	3.2077	6.2	2.13	0.0760
variety*rollgap	C	0.04	5.9965	3.2077	6.2	1.87	0.1091
variety*rollgap	C	0.06	6.1354	3.2077	6.2	1.91	0.1027
variety*rollgap	C	0.08	6.6941	3.2077	6.2	2.09	0.0804
moist*rollgap		0.02	0.9518	0.2475	6.1	3.85	0.0081
moist*rollgap		0.04	0.7703	0.2475	6.1	3.11	0.0201
moist*rollgap		0.06	0.5218	0.2475	6.1	2.11	0.0784
moist*rollgap		0.08	0.1762	0.2475	6.1	0.71	0.5027

Effect	NumDF	DenDF	FValue	ProbF
variety*rollgap	12	10.3	3.17	0.0368
moist*rollgap	4	12.4	8.33	0.0017

Table 15.7 contains the pairwise comparisons among the levels of variety at each roll gap at one moisture level using a LSD type method. These means were compared at just one moisture (any one) level since the slopes of the regression lines are not a function of variety, i.e., the regression lines are parallel for the different varieties. The means of varieties A and B are never significantly different ($p = 0.05$). The mean of variety C is significantly larger than the mean of variety A at roll gaps of 0.02, 0.04, and 0.06. The mean of variety C is significantly larger than the mean of variety B at roll gaps 0.04 and 0.06. The means of the varieties are not significantly different at roll gap of 0.08. Table 15.8 contains the LSD type pairwise comparisons of the roller gap means within each variety and at each level of moisture. All pairs of roller gap means are significantly different; thus the individual significance levels were not included in the table. Since the slopes of the regression lines are functions of

TABLE 15.6
Adjusted Means at Three Values of Moisture from the LSMEANS Statements

```
lsmeans variety*rollgap/diff at moist=12;
lsmeans variety*rollgap/diff at moist=14;
lsmeans variety*rollgap/diff at moist=16;
```

		Moist=12%		Moist=14%		Moist=16%	
variety	rollgap	Estimate	StdErr	Estimate	StdErr	Estimate	StdErr
A	0.02	17.296	0.446	19.200	0.354	21.103	0.736
A	0.04	13.999	0.446	15.539	0.354	17.080	0.736
A	0.06	11.555	0.446	12.599	0.354	13.642	0.736
A	0.08	8.471	0.446	8.823	0.354	9.176	0.736
B	0.02	17.885	0.406	19.788	0.386	21.692	0.789
B	0.04	14.345	0.406	15.886	0.386	17.426	0.789
B	0.06	11.410	0.406	12.454	0.386	13.497	0.789
B	0.08	8.379	0.406	8.731	0.386	9.083	0.789
C	0.02	18.243	0.381	20.147	0.412	22.050	0.827
C	0.04	15.240	0.381	16.781	0.412	18.321	0.827
C	0.06	12.397	0.381	13.441	0.412	14.484	0.827
C	0.08	8.808	0.381	9.160	0.412	9.513	0.827

TABLE 15.7
Pairwise Comparisons of the Variety Means within Each Level of Roller Gap and for One Value of Moisture

rollgap	variety	_variety	Estimate	StdErr	df	tValue	Probt
0.02	A	B	−0.5887	0.3174	5.8	−1.85	0.1149
0.02	A	C	−0.9474	0.3275	5.8	−2.89	0.0286
0.02	B	C	−0.3586	0.3148	5.8	−1.14	0.2997
0.04	A	B	−0.3464	0.3174	5.8	−1.09	0.3185
0.04	A	C	−1.2415	0.3275	5.8	−3.79	0.0096
0.04	B	C	−0.8950	0.3148	5.8	−2.84	0.0307
0.06	A	B	0.1449	0.3174	5.8	0.46	0.6647
0.06	A	C	−0.8421	0.3275	5.8	−2.57	0.0435
0.06	B	C	−0.9870	0.3148	5.8	−3.13	0.0213
0.08	A	B	0.0922	0.3174	5.8	0.29	0.7815
0.08	A	C	−0.3371	0.3275	5.8	−1.03	0.3442
0.08	B	C	−0.4294	0.3148	5.8	−1.36	0.2234

the levels of roller gap, analyses were carried out to provide overall tests of equality of the regression lines at the three moisture levels. Three new variables were computed: m12 = moisture − 12, m14= moisture − 14, and m16 = moisture − 16. The

TABLE 15.8
Pairwise Comparisons between the Levels of Roller Gap within Each Variety for Three Values of Moisture Content

```
all df=15; all comparisons significant at p < .0001
```

variety	rollgap	_rollgap	Moist=12%		Moist=14%		Moist=16%	
			Estimate	StdErr	Estimate	StdErr	Estimate	StdErr
A	0.02	0.04	3.297	0.272	3.660	0.214	4.023	0.453
A	0.02	0.06	5.741	0.272	6.601	0.214	7.461	0.453
A	0.02	0.08	8.825	0.272	10.376	0.214	11.927	0.453
A	0.04	0.06	2.444	0.272	2.941	0.214	3.438	0.453
A	0.04	0.08	5.528	0.272	6.716	0.214	7.904	0.453
A	0.06	0.08	3.084	0.272	3.775	0.214	4.467	0.453
B	0.02	0.04	3.540	0.247	3.903	0.234	4.266	0.486
B	0.02	0.06	6.475	0.247	7.335	0.234	8.195	0.486
B	0.02	0.08	9.506	0.247	11.057	0.234	12.608	0.486
B	0.04	0.06	2.935	0.247	3.432	0.234	3.929	0.486
B	0.04	0.08	5.966	0.247	7.155	0.234	8.343	0.486
B	0.06	0.08	3.031	0.247	3.723	0.234	4.414	0.486
C	0.02	0.04	3.003	0.231	3.366	0.251	3.729	0.509
C	0.02	0.06	5.846	0.231	6.706	0.251	7.566	0.509
C	0.02	0.08	9.435	0.231	10.986	0.251	12.538	0.509
C	0.04	0.06	2.843	0.231	3.340	0.251	3.837	0.509
C	0.04	0.08	6.432	0.231	7.620	0.251	8.808	0.509
C	0.06	0.08	3.589	0.231	4.280	0.251	4.971	0.509

three sets of PROC MIXED code in Table 15.9 provide the analyses where m12, m14, and m16 were each used as the covariate. The analysis using m12 is for a model with intercepts at 12% moisture. The test statistic corresponding to rollgap provides a test of the equality of the roller gap models evaluated at 12% moisture. Thus the analyses using m12, m14, and m16 provide tests of the equality of the roller gap models at a value of moisture of 12, 14, and 16%, respectively. The values of the F-test change as the moisture content increases. The test statistics for the other terms in the three analyses do not depend on the level of moisture; thus the test statistics are the same for all three analyses.

Generally a graphical presentation uses the covariate for the horizontal axis, but for this example, the horizontal axes are the value of the roller gap. Figures 15.4 to 15.6 are graphs of the variety means across the roller gap values for each of the three moisture levels. There is not a lot of difference between the variety means evaluated at a given moisture level within a roller gap value, though some are significantly different as displayed in Table 15.7. Figures 15.7 to 15.9 contain graphs of the models evaluated at the three moisture levels across the roller gap values for each variety. The models are diverging as the roller gap becomes smaller.

This example involved a whole plot covariate and the analysis demonstrated that the degrees of freedom for estimating the covariance parameters are deducted from

Analysis of Covariance for Split-Plot and Strip-Plot Design Structures

TABLE 15.9
PROC MIXED Code for Fitting Models with Three Different Values of the Covariate to Provide Comparisons of the Roll Gap Means at Those Values

```
proc mixed covtest cl data=mill;
class day variety rollgap;
model yield=variety|rollgap m12 m12*rollgap/ddfm=kr;
random day day*variety;

proc mixed covtest cl data=mill;
class day variety rollgap;
model yield=variety|rollgap m14 m14*rollgap/ddfm=kr;
random day day*variety;

proc mixed covtest cl data=mill;
class day variety rollgap;
model yield=variety|rollgap m16 m16*rollgap/ddfm=kr;
random day day*variety;
```

CovParm	Estimate	StdErr	Lower	Upper
day	0.1405	0.1817	0.0303	48.6661
day*variety	0.0943	0.0850	0.0279	2.0664
Residual	0.0518	0.0189	0.0283	0.1241

Effect	NumDF	DenDF	FValue	ProbF
variety	2	3.2	5.04	0.1018
rollgap	3	15	791.03	0.0000
variety*rollgap	6	15	3.45	0.0241
m12	1	4.1	6.97	0.0560
m12*rollgap	3	15	9.64	0.0009

Effect	NumDF	DenDF	FValue	ProbF
variety	2	3.2	5.04	0.1018
rollgap	3	15	1371.48	0.0000
variety*rollgap	6	15	3.45	0.0241
m14	1	4.1	6.97	0.0560
m14*rollgap	3	15	9.64	0.0009

Effect	NumDF	DenDF	FValue	ProbF
variety	2	3.2	5.04	0.1018
rollgap	3	15	265.02	0.0000
variety*rollgap	6	15	3.45	0.0241
m16	1	4.1	6.97	0.0560
m16*rollgap	3	15	9.64	0.0009

different error terms depending on where the covariate term resides in the model. A graphical display was presented where the covariate was not used on the horizontal axis, but the information presented was very meaningful. The next example is a continuation of the cookie baking example described in Section 15.4 where the covariate is measured on the sub-plot or small size experimental unit.

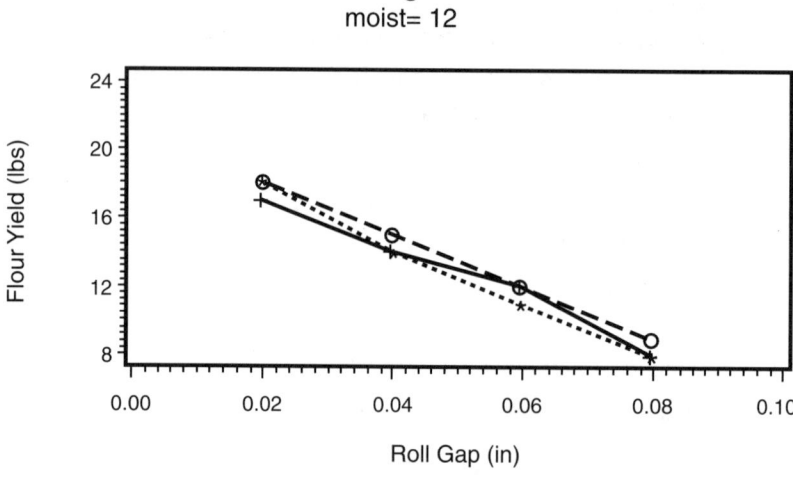

FIGURE 15.4 Variety by roll gap means evaluated at moisture = 12%.

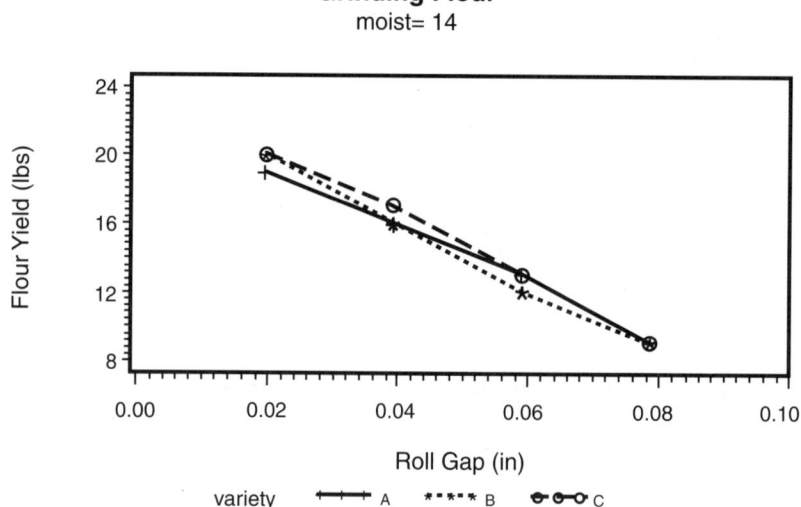

FIGURE 15.5 Variety by roll gap means evaluated at moisture = 14%.

15.8 EXAMPLE: COOKIE BAKING

The data in Table 15.10 are the diameters of cookies of three cookie types baked at one of three different temperatures, thus generating a two-way treatment structure. The levels of temperature are assigned to the ovens or whole plots or large size

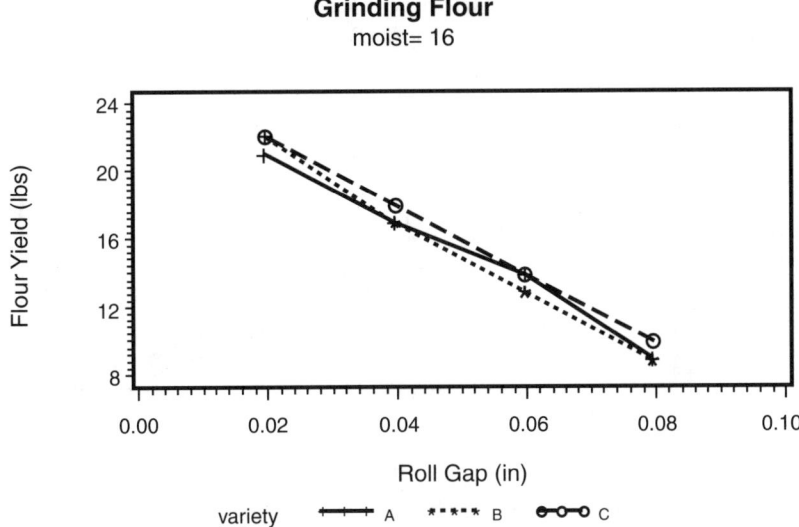

FIGURE 15.6 Variety by roll gap means evaluated at moisture = 16%.

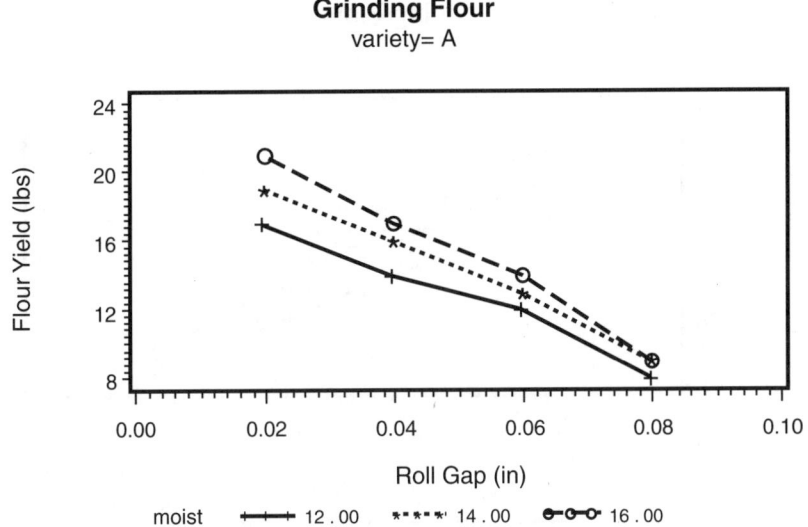

FIGURE 15.7 Roll gap means evaluated at three moisture levels for Variety A.

experimental units completely at random, forming a completely randomized design oven or whole plot design structure. One slice of cookie dough of each type of the three types of refrigerator cookie dough (prepackaged) was placed into each oven. The thickness of the slice of cookie dough was measured as a possible covariate. The schematic in Figure 15.2 is a graphical representation of this split-plot design

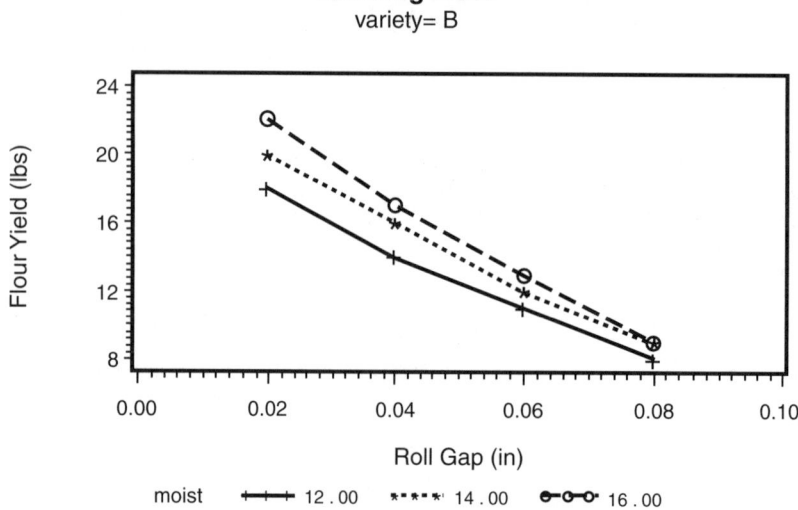

FIGURE 15.8 Roll gap means evaluated at three moisture levels for Variety B.

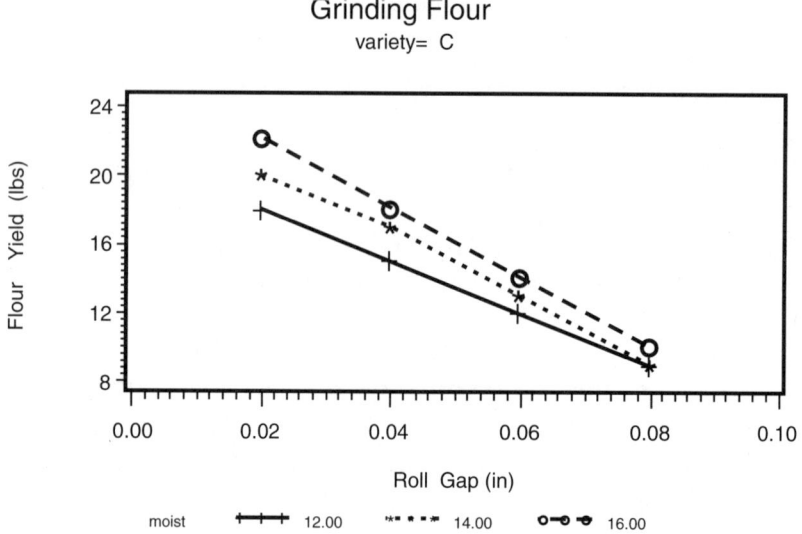

FIGURE 15.9 Roll gap means evaluated at three moisture levels for Variety C.

structure. The vertical rectangles represent the ovens (only one oven was used, so they really represent the run order for using a single oven) and the levels of temperature are randomly assigned to the ovens in a completely random fashion; thus there was no blocking at the oven level of the experiment. The oven is the experimental unit for the levels of temperature. The circles within each of the vertical

TABLE 15.10
Cookie Diameters (mm) of Three Different Cookie Types Baked at Three Different Temperatures with Cookie Dough Slice Thickness (mm) as Possible Covariate

		Chocolate Chip		Peanut Butter		Sugar	
rep	temp	diam	thick	diam	thick	diam	thick
1	300	53	3	62	7	59	7
1	350	65	6	68	6	71	7
1	400	64	6	72	5	66	6
2	300	67	7	75	9	66	7
2	350	68	6	68	6	64	5
2	400	79	9	82	8	72	5
3	300	65	7	67	5	63	6
3	350	77	8	81	9	63	3
3	400	72	7	76	5	67	4
4	300	57	5	60	5	52	4
4	350	75	7	67	5	67	5
4	400	60	4	79	9	74	9

TABLE 15.11
Analysis of Variance Table Without Covariate Information for the Baking Cookie Experiment

Source	df	EMS
Temperature	2	$\sigma_\varepsilon^2 + 3\sigma_a^2 + \phi^2(\text{temp})$
Error(oven) = Rep(Temperature)	9	$\sigma_\varepsilon^2 + 3\sigma_a^2$
Cookie	2	$\sigma_\varepsilon^2 + \phi^2(\text{cookie})$
Temperature*Cookie	4	$\sigma_\varepsilon^2 + \phi^2(\text{cookie} + \text{temp})$
Error(cookie)	18	σ_ε^2

rectangles represent the position on the baking pan to which a slice of cookie dough from each of the three cookie types is randomly assigned. The cookie is the experimental unit for the levels of cookie type and each oven forms a block of size three. Thus the cookie design structure is a randomized complete block with twelve blocks of size three. Since this design consists of twelve blocks of size three, there is considerable information about the models parameters contained in the between block comparisons; therefore, a mixed models analysis can be used to provide the combined within block and between block estimates of the models parameters. The usual analysis of variance table (without the covariate information) is in Table 15.11 (Milliken and Johnson, 1992). The term rep(temperature) is a measure of the variability among ovens or oven runs treated alike [Error(oven)] which is based on

TABLE 15.12
PROC MIXED Code and Analysis to Fit a Model with Factorial Effects for the Slopes and Intercepts for the Cooking Baking Data

```
proc mixed data=cookies cl covtest;
class cookie temp rep;
model diam=cookie temp cookie*temp thick thick*cookie thick*temp
  cookie*temp*thick/ddfm=kr;
random rep(temp);
```

CovParm	Estimate	StdErr	ZValue	ProbZ	Alpha	Lower	Upper
rep(temp)	9.5425	4.9174	1.94	0.0262	0.05	4.2713	36.9311
Residual	1.3924	0.6513	2.14	0.0163	0.05	0.6619	4.5847

Effect	NumDF	DenDF	FValue	ProbF
cookie	2	10.7	5.29	0.0252
temp	2	17.6	5.18	0.0171
cookie*temp	4	10.4	3.42	0.0503
thick	1	10.6	156.49	0.0000
thick*cookie	2	10.8	0.95	0.4164
thick*temp	2	10.5	4.85	0.0322
thick*cookie*temp	4	10.4	1.76	0.2098

9 degrees of freedom. The Error(cookie) term measures the variability of cookies within an oven, computed as cookie*rep(temp), and is based on 18 degrees of freedom. When the covariate part of the model is included, each error term can lose degrees of freedom depending on the form of the model. Table 15.12 contains the PROC MIXED code to fit Model 15.6 to cookie baking data. The model has two-way factorial effects for both the intercepts and slopes. To simplify the covariate part of the model, delete the highest-order interaction term involving the covariate that is not significant. In this case, delete the thick*cookie*temp term and refit the model. The thick*cookie term was deleted next, leaving the model with unequal slopes for the levels of temperature (analysis not shown). This is a balanced data set, but including the covariate causes the model to become unbalanced. The denominator degrees of freedom should be either 9 or 18, but the unbalancing causes the use of a combination of the batch error term with the cookie error term, resulting in using an approximate number of degrees of freedom. Thus the degrees of freedom are between 9 and 18 because nine degrees of freedom that were associated with one or the other of the error terms were used to estimate the slopes. Table 15.13 contains the PROC MIXED code to fit the final model with two-way factorial effects for the intercepts and a one-way effects for the slopes, or model

$$D_{ijk} = \mu + \tau_i + \theta T_{ijk} + \phi_i T_{ijk} d_j + a_{ij} \} \quad \text{oven part of model}$$
$$+ \rho_k + (\tau\rho)_{ik} + \varepsilon_{ijk} \} \quad \text{cookie part of model}$$
(15.37)

TABLE 15.13
PROC MIXED Code and Analysis to Fit the Final Model to the Cookie Baking Data with Factorial Effects for the Intercepts and a Reduced Model for the Slopes

```
proc mixed data=cookies cl covtest;
class cookie temp rep;
model diam=cookie temp cookie*temp thick thick*temp/ddfm=kr;
random rep(temp);
```

CovParm	Estimate	StdErr	ZValue	ProbZ	Alpha	Lower	Upper
rep(temp)	10.8857	5.4785	1.99	0.0235	0.05	4.9462	40.4047
Residual	1.6021	0.5881	2.72	0.0032	0.05	0.8719	3.8591

Effect	NumDF	DenDF	FValue	ProbF
cookie	2	14.9	35.69	0.0000
temp	2	22.5	3.08	0.0656
cookie*temp	4	14.9	10.97	0.0002
thick	1	15.6	196.56	0.0000
thick*temp	2	15.6	5.73	0.0137

The significance level corresponding to thick*temp tests the equality of the slopes for the levels of temperature. The corresponding significance level is 0.0137, indicating there is sufficient evidence to conclude the slopes are unequal. The full rank model with intercepts for each combination of cookie type and temperature and slopes for each level of temperature is

$$D_{ijk} = \mu_{ik} + \phi_i T_{ijk} + d_j + a_{ij} + \varepsilon_{ijk} \tag{15.38}$$

The PROC MIXED code in Table 15.14 fits Model 15.38 to the cookie data set. The second part of Table 15.14 contains the REML estimates of the variance components where the estimates of the oven and cookie variance components are 10.8857 and 1.6021, respectively. Table 15.15 contains the LSMEANS statements to compute adjusted means at three values of cookie dough thickness. The adjusted means and the estimated standard errors are provided for each combination of cookie type and oven temperature evaluated at cookie thicknesses of 3.125, 6.25, and 12.5 mm. Figures 15.10 to 15.12 are plots of the adjusted means for the three temperatures vs. the cookie thickness values for each of the cookie types. These sets of lines are not parallel as the slopes are functions of the levels of temperature. Figures 15.13 to 15.15 are graphs of the adjusted means for the levels of cookie type across the cookie thickness values for each of the temperatures. These sets of lines are parallel since the slopes are not functions of the cookie types. Pairwise comparisons among the levels of cookie type within each temperature evaluated at one cookie thickness (any one since the lines are parallel) are included in Table 15.16. The mean diameter of peanut butter cookies are significantly larger than the mean diameter of the sugar

TABLE 15.14
PROC MIXED Code and Analysis for a Means Model with a Two-Way Means Representation for the Intercepts and a One-Way Representation for the Slopes for the Cookie Baking Data

```
proc mixed data=cookies cl covtest; where rep<5;
class cookie temp rep;
model diam=cookie*temp thick*temp/noint ddfm=kr solution;
random rep(temp);
```

CovParm	Estimate	StdErr	ZValue	ProbZ	Alpha	Lower	Upper
rep(temp)	10.8857	5.4785	1.99	0.0235	0.05	4.9462	40.4047
Residual	1.6021	0.5881	2.72	0.0032	0.05	0.8719	3.8591

Effect	cookie	temp	Estimate	StdErr	df	tValue	Probt
cookie*temp	c_chip	300	50.8815	2.5561	23.1	19.91	0.0000
cookie*temp	c_chip	350	49.7046	2.6993	23.9	18.41	0.0000
cookie*temp	c_chip	400	54.5539	2.3251	21.4	23.46	0.0000
cookie*temp	p_butter	300	54.6327	2.8084	23.9	19.45	0.0000
cookie*temp	p_butter	350	50.2526	2.6427	23.8	19.02	0.0000
cookie*temp	p_butter	400	62.5079	2.3633	21.9	26.45	0.0000
cookie*temp	sugar	300	49.5071	2.6800	23.7	18.47	0.0000
cookie*temp	sugar	350	50.2905	2.3253	21.6	21.63	0.0000
cookie*temp	sugar	400	56.6459	2.2513	20.4	25.16	0.0000
thick*temp		300	1.7488	0.3358	16.2	5.21	0.0001
thick*temp		350	3.1919	0.3023	15.0	10.56	0.0000
thick*temp		400	2.1840	0.2325	15.3	9.39	0.0000

Effect	NumDF	DenDF	FValue	ProbF
cookie*temp	9	17.9	177.26	0.0000
thick*temp	3	15.5	75.60	0.0000

and chocolate chip cookies baked at 300°F. The mean diameters of the three cookie types are not significantly different when baked at 350°F. The mean diameters of peanut butter cookies are significantly larger than the mean diameters of the sugar cookies which are significantly larger than the mean diameters of the chocolate chip cookies when baked at 400°F. Table 15.17 contains the pairwise comparisons among the baking temperatures within each cookie type for three thicknesses. There are several significantly different sets of means in the table. Temperature mean diameters for 300 and 350°F are not significantly different for all cookie types at 3.125-mm thickness and for peanut butter at 6.25-mm thickness. Temperature mean diameters for 300 and 400°F are not significantly different for chocolate chips at 3.125-mm thickness. Temperature mean diameters for 350 and 400°F are not significantly different for chocolate chips at 3.125- and 6.25-mm thickness, for peanut butter at 6.25-mm thickness, and for sugar cookies at all three thicknesses.

Figure 15.16 contains the JMP® data table for the cookie baking experiment. Variable rep, temp, and cookie are declared to be nominal and thick and diam as

TABLE 15.15
Adjusted Mean Diameters for the Cookie Type by Temperature Combinations Evaluated at Three Values of Cookie Dough Thickness

```
lsmeans cookie*temp/at thick=6.25 diff;
lsmeans cookie*temp/at thick=3.125 diff;
lsmeans cookie*temp/at thick=12.5 diff;
```

		Thickness=6.25 mm		Thickness=3.125 mm		Thickness=12.5 mm	
cookie	temp	Estimate	StdErr	Estimate	StdErr	Estimate	StdErr
c_chip	300	61.81	1.78	56.35	1.94	72.74	2.94
c_chip	350	69.65	1.77	59.68	2.08	89.60	2.48
c_chip	400	68.20	1.77	61.38	1.93	81.85	2.25
p_butter	300	65.56	1.77	60.10	2.10	76.49	2.68
p_butter	350	70.20	1.77	60.23	2.04	90.15	2.53
p_butter	400	76.16	1.77	69.33	1.96	89.81	2.22
sugar	300	60.44	1.77	54.97	2.01	71.37	2.81
sugar	350	70.24	1.81	60.27	1.86	90.19	2.87
sugar	400	70.30	1.77	63.47	1.89	83.95	2.33

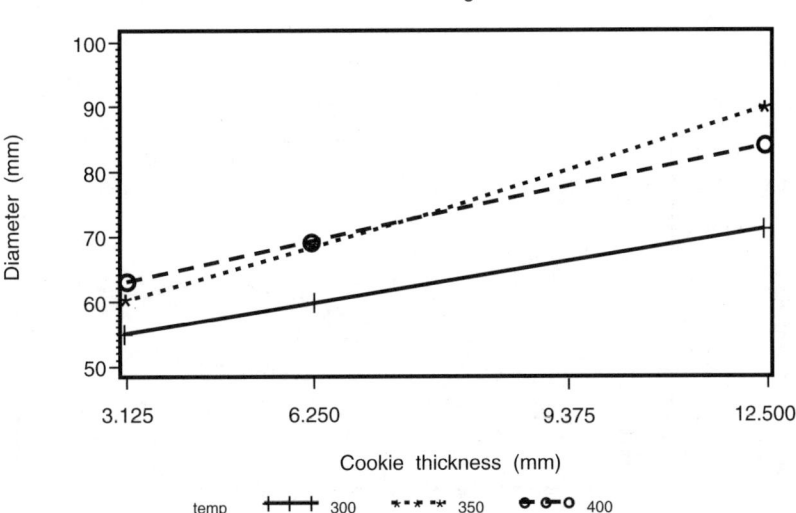

FIGURE 15.10 Plot of temperature least squares means for three slice thicknesses for sugar cookies.

continuous. The model is specified in the fit model screen of Figure 15.17. The terms temp, cookie, temp*cookie, thick, and thick*temp were included as fixed effects and rep(temp) was declared to be a random effect by using the attributes menu. Use

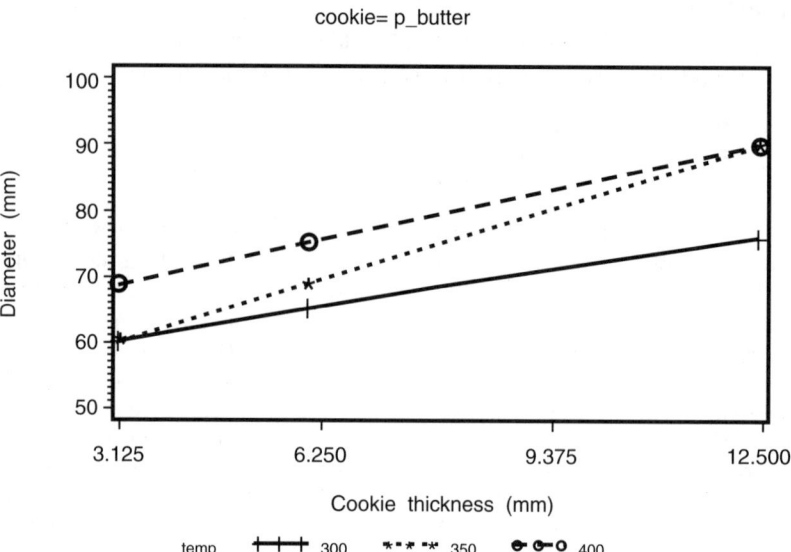

FIGURE 15.11 Plot of temperature least squares means for three slice thicknesses for peanut butter cookies.

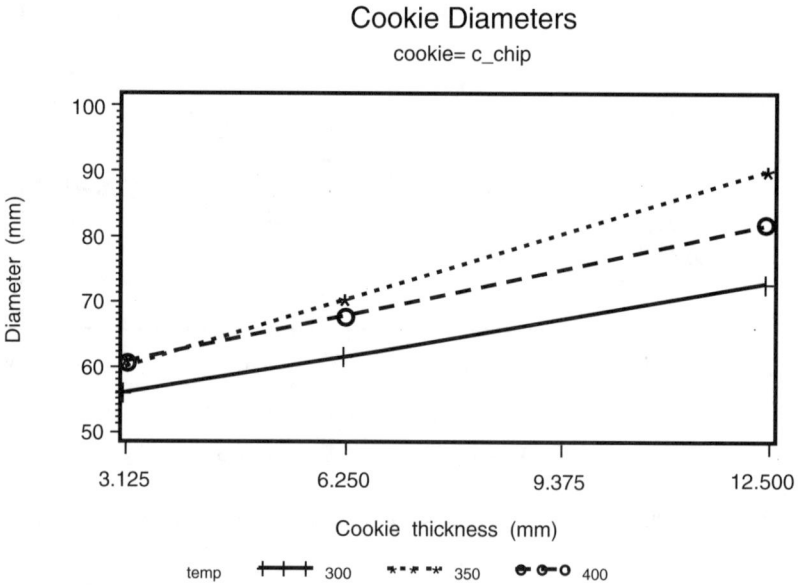

FIGURE 15.12 Plot of temperature least squares means for three slice thicknesses for chocolate chip cookies.

Analysis of Covariance for Split-Plot and Strip-Plot Design Structures

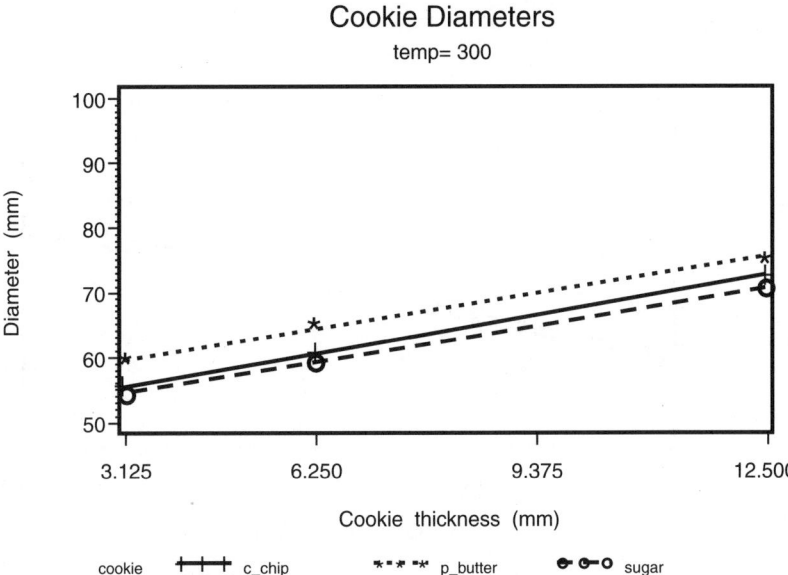

FIGURE 15.13 Plot of cookie type least squares means for three slice thickensses of cookies baked at temperature = 300°F.

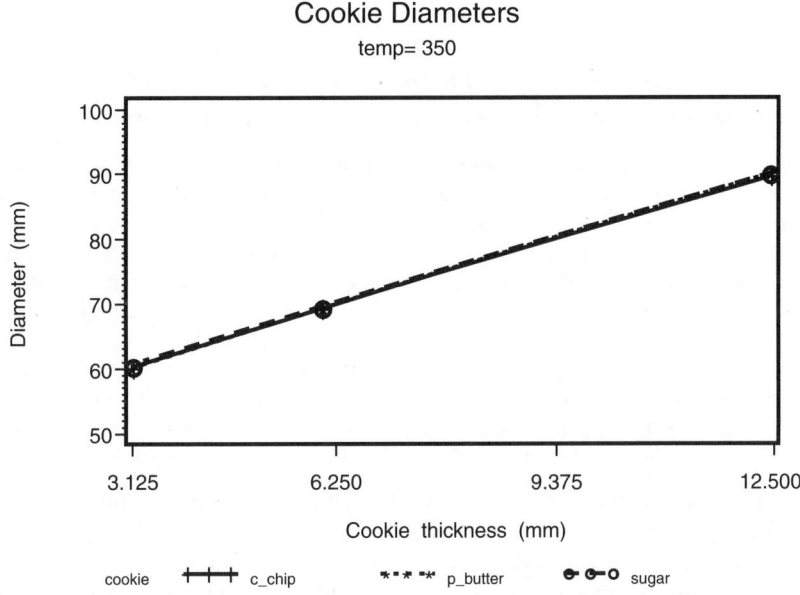

FIGURE 15.14 Plot of cookie type least squares means for three slice thicknesses of cookies baked at temperature = 350°F.

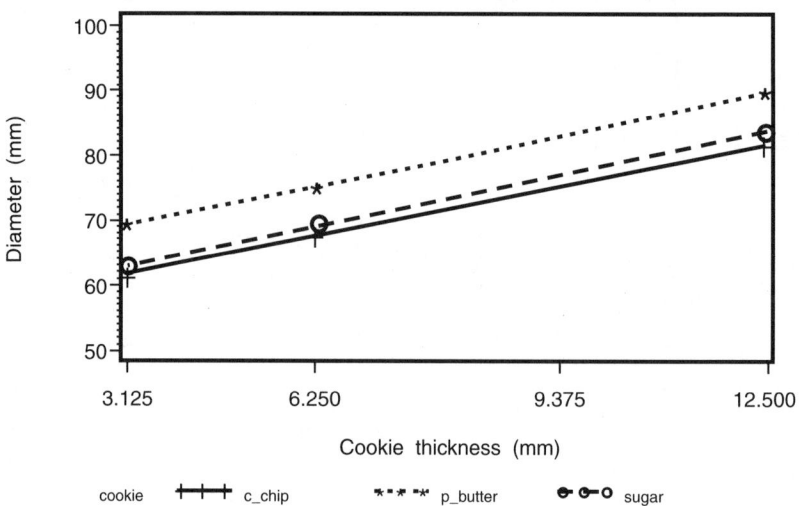

FIGURE 15.15 Plot of cookie type least squares means for three slice thicknesses of cookies baked at temperature = 400°F.

TABLE 15.16
Pairwise Comparisons among the Cookie Types within Each Level of Temperature for a Single Value of Cookie Dough Thickness

Effect	temp	cookie	_cookie	Estimate	StdErr	df	tValue	Probt
cookie*temp	300	c_chip	p_butter	−3.75	0.96	15.0	−3.92	0.0014
cookie*temp	300	c_chip	sugar	1.37	0.91	14.9	1.51	0.1521
cookie*temp	300	p_butter	sugar	5.13	0.91	14.9	5.63	0.0000
cookie*temp	350	c_chip	p_butter	−0.55	0.90	14.8	−0.61	0.5510
cookie*temp	350	c_chip	sugar	−0.59	1.04	14.9	−0.56	0.5815
cookie*temp	350	p_butter	sugar	−0.04	1.00	14.9	−0.04	0.9704
cookie*temp	400	c_chip	p_butter	−7.95	0.90	14.8	−8.87	0.0000
cookie*temp	400	c_chip	sugar	−2.09	0.90	14.9	−2.32	0.0351
cookie*temp	400	p_butter	sugar	5.86	0.91	14.9	6.43	0.0000

the model specification menu to deselect the center polynomials option. Select the run model option to have JMP® fit the model to the data. Figure 15.18 contains the estimates of the variance components and the analysis of variance with tests for the effects in the model. The results are similar to those in Table 15.13, but the differences are mainly due to the use of different denominator degrees of freedom approximations. The custom test menu can be used to provide adjusted means at selected values of the thickness of cookie dough. That process is not displayed here.

TABLE 15.17
Pairwise Comparisons among the Levels of Temperature within Each Cookie Type Evaluated at Three Values of Cookie Dough Thickness

			Thickness=3.125 mm		Thickness=6.25 mm		Thickness=12.5 mm	
cookie	**temp**	**_temp**	**Estimate**	**Probt**	**Estimate**	**Probt**	**Estimate**	**Probt**
c_chip	300	350	−3.33	0.2584	−7.84	0.0102	−16.86	0.0002
c_chip	300	400	−5.03	0.0872	−6.39	0.0280	−9.11	0.0216
c_chip	350	400	−1.70	0.5579	1.45	0.5748	7.75	0.0304
p_butter	300	350	−0.13	0.9652	−4.64	0.0921	−13.66	0.0011
p_butter	300	400	−9.24	0.0054	−10.60	0.0016	−13.32	0.0009
p_butter	350	400	−9.11	0.0055	−5.96	0.0376	0.34	0.9196
sugar	300	350	−5.29	0.0736	−9.80	0.0026	−18.82	0.0001
sugar	300	400	−8.50	0.0079	−9.86	0.0026	−12.58	0.0021
sugar	350	400	−3.21	0.2480	−0.06	0.9827	6.24	0.1044

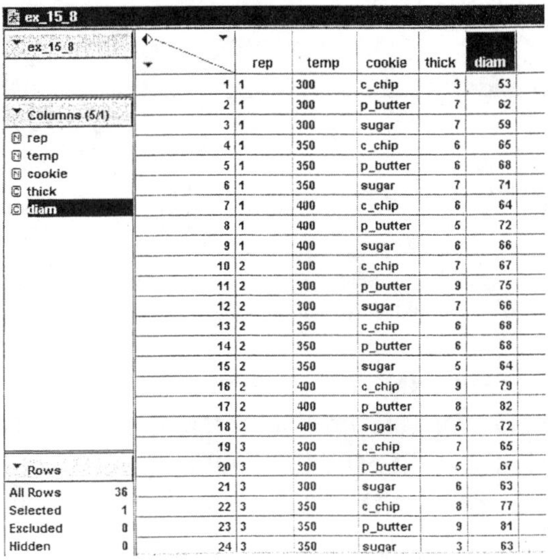

	rep	temp	cookie	thick	diam
1	1	300	c_chip	3	53
2	1	300	p_butter	7	62
3	1	300	sugar	7	59
4	1	350	c_chip	6	65
5	1	350	p_butter	6	68
6	1	350	sugar	7	71
7	1	400	c_chip	6	64
8	1	400	p_butter	5	72
9	1	400	sugar	6	66
10	2	300	c_chip	7	67
11	2	300	p_butter	9	75
12	2	300	sugar	7	66
13	2	350	c_chip	6	68
14	2	350	p_butter	6	68
15	2	350	sugar	5	64
16	2	400	c_chip	9	79
17	2	400	p_butter	8	82
18	2	400	sugar	5	72
19	3	300	c_chip	7	65
20	3	300	p_butter	5	67
21	3	300	sugar	6	63
22	3	350	c_chip	8	77
23	3	350	p_butter	9	81
24	3	350	sugar	3	63

FIGURE 15.16 JMP® table for the baking cookie experiment.

This example consisted of a split-plot design where the covariate is measured on the sub-plot or small size experimental unit. The analysis is really not different from that used in Section 15.7 where the covariate was measured on the whole plot. Hence, there is one process to follow for carrying out the analysis of covariance and it does not depend on which experimental unit of the covariate was measured. The next example is a combination of the examples in Sections 15.7 and 15.8 where there are two covariates, one measured on the large experimental unit and one measured on the small experimental unit.

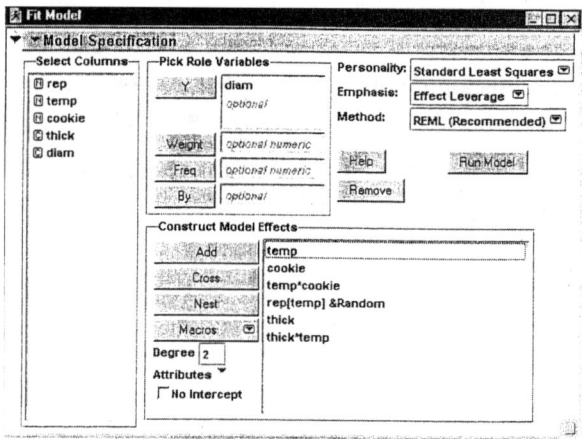

FIGURE 15.17 JMP® fit model screen to define final model for the baking cookie experiment.

REML Variance Component Estimates					
Random Effect	Var Ratio	Var Component	Std Error	95% Lower	95% Upper
rep[temp]&Random	6.7953736	10.886376	6.9565599	4.2104628	67.25443
Residual		1.6020275			
Total		12.488403			
-2 LogLikelihood = 143.79489					

Effect Tests							
Source	Nparm	DF	DFDen	Sum of Squares	F Ratio	Prob > F	
temp	2	2	9	9.92713	3.0983	0.0947	
cookie	2	2	15	114.51578	35.7409	<.0001	
temp*cookie	4	4	15	70.31600	10.9730	0.0002	
rep[temp]&Random	12	9	15	242.43318	16.8143	<.0001	Shru
thick	1	1	15	320.47411	200.0428	<.0001	
thick*temp	2	2	15	18.58424	5.8002	0.0136	

FIGURE 15.18 JMP® REML estimates of the variance components and test of the effects in the model for the baking cookie experiment.

15.9 EXAMPLE: TEACHING METHODS WITH ONE COVARIATE MEASURED ON THE LARGE SIZE EXPERIMENTAL UNIT AND ONE COVARIATE MEASURED ON THE SMALL SIZE EXPERIMENTAL UNIT

The data in Table 15.18 are from a study carried out by a school district to evaluate three methods of teaching mathematics as described in Section 15.5. The two factors in the study are the teaching methods and the sex of the students in the classroom. The design is a split-plot where the teacher or classroom is the large experimental unit to which the teaching methods were assigned and the student within the class room is the experimental unit to which the level of sex was assigned (at least conceptually). Figure 15.3 is a graphical representation of the assignment process. The squares represent the classrooms of students with four students shown per class (there should be 10 or more of each sex). The levels of teaching method were randomly assigned to the classrooms with each method assigned to three classes. The classroom is the experimental unit for the levels of teaching method with the experimental design being a one-way treatment structure in a completely randomized design structure. The levels of sex were conceptually randomly assigned to the students within each class; thus the student is the experimental unit for levels of sex. The student experimental design is a one-way treatment structure in a randomized complete block design structure where there are multiple students of each sex within each class or block. The data in Table 15.18 shows that there were 8 to 13 students of each sex within each classroom. Table 15.19 contains an analysis of variance table for this split-plot design without using any of the covariate information. The teacher error term is based on 6 degrees of freedom and the student error term is based on 181 degrees of freedom. Thus the approximate number of degrees of freedom associated with the denominator for the various tests from the mixed model can be between 2 and 183, depending on the number of parameters in the covariate part of the model. Pre-tests were given to each student to measure the initial knowledge of mathematics and the number of years of teaching experience for each teacher were used as possible covariates. A model with factorial effects for the intercepts, the slopes for years of experience and the slopes for pre-test scores is in Equation 15.8. The results of fitting Model 15.8 to the teaching method and the PROC MIXED code to do so are in Table 15.20. There are several terms in the covariate part of the model with coefficients that are not significantly different from zero; thus some model building needed to be carried out. The first term deleted was pre*method*sex as it has the largest significance level — 0.9131. The additional terms deleted using a backward or stepwise deletion process were pre*sex, yrs*method*sex, and yrs*method. The resulting reduced model is

$$Sc_{ijkm} = \mu + \rho_i + \theta\, Yrs_{ij} + \delta\, Pr\,e_{ijkm} + \kappa_i\, Pr\,e_{ijkm} + a_{ij} \} \text{ teacher} \tag{15.39}$$
$$+ \tau_k + (\rho\tau)_{ik} + \lambda_k\, Yrs_{ij} + \varepsilon_{ijkm} \} \text{ student}$$

The PROC MIXED code in Table 15.21 enables PROC MIXED to fit Model 15.39 to the teaching method data. The significance levels corresponding to pre*method and yrs*sex are 0.0098 and 0.0494, respectively, indicating there is sufficient evidence to conclude that the slopes for years of experience are different for each sex and the slopes for pre-test scores are different for the teaching methods. The full rank model for the intercepts, the slopes for years of experience, and the slopes for pre-test scores is

$$Sc_{ijkm} = \mu_{ik} + \kappa_i\, Pr\,e_{ijkm} + \lambda_k\, Yrs_{ij} + a_{ij} + \varepsilon_{ijkm} \tag{15.40}$$

The PROC MIXED code in Table 15.22 is used to fit the full rank model, Equation 15.40. The REML estimates of the variance components and combined estimates of the intercepts and slopes are in Table 15.22. The significance levels corresponding to method*sex, pre*method, and yrs*sex are very small, indicating there is sufficient evidence to conclude that all of the parameters are not equal to zero. The adjusted means evaluated at four values of pre-test scores, 77.67 (mean

TABLE 15.18
Test Scores for Male and Female Students Taught by One of Three Teaching Methods Where a Pre-Test Score and Years of Teaching Experience are Possible Covariates

		Teacher 1 with 7 years of exp				Teacher 2 with 4 years of exp				Teacher 3 with 4 years of exp			
		Male		Female		Male		Female		Male		Female	
Method	ID	Score	Pre Sc	Score	Pre Sc	Score	Pre Sc	Score	Pre Sc	Score	Pre Sc	Score	Pre Sc
I	1	82	94	76	69	79	67	77	66	77	67	79	90
I	2	80	92	81	79	80	74	78	74	76	93	76	92
I	3	75	81	74	79	78	92	78	81	81	79	71	66
I	4	77	72	81	79	79	72	77	64	79	72	76	85
I	5	80	90	76	70	81	75	77	75	77	62	72	61
I	6	82	86	76	77	73	60	76	64	79	82	79	95
I	7	77	65	78	80	77	85	79	88	74	62	77	91
I	8	75	65	77	63	79	76	79	62	79	88	74	95
I	9	79	65	77	84	76	81	77	75	76	63		
I	10	77	80	78	70	80	85			83	91		
I	11			79	83					78	62		
I	12			76	87					76	93		

		Teacher 4 with 7 years of exp				Teacher 5 with 8 years of exp				Teacher 6 with 3 years of exp			
		Male		Female		Male		Female		Male		Female	
		Score	Pre Sc	Score	Pre Sc	Score	Pre Sc	Score	Pre Sc	Score	Pre Sc	Score	Pre Sc
II	1	80	90	79	66	75	81	79	65	72	89	76	82
II	2	75	76	85	67	81	80	82	63	75	63	76	85
II	3	78	84	85	84	76	74	86	94	70	80	74	64

		Male		Female		Male		Female		Male		Female	
		Score	Pre Sc	Score	Pre Sc	Score	Pre Sc	Score	Pre Sc	Score	Pre Sc	Score	Pre Sc
II	4	76	76	85	76	77	71	83	75	73	84	77	83
II	5	78	94	80	68	78	74	83	65	74	69	77	74
II	6	71	85	79	66	76	84	84	86	74	83	75	90
II	7	70	87	81	91	78	86	86	87	76	91	78	78
II	8	78	89	80	84	80	86	85	68	67	70	76	73
II	9	78	91	81	90	78	83	82	72	73	72	80	75
II	10	77	90	79	90	74	70	81	69	65	62	76	80
II	11	74	63			78	64						
II	12	76	86										

Teacher 7 with 8 years of exp

		Male		Female	
		Score	Pre Sc	Score	Pre Sc
III	1	85	89	89	93
III	2	81	66	94	75
III	3	84	76	85	75
III	4	91	92	84	72
III	5	88	86	85	63
III	6	86	93	84	65
III	7	82	82	91	78
III	8	81	66	88	73
III	9	82	76	88	69
III	10	81	62	89	68
III	11	80	68	88	62
III	12	83	65	84	66
III	13			83	66

Teacher 8 with 6 years of exp

		Male		Female	
		Score	Pre Sc	Score	Pre Sc
III	1	83	63	84	79
III	2	80	64	84	66
III	3	78	61	88	94
III	4	77	65	84	90
III	5	82	74	85	67
III	6	88	89	84	74
III	7	83	83	88	75
III	8	82	87	89	78
III	9	88	77	90	92
III	10	84	75	85	75
III	11	77	65		

Teacher 9 with 6 years of exp

		Male		Female	
		Score	Pre Sc	Score	Pre Sc
III	1	86	66	99	88
III	2	88	85	99	89
III	3	89	86	89	78
III	4	87	86	95	75
III	5	90	94	90	76
III	6	90	91	91	82
III	7	86	83	92	90
III	8	91	90	92	74
III	9	86	74	92	75
III	10	90	89	90	70
III	11			94	92
III	12			90	78
III	13			92	89

TABLE 15.19
Analysis of Variance Table without Covariate Information for the Teaching Method Study

Source	df	EMS
Method	2	$\sigma_\varepsilon^2 + 21.293\,\sigma_a^2 + \phi^2(\text{method})$
Error(teacher) = Class(method)	6	$\sigma_\varepsilon^2 + 21.333\,\sigma_a^2$
Sex	1	$\sigma_\varepsilon^2 + \phi^2(\text{sex})$
Method*sex	2	$\sigma_\varepsilon^2 + \phi^2(\text{method}*\text{sex})$
Error(student)	181	σ_ε^2

TABLE 15.20
PROC MIXED Code and Results for Fitting a Model with Two-Way Factorial Effects for the Intercepts and Slopes for the Teaching Method Data Set

```
proc mixed data=score method=reml cl covtest;
class method teacher sex;
model score=method|sex pre pre*method pre*sex Pre*sex*method;
yrs yrs*method yrs*sex yrs*sex*method/ddfm=kr;
random teacher(method);
```

CovParm	Estimate	StdErr	ZValue	ProbZ	Alpha	Lower	Upper
Teacher(method)	0.2385	0.4422	0.54	0.2948	0.05	0.0371	32153.4762
Residual	5.5977	0.6039	9.27	0.0000	0.05	4.5798	6.9990

Effect	NumDF	DenDF	FValue	ProbF
method	2	22.4	2.28	0.1258
sex	1	174.7	1.84	0.1763
method*sex	2	174.1	0.58	0.5631
pre	1	172.0	44.65	0.0000
pre*method	2	172.0	4.69	0.0104
pre*sex	1	174.8	1.84	0.1766
pre*method*sex	2	174.4	0.09	0.9131
yrs	1	2.7	32.84	0.0134
yrs*method	2	2.8	4.16	0.1471
yrs*sex	1	172.3	3.65	0.0577
yrs*method*sex	2	172.2	0.12	0.8852

pre-test score), 70, 80, and 90, and the mean number of years of experience (6.59 years) are in Table 15.23. The adjusted means evaluated at three values of years of experience, 0, 5, and 10 years, and the mean pre-test score (77.67) are in Table 15.24. The adjusted means evaluated at 0 years of experience correspond to the expected mean test scores for new teachers. Since the slopes for years of experience are a function of sex, the female and male students need to be compared

TABLE 15.21
PROC MIXED Code and Results for Fitting a Model with Two-Way Factorial Effects for the Intercepts and One-Way Effects for the Slopes Corresponding to Pre-Test Scores and for Years of Teaching Experience

```
proc mixed data=score method=reml cl covtest;
class method teacher sex;
model score=method|sex pre pre*method
yrs yrs*sex/ddfm=kr;
random teacher(method);
```

CovParm	Estimate	StdErr	ZValue	ProbZ	Alpha	Lower	Upper
Teacher(method)	0.9198	0.7554	1.22	0.1117	0.05	0.2938	13.1064
Residual	5.5054	0.5853	9.41	0.0000	0.05	4.5169	6.8600

Effect	NumDF	DenDF	FValue	ProbF
method	2	150.4	0.40	0.6685
sex	1	177.5	0.13	0.7212
method*sex	2	177.3	28.64	0.0000
pre	1	181.2	49.17	0.0000
pre*method	2	181.1	4.75	0.0098
yrs	1	5.0	23.67	0.0047
yrs*sex	1	177.6	3.92	0.0494

at three or more values. Figures 15.19 and 15.20 contain plots of the teaching method by sex means in Tables 15.23 and 15.24. The female and male regression lines within a teaching method are parallel when graphed over the levels of pre-test scores. The regression models for the three teaching methods are parallel within each level of sex when graphed over the years of experience. Table 15.25 contains the pairwise comparisons of the female and male mean test scores within each level of teaching method for the for 0, 5, 10, and 6.59 years of experience and a pre-test score of 77.67. The means of the females and the means of the males are not significantly different ($p = 0.05$) for teaching method I with 6.59 and 10 years of experience and for teaching method III with 0 years of experience. The slopes for pre-test scores are a function of teaching method, the teaching methods need to be compared at three or more values of pre-test scores. Tables 15.26 and 15.27 are comparisons of the means of the three teaching methods within females and within males evaluated at 6.59 years of experience and pre-test scores of 70, 80, 90, and 77.67. The means of teaching method III are significantly larger than the means of teaching method I which are larger than the means of teaching method II for all comparisons. The above are LSD type comparisons, but other multiple comparison techniques could be used in making this set of comparisons.

This example involves two covariates where one is measured on the whole plot or large experimental unit and one is measured on the small experimental unit, a combination of the examples in Sections 15.7 and 15.8. As long as the computations are carried out using a mixed models approach, there is really no difference in the analysis of this split-plot design than carrying out an analysis of covariance with a

TABLE 15.22
PROC MIXED Code to Fit a Full Rank Means Model for the Intercepts and Slopes

```
proc mixed data=score method=reml cl covtest;
class method teacher sex;
model score=method*sex pre*method.yrs*sex/noint ddfm=kr solution;
random teacher(method);
```

CovParm	Estimate	StdErr	ZValue	ProbZ	Alpha	Lower	Upper
Teacher(method)	0.9198	0.7554	1.22	0.1117	0.05	0.2938	13.1064
Residual	5.5054	0.5853	9.41	0.0000	0.05	4.5169	6.8600

Effect	method	sex	Estimate	StdErr	df	tValue	Probt
method*sex	I	F	64.607	2.569	82.7	25.15	0.0000
method*sex	I	M	67.829	2.520	79.7	26.91	0.0000
method*sex	II	F	66.854	2.956	81.6	22.62	0.0000
method*sex	II	M	63.521	2.938	80.7	21.62	0.0000
method*sex	III	F	63.809	2.972	36.2	21.47	0.0000
method*sex	III	M	62.608	2.872	32.7	21.80	0.0000
pre*method	I		0.086	0.029	179.5	3.00	0.0031
pre*method	II		0.089	0.033	180.8	2.69	0.0077
pre*method	III		0.202	0.031	182.0	6.58	0.0000
yrs*sex		F	1.108	0.210	7.3	5.28	0.0010
yrs*sex		M	0.750	0.213	7.7	3.52	0.0084

Effect	NumDF	DenDF	FValue	ProbF
method*sex	6	109.7	227.84	0.0000
pre*method	3	181.0	19.63	0.0000
yrs*sex	2	12.2	13.26	0.0009

two-way treatment structure in a completely randomized or randomized complete block design structure. The next example involves a strip-plot design with three sizes of experimental units.

15.10 EXAMPLE: COMFORT STUDY IN A STRIP-PLOT DESIGN WITH THREE SIZES OF EXPERIMENTAL UNITS AND THREE COVARIATES

An environmental engineer designed an experiment to evaluate the affect of three environmental conditions on the comfort of female and male human subjects. On a given day, one female and one male were subjected to one of the three environments where both persons were put into an environmental chamber at the same time. On two other days, the same two persons were subjected to the other two environmental conditions; thus each person is subjected to all three environmental conditions on different days. After 1 hour in the environmental chamber, each person was given a set of questions which were used to compute a comfort score where the larger score indicates the person is warmer and the smaller score indicates the person is

TABLE 15.23
LSMEANS Code to Provide Adjusted method*sex Means Evaluated at the Mean Number of Years of Teaching Experience and Four Values of Pre-Test Scores

```
lsmeans method*sex/at means diff;
lsmeans method*sex/at (pre)=(70) diff;
lsmeans method*sex/at (pre)=(80) diff;
lsmeans method*sex/at (pre)=(90) diff;
```

Effect	method	sex	Years	Pre=77.76 Estimate	Pre=70 Estimate	Pre=80 Estimate	Pre=90 Estimate
method*sex	I	F	6.59	78.57	77.91	78.77	79.63
method*sex	I	M	6.59	79.43	78.77	79.63	80.49
method*sex	II	F	6.59	81.04	80.36	81.25	82.14
method*sex	II	M	6.59	75.35	74.67	75.56	76.44
method*sex	III	F	6.59	86.82	85.26	87.29	89.31
method*sex	III	M	6.59	83.26	81.70	83.73	85.75

TABLE 15.24
LSMEANS Code to Provide Adjusted method*sex Means Evaluated at the Mean Number of Pre-Test Score and for Three Values of Years of Teaching Experience

```
lsmeans method*sex/at (yrs)=(0) diff;
lsmeans method*sex/at (yrs)=(5) diff;
lsmeans method*sex/at (yrs)=(10) diff;
```

Effect	method	sex	pre	yr=0 Estimate	yr=5 Estimate	yr=10 Estimate
method*sex	I	F	77.67	71.27	76.81	82.35
method*sex	I	M	77.67	74.49	78.24	81.99
method*sex	II	F	77.67	73.74	79.28	84.83
method*sex	II	M	77.67	70.41	74.16	77.91
method*sex	III	F	77.67	79.52	85.06	90.60
method*sex	III	M	77.67	78.32	82.07	85.82

cooler (Table 15.28). Figure 15.21 is a graphical representation of the randomization process for this experiment. The picture depicts just 3 of the 12 blocks where 1 block consists of 3 horizontal rectangles and 2 vertical rectangles. The two vertical rectangles represent two persons and the three horizontal rectangles represent the days on which the two persons were observed within an environmental chamber set to one of the environmental conditions. The randomization process is to randomly assign sex of person to the two persons within each block (conceptually at least). Thus the person is the experimental unit for sex of person. If day is ignored, the person design is a one-way treatment structure (2 sexes) in a randomized complete block design structure (12 blocks). The error term for persons is computed as the

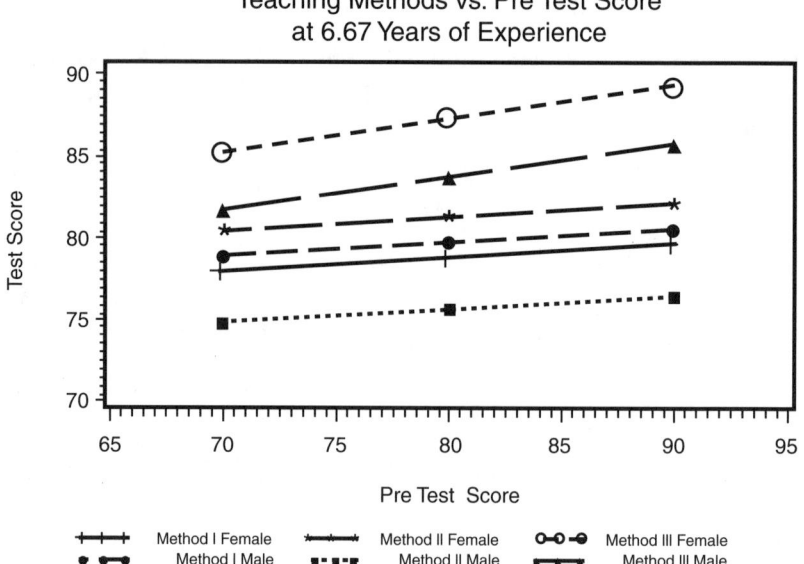

FIGURE 15.19 Graph of least squares means for method by sex combinations against pre-test score evaluated at years of experience of 6.59.

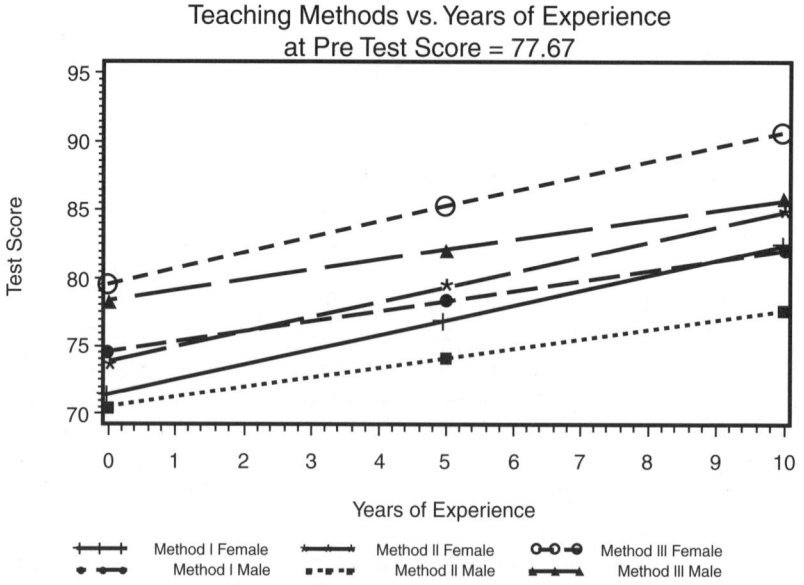

FIGURE 15.20 Graph of least squares means for method by sex combinations against years of experience evaluated at pre-test score of 77.67.

TABLE 15.25
Pairwise Comparisons of Female and Male Means within Each Teaching Method Evaluated at a Pre-Test Score of 77.67 and 0, 5, 10, and 6.59 Years of Teaching Experience

method	sex	_sex	pre	years	Estimate	StdErr	df	tValue	Probt
I	F	M	77.67	0.00	−3.222	1.102	178	−2.92	0.0039
II	F	M	77.67	0.00	3.333	1.242	177	2.68	0.0080
III	F	M	77.67	0.00	1.201	1.615	177	0.74	0.4583
I	F	M	77.67	5.00	−1.430	0.605	178	−2.36	0.0192
II	F	M	77.67	5.00	5.124	0.625	177	8.19	0.0000
III	F	M	77.67	5.00	2.992	0.832	177	3.60	0.0004
I	F	M	77.67	10.00	0.361	1.075	178	0.34	0.7372
II	F	M	77.67	10.00	6.916	0.937	177	7.38	0.0000
III	F	M	77.67	10.00	4.784	0.641	178	7.46	0.0000
I	F	M	77.67	6.59	−0.862	0.663	178	−1.30	0.1950
II	F	M	77.67	6.59	5.692	0.607	177	9.38	0.0000
III	F	M	77.67	6.59	3.560	0.652	177	5.46	0.0000

TABLE 15.26
Pairwise Comparisons of the Teaching Method Means for Female Students Evaluated at 6.59 Years of Teaching Experience and Pre-Test Scores of 70, 80, 90, and 77.67

sex	method	_method	pre	years	Estimate	StdErr	df	tValue	Probt
F	I	II	70.00	6.59	−2.451	1.060	9.1	−2.31	0.0459
F	I	III	70.00	6.59	−7.354	1.231	7.9	−5.97	0.0003
F	II	III	70.00	6.59	−4.903	1.133	8.3	−4.33	0.0023
F	I	II	80.00	6.59	−2.481	1.020	7.8	−2.43	0.0419
F	I	III	80.00	6.59	−8.519	1.210	7.4	−7.04	0.0002
F	II	III	80.00	6.59	−6.038	1.109	7.6	−5.44	0.0007
F	I	II	90.00	6.59	−2.510	1.154	12.5	−2.18	0.0494
F	I	III	90.00	6.59	−9.683	1.326	10.5	−7.30	0.0000
F	II	III	90.00	6.59	−7.174	1.254	12.2	−5.72	0.0001
F	I	II	77.67	6.59	−2.474	1.013	7.6	−2.44	0.0420
F	I	III	77.67	6.59	−8.248	1.202	7.2	−6.86	0.0002
F	II	III	77.67	6.59	−5.774	1.099	7.4	−5.25	0.0010

sex*block interaction. The two persons assigned to a given block will be subjected to the three environmental conditions on three different days. Next, randomly assign the levels of environment to the days within each block so that one environmental condition is used on a given day. Thus, day is the experimental unit for the levels of environment and the day design (ignoring persons) is a one-way treatment structure (three levels of environment) in a randomized complete block design structure where the day error is computed by the environment*day interaction. The time period

TABLE 15.27
Pairwise Comparisons of the Teaching Method Means for Male Students Evaluated at 6.59 Years of Teaching Experience and Pre-Test Scores of 70, 80, 90, and 77.67

sex	method	_method	pre	years	Estimate	StdErr	df	tValue	Probt
M	I	II	70.00	6.59	4.103	1.066	9.3	3.85	0.0037
M	I	III	70.00	6.59	−2.932	1.231	7.9	−2.38	0.0447
M	II	III	70.00	6.59	−7.035	1.148	8.7	−6.13	0.0002
M	I	II	80.00	6.59	4.074	1.006	7.4	4.05	0.0044
M	I	III	80.00	6.59	−4.096	1.213	7.5	−3.38	0.0107
M	II	III	80.00	6.59	−8.170	1.090	7.1	−7.49	0.0001
M	I	II	90.00	6.59	4.045	1.123	11.3	3.60	0.0040
M	I	III	90.00	6.59	−5.261	1.331	10.7	−3.95	0.0024
M	II	III	90.00	6.59	−9.305	1.206	10.5	−7.72	0.0000
M	I	II	77.67	6.59	4.081	1.003	7.3	4.07	0.0044
M	I	III	77.67	6.59	−3.825	1.204	7.3	−3.18	0.0148
M	II	III	77.67	6.59	−7.906	1.088	7.1	−7.27	0.0002

a person spends within an environment is the experimental unit for the interaction between the levels of sex and the levels of environment. Therefore, this experiment involves three different sizes of experimental units. Table 15.29 contains the analysis of variance table corresponding to this design using the model

$$\begin{aligned} Sc_{ijk} &= \mu + b_i \} \quad \text{block} \\ &+ \rho_j + p_{ij} \} \quad \text{person} \\ &+ \delta_k + d_{ik} \} \quad \text{day} \\ &+ (\rho\delta)_{jk} + \varepsilon_{ijk} \} \quad \text{day} * \text{person} \end{aligned} \quad (15.41)$$

The person error term is based on 11 degrees of freedom, the day error term is based on 22 degrees of freedom, and the day*person error term is based on 22 degrees of freedom. The approximate denominator degrees of freedom associated with the tests from the mixed model analysis can be from 2 to 55 depending on the number of covariance parameters in the model.

Three variables were measured to be considered as possible covariates for this analysis: the age of each person was determined, which is a covariate measured on the person experimental unit; the outdoor temperature (°F) at 10 a.m. in the morning of the day two persons were subjected to an environmental condition, which is a covariate measured on the day experimental unit; and finally, the baseline skin temperature (°F) of each person was measured at the beginning of the test, which is a covariate measured on the day*person experimental unit. A model with factorial effects for the intercepts and for the slopes of each of the three covariates is

TABLE 15.28
Comfort Scores for Males and Females Exposed to Three Different Environments Where the Person's Age, the Temperature of the Day, and Baseline Skin Temperature are Possible Covariates. Prefix F_ is for females and M_ is for males.

Block	Env	temp	F_score	F_age	F_baseSkT	M_score	M_age	M_baseSkT
1	I	53	10	24	86	8.9	23	90
1	II	52	10	24	88	9.4	23	92
1	III	54	13	24	94	11.0	23	89
2	I	54	8	26	87	8.4	25	91
2	II	54	9	26	90	8.5	25	87
2	III	53	11	26	88	10.0	25	92
3	I	51	9	18	90	8.0	20	89
3	II	54	10	18	91	8.9	20	90
3	III	54	12	18	90	10.0	20	90
4	I	58	8	26	92	8.6	25	90
4	II	59	9	26	92	9.0	25	87
4	III	64	10	26	88	11.2	25	93
5	I	64	8	24	91	8.6	18	93
5	II	67	9	24	91	9.9	18	91
5	III	69	11	24	90	11.0	18	94
6	I	69	8	25	93	7.8	23	93
6	II	69	9	25	94	9.3	23	93
6	III	67	10	25	89	9.8	23	94
7	I	66	8	19	90	7.1	24	91
7	II	65	9	19	88	8.5	24	86
7	III	65	11	19	93	9.2	24	94
8	I	63	8	20	94	8.3	24	87
8	II	64	10	20	90	9.8	24	89
8	III	61	11	20	93	10.5	24	87
9	I	59	8	20	93	7.6	22	93
9	II	56	9	20	94	8.3	22	90
9	III	57	11	20	90	9.9	22	91
10	I	54	9	21	89	8.4	22	92
10	II	54	10	21	93	9.1	22	91
10	III	51	11	21	93	9.8	22	88
11	I	51	8	21	88	7.0	21	93
11	II	53	10	21	92	8.8	21	88
11	III	51	11	21	86	9.4	21	91
12	I	51	8	19	94	7.4	19	86
12	II	47	10	19	87	8.5	19	91
12	III	51	11	19	90	9.6	19	92

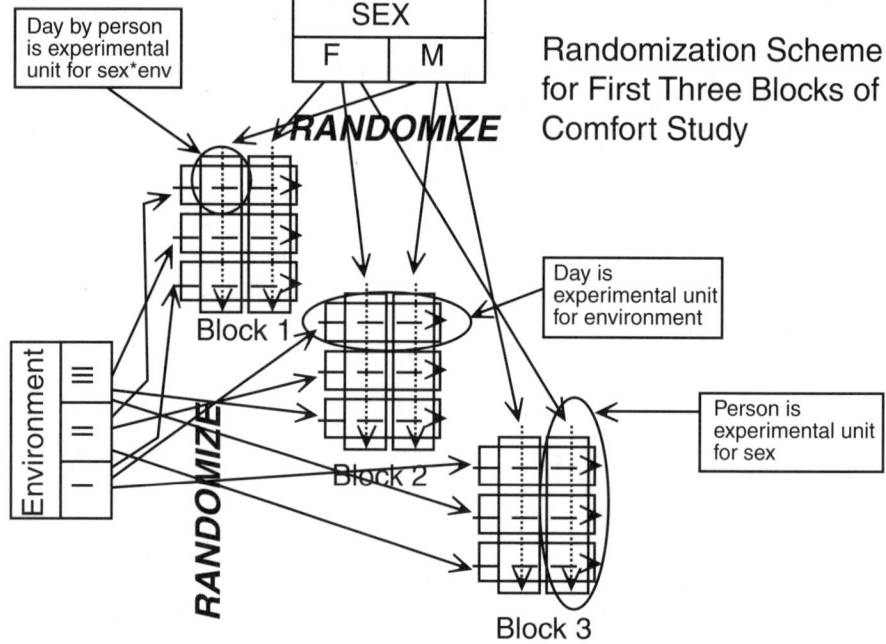

FIGURE 15.21 Graphical representation of the randomization scheme for the comfort study where the vertical rectangles represent persons and the horizontal rectangles represent days.

TABLE 15.29
Analysis of Variance Table without Covariate Information for the Comfort Study in a Strip-Plot Design

Source	df	EMS
Block	11	$\sigma_\varepsilon^2 + 2\sigma_d^2 + 3\sigma_p^2 + 6\sigma_b^2$
Sex	1	$\sigma_\varepsilon^2 + 3\sigma_p^2 + \phi^2(\text{sex})$
Error(person) = Sex*Block	11	$\sigma_\varepsilon^2 + 3\sigma_p^2$
Environment	2	$\sigma_\varepsilon^2 + 3\sigma_d^2 + \phi^2(\text{Env})$
Error(day) = Env*Block	22	$\sigma_\varepsilon^2 + 3\sigma_d^2$
Env*Sex	2	$\sigma_\varepsilon^2 + \phi^2(\text{Env}*\text{sex})$
Error(day*person)	22	σ_ε^2

TABLE 15.30
PROC MIXED Code and Results for Fitting a Model with a Two-Way Factorial Effects Model for the Intercepts and for the Slopes for Age, Day Temperature, and Baseline Skin Temperature

```
proc mixed method=reml cl covtest data=comfort;
class block env sex;
model score=env|sex age age*sex age*env age*sex*env
temp sex*temp temp*env temp*sex*env
baseSkt baseskt*sex baseskt*env baseskt*sex*env/solution ddfm=kr;
random block block*sex block*env;
```

CovParm	Estimate	StdErr	ZValue	ProbZ	Alpha	Lower	Upper
Block	0.1957	0.1357	1.44	0.0747	0.05	0.0713	1.5227
Block*sex	0.0834	0.0430	1.94	0.0261	0.05	0.0373	0.3224
Block*Env	0.1007	0.0384	2.62	0.0044	0.05	0.0537	0.2530
Residual	0.0126	0.0060	2.10	0.0179	0.05	0.0059	0.0425

Effect	NumDF	DenDF	FValue	ProbF
Env	2	20.9	1.03	0.3746
sex	1	19.7	9.53	0.0059
Env*sex	2	16.3	1.60	0.2320
age	1	14.3	1.45	0.2481
age*sex	1	10.2	4.28	0.0648
age*Env	2	13.9	1.80	0.2011
age*Env*sex	2	10.4	0.10	0.9022
temp	1	16.9	0.29	0.5990
temp*sex	1	20.1	10.77	0.0037
temp*Env	2	17.6	0.43	0.6580
temp*Env*sex	2	12.1	2.33	0.1391
baseSkT	1	10.9	13.84	0.0034
baseSkT*sex	1	14.4	0.62	0.4423
baseSkT*Env	2	13.2	0.57	0.5792
baseSkT*Env*sex	2	16.5	1.57	0.2385

$$\begin{aligned} Sc_{ijk} = \mu &+ \tau\, age_{ij} + \alpha\, temp_{ik} + \beta\, baseSkT_{ijk} + b_i\} \text{ block} \\ &+ \rho_j + \gamma_j\, age_{ij} + \psi_j\, temp_{ik} + \theta_j\, baseSkT_{ijk} + p_{ij}\} \text{ person} \\ &+ \delta_k + \zeta_k\, age_{ij} + \delta_k\, temp_{ik} + \phi_k\, baseSkT_{ijk} + d_{ik}\} \text{ day} \\ &+ (\rho\delta)_{jk} + \eta_{jk}\, age_{ij} + \nu_{jk}\, temp_{ik} + \omega_{jk}\, baseSkT_{ijk} + \varepsilon_{ijk}\} \text{ day} * \text{person} \end{aligned} \quad (15.42)$$

The PROC MIXED code in Table 15.30 is used to fit Model 15.42 to the comfort study data set. The REML estimates of the variance components are in the second part of Table 15.30 and the lower part of the table contains the tests statistics for

TABLE 15.31
PROC MIXED Code and Results for Fitting a Model with a Two-Way Factorial Effects Model for the Intercepts and One-Way Effects Model for the Slopes of Day Temperature, and a Common Slope for Baseline Skin Temperature

```
proc mixed method=reml cl covtest data=comfort;
class block env sex;
model score=env|sex temp sex*temp baseSkt/ddfm=kr;
random block block*sex block*env;
```

CovParm	Estimate	StdErr	ZValue	ProbZ	Alpha	Lower	Upper
Block	0.1445	0.1237	1.17	0.1213	0.05	0.0446	2.4677
Block*sex	0.1511	0.0682	2.22	0.0133	0.05	0.0734	0.4714
Block*Env	0.0730	0.0264	2.77	0.0028	0.05	0.0401	0.1729
Residual	0.0186	0.0058	3.18	0.0007	0.05	0.0109	0.0386

Effect	NumDF	DenDF	FValue	ProbF
Env	2	19.9	215.51	0.0000
sex	1	29.4	13.32	0.0010
Env*sex	2	20.4	42.09	0.0000
temp	1	18.6	0.13	0.7263
temp*sex	1	29.8	10.54	0.0029
baseSkT	1	23.0	10.68	0.0034

the various factorial effects in the model. Several of the terms corresponding to the covariate part of the model have significance levels that are larger than 0.05, indicating that some of the model's parameters are not significantly different from zero. So, some model building is in order. Using the stepwise deletion process, the first term to be deleted was age*env*sex ($p = 0.9022$). Continuing the process, the following terms were deleted in order: age*sex*env, baseskt*sex*env, baseskt*env, age*env, temp*sex*env, temp*env, age*sex, and age. The model, after reducing the covariate part by stepwise deletion, is

$$Sc_{ijk} = \mu + \alpha\ temp_{ik} + \beta\ baseSkT_{ijk} + b_i \} \quad \text{block}$$
$$+ \rho_j + \psi_j\ temp_{ik} + p_{ij} \} \quad \text{person} \quad (15.43)$$
$$+ \delta_k + d_{ik} \} \quad \text{day}$$
$$+ (\rho\delta)_{jk} + \varepsilon_{ijk} \} \quad \text{day} * \text{person}$$

The PROC MIXED code in Table 15.31 fits the reduced model to the data set. The REML estimates of the variance components in Table 15.31 are not too different than the REML estimates variance components for the full model (from Table 15.30).

TABLE 15.32
PROC MIXED Code to Fit the Full Rank Means Model for the Intercepts and Temperature Slopes

```
proc mixed method=reml cl covtest data=comfort;
class block env sex;
model score=env*sex  baseSkt sex*temp/noint solution
  ddfm=kr;
random block block*sex block*env;
```

CovParm	Estimate	StdErr	ZValue	ProbZ	Alpha	Lower	Upper
Block	0.1445	0.1237	1.17	0.1213	0.05	0.0446	2.4677
Block*sex	0.1511	0.0682	2.22	0.0133	0.05	0.0734	0.4714
Block*Env	0.0730	0.0264	2.77	0.0028	0.05	0.0401	0.1729
Residual	0.0186	0.0058	3.18	0.0007	0.05	0.0109	0.0386

Effect	Env	sex	Estimate	StdErr	DF	tValue	Probt
Env*sex	I	F	6.4634	1.5620	42.2	4.14	0.0002
Env*sex	I	M	2.8864	1.5464	41.3	1.87	0.0691
Env*sex	II	F	7.6986	1.5649	42.2	4.92	0.0000
Env*sex	II	M	3.9108	1.5420	41.1	2.54	0.0151
Env*sex	III	F	9.2540	1.5666	42.0	5.91	0.0000
Env*sex	III	M	4.9631	1.5555	41.3	3.19	0.0027
baseSkT			0.0331	0.0101	23.0	3.27	0.0034
temp*sex		F	−0.0222	0.0224	26.1	−0.99	0.3291
temp*sex		M	0.0368	0.0224	26.2	1.64	0.1125

Effect	NumDF	DenDF	FValue	ProbF
Env*sex	6	37.3	86.53	0.0000
baseSkT	1	23.0	10.30	0.0039
temp*sex	2	29.9	5.22	0.0114

The tests for the remaining terms of the covariate part of the model have significance levels less than 0.01. The exception is the coefficient for temp, but since there is a significant difference between the slopes for males and for females, temp was retained in the model. A means model with a two-way means model effects representation for the intercepts, a one-way means model effects representation for the temp slopes for sex, and a common slope for baseSkt is

$$Sc_{ijk} = \mu_{ik} + \beta \text{ baseSkT}_{ijk} + \psi_j \text{ temp}_{ik} + b_i + p_{ij} + d_{ik} + \varepsilon_{ijk} \qquad (15.44)$$

The PROC MIXED code and results from fitting Model 15.44 to the comfort data are in Table 15.32. The estimate of the slopes for temp are −0.0222 and 0.0368 for females and males, respectively. Neither temperature slope is significantly different from zero (significance levels of 0.3291 and 0.1125), but the two slopes are significantly different from each other; thus temperature with different slopes for the two sexes was included in the model. The REML estimates of the variance components

TABLE 15.33
LSMEANS Statements to Provide Adjusted Means for the Environment by Sex Combinations Evaluated at the Baseline Skin Temperature of 90.54 and Three Day Temperatures

```
lsmeans env*sex/diff at temp=50;
lsmeans env*sex/diff at temp=60;
lsmeans env*sex/diff at temp=70;
```

Env	sex	baseSkT	temp	Estimate	StdErr	DF	tValue	Probt
I	F	90.54	50.00	8.35	0.25	23	33.44	0.0000
I	M	90.54	50.00	7.72	0.25	23	30.92	0.0000
II	F	90.54	50.00	9.58	0.25	23	38.20	0.0000
II	M	90.54	50.00	8.74	0.25	23	34.74	0.0000
III	F	90.54	50.00	11.14	0.25	23	43.69	0.0000
III	M	90.54	50.00	9.80	0.25	23	38.45	0.0000
I	F	90.54	60.00	8.12	0.19	21	43.55	0.0000
I	M	90.54	60.00	8.09	0.19	21	43.34	0.0000
II	F	90.54	60.00	9.36	0.19	21	50.29	0.0000
II	M	90.54	60.00	9.11	0.19	21	48.95	0.0000
III	F	90.54	60.00	10.91	0.18	20	59.10	0.0000
III	M	90.54	60.00	10.16	0.18	21	54.96	0.0000
I	F	90.54	70.00	7.90	0.33	24	24.12	0.0000
I	M	90.54	70.00	8.45	0.33	24	25.78	0.0000
II	F	90.54	70.00	9.14	0.33	24	28.02	0.0000
II	M	90.54	70.00	9.48	0.33	24	29.09	0.0000
III	F	90.54	70.00	10.69	0.32	24	33.28	0.0000
III	M	90.54	70.00	10.53	0.32	24	32.70	0.0000

in Table 15.32 are identical to those in Table 15.31, indicating that the two models are fitting the data identically. Since the regression models are different for the two levels of sex, the models need to be evaluated at three or more values of temp. Table 15.33 contains the LSMEANS statements and results for computing the adjusted means at the mean baseSkt value (90.54°F) and at temp values of 50, 60, and 70°F. Pairwise comparisons using the LSD approach between the female means and the male means within each environment at temp values of 50, 60, and 70°F are in Table 15.34. Females are significantly warmer than males for environments I, II, and III at temp = 50°F and for environment III at 60°F. Table 15.35 contains the pairwise comparisons of the levels of environment within each sex evaluated at just one value of temp, 60°F. Persons exposed to environment I are significantly cooler than those exposed to environment II which are significantly cooler than those in environment III. Figure 15.22 is a plot of the adjusted means in Table 15.33 over the values of temp. The three environment models within each sex are parallel, but models from different levels of sex are not parallel.

Figure 15.23 contains the JMP® data table with the comfort data set. The variables block, env, and sex are declared to be nominal and the other variables are declared to be continuous. The fit model table is in Figure 15.24. The terms block,

TABLE 15.34
Comparisons of Female and Male Means within Each Environment Evaluated at the Baseline Skin Temperature of 90.54 and Three Day Temperatures

Env	sex	_sex	baseSkT	temp	Estimate	StdErr	df	tValue	Probt
I	F	M	90.54	50	0.63	0.22	18.3	2.86	0.0103
I	F	M	90.54	60	0.04	0.17	13.0	0.21	0.8355
I	F	M	90.54	70	−0.55	0.28	22.8	−1.98	0.0595
II	F	M	90.54	50	0.84	0.22	18.5	3.79	0.0013
II	F	M	90.54	60	0.25	0.17	13.1	1.43	0.1763
II	F	M	90.54	70	−0.34	0.28	22.8	−1.23	0.2301
III	F	M	90.54	50	1.34	0.22	18.7	6.00	0.0000
III	F	M	90.54	60	0.75	0.17	12.9	4.36	0.0008
III	F	M	90.54	70	0.16	0.27	22.6	0.58	0.5646

TABLE 15.35
Pairwise Comparisons of Environment Means within Males and Females Evaluated at the Baseline Skin Temperature of 90.54 and Three Day Temperatures

sex	Env	_Env	baseSkT	temp	Estimate	StdErr	df	tValue	Probt
F	I	II	90.54	60	−1.24	0.12	24.3	−9.99	0.0000
F	I	III	90.54	60	−2.79	0.12	24.4	−22.54	0.0000
F	II	III	90.54	60	−1.56	0.12	24.4	−12.56	0.0000
M	I	II	90.54	60	−1.02	0.12	24.6	−8.26	0.0000
M	I	III	90.54	60	−2.08	0.12	24.5	−16.76	0.0000
M	II	III	90.54	60	−1.05	0.12	25.1	−8.43	0.0000

sex*block, and env*block are declared to be random effects by using the attributes selection menu. The remaining terms in the model are the factorial effects for the intercepts, the one-way effects model for the slopes of temp, and a common slope for baseSkt, the same model fit by PROC MIXED to produce the results in Table 15.31. Deselect the center polynomials option from the model specification menu of the JMP® fit model screen. The analysis of variance table with the error mean square is in Figure 15.25. This error term provides an estimate of the day*person variance component. The F-ratio corresponding to the source "MODEL" is not interpretable since this model sum of squares combines all of the fixed effects and all of the random effects in one test. Figure 15.26 contains the REML estimates of the variance components and tests for the fixed and random effects in the model. The results in Figure 15.26 are a little different than the PROC MIXED results in Table 15.31, but differences are mainly due to different approximations to the denominator degrees of freedom. Adjusted means could be obtained from JMP® by using the custom test menu, as was done in Chapter 3, but those results were not included here.

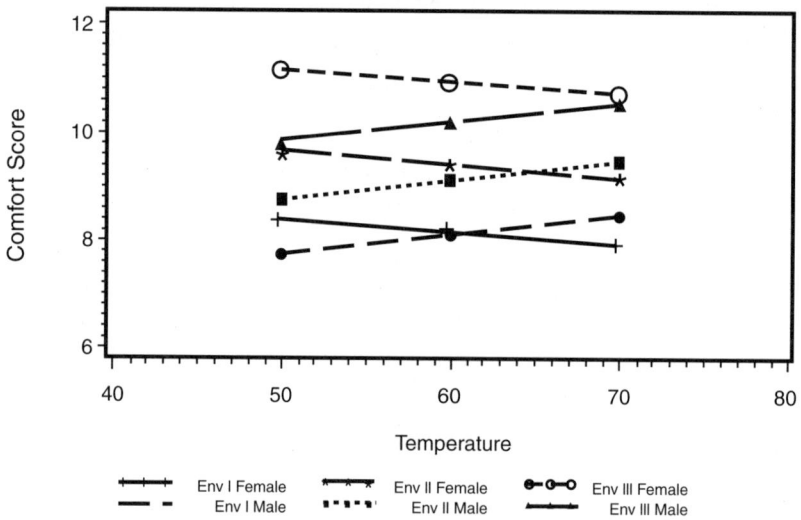

FIGURE 15.22 Comfort study. Plot of environment by sex least squares as a function of temperature.

ex_15_10

Columns (7/0): Block, Env, temp, sex, score, baseSkT, age

Rows: All Rows 72, Selected 0, Excluded 0, Hidden 0, Labelled 0

	Block	Env	temp	sex	score	baseSkT	age
1	1	I	53	F	9.7	86	24
2	1	I	53	M	8.9	90	23
3	1	II	52	F	10.2	88	24
4	1	II	52	M	9.4	92	23
5	1	III	54	F	12.5	94	24
6	1	III	54	M	11	89	23
7	2	I	54	F	7.9	87	26
8	2	I	54	M	8.4	91	25
9	2	II	54	F	8.7	90	26
10	2	II	54	M	8.5	87	25
11	2	III	53	F	10.6	88	26
12	2	III	53	M	10	92	25
13	3	I	51	F	9.4	90	18
14	3	I	51	M	8	89	20
15	3	II	54	F	10.3	91	18
16	3	II	54	M	8.9	90	20
17	3	III	54	F	12	90	18
18	3	III	54	M	10	90	20
19	4	I	58	F	7.8	92	26
20	4	I	58	M	8.6	90	25

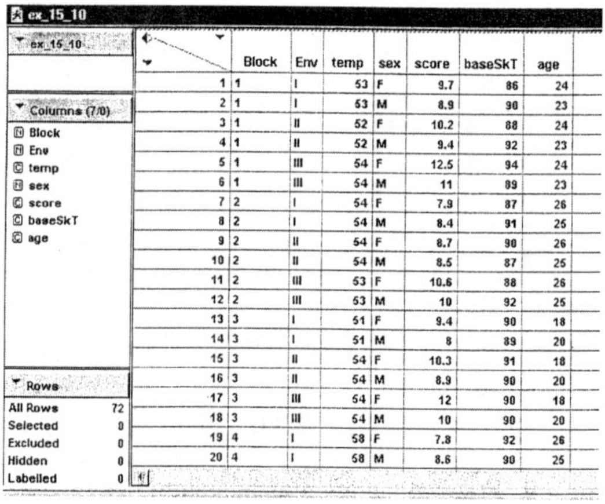

FIGURE 15.23 JMP® data table for comfort study.

15.11 CONCLUSIONS

The examples discussed in the above sections demonstrate that the covariate can be measured on any size of experimental unit. The main concern is that the values of the covariate must not be affected by the application of the treatment combinations. In the comfort experiment, there might be a remote chance that the baseline skin temperature could be affected by the person being in the previous environmental

Analysis of Covariance for Split-Plot and Strip-Plot Design Structures

FIGURE 15.24 JMP® model specification table to fit strip-plot model to the comfort data set.

Analysis of Variance

Source	DF	Sum of Squares	Mean Square	F Ratio
Model	52	102.37288	1.96871	105.8655
Error	19	0.35333	0.01860	Prob > F
C. Total	71	103.54444		<.0001

FIGURE 15.25 Overall analysis of variance table where the error term is the error for the person*day experimental unit.

Response score

Parameter Estimates

Term	Estimate	Std Error	t Ratio	Prob>|t|
sex[r]*temp	-0.029303	0.008511	-3.47	0.0025

REML Variance Component Estimates

Random Effect	Var Ratio	Var Component	Std Error	95% Lower	95% Upper	Pct of Total
Block&Random	7.7720046	0.1445307	0.1314405	0.0424626	3.3150504	37.327
sex*Block&Random	8.1238116	0.151073	0.0839662	0.0643622	0.6760258	39.017
Env*Block&Random	3.9253167	0.0729965	0.0370412	0.0329954	0.2748681	18.853
Residual		0.0185963				4.803
Total		0.3871965				100.000

-2 LogLikelihood = 97.650464

Effect Tests

Source	Nparm	DF	DFDen	Sum of Squares	F Ratio	Prob > F	
Env	2	2	22	8.0191452	215.8111	<.0001	
sex	1	1	11	0.2814696	15.1358	0.0025	
sex*Env	2	2	19	1.5674594	42.1443	<.0001	
Block&Random	12	11	11	0.2542865	1.2431	0.3623	Shrunk
sex*Block&Random	24	11	19	4.7868157	23.4006	<.0001	Shrunk
Env*Block&Random	36	22	19	2.7961889	6.8347	<.0001	Shrunk
baseSkT	1	1	19	0.2039336	10.9663	0.0037	
temp	1	1	19	0.0027134	0.1459	0.7067	
sex*temp	1	1	19	0.2234396	12.0153	0.0026	

Tests on Random effects refer to shrunken predictors rather than traditional estimates.

FIGURE 15.26 REML estimates of the variance components and the analysis of variance table for the effects of the split-plot design for the comfort data set.

condition, but that is not likely here. By using a mixed models approach to the analysis, all of the information from the between day model, the between person model and the within day*person model for the models' fixed effect parameters are combined into more efficient estimators than if a non-mixed models approach were used. Using the mixed models approach for the analysis of the split-plot and strip-plot designs depends on being able to appropriately describe the error terms, but the analysis of covariance strategy is the same as for the two-way or higher order treatment structure models discussed in Chapter 5.

REFERENCES

Kempthorne, O. (1952) *The Design and Analysis of Experiments,* John Wiley & Sons, New York.

Milliken, G. A. and Johnson, D. E. (1992) *Analysis of Messy Data, Volume I: Design Experiments,* Chapman & Hall, London.

EXERCISES

EXERCISE 15.1: In a study of the times required to dissolve a piece of a type of chocolate candy, each person dissolved three of six candy types: red, blue, and small M&M®s, buttons, chocolate chips, and Sno-Caps®. The study involved 10 females and 10 males. The person's age and the time required to dissolve a butterscotch chip (base) were measured as possible covariates. Use the following data set to carryout an analysis of covariance and make all meaningful comparisons.

Data for Exercise 15.1

Person	Sex	Age	Base	Blue M&M	Buttons	Choc Chip	Red M&M	Small M&M	Sno-Caps
1	F	21	16	31	46	37			
2	F	20	22	43	50				25
3	F	19	15	31		33	32		
4	F	22	29	50		48			39
5	F	19	29	45			41	24	
6	F	21	14		45	31	31		
7	F	19	28		54	44		24	
8	F	19	22		49		31		23
9	F	19	27			43		24	32
10	F	19	23				37	21	27
1	M	22	33	87	95	83			
2	M	19	32	79	86				64
3	M	19	24	66		67	64		
4	M	18	20	58		59			46
5	M	21	22	64			62	43	
6	M	19	18		73	60	55		
7	M	22	19		66	55		40	
8	M	18	23		75		58		47
9	M	19	22			60		41	51
10	M	21	21				57	38	48

EXERCISE 15.2: An animal scientist and a meat scientist worked together to study the effect of fat level in the diet of beef animals and different post-mortem treatments of the carcasses on the tenderness of two types of steaks. The three rations were control, high fat, and low fat. The animals of each ration were fed until at least all but one reached market weight, at which time all animals of a given ration were slaughtered. The final weight of each animal was measured as a possible covariate. At slaughter, each carcass was split into two halves where one half was assigned to the control or usual post-mortem treatment and the other half was assigned to the electric stimulated post-mortem treatment. There were some differences in the quality of the two halves, so the yield grade of each half of carcass was measured as a possible covariate. Two steaks were cut from each half of carcass, one from the loin and one from the round. Eight cores were punched in each steak and the shear force was determined for each. The mean force of the eight cores was used as the measure of the tenderness of a steak. The thickness of the fat layer around the outside of each steak was measured as a possible covariate. Provide a thorough analysis of the following data set.

Data for Exercise 15.2

				Post-Mortem Control				Post-Mortem Electric Stimulate				
				Round		Loin			Round		Loin	
Ration	Animal	Weight	Grade	Fat	Shear	Fat	Shear	Grade	Fat	Shear	Fat	Shear
Control	1	1144	3	18	6.3	13	6.0	3	14	6.3	15	5.7
Control	2	1067	4	12	7.9	19	7.1	4	14	7.7	15	7.2
Control	3	1163	4	16	6.3	16	6.0	3	13	6.3	13	5.6
Control	4	977	4	19	5.1	14	5.3	5	12	5.9	21	5.1
Control	5	1128	5	16	5.6	15	5.0	5	11	5.5	15	4.9
Control	6	1142	5	15	7.4	12	6.9	5	18	6.6	20	6.0
Control	7	1182	4	15	4.4	11	4.0	4	11	3.9	19	2.9
High Fat	1	1186	4	15	7.5	15	6.9	3	14	7.0	14	6.5
High Fat	2	1057	4	22	7.7	16	7.5	5	19	6.1	11	6.0
High Fat	3	1126	5	11	6.6	15	5.8	4	16	6.1	18	5.4
High Fat	4	1077	4	12	6.5	11	6.3	3	21	6.3	18	5.8
High Fat	5	1225	4	16	7.1	11	6.4	4	10	6.5	15	5.9
Low Fat	1	1119	5	11	6.2	11	5.5	5	14	5.7	14	5.0
Low Fat	2	1082	5	14	7.5	19	6.7	5	15	6.7	11	6.8
Low Fat	3	968	4	16	7.4	16	7.1	5	12	7.0	12	6.5
Low Fat	4	1171	5	17	4.5	13	4.3	5	15	3.9	18	3.2
Low Fat	5	1216	2	16	7.5	16	6.9	3	12	6.9	19	5.7
Low Fat	6	1069	5	16	6.8	13	6.5	5	12	6.8	20	5.6

EXERCISE 15.3: The following data are from a complex strip plot design from an experiment setup to study the effect of temperature (TEMP at three levels, 900, 1000, 1100°F) on the deposition rate (Delta) of a layer of polysilicon in the fabrication of wafers in the semiconductor industry. The experiment includes two wafer

types (A and B) in order to study the possibility of interaction between temperature and wafer type. The experiment consists of putting two wafers of each type into a cassette and then inserting the cassette into a furnace for treatment at a given level of temperature. Wafers of a given type are produced in groups of six wafers called lots, and four lots of each type of wafer were available for use in the experiment. The six wafers from one randomly selected lot of type A wafers were randomly divided into three groups of two wafers. These three groups of two type A wafers were randomly assigned to the three furnaces in block 1 of the experiment. Similarly, the six wafers from one randomly selected lot of type B wafers were randomly divided into three groups of two wafers. These three groups of two type B wafers were randomly assigned to the three furnaces in block 1 of the experiment. This process was repeated 4 times using a total of 4 lots of type A wafers, 4 lots of type B wafers, and 12 furnaces (actually, the same furnace was used each time) providing 4 blocks. The experiment was conducted by randomly assigning the three temperatures to the three furnaces in each block. The measurements are the amount of deposited material at three randomly chosen sites on each wafer. It was thought that the wafer thickness before the deposition process was applied may have an effect on the deposition rate. Therefore, the average thickness of each wafer was determined and used as a possible covariate. The diagram in Figure 15.27 illustrates the assignment of the factors to the different experimental units for each block. Carry out a thorough analysis of this data set and make all necessary comparisons.

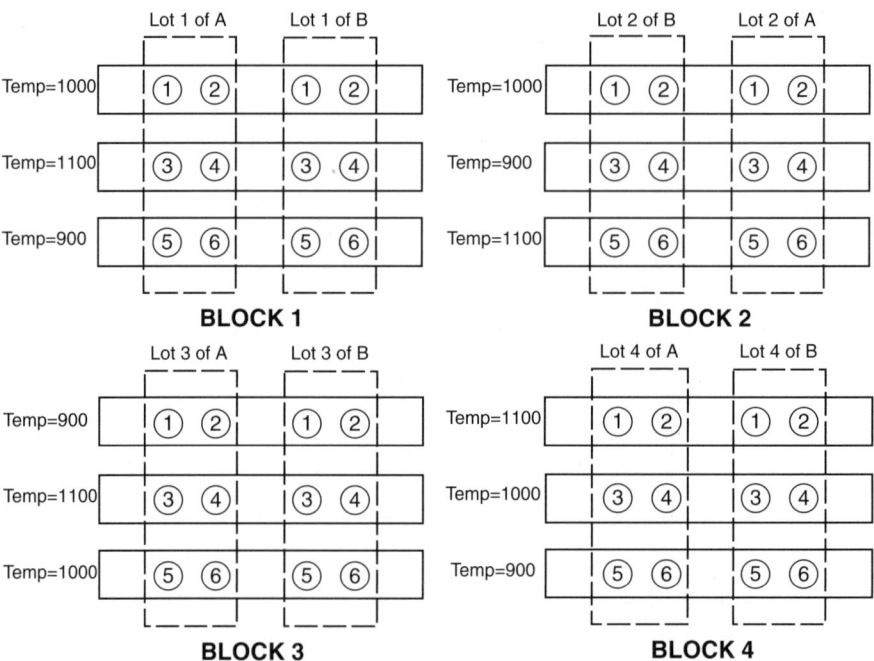

FIGURE 15.27 Randomization scheme for the data from Exercise 15.3.

Data Set – Wafers for Exercise 15.3

Temperature (°F)	Site	Wafer Type A				Wafer Type B			
		Wafer 1		Wafer 2		Wafer 1		Wafer 2	
		Delta	Thick	Delta	Thick	Delta	Thick	Delta	Thick
Block 1									
900	1	291	1919	318	2113	349	1965	332	1829
	2	295	1919	315	2113	348	1965	334	1829
	3	294	1919	315	2113	345	1965	331	1829
1000	1	319	2098	290	1823	358	2059	365	2145
	2	315	2098	289	1823	357	2059	367	2145
	3	321	2098	292	1823	362	2059	367	2145
1100	1	264	1846	276	2028	352	2086	330	1899
	2	266	1846	280	2028	353	2086	330	1899
	3	268	1846	278	2028	350	2086	334	1899
Block 2									
900	1	306	1841	342	2170	342	1981	366	2190
	2	302	1841	341	2170	341	1981	363	2190
	3	305	1841	336	2170	340	1981	361	2190
1000	1	299	1915	329	2161	348	2072	350	2082
	2	296	1915	330	2161	346	2072	346	2082
	3	297	1915	332	2161	346	2072	347	2082
1100	1	285	1854	306	2046	357	2062	361	2055
	2	292	1854	303	2046	360	2062	361	2055
	3	289	1854	304	2046	359	2062	360	2055

Data Set – Wafers for Exercise 15.3 (Continued)

Temperature (°F)	Site	Wafer Type A				Wafer Type B			
		Wafer 1		Wafer 2		Wafer 1		Wafer 2	
		Delta	Thick	Delta	Thick	Delta	Thick	Delta	Thick
Block 3									
900	1	318	2019	307	872	372	2182	348	1973
	2	323	2019	308	872	371	2182	349	1973
	3	323	2019	308	1872	370	2182	352	1973
1000	1	264	1828	274	1827	332	2109	322	2003
	2	265	1828	268	1827	337	2109	326	2003
	3	265	1828	275	1827	335	2109	321	2003
1100	1	273	1925	276	1942	333	1893	349	2170
	2	275	1925	273	1942	332	1893	350	2170
	3	276	1925	273	1942	332	1893	352	2170
Block 4									
900	1	295	1862	326	2149	322	1888	335	1998
	2	297	1862	326	2149	325	1888	332	1998
	3	296	1862	328	2149	327	1888	334	1998
1000	1	258	1815	280	1981	319	2012	311	1892
	2	260	1815	276	1981	322	2012	313	1892
	3	260	1815	278	1981	317	2012	313	1892
1100	1	282	2083	271	2036	335	2174	304	1802
	2	282	2083	271	2036	339	2174	303	1802
	3	279	2083	270	2036	338	2174	303	1802

16 Analysis of Covariance for Repeated Measures Designs

16.1 INTRODUCTION

Repeated measures designs have the same structure as split-plot designs with the main similarity being that they both involve more than one size of experimental unit. The difference between the two types of designs is that the levels of all of the factors associated with a split-plot design are randomly assigned to the respective experimental units, whereas for the repeated measures designs, there are one or more factors whose levels cannot be randomly assigned to their experimental units (Milliken and Johnson, 1992). Because of the non-random assignment process of the levels of a factor to one of the sizes of experimental units, the correlation among those experimental units may be different than the uncorrelated assumption used for the analysis of split-plot models. Thus, this chapter emphasizes specifying or selecting an adequate correlation structure associated with the repeated measurements instead of spending time on specifying the form of the slopes and intercepts, since the forms of the regression models for the repeated measures designs are identical to the forms of the regression models for the split-plot design and was discussed in detail in Chapter 15.

Repeated measures occur most often when time of exposure is one of the factors in the treatment structure (Littell et al., 1996). For example, for a study of blood pressure medications, persons were randomly assigned to one of four treatment regimes. The initial blood pressure of each person was determined to be used as a possible covariate. It is of interest to determine if there are different responses to the levels of the treatment regime caused by the length of time persons are exposed to the regime. In this case, suppose each person's blood pressure is measured each week for 6 weeks and time with 6 levels becomes part of the treatment structure of the experiment. The treatment structure for this experiment is a two-way factorial arrangement with the four levels of treatment regime crossed with the six time intervals. This experiment involves two sizes of experimental units. The person is the entity to which the levels of treatment regime were randomly assigned. A 1-week interval of time of exposure to a level of the treatment regime is the experimental unit for the levels of time. The levels of time cannot be randomly assigned to the time intervals, i.e., time 2 cannot occur before time 1 (times must occur in sequence), thus this is a repeated measures design. The person is the large size experimental unit and a 1-week interval of time of a person is the small size experimental unit. The covariate is measured on the person or the large size experimental unit. A model to describe this data using unequal slopes for each of the treatment by time combinations is

$$BP_{ijk} = \mu_{ik} + \beta_{ik} IBP_{ij} + p_{ij} + \varepsilon_{ijk},$$

$$i = 1, 2, 3, 4, \quad j = 1, 2, \ldots, n_i, \quad k = 1, 2, \ldots, 6$$

(16.1)

The terms BP_{ijk} and IBP_{ij} denote the blood pressure and initial blood pressure readings, respectively, μ_{ik} and β_{ik} are the intercepts and slopes for the regression model for the ith drug and k^{th} time. The p_{ij} term represents the random effect corresponding to the jth person within the ith treatment regime and it is assumed these random effects are distributed iid $N(0, \sigma_p^2)$. Unless persons are genetically related, the uncorrelated assumption among the persons within a treatment regime is reasonable. There is the possibility that the application of the treatment regime could have an effect on the variability among the persons, so there could be unequal treatment regime variances. The assumption made about the distribution of the ε_{ijk} in Chapter 15 was that they were distributed iid $N(0, \sigma_\varepsilon^2)$. This is likely not a reasonable assumption for the distribution of the ε_{ijk} within a person. In fact, the ε_{ijk} are most likely correlated. Let ε_{ij} denote the vector of time interval errors for the jth person from the ith treatment regime; then the covariance matrix of the six time measurements can be expressed as

$$\text{Var}(\varepsilon_{ij}) = \text{Var}\begin{bmatrix} \varepsilon_{ij1} \\ \varepsilon_{ij2} \\ \varepsilon_{ij3} \\ \varepsilon_{ij4} \\ \varepsilon_{ij5} \\ \varepsilon_{ij6} \end{bmatrix} = \begin{bmatrix} \sigma_1^2 & \sigma_{12} & \sigma_{13} & \sigma_{14} & \sigma_{15} & \sigma_{16} \\ \sigma_{12} & \sigma_2^2 & \sigma_{23} & \sigma_{24} & \sigma_{25} & \sigma_{26} \\ \sigma_{13} & \sigma_{23} & \sigma_3^2 & \sigma_{34} & \sigma_{35} & \sigma_{36} \\ \sigma_{14} & \sigma_{24} & \sigma_{34} & \sigma_4^2 & \sigma_{45} & \sigma_{46} \\ \sigma_{15} & \sigma_{25} & \sigma_{35} & \sigma_{45} & \sigma_5^2 & \sigma_{56} \\ \sigma_{16} & \sigma_{26} & \sigma_{36} & \sigma_{46} & \sigma_{56} & \sigma_6^2 \end{bmatrix} = \mathbf{R} \quad (16.2)$$

The new challenge of the repeated measures analysis of covariance is to determine an appropriate parameterization for $\mathbf{R}$ and then estimate the resulting parameters. The fixed effects of the repeated measures analysis of covariance model in terms of intercepts and slopes are identical to those of the split-plot models discussed in Chapter 15, so the new dimension for the repeated measures design is to determine the form of the covariance structure of the repeated measurements within each subject. Thus, the repeated measures analysis of covariance model has a covariate part of the model and a covariance part of the model (be careful and keep the definitions of these two terms straight).

The basic strategy recommended is to select a more than adequate form for the covariate part of the model, such as linear or quadratic. The more than adequate covariate part of the model will have unequal slopes for all of the treatment combinations, i.e., do not do any model building at this time (model building will be done later). Including the nonreduced covariate part of the model, determine the simplest adequate form of the covariance part of the model or covariance structure of the repeated measures within each person, i.e., select the simplest adequate

Analysis of Covariance for Repeated Measures Designs 453

parameterization for **R**. After **R** has been selected, use that form of the covariance structure and carry out the model building process to simplify the covariate part of the model as was done for the examples in Chapter 15. After the covariate part of the model is simplified, then the final step in the analysis of covariance is to compare the resulting regression models.

The next section consists of a discussion of some of the covariance structures that are possible choices for **R** and a description of how to use the repeated statement of PROC MIXED to enable the parameters to be estimated. The section continues with a discussion of the joint use of the repeated and random statements, i.e., when to use both and when to use only the repeated statement.

16.2 THE COVARIANCE PART OF THE MODEL — SELECTING R

Selecting an adequate covariance structure of the small size experimental units within each large experimental unit is a very important part of the repeated measures analysis of covariance (as well as analysis of variance). In fact, selecting the most appropriate covariance structure with the fewest number of parameters should provide more degrees of freedom for each of the estimated variances and the resulting denominator degrees of freedom for the tests and estimates should be larger. One could just use the unstructured covariance in Model 16.2 which, for this example, contains 21 parameters, but if a simpler structure would be just as adequate, then using the simpler structure should provide more powerful tests for the fixed effects. The larger the number of degrees of freedom associated with the denominator, the more powerful the tests. There are several possible structures for **R**; some are described below. The first structure is called compound symmetry (Box, 1954) where all of the variances are equal and all of the covariances or correlations are equal. **R** has compound symmetry form when

$$\mathbf{R} = \begin{bmatrix} \sigma_0^2+\sigma_1 & \sigma_1 & \sigma_1 & \sigma_1 & \sigma_1 & \sigma_1 \\ \sigma_1 & \sigma_0^2+\sigma_1 & \sigma_1 & \sigma_1 & \sigma_1 & \sigma_1 \\ \sigma_1 & \sigma_1 & \sigma_0^2+\sigma_1 & \sigma_1 & \sigma_1 & \sigma_1 \\ \sigma_1 & \sigma_1 & \sigma_1 & \sigma_0^2+\sigma_1 & \sigma_1 & \sigma_1 \\ \sigma_1 & \sigma_1 & \sigma_1 & \sigma_1 & \sigma_0^2+\sigma_1 & \sigma_1 \\ \sigma_1 & \sigma_1 & \sigma_1 & \sigma_1 & \sigma_1 & \sigma_0^2+\sigma_1 \end{bmatrix}$$

$$= \sigma_\varepsilon^2 \begin{bmatrix} 1 & \rho & \rho & \rho & \rho & \rho \\ \rho & 1 & \rho & \rho & \rho & \rho \\ \rho & \rho & 1 & \rho & \rho & \rho \\ \rho & \rho & \rho & 1 & \rho & \rho \\ \rho & \rho & \rho & \rho & 1 & \rho \\ \rho & \rho & \rho & \rho & \rho & 1 \end{bmatrix}$$

$$\sigma_0^2 \mathbf{I}_6 + \sigma_1 \mathbf{J}_6$$

where $\sigma_\varepsilon^2 = \sigma_0^2 + \sigma_1$, $\rho = \sigma_1/(\sigma_0^2 + \sigma_1)$, $\mathbf{I}_6$ is a 6×6 identity matrix and $\mathbf{J}_6$ is a 6×6 matrix of 1's. For equally spaced, either by time or by some other clock such as a biological clock, the first-order auto-regressive covariance structure is

$$\mathbf{R} = \sigma_\varepsilon^2 \begin{bmatrix} 1 & \rho & \rho^2 & \rho^3 & \rho^4 & \rho^5 \\ \rho & 1 & \rho & \rho^2 & \rho^3 & \rho^4 \\ \rho^2 & \rho & 1 & \rho & \rho^2 & \rho^3 \\ \rho^3 & \rho^2 & \rho & 1 & \rho & \rho^2 \\ \rho^4 & \rho^3 & \rho^2 & \rho & 1 & \rho \\ \rho^5 & \rho^4 & \rho^3 & \rho^2 & \rho & 1 \end{bmatrix}$$

where the correlation between two observations k time intervals apart is ρ^k. The first-order auto-regressive moving-average covariance matrix has the structure

$$\mathbf{R} = \sigma_\varepsilon^2 \begin{bmatrix} 1 & \gamma & \gamma\rho & \gamma\rho^2 & \gamma\rho^3 & \gamma\rho^4 \\ \gamma & 1 & \gamma & \gamma\rho & \gamma\rho^2 & \gamma\rho^3 \\ \gamma\rho & \gamma & 1 & \gamma & \gamma\rho & \gamma\rho^2 \\ \gamma\rho^2 & \gamma\rho & \gamma & 1 & \gamma & \gamma\rho \\ \gamma\rho^3 & \gamma\rho^2 & \gamma\rho & \gamma & 1 & \gamma \\ \gamma\rho^4 & \gamma\rho^3 & \gamma\rho^2 & \gamma\rho & \gamma & 1 \end{bmatrix}$$

A covariance structure with equal covariances for observations that are k time intervals apart is the Toplitz structure which provides

$$\mathbf{R} = \begin{bmatrix} \sigma_0^2 & \sigma_1 & \sigma_2 & \sigma_3 & \sigma_4 & \sigma_5 \\ \sigma_1 & \sigma_0^2 & \sigma_1 & \sigma_2 & \sigma_3 & \sigma_4 \\ \sigma_2 & \sigma_1 & \sigma_0^2 & \sigma_1 & \sigma_2 & \sigma_3 \\ \sigma_3 & \sigma_2 & \sigma_1 & \sigma_0^2 & \sigma_1 & \sigma_2 \\ \sigma_4 & \sigma_3 & \sigma_2 & \sigma_1 & \sigma_0^2 & \sigma_1 \\ \sigma_5 & \sigma_4 & \sigma_3 & \sigma_2 & \sigma_1 & \sigma_0^2 \end{bmatrix}$$

Another covariance structure used for time interval data is the first-order ante-dependence structure (a 4×4 matrix is shown) where

Analysis of Covariance for Repeated Measures Designs

$$\mathbf{R} = \begin{bmatrix} \sigma_1^2 & \sigma_1\sigma_2\rho_1 & \sigma_1\sigma_3\rho_1\rho_2 & \sigma_1\sigma_4\rho_1\rho_2\rho_3 \\ \sigma_1\sigma_2\rho_1 & \sigma_2^2 & \sigma_2\sigma_3\rho_2 & \sigma_2\sigma_4\rho_2\rho_3 \\ \sigma_1\sigma_3\rho_1\rho_2 & \sigma_2\sigma_3\rho_2 & \sigma_3^2 & \sigma_3\sigma_4\rho_3 \\ \sigma_1\sigma_4\rho_1\rho_2\rho_3 & \sigma_2\sigma_4\rho_2\rho_3 & \sigma_3\sigma_4\rho_3 & \sigma_4^2 \end{bmatrix}$$

There are heterogeneous variance forms of the compound symmetry, first-order autoregressive, and Toeplitz covariance structures. The heterogeneous compound symmetry covariance structure is

$$\mathbf{R} = \begin{bmatrix} \sigma_1^2 & \sigma_1\sigma_2\rho & \sigma_1\sigma_3\rho & \sigma_1\sigma_4\rho & \sigma_1\sigma_5\rho & \sigma_1\sigma_6\rho \\ \sigma_1\sigma_2\rho & \sigma_2^2 & \sigma_2\sigma_3\rho & \sigma_2\sigma_4\rho & \sigma_2\sigma_5\rho & \sigma_2\sigma_6\rho \\ \sigma_1\sigma_3\rho & \sigma_2\sigma_3\rho & \sigma_3^2 & \sigma_3\sigma_4\rho & \sigma_3\sigma_5\rho & \sigma_3\sigma_6\rho \\ \sigma_1\sigma_4\rho & \sigma_2\sigma_4\rho & \sigma_3\sigma_4\rho & \sigma_4^2 & \sigma_4\sigma_5\rho & \sigma_4\sigma_6\rho \\ \sigma_1\sigma_5\rho & \sigma_2\sigma_5\rho & \sigma_3\sigma_5\rho & \sigma_4\sigma_5\rho & \sigma_5^2 & \sigma_5\sigma_6\rho \\ \sigma_1\sigma_6\rho & \sigma_2\sigma_6\rho & \sigma_3\sigma_6\rho & \sigma_4\sigma_6\rho & \sigma_5\sigma_6\rho & \sigma_6^2 \end{bmatrix}$$

The heterogeneous first-order auto-regressive structure is

$$\mathbf{R} = \begin{bmatrix} \sigma_1^2 & \sigma_1\sigma_2\rho & \sigma_1\sigma_3\rho^2 & \sigma_1\sigma_4\rho^3 & \sigma_1\sigma_5\rho^4 & \sigma_1\sigma_6\rho^5 \\ \sigma_1\sigma_2\rho & \sigma_2^2 & \sigma_2\sigma_3\rho & \sigma_2\sigma_4\rho^2 & \sigma_2\sigma_5\rho^3 & \sigma_2\sigma_6\rho^4 \\ \sigma_1\sigma_3\rho^2 & \sigma_2\sigma_3\rho & \sigma_3^2 & \sigma_3\sigma_4\rho & \sigma_3\sigma_5\rho^2 & \sigma_3\sigma_6\rho^3 \\ \sigma_1\sigma_4\rho^3 & \sigma_2\sigma_4\rho^2 & \sigma_3\sigma_4\rho & \sigma_4^2 & \sigma_4\sigma_5\rho & \sigma_4\sigma_6\rho^2 \\ \sigma_1\sigma_5\rho^4 & \sigma_2\sigma_5\rho^3 & \sigma_3\sigma_5\rho^2 & \sigma_4\sigma_5\rho & \sigma_5^2 & \sigma_5\sigma_6\rho \\ \sigma_1\sigma_6\rho^5 & \sigma_2\sigma_6\rho^4 & \sigma_3\sigma_6\rho^3 & \sigma_4\sigma_6\rho^2 & \sigma_5\sigma_6\rho & \sigma_6^2 \end{bmatrix}$$

The heterogeneous Toeplitz structure is

$$\mathbf{R} = \begin{bmatrix} \sigma_1^2 & \sigma_1\sigma_2\rho_1 & \sigma_1\sigma_3\rho_2 & \sigma_1\sigma_4\rho_3 & \sigma_1\sigma_5\rho_4 & \sigma_1\sigma_6\rho_5 \\ \sigma_1\sigma_2\rho_1 & \sigma_2^2 & \sigma_2\sigma_3\rho_1 & \sigma_2\sigma_4\rho_2 & \sigma_2\sigma_5\rho_3 & \sigma_2\sigma_6\rho_4 \\ \sigma_1\sigma_3\rho_2 & \sigma_2\sigma_3\rho_1 & \sigma_3^2 & \sigma_3\sigma_4\rho_1 & \sigma_3\sigma_5\rho_2 & \sigma_3\sigma_6\rho_3 \\ \sigma_1\sigma_4\rho_3 & \sigma_2\sigma_4\rho_2 & \sigma_3\sigma_4\rho_1 & \sigma_4^2 & \sigma_4\sigma_5\rho_1 & \sigma_4\sigma_6\rho_2 \\ \sigma_1\sigma_5\rho_4 & \sigma_2\sigma_5\rho_3 & \sigma_3\sigma_5\rho_2 & \sigma_4\sigma_5\rho_1 & \sigma_5^2 & \sigma_5\sigma_6\rho_1 \\ \sigma_1\sigma_6\rho_5 & \sigma_2\sigma_6\rho_4 & \sigma_3\sigma_6\rho_3 & \sigma_4\sigma_6\rho_2 & \sigma_5\sigma_6\rho_1 & \sigma_6^2 \end{bmatrix}$$

Heterogeneous variance covariance structures are particularly useful when the repeated measures corresponds to time and the observed process is growing over time. When it is not reasonable to simplify the covariance structure, then use the one in Model 16.2, which is called unstructured.

16.3 COVARIANCE STRUCTURE OF THE DATA

Model 16.1 is a mixed model that can be expressed as $y = X\beta + Zu + \varepsilon$, where $X\beta$ denotes the fixed effects of the model, which in this case consists of the intercepts and slopes μ_{ik} and β_{ik}. The Zu part of the model denotes the random effects where the distribution of u is assumed to $N(0,G)$ and then ε denotes the residual or the within subject errors where the distribution is assumed to be $N(0, R)$, where R is a matrix with diagonal partitions, one for each person, as

$$R = \begin{pmatrix} R_{1(1)} & 0 & \cdots & 0 \\ 0 & R_{2(1)} & \cdots & 0 \\ \vdots & \vdots & \ddots & \vdots \\ 0 & 0 & \cdots & R_{n_4(4)} \end{pmatrix}$$

The covariance structure of the data is $\text{Var}(y) = \Sigma = ZGZ' + R$. For Model 16.1 the Σ matrix is

$$\Sigma = \begin{pmatrix} \Sigma_{1(1)} & 0 & \cdots & 0 \\ 0 & \Sigma_{2(1)} & \cdots & 0 \\ \vdots & \vdots & \ddots & \vdots \\ 0 & 0 & \cdots & \Sigma_{n_4(4)} \end{pmatrix}$$

$$= \begin{pmatrix} \sigma_p J_6 + R_{1(1)} & 0 & \cdots & 0 \\ 0 & \sigma_p J_6 + R_{2(1)} & \cdots & 0 \\ \vdots & \vdots & \ddots & \vdots \\ 0 & 0 & \cdots & \sigma_p J_6 + R_{n_4(4)} \end{pmatrix}$$

The covariance matrix for the observations made on the jth person assigned to the ith treatment regime is

$$\Sigma_{j(I)} = \sigma_p^2 \begin{pmatrix} 1 & 1 & 1 & 1 & 1 & 1 \\ 1 & 1 & 1 & 1 & 1 & 1 \\ 1 & 1 & 1 & 1 & 1 & 1 \\ 1 & 1 & 1 & 1 & 1 & 1 \\ 1 & 1 & 1 & 1 & 1 & 1 \\ 1 & 1 & 1 & 1 & 1 & 1 \end{pmatrix} + \begin{pmatrix} \sigma_1^2 & \sigma_{12} & \sigma_{13} & \sigma_{14} & \sigma_{15} & \sigma_{16} \\ \sigma_{12} & \sigma_2^2 & \sigma_{23} & \sigma_{24} & \sigma_{25} & \sigma_{26} \\ \sigma_{13} & \sigma_{23} & \sigma_3^2 & \sigma_{34} & \sigma_{35} & \sigma_{36} \\ \sigma_{14} & \sigma_{24} & \sigma_{34} & \sigma_4^2 & \sigma_{45} & \sigma_{46} \\ \sigma_{15} & \sigma_{25} & \sigma_{35} & \sigma_{45} & \sigma_5^2 & \sigma_{56} \\ \sigma_{16} & \sigma_{26} & \sigma_{36} & \sigma_{46} & \sigma_{57} & \sigma_6^2 \end{pmatrix}$$

where $\mathbf{R}$ is expressed as the unstructured covariance matrix. If instead, $\mathbf{R}$ has the compound symmetry structure, $\mathbf{R}_{j(i)} = \sigma_\varepsilon^2 (1-\rho) \mathbf{I}_6 + \sigma_\varepsilon \rho \mathbf{J}_6$, then $\Sigma_{j(i)} = \sigma_p^2 \mathbf{J}_6 + \sigma_\varepsilon^2 (1-\rho) \mathbf{I}_p + \sigma_\varepsilon^2 \rho \mathbf{J}_6 = \sigma_\varepsilon^2 (1-\rho) \mathbf{I}_6 + (\sigma_p^2 + \sigma_\varepsilon^2 \rho) \mathbf{J}_6 = \sigma_0^2 \mathbf{I}_6 + \sigma_1 \mathbf{J}_6$ where $\sigma_0^2 = \sigma_p^2 (1-\rho)$ and $\sigma_1 = (\sigma_p^2 + \sigma_\varepsilon^2 \rho)$. Thus the covariance structure effectively has only two parameters, σ_0^2 and σ_1, instead of the three parameters specified, σ_p^2, σ_ε^2, and ρ. For this case, one only needs to specify that $\mathbf{R}$ satisfies the compound symmetry structure and does not need to include the fact that the persons are a random effect, or just specify that the persons are a random effect and do not specify that $\mathbf{R}$ has the compound symmetry structure. In general if the structure specified for $\mathbf{R}$ includes in its structure (or column space) $\mathbf{J}$, then only the $\mathbf{R}$ structure needs to be specified and the persons within the treatment regime does not need to be included in the specification of the model.

16.4 SPECIFYING THE RANDOM AND REPEATED STATEMENTS FOR PROC MIXED OF THE SAS® SYSTEM

To analyze the mixed model, $\mathbf{y} = \mathbf{X}\boldsymbol{\beta} + \mathbf{Zu} + \boldsymbol{\varepsilon}$ where the distribution of $\mathbf{u}$ is assumed to $N(\mathbf{0},\mathbf{G})$ and the distribution of $\boldsymbol{\varepsilon}$ is assumed to be $N(\mathbf{0}, \mathbf{R})$ with PROC MIXED, the random statement is used to specify the form of $\mathbf{G}$ and the repeated statement is used to specify the form of $\mathbf{R}$. The specification of the random statements have been discussed in some detail in Chapters 12 and 13, so the specification of the repeated statement is discussed in this section. The syntax of the repeated statement is

```
REPEATED<experimental unit for the repeated measures
variable>/type=<covariance structure> subject=
<experimental unit being measured repeatedly> Group=
<grouping with unequal covariances matrices>.
```

For the blood pressure example, the repeated measures experimental units are the 1-week time intervals to which the person is exposed or subjected to the treatment regime. The type=option specifies the form of the covariance structure, such as CS, CSH, AR(1), ARH(1), TOEP, ANTE(1), ARMA(1,1), and UN (see SAS Institute Inc.,

1999, for a detailed description of the covariance structures available). The subject=option specifies the experimental unit on which the repeated measures have been made. In the blood pressure example, the subject is the person whose blood pressure is measured each of 6 weeks. The group=option specifies groups of observations where between groups heterogeneous covariance matrices exist. For the blood pressure example, the covariance structure of the repeated measurements could be affected by the application of the treatment regime, i.e., the covariance structures could be different for each level of the treatment regime. Thus, using group=treatment, causes PROC MIXED to fit a covariance structure with different parameter estimates for each level of the treatment regime. The following PROC MIXED code fits Model 16.1 to the data set where first-order auto-regressive is specified for the repeated measures covariance structure:

```
PROC MIXED CL IC Covtest;
Class Treatment Time Person;
Model BP = Treatment*Time IBP*Treatment*Time/Noint
DDFM=KR;
Random Person(Treatment);
Repeated Time/Subject = Person Type = AR(1);
```

When specifying covariance structures that contain the **J** matrix as a special case, then the random statement is not needed. Such covariance structures are CS, CSH, and UN.

16.5 SELECTING AN ADEQUATE COVARIANCE STRUCTURE

There are several criteria that can be used to select adequate covariance structures. The criteria that is recommended most often (SAS Institute Inc., 1999) is Akaike's Information Criterion (AIC) (Akaike, 1974) which has the value of AIC = $-2\text{LH}(\hat{\beta}, \hat{\Sigma}) - 2q$ where $\text{LH}(\hat{\beta}, \hat{\Sigma})$ is the value of the likelihood function evaluated at $(\hat{\beta}, \hat{\Sigma})$ and q is the effective number of parameters in Σ. The value of AIC can be used to compare two models with different covariance structure, but identical fixed effects. Select the model with the smallest AIC as the better model.

A likelihood ratio test statistic can be computed for testing H_0: $\mathbf{R}_1$ is as adequate as $\mathbf{R}_2$ vs. H_a: $\mathbf{R}_1$ is not as adequate as $\mathbf{R}_2$ when $\mathbf{R}_1$ is a special case of $\mathbf{R}_2$ (a nested hypothesis where $\mathbf{R}_2$ has more parameters than $\mathbf{R}_1$). Let $\text{LH}(\hat{\beta}_1, \hat{\Sigma}_1)$ and $\text{LH}(\hat{\beta}_2, \hat{\Sigma}_2)$ be the values of the likelihood function evaluated with parameter estimates from specifying $\mathbf{R}_1$ and $\mathbf{R}_2$, respectively, where $\mathbf{R}_1$ has q_1 parameters and where $\mathbf{R}_2$ has q_2 parameters. The test statistic is $Q = -2\,(\text{LH}(\hat{\beta}_1, \hat{\Sigma}_1) - \text{LH}(\hat{\beta}_2, \hat{\Sigma}_2))$ which is asymptotically distributed as a chi-square random variable based on $q_2 - q_1$ degrees of freedom. When $Q > \chi^2_{\alpha/2}$, $q_2 - q_1$ conclude that $\mathbf{R}_1$ is not as good as $\mathbf{R}_2$, so $\mathbf{R}_2$ is the better covariance structure. When $Q > \chi^2_{\alpha/2}$, $q_2 - q_1$ then conclude that $\mathbf{R}_1$ was not shown to be worse than $\mathbf{R}_2$, so use $\mathbf{R}_1$.

16.6 EXAMPLE: SYSTOLIC BLOOD PRESSURE STUDY WITH COVARIATE MEASURED ON THE LARGE SIZE EXPERIMENTAL UNIT

A medical team designed an experiment to study the effect of a drug and exercise program on a person's systolic blood pressure. Thirty-two persons were selected with marginal to high systolic blood pressure and then they were randomly assigned to one of four combinations of exercise and drug regimes: (Exercise=No, Drug=No), (Exercise=No, Drug=Yes), (Exercise=Yes, Drug=No) and (Exercise=Yes, Drug=Yes). Each person returned for a blood pressure measurement each week for 6 weeks after starting the treatment. It was thought that the initial systolic blood pressure (mmHg) could possibly have an impact on the person's response to the treatment combinations, so a blood pressure measurement was taken before the start of the study and is denoted by IBP. Table 16.1 contains the data for eight different persons assigned to each of the exercise by drug combinations. There are two sizes of experimental units for this study. The large size experimental unit is a person, the entity assigned to the levels of exercise and drug. Thus the person design is a two-way treatment structure (levels of exercise by levels of drug) in a completely randomized design structure. Each person's time on the study was split into six 1-week intervals. It was of interest to determine the effect of exposure time to the treatment regime when exposure time was measured for 6 consecutive weeks. Thus, there are six levels of exposure time and the levels of exposure time are observed on the 1-week intervals within a person. The 1-week interval is the small size experimental unit to which the levels of exposure time would be randomly assigned, but the randomization process is not possible (5 weeks of exposure cannot occur before 2 weeks of exposure). This nonrandom assignment causes the design of this experiment to be called a repeated measures design. A model to describe the systolic blood pressure measurement made on the mth time of the kth person assigned to the jth level of drug combined with the ith level of exercise is

$$BP_{ijkm} = \mu + E_i + D_j + (ED)_{ij} + p_{ijk} \text{ (person part of model)} \quad (16.3)$$

$$+ T_m + (TE)_{im} + (TD)_{jm} + (TED)_{ijm} + \varepsilon_{ijkm} \text{ (week interval part of model)}$$

for $i = 1, 2$, $j = 1, 2$, $k = 1, 2, ..., 8$, $m = 1, 2, ..., 6$, E_i denotes the effect of the ith level of exercise, D_j denotes the effect of the jth level of drug, T_m denotes the mth level of exposure time, p_{ijk} denotes the random person effect, and ε_{ijkm} denotes the week interval error. It is reasonable to assume that the p_{ijk} are independently distributed (unless the persons are genetically related) with variance component σ_p^2. The split-plot assumptions for the week interval errors are that they are independent with variance component σ_ε^2. Under those assumptions, a split-plot analysis of variance table can be generated as the one in Table 16.2. There are two error terms in Table 16.2 where the person error term is based on 28 degrees of freedom and the week interval is based on 140 degrees of freedom. Thus approximate degrees of freedom associated with the denominators of various effects, estimates, or tests

TABLE 16.1
Data for the Blood Pressure Study Where IBP is the Initial Systolic Blood Pressure and Week 1 to Week 6 are the Weekly Systolic Blood Pressure Measurements (in mmHg)

Drug	Exercise	Person	IBP	Week 1	Week 2	Week 3	Week 4	Week 5	Week 6
No	No	1	133	146	144	142	141	142	143
No	No	2	137	136	135	134	134	134	137
No	No	3	148	148	148	148	143	143	144
No	No	4	136	139	137	138	139	139	142
No	No	5	140	141	143	145	147	147	147
No	No	6	139	140	136	135	135	132	135
No	No	7	154	170	166	168	167	165	166
No	No	8	152	146	147	146	145	146	143
Yes	No	1	150	145	134	136	134	132	134
Yes	No	2	147	147	139	139	132	134	135
Yes	No	3	142	133	121	122	120	121	116
Yes	No	4	144	139	128	128	122	119	116
Yes	No	5	140	132	121	121	115	117	116
Yes	No	6	137	133	122	122	118	119	121
Yes	No	7	143	144	133	131	124	123	124
Yes	No	8	137	122	111	111	104	101	101
No	Yes	1	150	143	142	139	137	134	128
No	Yes	2	151	144	137	134	131	128	121
No	Yes	3	151	155	149	147	140	136	131
No	Yes	4	142	149	144	140	140	137	129
No	Yes	5	149	142	140	136	132	126	126
No	Yes	6	132	133	127	123	122	117	116
No	Yes	7	134	136	129	125	122	123	115
No	Yes	8	151	150	145	141	137	131	130
Yes	Yes	1	131	119	109	102	100	100	103
Yes	Yes	2	148	138	130	124	120	122	123
Yes	Yes	3	151	146	134	129	130	134	134
Yes	Yes	4	150	137	126	119	119	120	121
Yes	Yes	5	144	129	123	118	117	118	117
Yes	Yes	6	151	140	138	134	131	126	129
Yes	Yes	7	152	149	141	138	132	132	131
Yes	Yes	8	135	123	115	109	107	111	109

would be between 28 and 168, where those numbers will be reduced depending on the form of the covariance matrix and the form of the covariate part of the model. It is not reasonable to just assume that the split-plot assumptions are adequate without first looking at the fit of several of the possible covariance structures. In order to pursue the investigation into an adequate covariance structure, the covariate part of the model must be addressed. Plots of the measured blood pressure (BP) values

TABLE 16.2
Analysis of Variance Table without Covariate Information for the Blood Pressure Study

Source	df	EMS
Drug	1	$\sigma_\varepsilon^2 + 6\sigma_p^2 + \phi^2(\text{Drug})$
Exercise	1	$\sigma_\varepsilon^2 + 6\sigma_p^2 + \phi^2(\text{Exercise})$
Drug*Exercise	1	$\sigma_\varepsilon^2 + 6\sigma_p^2 + \phi^2(\text{Drug} * \text{Exercise})$
Error(person) = Person(Drug*Exercise)	28	$\sigma_\varepsilon^2 + 6\sigma_p^2$
Time	5	$\sigma_\varepsilon^2 + \phi^2(\text{Time})$
Time*Drug	5	$\sigma_\varepsilon^2 + \phi^2(\text{Time} * \text{Drug})$
Time*Exercise	5	$\sigma_\varepsilon^2 + \phi^2(\text{Time} * \text{Exercise})$
Time*Drug*Exercise	5	$\sigma_\varepsilon^2 + \phi^2(\text{Time} * \text{Drug} * \text{Exercise})$
Error(week interval) = Time*Person(Drug*Exercise)	140	σ_ε^2

against the initial blood pressure (IBP) values for each combination of exercise by drug by time indicates that a linear relationship is adequate to describe the relationship between BP and IBP (plots not shown). The form of the model used to investigate the selection of the covariance structure for the repeated measurements uses different intercepts and different slopes for each of the combinations of exercise by drug by time as:

$$BP_{ijkm} = \alpha_{ijm} + \beta_{ijm} IBP_{ijk} + p_{ijk} + \varepsilon_{ijkm} \tag{16.4}$$

The random person effects are specified by the random statement in PROC MIXED as "Random person(exercise drug);" and repeated statement is used to specify the covariance structure of the repeated measurements as "Repeated Time/type=xxx subject=person(exercise drug);" where xxx is one of the covariance structures that can be selected when using PROC MIXED. As discussed in Section 16.3, the inclusion of the random statement depends on which covariance structure has been selected to fit the data. Tables 16.3 through 16.10 contain the PROC MIXED code and results for fitting eight different covariance structures to the repeated measurements. The eight covariance structures are split-plot (Table 16.3), compound symmetry (Table 16.4), heterogeneous compound symmetry (Table 16.5), first-order auto-regressive (Table 16.6), heterogeneous first-order auto-regressive (Table 16.7), ante-dependence (Table 16.8), Toeplitz (Table 16.9), and unstructured (Table 16.10). The random statement was not needed for compound symmetry, heterogeneous compound symmetry, ante-dependence, Toeplitz, and unstructured, as discussed in Section 16.3. Each of the tables contains the PROC MIXED code, estimates of the covariance structure parameters, tests that the intercepts are all equal, tests that the

TABLE 16.3
Analysis of Variance Table for the Blood Pressure Data Using the Split-Plot Assumptions for the Error Terms

```
Proc Mixed cl ic covtest DATA=E165;
Class Drug Exercise person Time;
Model BP= Drug*Exercise*Time IBP*Drug*Exercise*Time/ddfm=kr;
Random person(Drug*Exercise);
```

Neg2LogLike	Parameters	AIC	AICC	HQIC	BIC	CAIC
896.94	2	900.94	901.03	901.91	903.87	905.87

CovParm	Estimate	StdErr	ZValue	ProbZ
person(drug*exercis)	28.4266	8.4131	3.38	0.0004
Residual	4.2917	0.5541	7.75	0.0000

Effect	NumDF	DenDF	FValue	ProbF
drug*exercise*TIME	23	123.2	1.09	0.3626
IBP*drug*exercise*TIME	24	121.7	3.23	0.0000

TABLE 16.4
Analysis of Variance Table with Covariance Parameter (CovParm) Estimates Using Compound Symmetry Covariance Structure

```
Proc Mixed cl ic covtest DATA=E165;
Class Exercise Drug person Time;
Model BP= Exercise*Drug*Time IBP*Exercise*Drug*Time/ddfm=kr;
repeated Time/type=cs subject=person(Exercise*Drug);
```

Neg2LogLike	Parameters	AIC	AICC	HQIC	BIC	CAIC
896.94	2	900.94	901.03	901.91	903.87	905.87

CovParm	Subject	Estimate	StdErr	ZValue	ProbZ
CS	person(exercise*drug)	28.4266	8.4131	3.38	0.0007
Residual		4.2917	0.5541	7.75	0.0000

Effect	NumDF	DenDF	FValue	ProbF
exercise*drug*TIME	23	123.2	1.09	0.3626
IBP*exercise*drug*TIME	24	121.7	3.23	0.0000

slopes are all equal to zero, and a list of information criteria. For the information criteria, let Q= −2 Log(likelihood), then AIC = Q + 2d (Akaike, 1974), AICC = Q + 2dn/(n − d − 1) (Burnham and Anderson, 1998), HQIC = Q + 2d log(log(n)) (Hannan and Quinn, 1979), BIC = Q + d log(n) (Schwarz, 1978), and CAIC = q + d (log(n) + 1) (Bozdogan, 1987) where d is the effective number of parameters in the covariance structure, n is the number of observations for maximum likelihood estimation, and n–p where p is the rank of the fixed effects part of the model for REML estimation (SAS Institute Inc., 1999). Using this form of the information

Analysis of Covariance for Repeated Measures Designs

TABLE 16.5
Analysis of Variance Table with Covariance Parameter (CovParm) Estimates Using Heterogeneous Variance Compound Symmetry Covariance Structure

```
Proc Mixed cl ic covtest DATA=E165;
Class Exercise Drug person Time;
Model BP= Exercise*Drug*Time IBP*Exercise*Drug*Time/ddfm=kr;
repeated Time/type=csh subject=person(Exercise*Drug);
```

Neg2LogLike	Parameters	AIC	AICC	HQIC	BIC	CAIC
893.07	7	907.07	907.89	910.47	917.33	924.33

CovParm	Subject	Estimate	StdErr	ZValue	ProbZ
Var(1)	person(exercise*drug)	33.8673	9.8103	3.45	0.0003
Var(2)	person(exercise*drug)	27.8799	8.0000	3.49	0.0002
Var(3)	person(exercise*drug)	30.6665	8.7844	3.49	0.0002
Var(4)	person(exercise*drug)	29.6744	8.5013	3.49	0.0002
Var(5)	person(exercise*drug)	37.7676	10.9035	3.46	0.0003
Var(6)	person(exercise*drug)	36.9608	10.6445	3.47	0.0003
CSH	person(exercise*drug)	0.8725	0.0360	24.21	0.0000

Effect	NumDF	DenDF	FValue	ProbF
exercise*drug*TIME	23	85.7	0.95	0.5378
IBP*exercise*drug*TIME	24	81.2	3.68	0.0000

TABLE 16.6
Analysis of Variance Table with Covariance Parameter (CovParm) Estimates Using First-Order Auto-Regressive Covariance Structure

```
PROC MIXED cl ic covtest DATA=E165;
Class Exercise Drug person Time;
Model BP= Exercise*Drug*Time IBP*Exercise*Drug*Time/ddfm=kr;
Random person(Exercise*Drug);
repeated Time/type=ar(1) subject=person(Exercise*Drug);
```

Neg2LogLike	Parameters	AIC	AICC	HQIC	BIC	CAIC
864.01	3	870.01	870.19	871.47	874.41	877.41

CovParm	Subject	Estimate	StdErr	ZValue	ProbZ
person(exercise*drug)		25.0939	8.9805	2.79	0.0026
AR(1)	person(exercise*drug)	0.7104	0.1295	5.49	0.0000
Residual		8.3096	3.6323	2.29	0.0111

Effect	NumDF	DenDF	FValue	ProbF
exercise*drug*TIME	23	111.7	0.87	0.6341
IBP*exercise*drug*TIME	24	111.2	2.91	0.0001

TABLE 16.7
Analysis of Variance Table with Covariance Parameter (CovParm) Estimates Using Heterogeneous First-Order Auto-Regressive Covariance Structure

```
PROC MIXED cl ic covtest DATA=E165;
Class Exercise Drug person Time;
Model BP= Exercise*Drug*Time IBP*Exercise*Drug*Time/ddfm=kr;
Random person(Exercise*Drug);
repeated Time/type=arh(1) subject=person(Exercise*Drug);
```

Neg2LogLike	Parameters	AIC	AICC	HQIC	BIC	CAIC
847.41	8	863.41	864.48	867.30	875.13	883.13

CovParm	Subject	Estimate	StdErr	ZValue	ProbZ
person(exercise*drug)		28.6767	8.7041	3.29	0.0005
Var(1)	person(exercise*drug)	10.4832	3.7663	2.78	0.0027
Var(2)	person(exercise*drug)	1.7199	1.5233	1.13	0.1294
Var(3)	person(exercise*drug)	3.1345	1.4346	2.18	0.0144
Var(4)	person(exercise*drug)	8.7116	3.2948	2.64	0.0041
Var(5)	person(exercise*drug)	14.4857	5.3342	2.72	0.0033
Var(6)	person(exercise*drug)	11.9145	4.2686	2.79	0.0026
ARH(1)	person(exercise*drug)	0.7486	0.0972	7.70	0.0000

Effect	NumDF	DenDF	FValue	ProbF
exercise*drug*TIME	23	74.7	1.07	0.3983
IBP*exercise*drug*TIME	24	75.9	3.08	0.0001

criteria, the better model is the one that has the smaller value within a criteria. Each of the criteria penalize the value of the likelihood function in different ways by a function of the number of parameters in the covariance matrix and the sample size. The AIC has been used by many modelers, so the AIC value is used here as the criteria to select the most adequate covariance matrix, i.e, covariance matrix 1 is more adequate than covariance matrix 2 if the AIC from the fit of 1 is less than the AIC from the fit of 2. The split-plot analysis in Table 16.3 and the analysis in Table 16.4 with the compound symmetry covariance structure provide identical analyses, but the compound symmetry model is a little more general since the CS parameter can be estimated with a negative value where as the person(drug*exercise) variance component cannot take on negative values since it is a variance. The heterogeneous compound symmetry covariance structure (see Table 16.5) has seven parameters and the corresponding AIC is larger than the one for the compound symmetry structure, i.e., 900.94 for CS and 907.07 for CSH. The AR(1) covariance structure is fit to the data by using the PROC MIXED code in Table 16.6. The repeated measures covariance structure has two parameters along with the person(drug*exercise) variance component. The estimate of the autocorrelation coefficient is 0.7104. The AIC for the AR(1) structure is 870.01, which is considerably smaller than the AIC for the CS and CSH structures. The estimate of the autocorrelation coefficient from fitting the ARH(1) model is 0.7486, as shown in Table 16.7. This model has eight parameters in the covariance structure, including

TABLE 16.8
Analysis of Variance Table with Covariance Parameter (CovParm) Estimates Using First-Order Ante-Dependence Covariance Structure

```
Proc Mixed cl ic covtest DATA=E165;
Class Exercise Drug person Time;
Model BP= Exercise*Drug*Time IBP*Exercise*Drug*Time/ddfm=kr;
repeated Time/type=ANTE(1) subject=person(Exercise*Drug);
```

Neg2LogLike	Parameters	AIC	AICC	HQIC	BIC	CAIC
848.70	11	870.70	872.70	876.04	886.82	897.82

CovParm	Subject	Estimate	StdErr	ZValue	ProbZ
Var(1)	person(exercise*drug)	32.0944	9.2648	3.46	0.0003
Var(2)	person(exercise*drug)	28.4515	8.2133	3.46	0.0003
Var(3)	person(exercise*drug)	31.7284	9.1592	3.46	0.0003
Var(4)	person(exercise*drug)	30.6694	8.8535	3.46	0.0003
Var(5)	person(exercise*drug)	36.7300	10.6030	3.46	0.0003
Var(6)	person(exercise*drug)	36.6363	10.5760	3.46	0.0003
Rho(1)	person(exercise*drug)	0.9062	0.0365	24.83	0.0000
Rho(2)	person(exercise*drug)	0.9782	0.0088	111.02	0.0000
Rho(3)	person(exercise*drug)	0.9253	0.0293	31.54	0.0000
Rho(4)	person(exercise*drug)	0.9245	0.0297	31.15	0.0000
Rho(5)	person(exercise*drug)	0.9073	0.0361	25.13	0.0000

Effect	NumDF	DenDF	FValue	ProbF
exercise*drug*TIME	23	65.1	1.04	0.4281
IBP*exercise*drug*TIME	24	62.6	2.83	0.0005

the person(drug*exercise) variance component, six residual variances, one for each time period, and the auto-correlation coefficient. The AIC value for ARH(1) is 863.41, which is the smallest value for the covariance structures discussed so far. The analyses in Tables 16.8, 16.9, and 16.10 fit the ANTE(1), TOEP, and UN covariance structures, respectively. The AIC values are all larger than that for the ARH(1) structure. Table 16.11 contains the values of the AIC from each of the covariance structures evaluated where the column of number of parameters contains of the number of covariance parameters estimated that have nonzero values (values not on the boundary). The covariance structure ARH(1) has the smallest AIC value; thus, based on AIC, the ARH(1) was selected to be the most adequate covariance structure among the set of covariance structures evaluated. The next step in the process is to express the slopes as factorial effects and attempt to simplify the covariate part of the model while using ARH(1) as the repeated measures covariance structure. The model with factorial effects for both the intercepts and slopes is

$$BP_{ijkm} = \mu + E_i + D_j + (ED)_{ij} + \left(\theta + \zeta_i + \delta_j + \phi_{ij}\right) IBP_{ijk} + p_{ijk} \qquad (16.5)$$

$$+ T_m + (TE)_{im} + (TD)_{jm} + (TED)_{ijm} + \left(\lambda_m + \kappa_{im} + \eta_{jm} + \omega_{ijm}\right) IBP_{ijk} + \varepsilon_{ijkm}$$

TABLE 16.9
Analysis of Variance Table with Covariance Parameter (CovParm) Estimates Using Toplitz Covariance Structure

```
Proc Mixed cl ic covtest DATA=E165;
Class Exercise Drug person Time;
Model BP= Exercise*Drug*Time IBP*Exercise*Drug*Time/ddfm=kr;
repeated Time/type=TOEP subject=person(Exercise*Drug);
```

Neg2LogLike	Parameters	AIC	AICC	HQIC	BIC	CAIC
863.21	6	875.21	875.82	878.12	884.00	890.00

CovParm	Subject	Estimate	StdErr	ZValue	ProbZ
TOEP(2)	person(exercise*drug)	30.9604	8.5780	3.61	0.0003
TOEP(3)	person(exercise*drug)	29.1006	8.5684	3.40	0.0007
TOEP(4)	person(exercise*drug)	27.8108	8.5641	3.25	0.0012
TOEP(5)	person(exercise*drug)	27.3703	8.5581	3.20	0.0014
TOEP(6)	person(exercise*drug)	26.8067	8.5861	3.12	0.0018
Residual		33.3667	8.5835	3.89	0.0001

Effect	NumDF	DenDF	FValue	ProbF
exercise*drug*TIME	23	87.5	0.91	0.5908
IBP*exercise*drug*TIME	24	88.4	2.82	0.0002

where E_i, D_j, $(ED)_{ij}$, T_m, $(TE)_{im}$, $(TD)_{jm}$ and $TED)_{ijk}$ denote the fixed effects for exercise, drug, time, and their interactions, and $\theta + \zeta_i + \delta_j + \phi_{ij}$ and $\lambda_m + \kappa_{im} + \eta_{jm} + \omega_{ijm}$ are the factorial effects for the slopes for the person experimental unit and the within person experimental unit, respectively. Table 16.12 contains the PROC MIXED code to fit Model 16.5 using the ARH(1) structure for the covariance part of the model. Several of the significance levels corresponding to the terms involving IBP have significance levels that are quite large, which indicates the model can be simplified. Since the IBP*exercise*drug*time term has a significance level of 0.9162, it was the first term to be deleted. The terms deleted and the order they were deleted are IBP*Exercise*Drug*Time, IBP*Time*Exercise, IBP*Time*Drug, IBP*Drug*Exercise, IBP*Time, and IBP*Exercise, leaving the covariate part of the model with just IBP and IBP*drug. The final model after reducing the covariate part is

$$BP_{ijkm} = \mu + E_i + D_j + (ED)_{ij} + \theta\, IBP_{ijk} + \delta_j\, IBP_{ijk} + p_{ijk}$$
$$+ T_m + (TE)_{im} + (TD)_{jm} + (TED)_{ijm} + \varepsilon_{ijkm}$$
(16.6)

Table 16.13 contains the PROC MIXED code to fit Model 16.6 using the ARH(1) covariance structure. The significance level corresponding to IBP*drug in Table 16.13 is 0.1045, so there is some indication that the slopes may not be equal. If a common slope model is fit to the data set, the additional variability causes PROC MIXED to fail to converge. Thus the model with unequal slopes for the two levels of drug was selected for further analyses. A full rank expression of Model 16.6 with intercepts

TABLE 16.10
Analysis of Variance Table with Covariance Parameter (CovParm) Estimates Using Unstructured Covariance Matrix

```
Proc Mixed cl ic covtest DATA=E165;
Class Exercise Drug person Time;
Model BP= Exercise*Drug*Time IBP*Exercise*Drug*Time/ddfm=kr;
repeated Time/type=UN subject=person(Exercise*Drug);
```

Neg2LogLike	Parameters	AIC	AICC	HQIC	BIC	CAIC
836.05	21	878.05	885.62	888.25	908.83	929.83

CovParm	Subject	Estimate	StdErr	ZValue	ProbZ
UN(1,1)	person(exercise*drug)	32.0944	9.2648	3.46	0.0003
UN(2,1)	person(exercise*drug)	27.3839	8.3242	3.29	0.0010
UN(2,2)	person(exercise*drug)	28.4515	8.2133	3.46	0.0003
UN(3,1)	person(exercise*drug)	27.9428	8.6581	3.23	0.0012
UN(3,2)	person(exercise*drug)	29.3897	8.5792	3.43	0.0006
UN(3,3)	person(exercise*drug)	31.7284	9.1592	3.46	0.0003
UN(4,1)	person(exercise*drug)	25.6162	8.2677	3.10	0.0019
UN(4,2)	person(exercise*drug)	26.2251	8.0631	3.25	0.0011
UN(4,3)	person(exercise*drug)	28.8656	8.6754	3.33	0.0009
UN(4,4)	person(exercise*drug)	30.6694	8.8535	3.46	0.0003
UN(5,1)	person(exercise*drug)	26.9072	8.9042	3.02	0.0025
UN(5,2)	person(exercise*drug)	26.2718	8.5030	3.09	0.0020
UN(5,3)	person(exercise*drug)	28.6413	9.0960	3.15	0.0016
UN(5,4)	person(exercise*drug)	31.0279	9.3301	3.33	0.0009
UN(5,5)	person(exercise*drug)	36.7300	10.6030	3.46	0.0003
UN(6,1)	person(exercise*drug)	27.4034	8.9600	3.06	0.0022
UN(6,2)	person(exercise*drug)	27.2981	8.6302	3.16	0.0016
UN(6,3)	person(exercise*drug)	29.2933	9.1754	3.19	0.0014
UN(6,4)	person(exercise*drug)	30.8522	9.2994	3.32	0.0009
UN(6,5)	person(exercise*drug)	33.2808	10.1104	3.29	0.0010
UN(6,6)	person(exercise*drug)	36.6363	10.5760	3.46	0.0003

Effect	NumDF	DenDF	FValue	ProbF
exercise*drug*TIME	23	36.8	0.89	0.6045
IBP*exercise*drug*TIME	24	36.3	2.47	0.0067

for each combination of exercise by drug by time and with slopes for each level of drug is

$$BP_{ijkm} = \mu_{ijm} + \delta_j IBP_{ijk} + p_{ijk} + \varepsilon_{ijkm} \quad (16.7)$$

Table 16.14 contains the PROC MIXED code needed to fit model 16.7 with the ARH(1) covariance structure to the data. Table 16.15 contains the estimates of the models intercepts and slopes, where the slope for those using no drug is estimated to be 0.9103 and for those using drug is 1.4138. Table 16.16 contains the adjusted means for each combination of exercise by drug by time evaluated at three values

TABLE 16.11
Summary of the Akiake Information Criterion (AIC) for Eight Covariance Structures

Covariance Structure	Number of Parameters	AIC
Split-plot	2	900.94
CS	2	900.94
CSH	7	907.07
AR(1)	3	870.01
ARH(1)	8	863.41
ANTE(1)	11	870.07
TOEP	6	875.21
UN	21	878.05

of IBP: 140, 150, and 160 mmHg. Since the slopes of the regression lines are not dependent on the levels of exercise, the levels of exercise can be compared within each level of drug and exposure time at a single value of IBP. The pairwise comparisons of the levels of exercise, where the simulate adjustment for multiple comparisons was used, are in Table 16.17. There are significantly lower mean systolic blood pressure values for those persons using an exercise program than those not using an exercise program for times of 3 weeks and beyond when drug was not in their regime. There are no significant differences between exercise and no exercise at any time when the persons included the drug in their regime. Pairwise comparisons between the drug and no-drug regimes at each time for exercise and no exercise at the three values of IBP are displayed in Table 16.18. All denominator degrees of freedom for each test statistic were between 27.9 and 31.2, and thus were not included in the table. Those including the drug in their regime had significantly lower mean systolic blood pressure values for (1) both no exercise and exercise when IBP = 140 mmHg at times 1 through 5, (2) for no exercise when IBP = 150 mmHg for times 2 through 6, (3) for exercise when IBP = 150 mmHg at time 3, and (4) for no exercise when IBP = 150 at times 6 for IBP = 140 mmHg. Figures 16.1 through 16.3 are graphs of the adjusted systolic blood pressure means for the exercise by drug combinations against the values of IBP at times 1, 3, and 6. On each graph, two lines with the same level of drug are parallel and lines with different levels of drug are not parallel.

The analysis of covariance strategy for this repeated measures design is identical to that for the split-plot or any other design that involves a factorial treatment structure, except the starting point is to model the covariance structure of the repeated measures part of the model. Eight different structures were considered here, but there are others that can be very meaningful covariance structures. In particular, the parameters for a given covariance structure could be unequal for different levels of one or more of the factors in the treatment structure. For example, the AR(1) structure might be different for the two levels of drug. To specify such a structure in PROC

Analysis of Covariance for Repeated Measures Designs

TABLE 16.12
Analysis of Variance Table with the Factorial Effects for the Intercepts and the Slopes with Covariance Parameter (CovParm) Estimates Using ARH(1) Covariance Structure for the Repeated Measurement Errors

```
Proc Mixed cl ic covtest DATA=E165;
Class Exercise Drug person Time;
Model BP= Exercise|Drug|Time IBP IBP*Drug IBP*Exercise
IBP*Drug*Exercise IBP*Time IBP*Time*Drug IBP*Time*Exercise
IBP*Exercise*Drug*Time/ddfm=kr;
Random person(Exercise*Drug);
repeated Time/type=arh(1) subject=person(Exercise*Drug);
```

Neg2LogLike	Parameters	AIC	AICC	HQIC	BIC	CAIC
847.41	8	863.41	864.48	867.30	875.13	883.13

CovParm	Subject	Estimate	StdErr	ZValue	ProbZ
person(exercise*drug)		28.6767	8.7041	3.29	0.0005
Var(1)	person(exercise*drug)	10.4832	3.7663	2.78	0.0027
Var(2)	person(exercise*drug)	1.7199	1.5233	1.13	0.1294
Var(3)	person(exercise*drug)	3.1345	1.4346	2.18	0.0144
Var(4)	person(exercise*drug)	8.7116	3.2948	2.64	0.0041
Var(5)	person(exercise*drug)	14.4857	5.3342	2.72	0.0033
Var(6)	person(exercise*drug)	11.9145	4.2686	2.79	0.0026
ARH(1)	person(exercise*drug)	0.7486	0.0972	7.70	0.0000

Effect	NumDF	DenDF	FValue	ProbF
exercise	1	26.2652	0.75	0.3953
drug	1	26.2652	6.14	0.0200
exercise*drug	1	26.2652	0.17	0.6824
Time	5	41.9903	1.97	0.1036
exercise*Time	5	41.9903	0.20	0.9600
drug*Time	5	41.9903	0.73	0.6037
exercise*drug*Time	5	41.9903	0.45	0.8092
IBP	1	26.2652	46.45	0.0000
IBP*drug	1	26.2652	4.38	0.0461
IBP*exercise	1	26.2652	1.02	0.3210
IBP*exercise*drug	1	26.2652	0.11	0.7459
IBP*Time	5	41.9903	1.45	0.2266
IBP*drug*Time	5	41.9903	0.69	0.6364
IBP*exercise*Time	5	41.9903	0.14	0.9812
IBP*exercise*drug*Time	5	41.9903	0.29	0.9162

MIXED, use the repeated statement "Repeated time/type=AR(1) subject=person(exercise*drug) group=drug;". The group = drug option is the specification that the covariance structures for the two levels of drug could have different values. However, one thing to keep in mind is that with more parameters in the covariance structure, it is more difficult computationally to obtain the maximum likelihood or REML estimators of the parameters.

TABLE 16.13
Analysis of Variance Table for the Reduced Model with Factorial Effects for the Slopes and Intercepts Using the ARH(1) Covariance Structure for the Repeated Measurement Errors

```
Proc Mixed cl ic covtest DATA=E165;
Class Exercise Drug person Time;
Model BP= Exercise|Drug|Time IBP IBP*Drug/ddfm=kr;
Random person(Exercise*Drug);
repeated Time/type=arh(1) subject=person(Exercise*Drug);
```

Neg2LogLike	Parameters	AIC	AICC	HQIC	BIC	CAIC
814.34	8	830.34	831.26	834.23	842.07	850.07

CovParm	Subject	Estimate	StdErr	ZValue	ProbZ
person(exercise*drug)		26.9435	8.0844	3.33	0.0004
Var(1)	person(exercise*drug)	10.5648	4.1654	2.54	0.0056
Var(2)	person(exercise*drug)	2.3860	2.0324	1.17	0.1202
Var(3)	person(exercise*drug)	3.7448	1.7661	2.12	0.0170
Var(4)	person(exercise*drug)	8.8289	3.7106	2.38	0.0087
Var(5)	person(exercise*drug)	14.3004	5.8853	2.43	0.0076
Var(6)	person(exercise*drug)	12.4418	4.9743	2.50	0.0062
ARH(1)	person(exercise*drug)	0.7642	0.1052	7.27	0.0000

Effect	NumDF	DenDF	FValue	ProbF
exercise	1	26.2	19.23	0.0002
drug	1	24.3	4.39	0.0467
exercise*drug	1	26.2	3.57	0.0701
Time	5	47.4	68.82	0.0000
exercise*Time	5	47.4	22.21	0.0000
drug*Time	5	47.4	14.90	0.0000
exercise*drug*Time	5	47.4	12.59	0.0000
IBP	1	24.3	60.61	0.0000
IBP*drug	1	24.3	2.84	0.1045

16.7 EXAMPLE: OXIDE LAYER DEVELOPMENT EXPERIMENT WITH THREE SIZES OF EXPERIMENTAL UNITS WHERE THE REPEATED MEASURE IS AT THE MIDDLE SIZE OF EXPERIMENTAL UNIT AND THE COVARIATE IS MEASURED ON THE SMALL SIZE EXPERIMENTAL UNIT

One of the many steps in the fabrication of semiconductors involves putting a wafer of silicon into a furnace set to a specific temperature to enable a layer of oxide to accumulate. A chemical engineer ran an experiment to study the effect of temperature of the furnace, position within the furnace, and wafer type on the thickness of the resulting layer of oxide for each wafer. The process was to measure the thickness of each wafer before putting the wafer into the furnace and then measuring the thickness of the wafer after the furnace run. The difference, or delta, between the

Analysis of Covariance for Repeated Measures Designs

TABLE 16.14
PROC MIXED Code to Fit the Final Model with Means Expression for Both the Intercepts and the Slopes Using the ARH(1) Covariance Structure for the Repeated Measurement Errors

```
Proc Mixed cl ic covtest DATA=E165;
Class Exercise Drug person Time;
Model BP= Exercise*Drug*Time IBP*Drug/ddfm=kr Solution noint;
Random person(Exercise*Drug);
repeated Time/type=arh(1) subject=person(Exercise*Drug);
```

Neg2LogLike	Parameters	AIC	AICC	HQIC	BIC	CAIC
814.34	8	830.34	831.26	834.23	842.07	850.07

CovParm	Subject	Estimate	StdErr	ZValue	ProbZ
person(exercise*drug)		26.9435	8.0844	3.33	0.0004
Var(1)	person(exercise*drug)	10.5648	4.1654	2.54	0.0056
Var(2)	person(exercise*drug)	2.3860	2.0324	1.17	0.1202
Var(3)	person(exercise*drug)	3.7448	1.7661	2.12	0.0170
Var(4)	person(exercise*drug)	8.8289	3.7106	2.38	0.0087
Var(5)	person(exercise*drug)	14.3004	5.8853	2.43	0.0076
Var(6)	person(exercise*drug)	12.4418	4.9743	2.50	0.0062
ARH(1)	person(exercise*drug)	0.7642	0.1052	7.27	0.0000

Effect	NumDF	DenDF	FValue	ProbF
exercise*drug*Time	24	88.0	25.01	0.0000
IBP*drug	2	24.3	32.32	0.0000

after and before measurements was used as the thickness of the oxide layer. The experiment used two furnace temperatures, 950 and 1100°F. The furnaces used in this step of the process are vertical furnaces that can hold 100 wafers from top to bottom. It was thought that the position within a furnace might have an effect on the resulting thickness of the oxide layer. Each furnace was partitioned into four sections or positions: top, middle top, middle bottom, and bottom, denoted by Ttop, Mtop, Mbot, and Bbot. Two furnaces were available for the experiment and one run from each furnace could be obtained within one shift. Thus, the day was considered as a block and the two levels of temperature were randomly assigned to the two furnaces within a day. The furnace is the experimental unit for the levels of temperature and the furnace design is a one-way treatment structure in a randomized complete block design structure. Each furnace was partitioned into four sections and the four levels of position were assigned to those sections, but the levels of position could not be randomly assigned to the sections. Hence, this overall design involves repeated measures at the section level. The section is the experimental unit for the levels of position (section and position are synonymous, but the two terms are used here so that section is the experimental unit and position is the factor with four levels) and the section design is a one-way treatment structure in a randomized complete block design structure with furnaces as blocks. Finally, one wafer from each of two wafer types were placed into slots within each of the positions of each

TABLE 16.15
Estimates of the Intercepts for the Exercise by Drug by Time Combinations and of the Slopes for Each Level of Drug

Effect	Exercise	Drug	Time	Estimate	StdErr	df	tValue	Probt
exercise*drug*Time	No	No	1	16.1438	26.2544	24.4	0.61	0.5443
exercise*drug*Time	No	No	2	14.8938	26.2349	24.3	0.57	0.5754
exercise*drug*Time	No	No	3	14.8938	26.2381	24.3	0.57	0.5755
exercise*drug*Time	No	No	4	14.2688	26.2502	24.4	0.54	0.5917
exercise*drug*Time	No	No	5	13.8938	26.2632	24.4	0.53	0.6016
exercise*drug*Time	No	No	6	15.0188	26.2588	24.4	0.57	0.5726
exercise*drug*Time	No	Yes	1	−64.5936	31.8888	24.4	−2.03	0.0539
exercise*drug*Time	No	Yes	2	−75.3436	31.8728	24.3	−2.36	0.0264
exercise*drug*Time	No	Yes	3	−75.2186	31.8755	24.3	−2.36	0.0266
exercise*drug*Time	No	Yes	4	−80.3436	31.8854	24.4	−2.52	0.0187
exercise*drug*Time	No	Yes	5	−80.7186	31.8961	24.4	−2.53	0.0182
exercise*drug*Time	No	Yes	6	−81.0936	31.8925	24.4	−2.54	0.0178
exercise*drug*Time	Yes	No	1	12.0042	26.7351	24.4	0.45	0.6574
exercise*drug*Time	Yes	No	2	7.1292	26.7160	24.3	0.27	0.7918
exercise*drug*Time	Yes	No	3	3.6292	26.7192	24.3	0.14	0.8931
exercise*drug*Time	Yes	No	4	0.6292	26.7311	24.4	0.02	0.9814
exercise*drug*Time	Yes	No	5	−2.9958	26.7439	24.4	−0.11	0.9117
exercise*drug*Time	Yes	No	6	−7.4958	26.7395	24.4	−0.28	0.7816
exercise*drug*Time	Yes	Yes	1	−70.2316	32.5014	24.4	−2.16	0.0407
exercise*drug*Time	Yes	Yes	2	−78.3566	32.4857	24.3	−2.41	0.0238
exercise*drug*Time	Yes	Yes	3	−83.7316	32.4883	24.3	−2.58	0.0164
exercise*drug*Time	Yes	Yes	4	−85.8566	32.4981	24.4	−2.64	0.0142
exercise*drug*Time	Yes	Yes	5	−84.9816	32.5086	24.4	−2.61	0.0151
exercise*drug*Time	Yes	Yes	6	−84.4816	32.5050	24.4	−2.60	0.0156
IBP*drug		No		0.9103	0.1838	24.3	4.95	0.0000
IBP*drug		Yes		1.4138	0.2233	24.3	6.33	0.0000

furnace. The wafer types are randomly assigned to the two slots within the section. The silicon wafer is the experimental unit for wafer types (or slot in the section of the furnace) and the wafer design is a one-way treatment structure in a randomized complete block design structure where each furnace section is a block. A furnace run involves 8 wafers, 2 in each of the 4 positions, where the remainder of the 100 slots are filled with dummy wafers to simulate an actual furnace run. This experiment involves three sizes of experimental units; the furnace, the section within the furnace, and the slot or wafer within the section within the furnace. Finally, it was thought that the initial thickness of the wafer could possibly have an effect on the thickness of the resulting oxide layer, so its value was retained as a possible covariate. The wafer thickness is a covariate measured on the small size experimental unit. The data in Table 16.19 consist of five blocks of the three-way treatment structure, i.e.,

TABLE 16.16
PROC MIXED Code to Produce Adjusted Mean Systolic Blood Pressure (mmHg) for the Exercise by Drug by Time Treatment Combinations Evaluated at IBP=140, 150, and 160 mmHg

```
lsmeans  Exercise*Drug*Time/Diff  AT  IBP=140  adjust=simulate;
lsmeans  Exercise*Drug*Time/Diff  AT  IBP=150  adjust=simulate;
lsmeans  Exercise*Drug*Time/Diff  AT  IBP=160  adjust=simulate;
```

			IBP=140		IBP=150		IBP=160	
Exercise	Drug	Time	Estimate	StdErr	Estimate	StdErr	Estimate	StdErr
No	No	1	143.59	2.21	152.69	2.58	161.79	3.90
No	No	2	142.34	1.96	151.44	2.37	160.54	3.76
No	No	3	142.34	2.01	151.44	2.41	160.54	3.79
No	No	4	141.71	2.16	150.82	2.54	159.92	3.87
No	No	5	141.34	2.31	150.44	2.67	159.54	3.96
No	No	6	142.46	2.26	151.57	2.62	160.67	3.93
No	Yes	1	133.34	2.24	147.48	2.74	161.62	4.47
No	Yes	2	122.59	1.99	136.73	2.54	150.87	4.35
No	Yes	3	122.72	2.04	136.85	2.58	150.99	4.37
No	Yes	4	117.59	2.19	131.73	2.70	145.87	4.44
No	Yes	5	117.22	2.34	131.35	2.82	145.49	4.52
No	Yes	6	116.84	2.29	130.98	2.78	145.12	4.49
Yes	No	1	139.45	2.35	148.55	2.35	157.65	3.51
Yes	No	2	134.57	2.12	143.68	2.12	152.78	3.36
Yes	No	3	131.07	2.16	140.18	2.16	149.28	3.38
Yes	No	4	128.07	2.31	137.18	2.31	146.28	3.47
Yes	No	5	124.45	2.45	133.55	2.45	142.65	3.57
Yes	No	6	119.95	2.40	129.05	2.40	138.15	3.54
Yes	Yes	1	127.70	2.46	141.84	2.41	155.98	3.94
Yes	Yes	2	119.58	2.25	133.72	2.19	147.85	3.81
Yes	Yes	3	114.20	2.28	128.34	2.23	142.48	3.83
Yes	Yes	4	112.08	2.42	126.22	2.37	140.35	3.91
Yes	Yes	5	112.95	2.56	127.09	2.51	141.23	4.00
Yes	Yes	6	113.45	2.51	127.59	2.46	141.73	3.97

two temperatures (temp) by two wafer types (waftype) by four positions (posit) where delta is the thickness of the resulting oxide layer and thick is the thickness of the wafer before processing. A model to describe this data set with different intercepts and slopes for each of the temp by waftype by posit combinations is

$$\text{Delta}_{ijkm} = \mu_{jkm} + \beta_{jkm}\text{Thick} + b_i + f_{ij} + s_{ijk} + w_{ijkm}$$
$$\text{for } i = 1, 2, \ldots, 5, \quad j = 1, 2, \quad k = 1, 2, 3, 4, \text{ and } m = 1, 2$$

(16.8)

TABLE 16.17
Pairwise Comparisons of Exercise Means within Each Combination of Drug and Time

Drug	Time	Exercise	_exercise	Estimate	StdErr	df	tValue	Adjp
No	1	No	Yes	4.14	3.10	32	1.34	0.9976
No	2	No	Yes	7.76	2.75	24	2.82	0.2520
No	3	No	Yes	11.26	2.81	27	4.01	0.0088
No	4	No	Yes	13.64	3.03	30	4.50	0.0014
No	5	No	Yes	16.89	3.25	32	5.20	0.0001
No	6	No	Yes	22.51	3.17	32	7.09	0.0000
Yes	1	No	Yes	5.64	3.12	32	1.81	0.9218
Yes	2	No	Yes	3.01	2.78	24	1.09	0.9999
Yes	3	No	Yes	8.51	2.84	27	3.00	0.1709
Yes	4	No	Yes	5.51	3.05	30	1.81	0.9216
Yes	5	No	Yes	4.26	3.27	32	1.30	0.9983
Yes	6	No	Yes	3.39	3.20	32	1.06	1.0000

Note: Slopes of regression lines are not dependent on the levels of exercise.

where μ_{jkm} and β_{jkm} are the intercepts and slopes for the three-way treatment combinations, b_i denotes the random block or day effect, f_{ij} denotes the random furnace effect within a block, s_{ijk} denotes the random section effect within a furnace within a day, and w_{ijkm} denotes the residual or random wafer effect. The split-split-plot type assumptions about the distributions of the random effects are to assume that the $b_i \sim$ iid $N(0, \sigma_b^2)$, the $f_{ij} \sim$ iid $N(0, \sigma_f^2)$, the $s_{ijk} \sim$iid $N(0, \sigma_s^2)$, and the w_{ijkm} iid $N(0, \sigma_w^2)$. The analysis of variance table corresponding to these split-split-plot model assumptions without the covariate information is in Table 16.20. There are three error terms in the table: error(furnace), error(segment), and error(wafer) based on 4, 24, and 32 degrees of freedom, respectively. The split-split-plot assumptions are reasonable for b_i, f_{ij}, and w_{ijkm}, but are not reasonable for the distribution of s_{ijk} since the levels of position cannot be randomly assigned to the sections. Let the covariance structure of the four s_{ijk} variables within each furnace be represented by

$$\mathbf{G}_s = \mathrm{Var}\begin{bmatrix} s_{ij1} \\ s_{ij2} \\ s_{ij3} \\ s_{ij4} \end{bmatrix} = \begin{bmatrix} \sigma_{s1}^2 & \sigma_{s12} & \sigma_{s13} & \sigma_{s14} \\ \sigma_{s12} & \sigma_{s2}^2 & \sigma_{s23} & \sigma_{s24} \\ \sigma_{s13} & \sigma_{s23} & \sigma_{s3}^2 & \sigma_{s34} \\ \sigma_{s14} & \sigma_{s24} & \sigma_{s34} & \sigma_{s4}^2 \end{bmatrix}.$$

The PROC MIXED code to fit a model with three-way factorial effects for both the intercepts and the slopes is

TABLE 16.18
Pairwise Comparisons between the Level of Drug at Combinations of Time and Exercise and Three Levels of IBP

				Exercise = NO			Exercise = Yes		
Time	Drug	_drug	IBP	Estimate	StdErr	Adjp	Estimate	StdErr	Adjp
1	No	Yes	140	10.52	2.93	0.0360	12.32	3.19	0.0140
1	No	Yes	150	4.34	3.55	0.9995	6.14	3.16	0.8846
1	No	Yes	160	−1.84	5.70	1.0000	−0.04	5.05	1.0000
2	No	Yes	140	20.02	2.93	0.0000	15.57	3.19	0.0001
2	No	Yes	150	13.84	3.55	0.0140	9.39	3.16	0.1892
2	No	Yes	160	7.66	5.70	0.9979	3.21	5.05	1.0000
3	No	Yes	140	19.89	2.93	0.0000	17.44	3.19	0.0000
3	No	Yes	150	13.71	3.55	0.0154	11.26	3.16	0.0392
3	No	Yes	160	7.53	5.70	0.9983	5.08	5.05	1.0000
4	No	Yes	140	24.39	2.93	0.0000	16.57	3.19	0.0000
4	No	Yes	150	18.21	3.55	0.0002	10.39	3.16	0.0822
4	No	Yes	160	12.03	5.70	0.7679	4.21	5.05	1.0000
5	No	Yes	140	24.39	2.93	0.0000	12.07	3.19	0.0179
5	No	Yes	150	18.21	3.55	0.0002	5.89	3.16	0.9176
5	No	Yes	160	12.03	5.70	0.7679	−0.29	5.05	1.0000
6	No	Yes	140	25.89	2.93	0.0000	7.07	3.19	0.7193
6	No	Yes	150	19.71	3.55	0.0000	0.89	3.16	1.0000
6	No	Yes	160	13.53	5.70	0.5653	−5.29	5.05	1.0000

Note: The denominator df for each test was between 27.9 and 31.2.

```
PROC MIXED cl ic covtest;
Class block temp posit waftype;
Model Delta = temp|posit|waftype|thick;
Random block temp*block;
Random posit/type=xxx subject=block*temp;
```

The last random statement selects the specific covariance structure for G_s by specifying xxx. The above model was fit to the data set using seven covariance structure types: CS, AR(1), CSH, ARH(1), TOEP, ANTE(1), and UN. For those covariance structures that contain the J_4 matrix as a special case, the first random statement needs to be just "Random block;". This is the case for CS, CSH, TOEP, ANTE(1), and UN. Using the AIC as a guide, the TOEP covariance structure was determined to be the most adequate of the structures tested. The wafers were assumed to be independent so a repeated statement was not needed in the model specification.

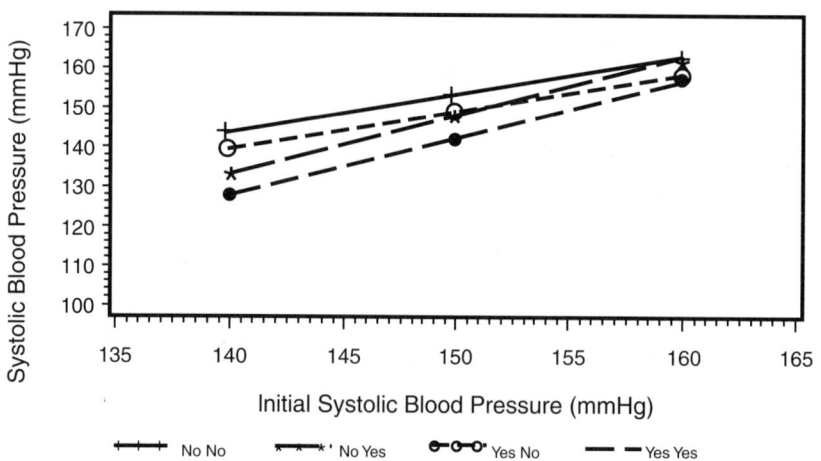

FIGURE 16.1 Plot of the exercise by drug adjusted systolic blood pressure means for (Exercise, Drug) combinations after week one (time = 1).

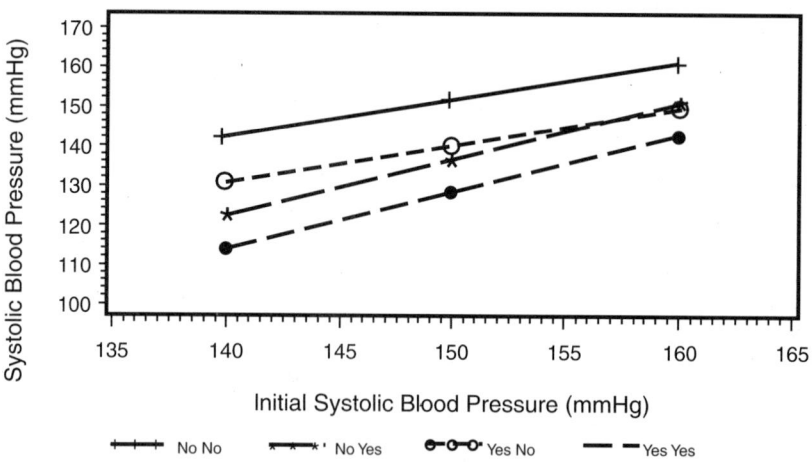

FIGURE 16.2 Plot of the exercise by drug adjusted systolic blood pressure means for (Exercise, Drug) combinations after week three (time = 3).

Table 16.21 contains the PROC MIXED code to fit the model with the three-way factorial effects for the intercepts and the slopes using the TOPE covariance structure for the repeated measures. The estimate of the block variance component was set

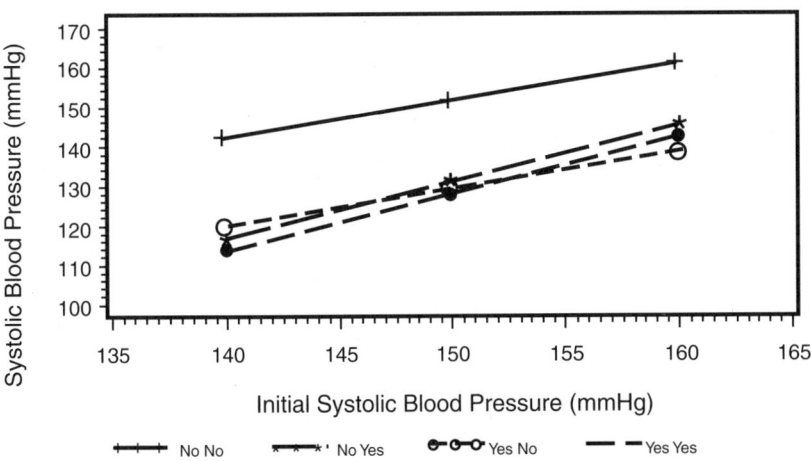

FIGURE 16.3 Plot of the exercise by drug adjusted systolic blood pressure means for (Exercise, Drug) combinations after week six (time = 6).

to zero (on the boundary) by PROC MIXED, so the information criteria indicates that there are effectively five parameters instead of six in the covariance structure of the model. Some of the significance levels of the effects involving thick are much larger than 0.05, indicating the model can be simplified by using the deletion process. The stepwise deletion process deleted the following terms in the order: thick*waftype*temp*posit, thick*waftype*posit, thick*posit*temp, thick*waftype*temp, thick*temp, and thick*posit. The final model with the reduced form of the covariate part of the model consists of thick and thick*waftype, as shown in Table 16.22. Thus, the adequate covariate part of the model involves different slopes for each wafer type. The model with factorial effects for the intercepts and two slopes, one for each wafer type, is

$$\text{Delta}_{ijkm} = \mu + b_i + T_j + f_{ij} + P_k + (TP)_{jk} + s_{ijk}$$
$$+ W_m + (WT)_{jm} + (WP)_{km} + (WTP)_{jkm} + \beta_m \text{Thick}_{ijkm} + w_{ijkm} \quad (16.9)$$

$$\text{for } i = 1, 2, \ldots, 5, \quad j = 1, 2, \quad k = 1, 2, 3, 4, \text{ and } m = 1, 2$$

and the PROC MIXED code to fit this model using the TOEP covariance structure for the section random effects is in Table 16.23. The estimate of the block variance component from this reduced model is still zero. The estimates of the slopes for wafer types 1 and 2 are 0.00992 and 0.00563, respectively. The significance level associated with waftype*temp*posit is 0.0360, indicating there is a significant interaction effect for the intercepts of the model. The PROC MIXED code in Table 16.24

TABLE 16.19
Data from the Oxide Layer Data Set Where Delta is the Thickness of the Oxide Layer and Thick is the Thickness of the Wafer

			Top		Middle Top		Middle Bottom		Bottom	
Block	temp	waftype	Delta	Thick	Delta	Thick	Delta	Thick	Delta	Thick
1	950	1	40	1804	39	1865	39	2055	36	2005
1	950	2	26	1959	29	2032	26	1892	29	2088
1	1100	1	71	2083	62	1802	58	1814	54	1848
1	1100	2	53	1986	52	1844	48	1784	47	1764
2	950	1	56	1901	59	1924	60	1935	49	1874
2	950	2	41	1856	48	2014	48	1843	44	2131
2	1100	1	66	1832	64	2032	61	1765	56	1967
2	1100	2	51	2008	53	2149	51	1767	50	2125
3	950	1	48	1834	47	2031	44	2135	43	2085
3	950	2	33	1932	34	1877	30	1776	33	1756
3	1100	1	70	2010	66	1911	67	1903	56	1771
3	1100	2	52	2032	54	2022	56	1839	53	2062
4	950	1	46	1901	40	1838	42	1868	42	1961
4	950	2	30	1900	29	2063	30	1947	36	2085
4	1100	1	68	2145	59	1956	56	1869	44	1910
4	1100	2	50	2116	49	2117	46	1790	39	2103
5	950	1	41	1902	40	1959	44	1793	48	1862
5	950	2	24	1900	27	1868	34	2065	42	1922
5	1100	1	67	1984	67	2059	63	1870	55	1792
5	1100	2	49	1908	54	1850	51	1773	49	1987

requests that adjusted means be computed for each of the 16 treatment combinations evaluated at 5 values of thickness. The 5 values of thickness were selected as the lower 10 percentile, lower 25 percentile, the median, the upper 25 percentile, and the upper 10 percentile of the distribution of the 80 observed wafer thickness values. Table 16.25 contains the pairwise comparisons of the position means within each wafer type and temperature combination using the simulate adjustment for multiple comparisons. Since the slopes are not a function of the levels of position, the comparisons are the same for each value of thickness; thus just one set is presented. There is a significantly thinner layer of oxide in the Bbot section of the furnace with wafer type 1 and temperature 1100°F than at the other three positions. There were no other pairs of temperature means that were significantly different. Table 16.26 provides the comparisons of the two levels of temperature within each wafer type and position. The 950°F temperature produced mean oxide thicknesses that were significantly less than those produced at 1100°F, except in the Bbot position for both wafer types ($p \leq 0.05$). The wafer type means needed to be compared at three or more values of thickness since the slopes are different for the two wafer types. Table 16.27 contains the pairwise differences between the two wafer type oxide layer

TABLE 16.20
Analysis of Variance Table for Split-Split-Plot Design Assumptions without Covariate Information for the Oxide Deposition Study

Source	df	EMS
Block	4	$\sigma_w^2 + 2\sigma_s^2 + 8\sigma_f^2 + 16\sigma_{blk}^2$
Temp	1	$\sigma_w^2 + 2\sigma_s^2 + 8\sigma_f^2 + \phi^2(\text{temp})$
Error(furnace) = Block*Temp	4	$\sigma_w^2 + 2\sigma_s^2 + 8\sigma_f^2$
Position	3	$\sigma_w^2 + 2\sigma_s^2 + \phi^2(\text{posit})$
Temp*Position	3	$\sigma_w^2 + 2\sigma_s^2 + \phi^2(\text{posit}*\text{temp})$
Error(segment) = Position*Block(Temp)	24	$\sigma_w^2 + 2\sigma_s^2$
WafType	1	$\sigma_w^2 + \phi^2(\text{waf})$
WafType*Temp	1	$\sigma_w^2 + \phi^2(\text{waf}*\text{temp})$
WafType*Posit	3	$\sigma_w^2 + \phi^2(\text{waf}*\text{posit})$
WafType*Posit*Temp	3	$\sigma_w^2 + \phi^2(\text{waf}*\text{posit}*\text{temp})$
Error(wafer) = waftype*block(posit temp)	32	σ_w^2

thickness means evaluated at the five wafer thickness values. In all cases, wafer type 1 had a mean oxide layer thickness that was significantly larger than that for wafer type 2. All of the adjusted significance levels were 0.0000. Finally, the wafer type by position adjusted means are displayed in a graph for each of the two temperatures in Figures 16.4 and 16.5. Regression lines for different wafer types are not parallel while lines of the same wafer type are parallel. The slopes are quite similar and it is difficult to visually see which lines are parallel and which lines are not.

This example presented a situation where the repeated measurement part of the model did not occur on the smallest size experimental unit. In this case, a Random statement was required to specify the form of the covariance structure for PROC MIXED rather than the Repeated statement that was used in Section 16.6.

16.8 CONCLUSIONS

Repeated measures designs involve factorial treatment structures and thus the covariate part of the models are identical to those in the split-plot type designs which are identical to those with factorial treatment structures in any design structure (such as discussed in Chapter 5). The additional complication provided with the repeated measures design is that the covariance structure of the repeated measures part of the model needs to be selected. The value of AIC was used to provide an index for the selection of the most adequate covariance structure among those evaluated. Two examples were discussed where one involved the repeated measures on the small

TABLE 16.21
PROC MIXED Code and Analysis with Factorial Effects for the Slopes and the Intercepts to the Oxide Wafer Data Using the Toplitz Covariance Structure for the Furnace Position Errors

```
proc mixed cl ic covtest data=oxide3;
class waftype block temp posit;
model delta= waftype|Temp|posit thick thick*waftype thick*temp
 thick*posit
thick*waftype*posit thick*posit*temp
thick*waftype*temp thick*waftype*temp*posit/ddfm=kr;
random block;
random posit/type=toep subject=block(temp);
```

Neg2LogLike	Parameters	AIC	AICC	HQIC	BIC	CAIC
392.74	5	402.74	404.16	397.49	400.78	405.78

CovParm	Subject	Estimate	StdErr	ZValue	ProbZ
block		0.0000			
Variance	block(temp)	30.1005	10.1721	2.96	0.0015
TOEP(2)	block(temp)	23.5908	8.7553	2.69	0.0071
TOEP(3)	block(temp)	10.9850	6.9174	1.59	0.1123
TOEP(4)	block(temp)	−1.6323	9.5859	−0.17	0.8648
Residual		0.3025	0.1036	2.92	0.0017

Effect	NumDF	DenDF	FValue	ProbF
waftype	1	27.3	1.39	0.2478
temp	1	31.3	0.82	0.3728
waftype*temp	1	27.3	3.49	0.0725
posit	3	23.0	0.90	0.4560
waftype*posit	3	26.8	1.68	0.1944
temp*posit	3	23.0	0.73	0.5435
waftype*temp*posit	3	26.8	1.50	0.2371
thick	1	24.3	23.89	0.0001
thick*waftype	1	27.2	4.95	0.0346
thick*temp	1	24.3	1.68	0.2075
thick*posit	3	20.5	0.51	0.6814
thick*waftype*posit	3	26.8	1.80	0.1708
thick*temp*posit	3	20.5	0.70	0.5627
thick*waftype*temp	1	27.2	3.60	0.0683
thic*waft*temp*posit	3	26.8	1.47	0.2445

size of the experimental unit and the Repeated statement was used to specify the covariance structure and the other example involved the repeated measures on an intermediate size experimental unit and the Random statement was used to specify the covariance structure. The two examples can be extended to other situations by following the approaches outlined above. None of the examples were analyzed using JMP® (ver 4) since it does no have the capability corresponding to the type=options of PROC MIXED.

TABLE 16.22
PROC MIXED Code to Fit the Final Effects Model to the Oxide Deposition Data Using the Toplitz Covariance Structure for the Furnace Position Errors

```
proc mixed cl ic covtest data=oxide3;
class waftype block temp posit;
model delta= waftype|Temp|posit thick thick*waftype/ddfm=kr;
random block;
random posit/type=toep subject=block(temp);
```

Neg2LogLike	Parameters	AIC	AICC	HQIC	BIC	CAIC
287.08	5	297.08	298.15	291.84	295.13	300.13

CovParm	Subject	Estimate	StdErr	ZValue	ProbZ
block		0.0000			
Variance	block(temp)	29.9597	10.1192	2.96	0.0015
TOEP(2)	block(temp)	24.0139	9.2193	2.60	0.0092
TOEP(3)	block(temp)	14.1059	8.3918	1.68	0.0928
TOEP(4)	block(temp)	4.3006	11.2715	0.38	0.7028
Residual		0.2514	0.0650	3.87	0.0001

Effect	NumDF	DenDF	FValue	ProbF
waftype	1	31.4	0.74	0.3961
temp	1	11.7	33.30	0.0001
waftype*temp	1	29.9	0.90	0.3516
posit	3	12.5	2.39	0.1178
waftype*posit	3	30.1	306.45	0.0000
temp*posit	3	12.4	2.73	0.0884
waftype*temp*posit	3	30.1	3.23	0.0360
thick	1	31.1	91.67	0.0000
thick*waftype	1	31.4	4.89	0.0345

TABLE 16.23
PROC MIXED Code and Analysis for the Final Model with Factorial Effects for the Intercepts and a Slope for Each Wafer Type

```
proc mixed cl ic covtest data=oxide3;
class waftype block temp posit;
model delta= waftype|Temp|posit thick*waftype/ddfm=kr solution;
random block;
random posit/type=toep subject=block(temp);
```

Neg2LogLike	Parameters	AIC	AICC	HQIC	BIC	CAIC
287.08	5	297.08	298.15	291.84	295.13	300.13

CovParm	Subject	Estimate	StdErr	ZValue	ProbZ
block		0.0000			
Variance	block(temp)	29.9597	10.1192	2.96	0.0015
TOEP(2)	block(temp)	24.0139	9.2193	2.60	0.0092
TOEP(3)	block(temp)	14.1059	8.3918	1.68	0.0928
TOEP(4)	block(temp)	4.3006	11.2715	0.38	0.7028
Residual		0.2514	0.0650	3.87	0.0001

Effect	waftype	Estimate	StdErr	df	tValue	Probt
thick*waftype	1	0.00992	0.00136	31.3	7.30	0.0000
thick*waftype	2	0.00563	0.00117	31.3	4.83	0.0000

Effect	NumDF	DenDF	FValue	ProbF
waftype	1	31.4	0.74	0.3961
temp	1	11.7	33.30	0.0001
waftype*temp	1	29.9	0.90	0.3516
posit	3	12.5	2.39	0.1178
waftype*posit	3	30.0	306.45	0.0000
temp*posit	3	12.3	2.73	0.0884
waftype*temp*posit	3	30.1	3.23	0.0360
thick*waftype	2	31.2	46.11	0.0000

TABLE 16.24
Adjusted Means for the Wafer Type by Temperatue by Position Combinations Evaluated at Five Values of Thickness, Lower 10 Percentile, Lower 25 Percentile, Median, Upper 25 Percentile and the Upper 10 Percentile of the Wafer Thickness Values

```
lsmeans waftype*posit*temp/diff at thick=1769
 adjust=simulate;**low 10%tile;
lsmeans waftype*posit*temp/diff at thick=1846
 adjust=simulate;**low 25% tile;
lsmeans waftype*posit*temp/diff at thick=1911
 adjust=simulate;**median;
lsmeans waftype*posit*temp/diff at thick=2031
 adjust=simulate;**up 25% tile;
lsmeans waftype*posit*temp/diff at thick=2128
 adjust=simulate;**up 10% tile;
```

			Wafer Thickness				
waftype	temp	posit	1769	1846	1911	2031	2128
1	950	BBot	41.73	42.50	43.14	44.33	45.29
1	950	MBot	43.93	44.70	45.34	46.53	47.49
1	950	MTop	43.47	44.23	44.88	46.07	47.03
1	950	TTop	45.21	45.98	46.62	47.81	48.77
1	1100	BBot	52.12	52.88	53.53	54.72	55.68
1	1100	MBot	60.25	61.02	61.66	62.85	63.81
1	1100	MTop	61.79	62.55	63.19	64.38	65.35
1	1100	TTop	66.00	66.77	67.41	68.60	69.56
2	950	BBot	35.52	35.95	36.32	36.99	37.54
2	950	MBot	32.84	33.27	33.64	34.31	34.86
2	950	MTop	32.26	32.70	33.06	33.74	34.28
2	950	TTop	30.01	30.44	30.81	31.48	32.03
2	1100	BBot	46.25	46.69	47.05	47.73	48.27
2	1100	MBot	50.28	50.71	51.08	51.75	52.30
2	1100	MTop	51.12	51.55	51.92	52.59	53.14
2	1100	TTop	49.64	50.08	50.44	51.12	51.66

Note: estimated standard errors were from 2.46 to 2.48, thus are not included.

TABLE 16.25
Pairwise Comparisons between the Furnace Positions within Each Wafer Type by Temperature Combination

waftype	temp	posit	_posit	Estimate	StdErr	df	tValue	Adjp
1	950	BBot	MBot	−2.20	1.57	19.8	−1.40	0.9263
1	950	BBot	MTop	−1.74	2.54	11.5	−0.68	0.9999
1	950	BBot	TTop	−3.48	3.22	7.7	−1.08	0.9887
1	950	MBot	MTop	0.46	1.58	19.8	0.30	1.0000
1	950	MBot	TTop	−1.28	2.54	11.5	−0.50	1.0000
1	950	MTop	TTop	−1.75	1.58	19.9	−1.11	0.9861
1	1100	BBot	MBot	−8.13	1.57	19.8	−5.17	0.0007
1	1100	BBot	MTop	−9.66	2.54	11.5	−3.80	0.0168
1	1100	BBot	TTop	−13.88	3.23	7.8	−4.30	0.0044
1	1100	MBot	MTop	−1.53	1.58	20.1	−0.97	0.9956
1	1100	MBot	TTop	−5.75	2.55	11.6	−2.26	0.4260
1	1100	MTop	TTop	−4.22	1.58	19.9	−2.67	0.2134
2	950	BBot	MBot	2.68	1.58	19.9	1.70	0.7845
2	950	BBot	MTop	3.26	2.54	11.5	1.28	0.9585
2	950	BBot	TTop	5.51	3.22	7.7	1.71	0.7775
2	950	MBot	MTop	0.57	1.58	19.9	0.36	1.0000
2	950	MBot	TTop	2.83	2.54	11.5	1.11	0.9855
2	950	MTop	TTop	2.25	1.58	19.8	1.43	0.9148
2	1100	BBot	MBot	−4.02	1.59	20.8	−2.52	0.2776
2	1100	BBot	MTop	−4.87	2.54	11.5	−1.92	0.6473
2	1100	BBot	TTop	−3.39	3.22	7.7	−1.05	0.9904
2	1100	MBot	MTop	−0.84	1.59	20.7	−0.53	1.0000
2	1100	MBot	TTop	0.63	2.55	11.7	0.25	1.0000
2	1100	MTop	TTop	1.48	1.57	19.8	0.94	0.9971

Note: The slopes of the regression lines are not functions of the positions, thus the comparisons need be made at only one value of wafer thickness.

TABLE 16.26
Pairwise Comparisons between the Levels of Temperature for Each Wafer Type and Position

waftype	posit	temp	_temp	Estimate	StdErr	df	tValue	Adjp
1	BBot	950	1100	−10.39	3.48	17.9	−2.99	0.1126
1	MBot	950	1100	−16.32	3.48	17.9	−4.69	0.0017
1	MTop	950	1100	−18.32	3.48	17.8	−5.27	0.0006
1	TTop	950	1100	−20.79	3.48	17.9	−5.97	0.0000
2	BBot	950	1100	−10.73	3.48	17.8	−3.09	0.0911
2	MBot	950	1100	−17.44	3.48	17.9	−5.01	0.0009
2	MTop	950	1100	−18.86	3.48	17.8	−5.42	0.0004
2	TTop	950	1100	−19.63	3.48	17.9	−5.64	0.0001

Note: They were evaluated at only one value of wafer thickness as the slopes do not depend on the levels of temperature.

TABLE 16.27
Pairwise Comparisons between the Mean Thickness of the Oxide Layer for the Two Wafer Types for Each Temperature and Furnace Position and Five Values of Wafer Thickness

				Wafer Thickness				
temp	posit	waftype	_waftype	1769	1846	1911	2031	2128
950	BBot	1	2	6.21	6.54	6.82	7.34	7.75
950	MBot	1	2	11.10	11.43	11.71	12.22	12.64
950	MTop	1	2	11.20	11.53	11.81	12.33	12.74
950	TTop	1	2	15.20	15.53	15.81	16.33	16.74
1100	BBot	1	2	5.87	6.20	6.48	6.99	7.41
1100	MBot	1	2	9.98	10.31	10.59	11.10	11.52
1100	MTop	1	2	10.66	10.99	11.27	11.79	12.21
1100	TTop	1	2	16.36	16.69	16.97	17.48	17.90

Note: All adjusted significance levels were 0.0000, indicating that all comparisons have significantly different means.

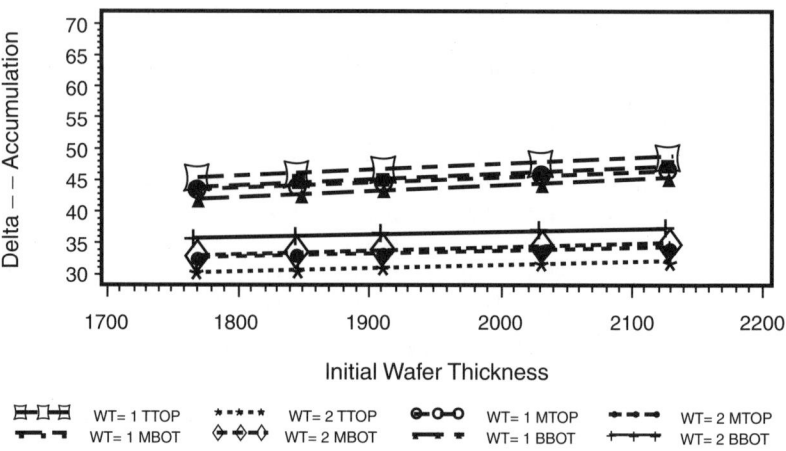

FIGURE 16.4 Plot of the regression lines for each wafer by position combination for temperature = 950°F.

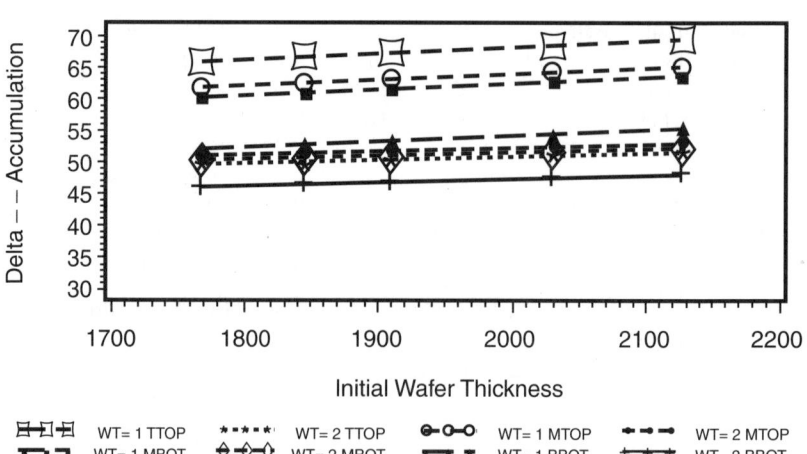

FIGURE 16.5 Plot of the regression lines for each wafer by position combination for temperature = 1100°F.

REFERENCES

Akaike, H. (1974) A New Look at the Statistical Model Identification, *IEEE Transactions on Automatic Control,* AC-19:716-723.

Box, G. E. P. (1954) Some Theorens on Quadratic Forms Applied in the Study of Analysis of Variance Problems, *Annals of Mathematical Statistics 25*:290-302.

Bozdogan, H. (1987) Model Selection and Akaike's Information Criterion (AIC): The General Theory and Its Analytical Extensions, *Psychometrika* 52:345-370.

Burnham, K. P. and Anderson, D. R. (1998) *Model Selection and Inference: A Practical Information-Theoretic Approach,* Spring Verlag, New York.

Hannan, E. J and Quinn, B. G. (1979) The Determination of the Order of an Autoregression, *Journal of Royal Statistical Society, Ser. B,* 41:190-195.

Littell, R., Milliken, G. A., Stroup, W., Wolfinger, R. (1996) *SAS System for Mixed Models.* SAS Institute Inc., Cary, NC.

Milliken, G. A. and Johnson, D. E. (1992) *Analysis of Messy Data, Volume I: Design Experiments*, Chapman & Hall, London.

SAS Institute Inc. (1999) *SAS/STAT® User's Guide,* Version 8, SAS Institute Inc., Cary, NC.

Schwarz, G. (1978) Estimating the Dimension of a Model, *Annals of Statistics,* 6:461-464.

EXERCISES

EXERCISE 16.1: The data in the following table are from a study evaluating the effectiveness of two different teaching techniques and the use or not of a computer to teach geography to elementary school students. The information was presented in two different ways: (1) a teacher presented the material and (2) the material was presented via video tape. Additional exercises were presented using a computer and using hard copy. Thus the basic treatment structure involves two levels of method of teaching (MD) (teacher or video) and two levels of using the computer (NO and YES). Sixty fourth grade students were randomly selected from the population of fourth grade students attending an elementary school and they were randomly divided into 12 classes of 5 students each. The four combinations of teaching method by computer usage were each randomly assigned to three of the classes. The students were given standard exams at the end of each trimester. Class attendance is important to learning; thus the number of days a student was absent from class during a given trimester was considered as a possible covariate. In the table, the headings S1 and D1 are the score and days missing class for each student during the first trimester, S2 and D2 are for the second trimester; and S3 and D3 are for the third trimester. The three major columns denote the three classes assigned to the method by computer combination and the five rows within each rectangle correspond to the five students within each class. Some students were absent on the days exams were given and thus have missing data. Identify the different sizes of experimental units and specify which experimental unit is being repeatedly measured. Carry out a thorough analysis of covariance using this data set.

Data for the Exercise 16.1

Method	Computer	Class 1						Class 2						Class 3					
		S1	D1	S2	D2	S3	D3	S1	D1	S2	D2	S3	D3	S1	D1	S2	D2	S3	D3
Teacher	No	51	0	62	1	75	1	—	12	71	7	—	16	69	0	84	0	88	0
		70	0	78	0	86	0	66	0	73	3	80	2	66	0	78	0	84	0
		42	5	50	4	64	4	58	3	70	0	84	0	77	0	88	0	99	0
		47	3	59	0	65	0	51	1	56	6	66	8	59	8	58	16	74	9
		56	0	67	0	81	0	62	0	70	0	79	0	—	12	76	8	—	14
Teacher	Yes	56	0	74	0	85	0	73	0	85	0	98	0	78	0	99	0	100	4
		60	2	76	0	88	0	63	9	80	6	78	16	85	0	100	0	100	0
		59	0	77	0	89	0	55	10	—	7	72	16	52	13	81	0	95	0
		60	6	81	7	91	4	70	8	89	5	97	8	77	10	100	4	100	4
		72	1	92	0	100	0	76	0	86	0	94	2	71	0	86	0	96	0
Video	No	54	0	63	0	59	4	88	0	97	0	100	0	40	11	74	3	71	7
		64	5	64	7	—	14	61	0	73	0	79	0	42	0	58	0	—	2
		61	0	66	2	64	6	50	6	55	8	71	2	68	0	78	0	86	0
		67	0	76	0	82	0	64	0	72	0	80	0	27	12	—	14	52	9
		74	0	81	0	85	0	38	11	47	10	—	7	70	0	87	0	86	0
Video	Yes	78	0	94	0	100	0	87	3	95	0	100	0	64	0	74	0	85	0
		77	0	81	2	90	4	55	0	72	0	—	2	51	0	60	0	73	0
		79	0	93	0	100	0	61	9	—	14	85	4	50	0	56	0	73	2
		75	0	82	2	98	0	77	0	87	1	97	2	42	0	54	0	65	3
		88	0	99	0	100	0	79	0	85	2	98	5	—	2	62	5	79	5

Analysis of Covariance for Repeated Measures Designs

EXERCISE 16.2: The data in the table below are from an experiment evaluating the effect of three treatment regimes on reducing systolic blood pressure of males (human). There is evidence that the amount of salt in one's diet can have an effect on blood pressure. Thus each person kept track of the foods eaten each day and the amount of salt consumed per day was determined. Each persons blood pressure was measured after 2 weeks, 4 weeks, 6 weeks, and 8 weeks on the treatment regime. There were 21 persons enrolled in the study and 7 were randomly assigned to each treatment regime. The variables in the table are bp1 and salt1: the systolic blood pressure readings and the salt intake for the ith time period. Identify the different sizes of experimental units and specify which experimental unit is being repeatedly measured. Carry out a thorough analysis of covariance using this data set.

Data for Exercise 16.2

trt	sub	bp1	salt1	bp2	salt2	bp3	salt3	bp4	salt4
1	1	168	5660	166	6240	163	5480	159	4880
1	2	147	5190	142	4070	138	3210	136	3790
1	3	148	3520	145	3840	144	5250	139	3950
1	4	170	5460	166	4910	162	3960	160	4600
1	5	159	4410	156	4630	155	5980	151	5630
1	6	147	6120	140	3400	141	5940	135	3500
1	7	155	3880	153	3960	151	5030	147	4400
2	1	165	3360	159	6080	152	2800	146	2460
2	2	159	2860	154	5740	148	5400	142	2800
2	3	152	4890	147	6280	139	4930	134	5440
2	4	157	5980	152	4420	145	5000	140	4880
2	5	151	3650	147	2970	141	3420	135	5560
2	6	161	4310	156	5130	149	3400	144	5710
2	7	150	5240	142	4970	138	4670	131	3300
3	1	155	4180	142	5510	136	4040	126	3970
3	2	149	3330	136	5450	132	2960	123	3540
3	3	158	3000	149	3730	138	4200	128	6240
3	4	141	5610	132	6110	123	6390	116	5160
3	5	166	3180	154	3500	143	5440	135	5240
3	6	154	3690	144	3920	136	4510	132	2470
3	7	164	4860	155	4870	145	4770	135	5360

EXERCISE 16.3: Women (32) were selected to participate in a study to evaluate the effect of 4 different drugs on their heart rate. The 32 women were randomly divided into 4 groups of 8 each. Each group of eight women was then randomly assigned to one of the four levels of drug. At the beginning of the study, each woman's heart rate was measured and then was measured every hour for 5 hours after taking the drug. Not all of the women finished the trial, leaving unequal number of observations per treatment group. The initial heart rate was used as a possible covariate. The data are in the table where IHR is the initial heart rate and Hr1 is the heart rate measured on the ith hour. Identify the different sizes of experimental units and specify which experimental unit is being repeatedly measured. Carry out a thorough analysis of covariance for this data set.

Data for Exercise 16.3

Drug	Subject	IHR	HR1	HR2	HR3	HR4	HR5
WW3	1	61	66	75	74	45	46
WW3	2	60	78	89	89	73	68
WW3	3	67	64	68	90	74	78
WW3	4	66	55	68	90	63	61
WW3	5	58	73	95	104	89	82
WW3	6	65	80	82	89	77	75
DD4	1	74	78	99	102	100	96
DD4	2	72	76	89	86	90	93
DD4	3	72	82	91	94	106	98
DD4	4	71	93	100	102	104	112
DD4	5	65	76	95	92	86	85
DD4	6	60	74	78	76	88	82
DD4	7	58	82	94	98	98	102
DD4	8	68	81	91	93	94	101
CC5	1	55	74	72	54	55	60
CC5	2	55	75	73	71	73	75
CC5	3	71	87	82	70	62	73
CC5	4	68	98	103	94	104	91
CC5	5	60	69	64	60	63	56
CC5	6	69	75	78	68	71	73
CC5	7	68	76	78	73	73	71
AA3	1	55	93	93	73	58	59
AA3	2	56	96	96	84	60	63
AA3	3	65	121	116	99	86	79
AA3	4	69	103	104	85	67	69
AA3	5	60	114	118	94	77	74
AA3	6	64	101	103	81	74	66
AA3	7	74	107	100	74	72	76

Analysis of Covariance for Repeated Measures Designs

EXERCISE 16.4: Rolled oats are produced by steam treating the oat grain and then processing the grain through a set of rollers. An experiment was set up to study the effect of three steam temperatures on the quality of the rolled oats produced from four varieties of oats. The process was to first generate steam at the lower temperature, process the four varieties, then increase to the next temperature and process the four varieties, and finally increase to the next temperature and process the four varieties. The order of processing the varieties was randomized within each of the temperature runs. The percent of the batch that was ground into fine particles, was used as the measure of quality. The lower the percent of fine particles the better the quality of the end product. This process of ramping up the temperature is carried out in order to save time in conducting the experiment. This process was repeated on four different days, where day is a block. The moisture content of each batch of oats was determined prior to processing to be used as a possible covariate. The data are in the table below. Identify the different sizes of experimental units and specify which experimental unit is being repeatedly measured. Carry out a thorough analysis of covariance for this data set.

Data for Exercise 16.4

		Variety Kan1		Variety Colo		Variety Nebr	
Day	temp	Percent	Moisture	Percent	Moisture	Percent	Moisture
1	High	7.0	16.0	8.4	12.3	9.1	15.6
1	Low	11.0	14.7	14.1	11.4	12.8	15.7
1	Mid	11.0	14.6	11.0	15.9	12.3	13.5
2	High	11.1	13.8	12.5	14.3	12.7	14.9
2	Low	11.1	16.0	13.3	13.9	14.4	13.6
2	Mid	13.3	11.3	14.2	14.5	14.6	14.5
3	High	4.2	14.1	5.7	12.5	6.0	12.6
3	Low	9.8	12.7	9.8	14.3	11.7	11.8
3	Mid	9.0	14.9	10.0	15.4	11.5	14.7
4	High	11.6	14.1	11.7	14.5	14.0	12.6
4	Low	13.0	16.0	14.9	13.1	17.2	12.5
4	Mid	12.6	14.4	14.7	12.1	15.1	16.0

17 Analysis of Covariance for Nonreplicated Experiments

17.1 INTRODUCTION

A method is described that incorporates covariate information into the analysis of nonreplicated experiments with factorial and fractional factorial treatment structures. The data are used to estimate the slope corresponding to the covariate (or slopes for multiple covariates) and then the estimates of the effects are adjusted for differing values of the covariate or covariates, i.e., the methodology provides estimates of the factorial effects adjusted for the covariate or covariates or after removing the confounding effects of the covariates on the factorial effects. Regression analyses are used to provide estimates of the factorial effects and of the sampling variance. The half-normal probability plot is one method for analyzing nonreplicated experiments without covariates and that method is used to compare the results of the covariance analysis. Five examples are presented to demonstrate the methodology. Three examples involve a single covariate, one example involves two covariates, and one example involves one covariate with unequal slopes for the levels of one of the factors. Because of the limited number of data points in nonreplicated experiments, there is a limit as to the number of covariate parameters that can be estimated.

The data in Table 17.1 are from an experiment designed to evaluate the effect of three factors each at two levels on the amount of flour milled in a 10-min test run (Y denotes the pounds of flour milled). The three factors are A, the feed rate; B, the type of grinding screen; and C, the pressure of the grinding surface. Only one replicate of each treatment combination was run, thus producing a nonreplicated experiment. The objective is to determine which of the three factors and/or interactions have an influence on the amount of flour milled in a 10-min run. However, the batches of wheat used for each run were quite variable, so characteristics of a sample taken from each batch were determined for possible use as covariates. In this case, only the percent of undersized kernels (X) in the sample was considered. There are eight observations in the data set and there are the mean and seven factorial effects that need to be estimated. Thus there are no degrees of freedom left over to provide an estimate of the sample variance. In addition, the covariate information is confounded with the factorial effects and to utilize this information, additional parameters need to be estimated. Thus, in order to extract information from this data set, some decisions need to be made as to which of the factorial effects are possibly real and which are possibly not real. The parameters corresponding to the unreal effects are set to zero, which enables parameters associated with the covariates and the sample variance to be estimated. Once the slope parameters are estimated, then the confounding effects of the covariates can be removed from the factorial effects.

In general, the analysis of nonreplicated experiments involves the identification of important or significant effects. Normal and half-normal probability plots (Daniel, 1959, and Milliken and Johnson, 1989) are two such techniques for identifying possibly insignificant effects and using them to provide an estimate of experimental error. The techniques have been successfully used by many researchers involved with projects in which experimental units are expensive or it is impossible to obtain true replications. Since there are techniques for analyzing data from experiments without replication, it is often prudent as a cost- and time-saving process to use nonreplicated experiments. For the wheat-to-flour experiment, the estimates of the factorial effects for Y are in Table 17.4 and the half-normal probability plot of those effects is in Figure 17.1. The half-normal plot indicates that the main effect of C is the only important effect in the data set. Thus a regression analysis is used to provide estimates of the parameters of a model that only has C as a main effect, i.e., there are no other effects. Such a strategy provides six degrees of freedom for the residual sum of squares which can be used as a possible estimate of the variance in the system. Table 17.2 consists of the SAS® system code using PROC REG to fit the model and the pertinent results. The estimate of the effect of C is −12.525 with estimated standard error 3.964, and the estimate of the variance is 125.7 which is based on six degrees of freedom.

As in this wheat-to-flour experiment, often the experimental units are not identical and one or more covariates reflect the possible affect the variability has on the levels of the factors assigned to the respective experimental units. These possibly influential covariates cannot be controlled, but their values can be measured. Table 17.3 contains the SAS® system code to fit a model with the main effect of C and the covariate X and the results of that analysis. The estimate of the main effect of C is −12.345 with estimate standard error of 5.603, the estimate of the slope corresponding to X is 0.038 with estimated standard error of 0.751, and the estimate of the sample variance is 150.776. This analysis indicates that the variability cannot be reduced by taking the covariate information into account and the loss of the one degree of freedom used to fit the slope for the covariate inflates the estimate of the sample variance. The main problem with this strategy is that the effect of the covariate is evaluated after decisions about the factorial effects have been made. A better approach is to jointly evaluate the effects of the covariates and the factorial effects.

The objective of this chapter is to describe a method that jointly incorporates covariates and factorial effects into the analysis of nonreplicated experiments, thus removing the confounding of the covariates on the factorial effects. The methods involve using a simple plotting strategy followed by regression analysis. The wheat-to-flour example is reanalyzed in Section 17.6 using the described strategy and the decisions are greatly changed. Section 17.2 describes the basic method for the case where there is a single covariate and a common slopes model, and the multiple covariate case is discussed in Section 17.3. Section 17.4 presents the graphical method of selecting possible significant or important factorial effects while taking the covariate information into account, and Section 17.5 shows how to do the

Analysis of Covariance for Nonreplicated Experiments

computations using a regression analysis computer package. Sections 17.6 through 17.10 contain five examples to demonstrate various aspects of the methodology.

17.2 EXPERIMENTS WITH A SINGLE COVARIATE

Assume an experiment consists of n treatment combinations from some treatment structure observed on experimental units in a completely randomized design structure. Possible treatment structures are 2^n or 3^n factorial arrangements, 2^{n-k} fractional factorial arrangements, other fractional factorial arrangements, and mixed factorials. Let μ_i denote the mean response of a population of experimental units subjected to the ith treatment combination when the covariates are not included in the model (or when the slopes of the regression lines are zero). The analysis of covariance model can be expressed as a set of linear (or nonlinear, not discussed here) regression models where the intercepts of the models are the means of the treatment combinations and the covariate part of the models are linear functions of the covariates with possibly different slopes for each treatment combination. The analysis of covariance is the process of making decisions about the form of the covariate part of the models and then comparing the treatment combinations through the respective regression models. An important decision concerns the form of the covariate, i.e., are the slopes equal for all treatment combinations, are the slopes zero, or the slopes not equal? The first part of this discussion assumes the slopes of the regression models are equal, providing a set of parallel lines.

One model relating the dependent variable Y_i to its mean and the covariate X_i is

$$y_i = \mu_i + \beta x_i + \varepsilon_i, \quad i = 1, 2, \ldots, n, \tag{17.1}$$

where i denotes the ith treatment combination, μ_i denotes the mean response of the ith treatment combination when the value of x is zero, β denotes the slope of the family of parallel regression lines, and ε_i denotes the random error, where the error terms are assumed to be iid $N(0, \sigma^2)$.

Model 17.1 can be generalized in at least two ways. First the model could have unequal slopes. Examples are (1) the slopes could be unequal for the levels of one of the factors in the experiment providing a model with two slopes (e.g., where the factor has two levels) or (2) the slopes could be unequal for all treatment combinations providing a model with n slopes, one for each treatment combination. For replicated experiments, there are often sufficient numbers of observations for each of the treatment combinations to fit the unequal slopes models and construct statistics to test the slopes equal to zero and the equal slopes hypotheses. For nonreplicated experiments, there is an extreme limit on the number of observations and thus a severe limit on the number of parameters that can be estimated from the data set. The model with unequal slopes for each of the treatment combinations is

$$y_i = \mu_i + \beta_i x_i + \varepsilon_i, \quad i = 1, 2, \ldots, n, \tag{17.2}$$

which is a model with n observations and 2n parameters. All of the parameters cannot be estimated from the data set, thus some simplification of the model must be accomplished. The second generalization is to include more covariates, a topic discussed in Section 17.3.

The first step in the analysis of factorial or fractional factorial treatment structures is to estimate the estimable effects in the model. For 2^n and 2^{n-k} factorial treatment structures, the estimates of the factorial effects for a model without covariates are computed by orthogonal linear combinations of the observations as follows. Let C_{ji}, $j = 1, 2, \ldots, n-1$, and $i = 1, 2, \ldots, n$ be $n-1$ sets of coefficients of orthogonal contrasts such that

$$\sum_{i=1}^{n} C_{ji} = 0, \quad \sum_{i=1}^{n} C_{ji}^2 = c, \quad \sum_{i=1}^{n} C_{ji} C_{j'i} = 0$$

for $j = 1, 2, \ldots, n-1$, $j' = 1, 2, \ldots, n-1$, $j \neq j'$

Compute contrasts of the observations as $v_j = \sum_{i=1}^{n} C_{ji} y_i$, $j = 1, 2, \ldots, n-1$. The models corresponding to the computed contrasts are

$$v_j = \sum_{i=1}^{n} C_{ji} y_i = \sum_{i=1}^{n} C_{ji} \mu_i + \beta \sum_{i=1}^{n} C_{ji} x_i + \sum_{i=1}^{n} C_{ji} \varepsilon_i, \quad j = 1, 2, \ldots, n-1.$$

Let

$$\theta_j = \sum_{i=1}^{n} C_{ji} \mu_i \quad \text{and} \quad U_j = \sum_{i=1}^{n} C_{ji} x_i$$

then the v_j are independently distributed as $v_j \sim N(\theta_j + \beta U_j, c\sigma^2)$. If the C_{ji} are selected such that the corresponding contrasts relate to the desired factorial effects, then the θ_j represent the factorial effect parameters. If the slope, β, is zero, the objective of such a study is to determine which θ_j are nonzero and which are negligible or zero. Then separate the v_j into two sets, one set containing the non-null effects (corresponding to $\theta_j \neq 0$) and one set containing the null effects (corresponding to $\theta_j = 0$). When the slope is zero or when there is no available covariate information, decisions can be made about the θ_j by using probability plotting techniques such as a half-normal probability plot (Daniel, 1959) for the v_j. However, when there a relationship between the v_j and the covariates, a half-normal plot of the v_j provides confusing information about the θ_j. Suppose that the first S_1 values of θ_j are non-null (denoted by θ_{1j}) and the remaining S_2 values of θ_j are zero, where $S_1 + S_2 = n - 1$. A procedure for determining this partition of the effects is described in Section 17.4. With this assumption, the models for the contrasts can be separated into two sets as

Analysis of Covariance for Nonreplicated Experiments

$$v_{1j} = \theta_{1j} + \beta U_{1j} + e_{1j}, \quad j = 1, 2, \ldots, S_1$$

$$v_{2j} = \beta U_{2j} + e_{2j}, \quad j = 1, 2, \ldots, S_2$$

where

$$e_{1j} = \sum_{i=1}^{n} C_{1ji} \varepsilon_i \sim N[0, c\sigma^2], \quad j = 1, 2, \ldots, S_1$$

$$e_{2j} = \sum_{i=1}^{n} C_{2ji} \varepsilon_i \sim N[0, c\sigma^2], \quad j = 1, 2, \ldots, S_2$$

The additional subscripts "1" and "2" are used to delineate between the non-null set ("1") and the null set ("2"). The contrasts, v_{1j}, correspond to the non-null effects and the contrasts, v_{2j}, correspond to the null effects. Since the error terms for Model 17.1 are iid normal random variables and the contrasts are orthogonal, all of the elements of the v_{1j} and the v_{2j} are independent random variables. As in the usual half-normal plot methodology, the estimate of σ is obtained from the v_{2j} information. For the analysis of covariance model, the estimate of β is also computed from the v_{2j} data. The least squares estimate (also maximum likelihood under normality) of β based on the v_{2j} data is

$$\hat{\beta} = \frac{\sum_{j=1}^{S_2} v_{2j} U_{2j}}{\sum_{j=1}^{S_2} U_{2j}^2}.$$

The sampling distribution of $\hat{\beta}$ is

$$N\left[\beta, c\sigma^2 \left[\sum_{j=1}^{S_2} U_{2j}^2\right]^{-1}\right].$$

The estimate of σ^2 is

$$\hat{\sigma}^2 = \frac{1}{c} \frac{\sum_{j=1}^{S_2}\left(v_{2j} - \hat{\beta} U_{2j}\right)^2}{S_2 - 1} \quad \text{or} \quad c\hat{\sigma}^2 = \frac{\sum_{j=1}^{S_2}\left(v_{2j} - \hat{\beta} U_{2j}\right)^2}{S_2 - 1}$$

Under the normality assumption, the sampling distribution of

$$\frac{(S_2-1)\hat{\sigma}^2}{\sigma^2} = \frac{(S_2-1)c\hat{\sigma}^2}{c\sigma^2} \text{ is } \chi^2_{(S_2-1)}.$$

The next step is to adjust or filter or un-confound the effects of the covariate from the non-null factorial effects in order to provide unbiased estimates of the θ_{1j}. The adjusted effects are

$$\hat{\theta}_{1j} = v_{1j} - \hat{\beta} U_{1j}, \quad j = 1, 2, \ldots, S_1$$

which have sampling distributions

$$\hat{\theta}_{1j} \sim N\left[\theta_{1j}, c\sigma^2\left(1 + U_{1j}^2\left[\sum_{j=1}^{S_2} U_{2j}^2\right]^{-1}\right)\right].$$

The residuals from the null effects are $r_{2j} = v_{2j} - \hat{\beta} U_{2j}$, $j = 1, 2, \ldots, S_2$ which, under the assumption that the null effects have been correctly identified, have sampling distributions

$$r_{2j} \sim N\left[0, c\sigma^2\left(1 - U_{2j}^2\left[\sum_{j=1}^{S_2} U_{2j}^2\right]^{-1}\right)\right]$$

and $(1 - \alpha)$ 100% confidence intervals about the non-null effects are

$$\hat{\theta}_{1j} \pm t_{\frac{\alpha}{2}, S_2-1} \sqrt{c\hat{\sigma}^2_{\hat{\theta}_{1j}}}$$

where

$$c\hat{\sigma}^2_{\hat{\theta}_{1j}} = c\hat{\sigma}^2\left(1 + U_{1j}^2\left[\sum_{j=1}^{S_2} U_{2j}^2\right]^{-1}\right).$$

A set of $(1 - \alpha)$ 100% prediction intervals about the null effects are

$$r_{2j} \pm t_{\frac{\alpha}{2}, S_2-1} \sqrt{c\hat{\sigma}^2_{\hat{\theta}_{1j}}}$$

where

$$c\hat{\sigma}^2_{r_2j} = c\hat{\sigma}^2\left(1 - U^2_{2j}\left[\sum_{j=1}^{S_2} U^2_{2j}\right]^{-1}\right).$$

These interval estimates can be used to help determine if an appropriate division of the factorial effects into null and non-null sets has been achieved. The computations described can easily be carried out by a standard regression analysis computer program, as described in Section 17.5.

17.3 EXPERIMENTS WITH MULTIPLE COVARIATES

Assume the experiment consists of n treatment combinations from some treatment structure observed on experimental units in a completely randomized design structure as described in Section 17.2, where h covariates are measured on each experimental unit. A model with intercepts for the treatment effects and linear relationships for the covariates with equal slopes for each of the treatment combinations is:

$$y_i = \mu_i + \beta_1 x_{i1} + \beta_2 x_{i2} + \ldots + \beta_h x_{ih} + \varepsilon_i, \quad i = 1, 2, \ldots, n \qquad (17.3)$$

where i denotes the ith treatment combination, μ_i denotes the mean response of the ith treatment combination when the values of all of the x_{ik}'s are zero, β_k $k = 1, 2, \ldots, h$ are the slopes of regression planes in the direction of the respective covariates, and ε_i denotes the random error, where the error terms are assumed to be iid $N(0, \sigma^2)$.

As in Section 17.2, the first step in the analysis of factorial or fractional factorial treatment structures is to estimate the estimable factorial effects in the model. Let C_{ji}, $j = 1, 2, \ldots, n-1$, and $i = 1, 2, \ldots, n$ be $n-1$ sets of coefficients of orthogonal contrasts as defined in section 17.2. Compute contrasts of the observations as $v_j = \sum_{i=1}^{n} C_{ji} Y_i$, $j = 1, 2, \ldots, n-1$, then the models corresponding to the computed contrasts are

$$v_j = \sum_{i=1}^{n} C_{ji} y_i = \sum_{i=1}^{n} C_{ji} \mu_i + \beta_1 \sum_{i=1}^{n} C_{ji} x_{i1} + \beta_2 \sum_{i=1}^{n} C_{ji} x_{i2}$$

$$+ \ldots + \beta_h \sum_{i=1}^{n} C_{ji} x_{ih} + \sum_{i=1}^{n} C_{ji} \varepsilon_i, \quad j = 1, 2, \ldots n-1$$

Let $\theta_j = \sum_{i=1}^{n} C_{ji} U_{jk}$ and $U_{jk} = \sum_{i=1}^{n} C_{ji} y_{ik}$, then the v_j are independently distributed as $v_j \sim N(\theta_j + \beta_1 U_{j1} + \beta_2 U_{j2} + \ldots + \beta_h U_{jh}, c\sigma^2)$. If the C_{ji} are selected such that the corresponding contrasts relate to the desired factorial effects, then the θ_j represent the factorial effect parameters. If the slopes, β_k $k = 1, 2, \ldots, h$, are zero, the objective of such a study is to determine which θ_j are nonzero and which are negligible or zero. Then separate the v_j into two sets: one set containing the non-null effects (corresponding to $\theta_j \neq 0$) and one set containing the null effects (corresponding to $\theta_j = 0$). When the slopes are zero or when there is no available covariate information, decisions can be made about the θ_j by using probability plotting techniques such as a half-normal plot (Daniel, 1959) for the v_j. However, when there is covariate information, a half-normal plot of the v_j provides confusing information about the θ_j, i.e., the effects of the θ_j are confused with the effects of the covariates. As in Section 17.2, suppose that the first S_1 values of θ_j are non-null (denoted by θ_{1j}) and the remaining S_2 values of $\theta_j = 0$, where $S_1 + S_2 = n - 1$. A method of separating the effects into the null set and the non-null set is described in Section 17.4. Then the contrast models can be separated into two sets as

$$v_{1j} = \theta_{1j} + \beta_1 U_{1j1} + \beta_2 U_{1j2} + \ldots + \beta_h U_{1jh} + e_{1j}, \quad j = 1, 2, \ldots, S_1$$

$$v_{2j} = \beta_1 U_{2j1} + \beta_2 U_{2j2} + \ldots + \beta_h U_{2jh} + e_{2j}, \quad j = 1, 2, \ldots, S_2$$

where

$$e_{1j} = \sum_{i=1}^{n} C_{1ji} \varepsilon_i \sim N[0, c\sigma^2], \quad j = 1, 2, \ldots, S_1$$

$$e_{2j} = \sum_{i=1}^{n} C_{2ji} \varepsilon_i \sim N[0, c\sigma^2], \quad j = 1, 2, \ldots, S_2$$

The additional subscripts "1" and "2" are used to delineate between the non-null set ("1") and the null set ("2"). The contrasts, v_{1j}, correspond to the non-null effects and the contrasts, v_{2j}, correspond to the null effects. Since the error terms for Model 17.3 are iid normal random variables and the contrasts are orthogonal, all of the elements of the v_{1j} and the v_{2j} are independent random variables. As in the usual half-normal plot methodology, the estimate of σ is obtained from the v_{2j} information, but for the analysis of covariance model, the estimates of the slopes are also computed from the v_{2j} data.

The models for the v_{2j}'s can be expressed in matrix form as

$$\mathbf{v}_2 = \begin{bmatrix} v_{21} \\ v_{22} \\ \vdots \\ v_{2S_2} \end{bmatrix} = \begin{bmatrix} U_{211} & U_{212} & \cdots & U_{21h} \\ U_{221} & U_{222} & \cdots & U_{22h} \\ \vdots & \vdots & \ddots & \vdots \\ U_{2S_21} & U_{2S_22} & \cdots & U_{2S_2h} \end{bmatrix} \begin{bmatrix} \beta_1 \\ \beta_2 \\ \vdots \\ \beta_h \end{bmatrix} + \begin{bmatrix} e_{21} \\ e_{22} \\ \vdots \\ e_{2S_2} \end{bmatrix} = \mathbf{U}_2 \boldsymbol{\beta} + \mathbf{e}_2.$$

Analysis of Covariance for Nonreplicated Experiments

The least squares estimate of the vector of slopes based on the v_2 data is $\hat{\beta} = (U_2' U_2)^{-1} U_2' v_2$.

The sampling distribution of $\hat{\beta}$ is $N[\beta, c\sigma^2 (U_2' U_2)^{-1}]$. The estimate of $c\sigma^2$ is

$$c\hat{\sigma}^2 = \frac{\sum_{j=1}^{S_2}\left(v_{2j} - \hat{\beta}_1 U_{2j1} - \hat{\beta}_2 U_{2j2} - \ldots - \hat{\beta}_h U_{2jh}\right)}{S_2 - h} = \frac{v_2'\left(I_{S_2} - U_2(U_2' U_2)^{-1} U_2'\right)v_2}{S_2 - h}.$$

In order to obtain an estimate of σ^2, there must be more than "h" elements in the null set or $S_2 > h$. Under the normality assumption, the sampling distribution of $(S_2 - h) c \hat{\sigma}^2 / c \sigma^2$ is $\chi^2_{(S_2-h)}$.

The next step is to adjust or filter or un-confound the effects of the covariate from the non-null effects to provide unbiased estimates of the θ_{1j}. The estimates of the adjusted effects are $\hat{\theta}_{1j} = v_{1j} - \hat{\beta}' u_{1j}$, $j = 1, 2, \ldots, S_1$ where $u_{1j} = (U_{1j1} U_{1j2} \ldots U_{1jh})'$ or the vector of adjusted estimates of the non-null factorial effects is

$$\hat{\theta}_1 = v_1 - U_1 \hat{\beta} \text{ where}$$

$$U_1 = \begin{bmatrix} U_{111} & U_{112} & \cdots & U_{11h} \\ U_{121} & U_{122} & \cdots & U_{12h} \\ \vdots & \vdots & \ddots & \vdots \\ U_{1S_1 1} & U_{1S_1 2} & \cdots & U_{1S_1 h} \end{bmatrix}$$

with sampling distribution $N[\theta_1, c\sigma^2 (I_{S_1} + U_1(U_2' U_2)^{-1} U_1')]$. The adjusted null effects or residuals for the null effects model are $r_{2j} = v_{2j} - \hat{\beta}' u_{2j}$, $j = 1, 2, \ldots, S_2$, where $u_{2j} = (U_{2j1} U_{2j2} \ldots U_{2jh})'$, or the vector of residuals is $r_2 = v_2 - U_2\hat{\beta}$ with sampling distribution $N[0, c\sigma^2 (I_{S_2} - U_2(U_2' U_2)^{-1} U_2')]$.

A multiple covariate model can occur where there is only one covariate. The multiple covariate model comes about when the slopes are unequal for the levels of one or more of the factors in the experiment. That situation is demonstrated by example in Section 17.10.

17.4 SELECTING NON-NULL AND NULL PARTITIONS

The key to successfully applying this method is being able to partition v into the non-null effects denoted by v_1 and the null effects denoted by v_2. When there is no covariate information, a partitioning method is not fool-proof since there is no initial estimate of $c\sigma^2$ with which to judge the magnitude of any estimated effect. Then throw the covariate information into the brew and the partitioning becomes more complex. In any case, the partitioning method becomes an iterative process. One method that helps carry out the partitioning is to plot the elements of v against each column of U where

$$\mathbf{U} = \begin{bmatrix} U_{111} & U_{112} & \cdots & U_{11h} \\ U_{121} & U_{122} & \cdots & U_{12h} \\ \vdots & \vdots & \ddots & \vdots \\ U_{1n1} & U_{1n2} & \cdots & U_{1nh} \end{bmatrix}.$$

In a plot of $\mathbf{v}$ against the columns of $\mathbf{U}$, those v_j corresponding to the null effects should cluster about a plane through the origin with slopes $\beta_1, \beta_2, \ldots, \beta_h$ where the non-null effects will most likely not lie on the plane. Since the null effects are used to provide the estimate of σ^2, it is more important for $\mathbf{v}_2$ to contain only null effects than for $\mathbf{v}_1$ to consist only of non-null effects. It is also important that the number of null effects be large enough to provide sufficient degrees of freedom for estimating the sample variance. If there are just a few degrees of freedom (say less than 4) available for estimating the sample variance, all of the confidence intervals will be extremely wide and the t-statistics will have to be very large before effects can be considered significant. If it becomes apparent that an incorrect partitioning has been used, go back and repartition the factorial effects into new sets of null and non-null effects and re-estimate the parameters of the model.

For the wheat-to-flour experiment, the factorial effects of Y and of X have been computed and they are listed in Table 17.4. Figure 17.3 is a bivariate scatter plot of these treatment effects, indicating that AC and B are the non-null effects and all other effects are considered as null effects. In this case C is very close to the estimated regression line. Figure 17.1 is a half-normal probability plot where C is denoted as a non-null effect and all other effects are considered as the null effects. So, the half-normal probability plot can provide one signal about the possible non-null effects and the use of the covariate information can provide another signal about the non-null effects. The examples in Sections 17.6 to 17.10 are used to demonstrate this partitioning process.

17.5 ESTIMATING THE PARAMETERS

Once the non-null effects have been determined from the graph of the response variable factorial effects with the covariate factorial effects, then use a regression code to fit a model that includes terms for the selected non-null effects and the covariate. If a selected term in the model is really not non-null, the significance level corresponding to that term from the regression analysis can be used as a guide as to whether to keep a term in the model or delete the term from the model. If the process has failed to identify a non-null effect that should be in the model, use a model building procedure that forces the identified non-null effects into the model and then use the stepwise process (SAS Institute, 1989) to determine if additional variables need to be included in the model.

These modeling ideas are used in the following examples. Five examples are presented to demonstrate the described methodology. The first two examples involve one covariate with a positive slope, the third example has one covariate with a negative slope, the fourth example has two covariates, and the fifth example has one covariate with unequal slopes for the levels of one of the factors. This collection of examples is provided to demonstrate the possible types of analyses. Graphics provide views of the different data sets to help partition the estimated effects into null and non-null sets, as well as exhibit the half-normal probability plots when the covariate information is ignored. In some of the examples, half-normal probability plots of effects without the covariate adjustment indicate that some factorial effects are non-null whereas other factorial effects become non-null after adjustment for the covariate.

17.6 EXAMPLE: MILLING FLOUR USING THREE FACTORS EACH AT TWO LEVELS

The data in Table 17.1 are pounds of flour milled in a 10-min test run from a pilot mill (Y) and the percent of kernels from a sample of wheat from the run that were undersized (X) from a three-way factorial experiment where the factors were (A) the feed rate, (B) the type of grinding screen, and (C) the pressure of the grinding surface. Figure 17.1 contains a half-normal probability plot of the factorial effects computed from the flour milled data. The graph indicates that the main effect of C is the only important effect in the data. The PROC REG code in Table 17.2 fits a model to the data set with only the main effect of C. The estimate of the variance is 125.7117. Including the percent undersized kernels as a covariate along with the main effect of C is accomplished by the PROC REG code in Table 17.3. The estimate of the slope is 0.0382, with significance level of 0.9614, corresponding to the testing that the slope is zero. This analysis indicates that there is not a linear relationship between the mean of the pounds of flour produced and the percent small kernels when just C is in the model. The next step is to include the information about the covariate in the analysis. Figure 17.2 is a graph of the pounds of flour milled vs. percent of undersized kernels. A graphical observation is that there seems to be a slight increase in the pounds of flour milled as the percent of undersized kernels increases, although that observation is confounded with the treatment combinations. To remove some of the confounding, the PROC GLM code in Table 17.4 is used to

TABLE 17.1
Treatment Combinations with Pounds of Flour Milled as Response Variable and Percent Small Kernels as a Covariate

a	b	c	Flour	% Small Kernels
−1	−1	−1	155.3	43.5
−1	−1	1	121.4	39.5
−1	1	−1	149.4	50.0
−1	1	1	120.9	50.3
1	−1	−1	145.0	48.3
1	−1	1	145.3	39.5
1	1	−1	147.1	58.4
1	1	1	109.0	33.2

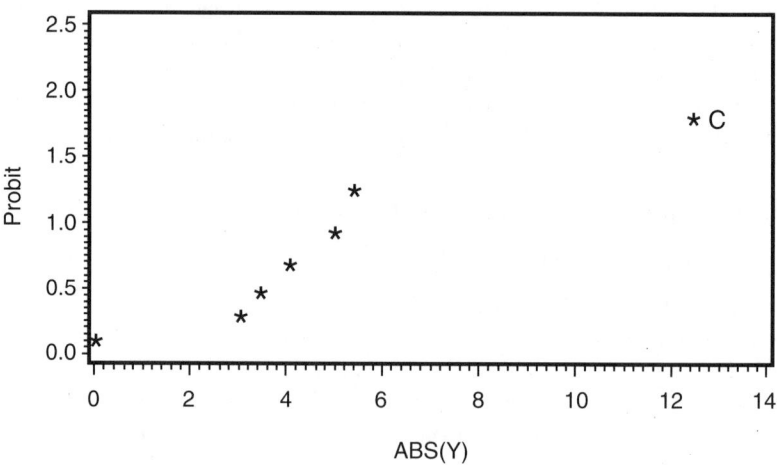

FIGURE 17.1 Half-normal plot of the milled flour data without the covariate.

TABLE 17.2
PROC REG Code to Fit Model with Main Effect for C with Analysis of Variance and Parameter Estimates

```
proc reg data=values;
model y=c;
```

Source	df	SS	MS	FValue	ProbF
Model	1	1255.0050	1255.0050	9.98	0.0196
Error	6	754.2700	125.7117		
Corrected Total	7	2009.2750			

Variable	df	Estimate	StdErr	tValue	Probt
Intercept	1	136.6750	3.9641	34.48	0.0000
c	1	−12.5250	3.9641	−3.16	0.0196

fit the full three-way factorial effects model to the pounds of flour data (y) and the percent small kernels data (x). The estimates of the factorial effects for each of the variables are in lower part of Table 17.4. Figure 17.3 is a scatter plot of the factorial effects of pounds of flour milled by the factorial effects of the percent undersized kernels. The observation from Figure 17.3 is that B and AC are possible non-null effects, while C is not a possible non-null effect since its point is on the line drawn through the other null effects points. Thus, B and AC are identified as possible non-null effects and the PROC GLM code in Table 17.5 fits a model with the two non-null effects and the covariate x (percent of undersized kernels) to the data set. The estimate of the variance for this model is 7.0329 and the significance levels corresponding to

TABLE 17.3
PROC REG Code to Fit Model with Main Effect of C and Covariate with Analysis of Variance Table and Parameter Estimates

```
proc reg data=values;
model y=c x;
```

Source	df	SS	MS	FValue	ProbF
Model	2	1255.3943	627.6972	4.16	0.0862
Error	5	753.8807	150.7761		
Corrected Total	7	2009.2750			

Variable	df	Estimate	StdErr	tValue	Probt
Intercept	1	134.9436	34.3479	3.93	0.0111
c	1	−12.3450	5.6027	−2.20	0.0788
x	1	0.0382	0.7515	0.05	0.9614

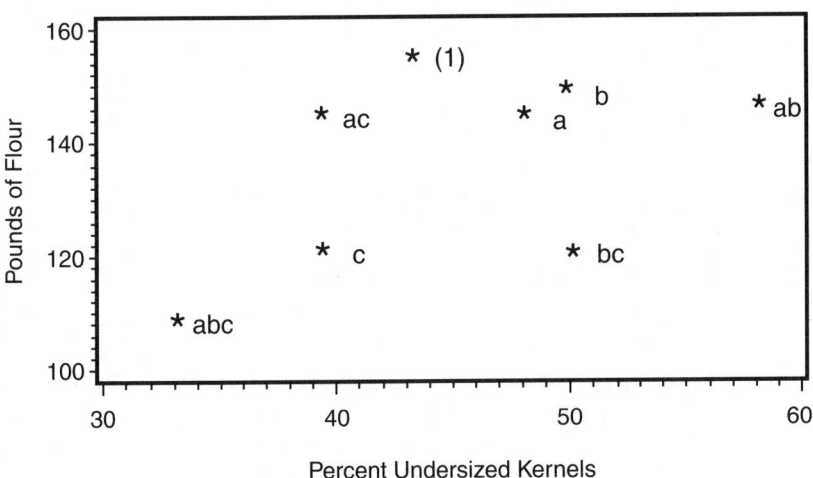

FIGURE 17.2 Plot of pounds of flour by percent undersized kernels.

the three terms in the model are all less than 0.0004, indicating all terms are needed in the model. The estimate of the slope is 2.1798, indicating the model predicts that there is a 2.1798-lb increase in the amount of flour milled for each 1% increase in the percent of undersized kernels. A stepwise regression process was used to determine if there are any other possible non-null effects. Using the stepwise method, no other variables would be entered in the model with a significance level of 0.15 or less. Thus the use of this analysis of covariance process discovered the important effects in the model are B and AC instead of C, the effect that would be used if the

TABLE 17.4
PROC GLM Code to Compute the Estimates of the Effects for the Response Variable and the Covariate

```
proc glm data=values;
model y x=a b c a*b a*c b*c a*b*c/solution;
```

Parameter	Estimate from y	Estimate from x
Intercept	136.6750	45.3375
a	−0.0750	−0.4875
b	−5.0750	2.6375
c	−12.5250	−4.7125
a*b	−3.4750	−1.6875
a*c	3.0750	−3.7875
b*c	−4.1250	−1.5125
a*b*c	−5.4750	−2.5875

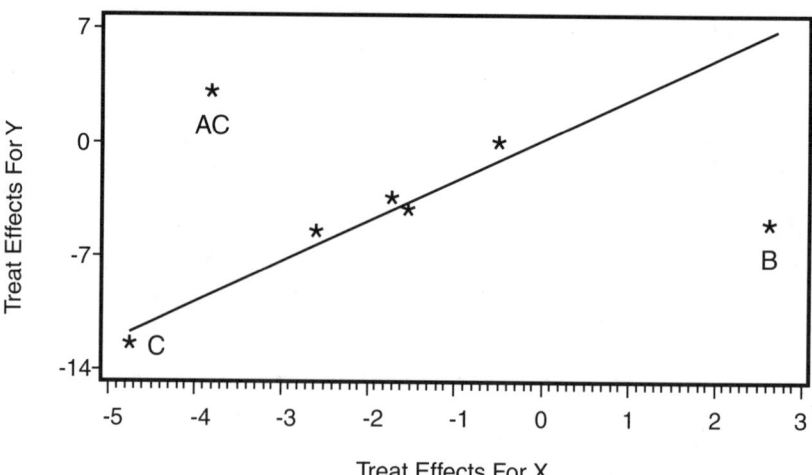

FIGURE 17.3 Plot of the factorial effects of pounds of flour (y) by the effects of percent undersized kernels (x).

covariate information were to be ignored. Table 17.6 consists of the PROC GLM code where C is included with the non-null effects and the covariate. The significance level corresponding to C is 0.1766, indicating that C is not likely to be a non-null effect. The estimate of the variance using the covariate (7.0329) is considerably less than the estimate of the variance from the model without the covariate (125.7117). Thus the covariate is important in describing the mean of the number of pounds milled in the presence of the factorial effects.

TABLE 17.5
PROC GLM Code to Fit the Model with the Main Effect of B, the A*C Interaction Term, and the Covariate with the Analysis of Variance Table and Parameter Estimates

```
proc glm data=values; model y=b a*c x/solution;
```

Source	df	SS	MS	FValue	ProbF
Model	3	1981.1435	660.3812	93.90	0.0004
Error	4	28.1315	7.0329		
Corrected Total	7	2009.2750			

Source	df	SS(III)	MS	FValue	ProbF
b	1	901.2185	901.2185	128.14	0.0003
a*c	1	881.7723	881.7723	125.38	0.0004
x	1	1699.4535	1699.4535	241.64	0.0001

Parameter	Estimate	StdErr	tValue	Probt
Intercept	23.8063	7.3211	3.25	0.0313
b	−11.6411	1.0284	−11.32	0.0003
a*c	12.5041	1.1167	11.20	0.0004
x	2.4895	0.1602	15.54	0.0001

TABLE 17.6
PROC GLM Code to Fit the Model with the Main Effects of B and C, the A*C Interaction Term, and the Covariate with the Analysis of Variance Table and Parameter Estimates

```
proc glm data=values; model y=b c a*c x/solution;
```

Source	df	SS	MS	FValue	ProbF
Model	4	1995.4365	498.8591	108.15	0.0014
Error	3	13.8385	4.6128		
Corrected Total	7	2009.2750			

Source	df	SS(III)	MS	FValue	ProbF
b	1	594.5822	594.5822	128.90	0.0015
c	1	14.2930	14.2930	3.10	0.1766
c*a	1	469.2961	469.2961	101.74	0.0021
x	1	458.7415	458.7415	99.45	0.0021

Parameter	Estimate	StdErr	tValue	Probt
Intercept	37.8477	9.9391	3.81	0.0318
b	−10.8243	0.9534	−11.35	0.0015
c	−2.2526	1.2797	−1.76	0.1766
c*a	11.3310	1.1234	10.09	0.0021
x	2.1798	0.2186	9.97	0.0021

TABLE 17.7
Data on Loaf Volume from a Four-Way Factorial Treatment Structure with One Covariate, Yeast Viability

a	b	c	d	Loaf Volume	Yeast Viability
−1	−1	−1	−1	160.2	16.4
−1	−1	−1	1	187.4	29.0
−1	−1	1	−1	207.0	39.8
−1	−1	1	1	189.1	26.5
−1	1	−1	−1	162.5	12.5
−1	1	−1	1	178.3	13.1
−1	1	1	−1	195.3	17.0
−1	1	1	1	185.5	21.3
1	−1	−1	−1	196.0	35.3
1	−1	−1	1	201.0	23.8
1	−1	1	−1	200.0	18.5
1	−1	1	1	190.0	30.1
1	1	−1	−1	191.1	15.2
1	1	−1	1	199.0	19.1
1	1	1	−1	203.7	39.0
1	1	1	1	184.8	14.3

17.7 EXAMPLE: BAKING BREAD USING FOUR FACTORS EACH AT TWO LEVELS

The data in Table 17.7 are from a baking experiment where loaf volume (y) in cm^3 is the response variable and yeast viability (x) of dough used to make a loaf of bread is a covariate. The treatment structure consists of four factors each at two levels where a, b, c, and d denote mixing time, rising time, mixing temperature, and rising temperature, respectively. The half-normal probability plot of the factorial effects computed from loaf volume is in Figure 17.4. The half-normal probability plot indicates that A, C, AC, and CD are the important factorial effects in the data. The scatter plot in Figure 17.5 provides a visual suggestion that there is an increase in the loaf volume as the level of yeast viability increases, although this conclusion is made with the knowledge that the factorial effects are confounded with the level of yeast viability. The full four-way factorial model is fit to the loaf volume and viability data by using the PROC GLM code in Table 17.8. The lower part of Table 17.8 contains the estimates of the factorial effects. The scatter plot of the factorial effects of loaf volume and the factorial effects of yeast viability is in Figure 17.6. The observation from the scatter plot is that A, C, AC, and CD are possible non-null effects, the same as identified by the half-normal probability plot in Figure 17.4. Table 17.9 contains the PROC GLM code to fit the model with the terms A, C, AC, CD, and yeast viability (x). The estimate of the variance is 9.2739, while the estimate

Analysis of Covariance for Nonreplicated Experiments 509

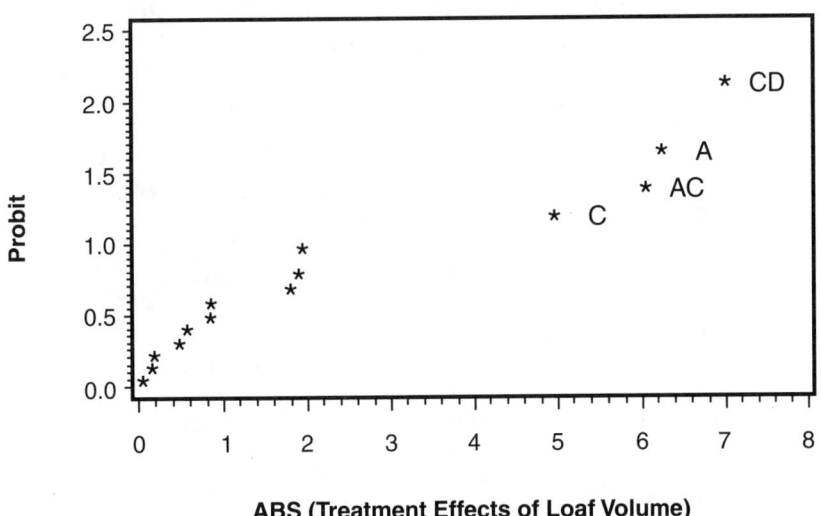

FIGURE 17.4 Half-normal plot of the treatment effects for the loaf volume data without the covariate.

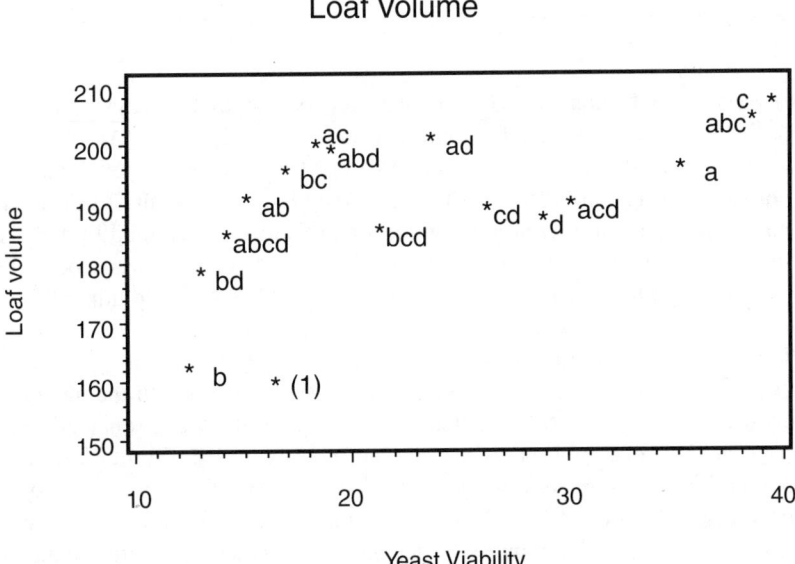

FIGURE 17.5 Plot of the loaf volume and yeast viability data.

TABLE 17.8
PROC GLM Code to Fit the Full Factorial Effects Model to the Loaf Volume and Yeast Viability Data

```
proc glm data=values;
model y x=a b c a*b a*c b*c a*b*c d a*d b*d
  a*b*d c*d a*c*d b*c*d a*b*c*d/solution;
```

Parameter	Estimate for Loaf Volume	Estimate for Yeast Viability
Intercept	189.4313	23.1813
a	6.2688	1.2313
b	−1.9063	−4.2438
c	4.9938	2.6313
a*b	0.8562	1.7313
a*c	−6.0688	−1.5688
b*c	−0.1938	1.3313
a*b*c	0.8688	2.3563
d	−0.0437	−1.0313
a*d	−1.9563	−1.5563
b*d	−0.5812	−0.9563
a*b*d	−0.1687	−1.6563
c*d	−7.0313	−1.7313
a*c*d	1.8063	1.0438
b*c*d	0.4813	−1.3813
a*b*c*d	−1.9563	−5.0813

Note: Results are the estimates of the factorial effects.

of the variance for an analysis without the covariate is 24.2552 (analysis not shown). The estimate of the slope is 0.4114 cm^3 per unit of viability score. All of the terms in the model have significance levels of 0.0015 or less, indicating the selected terms are non-null effects. The PROC REG code in Table 17.10 was used to investigate the possibility of additional non-null effects that were not identified by the scatter plot in Figure 17.6. The options indicate that the first five terms are to be included in the model and the other terms are to be considered as possible non-null effects if one or more can be entered with a significance level of 0.15 or less and can stay in the model if the resulting significance level is less than 0.05, using the variable selection procedure stepwise. No other variables were included in the model, indicating that the non-null effects selected from the scatter plot are adequate to describe the data set. This example shows that the half-normal probability plot can identify the non-null effects to be used in the analysis of covariance, but using the covariate in the analysis substantially reduces the estimate of the variance. This reduced variance could be used to compare adjusted means computed for the important effects in the model. The adjusted means are not computed for this example, but the process is demonstrated in the last two examples in this chapter.

Analysis of Covariance for Nonreplicated Experiments

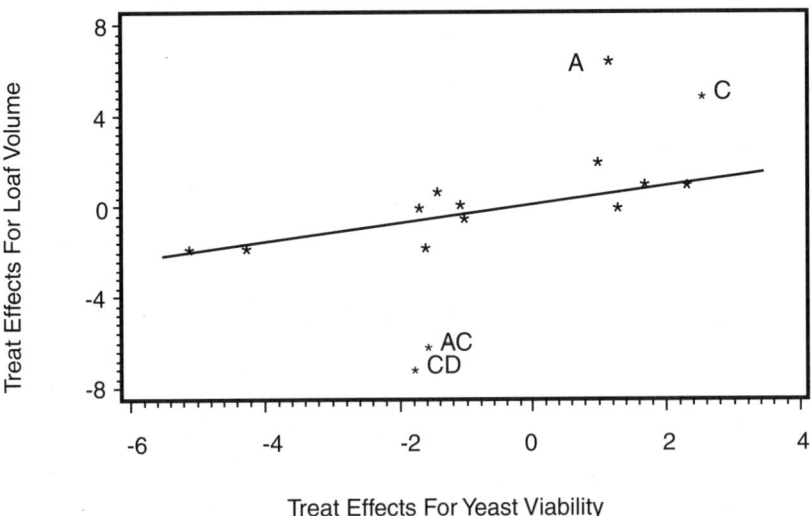

FIGURE 17.6 Plot of the treatment effects for loaf volume by treatment effects for yeast viability.

17.8 EXAMPLE: HAMBURGER PATTIES WITH FOUR FACTORS EACH AT TWO LEVELS

The data in Figure 17.7 displayed in a JMP® table are the acceptability scores of hamburger patties made several different ways when the fat content of the batch of hamburger from which a patty was made is considered as a possible covariate. The different ways the hamburger patties were made are from a four-way factorial treatment structure where A is cooking method, B is grind size, C is filler amount, and D is patty size. The half-normal probability plot in Figure 17.8 indicates that when the information about the fat content is ignored, ABD is the only important factorial effect. The scatter plot in Figure 17.9 indicates that the acceptability scores generally decrease when the level of fat increases, although the fat content is confounded with the factorial effects. Figure 17.10 is the fit model screen used to fit the full factorial effects model to the acceptance scores and the fat content data. Figure 17.11 is a data table display of the estimated factorial effects for both variables, which also includes their scatter plot. The scatter plot in Figure 17.12 is an enlargement of the scatter plot in Figure 17.11 where the effects not on the common line are denoted. Thus the possible non-null effects are A, C, AC, and CD. The fit model screen in Figure 17.13 contains the model specification to fit the fat content as a covariate along with the factorial effects A, C, AC, and CD. The results of the least squares fit of the model to the acceptance scores are in Figure 17.14. The significance levels of the effects in the model are all less than 0.0021, indicating the

TABLE 17.9
PROC GLM Code to Fit the Final Model to the Loaf Volume Data Using Yeast Viability as the Covariate

```
proc glm data=values; model y=a c a*c c*d
  x/solution;
```

Source	df	SS	MS	FValue	ProbF
Model	5	2582.1152	516.4230	55.69	0.0000
Error	10	92.7392	9.2739		
Corrected Total	15	2674.8544			

Source	df	SS(III)	MS	FValue	ProbF
a	1	519.0192	519.0192	55.97	0.0000
c	1	220.9805	220.9805	23.83	0.0006
a*c	1	453.2654	453.2654	48.88	0.0000
c*d	1	610.4328	610.4328	65.82	0.0000
x	1	174.0677	174.0677	18.77	0.0015

Parameter	Estimate	StdErr	tValue	Probt
Intercept	179.8953	2.3290	77.24	0.0000
a	5.7623	0.7703	7.48	0.0000
c	3.9113	0.8013	4.88	0.0006
a*c	−5.4234	0.7758	−6.99	0.0000
c*d	−6.3191	0.7789	−8.11	0.0000
x	0.4114	0.0950	4.33	0.0015

TABLE 17.10
PROC REG Code to Use Stepwise Regression to Search for Additional Non-Null Effects

```
proc reg data=check; model y =a c ac cd x b ab bc abc d ad bd abd
acd bcd abcd/selection=stepwise include=5 slstay=0.05 slentry=0.15;
```

selected effects are non-null. The estimate of the slope is −1.94 acceptance score units per one unit of fat content. The evaluation of possible other non-null effects is left as an exercise for the reader.

17.9 EXAMPLE: STRENGTH OF COMPOSITE MATERIAL COUPONS WITH TWO COVARIATES

This example involves two covariates and thus demonstrates the process needed to be followed to evaluate the effect of more than one covariate. The same process can be used when it is suspected that the slopes are not equal for the levels of one of the factors, as is demonstrated in Section 17.10. The data in Table 17.11 are strength of a coupon made of composite material (Y), average thickness of a coupon (X),

Analysis of Covariance for Nonreplicated Experiments

FIGURE 17.7 Data table for acceptability scores of hamburger patties from a four-way treatment structure and fat content as a possible covariate.

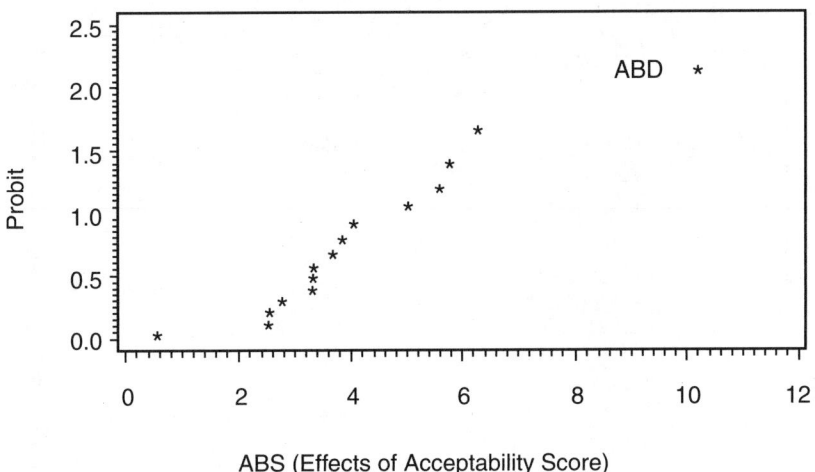

FIGURE 17.8 Half-normal probability plot of the estimated effects of the hamburger acceptability scores.

and the average thickness of the outer coating of the coupon (Z). The coupons were constructed using the settings from a four-way treatment structure where A, B, C, and D represent two levels of heat, two times of processing, two pressures, and two coupon compositions, respectively. Figure 17.15 is the half-normal probability plot of the strength values and there is evidence that D, BC, AB, and ABC are the non-null

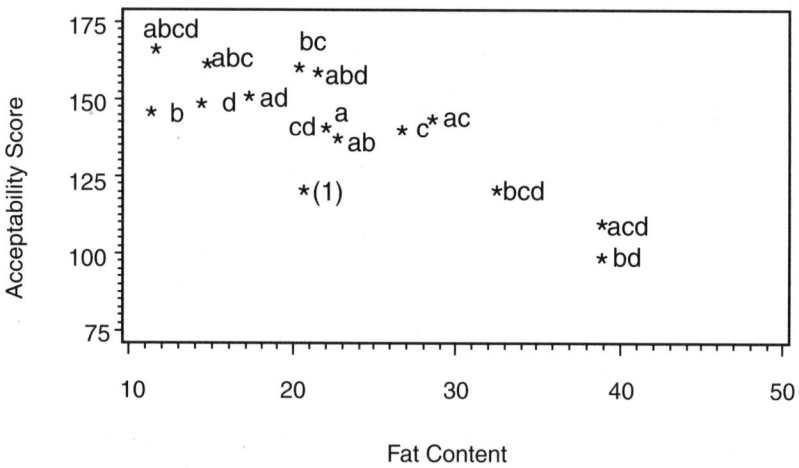

FIGURE 17.9 Plot of the hamburger acceptability scores data by the fat content data for each of the treatment combinations.

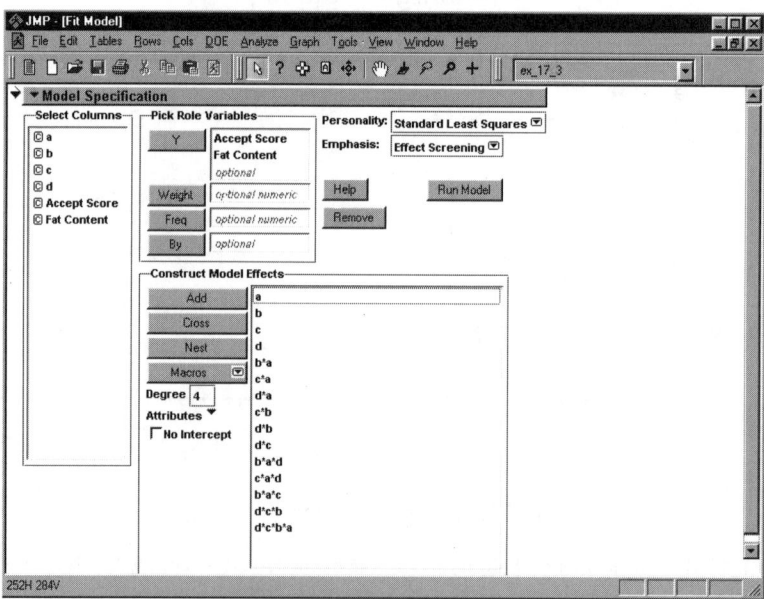

FIGURE 17.10 Fit model display for JMP® to fit the full four-way factorial treatment structure model to the hamburger acceptability scores and the fat content data.

effects when the information about the two covariates is ignored. The three-dimensional scatter plot of the data is in Figure 17.16, where the y, x, and z axes represent the strength, coupon thickness, and coating thickness, respectively. The type of

Analysis of Covariance for Nonreplicated Experiments

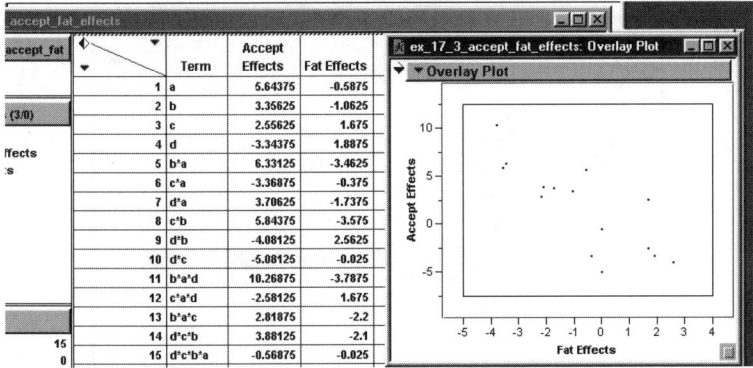

FIGURE 17.11 Estimates of the factorial effects for acceptability scores and fat content data with plot.

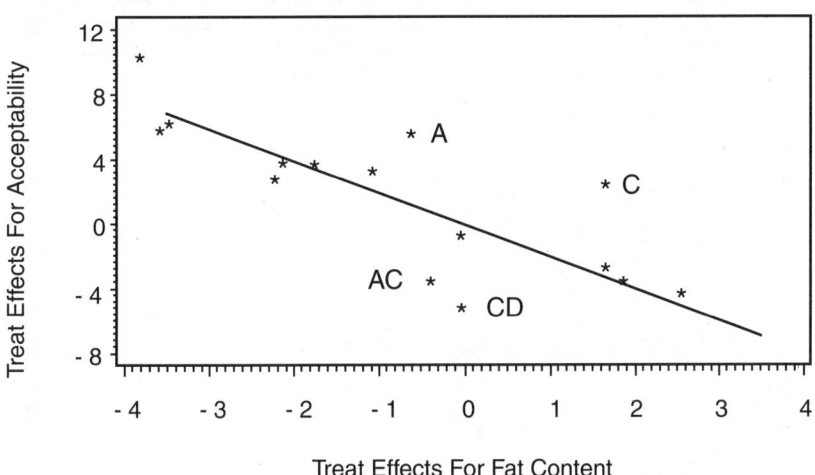

FIGURE 17.12 Plot of the factorial effects for acceptability scores and fat content data of hamburger patties.

relationship among the variables is not obvious because of the confounding of the covariates with the factorial effects. The next step is to fit the full factorial effects model to each of the three variables, using the PROC GLM code in Table 17.12. The lower part of Table 17.12 contains the estimates of the factorial effects for strength, coupon thickness, and coating thickness. The three-dimensional scatter plot of the factorial effects is in Figure 17.17 where D, AB, BC, and ABC are possible non-null effects (the same as identified via the half-normal probability plot). The PROC GLM code to fit the model with both covariates and the identified non-null effects is in Table 17.13. The results in Table 17.13 indicate that the effects D, AB,

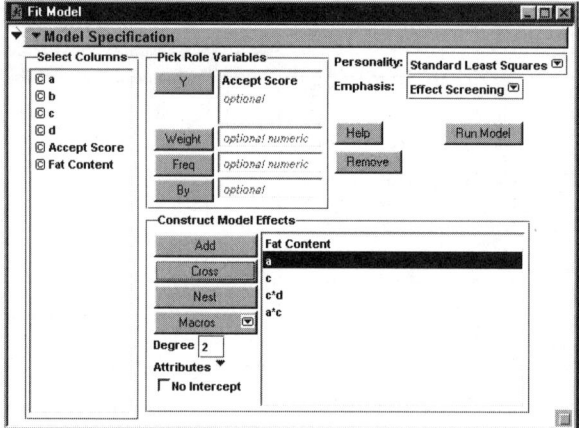

FIGURE 17.13 Fit model display to fit the final model to the hamburger data.

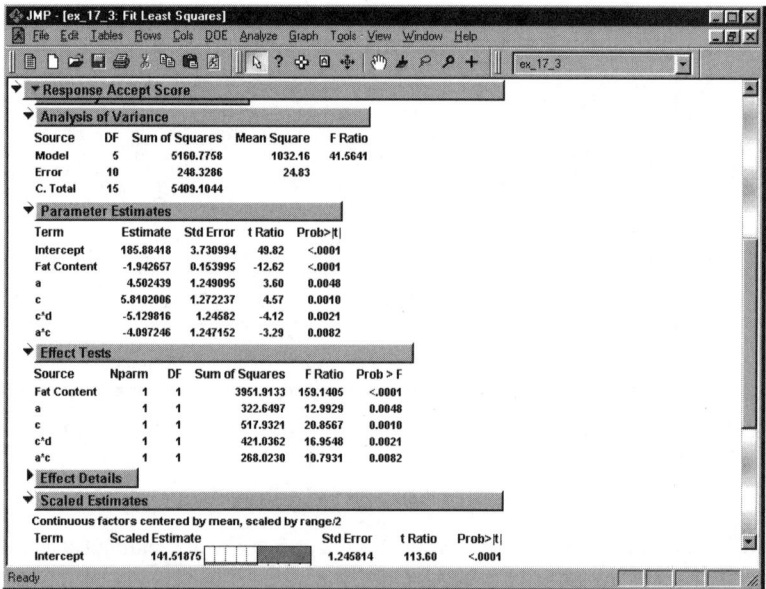

FIGURE 17.14 Results of fitting the final model to the hamburger acceptability score data.

and BC are important effects, while the significance level for ABC is 0.1814, indicating that it is possibly not a non-null effect. Also the significance levels corresponding to coupon thickness (x) and coating thickness (z) are 0.2652 and 0.1459, respectively, indicating they are possibly not needed in the model. But for the rest of the analysis here, the model fit in Table 17.13 is used to describe the data. The estimate statements in Table 17.14 are used to provide estimates of the means and comparisons of the means for the factorial effects. The model used to describe

Analysis of Covariance for Nonreplicated Experiments

TABLE 17.11
Coupon Strength Data from Four-Way Factorial Treatment Structure

a	b	c	d	Strength	Coupon Thickness	Coating Thickness
-1	-1	-1	-1	69.6	30.3	0.60
-1	-1	-1	1	61.5	31.1	0.77
-1	-1	1	-1	74.4	21.2	1.68
-1	-1	1	1	61.8	12.5	1.36
-1	1	-1	-1	62.6	35.1	1.33
-1	1	-1	1	53.9	13.4	0.62
-1	1	1	-1	60.1	15.9	1.46
-1	1	1	1	54.2	27.7	0.53
1	-1	-1	-1	49.3	34.9	1.77
1	-1	-1	1	40.4	13.2	1.46
1	-1	1	-1	71.8	22.2	0.71
1	-1	1	1	52.3	21.9	0.84
1	1	-1	-1	83.7	13.5	0.96
1	1	-1	1	77.0	18.0	1.13
1	1	1	-1	67.3	21.2	1.50
1	1	1	1	69.3	10.3	0.68

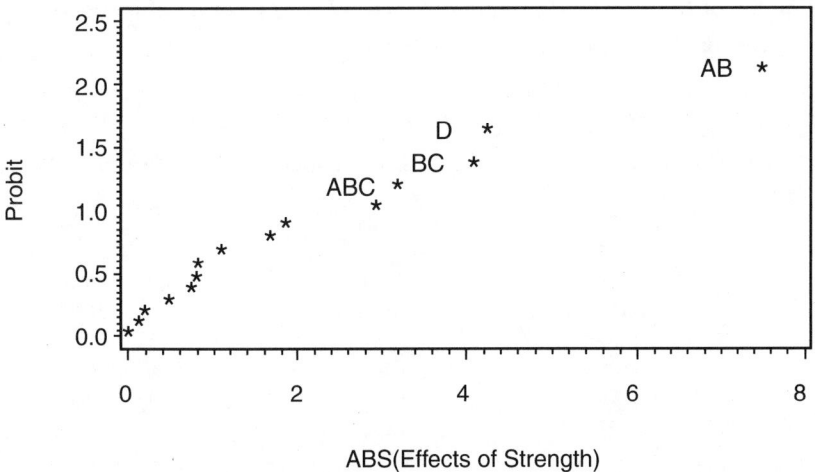

FIGURE 17.15 Half-normal probability plot of the strength effects.

the mean strength as a function of the non-null effects and the two covariates is $y_{ijkm} = \alpha + d_m + ab_{ij} + bc_{jk} + abc_{ijk} + \beta_x x_{ijkm} + \beta_z z_{ijkm} + \varepsilon_{ijkm}$, so the estimate statements are constructed using -1 and $+1$ for the two levels of each of the factorial effects and $x = 21.4$ and $z = 1.0875$, the means of the two covariates. The first two estimate

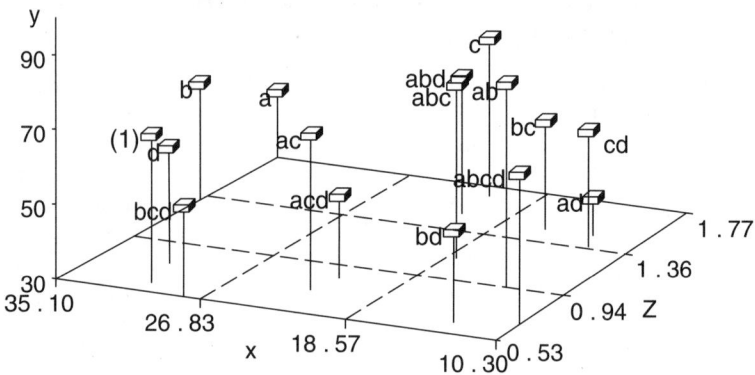

FIGURE 17.16 Plot of the strength (y), coupon thickness (x), and coating thickness (z) for the factorial effects.

TABLE 17.12
PROC GLM Code to Fit a Full Factorial Model to the Data Set to Provide Estimates of the Factorial Effects for Strength (y), Coupon Thickness (x), and Coating Thickness (z).

```
proc glm data=values; model y x z=a b c d a*b a*c a*d b*c b*d c*d
   a*b*c a*b*d a*c*d b*c*d a*b*c*d/solution;
```

Parameter	Effects of Strength	Effects of Coupon Thickness	Effects of Coating Thickness
Intercept	63.0750	21.4000	1.0875
a	0.8125	−2.0000	0.0438
b	2.9375	−2.0125	−0.0613
c	0.8250	−2.2875	0.0075
d	−4.2750	−2.8875	−0.1638
a*b	7.5000	−1.6375	−0.0025
a*c	0.4625	1.7875	−0.2063
a*d	0.1375	−0.6625	0.0600
b*c	−4.1125	1.6750	0.0087
b*d	1.8625	0.8500	−0.1225
c*d	−0.2250	1.8750	−0.0788
a*b*c	−3.2000	−1.1750	0.2125
a*b*d	1.1000	1.1000	0.0638
a*c*d	−0.0125	−1.1250	0.0100
b*c*d	1.6625	0.3875	−0.0725
a*b*c*d	0.7500	−4.9875	−0.1063

Analysis of Covariance for Nonreplicated Experiments 519

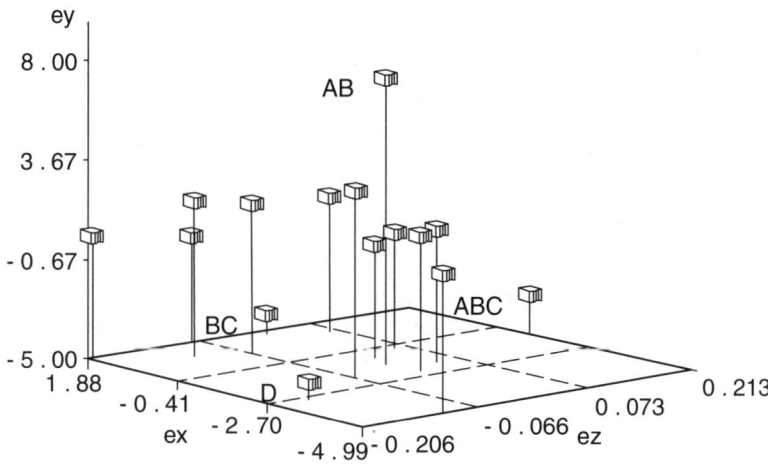

FIGURE 17.17 Graph of the estimated factorial effects for strength (ey), coupon thickness (ex), and coating thickness (ez).

statements provide least squares means for the two levels of D evaluated at the average levels of A, B, C, and the two covariates. The third estimate statement provides the estimate of the difference between the two levels of D. The results of the estimate statements are in Table 17.15. The adjusted mean strengths for the two types of composite materials are 57.1953 and 68.9547 with a difference of –11.7593 where the significance level indicates that the material coded –1 is stronger than the material coded +1. Since the levels of D are qualitative, it does not make sense to evaluate the means of A × B × C at the average level of D. The next 16 estimate statements provide estimates of the ABC combination means for each of the levels of D or for each of the composite materials. The last 12 estimate statements provide comparisons of the levels of A at each combination of B and C, comparisons of the levels of B at each combination of the levels of A and C, and comparisons of levels of C at each combination of the levels of A and B. Since there is no interaction between the levels of D and the levels of the other factors, the comparisons of the levels of A, B, and C do not depend on the levels of D. The combination of a = 1, b = 1, and c = –1 provide the maximum strengths for both types of materials. The means of a = 1 are significantly larger than the means of a = –1 for b = 1, but the trend is reversed for b = –1. The means of b = 1 are significantly smaller than the means of b = –1 for a = –1, but the mean of b = 1 is significantly larger than the mean of b = –1 for a = 1 and c = –1. Finally the mean of c = 1 is signifcantly larger than the mean of c = –1 at a = 1 and b = –1, but the trend is reversed for a = 1 and b = 1.

The analysis in Table 17.13 indicates that ABC is possibly a null effect and that the covariates are possibly not needed. The analysis was carried out with these terms in the model to demonstrate the process of using two covariates and of computing and comparing adjusted means. Refining the analysis is reserved as an exercise.

TABLE 17.13
PROC GLM Code to Fit the Proposed Model to the Coupon Strength Data

```
proc glm data=values;
model y=d a*b b*c a*b*c x z/solution;
```

Source	df	SS	MS	FValue	ProbF
Model	6	1713.7021	285.6170	12.52	0.0006
Error	9	205.2879	22.8098		
Corrected Total	15	1918.9900			

Source	df	SS(III)	MS	FValue	ProbF
d	1	375.0679	375.0679	16.44	0.0029
a*b	1	775.2901	775.2901	33.99	0.0002
b*c	1	208.7569	208.7569	9.15	0.0144
a*b*c	1	47.8525	47.8525	2.10	0.1814
x	1	32.1949	32.1949	1.41	0.2652
z	1	57.7972	57.7972	2.53	0.1459

Parameter	Estimate	StdErr	tValue	Probt
Intercept	74.1844	5.8349	12.71	0.0000
d	−5.8797	1.4500	−4.06	0.0029
a*b	7.1512	1.2266	5.83	0.0002
b*c	−3.7173	1.2288	−3.03	0.0144
a*b*c	−2.1193	1.4632	−1.45	0.1814
x	−0.2035	0.1713	−1.19	0.2652
z	−6.2109	3.9018	−1.59	0.1459

17.10 EXAMPLE: EFFECTIVENESS OF PAINT ON BRICKS WITH UNEQUAL SLOPES

There is not enough data to enable one to evaluate the possibility of a different slope for each of the treatment combinations, but there are generally sufficient data to fit a model with unequal slopes for one or two of the factors. The process of analysis is similar to that used for the multiple covariate problem described in Section 17.9. The data in Table 17.16 are from a four-way factorial treatment structure in a CR design structure experiment set up to evaluate the effectiveness of painting a brick's surface to withstand abrasion. The dependent measure is the average time in minutes for a fine surface grinder to wear through the paint. The time to wear through the paint surface was recorded at seven locations on the surface of each brick. The covariate is a measure of the average porosity of the brick's surface measured at the seven locations used for the abrasion test. The four factors are type of applicator (A, spray or brush), type of paint (B, epoxy or enamel), drying agent (C, no or yes), and bonding agent (D, no or yes). Figure 17.18 contains the half-normal probability plot of the factorial effects and the indication is that without including information about the covariate, the non-null effects are A, B, D, and AC. The scatter plot of the

TABLE 17.14
Estimate Statements for PROC GLM to Provide Estimates of the Adjusted Means and Comparison of the Adjusted Means

```
estimate 'd=1'  intercept 1 d  1 x 21.4 z 1.0875;
estimate 'd=-1' intercept 1 d -1 x 21.4 z 1.0875;
estimate 'd=1 minus d=-1' d 2;
estimate 'a=-1,b=-1,c=-1 d=-1' intercept 1 d -1 a*b  1 b*c  1 a*b*c
 -1 x 21.4 z 1.0875;
estimate 'a=-1,b=-1,c=1 d=-1' intercept 1 d -1 a*b  1 b*c -1 a*b*c
  1 x 21.4 z 1.0875;
estimate 'a=-1,b=1,c=-1 d=-1' intercept 1 d -1 a*b -1 b*c -1 a*b*c
  1 x 21.4 z 1.0875;
estimate 'a=-1,b=1,c= 1 d=-1' intercept 1 d -1 a*b -1 b*c  1 a*b*c
 -1 x 21.4 z 1.0875;
estimate 'a=-1,b=-1,c=-1 d=1' intercept 1 d  1 a*b  1 b*c  1 a*b*c
 -1 x 21.4 z 1.0875;
estimate 'a=-1,b=-1,c=1 d=1' intercept 1 d  1 a*b  1 b*c -1 a*b*c  1
  x 21.4 z 1.0875;
estimate 'a=-1,b=1,c=-1 d=1' intercept 1 d  1 a*b -1 b*c -1 a*b*c
  1 x 21.4 z 1.0875;
estimate 'a=-1,b=1,c= 1 d=1' intercept 1 d  1 a*b -1 b*c  1 a*b*c
 -1 x 21.4 z 1.0875;
estimate 'a=1,b=-1,c=-1 d=-1' intercept 1 d -1 a*b -1 b*c  1 a*b*c
  1 x 21.4 z 1.0875;
estimate 'a=1,b=-1,c=1 d=-1' intercept 1 d -1 a*b -1 b*c -1 a*b*c
 -1 x 21.4 z 1.0875;
estimate 'a=1,b=1,c=-1 d=-1' intercept 1 d -1 a*b  1 b*c -1 a*b*c
 -1 x 21.4 z 1.0875;
estimate 'a=1,b=1,c= 1 d=-1' intercept 1 d -1 a*b  1 b*c  1 a*b*c
  1 x 21.4 z 1.0875;
estimate 'a=1,b=-1,c=-1 d=1' intercept 1 d  1 a*b -1 b*c  1 a*b*c  1
  x 21.4 z 1.0875;
estimate 'a=1,b=-1,c=1 d=1' intercept 1 d  1 a*b -1 b*c -1 a*b*c
 -1 x 21.4 z 1.0875;
estimate 'a=1,b=1,c=-1 d=1' intercept 1 d  1 a*b  1 b*c -1 a*b*c
 -1 x 21.4 z 1.0875;
estimate 'a=1,b=1,c= 1d=1' intercept 1 d  1 a*b  1 b*c  1 a*b*c
  1 x 21.4 z 1.0875;
estimate 'a=1 minus a=-1,@b=-1,c=-1' a*b -2 a*b*c  2;
estimate 'a=1 minus a=-1,@b=-1,c= 1' a*b -2 a*b*c -2;
estimate 'a=1 minus a=-1,@b= 1,c=-1' a*b  2 a*b*c -2;
estimate 'a=1 minus a=-1,@b= 1,c= 1' a*b  2 a*b*c  2;
estimate 'b=1 minus b=-1,@a=-1,c=-1' a*b -2 b*c -2 a*b*c  2;
estimate 'b=1 minus b=-1,@a=-1,c= 1' a*b -2 b*c  2 a*b*c -2;
estimate 'b=1 minus b=-1,@a= 1,c=-1' a*b  2 b*c -2 a*b*c -2;
estimate 'b=1 minus b=-1,@a= 1,c= 1' a*b  2 b*c  2 a*b*c  2;
estimate 'c=1 minus c=-1,@a=-1,b=-1' b*c -2 a*b*c  2;
estimate 'c=1 minus c=-1,@a=-1,b= 1' b*c  2 a*b*c -2;
estimate 'c=1 minus c=-1,@a= 1,b=-1' b*c -2 a*b*c -2;
estimate 'c=1 minus c=-1,@a= 1,b= 1' b*c  2 a*b*c  2;
```

Note: Results are in Table 17.15.

TABLE 17.15
Results from the Estimate Statements in Table 17.14 with Adjusted Means and Comparisons of Adjusted Means for the Levels of a, b, c, and d

Parameter	Estimate	StdErr	tValue	Probt
d=1	57.1953	1.8783	30.45	0.0000
d=−1	68.9547	1.8783	36.71	0.0000
d=1 minus d=−1	−11.7593	2.8999	−4.06	0.0029
a=−1,b=−1,c=−1 d=−1	74.5079	2.7640	26.96	0.0000
a=−1,b=−1,c=1 d=−1	77.7039	3.0334	25.62	0.0000
a=−1,b=1,c=−1 d=−1	63.4014	3.0538	20.76	0.0000
a=−1,b=1,c= 1 d=−1	60.2054	2.9549	20.37	0.0000
a=−1,b=−1,c=−1 d=1	62.7486	3.0538	20.55	0.0000
a=−1,b=−1,c=1 d=1	65.9446	2.9549	22.32	0.0000
a=−1,b=1,c=−1 d=1	51.6421	2.7640	18.68	0.0000
a=−1,b=1,c= 1 d=1	48.4461	3.0334	15.97	0.0000
a=1,b=−1,c=−1 d=−1	55.9668	3.2021	17.48	0.0000
a=1,b=−1,c=1 d=−1	67.6400	2.7637	24.47	0.0000
a=1,b=1,c=−1 d=−1	81.9425	2.6827	30.54	0.0000
a=1,b=1,c= 1 d=−1	70.2693	3.0788	22.82	0.0000
a=1,b=−1,c=−1 d=1	44.2075	2.6827	16.48	0.0000
a=1,b=−1,c=1 d=1	55.8807	3.0788	18.15	0.0000
a=1,b=1,c=−1 d=1	70.1832	3.2021	21.92	0.0000
a=1,b=1,c= 1d=1	58.5100	2.7637	21.17	0.0000
a=1 minus a=−1,@b=−1,c=−1	−18.5411	3.7769	−4.91	0.0008
a=1 minus a=−1,@b=−1,c= 1	−10.0639	3.8599	−2.61	0.0284
a=1 minus a=−1,@b= 1,c=−1	18.5411	3.7769	4.91	0.0008
a=1 minus a=−1,@b= 1,c= 1	10.0639	3.8599	2.61	0.0284
b=1 minus b=−1,@a=−1,c=−1	−11.1065	4.4519	−2.49	0.0342
b=1 minus b=−1,@a=−1,c= 1	−17.4984	4.6642	−3.75	0.0045
b=1 minus b=−1,@a= 1,c=−1	25.9756	4.5595	5.70	0.0003
b=1 minus b=−1,@a= 1,c= 1	2.6293	4.4858	0.59	0.5722
c=1 minus c=−1,@a=−1,b=−1	3.1960	3.8430	0.83	0.4271
c=1 minus c=−1,@a=−1,b= 1	−3.1960	3.8430	−0.83	0.4271
c=1 minus c=−1,@a= 1,b=−1	11.6732	3.7997	3.07	0.0133
c=1 minus c=−1,@a= 1,b= 1	−11.6732	3.7997	−3.07	0.0133

time and porosity data for each of the treatment combinations is in Figure 17.19. This plot does not provide any indication that there is a relationship between the abrasion time and porosity, but that relationship is confounded with the factorial treatment effects. From prior experience, the relationship between resistance to abrasion and porosity can depend on the method of application, i.e., there may be different slopes for spray and brush. The possible model that uses a linear relationship between the mean of the abrasion time and porosity is

$$y_{ijkm} = \mu_{ijkm} + \beta_i x_{ijkm} + \varepsilon_{ijkm} \text{ for } i = -1, 1, \ j = -1, 1, \ k = -1, 1, \ m = -1, 1,$$

Analysis of Covariance for Nonreplicated Experiments

TABLE 17.16
Times to Wear through the Paint Layer on Bricks from a Four-Way Factorial Treatment Structure with Porsity of the Brick as a Possible Covariate

a	b	c	d	Time	Porosity	Porosity a = −1 (xx)	Porosity a = 1 (zz)
−1	−1	−1	−1	13.11	0.10	0.10	0
−1	−1	−1	1	11.13	0.27	0.27	0
−1	−1	1	−1	8.32	0.10	0.10	0
−1	−1	1	1	5.40	0.10	0.10	0
−1	1	−1	−1	16.42	0.01	0.01	0
−1	1	−1	1	13.17	0.15	0.15	0
−1	1	1	−1	10.71	0.10	0.10	0
−1	1	1	1	10.64	0.16	0.16	0
1	−1	−1	−1	7.41	0.14	0	0.14
1	−1	−1	1	3.70	0.02	0	0.02
1	−1	1	−1	12.21	0.29	0	0.29
1	−1	1	1	8.34	0.21	0	0.21
1	1	−1	−1	9.76	0.18	0	0.18
1	1	−1	1	8.63	0.20	0	0.20
1	1	1	−1	14.01	0.09	0	0.09
1	1	1	1	13.88	0.19	0	0.19

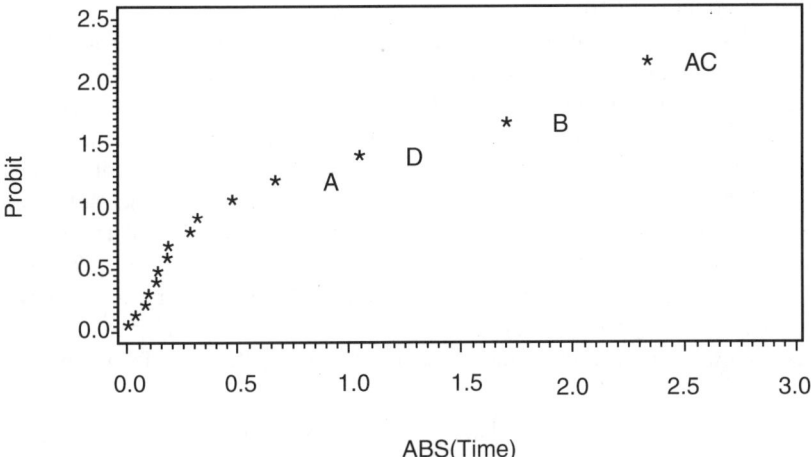

FIGURE 17.18 Half-normal probability plot of the grinding time data.

where i denotes the level of A; j, the level of B; k, the level of C; m, the level of D; and β_i is the slope for the ith level of A. The last two columns of Table 17.16 are the values of the covariate (porosity) separated into two columns so that the model

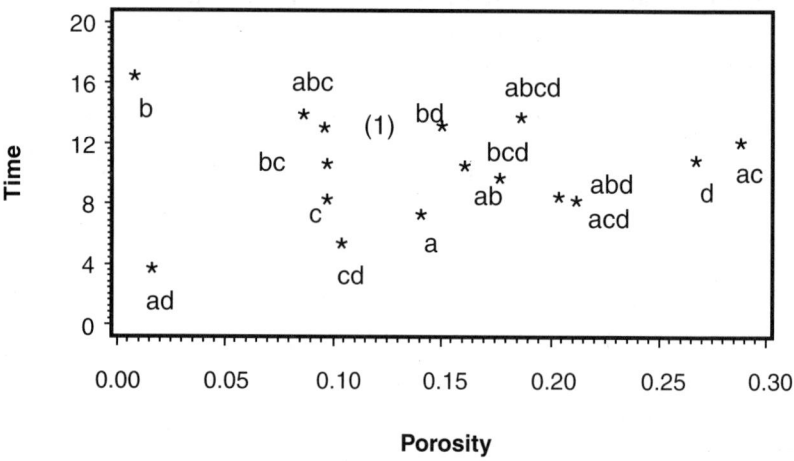

FIGURE 17.19 Plot of the grinding time and porosity data factorial effects.

can be expressed as with two covariates (as in Section 17.9) so there are different slopes for spray and brush. The possible model is

$$y_{ijkm} = \mu_{ijkm} + \beta_{-1} x_{-1jkm} + \beta_1 x_{1jkm} + \varepsilon_{ijkm} \text{ for } i = -1, 1, \quad j = -1, 1, \quad k = -1, 1, \quad m = -1, 1$$

where x_{-1jkm} and x_{1jkm} are the porosity values for $a = -1$ and $a = 1$, respectively as displayed in Table 17.16.

The first step in the analysis is to compute the estimates of the factorial effects for three variables: time(y), porosity for $a = -1$(xx), and porosity for $a = 1$(zz). Table 17.17 contains the PROC GLM code to fit the full four-way factorial treatment structure model to each of the three variables and the respective estimates of the factorial effects are displayed in the lower portion of the table. Since the two covariates are constructed from a single covariate, two-dimensional scatter plots were used to look for the possible non-null effects. Figure 17.20 contains the scatter plot for $a = -1$ and Figure 17.21 contains the scatter plot for $a = 1$. The visual indication from each of the scatter plots is that B, D, and AC are non-null effects with possibly A also being considered as non-null. Using the possible non-null effects and the two covariates, the PROC GLM code in Table 17.18 fits the model

$$y_{ijkm} = \alpha + A_i + B_j + D_m + AC_{ik} + \beta_{-1} xx_{ijkm} + \beta_1 zz_{ijkm} + \varepsilon_{ijkm}$$

to the data set. The estimate of the variance is 0.7984. The significance levels corresponding to testing the hypothesis that the factorial effects are zero, i.e., A = 0, B = 0, D = 0, and AC = 0, are 0.0893, 0.0000, 0.0020, and 0.0000, respectively. The significance levels corresponding testing the slopes are zero, i.e., $\beta_{-1} = 0$ and $\beta_1 = 0$, are 0.5984 and 0.1783, indicating the covariates are possibly not needed in the

TABLE 17.17
PROC GLM Code to Fit Complete Four-Way Factorial Effects Model and Estimates for Time (y), the Porosity for a = −1 (xx) and Porosity for a = 1 (zz)

```
proc glm data=values;
model y xx zz=a b c a*b a*c b*c a*b*c d a*d b*d a*b*d c*d
    a*c*d b*c*d a*b*c*d/solution;
```

Parameter	Estimate (Time)	Estimate (Porosity a = −1)	Estimate (Porosity a = 1)
Intercept	10.4275	0.0613	0.0819
a	−0.6850	−0.0613	0.0819
b	1.7250	−0.0093	−0.0002
c	0.0113	−0.0039	0.0147
a*b	0.1025	0.0093	−0.0002
a*c	2.3563	0.0039	0.0147
b*c	0.1463	0.0164	−0.0282
a*b*c	−0.1388	−0.0164	−0.0282
d	−1.0663	0.0241	−0.0046
a*d	−0.0388	−0.0241	−0.0046
b*d	0.4938	0.0017	0.0206
a*b*d	0.2963	−0.0017	0.0206
c*d	0.1925	−0.0152	0.0077
a*c*d	−0.0875	0.0152	0.0077
b*c*d	0.3300	0.0054	0.0016
a*b*c*d	−0.1850	−0.0054	0.0016

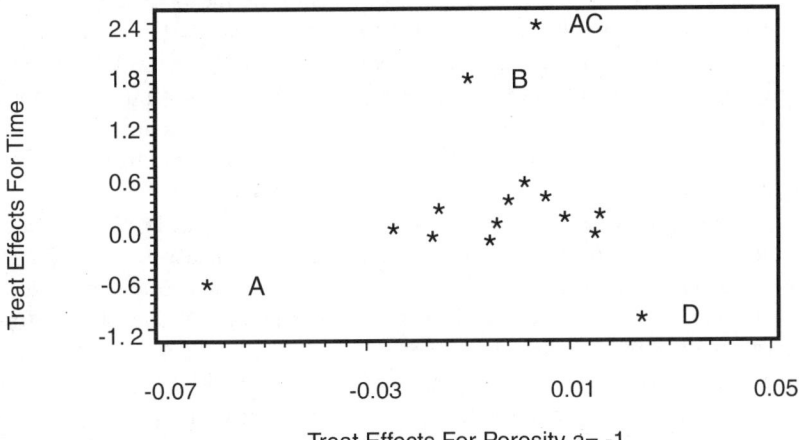

FIGURE 17.20 Plot of the factorial effects of grinding time by factorial effects of porosity at level a = −1.

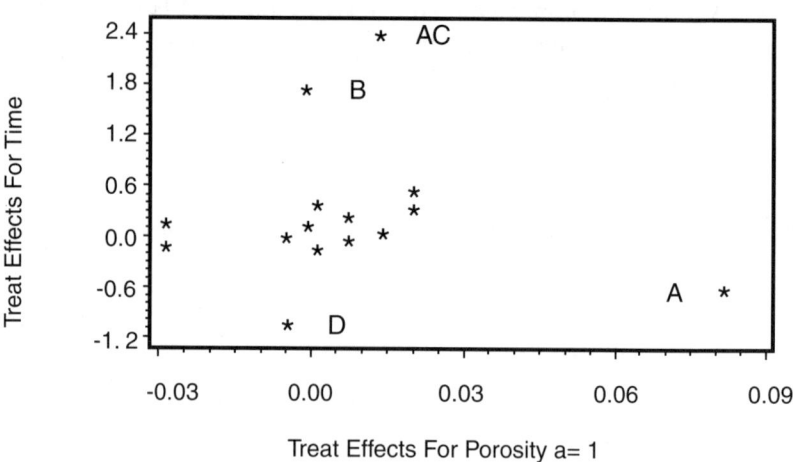

FIGURE 17.21 Plot of the factorial effects of grinding time by factorial effects of porosity at level a = 1.

analysis. Using the stepwise regression approach, no additional effects were identified as non-null. The model was not simplified for the remainder of the analysis in this section, but the covariate part of the model should be simplified and then evaluate the possibility of declaring A as a null effect should be evaluated. To continue with the analysis, estimate statements in Table 17.19 were used to provide adjusted means for the factorial effects remaining in the model. The first line provides a test of the equal slopes hypothesis for the levels of A. Since the covariate has been split into two parts, porosity at a = –1 or xx is used when evaluating means that involve a = –1 and porosity at a = 1 or zz is used when evaluating the means that involve a = 1. The next two lines provide estimates of the effect of A at b = 1, c = –1, and d = 1 at a porosity of 0.1. The first statement includes "zz 0.1" for level a = 1 and the second statement includes "xx 0.1" for level a = –1. The third estimate statement is a comparison of the above two means or a comparison of a = 1 and a = –1 at b = 1, c = –1, and d = 1 which includes "zz .1 and xx – .1" as this estimate statement is constructed by subtracting the first two estimate statements. Since the slopes are different for the levels of A, the means for a = 1 and a = –1 need to be evaluated at three or more values of the covariate. In this case, the means were evaluated at porosity values of 0.1, 0.2, and 0.3. Also, since AC is included in the model, the means for the levels of A were evaluated and compared at each level of C for the three values of porosity where b = 1 and d = 1. The levels C were evaluated and compared at each level of A for a porosity value of 0.2 and b = 1 and d = 1. The levels of B were evaluated compared at a = 1, c = 1, d = 1, and porosity = 0.2 and the levels of D were evaluated and compared at a = 1, b = 1, c = 1, and porosity of 0.2. The results of the estimate statements are in Table 17.20 where the first partition

TABLE 17.18
PROC GLM Code to Fit the Final Model to the Time Data with Different Slopes for Each Level of a

```
proc glm data=values; model y=a b d a*c xx
   zz/solution;
```

Source	df	SS	MS	FValue	ProbF
Model	6	164.1078	27.3513	34.26	0.0000
Error	9	7.1853	0.7984		
Corrected Total	15	171.2931			

Source	df	SS(III)	MS	FValue	ProbF
a	1	2.8941	2.8941	3.62	0.0893
b	1	46.8610	46.8610	58.70	0.0000
d	1	14.6537	14.6537	18.35	0.0020
a*c	1	74.9340	74.9340	93.86	0.0000
xx	1	0.2379	0.2379	0.30	0.5984
zz	1	1.7015	1.7015	2.13	0.1783

Parameter	Estimate	StdErr	tValue	Probt
Intercept	9.7430	0.5216	18.68	0.0000
a	−1.0117	0.5314	−1.90	0.0893
b	1.7533	0.2289	7.66	0.0000
d	−1.1083	0.2587	−4.28	0.0020
a*c	2.2541	0.2327	9.69	0.0000
xx	2.9172	5.3446	0.55	0.5984
zz	6.1707	4.2268	1.46	0.1783

contains a test of the equality of the slopes, the second partition contains the results for evaluating and comparing the levels of A, the third partition contains the results for evaluating and comparing the levels of C, and the last two partitions contains the results for evaluating and comparing the levels of B and D, respectively. All of the comparisons of the pairs of means are significantly different. The estimates of the parameters of the model in Table 17.18 indicate that the maximum time is predicted to occur at $a = -1$, $b = 1$, $c = -1$, and $d = -1$ since the estimates of the slopes for a and d are negative and for b and ac are positive.

17.11 SUMMARY

A methodology is described that incorporates covariates into the analysis of non-replicated factorial or fractional factorial treatment structures. The process depends on being able to separate the non-null factorial effects from the null effects. The slopes corresponding to the covariates are estimated from the set of factorial effects declared to be null effects, computed for the dependent variables as well as for each covariate. A scatter plot of the estimated factorial effects from the response variable

TABLE 17.19
Estimate Statements to Provide Adjusted Means and Comparisons of Adjusted Means for Model with Unequal Slopes for the Covariate for Each Level of a

```
estimate 'compare slopes' xx 1 zz -1;
estimate 'a= 1 b=1 c=-1 d=1 x=.1' intercept 1 a 1 b 1 d 1 a*c -1 zz .1;
estimate 'a=-1 b=1 c=-1 d=1 x=.1' intercept 1 a -1 b 1 d 1 a*c 1 xx .1;
estimate 'a=1 minus a=-1 at c=-1 x=.1' a 2 a*c -2 zz .1 xx -.1;
estimate 'a= 1 b=1 c=-1 d=1 x=.2' intercept 1 a 1 b 1 d 1 a*c -1 zz .2;
estimate 'a=-1 b=1 c=-1 d=1 x=.2' intercept 1 a -1 b 1 d 1 a*c 1 xx .2;
estimate 'a=1 minus a=-1 at c=-1 x=.2' a 2 a*c -2 zz .2 xx -.2;
estimate 'a= 1 b=1 c=-1 d=1 x=.3' intercept 1 a 1 b 1 d 1 a*c -1 zz .3;
estimate 'a=-1 b=1 c=-1 d=1 x=.3' intercept 1 a -1 b 1 d 1 a*c 1 xx .3;
estimate 'a=1 minus a=-1 at c=-1 x=.3' a 2 a*c -2 zz .3 xx -.3;
estimate 'a= 1 b=1 c=1 d=1 x=.1' intercept 1 a 1 b 1 d 1 a*c 1 zz .1;
estimate 'a=-1 b=1 c=1 d=1 x=.1' intercept 1 a -1 b 1 d 1 a*c -1 xx .1;
estimate 'a=1 minus a=-1 at c=1 x=.1' a 2 a*c 2 zz .1 xx -.1;
estimate 'a= 1 b=1 c=1 d=1 x=.2' intercept 1 a 1 b 1 d 1 a*c 1 zz .2;
estimate 'a=-1 b=1 c=1 d=1 x=.2' intercept 1 a -1 b 1 d 1 a*c -1 xx .2;
estimate 'a=1 minus a=-1 at c=1 x=.2' a 2 a*c 2 zz .2 xx -.2;
estimate 'a= 1 b=1 c=1 d=1 x=.3' intercept 1 a 1 b 1 d 1 a*c 1 zz .3;
estimate 'a=-1 b=1 c=1 d=1 x=.3' intercept 1 a -1 b 1 d 1 a*c -1 xx .3;
estimate 'a=1 minus a=-1 at c=1 x=.3' a 2 a*c 2 zz .3 xx -.3;
estimate 'c=1 at a=1 b=1 d=1 x=.2' intercept 1 a 1 b 1 d 1 a*c 1 xx .2;
estimate 'c=-1 at a=1 b=1 d=1 x=.2' intercept 1 a 1 b 1 d 1 a*c -1 xx .2;
estimate 'c=1 minus c=-1@a=1' a*c 2;
estimate 'c=1 at a=-1 b=1 d=1 x=.2' intercept 1 a -1 b 1 d 1 a*c -1 zz .2;
estimate 'c=-1 at a=-1 b=1 d=1 x=.2' intercept 1 a -1 b 1 d 1 a*c 1 zz .2;
estimate 'c=1 minus c=-1@a=-1' a*c -2;
estimate 'b=1 at a=1 c=1 d=1 x=.2' intercept 1 a 1 b 1 d 1 a*c 1 xx .2;
estimate 'b=-1 at a=1 c=1 d=1 x=.2' intercept 1 a 1 b -1 d 1 a*c 1 xx .2;
estimate 'b=1 minus b=-1' b 2;
estimate 'd=1 at a=1 b=1 c=1 x=.2' intercept 1 a 1 b 1 d 1 a*c 1 xx .2;
estimate 'd=-1 at a=1 b=1 c=1  x=.2' intercept 1 a 1 b 1 d -1 a*c 1 xx .2;
estimate 'd=1 minus d=-1' d 2;
```

and the estimated factorial effects from each of the covariates is used to help identify the non-null effects and the regression analysis is used to estimate the parameters of the model corresponding to the non-null effects and the covariates. Stepwise regression can be used to determine if additional factorial effects should be included in the set of non-null effects. Adjusted means can be computed using the final model to provide a method of comparing the means of the factors contained in the non-null effects. Five examples were included to demonstrate possible cases for using analysis of covariance to analyze nonreplicated treatment structures. These examples show that the non-null effects can change from the unadjusted analysis to the covariate adjusted analysis, and the use of covariates in the analyses can greatly improve the estimate of the standard deviation.

TABLE 17.20
Results from the Estimate Statements for Model with the Covariate

Parameter	Estimate	StdErr	tValue	Probt
compare slopes	−3.2534	6.8889	−0.47	0.6480
a= 1 b=1 c=−1 d=1 x=.1	7.7392	0.5363	14.43	0.0000
a=−1 b=1 c=−1 d=1 x=.1	13.9457	0.5473	25.48	0.0000
a=1 minus a=−1 at c=−1 x=.1	−6.2064	0.6694	−9.27	0.0000
a= 1 b=1 c=−1 d=1 x=.2	8.3563	0.5555	15.04	0.0000
a=−1 b=1 c=−1 d=1 x=.2	14.2374	0.5918	24.06	0.0000
a=1 minus a=−1 at c=−1 x=.2	−5.8811	0.7864	−7.48	0.0000
a= 1 b=1 c=−1 d=1 x=.3	8.9734	0.8288	10.83	0.0000
a=−1 b=1 c=−1 d=1 x=.3	14.5291	0.9860	14.74	0.0000
a=1 minus a=−1 at c=−1 x=.3	−5.5557	1.3183	−4.21	0.0023
a= 1 b=1 c=1 d=1 x=.1	12.2475	0.5961	20.54	0.0000
a=−1 b=1 c=1 d=1 x=.1	9.4374	0.5374	17.56	0.0000
a=1 minus a=−1 at c=1 x=.1	2.8101	0.7498	3.75	0.0046
a= 1 b=1 c=1 d=1 x=.2	12.8646	0.5217	24.66	0.0000
a=−1 b=1 c=1 d=1 x=.2	9.7291	0.6176	15.75	0.0000
a=1 minus a=−1 at c=1 x=.2	3.1355	0.7799	4.02	0.0030
a= 1 b=1 c=1 d=1 x=.3	13.4816	0.7392	18.24	0.0000
a=−1 b=1 c=1 d=1 x=.3	10.0208	1.0224	9.80	0.0000
a=1 minus a=−1 at c=1 x=.3	3.4608	1.2662	2.73	0.0231
c=1 at a=1 b=1 d=1 x=.2	12.2139	1.3269	9.20	0.0000
c=−1 at a=1 b=1 d=1 x=.2	7.7056	1.2933	5.96	0.0002
c=1 minus c=−1@a=1	4.5083	0.4653	9.69	0.0000
c=1 at a=−1 b=1 d=1 x=.2	10.3798	1.2839	8.08	0.0000
c=−1 at a=−1 b=1 d=1 x=.2	14.8881	1.2218	12.19	0.0000
c=1 minus c=−1@a=−1	−4.5083	0.4653	−9.69	0.0000
b=1 at a=1 c=1 d=1 x=.2	12.2139	1.3269	9.20	0.0000
b=−1 at a=1 c=1 d=1 x=.2	8.7072	1.2557	6.93	0.0001
b=1 minus b=−1	3.5066	0.4577	7.66	0.0000
d=1 at a=1 b=1 c=1 x=.2	12.2139	1.3269	9.20	0.0000
d=−1 at a=1 b=1 c=1 x=.2	14.4305	1.5473	9.33	0.0000
d=1 minus d=−1	−2.2166	0.5174	−4.28	0.0020

REFERENCES

Daniel, C. (1959). Use of Half-Normal Plots in Interpreting Factorial Two-Level Experiments. *Technometrics* 1:311-341.

Milliken, G. A. and Johnson, D. E. (1989). *Analysis of Messy Data, Volume II: Nonreplicated Experiments.* Chapman & Hall, New York.

SAS Institute Inc. (1989). *SAS/STAT® User's Guide, Version 6, Fourth Edition, Volume 2,* Cary, NC.

EXERCISES

EXERCISE 17.1: Use the data in the following table to carry out the analysis of covariance on the single replication of the five-way treatment structure. Make any necessary comparisons among the non-null effects.

Data for Exercise 17.1 (a Five-Way Treatment Structure with y as the Response Variable and x as the Covariate)

a	b	c	d	e	x	y
−1	−1	−1	−1	−1	130	265
−1	−1	−1	−1	1	113	219
−1	−1	−1	1	−1	123	234
−1	−1	−1	1	1	112	186
−1	−1	1	−1	−1	124	254
−1	−1	1	−1	1	105	182
−1	−1	1	1	−1	138	284
−1	−1	1	1	1	147	300
−1	1	−1	−1	−1	143	283
−1	1	−1	−1	1	110	247
−1	1	−1	1	−1	108	199
−1	1	−1	1	1	127	255
−1	1	1	−1	−1	114	206
−1	1	1	−1	1	137	261
−1	1	1	1	−1	104	190
−1	1	1	1	1	123	264
1	−1	−1	−1	−1	123	213
1	−1	−1	−1	1	104	210
1	−1	−1	1	−1	118	251
1	−1	−1	1	1	125	264
1	−1	1	−1	−1	117	241
1	−1	1	−1	1	120	250
1	−1	1	1	−1	144	261
1	−1	1	1	1	112	248
1	1	−1	−1	−1	120	243
1	1	−1	−1	1	130	258
1	1	−1	1	−1	122	274
1	1	−1	1	1	124	248
1	1	1	−1	−1	116	259
1	1	1	−1	1	114	212
1	1	1	1	−1	120	233
1	1	1	1	1	134	252

Analysis of Covariance for Nonreplicated Experiments

EXERCISE 17.2: Carry out a thorough analysis of covariance using the following data set involving a four-way treatment structure. Make all necessary comparisons among the selected non-null effects.

Data for Exercise 17.2 (Four-Way Treatment Structure with Response Variable y and Covariate x)

a	b	c	d	x	y
−1	−1	−1	−1	1.18	24.66
−1	−1	−1	1	2.57	28.65
−1	−1	1	−1	4.51	35.56
−1	−1	1	1	3.85	37.16
−1	1	−1	−1	1.38	27.34
−1	1	−1	1	4.53	37.73
−1	1	1	−1	1.08	24.46
−1	1	1	1	1.37	23.85
1	−1	−1	−1	1.54	27.98
1	−1	−1	1	2.88	32.27
1	−1	1	−1	2.45	33.88
1	−1	1	1	4.16	39.25
1	1	−1	−1	1.33	28.86
1	1	−1	1	2.99	34.99
1	1	1	−1	2.53	30.62
1	1	1	1	2.73	30.45

EXERCISE 17.3: Compute adjusted means for the effects in the model of Section 17.7 and make necessary comparisons.

EXERCISE 17.4: Determine if there are other possible non-null effects for the example of Section 17.8.

EXERCISE 17.5: Refine the model used in Section 17.9 and make necessary comparisons among the adjusted means of the non-null effects remaining in the model.

EXERCISE 17.6: Re-analyze the data in Section 17.10 and include only those effects with coefficients that are significantly different from zero at $p \leq 0.05$. Are the conclusions from the reanalysis different than those reached using the analysis in Section 17.10?

18 Special Applications of Analysis of Covariance

18.1 INTRODUCTION

Previous chapters detailed most of the applications of analysis of covariance models, but there are some applications that do not exactly fit into any of those chapters. This chapter is a collection of examples where each is a specific application of analysis of covariance.

18.2 BLOCKING AND ANALYSIS OF COVARIANCE

Chapter 11 considered situations where the covariate was measured on a block or all experimental units within a block had identical values of the covariate. There are many areas where it is the normal operation procedure to rank the experimental units on one or more characteristics and then form blocks based on that ordering. For example, animal scientists often weigh animals before starting an experiment. When they have four treatments, they will form blocks of size four where the four heaviest animals are put into block 1, the next four heaviest animals are put into block 2, all the way down to the four lightest animals which are put into the last block. Then the four treatments are randomly assigned to the four animals within each block. This is similar to the situation described in Chapter 11, but not exactly since the animals within each block do not have identical values of the covariate or have identical starting weights. One concept about blocking that is generally ignored is that the factor or factors that are used to create the blocks must not interact with the factors in the treatment structure (Milliken and Johnson, 1992; elements of the design structure must not interact with elements of the treatment structure). For many experiments, using initial weight as a blocking factor is satisfactory because the treatments in the treatment structure do not interact with initial weight of the animals (or what ever characteristics are used to form blocks), but there are obviously many types of treatments that can possibly interact with animal size, and in those cases, using weight as a blocking factor is not reasonable. When the covariate was measured on the block (as in Chapter 11), all of the information available for estimating the slopes of the regression models came from the between block comparisons (there was no within block information about the slopes). When weight is used as a blocking factor (or some other characteristic), there will generally be considerable variability in the weights of the experimental units between the blocks and a little variability in the weights of the experimental units within blocks. Thus, most of the information about the slopes comes from the between block comparisons, but there is still some within block information about the slopes. The reason scientists use a characteristic to form blocks is that by doing so they think that they do not need to use that characteristic as a covariate in the analysis. However, as was discussed in Chapter 11, even though a characteristic is used as a blocking factor, there are still situations where it is beneficial to include the characteristic as a covariate in the model. This is the case when the treatment's slopes are unequal. The additional problem that can occur with this type of blocking process is that there can be great block to block differences in the variability among experimental units within each block. If the distribution of the characteristic used for blocking is similar to a normal distribution, then the weights of the four animals in the heaviest weight blocks and the weights of the animals in the lightest weight blocks will be much more variable than the weights of the animals within the middle weight blocks.

An animal nutritionist designed a study to evaluate the effect of four diets on the ability of Hereford heifers to gain weight. Forty heifers were purchased for the study and they were weighed (initial weight in pounds). Their weights were ranked from lowest to highest. The heifers were grouped in sets of four in which the lightest four heifers formed block 0, the next four heaviest heifers formed block 1, and so on to the four heaviest heifers which formed block 9. The four diets were randomly assigned to the four heifers within each block. The design of this experiment is a one-way treatment structure (four diets) in a randomized complete block design structure (ten blocks). The data are in Table 18.1, where iwt is the initial weight and adg is the average daily gain of a heifer computed as the number of pounds gained during the study divided by the number of days the study was conducted (pounds/day). The PROC GLM code in Table 18.2 provides an analysis of the initial weights where block is the effect in the model and the Levene's test for homogeneity of variances (see Chapter 14) is requested by the HOVTEST=Levene(Type=Abs)

TABLE 18.1
Data for the Average Daily Gain Study Where Initial Weight was Used as the Blocking Factor

block	Diet A		Diet B		Diet C		Diet D	
	adg	iwt	adg	iwt	adg	iwt	adg	iwt
0	1.43	586.3	1.81	599.6	1.80	599.5	1.71	597.2
1	1.96	618.6	1.99	603.9	1.97	610.7	2.06	606.8
2	2.12	619.2	2.33	620.2	2.25	625.1	2.35	627.0
3	2.22	628.7	2.37	628.3	2.33	638.8	2.52	632.6
4	2.08	639.6	2.66	645.9	2.46	642.4	2.54	645.3
5	2.50	649.2	2.79	653.9	2.59	652.2	2.80	648.3
6	2.22	655.7	2.96	654.6	2.38	655.4	2.97	657.9
7	2.46	660.9	2.95	659.1	2.66	662.0	2.72	661.1
8	2.31	666.7	3.37	674.5	2.69	673.7	2.72	669.3
9	2.73	695.8	3.48	707.8	2.96	685.6	3.04	681.8

TABLE 18.2
Levene's Test for Homogeneity of Initial Weight Block Variances for the Average Daily Gain Study (Block Means and Standard Deviations are Included)

```
proc glm data=design;
class block;
model iwt=block;
means block/hovtest=levene(type=abs);
```

Source	df	SS	MS	FValue	ProbF
block	9	201.596	22.400	3.69	0.0033
Error	30	182.314	6.077		

block	iwtmean	iwtstd
0	595.65	6.33
1	610.00	6.37
2	622.88	3.77
3	632.10	4.87
4	643.30	2.90
5	650.90	2.60
6	655.90	1.41
7	660.78	1.21
8	671.05	3.69
9	692.75	11.65

option on the means statement. The computed F value has a significance level of 0.003, an indication that the variances of the initial weights within the blocks are not equal. The lower part of Table 18.2 contains the means and standard deviations of the initial weights for each of the blocks. The means are increasing, as they should, since the blocks were formed by ranking the initial weights. However, the variances are not equal where the larger variances are for blocks 0, 1, and 9, a phenomenon that occurs when the characteristic used to form blocks has a distribution with tails, similar to those of a normal distribution (but, do not have to be symmetric).

A common step in the analysis of covariance is to determine if the randomization of the experimental units to the treatments was carried out in an appropriate manner. One way to accomplish that is to carry out an analysis of variance on the covariate. Table 18.3 contains the PROC MIXED code to request such an analysis of variance. The significance level corresponding to diet is 0.6620, which indicates that there were no significant differences between the diet means for the initial weight, hence the randomization process was adequately executed.

Since the initial weights were used to construct the blocks, then the first step (and only step by many analysts) is to ignore the covariate and carry out an analysis of variance on the adg data. A model that can be used for the analysis of variance is

TABLE 18.3
PROC MIXED Code and Results for Checking the Random Assignment by Analyzing the Initial Weight Values

```
proc mixed cl covtest data=design;
class Diet block;
model iwt=Diet;
random block;
```

CovParm	Estimate	StdErr	ZValue	ProbZ
block	843.510	401.158	2.10	0.0177
Residual	29.856	8.126	3.67	0.0001

Effect	NumDF	DenDF	FValue	ProbF
Diet	3	27	0.60	0.6220

$$ADG_{ij} = \mu + \rho_i + b_j + a_{ij}, \quad i = 1, 2, 3, 4, \text{ and } j = 0, 1, 2, \ldots, 9$$

$$\text{where } b_j \sim \text{iid } N(0, \sigma_b^2) \text{ and } a_{ij} \sim \text{iid } N(0, \sigma_a^2)$$

(18.1)

The parameter ρ_i denotes the diet effect, b_j is the random block effect, and a_{ij} is the random animal effect. The PROC MIXED code in Table 18.4 was used to fit Model 18.1 to the adg data, where diet is a fixed effect and block is a random effect. The estimates of the variance components are $\hat{\sigma}_b^2 = 0.1538$ and $\hat{\sigma}_a^2 = 0.0255$. The significance level corresponding the diet is 0.0000 which indicates the diet means are not equal. Table 18.5 contains the adjusted means for the diets as well as pairwise comparison among the diets where the simulate adjustment was used for multiple comparisons. The estimated standard errors for the adjusted means are 0.13 and for the differences are 0.07. The results indicate that the mean adg of ration A is

TABLE 18.4
PROC MIXED Code and Results for the Analysis of Variance of the Average Daily Gain Data

```
proc mixed cl covtest data=ex_18_2;
class block Diet;
model adg = Diet/ddfm=kr;
random block;
```

CovParm	Estimate	StdErr	ZValue	ProbZ
block	0.1538	0.0755	2.04	0.0209
Residual	0.0255	0.0069	3.67	0.0001

Effect	NumDF	DenDF	FValue	ProbF
Diet	3	27	15.71	0.0000

TABLE 18.5
Adjusted Means and Pairwise Comparisons of the Diet Means Using the Simulate Adjustment for Multiple Comparisons

```
lsmeans Diet/diff adjust=simulate;
```

Effect	Diet	Estimate	StdErr	df	tValue	Probt
Diet	A	2.20	0.13	11.2	16.45	0.0000
Diet	B	2.67	0.13	11.2	19.95	0.0000
Diet	C	2.41	0.13	11.2	17.99	0.0000
Diet	D	2.54	0.13	11.2	18.99	0.0000
Diet	_Diet	Estimate	StdErr	df	tValue	Adjp
A	B	−0.47	0.07	27	−6.56	0.0000
A	C	−0.21	0.07	27	−2.89	0.0368
A	D	−0.34	0.07	27	−4.76	0.0005
B	C	0.26	0.07	27	3.67	0.0060
B	D	0.13	0.07	27	1.79	0.3000
C	D	−0.13	0.07	27	−1.88	0.2632

significantly less than the means for the other three rations and the mean adg of diet B is significantly greater than the mean for diet C.

The next analysis carried out is an analysis of covariance with equal slopes using the model

$$\text{ADG}_{ij} = \mu + \rho_i + + \beta\, \text{IWT}_{ij} + b_j + a_{ij}, \quad i = 1, 2, 3, 4, \text{ and } j = 0, 1, 2, \ldots, 9 \tag{18.2}$$

$$\text{where } b_j \sim \text{iid}\, N(0, \sigma_b^2) \text{ and } a_{ij} \sim \text{iid}\, N(0, \sigma_a^2)$$

The common slope of the set of regression models is β. The PROC MIXED code in Table 18.6 provides the analysis for Model 18.2. The estimates of the variance components are $\hat{\sigma}_b^2 = 0.0009$ and $\hat{\sigma}_a^2 = 0.0198$. By including the covariate in the model, the estimate of the block variance component is greatly reduced (0.1938 to 0.0009), while the estimate of the animal variance component is reduced just a little (0.0255 to 0.0198). Thus including the covariate in the model has the major effect on the block variance component (as it should since the initial weights were used to form the blocks), but still has an effect of reducing the animal variance component as there is variation among animals within the blocks. The F statistic corresponding to diet has a significance level of 0.0000, which indicates that the set of parallel regression lines are not identical. Table 18.7 contains the adjusted diet means and pairwise comparisons of the diet means. The estimated standard errors for the adjusted means are 0.05 and for the differences are 0.06. These estimated standard errors are smaller than those in Table 18.5 with the main reduction in the estimated standard errors for the adjusted means (0.13 to 0.05). These estimated standard errors have different inference space interpretations. The estimated standard errors from Table 18.5 are to enable inferences to the population of Hereford heifers, while the

TABLE 18.6
PROC MIXED Code and Results for Fitting the Common Slope Analysis of Covariance Model to the Average Daily Gain Data

```
proc mixed cl covtest data=ex_18_2;
  class block Diet;
  model adg = Diet iwt/ddfm=kr;
  random block;
```

CovParm	Estimate	StdErr	ZValue	ProbZ
block	0.0009	0.0032	0.28	0.3910
Residual	0.0198	0.0054	3.67	0.0001

Effect	NumDF	DenDF	FValue	ProbF
Diet	3	27.0	17.97	0.0000
iwt	1	8.6	241.90	0.0000

TABLE 18.7
Adjusted Means Evaluated at iwt = 643.53 lb and Pairwise Comparisons Using the Simulate Adjustment for Multiple Comparisons

```
lsmeans Diet/diff adjust=simulate at means;
```

Effect	Diet	Estimate	StdErr	df	tValue	Probt
Diet	A	2.22	0.05	34.5	48.83	0.0000
Diet	B	2.65	0.05	34.5	58.31	0.0000
Diet	C	2.40	0.05	34.5	52.63	0.0000
Diet	D	2.55	0.05	34.5	56.12	0.0000

Diet	_Diet	Estimate	StdErr	df	tValue	Adjp
A	B	−0.43	0.06	27.1	−6.85	0.0000
A	C	−0.17	0.06	27.1	−2.74	0.0498
A	D	−0.33	0.06	27.0	−5.26	0.0002
B	C	0.26	0.06	27.0	4.11	0.0021
B	D	0.10	0.06	27.0	1.59	0.3998
C	D	−0.16	0.06	27.0	−2.52	0.0814

estimated standard errors in Table 18.7 are to enable inferences to the population of Hereford heifers that have a specified weight. For example, if one has a population of Hereford heifers that all have initial weights of 600 lb, then the estimated standard errors of the adjusted means can be used to infer as to a diet's performance. The pairwise comparisons provide the same conclusions as obtained from the analysis of variance in Table 18.5.

Special Applications of Analysis of Covariance

TABLE 18.8
PROC MIXED Code and Results for the Unequal Diet Slope Analysis of the Average Daily Gain Data

```
proc mixed cl covtest data=ex_18_2;
class block Diet;
model adg = Diet iwt iwt*Diet/ddfm=kr;
random block;
```

CovParm	Estimate	StdErr	ZValue	ProbZ
block	0.0026	0.0033	0.77	0.2195
Residual	0.0149	0.0043	3.46	0.0003

Effect	NumDF	DenDF	FValue	ProbF
Diet	3	24.3	2.92	0.0541
iwt	1	8.6	213.69	0.0000
iwt*Diet	3	24.3	3.74	0.0244

The unequal slopes model is

$$ADG_{ij} = \mu + \rho_i + \beta_i \ IWT_{ij} + b_j + a_{ij}, \quad i = 1, 2, 3, 4, \text{ and } j = 0, 1, 2, \ldots, 9 \quad (18.3)$$

$$\text{where } b_j \sim \text{iid N}(0, \sigma_b^2) \text{ and } a_{ij} \sim \text{iid N}(0, \sigma_a^2)$$

The β_i are the slopes of the regression lines. The results of using PROC MIXED to fit Model 18.3 are in Table 18.8. The block variance component is increased a little (0.0009 to 0.0026) and the animal variance component has decreased a little (0.0198 to 0.0149) as compared to the common slope analysis in Table 18.6. The significance level corresponding to iwt*diet is 0.0244, indicating that there is sufficient evidence to conclude that the slopes are different; thus the unequal slopes model fits the data better than the common slope model. The significance level corresponding to diet indicates that there is a significance difference among the intercepts (initial weight = 0, a meaningless place to compare the models). The adjusted means evaluated at iwt = 600, 643.54 (mean), and 700 lb are in Table 18.9. The estimated standard errors of the adjusted means are 0.04 at the mean, 0.07 to 0.08 at 600, and 0.08 to 0.10 at 700 lb. The lower part of Table 18.9 contains the pairwise comparisons between the diet means evaluated at the three values of initial weight. The estimated standard errors of the differences are 0.10 to 0.11 at 600, 0.05 at the mean, and 0.12 to 0.13 at 700 lb. Using the simulate method to provide adjustments for multiple comparisons within a value of initial weight, none of the diets have significantly different means when evaluated at iwt = 600 lb, diets B and D do not have significantly different means when evaluated at iwt = 643.53 lb, and the means of diets B, C, and D are not significantly different and the means of diets A and C are not significantly different at iwt = 700 lb. These conclusions are quite different than those reached for the analysis without the covariate (see Table 18.5) and for the analysis of covariance with the common slope (see Table 18.7). Taking into account the covariate

TABLE 18.9
Adjusted Means for the Diets Evaluated at iwt Values of 600, 643.54, and 700 lbs With Pairwise Comparions within Each Value of iwt

```
lsmeans Diet/diff at iwt=600 adjust=simulate;
lsmeans Diet/diff at means adjust=simulate;
lsmeans Diet/diff at iwt=700 adjust=simulate;
```

Effect	Diet	iwt	Estimate	StdErr	df	tValue	Probt
Diet	A	600	1.76	0.07	30.3	23.92	0.0000
Diet	B	600	1.94	0.07	30.3	26.56	0.0000
Diet	C	600	1.86	0.08	30.3	22.17	0.0000
Diet	D	600	1.94	0.08	30.3	23.72	0.0000
Diet	A	643.53	2.22	0.04	30.1	52.95	0.0000
Diet	B	643.53	2.65	0.04	30.1	63.29	0.0000
Diet	C	643.53	2.40	0.04	30.1	57.23	0.0000
Diet	D	643.53	2.55	0.04	30.1	61.02	0.0000
Diet	A	700	2.81	0.09	30.4	30.04	0.0000
Diet	B	700	3.57	0.08	30.4	42.04	0.0000
Diet	C	700	3.09	0.10	30.3	31.01	0.0000
Diet	D	700	3.36	0.10	30.4	32.67	0.0000

Diet	_Diet	iwt	Estimate	StdErr	df	tValue	Adjp
A	B	600	−0.18	0.10	24.1	−1.85	0.2692
A	C	600	−0.10	0.10	24.3	−0.94	0.7862
A	D	600	−0.17	0.10	24.3	−1.69	0.3467
B	C	600	0.08	0.10	24.3	0.79	0.8582
B	D	600	0.01	0.10	24.3	0.06	1.0000
C	D	600	−0.08	0.11	24.0	−0.70	0.8966
A	B	643.53	−0.43	0.05	24.0	−7.90	0.0000
A	C	643.53	−0.18	0.05	24.0	−3.26	0.0140
A	D	643.53	−0.34	0.05	24.0	−6.15	0.0000
B	C	643.53	0.25	0.05	24.0	4.65	0.0002
B	D	643.53	0.10	0.05	24.0	1.76	0.3202
C	D	643.53	−0.16	0.05	24.0	−2.89	0.0351
A	B	700	−0.76	0.12	24.2	−6.54	0.0000
A	C	700	−0.28	0.13	24.2	−2.25	0.1412
A	D	700	−0.55	0.13	24.2	−4.28	0.0015
B	C	700	0.48	0.12	24.4	3.95	0.0028
B	D	700	0.21	0.12	24.6	1.73	0.3336
C	D	700	−0.27	0.13	24.0	−2.01	0.2139

and using the unequal slopes model when the values of the covariate were used to form the blocks can still provide important information about the process that can be missed when just doing an analysis of variance because it was thought that when

TABLE 18.10
PROC MIXED Code and Results for Testing the Equality of the ADG Models at Initial Weight of 600, 643.54, and 700 lb

```
proc mixed cl covtest data=ex_18_2;
class block Diet;
model adg = Diet iwt600*Diet/ddfm=kr;
random block;
```

CovParm	Estimate	StdErr	ZValue	ProbZ
block	0.0026	0.0033	0.77	0.2195
Residual	0.0149	0.0043	3.46	0.0003

Effect	NumDF	DenDF	FValue	ProbF
Diet	3	24.2	1.44	0.2565
iwt600*Diet	4	23.0	57.28	0.0000

```
proc mixed cl covtest data=ex_18_2;
class block Diet;
model adg = Diet iwtmn*Diet/ddfm=kr;
random block;
```

Effect	NumDF	DenDF	FValue	ProbF
Diet	3	24.0	23.98	0.0000
iwtmn*Diet	4	23.0	57.28	0.0000

```
proc mixed cl covtest data=ex_18_2;
class block Diet;
model adg = Diet iwt700*Diet/ddfm=kr;
random block;
```

Effect	NumDF	DenDF	FValue	ProbF
Diet	3	24.3	15.59	0.0000
iwt700*Diet	4	23.0	57.28	0.0000

the blocking took place, the covariate was no longer needed in the analysis. Table 18.10 contains the PROC MIXED code to fit models where the covariate was translocated so the F statistic corresponding to diet provided a test of the equallity of the models (new intercepts) within the range of initial weight values. The variables iwt600 = iwt − 600, iwtmn = iwt − 643.54, and iwt 700 = iwt − 700, so the terms for diet in Table 18.10 provide tests of equality of models at the three values of initial weight. The significance level for comparing the models at iwt = 600 is 0.2195 (diet), which indicates the models are not providing significantly different responses at iwt = 600. On the other hand, the significance levels for F tests for diet at iwt = 643.54 and 700 are 0.0000, which indicate the regression models are providing significantly different responses. The pairwise comparisons in Table 18.9 provide the same type of information, except the pairwise comparisons provide the details as to which regression lines are providing similar responses and which are not. Figure 18.1 is a plot of the estimated regression lines for each of the diets with the

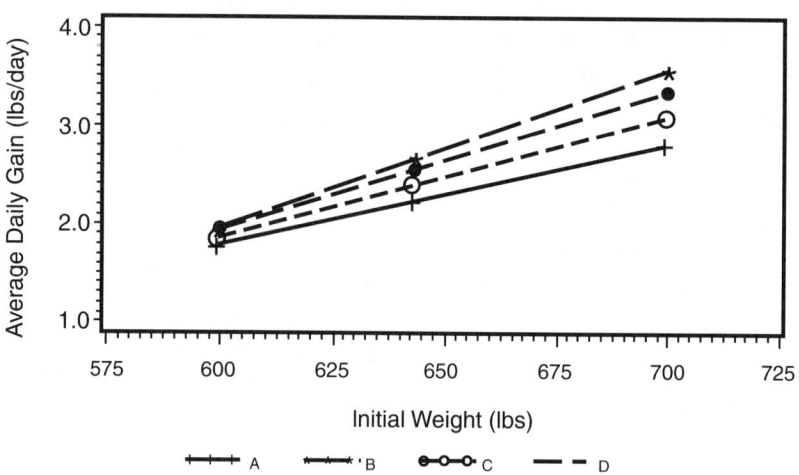

FIGURE 18.1 Plot of the average daily gain diet regression models against the initial weights.

symbols occurring at values of iwt where the adjusted means in Table 18.9 were computed.

The final step in the analysis of this data set is to assume that the animals were assigned to the diets completely at random; thus the design structure under this assumption is a completely randomized design. Under no circumstances would one ignore the blocking factor in the analysis after blocking has occurred because those degrees of freedom must be forfeited when the randomization has been restricted. So, this is just for comparison purposes. Table 18.11 contains the PROC MIXED code and analysis for this completely randomized block design. The estimate of the

TABLE 18.11
PROC MIXED Code to Fit Model with Completely Randomized Design Structure to Analyze the Average Daily Gain Data

```
proc mixed cl covtest data=ex_18_2;
class block Diet;
model adg = Diet iwt iwt*Diet/ddfm=kr;
```

CovParm	Estimate	StdErr	ZValue	ProbZ
Residual	0.0175	0.0044	4.00	0.0000

Effect	NumDF	DenDF	FValue	ProbF
Diet	3	32	2.40	0.0864
iwt	1	32	311.97	0.0000
iwt*Diet	3	32	3.08	0.0413

Special Applications of Analysis of Covariance   543

animal variance component is 0.0175, a value that is slightly larger than the 0.0149 in Table 18.8. The F test corresponding to iwt*diet has a significance level of 0.0413, which indicates that there is enough evidence to conclude that the slopes are not all equal. The adjusted means and pairwise comparisons of adjusted means evaluated at iwt = 600, 643.54, and 700 lb are in Table 18.12. The estimated standard errors of the adjusted means are the same (to two decimal places) as those in Table 18.9. The estimated standard errors for the pairwise differences are a little larger than for the model with blocks in Table 18.9, so the adjusted significance levels are a little larger in Table 18.12. However, the significance levels did not change enough to alter the conclusions drawn when using blocks in the model. This demonstration shows that if one would just use a completely randomized design structure for these types of experiments instead of using initial weight as a factor to form blocks, then the results from the analysis of covariance of the CRD most likely would be similar. In these cases, it is the recommendation to use a CRD design structure and do an analysis of covariance, rather than to use the covariate to form blocks and then ignore the covariate while doing the analysis.

18.3 TREATMENTS HAVE DIFFERENT RANGES OF THE COVARIATE

One of the assumptions for using the analysis of covariance is that the covariate should not be influenced by the application of the treatments, but it is not always possible to satisfy that assumption. This is particularly true when the experimental units are from different populations; then the covariates that are measured on one population may have a different mean and possibly a different variance than another population. Using experimental units of different sexes, different ages, different manufacturers, different breeds, etc. can result in covariates that have different distributions. When this occurs, the analysis can still be accomplished, but first some adjustment needs to be carried out (Urquhart, 1982). An example is used to demonstrate the adjustment process as well as the need for the adjustment process.

A dairy scientist carried out a study of how the addition of a supplemental drug to the diet of a dairy cow would effect milk production. After a calf was born, the cow was put on the control diet for the first 75 days. The mean number of pounds of milk produced per day during days 60 through 74 was to be used as a possible covariate. On day 75, half of the cows were assigned to a control diet and half were assigned a diet with a supplemental drug. Milk production was then measured during days 161 through 175 and expressed as the mean number of pounds of milk produced per day. The cows selected for the study were from two groups called heifers and cows. The heifer group consists of those cows that have had their first calf and thus their first time for milk production during the current study (called parity 1). The cow group is the type of cow that has had more than one calf or this is not their first time for milk production for the current milk production period (called parity 2). Cows generally produce much more milk than heifers. The data in Table 18.13 consist of milk production and pre-milk (lb/day) from cows and heifers from the control and supplemental drug diets. The cows and heifers were randomly assigned

TABLE 18.12
Adjusted Means and Pairwise Comparisons from Using the Completely Randomized Design Structure to Analyze the Average Daily Gain Data

```
lsmeans Diet/diff at iwt=600 adjust=simulate;
lsmeans Diet/diff at means adjust=simulate;
lsmeans Diet/diff at iwt=700 adjust=simulate;
```

Effect	Diet	iwt	Estimate	StdErr	df	tValue	Probt
Diet	A	600	1.76	0.07	32	23.92	0.0000
Diet	B	600	1.94	0.07	32	26.63	0.0000
Diet	C	600	1.86	0.08	32	22.17	0.0000
Diet	D	600	1.93	0.08	32	23.71	0.0000
Diet	A	643.53	2.22	0.04	32	52.95	0.0000
Diet	B	643.53	2.65	0.04	32	63.29	0.0000
Diet	C	643.53	2.40	0.04	32	57.23	0.0000
Diet	D	643.53	2.55	0.04	32	61.02	0.0000
Diet	A	700	2.81	0.09	32	30.15	0.0000
Diet	B	700	3.57	0.08	32	42.10	0.0000
Diet	C	700	3.10	0.10	32	31.12	0.0000
Diet	D	700	3.36	0.10	32	32.79	0.0000

Diet	_Diet	iwt	Estimate	StdErr	df	tValue	Adjp
A	B	600	−0.18	0.10	32	−1.76	0.3128
A	C	600	−0.10	0.11	32	−0.86	0.8287
A	D	600	−0.17	0.11	32	−1.55	0.4220
B	C	600	0.09	0.11	32	0.78	0.8619
B	D	600	0.01	0.11	32	0.11	0.9995
C	D	600	−0.07	0.12	32	−0.64	0.9177
A	B	643.53	−0.43	0.06	32	−7.30	0.0000
A	C	643.53	−0.18	0.06	32	−3.01	0.0229
A	D	643.53	−0.34	0.06	32	−5.67	0.0000
B	C	643.53	0.25	0.06	32	4.29	0.0006
B	D	643.53	0.10	0.06	32	1.63	0.3811
C	D	643.53	−0.16	0.06	32	−2.67	0.0535
A	B	700	−0.76	0.13	32	−6.00	0.0000
A	C	700	−0.28	0.14	32	−2.09	0.1802
A	D	700	−0.55	0.14	32	−3.97	0.0017
B	C	700	0.47	0.13	32	3.60	0.0044
B	D	700	0.21	0.13	32	1.54	0.4250
C	D	700	−0.27	0.14	32	−1.86	0.2665

to the two diets; thus the design of this experiment is a two-way treatment structure in a completely randomized design structure. A plot of the data for each of the treatment combinations reveals that there is a linear relationship between the milk

Special Applications of Analysis of Covariance

TABLE 18.13
Milk Production Data for Cows and Heifers Either on a Control Diet or a Diet Supplemented with a Drug Where Pre-Milk (lb/day) is the Possible Covariate and Milk (lb/day) is the Response Variable

	Cow				Heifer			
	Control		Drug		Control		Drug	
Animal	Milk	Pre_milk	Milk	Pre_milk	Milk	Pre_milk	Milk	Pre_milk
1	85.0	82.5	99.9	88.2	71.3	62.8	75.6	60.8
2	82.2	88.8	87.2	84.7	69.9	58.4	74.0	61.1
3	92.3	95.3	104.3	95.1	64.8	55.7	76.8	61.6
4	92.3	89.4	89.4	86.8	78.6	59.7	72.1	56.0
5	90.7	93.0	115.2	102.4	69.9	59.0	77.9	59.7
6	75.9	89.2	91.0	84.0	74.3	58.0	68.5	53.6
7	85.9	90.2	90.8	81.5	75.0	62.9	86.9	65.5
8	91.0	88.9	114.6	96.7	60.7	51.6	84.4	65.1
9	100.1	93.5	97.1	82.8	81.2	66.2	80.8	60.2
10			101.9	95.8	67.4	58.8		
11			98.5	89.7	67.4	55.4		
12			118.2	100.8	79.1	65.0		

production and the pre-milk data. An analysis of covariance model (see Chapter 5) for this data is

$$M_{ijk} = \mu + \delta_i + \tau_j + (\delta\tau)_{ij} + \beta_{ij} P_{ijk} + \varepsilon_{ijk}, \quad i=1, 2, \quad j=1, 2, \quad k=1, 2, \ldots, n_{ij} \quad (18.4)$$

M_{ijk} is the milk production during the test period, P_{ijk} is the milk production prior to the start of the trial (pre-milk), δ_i, τ_j, and $(\delta\tau)_{ij}$ are the fixed effects parameters for diet, type of cow, and their interaction, β_{ij} is the slope for diet i and cow type j, and ε_{ijk} is the random animal effect. During the analysis of covariance, it was discovered that the variances of the treatment combinations were different where the difference was due to the levels of type of cow (heifer or cow). Using AIC as a guide, the covariance model with different variances for each level of type had a smaller AIC than the other possibilities, equal variances, unequal variances for each diet, and unequal variances for each combination of diet and type of cow (analysis not shown). A model with unequal variances and unequal slopes expressed as factorial effects was fit to the data and the stepwise deletion process was used to simplify the model (analysis not shown). The resulting model has equal slopes; thus the model:

$$M_{ijk} = \mu + \delta_i + \tau_j + (\delta\tau)_{ij} + \beta P_{ijk} + \varepsilon_{ijk}, \quad i=1, 2, \quad j=1, 2, \quad k=1, 2, \ldots, n_{ij} \quad (18.5)$$

where $\varepsilon_{ijk} \sim N(0, \sigma_{\varepsilon j}^2)$ was used to compare the combinations of diet and type of cow. The PROC MIXED code in Table 18.14 was used to fit Model 18.5 to the data where the repeated statement is used to specify the unequal variances. The estimate of the

TABLE 18.14
PROC MIXED Code to Fit the Common Slope Model with Unequal Type Variances to the Milk Production Data

```
proc mixed cl covtest data=ex18_3;
class diet type;
model milk=diet type diet*type pre_milk/ddfm=kr;
repeated/group=type;
```

CovParm	Group	Estimate	StdErr	ZValue
Residual	type Cow	27.962	9.114	3.07
Residual	type Heifer	8.711	2.854	3.05

Effect	NumDF	DenDF	FValue	ProbF
diet	1	29.4	36.04	0.0000
type	1	36.5	20.42	0.0001
diet*type	1	29.3	6.68	0.0150
pre_milk	1	35.7	92.07	0.0000

TABLE 18.15
Adjusted Milk Yield (lb/day) Means for the Diet by Type Combinations Evaluated at Three Pre-Milk Values

```
lsmeans diet*type/diff at means;
lsmeans diet*type/diff at pre_milk=105;
lsmeans diet*type/diff at pre_milk=50;
```

Diet	Type	Pre Milk = 50 (lb/day)		Pre Milk = 75.15 (lb/day)		Pre Milk = 105 (lb/day)	
		Estimate	StdErr	Estimate	StdErr	Estimate	StdErr
Control	Cow	36.30	5.71	68.98	2.68	107.75	2.68
Control	Heifer	59.35	1.54	92.02	2.29	130.79	6.22
Drug	Cow	47.80	5.72	80.47	2.60	119.24	2.46
Drug	Heifer	63.94	1.72	96.61	2.23	135.38	6.12

cow and heifer variances are 27.962 and 8.711, respectively. There is a significant diet by type interaction; thus the interaction means need to be compared. The adjusted milk production means evaluated at pre-milk values of 50, 75.15 (mean), and 105 lb/day are in Table 18.15. The regression lines are parallel; thus a comparison among the means can occur at any of the pre-milk values. Table 18.16 contains some pairwise comparison. The top two comparisons compare the control diet to the drug-supplemented diet for each cow type. In both cases, the drug supplemented diet produces more milk than the control diet (significance levels of 0.0001 and 0.0024). The second two comparisons compare the two types of cows within each of the two diets. In both cases, heifers produce significantly more milk than cows. However, this is not a reasonable conclusion since it is known that cows produce more milk than heifers. A graph of the data and estimated regression models for each diet are in Figures 18.2 and 18.3. The problem is that the range of pre-milk values for the

TABLE 18.16
Pairwise Comparisons of the Levels of Diet within Each Type and of the Levels of Type within Each Level of Diet from a Model with Pre-Milk as the Covariate

Diet	Type	_diet	_type	Estimate	StdErr	df	tValue	Probt
Control	Cow	Drug	Cow	−11.49	2.33	19	−4.93	0.0001
Control	Heifer	Drug	Heifer	−4.59	1.31	19	−3.51	0.0024
Control	Cow	Control	Heifer	−23.04	4.59	37	−5.03	0.0000
Drug	Cow	Drug	Heifer	−16.14	4.49	37	−3.60	0.0009

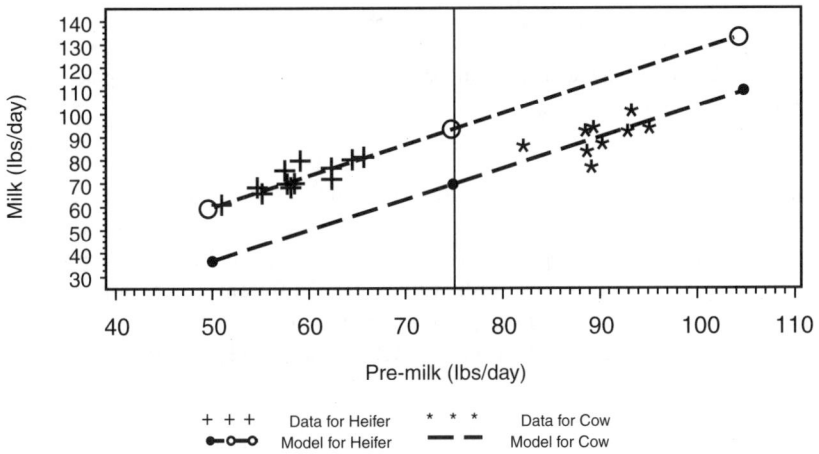

FIGURE 18.2 Plot of cow and heifer models and data for control diet using pre-milk as the covariate.

heifers and cows do not overlap and the regression model for heifers is above the regression model for cows. Thus, anytime a comparison is made between cows and heifers at a common value of pre-milk, the predicted value from the heifer model will be larger than the predicted value from the cow model. Since the values of the covariate do not overlap, it is not reasonable to compare cows and heifers at the same value of pre-milk. For example, the vertical lines on Figures 18.2 and 18.3 are at the mean value of pre-milk (75.15), a place where most adjusted means would be computed. However, for this data set no heifer has a pre-milk value as large as 75.15 and no cow has a pre-milk as small as 75.15. If the two types of cows are compared at the same value of pre-milk, the selected value will not be a reasonable value for one or both. The solution is to adjust the value of the covariate (Urquhart, 1982) and make comparisons at values of the covariate that are within the range of

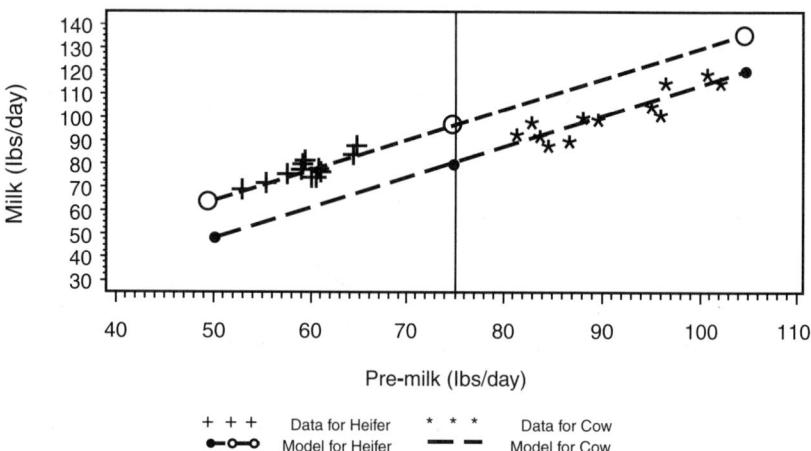

FIGURE 18.3 Plot of cow and heifer models and data for diet with drug supplement using pre-milk as the covariate.

each of the data sets (after adjustment). The suggested process is to subtract the subclass mean of the covariate from each value of the covariate within the subclass. This process sets the intercepts of the models to be evaluated at the subclass means. For this example, since the pre-milk values have different means for each of the levels of cow type, the cow type pre-milk means need to be subtracted from the respective cow type pre-milk data. The model with adjusted pre-milk as the covariate is

$$M_{ijk} = \mu + \delta_i + \tau_j + (\delta\tau)_{ij} + \beta\left(P_{ijk} - \bar{P}_{.k.}\right) + \varepsilon_{ijk}, \qquad (18.6)$$

$$i = 1, 2, \quad j = 1, 2, \quad k = 1, 2, \ldots, n_{ij}$$

where $\varepsilon_{ijk} \sim N(0, \sigma_{\varepsilon j}^2)$ and where $\mu + \delta_i + \tau_j + (\delta\tau)_{ij}$ are intercepts when the adjusted covariate is zero or when each cow type's model is evaluated at it's $\bar{P}_{.k.}$. Since there are unequal numbers of observations per treatment combination, there are at least two methods that can be used to carry out the adjustment. The first is as above: just compute the mean pre-milk of all of the heifers and the mean pre-milk of all of the cows. However, since there is another factor in the treatment structure, one could compute the mean pre-milk for the heifers for the control and for the drug supplemental diets and take the means of these two means (like least squares means Milliken and Johnson, 1992). This second approach was carried out here. One way to obtain these means to be used for the adjustment is to do an analysis of variance on the pre-milk values and then compute the least squares means, means that are

TABLE 18.17
PROC MIXED Code for the Analysis of the Pre-Milk Data to Provide Adjusted Means for the Two Cow Types

```
proc mixed cl covtest data=ex18_3;
class treat type;
model pre_milk=treat|type;
lsmeans type/diff;
```

CovParm	Estimate	StdErr	ZValue	ProbZ		
Residual	26.097	5.987	4.36	0.0000		

Effect	NumDF	DenDF	FValue	ProbF		
treat	1	38	0.24	0.6269		
type	1	38	365.92	0.0000		
treat*type	1	38	0.01	0.9200		

type	Estimate	StdErr	df	tValue	Probt
Cow	90.40	1.13	38	80.26	0.0000
Heifer	59.93	1.13	38	53.21	0.0000

computed as if all cells have the same number of observations. The PROC MIXED code in Table 18.17 was used to provide the means used for the adjustment process by fitting the model

$$P_{ijk} = \mu + \delta_i + \tau_j + (\delta\tau)_{ij} + \varepsilon_{ijk} \tag{18.7}$$

The results in Table 18.17 indicate there are significant differences between the cow type means (significance level is 0.0000). Using the estimates of the parameters from the model, adjusted or least squares means were computed. The least squares means are 90.40 and 59.93 lb/day for the cows and heifers, respectively. These least squares means were computed by obtaining the mean for each cow type for each of the diets and then taking the means of the cow types over the diets, i.e.,

$$\bar{P}_1 = \frac{\bar{P}_{11\cdot} + \bar{P}_{21\cdot}}{2} \text{ and } \bar{P}_2 = \frac{\bar{P}_{12\cdot} + \bar{P}_{12\cdot}}{2}$$

Next adjust the covariates by subtracting the respective least squares mean from either the cow or heifer data as $AdjP_{ijk} = P_{ijk} - \bar{P}_j$. The PROC MIXED code in Table 18.18 fits the common slope model

$$M_{ijk} = \mu + \delta_i + \tau_j + (\delta\tau)_{ij} + \beta\ AdjP_{ijk} + \varepsilon_{ijk},$$
$$i = 1, 2,\ \ j = 1, 2,\ \ k = 1, 2, \ldots, n_{ij} \tag{18.8}$$

TABLE 18.18
PROC MIXED Code to Fit the Common Slope Model with the Adjusted Pre-Milk Values as the Covariate

```
proc mixed cl covtest data=cex18_3;
class diet type;
model milk=diet type diet*type adj1_pre_milk/ddfm=kr;
repeated/group=type;
```

CovParm	Group	Estimate	StdErr	ZValue	ProbZ
Residual	type Cow	27.962	9.114	3.07	0.0011
Residual	type Heifer	8.711	2.854	3.05	0.0011

Effect	NumDF	DenDF	FValue	ProbF
diet	1	29.4	36.04	0.0000
type	1	29.3	224.10	0.0000
diet*type	1	29.3	6.68	0.0150
adj1_pre_milk	1	35.7	92.07	0.0000

TABLE 18.19
Adjusted Milk Production Means for the Diet by Type Combinations Evaluated at Four Values of Adjusted Pre-Milk

```
lsmeans diet|type/diff at means;
lsmeans diet|type/diff at adj1_pre_milk=0;
lsmeans diet|type/diff at adj1_pre_milk=-15;
lsmeans diet|type/diff at adj1_pre_milk=15;
```

Diet	Type	Adjusted Pre-milk = −0.11 lb/day		Adjusted Pre-milk = 0 lb/day		Adjusted Pre-milk = −15 lb/day		Adjusted Pre-milk = 15 lb/day	
		Estimate	StdErr	Estimate	StdErr	Estimate	StdErr	Estimate	StdErr
Control	Cow	88.77	1.76	88.78	1.76	69.30	2.66	108.26	2.72
Control	Heifer	72.23	0.85	72.24	0.85	52.76	2.14	91.73	2.26
Drug	Cow	100.26	1.53	100.27	1.53	80.79	2.57	119.76	2.51
Drug	Heifer	76.82	0.99	76.83	0.99	57.35	2.31	96.32	2.20

The estimates of the variance components in Table 18.18 are identical to those in Table 18.14 (the results from the model with pre-milk as the covariate). The tests for the intercepts are comparisons of the regression models evaluated at $\bar{P}_1$ for the cow models and the regression models evaluated at $\bar{P}_2$ for the heifer models. The adjusted means are computed using the LSMEANS statements in Table 18.19 evaluated at the mean of the pre-milk values, intercept or center and at ±15 lb/day from the center. Importantly, the adjusted means for the cows are larger than the adjusted means for the heifers (as was expected). The graphs of the regression models with the data points for heifers and cows evaluated for the control and drug are in Figures 18.4 and 18.5. Table 18.20 contains pairwise comparisons of the diets within each cow

Special Applications of Analysis of Covariance

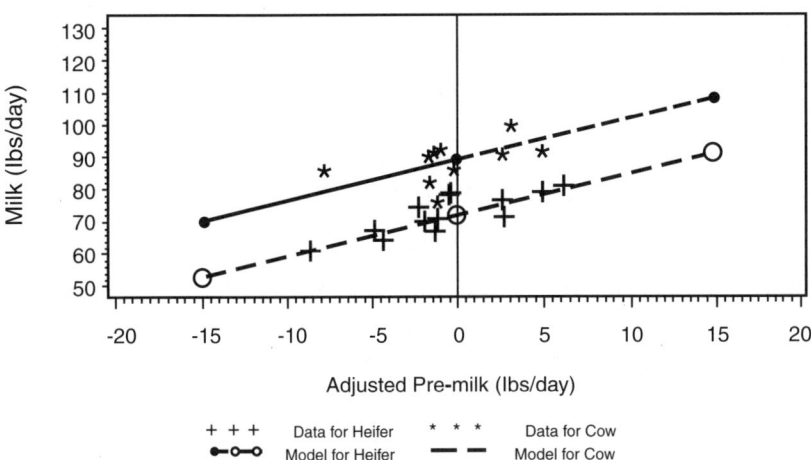

FIGURE 18.4 Plot of cow and heifer models and data for control diet using adjusted pre-milk as the covariate.

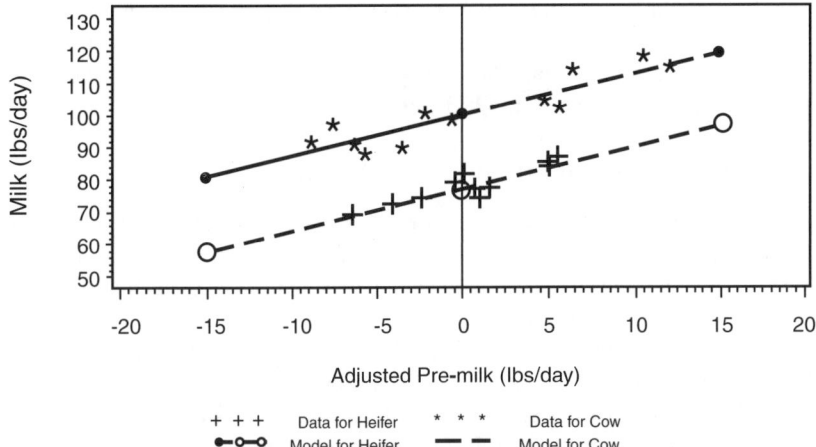

FIGURE 18.5 Plot of cow and heifer models and data for diet with drug supplement using adjusted pre-milk as the covariate.

type and the types of cows within each diet, evaluated at 0 adjusted pre-milk lb/day. The drug-supplemented diet increased milk yield 11.49 lb/day for cows and 4.59 lb/day for heifers.

TABLE 18.20
Pairwise Comparisons of the Levels of Diet within Each Type and of the Levels of Type within Each Level of Diet from a Model with Adjusted Pre-Milk as the Covariate

Diet	Type	_diet	_type	Estimate	StdErr	df	tValue	Probt
Control	Cow	Drug	Cow	−11.49	2.33	19	−4.93	0.0001
Control	Heifer	Drug	Heifer	−4.59	1.31	19	−3.51	0.0024
Control	Cow	Control	Heifer	16.54	1.96	27	8.45	0.0000
Drug	Cow	Drug	Heifer	23.44	1.82	32	12.91	0.0000

This experiment demonstrates the importance of correcting for unequal means of the covariate so that the conclusions are meaningful. This example also involves unequal variances and the analysis incorporates that fact into the modeling using PROC MIXED.

18.4 NONPARAMETRIC ANALYSIS OF COVARIANCE

The analyses discussed so far in this and previous chapters relied on the assumption that the distributions of the random effects associated with a model are approximately normally distributed. Conover and Imam (1982) proposed using the rank transform in the analysis of covariance where the response variable values are replaced with their ranks and the values of each of the covariates are also replaced with their ranks. The analysis of covariance is carried out using the ranks as the data. Using the ranks is the only thing that has changed; the analysis of covariance strategy is then followed to provide the analysis. The rank transformation has been used in completely randomized design structures; thus the following two examples involve a one-way and a two-way treatment structure in CR design structures.

18.4.1 Heart Rate Data from Exercise Programs

The data in Table 18.21 are from Section 3.3 where 24 persons were randomly assigned to 1 of 3 exercise programs. Before starting the study, the initial heart rate of each person was determined. After a person was on the assigned exercise program for the stated number of weeks, the heart rate was determined again. Table 18.21 contains heart rates (hr), ranks of the heart rates (Rank hr), initial heart rates (ihr), and ranks of the initial heart rates (Rank_ihr). The PROC MIXED code in Table 18.22 fits the unequal slopes model to the rank transformed heart rates and initial heart rates. The analysis indicates that the exercise slopes are unequal — the same conclusion made for the parametric analysis carried out in Section 3.3. Since the slopes are unequal, the regression models for the exercise programs need to be evaluated at three or more values of the ranks of the initial heart rates. In this case, using ranks of 1 (low), 12.5 (middle), and 24 (high) for the rank of the initial heart rate provides comparisons of the means of the ranks of the heart rate data. Table 18.23 contains the predicted values or adjusted means from the three models and significance

TABLE 18.21
Heart Rate (hr) and Initial Heart Rate (ihr) with the Rank Transform Values (Rank hr and Rank ihr)

Exercise Program 1				Exercise Program 2				Exercise Program 3			
hr	Rank hr	ihr	Rank ihr	hr	Rank hr	ihr	Rank ihr	hr	Rank hr	ihr	Rank ihr
118	1.0	56	1.5	148	5.0	60	5.0	153	8.0	56	1.5
138	2.0	59	4.0	159	13.0	62	7.5	150	6.0	58	3.0
142	3.0	62	7.5	162	16.0	65	10.0	158	12.0	61	6.0
147	4.0	68	12.0	157	11.0	66	11.0	152	7.0	64	9.0
160	14.5	71	13.0	169	21.0	73	15.0	160	14.5	72	14.0
166	20.0	76	18.0	164	17.5	75	16.5	154	9.0	75	16.5
165	19.0	83	20.0	179	24.0	84	21.0	155	10.0	82	19.0
171	22.0	87	23.0	177	23.0	88	24.0	164	17.5	86	22.0

TABLE 18.22
PROC GLM Code to Fit the Unequal Slopes Model to the Expercise Program Data Using the Rank Transformed Data

```
proc glm data=heart;
class Epro;
model rank_hr = Epro rank_ihr rank_ihr*Epro;
```

Source	df	SS	MS	FValue	ProbF
Model	5	959.445	191.889	18.22	0.0000
Error	18	189.555	10.531		
Corrected Total	23	1149.000			

Source	df	SS(III)	MS	FValue	ProbF
EPRO	2	102.519	51.260	4.87	0.0204
rank_ihr	1	658.644	658.644	62.54	0.0000
rank_ihr*EPRO	2	125.891	62.946	5.98	0.0102

levels for the pairwise comparisons. The conclusions are (1) EPRO 1 has significantly lower heart rates than EPROs 2 and 3 for low ranks, (2) EPROs 1 and 3 have significantly lower heart rates than EPRO 2 for middle ranks, and (3) EPRO 3 has significantly lower mean heart rates than EPROs 1 and 2 for high ranks. These are identical to the conclusions made from the parametric analysis in Section 3.3. One drawback from using the rank transform for carrying out the data analysis is that the regression models provided predicted ranks and not predicted values for heart rates. To work with this problem, a regression model was constructed to describe the relationship between the heart rates and the ranks of the heart rates. This relationship is not necessarily linear as displayed in Figure 18.6. A cubic regression model was used to model this relationship and the results are in Table 18.24. Using

TABLE 18.23
Adjusted Means Evaluated at Three Values of Rank_ihr with Significance Levels for Pairwise Comparions of the Expercise Program within a Value

```
lsmeans Epro/pdiff at rank_ihr=1;
lsmeans Epro/pdiff at rank_ihr=24;
lsmeans Epro/pdiff at rank_ihr=12.5;
```

Rank IHR	EPRO	LSMean	Row	_1	_2	_3
1	1	−1.81	1		0.0590	0.0077
	2	5.04	2	0.0590		0.5565
	3	7.03	3	0.0077	0.5565	
24	1	23.46	1		0.5460	0.0135
	2	25.38	2	0.5460		0.0040
	3	14.73	3	0.0135	0.0040	
12.5	1	10.82	1		0.0155	0.9751
	2	15.21	2	0.0155		0.0171
	3	10.88	3	0.9751	0.0171	

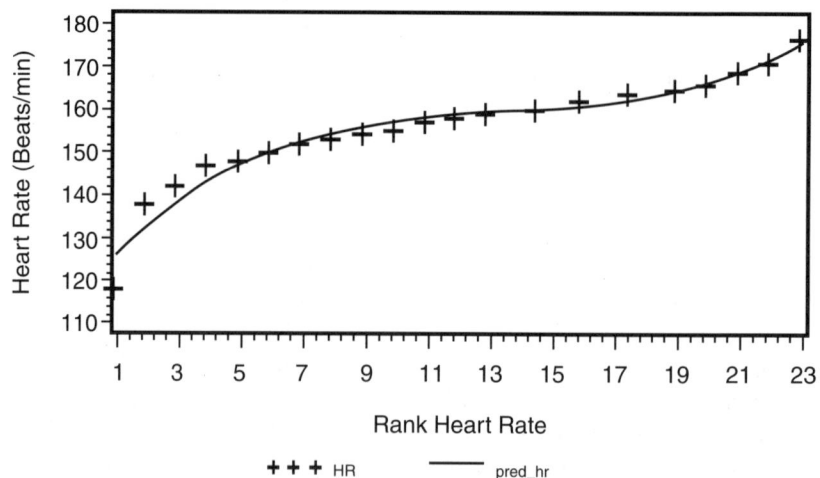

FIGURE 18.6 Plot of model to back transform the rank of heart rates to heart rates.

the regression model, the adjusted rank means were back transformed to heart rate means (new_hr). These back transformed means are quite similar to those in Table 3.22, thus indicating the success of this method. The new adjusted mean for EPRO 1 is lower than that in Table 3.22 which occurs because of having to extrapolate outside of the range of the data set.

TABLE 18.24
Model to Back Transform the Rank_hr Values to Heart Rates with the Back Transformed Adjusted Means

```
proc glm data=heart;
model hr=rank_hr rank_hr*rank_hr
  rank_hr*rank_hr*rank_hr;
```

Rank IHR	EPRO	LSMean	new_hr
1	1	−1.81	102.27
	2	5.04	147.03
	3	7.03	152.66
24	1	23.46	177.84
	2	25.38	188.21
	3	14.73	160.16
12.5	1	10.82	158.00
	2	15.21	160.43
	3	10.88	158.04

Parameter	Estimate
Intercept	118.829411
rank_hr	8.076811
rank_hr*rank_hr	−0.561608
rank_hr*rank_hr*rank_hr	0.013834

18.4.2 Average Daily Gain Data from a Two-Way Treatment Structure

The second example of the rank transform used the data from Section 5.4 which is from a two-way treatment structure. The experiment involves measuring the average daily gain (ADG) of male and female calves fed one of four diets where the diets consist four levels of a drug. Table 18.25 contains the ADG data, the ranks of the ADG data (radg), the birth weight data (birth wt), and the ranks of the birth weight data (rbw). A model with the slopes expressed as two-way factorial effects was used to describe the data and model building was carried out. The stepwise deletion process removed all but rbw, indicating that a common slope model was adequate to describe the relationship between the ranks of the ADG and the ranks of the birth weights (analyses not shown). The model fit to the rank transformed data is

$$\text{Radg}_{ijk} = \mu + \delta_i + \tau_j + (\delta\tau)_{ij} + \beta \text{ Rbw}_{ijk} + \varepsilon_{ijk},$$

(18.9)

$$i = 1, 2, 3, 4, \quad j = 1, 2, \quad k = 1, 2, \ldots, 8$$

TABLE 18.25
Average Daily Gain Data (adg) and Birth Weights (birth_wt) with Their Ranks (radg and rbw)

	Females				Males			
Drug	adg	radg	birth_wt	rbw	adg	radg	birth_wt	rbw
Control_(0mg)	2.58	23.5	69	20.0	2.43	15.0	76	40.5
Control_(0mg)	2.60	26.0	70	24.0	2.36	11.5	69	20.0
Control_(0mg)	2.13	3.0	74	35.0	2.93	51.0	70	24.0
Control_(0mg)	2.00	2.0	66	12.5	2.58	23.5	72	29.5
Control_(0mg)	1.98	1.0	74	35.0	2.27	7.0	60	1.0
Control_(0mg)	2.31	10.0	78	48.5	3.11	56.0	80	55.0
Control_(0mg)	2.30	9.0	78	48.5	2.42	13.5	71	27.0
Control_(0mg)	2.19	5.0	64	7.5	2.66	27.0	77	44.5
Med_(5.0mg)	3.01	54.0	77	44.5	3.14	60.0	69	20.0
Med_(5.0mg)	3.13	58.0	85	64.0	2.91	46.5	69	20.0
Med_(5.0mg)	2.75	34.5	77	44.5	2.53	21.0	71	27.0
Med_(5.0mg)	2.36	11.5	61	2.5	3.09	55.0	68	17.0
Med_(5.0mg)	2.71	30.0	76	40.5	2.86	43.0	63	5.0
Med_(5.0mg)	2.79	37.0	74	35.0	2.88	44.0	61	2.5
Med_(5.0mg)	2.84	39.0	75	37.5	2.75	34.5	64	7.5
Med_(5.0mg)	2.59	25.0	65	9.0	2.91	46.5	73	32.0
Low_(2.5mg)	2.92	49.5	78	48.5	2.52	20.0	71	27.0
Low_(2.5mg)	2.51	19.0	73	32.0	2.54	22.0	81	56.0
Low_(2.5mg)	2.78	36.0	83	59.5	2.95	53.0	66	12.5
Low_(2.5mg)	2.91	46.5	79	52.5	2.46	17.0	69	20.0
Low_(2.5mg)	2.18	4.0	66	12.5	3.13	58.0	70	24.0
Low_(2.5mg)	2.25	6.0	63	5.0	2.72	31.0	76	40.5
Low_(2.5mg)	2.91	46.5	82	57.0	3.41	63.0	79	52.5
Low_(2.5mg)	2.42	13.5	75	37.5	3.43	64.0	78	48.5
High_(7.5mg)	2.94	52.0	67	16.0	2.73	32.5	77	44.5
High_(7.5mg)	2.80	38.0	84	62.5	3.17	61.0	79	52.5
High_(7.5mg)	2.85	41.0	83	59.5	2.92	49.5	76	40.5
High_(7.5mg)	2.44	16.0	73	32.0	2.85	41.0	72	29.5
High_(7.5mg)	2.28	8.0	66	12.5	2.47	18.0	66	12.5
High_(7.5mg)	2.70	28.5	63	5.0	3.28	62.0	84	62.5
High_(7.5mg)	2.70	28.5	66	12.5	3.13	58.0	79	52.5
High_(7.5mg)	2.85	41.0	83	59.5	2.73	32.5	83	59.5

where $Radg_{ijk}$ and Rbw_{ijk} denote the ranks of the average daily gain values and of the birth weight values, and δ_i, τ_j, and $(\delta\tau)_{ij}$ are the fixed effects parameters for diet, sex, and diet by sex interaction, respectively. The PROC GLM code is in Table 18.26. The significance levels indicate there are important drug effects (p = 0.0001), sex effects (p = 0.0004), and rank birth weight effect (p = 0.0001), but there is no interaction between the levels of drug and the levels of sex (p = 0.8382). Table 18.27

TABLE 18.26
PROC GLM Code to Fit the Common Slope Model to the Rank Transform Data

```
Proc glm data=ex18_4_2;
class drug sex;
model radg=drug|sex rbw;
```

Source	df	SS	MS	FValue	ProbF
Model	8	10966.45	1370.81	6.94	0.0000
Error	55	10861.05	197.47		
Corrected Total	63	21827.50			

Source	df	SS(III)	MS	FValue	ProbF
drug	3	4797.88	1599.29	8.10	0.0001
sex	1	2792.71	2792.71	14.14	0.0004
drug*sex	3	167.07	55.69	0.28	0.8382
rbw	1	3489.45	3489.45	17.67	0.0001

TABLE 18.27
Adjusted Means and Pairwise Comparisons for Analysis of the Rank Transformed Data

```
lsmeans drug sex/pdiff;
```

Drug	LSMean	Row	_1	_2	_3	_4
Control_(0 mg)	19.05	1		0.0021	0.0097	0.0000
High_(7.5 mg)	35.40	2	0.0021		0.5637	0.1441
Low_(2.5 mg)	32.52	3	0.0097	0.5637		0.0440
Med_(5.0 mg)	43.03	4	0.0000	0.1441	0.0440	
Female	25.88		0.0004			
Male	39.12					

contains the adjusted means of the rank transformed ADG values for the levels of drug and sex evaluated at the average rank of the birth weights, which is 32.5. The significance levels indicate that the control has a mean rank ADG that is significantly less than the other three treatments and the low dose has a mean rank ADG that is significantly lower than the median dose. Females have a lower mean rank ADG than the males. A cubic polynomial regression model was used to describe the relationship between the ADG values and the rank ADG values, with the PROC GLM code and estimates of the regression coefficients in Table 18.28. Figure 18.7 is a plot of the data and the regression model using the coefficients in Table 18.28. Back transformed mean ADG values for the levels of drug and sex were computed using the regression model and the results (called new-adg) are in Table 18.28. These values are quite similar to the adjusted means computed using the parametric analysis

TABLE 18.28
Back Transformed ADG Values for the Levels of Drug and Sex with Coefficients for the Back Transform Models

Effect	LSMean	new_adg
Control_(0 mg)	19.05	2.53
High_(7.5 mg)	35.40	2.75
Low_(2.5 mg)	32.52	2.72
Med_(5.0 mg)	43.03	2.85
Female	25.88	2.63
Male	39.12	2.79

```
proc glm data=ex18_4_2;
model adg=radg radg*radg radg*radg*radg;
output out=pred p=p;
```

Parameter	Estimate
Intercept	1.97045986
radg	0.04408576
radg*radg	–0.00095782
radg*radg*radg	0.00000944

Plot of Model for ADG from Rank ADG

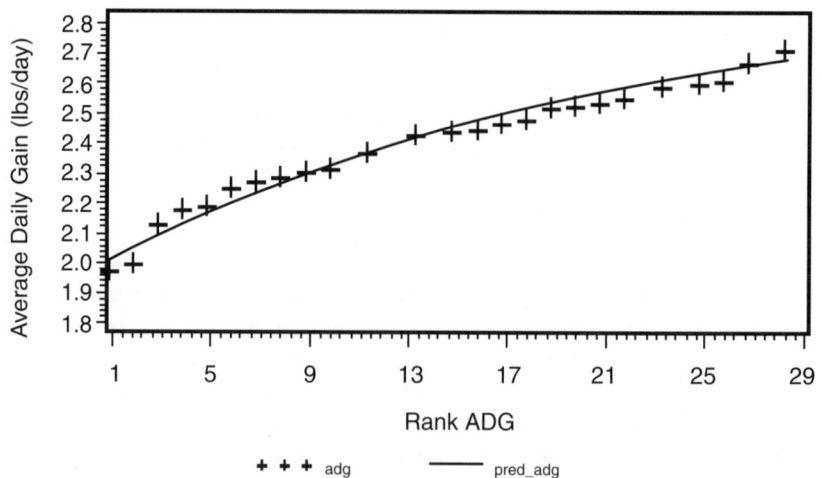

FIGURE 18.7 Plot of model used to back transform from the ranks of the ADG data to the ADG values.

in Tables 5.6 and 5.7. The new adjusted means could possibly be more accurate if a better relationship between the ADG data and the rank ADG was determined, but these results are very close.

Special Applications of Analysis of Covariance

The rank transform is one possible process to provide an analysis of covariance when the data are not normally distributed, and the two examples discussed above indicate that results similar to the parametric analyses were obtained.

18.5 CROSSOVER DESIGN WITH COVARIATES

There are many situations in which analysis of covariance can be applied and one of the more complex designs is the crossover design. The crossover design (Milliken and Johnson, 1992) is a design where the treatments are administered to each experimental unit in a sequence so that each treatment is applied to each experimental unit. The experimental unit's time in the study is split into time periods which become the experimental units for the treatments. More than one sequence of treatment must be used in order for the design to be called a crossover design. A nutritionist designed an experiment to evaluate the effect of two types of diet supplements on a person's cholesterol level. Three treatments were selected: A is the control, B is supplement 1, and C is supplement 2. The data in Table 18.29 are from the six possible sequences of A, B, and C. Thirty six persons were randomly assigned to one of the six sequences so that there were six persons per sequence. The population of persons considered for the study were those with a history of relatively high cholesterol levels; thus before the study started, a base line or initial cholesterol level was determined and is indicated by ichol. A person was given the supplement for the treatment for the first time period (trt1) and used that supplement for 1 month. At the end of the month a cholesterol measurement (chol1) was made. Additionally, each person kept a food diary for the month and the number of grams of fat per day (fat1) was calculated to be used as a possible covariate. After a 2-week break, the person was given the supplement for the treatment in the second time period (trt2), and after a month the cholesterol (chol2) and diet fat content (fat2) were measured. After another 2-week break, the supplement for the third time period (trt3) was provided, and after a month the cholesterol (chol3) and diet fat content (fat3) were measured.

This design is balanced with respect to first order carryover effects in that each treatment follows each other treatment and equal number of times. This enables one to determine if there are any carryover effects in the study as well as estimate the direct effects of the treatments (Milliken and Johnson, 1992). There are two possible covariates in this study: (1) the initial cholesterol level is measured on the person and (2) the fat content of the diet is measured on the time period. An analysis of variance on the initial cholesterol levels showed there were no first period treatment mean differences and a crossover analysis using the AR(1) covariance structure of the diet fat content showed no treatment mean differences; thus it was concluded that the covariates were not affected by the treatments (analyses not shown). This design enables one to provide estimates of the Sequence, Treatment, and Time effects, but not of their interactions (Milliken and Johnson, 1992) and estimates of contrasts of the carryover effects. A model can be constructed by including carryover effects for just two of the treatments as the corresponding parameter estimates are contrasts of the included effect and the not included effect. A model to describe this data set with unequal slopes for the two covariates for the levels of treatment and for the levels of time is

TABLE 18.29
Data for the Cholesterol Study Where trti, choli, and fati are the Treatment, Cholesterol Measurement, and Fat g/day for Time Period i and ichol is the Baseline Cholesterol Measurement

			Time Period 1			Time Period 2			Time Period 3		
seq	Person	ichol	trt1	chol1	fat1	trt2	chol2	fat2	trt3	chol3	fat3
1	1	235	A	221	66	B	201	59	C	191	63
1	2	219	A	228	93	B	201	94	C	193	93
1	3	231	A	246	88	B	220	88	C	212	90
1	4	235	A	259	96	B	231	102	C	222	106
1	5	226	A	251	110	B	216	108	C	212	107
1	6	206	A	207	104	B	174	99	C	170	103
2	1	225	A	248	45	C	226	40	B	230	46
2	2	213	A	213	75	C	189	71	B	187	76
2	3	217	A	219	76	C	194	81	B	197	84
2	4	210	A	206	50	C	180	55	B	184	53
2	5	228	A	219	68	C	192	68	B	193	63
2	6	213	A	243	73	C	217	69	B	215	69
3	1	215	B	181	82	A	197	78	C	173	81
3	2	240	B	229	62	A	238	62	C	214	56
3	3	225	B	227	92	A	249	99	C	214	97
3	4	211	B	198	47	A	206	42	C	187	47
3	5	214	B	201	117	A	226	123	C	202	126
3	6	215	B	220	89	A	232	85	C	203	83
4	1	227	B	242	120	C	240	121	A	263	128
4	2	233	B	208	57	C	198	60	A	217	65
4	3	234	B	213	81	C	201	75	A	215	76
4	4	237	B	226	51	C	209	45	A	227	50
4	5	216	B	196	47	C	183	47	A	199	46
4	6	210	B	199	116	C	199	121	A	220	124
5	1	241	C	232	101	A	249	97	B	228	102
5	2	222	C	209	90	A	224	96	B	195	95
5	3	230	C	212	68	A	225	64	B	207	65
5	4	214	C	188	117	A	210	119	B	175	118
5	5	226	C	198	65	A	212	67	B	192	70
5	6	218	C	190	58	A	205	63	B	184	57
6	1	235	C	217	78	B	217	81	A	234	87
6	2	228	C	221	117	B	213	119	A	233	114
6	3	212	C	201	57	B	204	52	A	218	60
6	4	200	C	197	83	B	192	86	A	211	88
6	5	222	C	230	114	B	219	111	A	238	107
6	6	222	C	189	71	B	188	72	A	200	73

Special Applications of Analysis of Covariance

$$C_{ijkm} = \mu + \delta_i + p_{ij} + \pi_k + \tau_m + \beta_k \, IChol_{ij} + \phi_m IChol_{ij} \quad (18.10)$$

$$+ \theta_k \, Fat_{ijkm} + \gamma_m Fat_{ijkm} + \lambda_A C_{ijkA} + \lambda_B C_{ijkB} + \varepsilon_{ijkm}$$

where $i = i, 2, \ldots, 6$, $j = i, 2, \ldots, 6$, $k = 1, 2, 3$ and $m = A, B, C$ is treatment

$$p_{ij} \sim iid \, N(0, \sigma_p^2) \text{ and } \left(\varepsilon_{ij1m_1}, \varepsilon_{ij2m_2}, \varepsilon_{ij3m_3} \right)' \sim iid \, N(\mathbf{0}, \mathbf{R}),$$

where δ_i, π_k, and τ_m denote the sequence, period, and treatment effects, β_k, ϕ_m, θ_k, and γ_m denote the slopes corresponding to the covariates Ichol and Fat, λ_A and λ_B denote the first order carry over effects of treatments A and B from one period to the next, p_{ij} denotes the random effect of the jth person assigned to the ith sequence, and ε_{ijk} denotes the time interval within a person error. The value of C_{ijkA} is "1" if treatment A was the previous treatment and zero if not (the same definition for C_{ijkB}). The $\mathbf{R}$ matrix is the covariance matrix of the three repeated measurements from each person. Using the strategy described in Chapter 16, several possible forms of $\mathbf{R}$ were investigated before simplifying the covariate part of the model. Using the AIC as an index, the first-order autoregressive covariance structure fit the data better than the other structures attempted [CS, CSH, ARH(1), UN, TOEP, and ANTE(1)] (analyses not shown). The next step in the analysis was to simplify the covariate part of Model 18.10 using the stepwise deletion process (analyses not shown). The final model has a common slope for the initial cholesterol values and unequal treatment slopes for the fat content in the diet, i.e., the model is

$$C_{ijkm} = \mu + \delta_i + p_{ij} + \pi_k + \tau_m + \beta \, IChol_{ij} + \theta \, Fat_{ijkm} + \gamma_m Fat_{ijkm} \quad (18.11)$$

$$+ \lambda_A C_{ijkA} + \lambda_B C_{ijkB} + \varepsilon_{ijkm}$$

where $i = i, 2, \ldots, 6$, $j = i, 2, \ldots, 6$, $k = 1, 2, 3$ and $m = A, B, C$ is treatment

$$p_{ij} \sim iid \, N(0, \sigma_p^2) \text{ and } \left(\varepsilon_{ij1m_1}, \varepsilon_{ij2m_2}, \varepsilon_{ij3m_3} \right)' \sim iid \, N(\mathbf{0}, \mathbf{R}),$$

The PROC MIXED code in Table 18.30 specifies Model 18.11, where AR(1) is the covariance structure of $\mathbf{R}$. The term fat*time was included in the model because when it was deleted, the resulting model was unstable. The significance levels associated with LamA and LamB are 0.5502 and 0.8968, indicating there are no important carryover effects. The significance levels associated with ichol, fat, and fat*treat are 0.0000, 0.0249, and 0.0001, indicating that the model depends linearly on the base line cholesterol level and the slopes for fat are different for each level of treatment. There is a significant treatment effect, indicating that the intercepts are significantly different (at fat = 0). Table 18.31 contains the adjusted treatment means evaluated at base line cholesterol of 222.4 (the mean from the base line values) and at five values of fat in the diet, 45, 65, 85, 105, and 125 g/day. The adjusted means

TABLE 18.30
Proc Mixed Code to Fit the Final Model to the Cholesterol Data with Unequal Treatment Slopes for Fat and AR(1) Covariance Structure

```
proc mixed cl covtest data=crosovr;
class treat time seq person;
model chol= seq treat time ichol fat fat*treat
  fat*time LamA LamB/ddfm=kr solution;
random person(seq);
repeated time/type=ar(1) subject=person(seq);
```

CovParm	Subject	Estimate	StdErr	ZValue
person(seq)		138.202	67.115	2.06
AR(1)	person(seq)	0.852	0.362	2.35
Residual		21.613	52.681	0.41

Effect	NumDF	DenDF	FValue	ProbF
seq	5	29.3	0.96	0.4575
treat	2	53.5	8.99	0.0004
time	2	42.9	0.65	0.5246
ichol	1	29.2	28.53	0.0000
fat	1	85.7	5.21	0.0249
fat*treat	2	51.5	11.24	0.0001
fat*time	2	43.3	0.30	0.7412
LamA	1	56.1	0.36	0.5502
LamB	1	56.7	0.02	0.8968

TABLE 18.31
Adjusted Treatment Means Evaluated at icho = 222.4 and Five Values of Diet Fat Content (g/day)

```
lsmeans treat/diff at fat=45 adjust=simulate;
lsmeans treat/diff at fat=65 adjust=simulate;
lsmeans treat/diff at fat=85 adjust=simulate;
lsmeans treat/diff at fat=105 adjust=simulate;
lsmeans treat/diff at fat=125 adjust=simulate;
```

	Fat = 45		Fat = 65		Fat = 85		Fat = 105		Fat = 125	
treat	Mean	Stde	Mean	Stde	Mean	Stde	Mean	Stde	Mean	Stde
A	210.1	5.5	218.4	3.2	226.7	2.2	235.0	3.8	243.2	6.2
B	199.4	5.3	202.8	3.1	206.2	2.2	209.6	3.9	213.0	6.3
C	190.8	5.4	197.4	3.1	204.0	2.2	210.6	3.8	217.2	6.3

are displayed in Figure 18.8. Table 18.32 contains the pairwise comparisons of the three treatment's mean cholesterol levels within each of the values of fat in the diet. All three treatments are significantly different for fat values of 45 and 65 g/day. Treatment A has significantly larger mean cholesterol levels than treatments B and

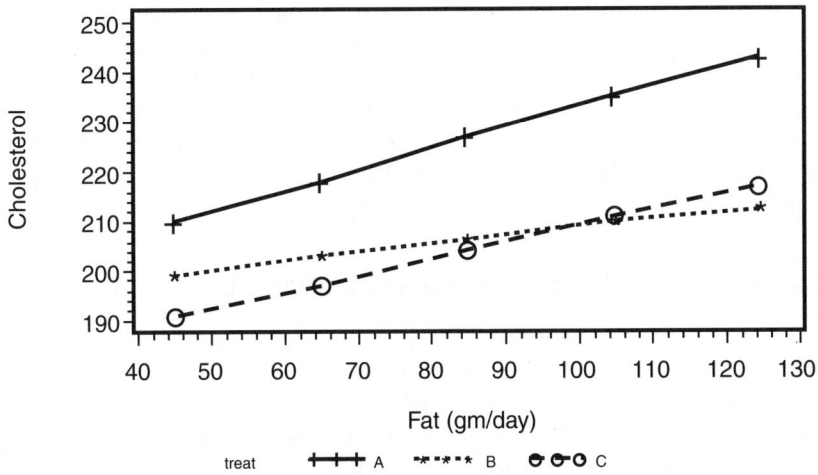

FIGURE 18.8 Plot of the treatment's regression lines vs. fat content of the diet (g/day).

TABLE 18.32
Pairwise Comparisons of the Adjusted Treatment Means within Each Value of Fat Content

Effect	treat	_treat	fat	Estimate	StdErr	Adjp
treat	A	B	45	10.7	2.4	0.0001
treat	A	C	45	19.3	2.2	0.0000
treat	B	C	45	8.6	2.4	0.0024
treat	A	B	65	15.6	1.7	0.0000
treat	A	C	65	21.0	1.6	0.0000
treat	B	C	65	5.4	1.7	0.0066
treat	A	B	85	20.5	1.4	0.0000
treat	A	C	85	22.7	1.4	0.0000
treat	B	C	85	2.2	1.5	0.2948
treat	A	B	105	25.4	1.9	0.0000
treat	A	C	105	24.4	1.8	0.0000
treat	B	C	105	−1.0	1.9	0.8559
treat	A	B	125	30.3	2.6	0.0000
treat	A	C	125	26.1	2.5	0.0000
treat	B	C	125	−4.2	2.7	0.2665

C for fat values of 85, 105, and 125 g/day, while the means of treatments B and C are not significantly different.

This example demonstrates the level of complication that can be achieved when doing analysis of variance and covariance for complex designs. The main idea of this example is that the analysis of covariance strategy is the same no matter what design and treatment structures are used in an experiment. The first step is to determine a more than adequate form of the covariate part of the model. The second step is to ascertain an adequate covariance structure for the repeated measures part of the model (if the model involves repeated measures) and to evaluate if the equality of variances is a reasonable assumption. The third step involves simplifying the covariate part of the model. The final step is to use the selected covariance structure and the simplified covariate part of the model to make comparisons among the resulting regression models. Using these four steps, one can be successful in applying the analysis of covariance strategy to complex designs.

18.6 NONLINEAR ANALYSIS OF COVARIANCE

A basic assumption used in the analysis of covariance is that there is a linear model relationship between the mean of the response variable and the covariates. A linear model relationship is one where the model is linear in the coefficients or slopes, but does not have to be linear in the covariates. For example, a linear model relationship between the mean of y and the covariate x could be expressed as

$$y_{ij} = \alpha_i + \beta_i x_{ij} + \delta_i X_{ij}^2 + \gamma_i x_{ij}^3 + \varepsilon_{ij} \tag{18.12}$$

Model 18.12 is a cubic function of x_{ij}, but it is a linear function of the parameters, α_i, β_i, δ_i, and γ_i. The graph of the model is not a straight line, but Model 18.12 is a linear model or a model that is linear in the parameters. The usual analysis of variance and covariance models discussed in the previous chapters are all linear models. The linear model can provide good approximations to the relationship between the mean of y and x, but there are times when a nonlinear model can be used to describe the relationship that has fewer parameters than the approximating linear model and the resulting parameters may have an interpretation in the setting of the study. One useful model is the exponential decay model

$$y_{ij} = \alpha_i + \beta_i e^{-\gamma_i x_{ij}} + \varepsilon_{ij} \tag{18.13}$$

Model 18.13 has a maximum of $\alpha_i + \beta_i$ at $x_{ij} = 0$ and asymptotes to α_i as $x_{ij} \to \infty$. The parameter γ_i is the decay rate of the ith model. The software packages PROC GLM and PROC MIXED cannot fit the Model 18.13, since a nonlinear estimation algorithm is required. PROC NLIN and PROC NLMIXED of the SAS® system are two procedures that can be used to obtain estimates of the parameters of a nonlinear model (SAS Institute Inc., 1999). A plant breeder and entomologist designed a study to evaluate how the level of insect infestation affected the yield of four varieties of corn. They planted large plots of each of the four varieties at ten different location in an area with known corn borer infestations; thus locations were considered as a block. The resulting design is a one-way treatment structure (four varieties) in a

Special Applications of Analysis of Covariance

TABLE 18.33
Data Set for Varieties with Yield (bu/acre) and Insects Per Plant for Nonlinear Analysis of Covariance

	Variety 1		Variety 2		Variety 3		Variety 4	
Block	Yield	Insect	Yield	Insect	Yield	Insect	Yield	Insect
1	127.3	5.7	164.9	1.4	124.4	2.1	146.4	3.4
2	120.1	5.8	124.1	3.5	96.6	4.0	129.0	4.9
3	196.9	1.1	123.1	3.7	94.0	5.4	142.0	4.0
4	147.4	3.2	126.1	3.2	104.5	2.8	122.5	5.0
5	170.7	2.2	152.2	1.8	118.9	2.3	164.8	2.2
6	185.2	1.5	125.2	3.4	95.3	4.4	165.3	2.1
7	142.5	4.1	146.8	2.2	93.5	4.5	125.2	5.4
8	158.9	2.9	117.6	3.9	102.6	3.3	123.4	5.3
9	170.1	2.1	97.9	6.1	106.2	3.1	123.6	5.4
10	191.6	1.2	112.7	4.4	93.1	4.9	134.1	4.5

randomized complete block design structure (ten blocks). Past experience indicated that an exponential decay model like the one in Model 18.13 should describe data from this experiment. The data set is in Table 18.33, where yield is in bushels per acre and the level of insect infestation is in insects per plant. At the time of the highest insect infestation, insect counts were made on 20 randomly selected plants throughout the plot and the mean number of insects per plant were recorded. Since the design structure is a RCB and the block effects are assumed to be a random effect, the model to describe the yield of corn as a function of the insect numbers per plant is

$$\text{yield}_{ij} = \alpha_i + \beta_i\, e^{-\gamma_i\, \text{insect}_{ij}} + b_j + \varepsilon_{ij}$$
$$\text{for } i = 1, 2, 3, 4, \quad j = 1, 2, \ldots, 10 \qquad (18.14)$$

where $b_i \sim \text{iid}\, N(0, \sigma_b^2)$ and $\varepsilon_{ij} \sim \text{iid}\, N(0, \sigma_\varepsilon^2)$

Table 18.34 contains the PROC NLMIXED code used to specify Model 18.14. PROC NLMIXED is an iterative estimation technique, so initial estimates or guesses of the parameters need to be provided using a parms statement. Almost any data step programming code can be used to specify computations in the model. The model statement is used to specify the distribution to be used to describe the dependent variable. In this case, the normal distribution is used (other distributions are available) and the model statement specifies the conditional mean and variance of the response variable given the random effects (blocks in this case). The code mu = (variety=1) * (a1+b1*exp(-c1*ins)) + (variety=2) * (a2+ b2*exp(-c2*ins)) + (variety=3) * (a3+b3*exp(-c3*ins)) + (variety=4) * (a4+b4*exp(-c4*ins)); specifies an exponential decay

TABLE 18.34
PROC NLMIXED Code to Fit Separate Nonlinear Models to Each Variety

```
proc nlmixed df=19 data=corn;
parms a1=80 a2=80 a3=80 a4=80 b1=140 b2=130 b3=115 b4= 125
c1=.25 c2=.27 c3=.4 c4=.25 s2e=3 sblk2=2;
mu=(variety=1)*(a1+b1*exp(-c1*ins))+(variety=2)*(a2+b2*exp(-c2*ins))
 +(variety=3)*(a3+b3*exp(-c3*ins))+(variety=4)*(a4+b4*exp(-c4*ins));
value=mu+blk;
model yield ~ normal(value,s2e);
random blk ~ normal(0,sblk2) subject=block;
bounds sblk2>0,s2e>0;
contrast 'eq a s' a1-a2, a1-a3, a1-a4;
contrast 'eq b s' b1-b2, b1-b3, b1-b4;
contrast 'eq c s' c1-c2, c1-c3,c1-c4;
```

Parameter	Estimate	StandardError	df	tValue	Probt	Lower	Upper
a1	80.767	9.006	19	8.97	0.0000	61.917	99.617
a2	69.244	7.101	19	9.75	0.0000	54.381	84.107
a3	84.143	3.033	19	27.74	0.0000	77.795	90.492
a4	82.593	18.336	19	4.50	0.0002	44.215	120.972
b1	142.457	6.734	19	21.15	0.0000	128.362	156.553
b2	131.953	4.549	19	29.01	0.0000	122.432	141.474
b3	119.088	16.413	19	7.26	0.0000	84.735	153.441
b4	128.534	10.979	19	11.71	0.0000	105.556	151.513
c1	0.211	0.027	19	7.89	0.0000	0.155	0.267
c2	0.250	0.030	19	8.23	0.0000	0.187	0.314
c3	0.547	0.091	19	5.98	0.0000	0.356	0.738
c4	0.209	0.064	19	3.24	0.0043	0.074	0.344
s2e	1.228	0.322	19	3.81	0.0012	0.554	1.902
sblk2	3.772	1.963	19	1.92	0.0698	−0.337	7.880

Descr	Value
−2 Log Likelihood	147.5
AIC (smaller is better)	175.5
AICC (smaller is better)	192.3
BIC (smaller is better)	179.7

Label	NumDF	DenDF	FValue	ProbF
eq a s	3	19	1.27	0.3120
eq b s	3	19	0.68	0.5749
eq c s	3	19	5.07	0.0095

model with unique parameters for each variety, as has been done in previous chapters using different intercepts and slopes for each of the levels of a factor in the treatment structure. There is no class statement available in PROC NLMIXED. The conditional mean yield of a given variety from a given block with a given value of insect infestation is described by the variable value = mu + blk. The variance of the

Special Applications of Analysis of Covariance

conditional distribution of the yield data is denoted by se2. The random statement is used to specify random effects in the model. For random effects, only the normal distribution is available. The random statement specifies that the distribution of the block effects is a normal distribution with mean 0 and variance sblk2. The subject=block indicates the variable being used as the random effect and the data must be sorted by the effect or effects used in the subject= specification. If more than one random effect is needed in the model, a vector of random effects is used in the random statement and a lower triangular portion of the covariance is specified (see the PROC NLMIXED documentation for the latest options and examples, SAS Institute Inc., 1999). The bounds statement is to require so that the estimates of the two variance components are greater than zero. The output provided by using the parms, model, and random statements is the set of estimates of the models parameters with estimated standard errors, t values for testing the respective parameter is equal to zero, and large sample size confidence intervals about each parameter. The first step in the analysis of covariance process is to see if the model can be simplified, which can be accomplished by individually testing the hypotheses

$$H_{o1}: \alpha_1 = \alpha_2 = \alpha_3 = \alpha_4 \text{ vs. } H_{a1}: (\text{not } H_{o1}:)$$

$$H_{o2}: \beta_1 = \beta_2 = \beta_3 = \beta_4 \text{ vs. } H_{a2}: (\text{not } H_{o2}:) \quad (18.15)$$

$$H_{o3}: \gamma_1 = \gamma_2 = \gamma_3 = \gamma_4 \text{ vs. } H_{a3}: (\text{not } H_{o3}:)$$

or by testing

$$H_{o1}: \alpha_1 - \alpha_2 = 0, \alpha_1 - \alpha_3 = 0, \alpha_1 - \alpha_4 = 0 \text{ vs. } H_{a1}: (\text{not } H_{o1}:)$$

$$H_{o2}: \beta_1 - \beta_2 = 0, \beta_1 - \beta_3 = 0, \beta_1 - \beta_4 = 0 \text{ vs. } H_{a2}: (\text{not } H_{o2}:) \quad (18.16)$$

$$H_{o3}: \gamma_1 - \gamma_2 = 0, \gamma_1 - \gamma_3 = 0, \gamma_1 - \gamma_4 = 0 \text{ vs. } H_{a3}: (\text{not } H_{o3}:)$$

or some other set of three linearly independent contrasts of the parameters within each hypothesis. The contrast statement can be used to provide a simultaneous test of a set of functions of the parameters where the syntax of the contrast statement is "contrast 'label' function 1, function 2, ...,function k;", where the statistic provides a test of the hypothesis H_o: function 1 = 0, function 2 = 0, ..., function k = 0 vs. H_o: (not H_o:). The contrast statements in Table 18.34 were used to obtain tests of the hypotheses in Equation 18.15 by using the contrasts in Equation 18.16. The results of the analysis indicate that the γ's are not equal ($p = 0.0095$), but the significance levels corresponding to testing the equality of the α's and equality of the β's are 0.3120 and 0.5749, respectively, indicating that the model can be simplified. Since the significance level corresponding to the test of equal β's hypothesis is the largest, a model was constructed with equal β's, but unequal α's and unequal γ's. The required model is

$$\text{yield}_{ij} = \alpha_i + \beta\, e^{-\gamma_i \operatorname{insect}_{ij}} + b_j + \varepsilon_{ij}$$

$$\text{for } i = 1,\ 2,\ 3,\ 4,\ \ j = 1,\ 2,\ \ldots,\ 10 \tag{18.17}$$

where $b_j \sim \text{iid } N(0, \sigma_b^2)$ and $\varepsilon_{ij} \sim \text{iid } N(0, \sigma_\varepsilon^2)$

Table 18.35 contains the PROC NLMIXED code to fit Model 18.17 to the data set. The estimate of the parameters are provided, but the main interest is in the results of the contrasts statements that provide tests of the equality of the α's and the equality of the γ's. The significance levels of the two tests are 0.0003 and 0.0000, respectively,

TABLE 18.35
PROC NLMIXED Code to Fit a Model with Common b Parameters But Different a and c Parameters for the Varieties

```
proc nlmixed df=22 data=corn;
parms a1=80 a2=80 a3=80 a4=80 b =140 c1=.25 c2=.27 c3=.4 c4=.25
  s2e=3 sblk2=2;
mu=(variety=1)*(a1+b *exp(-c1*ins))+(variety=2)*(a2+b *exp(-c2*ins))
  +(variety=3)*(a3+b *exp(-c3*ins))+(variety=4)*(a4+b *exp(-c4*ins));
value=mu+blk;
model yield ~ normal(value,s2e);
random blk ~ normal(0,sblk2) subject=block;
bounds sblk2>0,s2e>0;
contrast 'eq a s' a1-a2, a1-a3, a1-a4;
contrast 'eq c s' c1-c2, c1-c3,c1-c4;
```

Parameter	Estimate	StandardError	df	tValue	Probt	Lower	Upper
a1	92.198	4.443	22	20.75	0.0000	82.984	101.413
a2	68.473	5.837	22	11.73	0.0000	56.368	80.578
a3	86.060	1.490	22	57.74	0.0000	82.969	89.151
a4	76.342	5.884	22	12.97	0.0000	64.138	88.545
b	134.076	2.921	22	45.91	0.0000	128.019	140.133
c1	0.249	0.019	22	12.83	0.0000	0.208	0.289
c2	0.251	0.027	22	9.24	0.0000	0.195	0.307
c3	0.619	0.029	22	21.63	0.0000	0.560	0.679
c4	0.192	0.021	22	9.28	0.0000	0.149	0.235
s2e	1.355	0.363	22	3.74	0.0011	0.603	2.107
sblk2	3.030	1.550	22	1.96	0.0634	−0.184	6.245

Descr	Value
−2 Log Likelihood	148.6
AIC (smaller is better)	170.6
AICC (smaller is better)	180.0
BIC (smaller is better)	173.9

Label	NumDF	DenDF	FValue	ProbF
eq a s	3	22	9.38	0.0003
eq c s	3	22	41.05	0.0000

Special Applications of Analysis of Covariance

indicating that unequal α's and unequal γ's are needed in the model. The estimates of the variance components are maximum likelihood estimates and not REML estimates; thus they are likely underestimates of the corresponding parameters. The confidence intervals computed for the parameters are based on large sample results (but using t values instead of z values) and are computed as estimate $\pm\, t_{\alpha/2,\, df}$ st derr where the degrees of freedom are either specified in the PROC NLMIXED statement or the default number of degrees of freedom are computed as the sample size minus the number of random effects in the model. The degrees of freedom specified in the PROC NLMIXED statement in Table 18.34 were computed as the sample size minus the number of parameters in the model minus the degrees of freedom associated with the random effects or $40 - 12 - 9 = 19$. The number of degrees of freedom specified in the PROC NLMIXED statement in Table 18.35 is 22 since the reduced model has three fewer parameters. The confidence intervals for the variance components are most likely not going to be appropriate unless there are large numbers of degrees of freedom associated with the estimates. The Satterthwaite confidence intervals (Littell et al., 1996) could be constructed for the variance components using the chi-square distribution where the approximate degrees of freedom are computed df $= 2$ (tValue)^2. For the block variance component, the approximate number of degrees of freedom is 7.7, and for the plot variance component, the approximate number of degrees of freedom is 28.0. A $(1-\alpha)\,100\%$ confidence interval about a variance using the chi-square distribution can be constructed as

$$\frac{\text{d.f.}\,\hat{\sigma}^2}{\chi^2_{\alpha/2,\text{d.f.}}} < \sigma^2 < \frac{\text{d.f.}\,\hat{\sigma}^2}{\chi^2_{1-\alpha/2,\text{d.f.}}} \tag{18.18}$$

The corresponding 95% confidence intervals for the two variance components are $0.85 < \sigma_\varepsilon^2 < 2.48$ and $1.32 < \sigma_b^2 < 12.55$ where the number of degrees of freedom used to construct the confidence interval about σ_b^2 was seven. The next step in the analysis was to use the contrast statements to test the equality of the four variety models at selected values of the number of insects per plant or test

$$H_o: \alpha_1 + \beta\, e^{-\gamma_1 \text{insect}_0} = \alpha_2 + \beta\, e^{-\gamma_2 \text{insect}_0} =$$
$$\alpha_3 + \beta\, e^{-\gamma_3 \text{insect}_0} = \alpha_4 + \beta\, e^{-\gamma_4 \text{insect}_0} \text{ vs. } H_a: (\text{not } H_o:) \tag{18.19}$$

The two contrast statements in Table 18.36 were used to provide tests of the hypothesis in Equation 18.19 evaluated at ins = 1 and 7, but the results in the table provide tests of the equality of models at ins = 1, 2, ..., 7 (other contrast statement are not shown). The models are producing significantly different responses at each value of ins. Table 18.37 contains estimate statements to evaluate the models at ins = 0 and 1, but the results (estimates and estimated standard errors) are provided for the four varieties evaluated at ins = 0, 1, ..., 7. These predicted values are displayed in the Figure 18.9, which show the exponential decay aspect of the process. Finally, estimate statements were used to provide estimates of the differences between pairs of

TABLE 18.36
Contrast Statements to Test Equality of the Models Evaluated at 1, 2, ..., 7 Insects Per Plant

```
contrast 'models= at ins=1'
 (a1+b*exp(-c1*1))-(a2+b*exp(-c2*1)),
 (a1+b*exp(-c1*1))-(a3+b*exp(-c3*1)),
 (a1+b*exp(-c1*1))-(a4+b*exp(-c4*1));
contrast 'models= at ins=7'
 (a1+b*exp(-c1*7))-(a2+b*exp(-c2*7)),
 (a1+b*exp(-c1*7))-(a3+b*exp(-c3*7)),
 (a1+b*exp(-c1*7))-(a4+b*exp(-c4*7));
```

Label	NumDF	DenDF	FValue	ProbF
models= at ins=1	3	22	209.90	0.0000
models= at ins=2	3	22	1206.02	0.0000
models= at ins=3	3	22	2160.68	0.0000
models= at ins=4	3	22	2047.75	0.0000
models= at ins=5	3	22	966.75	0.0000
models= at ins=6	3	22	909.11	0.0000
models= at ins=7	3	22	667.24	0.0000

TABLE 18.37
Estimate Statements to Provide Estimates from the Variety Regression Models Evaluated at 0, 1, 2, ..., 7 Insects Per Plant

```
estimate 'm1 at 0' (a1+b*exp(-c1*0));
estimate 'm2 at 0' (a2+b*exp(-c2*0));
estimate 'm3 at 0' (a3+b*exp(-c3*0));
estimate 'm4 at 0' (a4+b*exp(-c4*0));
estimate 'm1 at 1' (a1+b*exp(-c1*1));
estimate 'm2 at 1' (a2+b*exp(-c2*1));
estimate 'm3 at 1' (a3+b*exp(-c3*1));
estimate 'm4 at 1' (a4+b*exp(-c4*1));
```

	Variety 1		Variety 2		Variety 3		Variety 4	
Insects	Est	Stde	Est	Stde	Est	Stde	Est	Stde
0	226.27	2.53	202.55	4.16	220.14	3.55	210.42	4.32
1	196.76	1.08	172.78	1.78	158.22	1.25	187.01	2.34
2	173.74	0.71	149.62	0.78	124.90	0.95	167.68	1.18
3	155.79	0.78	131.60	0.72	106.97	0.73	151.73	0.72
4	141.79	0.80	117.58	0.75	97.31	0.68	138.57	0.69
5	130.87	0.82	106.68	0.77	92.12	0.85	127.70	0.72
6	122.36	0.94	98.19	0.97	89.32	1.05	118.74	0.75
7	115.72	1.20	91.59	1.36	87.81	1.20	111.33	0.85

Special Applications of Analysis of Covariance

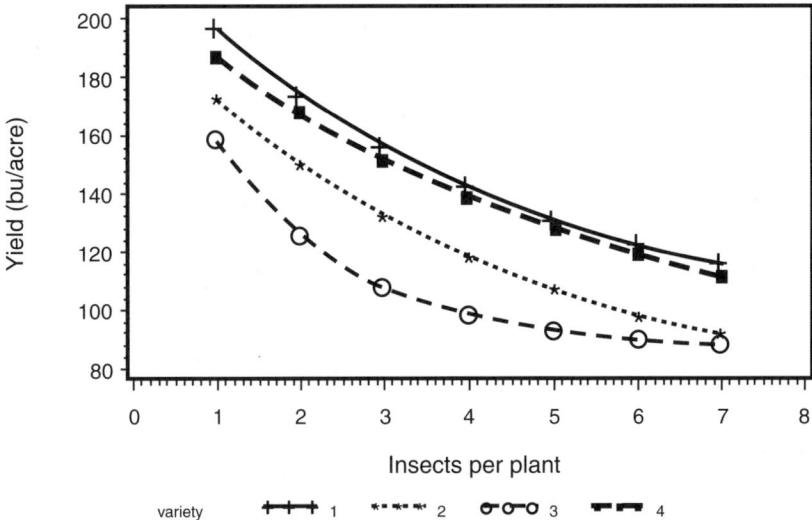

FIGURE 18.9 Graph of the regression models for each variety.

models at ins = 0, 1, 2, ..., 7, where the function in the estimate statement is of the form

$$\alpha_1 + \beta\, e^{-\gamma_1\, \text{insect}_0} - \alpha_2 + \beta\, e^{-\gamma_2\, \text{insect}_0} \tag{18.20}$$

which is a comparison of the models for variety 1 and variety 2 at a selected value of insect_0. The results are in Tables 18.38 and 18.39 where Di-i' denotes the difference between the models for variety i and i' and the "at ins" denotes the value of the number of insects per plant. Models 1 and 3, 2 and 4, and 3 and 4 are not significantly different at ins = 0 and Models 2 and 3 are not different at ins = 7. All other comparisons have very small significance levels. Figures 18.10 through 18.15 are graphs of the differences of each pair of models along with the upper and lower confidence limits which were used to construct confidence bands. When in the range of the differences and confidence bands included zero, a horizontal line on the yield (differences) axes at zero was included to provide a picture as to where the models are not significantly different and as to where they are by seeing where the horizontal line penetrates the interior of the confidence band. The curves for these varieties are quite diverse, so there are not many places where a pair of curves are similar.

This example of nonlinear analysis of covariate extends the analysis of variance to include both linear and nonlinear models. The development of PROC NLIN and NLMIXED enables many of these models to be fit to a data set and has greatly expanded the number of analysis of covariance models that can be used in practice.

TABLE 18.38
Estimate Statements to Provide Differences in the Variety Regression Models Evaluated at 0, 1, ..., 7 Insects Per Plant (Part I)

```
estimate 'D1-2 at 0' (a1+b*exp(-c1*0))-(a2+b*exp(-c2*0));
estimate 'D1-2 at 1' (a1+b*exp(-c1*1))-(a2+b*exp(-c2*1));
estimate 'D1-2 at 2' (a1+b*exp(-c1*2))-(a2+b*exp(-c2*2));
```

Label	Estimate	StError	df	tValue	Probt	Alpha	Lower	Upper
D1-2 at 0	23.73	4.80	22	4.95	0.0001	0.05	13.65	33.80
D1-2 at 1	23.98	2.01	22	11.93	0.0000	0.05	19.76	28.21
D1-2 at 2	24.12	0.66	22	36.82	0.0000	0.05	22.75	25.50
D1-2 at 3	24.19	0.67	22	36.13	0.0000	0.05	22.79	25.60
D1-2 at 4	24.21	0.80	22	30.32	0.0000	0.05	22.53	25.89
D1-2 at 5	24.20	0.70	22	34.38	0.0000	0.05	22.72	25.68
D1-2 at 6	24.17	0.56	22	42.79	0.0000	0.05	22.98	25.35
D1-2 at 7	24.13	0.66	22	36.40	0.0000	0.05	22.73	25.52
D1-3 at 0	6.14	4.75	22	1.29	0.2123	0.05	−3.83	16.11
D1-3 at 1	38.53	1.55	22	24.80	0.0000	0.05	35.27	41.80
D1-3 at 2	48.84	0.86	22	56.76	0.0000	0.05	47.03	50.65
D1-3 at 3	48.82	0.72	22	68.05	0.0000	0.05	47.32	50.33
D1-3 at 4	44.48	0.73	22	60.85	0.0000	0.05	42.94	46.01
D1-3 at 5	38.76	0.95	22	40.93	0.0000	0.05	36.77	40.75
D1-3 at 6	33.04	1.27	22	25.94	0.0000	0.05	30.36	35.72
D1-3 at 7	27.91	1.64	22	17.05	0.0000	0.05	24.47	31.34
D2-3 at 0	−17.59	6.38	22	−2.76	0.0129	0.05	−30.98	−4.19
D2-3 at 1	14.55	2.09	22	6.98	0.0000	0.05	10.17	18.93
D2-3 at 2	24.71	0.87	22	28.51	0.0000	0.05	22.89	26.53
D2-3 at 3	24.63	0.70	22	35.06	0.0000	0.05	23.16	26.11
D2-3 at 4	20.27	0.63	22	32.13	0.0000	0.05	18.94	21.59
D2-3 at 5	14.56	0.82	22	17.67	0.0000	0.05	12.83	16.29
D2-3 at 6	8.87	1.27	22	6.96	0.0000	0.05	6.20	11.55
D2-3 at 7	3.78	1.82	22	2.08	0.0521	0.05	−0.04	7.60

18.7 EFFECT OF OUTLIERS

The occurrence of outliers in a data set can have drastic effects on the resulting analysis. Outliers can occur in many ways with the most frequent cause being data recording and entry errors. As with the analysis of any data set, care must be used when declaring an observation to be an outlier and thus removable from the data set. Before an observation can be removed, there must be an established reason for why the outlier occurred. If a reason cannot be found, then the observation should not be removed from further analyses.

The data in Table 18.40 are from an experiment to evaluate the hotness of salsa products. The salsa products to be evaluated were of medium heat level. A mild

Special Applications of Analysis of Covariance 573

TABLE 18.39
Estimate Statements to Provide Differences in the Variety Regression Models Evaluated at 0, 1, ..., 7 Insects Per Plant (Part II)

Label	Estimate	StdError	df	tValue	Probt	Alpha	Lower	Upper
D1-4 at 0	15.86	4.61	22	3.44	0.0029	0.05	6.17	25.54
D1-4 at 1	9.75	2.41	22	4.05	0.0007	0.05	4.70	14.81
D1-4 at 2	6.06	1.14	22	5.30	0.0000	0.05	3.66	8.46
D1-4 at 3	4.06	0.69	22	5.92	0.0000	0.05	2.62	5.49
D1-4 at 4	3.22	0.71	22	4.57	0.0002	0.05	1.74	4.70
D1-4 at 5	3.17	0.74	22	4.28	0.0004	0.05	1.61	4.72
D1-4 at 6	3.62	0.72	22	5.05	0.0001	0.05	2.12	5.13
D1-4 at 7	4.39	0.71	22	6.15	0.0000	0.05	2.89	5.88
D2-4 at 0	−7.87	5.35	22	−1.47	0.1583	0.05	−19.10	3.36
D2-4 at 1	−14.23	2.66	22	−5.36	0.0000	0.05	−19.81	−8.65
D2-4 at 2	−18.07	1.16	22	−15.62	0.0000	0.05	−20.50	−15.64
D2-4 at 3	−20.14	0.62	22	−32.41	0.0000	0.05	−21.44	−18.83
D2-4 at 4	−20.99	0.64	22	−32.71	0.0000	0.05	−22.34	−19.64
D2-4 at 5	−21.03	0.68	22	−30.80	0.0000	0.05	−22.46	−19.59
D2-4 at 6	−20.54	0.73	22	−28.13	0.0000	0.05	−22.08	−19.01
D2-4 at 7	−19.74	0.89	22	−22.16	0.0000	0.05	−21.61	−17.87
D3-4 at 0	9.72	6.29	22	1.55	0.1395	0.05	−3.49	22.92
D3-4 at 1	−28.78	2.69	22	−10.71	0.0000	0.05	−34.43	−23.14
D3-4 at 2	−42.78	1.28	22	−33.44	0.0000	0.05	−45.47	−40.09
D3-4 at 3	−44.77	0.67	22	−66.99	0.0000	0.05	−46.17	−43.36
D3-4 at 4	−41.26	0.58	22	−71.01	0.0000	0.05	−42.48	−40.04
D3-4 at 5	−35.59	0.79	22	−45.24	0.0000	0.05	−37.24	−33.94
D3-4 at 6	−29.42	1.05	22	−28.11	0.0000	0.05	−31.61	−27.22
D3-4 at 7	−23.52	1.33	22	−17.73	0.0000	0.05	−26.31	−20.73

salsa was used as a warm-up sample and the panelist's ratings were used as a possible covariate. Each of the 20 panelists tasted each of the 4 salsa products where the order of the products was randomized for each panelist. The data set includes the responses for each of the four products (resp1, resp2, resp3, and resp4) as well as the order the respective product was tasted. It was thought that there could be an order effect as well as possibly an order by product interaction. Thus, the model to describe the data is

$$y_{ij} = \mu + O_i + \tau_{O_{ij}} + O\tau_{iO_{ij}} + \beta_{O_{ij}} M_j + p_j + \varepsilon_{ij}$$

$$p_j \sim iidN(0, \sigma^2_{person}) \text{ and } \varepsilon_{ij} \sim iidN(0, \sigma^2_\varepsilon)$$

where O_i denotes the effect of a product tasted in the ith order, $\tau_{O_{ij}}$ denotes the effect of the product tasted by the jth person during the ith order, $O\tau_{iO_{ij}}$ denotes the order by product interaction effect, $\beta_{O_{ij}}$ denotes the slope of the regression line for the

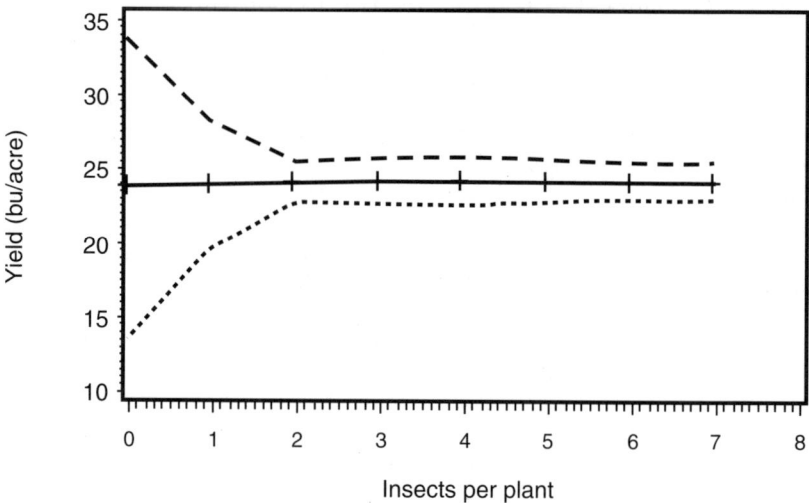

FIGURE 18.10 Graph of the difference between the models for variety 1 and variety 2 with 95% confidence bands.

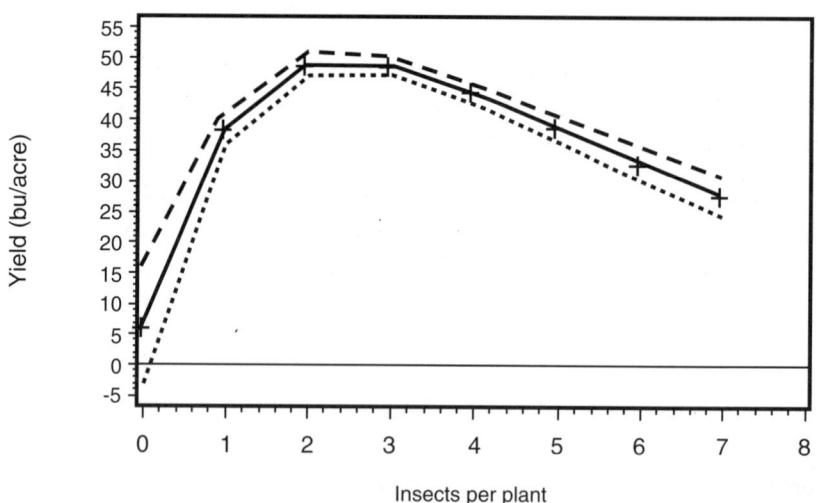

FIGURE 18.11 Graph of the difference between the models for variety 1 and variety 3 with 95% confidence bands.

Special Applications of Analysis of Covariance

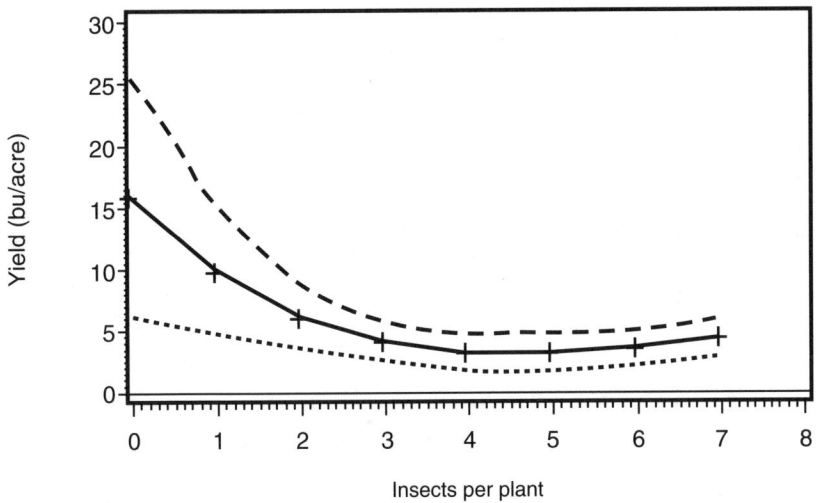

FIGURE 18.12 Graph of the difference between the models for variety 1 and variety 4 with 95% confidence bands.

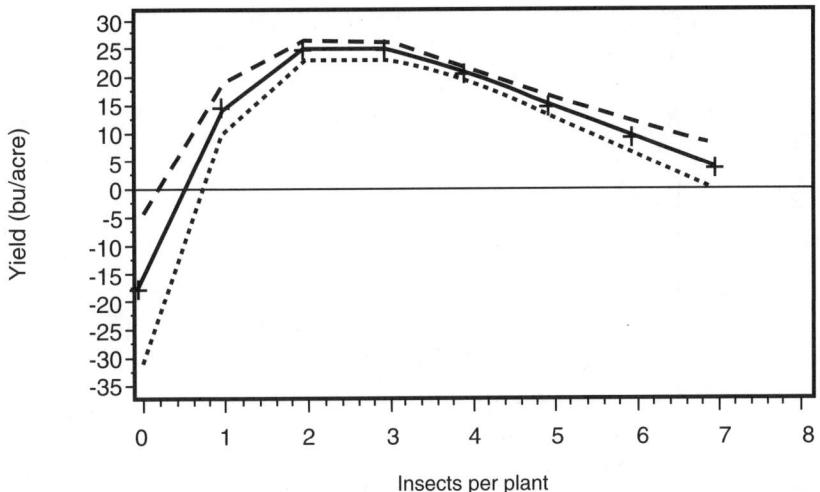

FIGURE 18.13 Graph of the difference between the models for variety 2 and variety 3 with 95% confidence bands.

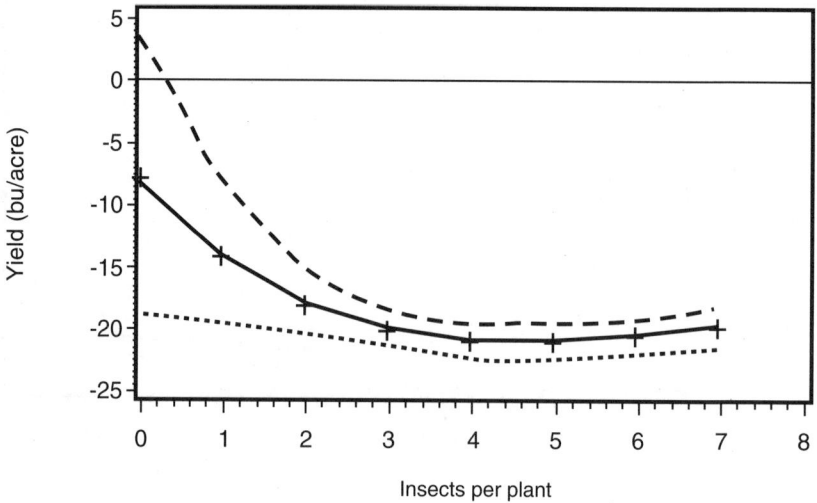

FIGURE 18.14 Graph of the difference between the models for variety 2 and variety 4 with 95% confidence bands.

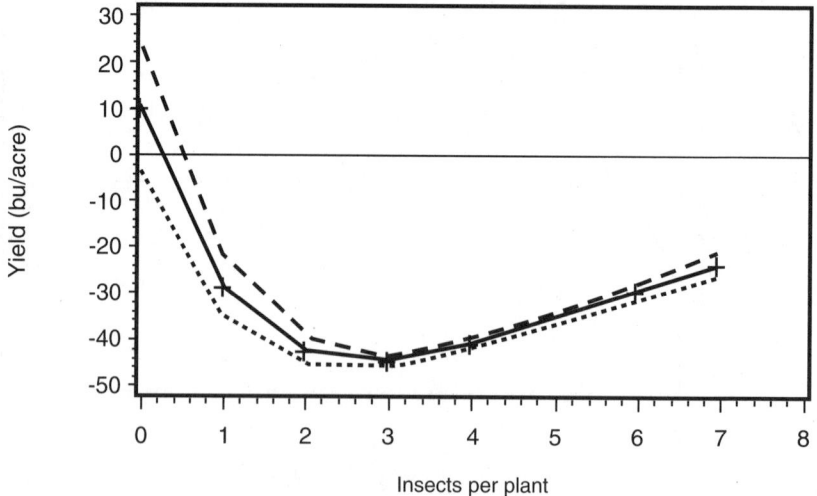

FIGURE 18.15 Graph of the difference between the models for variety 3 and variety 4 with 95% confidence bands.

TABLE 18.40
Data for Salsa Tasting Study Where Mild is Heat Rating of the Mild Salsa Used for Warm Up, resp1, resp2, resp3, and resp4 are the Heat Ratings for the Four Medium Salsa Products and order1, order2, order3, and order4 are the Orders the Respective Products were Tasted

Person	Mild	resp1	order1	resp2	order2	resp3	order3	resp4	order4
1	4.0	8.5	2	9.6	3	8.3	1	8.6	4
2	5.6	9.3	2	9.9	3	10.8	1	10.9	4
3	6.2	10.2	2	11.9	3	10.5	4	15.0	1
4	7.6	11.2	2	12.0	4	12.2	3	11.6	1
5	7.2	11.8	3	10.4	1	10.3	4	11.6	2
6	4.9	15.0	3	14.3	4	14.2	2	14.6	1
7	3.6	8.7	4	9.1	1	8.6	2	7.9	3
8	8.0	10.3	4	11.0	2	10.4	1	11.2	3
9	6.5	5.5	3	10.3	4	10.4	1	10.4	2
10	3.7	8.4	3	9.5	1	8.7	4	9.8	2
11	8.3	12.3	2	12.3	4	13.5	3	12.6	1
12	3.4	9.7	1	12.5	3	10.1	2	11.2	4
13	3.3	7.1	2	7.0	1	7.5	3	7.5	4
14	8.8	11.2	1	10.6	3	11.0	2	12.1	4
15	4.4	9.3	3	9.5	2	9.3	4	10.2	1
16	4.5	10.3	2	10.7	1	10.4	4	11.2	3
17	4.2	8.9	1	9.8	4	9.3	2	10.0	3
18	3.3	7.9	1	8.7	4	7.7	2	8.4	3
19	8.0	11.4	2	11.6	1	10.7	4	12.1	3
20	5.0	10.1	4	10.7	1	10.1	2	10.7	3

product tasted by the jth panelist, p_j is the random person effect, and ε_{ij} is the random error corresponding to the ith order of the jth panelist. This is a repeated measures design since each panelist tastes the four products during the four sessions denoted by order. From past experience, the ARH(1) residual covariance structure often describes the data instead of the independent errors model but the produced results that were not meaningful. The PROC MIXED code to fit the model with unequal product slopes and ARH(1) covariance structure is in Table 18.41. The problem with the analysis is that the F value corresponding to mild*prod is infinity, a result that is not meaningful. A plot of the resp by mild data indicates that a simple linear regression model is likely to describe the data. When trying to fit the UN(1) covariance to this data set, PROC MIXED failed to converge (analyses not shown).

As it turns out, outliers in the data set can have a major influence on the ability of PROC MIXED to fit some of the more complex covariance structures. Table 18.42 contains the PROC GLM code to provide the residuals by fitting the model with unequal product slopes and independent errors. The set of Box-plots in Table 18.42 are the plots of the residuals for each of the orders. The plot for order three indicates that a possible outlier exists. The outlying observation was identified as coming from the ninth person. That observation was deleted from the data set and PROC GLM

TABLE 18.41
PROC MIXED Code to Fit Unequal Slopes Model with ARH(1) Covariance Structure and the Results

```
proc mixed ic data=wide;
class person order prod;
model resp=prod order order*prod mild*prod/ddfm=kr;
random person;
repeated order/subject=person type=arh(1);
```

CovParm	Subject	Estimate		
person		1.635490		
Var(1)	person	0.806474		
Var(2)	person	0.255290		
Var(3)	person	1.744888		
Var(4)	person	0.000003		
ARH(1)	person	0.070666		
Effect	**NumDF**	**DenDF**	**FValue**	**ProbF**
prod	1	35.1	4.95	0.0326
order	2	18.3	0.40	0.6778
order*prod	4	30.4	1.85	0.1457
mild*prod	4	30.4	∞	0.0000

was again used to fit the model to the remaining data in order to compute the residuals. The PROC GLM code and resulting Box-plots are in Table 18.43. There is still one possible outlier in order one which was identified as coming from person three. Eliminating the observation from the first order from person three and refitting the model using PROC GLM provided the Box-plots of the residuals in Table 18.44. These plots do not indicate there are any additional possible outliers. The PROC MIXED code in Table 18.45 was used to fit the UN(1) model to the data set (it fit this time), so that the variances could be used as starting values for fitting the ARH(1) covariance structure. Using starting values greatly speeds up the execution time for fitting the ARH(1) covariance structure. The variances from Table 18.45 were used in the Parms (parameters) statement of Table 18.46 as starting values for the person variance and the order variances. A value of 0.1 was used as the initial value for the correlation coefficient. The "lowerb=" option was used to keep the estimates of the covariance structure parameters from going to zero. Sometimes the use of the lowerb option enables covariance parameters to be estimated so that the resulting tests for fixed effects are meaningful. The F value corresponding to mild*prod provides a test of the slopes equal to zero hypothesis and the corresponding significance level is 0.0790, indicating there is some doubt as to the need of the covariate in the model.

There is also the possibility of outliers occurring for all the other random effect terms in the model, which in this case is p_j. The "/solution" option on the random statement in Table 18.46 provides predictions of the random effects in the model (estimated best linear unbiased predictors or EBLUPs) and the ods statement writes

Special Applications of Analysis of Covariance

TABLE 18.42
PROC GLM Code to Fit the Unequal Slopes Model with Independent Errors to Provide Residuals Used to Construct the Box-Plot

```
proc glm data=wide;
class person order prod;
model resp=prod order order*prod mild*prod person;
output out=glmres r=glmr;
proc sort data=glmres; by order;
proc univariate data=glmres plot normal; var glmr; by
 order;
```

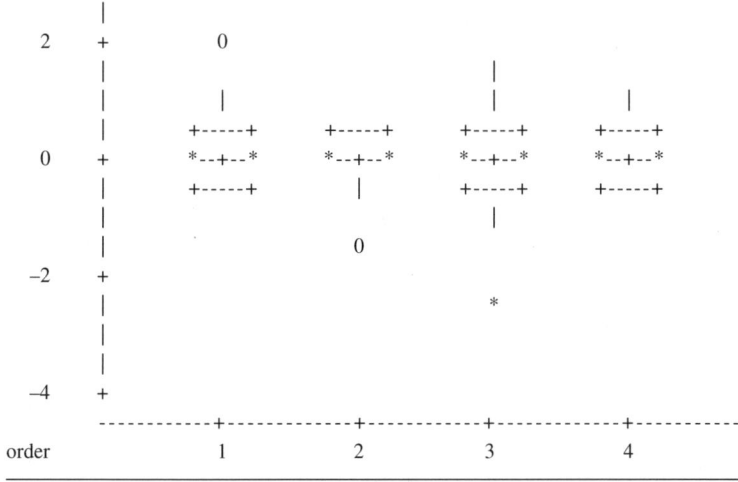

the predictions to a data set. The EBLUPs are listed in Table 18.47 and the Box-plot of the t-values corresponding the EBLUPs is in Table 18.48. The Box-plot indicates that there is a possible outlier, which corresponds to person or panelist six. The PROC MIXED code in Table 18.48 was used fit the UN(1) covariance structure to provide starting values for the variances with data of person six being removed. The variances in Table 18.49 were used in Table 18.50 as starting values for the ARH(1) structure. The estimate of the person variance component from Table 18.46 is 1.663316, but when person six was removed, the estimate of the person variance component is 0.689719. The other major change in the analysis of Table 18.50 is the significance level corresponding to mild*prod is 0.0015, which is strong evidence that the covariate is needed in the model (a result much different from that in Table 18.46). The EBLUPs of the person effects are in Table 18.51 and the Box-plot is in Table 18.52. The Box-plot does not indicate there are any additional outlying persons in the data set. Several covariance structures were fit to the data set and the ARH(1) structure had the smallest AIC value, so it was selected for the remaining analysis. The PROC MIXED code in Table 18.53 fits a model that includes

TABLE 18.43
PROC GLM Code to Fit the Model with the Data for Order Three of Person Nine Removed and the Resulting Box-Plot of the Residuals

```
data wide; set wide; if person=9 and order=3 then delete;
proc glm data=wide; class person order prod;
model resp=prod order order*prod mild*prod person;
output out=glmres r=glmr;
proc sort data=glmres; by order;
proc univariate data=glmres plot normal; var glmr;by order;
```

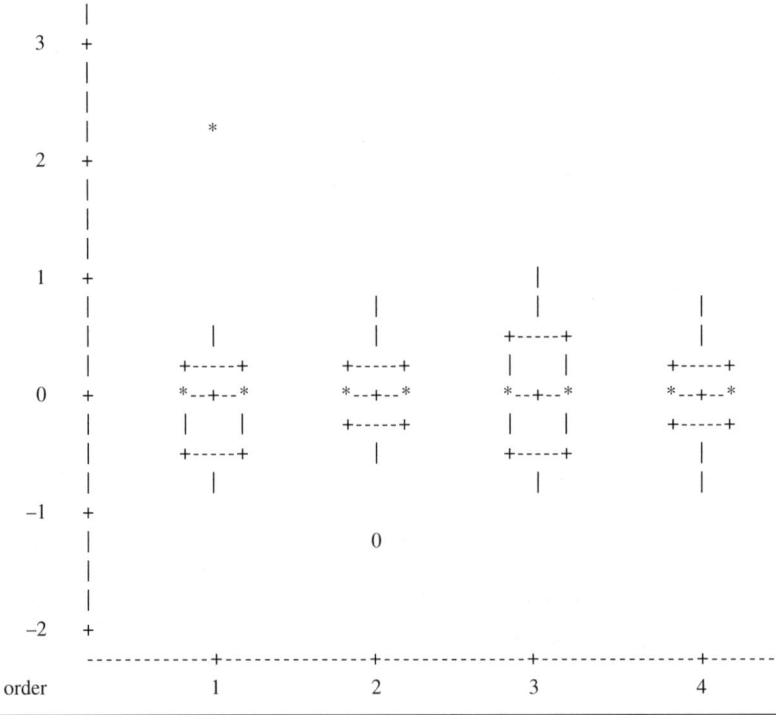

both mild and mild*prod so the F value corresponding to mild*prod provides a test of the equal slopes hypothesis. The significance level corresponding to mild*prod is 0.9488, indicating there is not sufficient evidence to conclude the equal slopes hypothesis is false. The PROC MIXED code in Table 18.54 was used to fit the common slope model with the ARH(1) covariance structure. The significance level corresponding to order*prod is 0.0602, indicating some evidence of a product by order interaction. A product by order interaction occurs when the mean rating of the product depends on the order it was rated. For the remainder of the analysis, assume the product by order interaction is not important. The adjusted means for the four products and pairwise comparisons of the products are in Table 18.55. The

TABLE 18.44
PROC GLM Code to Fit the Model with the Data for Order One of Person Three Also Removed and the Resulting Box-Plot of the Residuals

```
data wide; set wide; if person=3 and order=1 then delete;
proc glm data=wide; class person order prod;
model resp=prod order order*prod mild*prod person;
output out=glmres r=glmr;
proc sort data=glmres; by order;
proc univariate data=glmres plot normal; var glmr;by order;
```

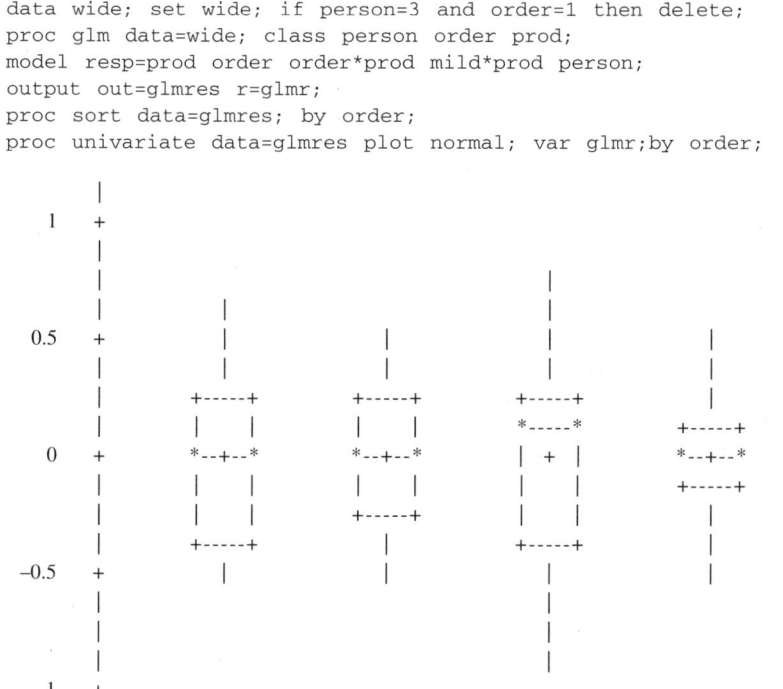

adjust=simulate option was used to control the experiment wise error rate. There are significance differences ($p = 0.05$) between products 1 and 2, 1 and 4, and 3 and 4.

This example demonstrates the drastic effects that outliers can have on the analysis. Additionally, outliers can occur for any size of experimental unit or random effect in the model. This example illustrated that residual outliers can have a disabling effect on the convergence for complex covariance structures as well as on the estimates of the covariance structure parameters. When outliers occur for other random effects, the estimates of the corresponding variance components are greatly effected as well as possibly having an impact on some of the tests for fixed effects. The data analyst needs to make sure the data set has been screened for possible outliers and when such observations or levels of random effects are identified, then the researcher needs to go back to the records to determine if there is an identifiable cause. If no cause can be identified, then there is no basis to remove the data from the analysis.

TABLE 18.45
PROC MIXED Code to Fit the Unequal Slopes Model with the UN(1)Covariance Structure to the Data with Two Outliers Removed and the Results to be Used as Starting Values for the ARH(1) Covariance Structure

```
proc mixed ic scoring=20 data=wide; class person
 order prod;
model resp=prod order order*prod
 mild*prod/ddfm=kr;
random person;
repeated order/subject=person type=un(1);
```

CovParm	Subject	Estimate		
person		1.680400		
UN(1,1)	person	0.088202		
UN(2,1)	person	0.000000		
UN(2,2)	person	0.200085		
UN(3,1)	person	0.000000		
UN(3,2)	person	0.000000		
UN(3,3)	person	0.764544		
UN(4,1)	person	0.000000		
UN(4,2)	person	0.000000		
UN(4,3)	person	0.000000		
UN(4,4)	person	0.014563		

Effect	NumDF	DenDF	FValue	ProbF
prod	3	29.1	1.95	0.1433
order	3	17.0	1.35	0.2904
order*prod	9	28.3	1.97	0.0811
mild*prod	4	26.1	2.61	0.0586

TABLE 18.46
PROC MIXED Code to Fit the Unequal Slopes Model with the ARH(1) Covariance Structure Using Starting Values from the UN(1) Covariance Structure to the Data with Two Outliers Removed and the Results

```
proc mixed ic data=wide; class person order prod;
**starting values from un(1);
model resp=prod order order*prod
 mild*prod/ddfm=kr outp=predicted;
random person/solution;
repeated order/subject=person type=arh(1);
parms 1.6804 0.0882 .2001 .7645 .01456 .1/lowerb =
 1e-4, 1e-4, 1e-4, 1e-4, 1e-4, 1e-4;
lsmeans prod/diff;
ods output solutionr=randomsol;
```

CovParm	Subject	Estimate		
person		1.663316		
Var(1)	person	0.041829		
Var(2)	person	0.463647		
Var(3)	person	1.631961		
Var(4)	person	0.180904		
ARH(1)	person	0.843573		

Effect	NumDF	DenDF	FValue	ProbF
prod	3	25.4	0.79	0.5127
order	3	13.8	0.57	0.6468
order*prod	9	20.8	1.73	0.1451
mild*prod	4	29.2	2.33	0.0790

TABLE 18.47
The EBLUPs of the Person Effects from the ARH(1) Model Fit to the Data Set with Outliers Removed

Person	Estimate	StdErrPred	df	tValue	Probt
1	−1.578	0.510	21	−3.10	0.0055
2	0.617	0.436	14	1.42	0.1791
3	0.139	0.519	29	0.27	0.7905
4	−0.181	0.550	19	−0.33	0.7454
5	−0.962	0.519	17	−1.85	0.0813
6	3.926	0.459	15	8.55	0.0000
7	−0.239	0.515	16	−0.46	0.6484
8	−1.244	0.583	20	−2.13	0.0456
9	−0.329	0.461	11	−0.71	0.4909
10	0.117	0.501	16	0.23	0.8176
11	0.263	0.620	20	0.42	0.6762
12	0.942	0.534	20	1.76	0.0933
13	−2.344	0.538	17	−4.35	0.0004
14	0.385	0.707	22	0.54	0.5915
15	0.097	0.477	18	0.20	0.8404
16	0.844	0.452	14	1.87	0.0831
17	−0.166	0.482	18	−0.34	0.7353
18	−0.619	0.554	19	−1.12	0.2782
19	−0.201	0.604	20	−0.33	0.7434
20	0.532	0.435	12	1.22	0.2460

TABLE 18.48
Box-Plot of the t-Values Corresponding to the EBLUPs in Table 18.47

```
proc univariate normal plot data=randomsol; var tvalue
```

Variable: tValue   (t Value)

```
   Stem   Leaf                    #    Boxplot
      8   5                       1       *
      6
      4
      2
      0   223452489               9    +--+--+
     -0   9175333                 7    *-----*
     -2   11                      2       |
     -4   4                       1       0
          ----+----+----+----+
Multiply Stem.Leaf by 10**-1
```

TABLE 18.49
PROC MIXED Code to Fit the Unequal Slopes Model with UN(1) Covariance Structure to the Data Set with Person Six Removed to Obtain Starting Values for the ARH(1) Covariance Structure

```
data wide; set wide; if person=6 then delete;
proc mixed ic scoring=20 data=wide; class person
 order prod;
model resp=prod order order*prod mild*prod/ddfm=kr;
random person;
repeated order/subject=person type=un(1);
```

CovParm	Subject	Estimate
person		0.686824
UN(1,1)	person	0.094153
UN(2,1)	person	0.000000
UN(2,2)	person	0.166396
UN(3,1)	person	0.000000
UN(3,2)	person	0.000000
UN(3,3)	person	0.754291
UN(4,1)	person	0.000000
UN(4,2)	person	0.000000
UN(4,3)	person	0.000000
UN(4,4)	person	0.029599

Effect	NumDF	DenDF	FValue	ProbF
prod	3	26.5	1.84	0.1647
order	3	15.1	0.96	0.4366
order*prod	9	25.4	1.72	0.1360
mild*prod	4	24.2	6.62	0.0010

TABLE 18.50
PROC MIXED Code to Fit the Unequal Slopes Model with ARH(1) Covariance Structure Using Starting Values from the UN(1) Structure in Table 18.49

```
proc mixed ic data=wide; class person order prod;
**starting values from un(1);
model resp=prod order order*prod
 mild*prod/ddfm=kr outp=predicted;
random person/solution;
repeated order/subject=person type=arh(1);
parms .6868 .09414 .1664 .7543 .0296 .1/lowerb =
 1e-4, 1e-4, 1e-4, 1e-4, 1e-4, 1e-4;
ods output solutionr=randomsol;
```

CovParm	Subject	Estimate
person		0.689719
Var(1)	person	0.096455
Var(2)	person	0.223257
Var(3)	person	0.840527
Var(4)	person	0.031573
ARH(1)	person	0.375293

Effect	NumDF	DenDF	FValue	ProbF
prod	3	26.3	1.18	0.3360
order	3	12.4	0.81	0.5135
order*prod	9	23.0	1.36	0.2609
mild*prod	4	25.2	6.02	0.0015

TABLE 18.51
EBLUPs from the Analysis without Person Six

Person	Estimate	StdErrPred	df	tValue	Probt
1	−1.167	0.354	15	−3.30	0.0047
2	0.305	0.329	12	0.93	0.3719
3	0.414	0.332	7.6	1.25	0.2494
4	0.432	0.383	19	1.13	0.2733
5	−0.311	0.366	18	−0.85	0.4058
7	0.026	0.392	24	0.07	0.9481
8	−0.888	0.415	24	−2.14	0.0424
9	−0.513	0.326	13	−1.57	0.1402
10	0.076	0.377	22	0.20	0.8414
11	0.496	0.421	21	1.18	0.2519
12	1.539	0.386	19	3.99	0.0008
13	−1.831	0.386	20	−4.74	0.0001
14	−0.149	0.477	23	−0.31	0.7577
15	0.243	0.340	14	0.71	0.4865
16	1.170	0.351	16	3.34	0.0041
17	0.123	0.371	19	0.33	0.7440
18	−0.461	0.415	24	−1.11	0.2780
19	−0.131	0.409	21	−0.32	0.7528
20	0.627	0.347	18	1.81	0.0874

TABLE 18.52
Box-Plot of the t-Values Corresponding to the EBLUPs for the Analysis without Person Six

```
proc univariate normal plot data=randomsol; var tvalue;
```

Variable: Resid

```
    Stem   Leaf              #     Boxplot
       6   8                 1       |
       4   6                 1       |
       2   22355             5     +-----+
       0   121               3     *--+--*
      -0   9328              4     |     |
      -2   40                2     +-----+
      -4   10                2       |
      -6                             |
      -8   7                 1       |
           ----+----+----+----+
Multiply Stem.Leaf by 10**-1
```

TABLE 18.53
PROC MIXED Code to Fit the Unequal Slopes Model with Test for Equal Slopes

```
proc mixed ic data=wide;
class person order prod;
**starting values from un(1);
model resp=prod order order*prod mild
 mild*prod/ddfm=kr;
random person;
repeated order/subject=person  type=arh(1);
parms .6868 .09414 .1664 .7543 .0296  .1/lowerb =
 1e-4, 1e-4, 1e-4, 1e-4, 1e-4, 1e-4;
```

CovParm	Subject	Estimate
person		0.689719
Var(1)	person	0.096455
Var(2)	person	0.223257
Var(3)	person	0.840527
Var(4)	person	0.031573
ARH(1)	person	0.375293

Effect	NumDF	DenDF	FValue	ProbF
prod	3	26.3	1.18	0.3360
order	3	12.4	0.81	0.5135
order*prod	9	23.0	1.36	0.2609
mild	1	16.2	22.92	0.0002
mild*prod	3	20.7	0.12	0.9499

TABLE 18.54
PROC MIXED Code to Fit the Equal Slopes Model with ARH(1) Covariance Structure to the Data Set with Outliers Removed

```
proc mixed ic data=wide;
class person order prod;
**starting values from un(1);
model resp=prod order order*prod mild /ddfm=kr
 outp=predicted;
random person/solution;
repeated order/subject=person  type=arh(1);
parms .6868 .09414 .1664 .7543 .0296 .1/lowerb =
 1e-4, 1e-4, 1e-4, 1e-4, 1e-4, 1e-4;
```

CovParm	Subject	Estimate
person		0.722087
Var(1)	person	0.000100
Var(2)	person	0.247060
Var(3)	person	1.113557
Var(4)	person	0.105590
ARH(1)	person	0.676322

Effect	NumDF	DenDF	FValue	ProbF
prod	3	27.1	8.96	0.0003
order	3	18.9	0.70	0.5661
order*prod	9	28.1	2.14	0.0602
mild	1	16.8	24.30	0.0001

TABLE 18.55
Adjusted Means and Pairwise Comparisons of the Means Using the Simulate Adjustment for Multiple Comparisons

```
lsmeans prod/diff adjust=simulate;
```

Effect	prod	Estimate	StdErr	df	tValue	Probt
prod	1	9.79	0.24	32.9	40.14	0.0000
prod	2	10.40	0.24	31.6	43.69	0.0000
prod	3	10.01	0.24	32.5	40.93	0.0000
prod	4	10.53	0.23	29.5	45.27	0.0000

prod	_prod	Estimate	StdErr	df	tValue	Probt	Adjp
1	2	−0.61	0.17	27.2	−3.61	0.0012	0.0056
1	3	−0.21	0.17	22.6	−1.25	0.2231	0.5866
1	4	−0.74	0.15	28.4	−4.96	0.0000	0.0003
2	3	0.40	0.17	28.3	2.35	0.0261	0.1057
2	4	−0.13	0.14	28.4	−0.92	0.3672	0.7887
3	4	−0.53	0.17	31.1	−3.13	0.0037	0.0184

REFERENCES

Conover, W. J. and Iman, R. L. (1982) Analysis of covariance using the rank transformation, *Biometrics* 38(3):715–724.

R. Littell, G. A. Milliken, W. Stroup, and R. Wolfinger (1996) *SAS System for Mixed Models*. SAS Institute Inc., Cary, NC.

Milliken, G. A. and Johnson, D. E. (1992) *Analysis of Messy Data, Volume I: Design Experiments*, Chapman & Hall, London.

SAS Institute Inc. (1999) *SAS/STAT® User's Guide, Version 8,* Cary, NC.

Urquhart, N. S. (1982) Adjustment in covariance when one factor affects the covariate, *Biometrics* 38(3):651–660.

EXERCISES

EXERCISE 18.1: A nutritionist designed a study to evaluate the ability of three diets (A, B, and C) to affect weight loss for both males and females. At the start of the study, the males and females were randomly assigned to the diets and each subject's initial weight and percent body fat were measured. The subjects were given the diet and were to return every month to measure their weight. They were also given a food diary in which they recorded all of the items they ate during the third week of the month and the number of grams of fat and the milligrams of salt on a daily basis were determined. Each subject was weighed after 1, 2, 3, and 4 months. The data in the table, salti, dfati, and wti, are the measurements after the ith month. Use the initial weight (iwt), percent body fat (bfat), diet salt (salti), and diet fat (dfati) as possible covariates and carry out an thorough analysis of this data set.

Data for Exercise 18.1

diet	sex	sub	lwt	bfat	salt1	dfat1	wt1	salt2	dfat2	wt2	salt3	dfat3	wt3	salt4	dfat4	wt4
A	F	1	175	27.1	1932	111	179	1929	105	177	1904	102	176	1828	98	174
A	F	2	185	32.4	2088	70	181	2011	77	180	2035	72	175	2149	77	173
A	F	3	172	30.5	2246	89	174	2081	96	171	2173	103	171	2261	103	172
A	F	4	174	27.2	2897	79	178	2887	86	176	2846	83	173	2956	79	172
A	M	1	200	27.4	2098	110	206	2081	112	204	2010	115	209	2043	119	210
A	M	2	223	32.4	2257	104	224	2322	110	222	2180	105	225	2182	103	221
A	M	3	193	25.2	2078	84	192	2153	86	186	2325	89	185	2326	88	184
A	M	4	194	23.7	2059	85	198	1851	83	190	1795	77	188	1980	72	185
A	M	5	236	35.7	2795	81	239	2997	85	233	2976	92	228	2941	87	226
B	F	1	172	27.8	2620	58	168	2628	65	167	2746	62	169	2774	65	162
B	F	2	157	26.8	2269	107	159	2173	108	161	2323	102	167	2211	96	162
B	F	3	158	26.8	2703	103	156	2811	102	159	2753	95	163	2833	99	156
B	F	4	161	30.2	2809	109	164	2882	116	164	2963	111	172	2868	109	166
B	F	5	178	31.1	2099	76	175	2127	80	173	2230	79	171	2248	84	160
B	F	6	169	29.1	2398	73	166	2491	70	164	2606	74	165	2733	70	157
B	M	1	230	30.8	2690	86	231	2714	92	229	2612	86	232	2585	82	226
B	M	2	230	34.1	2310	65	229	2328	58	225	2330	59	224	2217	58	210
B	M	3	203	29.9	2482	122	201	2550	122	203	2660	129	207	2518	125	197
B	M	4	204	27.7	1953	75	202	1953	80	197	1898	85	190	1880	81	172
B	M	5	209	30.6	2698	115	206	2707	110	209	2693	103	213	2659	106	197

Data for Exercise 18.1

diet	sex	sub	lwt	bfat	salt1	dfat1	wt1	salt2	dfat2	wt2	salt3	dfat3	wt3	salt4	dfat4	wt4
C	F	1	165	25.5	2576	116	168	2662	113	166	2608	107	164	2609	109	165
C	F	2	168	29.5	2487	96	170	2489	101	168	2539	96	161	2592	94	165
C	F	3	159	24.4	2247	81	160	2210	77	160	2276	82	157	2369	84	154
C	F	4	171	30.7	2395	73	169	2418	69	167	2351	67	161	2330	61	162
C	F	5	174	30.1	2487	70	173	2543	64	172	2583	61	167	2649	68	167
C	F	6	152	23.0	2740	128	152	2694	130	152	2658	138	147	2574	142	151
C	F	7	173	28.2	2327	108	174	2215	110	173	2297	109	166	2179	103	164
C	F	8	173	29.2	2304	117	172	2324	114	168	2217	115	163	2243	111	162
C	M	1	216	31.5	2411	117	214	2349	115	211	2306	117	205	2257	117	201
C	M	2	235	31.7	2178	111	233	2174	117	233	2137	122	224	2096	129	226
C	M	3	224	30.2	1876	111	220	1558	111	217	1681	116	209	1522	123	208
C	M	4	184	24.8	2164	60	189	2210	53	187	2167	55	176	2149	61	174
C	M	5	234	33.6	2394	122	234	2444	116	228	2301	119	216	2378	118	216

Special Applications of Analysis of Covariance

EXERCISE 18.2: Use the rank transformation and provide an analysis of the data in Exercise 3.1.

EXERCISE 18.3: Use the rank transformation and provide an analysis of the data in Exercise 4.1.

EXERCISE 18.4: Use the rank transformation and provide an analysis of the data in Section 5.5.

EXERCISE 18.5: Use the rank transformation and provide an analysis of the data in Section 5.7.

EXERCISE 18.6: The data in the table below are times to dissolve a piece of chocolate candy. A class of 48 students was available for the study. Each student was given a butterscotch chip and the time to dissolve the chip by mouth was recorded. The students were put into blocks of size six based on their time to dissolve the butterscotch chip (bst). Then the six students in each block were randomly assigned to one of the chocolate candy types and the time to dissolve the candy piece (time) by mouth was determined. Provide a complete analysis of this data set. Analyze the data without using the bst as a covariate. Reanalyze the data ignoring the blocking factor but use bst as a covariate. Compare the three analyses.

Data for exercise 18.6

	Choc Chip		Red M&M®		Small M&M®		Button		Blue M&M®		Snow Cap	
block	bst	time	bst	time	bst	time	bst	time	bst	time	bst	time
1	15	15	16	24	16	14	16	32	16	24	16	14
2	16	16	18	19	21	16	21	35	21	28	21	24
3	21	26	21	31	22	21	22	35	23	30	23	25
4	23	20	23	19	24	16	25	28	25	30	25	27
5	26	31	27	33	27	28	27	37	28	33	28	33
6	28	31	30	34	30	31	31	35	31	31	31	37
7	31	43	31	38	33	28	34	41	35	45	35	40
8	35	45	36	42	36	33	36	48	36	42	39	42

EXERCISE 18.7: The data for this example involved finding ten groups of three persons each and then randomly assigning one of the three shapes of hard candy to each person within a group. Each person was given a butterscotch chip to dissolve by mouth and the time was recorded. Then each person recorded the time to dissolve the assigned piece of hard candy. The groups form blocks as they were observed on different days, bst is the time to dissolve the butterscotch piece, and time is the time to dissolve the hard candy piece. Carry out a thorough analysis of this data set.

Data for Exercise 18.7

	Round		Square		Flat	
Block	bst	time	bst	time	bst	time
1	25	99	34	184	30	105
2	23	92	19	92	36	127
3	31	123	25	102	35	125
4	32	115	17	88	21	91
5	38	188	21	93	25	92
6	23	96	27	110	37	133
7	22	94	30	125	33	118
8	20	88	32	148	20	93
9	19	99	23	98	25	95
10	32	113	21	96	22	91

EXERCISE 18.8: Use the beta-hat model approach described in Chapter 6 to test each of the following hypotheses for the data in Table 18.33 and compare the results obtained by PROC NLMIXED.

$H_{o1}: \alpha_1 = \alpha_2 = \alpha_3 = \alpha_4$ vs. $H_{a1}:$ (not H_{o1}:)

$H_{o2}: \beta_1 = \beta_2 = \beta_3 = \beta_4$ vs. $H_{a2}:$ (not H_{o2}:)

$H_{o3}: \gamma_1 = \gamma_2 = \gamma_3 = \gamma_4$ vs. $H_{a3}:$ (not H_{o3}:)

$H_o: \alpha_1 + \beta\, e^{-\gamma_1 \text{insect}_0} = \alpha_2 + \beta\, e^{-\gamma_2 \text{insect}_0} =$
$\alpha_3 + \beta\, e^{-\gamma_3 \text{insect}_0} = \alpha_4 + \beta\, e^{-\gamma_4 \text{insect}_0}$ vs. $H_a:$ (not H_o:)

EXERCISE 18.9: Use the model comparison approach described in Chapter 8 to test each of the hypotheses in Exercise 18.8. Also provide a test of the equal model hypothesis.

Special Applications of Analysis of Covariance   595

EXERCISE 18.10: The data in in the table below are from an experiment similar to that in Section 16.5. Determine an appropriate covariance matrix and then provide a detailed analysis of the data set.

Data for Exercise 18.10 Where IBP Is Initial Blood Pressure and bp1, ..., bp6 are the Blood Pressure Readings for the Six Time Periods

Drug	Exercise	Person	IBP	bp1	bp2	bp3	bp4	bp5	bp6
No	No	1	133	146	144	142	141	142	143
No	No	2	137	136	135	134	134	134	137
No	No	3	148	148	148	148	143	143	144
No	No	4	136	139	137	138	139	139	142
No	No	5	140	141	143	145	147	147	147
No	No	6	139	140	136	135	135	132	135
No	No	7	154	185	181	183	182	180	181
No	No	8	152	146	147	146	145	146	143
Yes	No	1	150	145	134	136	134	132	134
Yes	No	2	147	147	139	139	132	134	135
Yes	No	3	142	133	121	122	120	165	116
Yes	No	4	144	139	128	128	122	119	116
Yes	No	5	140	132	121	121	115	117	116
Yes	No	6	137	133	122	122	118	119	121
Yes	No	7	143	144	133	131	124	123	124
Yes	No	8	137	122	111	111	104	101	101
No	Yes	1	150	143	142	139	137	134	128
No	Yes	2	151	144	137	134	131	128	121
No	Yes	3	151	155	149	147	140	136	131
No	Yes	4	142	149	101	140	140	137	129
No	Yes	5	149	142	140	136	132	126	126
No	Yes	6	132	133	127	123	122	117	116
No	Yes	7	134	136	129	125	122	123	115
No	Yes	8	151	150	145	141	137	131	130
Yes	Yes	1	131	95	85	78	76	76	79
Yes	Yes	2	148	138	130	124	120	122	123
Yes	Yes	3	151	146	134	129	130	134	134
Yes	Yes	4	150	137	126	119	119	120	121
Yes	Yes	5	144	129	123	118	117	118	117
Yes	Yes	6	151	140	138	134	131	126	129
Yes	Yes	7	152	149	141	138	132	132	131
Yes	Yes	8	135	123	115	109	107	111	109

Index

A

Adjusted means, 21, 28, 60–61, 72–73, 80–81, 99, 111, 140, 159, 215, 224, 419–420, 442–443, 510, 543, 580–581
  for common slope model, 264
  for measurement on the block, 264–265
  for non-parallel lines model, 264
Adjusted R^2 variable selection method, 179, 181
Adjusted significance levels, 336
`adjust=simulate` option, 336
Analysis of covariance
  covariate adjustment process, 1–7
  defined, 1
  general model and basic philosophy, 7–10
  strategy for, 8
ANTE(1) structure, 561
Approximate error degrees of freedom, 311
AR(1) covariance structure, 464
ARH(1) covariance structure, 464–466, 561, 577–580
Autoregressive covariance structure, first-order, 561

B

Backward variable selection method, 181
Balanced incomplete block (BIB) designs, 277–282
  example of four treatments, 278–282
Bartlett's test
  for equal variances, 355–356
  for unequal variances, 367
Baseline, analysis of change from, 70–74, 221
Basic strategy, exceptions to
  diet/cholesterol example, 70–74
  shoe tread design example, 74–78
Behrens-Fisher problem, 359, 360, 361
Best linear unbiased estimator (BLUE), 208, 237, 326
Best linear unbiased predictor (BLUP), 326
  estimated (EBLUP), 300–302, 578
Beta-hat models, 163–174
  for within block/between block comparison, 261–262
  for within block/between block estimates, 237
  for block designs, 207, 210, 213–214
  for within block estimates, 236–237
  for between block information estimates, 275–276
  for complex treatment structures, 166
  for equality of parameters, 165–166
  examples
    one-way treatment structure, 167–171
    two-way treatment structure, 171–173
  model and analysis, 163–165
  for nonparallel lines, 359–360
  for parallelism hypothesis, 238–239
  for slopes equal to zero hypothesis, 358–359
  for three-treatment incomplete block design structure, 239
Between block analysis, *see under* Block designs
Block designs
  block total model, 206
  complete, 203–231
    between block analysis, 206–207
    within block analysis, 204–206
    combined within block/between block designs, 207–209, 216–217, 218, 272–277, 329
    common slope model, 211–214
    comparing treatments, unequal slopes model, 215
    computations using SAS® system, 217–220
    confidence intervals about differences of two regression lines, 215–217
    within block analysis, 215–216
    example: drug effect on heart rate, 220–226
    model comparison method, 209–211
    model for, 203–204
  completely randomized, for two-way treatment in split-plot design, 400–406, *see also* Split-plot and strip-plot design structures
  incomplete
    examples
      balanced incomplete structure with four treatments, 247–251
      balanced incomplete structure with four treatments using JMP®, 251–254

597

five treatments in randomized design
structure, 240–247
for whole plot or large size of
experimental unit, 392–395
randomized, 175
randomized complete block, 233–257
one-way treatment structure, 233–234
whole plots or large size experimental
units, 415–425
randomized incomplete
between block design, 235
within block design, 234–235
combined within block/between block
information, 272–277, 417
combined within block/between block
treatment, 236–240
Block effect, 176
Blocking, 533–543
BLUE (best linear unbiased estimator), 208, 237, 326
Bonferroni type multiple comparison procedure, 20, 24, 28, 61–63, 77
Box-plots, 577–578

C

Cell means, 123, 157–158, 197
Centrality parameter, 15
Chocolate candy example, 41–54
  JMP® analysis, 50–54
  PROC GLM analysis, 42–47
  PROC MIXED analysis, 47–50
Cholesterol/diet example, 66–74
Coefficients, orthogonal polynomial, 387
Combined estimators of slopes, 272
Common slope model, 104, 125–126, 130–132
Common slopes, 133, 159
  in block designs, 211–214
  within large size of experimental unit, 403
  for measurement on the block, 263, 264
  within small size of experimental unit, 404
  with unequal between location variances, 385–386
Completely randomized design
  for heterogeneous errors, multilocation trials example, 381–386
  one-way treatments
    multiple covariates, 93–122
    one covariate, 11–40
      examples, 41–92
      unequal variance analysis, 353–354
  two-way treatments, 123–161
  two-way mixed effects treatment, 332–337
Compound symmetry, 453–454, 461, 462, 464

Conditional error, principle of, 14–16, 21–24
Confidence bands, 63, 191, 571
  about difference of two treatments, 25, 29–30
  Scheffé method, 298
Confidence intervals
  for block designs, 215–217
    within block designs, 215–216
    combined within block/between block designs, 216–217
  simultaneous, 30–31, 217
  Tukey simultaneous, 104
Confidence limits, 571
Confounding, 494
Contrast statements, 189, *see also* Model comparison
Covariance structure of data, 456–457
Covariate adjustment process, 1–7
Covariate by treatment interaction, 21–24
Covariate location, estimation of variance components and, 292–297
CP variable selection method, 179, 181
Crossover designs, 559–564
CS structure, 561

D

Degrees of freedom
  approximate error, 311
  pooled, 13
Distribution, student t, 20

E

EBLUP (estimated best linear unbiased predictor), 300–302, 326, 578
Effects models, 124, 130, 147, 158–159
Equal intercept hypothesis, 21
Equality of models, 189, 197–199, *see also* Model comparison
Equality (homogeneity) of slopes, 15
Equal slopes, 17–21, 27, 99
  in block designs, 214–215
  examples, equal slopes within treatment groups with unequal slopes between groups, 83–83
  model building, variable selection, 175–184
  one-way treatments, chocolate candy example, 41–54
    JMP® analysis, 50–54
    PROC GLM analysis, 42–47
    PROC MIXED analysis, 47–50
Equal slopes hypothesis, 247
Equal variance assumption, 12

Index

Error of prediction, estimated standard, 309
Error rate, experiment-wise, 20, 24, 25
Errors, heterogeneous, 353–388, *see also* Heterogeneous errors
Estimability, 146
Estimable functions, 146, 148, 329
Estimated best linear unbiased predictors (EBLUPS), 300–302, 578
Estimated regression model, 44
Estimated standard error of prediction, 309
Estimate of variance, pooled, 164
Estimates
    least squares, 95
    method of moments, 311
Estimate statements, 59, 132, 387, 517
Estimators
    best linear unbiased (BLUE), 208, 237, 326
    combined of slopes, 272
    least squares, 12–13, 16
    mixed models, 218–220
Exercise/initial heart rate example, 54–66
Experimental design, 8
Experimental unit, size of, *see* Size of experimental unit
Experiment-wise error rate, 20, 24, 25
Exponential decay model, 565

F

Factorial treatments
    five-way, 530
    four-way, 520
        at each of two levels, 508–511, 511–512
    fractional, 499–501
    scatter plots of, 504, 505, 506
    three-way at each of two levels, 503–508
    two-way, 418
F distribution, noncentral, 19–20, 22–24
First-order ante-dependence structure, 454–455, 461
First-order auto-regressive covariance matrix, 454
First-order auto-regressive covariance structure, 454, 561
Fisher Protected LSD comparison, 78, 178–179, 182
Fit model screen window, 129
Five-way factorial treatments, 530
Fixed effects treatment structure, 329–350
    fixed-effects estimation and small sample size approximations, 332–337
    fixed treatments with locations random, 331–332
    random locations with randomized complete block at each location, 337–350

Forward variable selection method, 179, 180–181
Four-way factorial treatments, at each of two levels, 508–511, 511–512
Fractional factorial treatments, 499–501
Full rank equal slopes model, 45
Full rank means model, 409–410
Full rank model, 35–36

H

Half-normal probability plot, 511
Hartley's F-Max test for equal variances, 355, 367
Heterogeneous compound symmetry covariance structure, 455, 461, 464
Heterogeneous errors, 353–388
    comparing models, 359–362
        nonparallel lines models, 359–361
        parallel lines models, 361–362
    computational issues, 362
    determining form of model, 358–359
    estimating parameters of regression model, 356–357
        least squares estimation, 356–357
        maximum likelihood methods, 357
    examples
        one-way treatment structure with unequal variances, 362–369
        treatments in multilocation trial, 381–388
        two-way treatment structure with unequal variances, 369–381
    tests for, 354–356
        Bartlett's test for equal variances, 355–356
        Harley's F-Max test for equal variances, 355, 367
        Levene's test for homogeneity of variances, 354–355, 363, 364, 367, 371, 534
        likelihood ratio test, 356
        unequal variance model, 353–354
Heterogeneous first-order auto-regressive structure, 455, 461
Heterogeneous Töplitz structure, 455
Homogeneity (equality) of slopes, 15
HOVTEST=Levene option, 535–536
Hyper-plane models, 93, 100

I

Incomplete block designs, *see* under Block designs
Information criteria, 365–367, 462, 477

Interaction comparisons, 127
Intercepts, of parallel regression lines, 17
Intervals, prediction, 309

J

JMP® system, 31–38
    attributes window, 421–422
    balanced incomplete block design structure with four treatments, 251
    chocolate candy example, 50–54
    data table, 420, 442
    fit model screen, 511
    fit model window, 100, 129, 421
    two-way treatments, 129–130

K

Kenward-Roger approximation, 216–217, 238, 358–359, 360, 361, 362

L

Least squares estimates, 95, 205–206, 356–357
Least squares estimators, 12–13, 16
Least squares means, 28, 33, 36, 52–54, 53, 57, 99, 106, 135–136, 246, 247
Less than full range equal slopes model, 51
Less than full rank model, 44–45
Levene's statistic, 371
Levene's test for homogeneity of variances, 354–355, 363, 364, 367, 371, 534
Likelihood ratio test, 356
Linear contrast, 20
Linear effects, 20
Linear regression models, 11–12
    multiple, 94, 205
    simple, 15, 212, 260–261, 362–365
Linear trend, 21
    significant, 387
LSD type multiple comparison procedure, 19–20, 24, 28, 371
    Fisher Protected LSD, 78, 178–179, 182
    pairwise, 409–412
LSMEANS statements, 140–141, 159, 220, 419

M

Main effects, with no interaction for intercepts, 126
Matrix forms, 97–98
    of block total model, 206–207
    of mixed model, 325–329
    of multiple regression model, 205
Matrix notation, 184–185
Maximum likelihood methods, 357
Maximum response, of drug, 136
Means
    adjusted, 21, 28, 60–61, 72–73, 80–81, 99, 104, 111, 140, 159, 215, 224, 419–420, 442–443, 510, 543, 580–581
    for common slope model, 264
    for measurement on the block, 264–265
    for non-parallel lines model, 264
    cell, 123, 157–158
    least squares, 28, 33, 36, 52, 53, 57, 99, 106, 135–136, 246, 247
    of y_i for given value of X, 11
Means models, 147
    full rank, 409–410
Mean squares, 311
Method of moments estimates, 311
Missing treatment combinations, 144–147
Mixed models, 325–352, 362
    examples
        fixed treatments and random locations with randomized complete block at each location, 337–350
        two-way mixed effects treatment in completely randomized design, 332–337
    fixed-effects estimation and small sample size approximations, 329–331
    fixed effects treatment structure, 329
    fixed treatments and locations random, 331–332
    matrix form, 325–329
    within small size of experimental unit, 404, 406
    within whole plot, 407–408
Mixed models estimate, 208
Mixed models estimator, 218–220
Model building, 555
    for nonreplicated experiments, 502–503
    with PROC GLM and PROC MIXED, 128–129
    variable selection in, 175–187
        example: one-way treatment structure with equal slopes, 177–184
        procedure for equal slopes, 175–177
Model comparison, 14–15, 19, 21–24, 97–99, 164, 189–202, 190
    for block designs, 209
    examples

Index

one-way treatment structure with one covariate, 193–195
one-way treatment structure with three covariates, 195–197
two-way treatment structure with one covariate, 197–201
one-way treatment structure, 190–191
two-way treatment structure, 191–193
Model selection, 176
Multiple comparisons, 336
   Bonferroni type, 20, 24, 28, 61–63, 77
   Fisher Protected LSD, 78, 178–179, 182
   LSD type, 19–20, 24, 28, 371
   pairwise, 47, 48, 50, 409–412
   procedure, 19–20
   Scheffé type, 20, 24, 25, 30, 63
   with simulate adjustment, 376–381, 387
   simulate adjustment for, 536, 539
   Tukey method, 67–69
Multiple covariates
   within large size of experimental unit, 405–406
   one-way treatments in completely randomized design, 93–122
      estimation, 95
      examples
         comparing response surface models, 112–120
         driving golf ball with different shafts, 95–99
         herbicides and soybean yield — three covariates, 99–105
         models that are quadratic functions of covariate, 105–112
   model, 93–94
Multiple linear regression models, 94
Multiple regression model, 205

N

No interaction assumption, 156
Noncentral F distribution, 22–24
Noncentrality parameter, 16–17
Nonlinear analysis of covariance, 564–572
Non-null effects, 501–502, 504, 505, 511, 517, 520
Nonparallel lines models, 16, 126–127, 139, 143, 211, 264
   for heterogeneous errors, 359–361
Nonparametric analysis, 552–559
Nonreplicated experiments, 493–531
   basic principles, 493–495
   estimating parameters, 502–503
   examples

four factors each at two levels, 508–511, 511–512
three factors each at two levels, 503–508
two covariates, 512–520
unequal slopes, 520–527
   with multiple covariates, 499–501
   selecting non-null and null partitions, 501–502
   with single covariate, 495–499
Null effects, 501–502, 504
Null hypothesis, model restricted by, 14

O

One-way treatments
   balanced, for random effects models, 299–304
   beta-hat test of, 167–171
   comparing treatments or regression lines, 17–23
      equal slopes model, 18–21
      unequal slopes model — covariate treatment by interaction, 21–24
   confidence bands about the difference of two treatments, 25
   estimation, 12–14
   examples, 41–91
      equal slopes (chocolate candy example), 41–54
         JMP® analysis, 50–54
         PROC GLM analysis, 42–47
         PROC MIXED analysis, 47–50
      equal slopes within treatment groups and unequal slopes between groups, 83–83
      exceptions to basic strategy
         cholesterol/diet example, 66–74
         shoe trade design data, 74–83
      unequal slopes (exercise/initial heart rate), 54–66
      unequal slopes and equal intercepts
         part 1, 83–85
         part 2, 85–89
   heterogeneous errors with unequal variances, 362–369
   model, 11–12
   model comparison, 190–191
      example with one covariate, 193–195
      example with three covariates, 195–197
   multiple covariates in completely randomized design, 93–122
      estimation, 95
      examples

comparing response surface models, 112–120
driving golf ball with different shafts, 95–99
herbicides and soybean yield — three covariates, 99–105
models that are quadratic functions of covariate, 105–112
model, 93–94
nonlinear analysis, 564–572
for random effects models
balanced, 299–304
unbalanced, 304–309
randomized complete block, 233–234, 269–277
measured on whole plot, 406–414
strategy for determining form of model, 14–17
summary of strategies, 25–26
unbalanced, for random effects models, 304–309
unequal variance analysis, 353–354
using JMP®, 31–38
using PROC GLM and PROC MIXED, 26–31
using SAS® system, 26
variable selection with equal slopes, 175–184
for whole plot or large size of experimental unit, 406–414
Orthogonal polynomial coefficients, 387
Outliers, 572–589

P

Pairwise comparisons, 76, 99, 111, 182–184, 369, 381, 419, 443, 468, 541–542, 543, 553, 562–563, 580–581
LSD type, 409–412
multiple, 47, 48, 50
Tukey method, 72–73
Parallelism, tests for, 402
Parallelism hypothesis, 98–99, 115–116, 125, 238
Parallel line comparisons, 139
Parallel lines models, for heterogeneous errors, 361–362
Parallel regression lines, 15–18, 26, 27
Parsimony principle, 8
Pooled degrees of freedom, 13
Pooled estimate of variance, 164
Pooled residual sum of squares, 13
Predicted regression models, 376
Prediction, estimated standard error of, 309
Prediction bands, 309

Prediction intervals, 309
Predictors
best linear unbiased (BLUP), 326
estimated best linear unbiased (EBLUP), 300–302, 326, 578
Principle of conditional error, 14–16, 21–24
Principle of parsimony, 8
Probability plot, half-normal, 511
PROC GLM/PROC MIXED, 26–31
two-way treatments, 128
PROC MIXED chocolate candy example, 47–50
PROC NLIN procedure, 564
PROC NLMIXED procedure, 564
p values, simulate-adjusted, 50

Q

Quadratic effects, 20
Quadratic function, 94, 105–112
one-way treatments of multiple covariates, 105–122
Quadratic regression model, 136
Quadratic response surface, 94, 112–120
Quadratic trends, 21

R

Random coefficient model, 287–292, 298
two-way, 311
Random effects, 287–324
changing covariate location, 297–299
estimation of variance components, 292–297
examples
balanced one-way treatment structure, 299–304
two-way treatment structure, 309–315
unbalanced one-way treatment structure, 304–309
in mixed model, 325
random coefficient model, 287–292, 298
Randomized block designs, 175
complete, 233–257
fixed treatments and locations random, 337–350
nonlinear analysis, 564–572
one-way treatment structure, 233–234
with unequal variances, 543–552
incomplete
between block design, 235
within block design, 234–235
combined within block/between block treatment, 236–240

Index

examples
 balanced incomplete structure with four treatments, 247–251
 balanced incomplete structure with four treatments using JMP®, 251–254
 five treatments in randomized design structure, 240–247
measurements on the block, 259–285
 adjusted means and treatment comparisons, 264–265
  common slope model, 264
  non-parallel lines model, 264–265
 between block model, 261
 within block model, 286–261
 combined within block/between block information, 261–263, 272–277
  common slope model, 263
 examples
  four treatments in balanced incomplete block structure, 278–282
  four treatments in randomized complete block, 269–279
  two treatments, 265–269
Regression
 forward, 179, 180–181
 stepwise, 176–177
Regression lines, 14, 94
 nonparallel, 16
 parallel, 15–18, 26
Regression models, 8, 60, 132
 estimating parameters for heterogeneous errors, 356–357
  least squares estimation, 356–357
  maximum likelihood methods, 357
 predicted, 376
 quadratic, 135–136, 136
Regression surfaces, 124
REML estimation, 224
repeated/group=method, 365
Repeated measures designs, 329, 451–491, *see also* Split-plot and strip-plot designs
 basic principles, 451–453
 covariance part of model — selecting **R**, 453–456
  AR(1) covariance structure, 464
  ARH(1) covariance structure, 464–466
  compound symmetry, 453–454, 461, 462, 464
  first-order ante-dependence structure, 454–455, 461
  first-order auto-regressive covariance matrix, 454
  first-order auto-regressive covariance structure, 454
  heterogeneous compound symmetry covariance structure, 455, 461, 464
  heterogeneous first-order auto-regressive structure, 455, 461
  heterogeneous Toplitz structure, 455
  Toplitz structure, 454, 461, 466
  unstructured covariance model, 456, 461
 covariance structure of data, 456–457
 examples
  covariate measured on large size experimental unit, 459–470
  three sizes of experimental units with repeated measure at middle size and covariate at small size, 470–479
 random and repeated statements for SAS® system, 457–458
 selecting adequate covariance structure, 458
 unstructured covariance matrix, 457, 467
Residuals, 176
Response surface models, one-way treatments of multiple covariates, 112–120
R^2 variable selection method, 179, 181

S

SAS® system, 26–31
 in block model computation, 217–220
 two-way treatments, 127
Satterthwaite approximation, 216–217, 238, 358–359, 360, 361, 362
Scatter plots
 four-dimensional, 504, 505, 506
 three-dimensional, 514, 515
Scheffé percentage points, 63
Scheffé type multiple comparison procedure, 20, 24, 25, 30, 63, 298
Shoe tread design example, 74–78
Significance levels, adjusted, 336
Significant linear trend, 387
Simple linear regression model, 15, 212, 260–261, 362–365
Simulate-adjusted p values, 50
Simulate adjustment, 376–381, 387, 468, 536, 539
Simulate multiple comparison method, 52, 54
Simultaneous confidence intervals, 30–31, 217
Size of experimental unit
 concepts related to, 392
 covariate measured on large and small size of experimental unit, 398–399

covariate measured on small size of experimental unit, 392–398
covariate measured on whole plot or large size of experimental unit, 392–395
examples
   combining large size and small size experimental units, 426–432
   cookie baking two-way treatment, 414–425
   covariate measured on whole plot, 406–414
   strip-plot design with three sizes of experimental units and three covariattes, 432–443
general representation, 399–406
covariate measured on large size of experimental unit,392–395, 401–403
covariate measured on small size of experimental unit, 398–399, 403–405
Slopes
   common, eee Common slopes
   equal, *see* Equal slopes
   unequal, *see* Unequal slopes
Small sample size approximations, with fixed-effects estimation, 329–331
Small size of experimental unit, Size of experimental unit
Special applications, 533–545
   blocking and analysis of covariance, 533–543
   crossover design with covariates, 559–564
   effect of outliers, 572–589
   nonlinear analysis of covariance, 564–572
   nonparametric analysis of covariance, 552–559
   treatments with different ranges of covariate, 543–552
Split-plot and strip-plot designs, 329, 391–450, 461, *see also* Repeated measures designs
   concepts related to, 392
   covariate measured on large and small size of experimental unit, 398–399
   covariate measured on small size of experimental unit, 392–398
   covariate measured on whole plot or large size of experimental unit, 392–395
   examples
      combining large size and small size experimental units, 426–432
      cookie baking two-way treatment, 414–425

covariate measured on whole plot, 406–414
strip-plot design with three sizes of experimental units and three covariates, 432–443
general representation, 399–406
covariate measured on large size of experimental unit, 392–395, 401–403
covariate measured on small size of experimental unit, 398–399, 403–405
split-split plot, 474–475
Standard error of prediction, estimated, 309
Stationary point, 136
Stepwise deletion, 104, 116, 427, 477, 502–503, 510, 545, 555, 561
Stepwise variable selection, 176–177, 179–180
Student t distribution, 20
Sub-plot versus whole-plot information, 392, *see also* Split-plot and strip-plot design structures
Sum of squares
   pooled residual, 13
   under principle of conditional error, 15
   in random effects models, 294–296
   Type I, 150, 152
   Type II, 150, 152
   Type III, 150, 153
   Type IV, 150, 153
Sum of squares lack of fit, 112

T

Three-treatment incomplete block design structure, 239
Three-way factorial effects, 476–477
Three-way treatments
   at each of two levels, 503–508
   in repeated measures designs, 472
TOPE covariance structure, 476–477, 561
Töplitz structure, 454, 455, 461, 466
Tukey method, 67–69, 72–73, 107
Tukey simultaneous confidence intervals, 104
2 × 2 table differences, 148, 200–201
Two-way factorial effect, 129, 418
Two-way treatments
   beta-hat test of, 171–173
   in completely randomized design structure, 123–161
      examples
         daily gains and birth weight — common slope, 130–136

Index

energy from wood of different trees — unequal slopes, 136–144
two-way treatment structure with missing cells, 147–158
extensions, 158–159
with JMP® system, 129–130
missing treatment combinations, 144–147
model, 123–127
with PROC GLM and PROC MIXED, 128
with SAS® system, 127
with different ranges of covariate, 543–552
heterogeneous errors with unequal variances, 369–381
mixed effects in completely randomized design, 332–337
model comparison, 191–193
example with one covariate, 197–201
with one covariate, 129
random effects, 289–292, 309–315
in split-plot design, 400–406, *see also* Split-plot and strip-plot design structures
whole plots or large size experimental units, 415–425
Type I sums of squares, 150, 152
Type II estimable functions, 146
Type II sums of squares, 150, 152
Type III estimable functions, 146
Type III sums of squares, 150, 153
Type IV estimable functions, 146
Type IV estimable hypotheses, 148–152
Type IV sums of squares, 150, 153

U

Unequal slopes, 21–24, 99, 100, 108
for block designs, 215
examples
equal intercepts
part 2, 85–89
part one, 83–85
equal slopes within treatment groups with unequal slopes between groups, 83–83
within large size of experimental unit, 402, 403
in nonreplicated experiments, 495–496
one-way treatments, exercise/initial heart rate example, 54–66
within small size of experimental unit, 404
with unequal between location variances, 384–385
variable selection, 185–186
Unequal variance model, 353–354, 367, *see also* Heterogeneous errors
Unequal variances, in two-way treatment, 543–552
UN structure, 561
Unstructured covariance matrix, 457, 467
Unstructured covariance model, 456, 461

V

Variable selection
matrix notation, 184–185
in model building, 175–187
example: one-way treatment structure with equal slopes, 177–184
procedure for equal slopes, 175–177
unequal slopes, 185–186

W

Weighted least squares, 208
Whole plot covariate, 412–413
Whole-plot versus sub-plot information, 392, *see also* Split-plot and strip-plot design structures
Within block analysis, *see* under Block designs

Z

Zero-slopes hypothesis, 14, 42